# Alfvén Waves Across Heliophysics

# Geophysical Monograph Series

235 **Electric Currents in Geospace and Beyond** *Andreas Keiling, Octav Marghitu, and Michael Wheatland (Eds.)*
236 **Quantifying Uncertainty in Subsurface Systems** *Celine Scheidt, Lewis Li, and Jef Caers (Eds.)*
237 **Petroleum Engineering** *Moshood Sanni (Ed.)*
238 **Geological Carbon Storage: Subsurface Seals and Caprock Integrity** *Stephanie Vialle, Jonathan Ajo-Franklin, and J. William Carey (Eds.)*
239 **Lithospheric Discontinuities** *Huaiyu Yuan and Barbara Romanowicz (Eds.)*
240 **Chemostratigraphy Across Major Chronological Eras** *Alcides N. Sial, Claudio Gaucher, Muthuvairavasamy Ramkumar, and Valderez Pinto Ferreira (Eds.)*
241 **Mathematical Geoenergy: Discovery, Depletion, and Renewal** *Paul Pukite, Dennis Coyne, and Daniel Challou (Eds.)*
242 **Ore Deposits: Origin, Exploration, and Exploitation** *Sophie Decree and Laurence Robb (Eds.)*
243 **Kuroshio Current: Physical, Biogeochemical and Ecosystem Dynamics** *Takeyoshi Nagai, Hiroaki Saito, Koji Suzuki, and Motomitsu Takahashi (Eds.)*
244 **Geomagnetically Induced Currents from the Sun to the Power Grid** *Jennifer L. Gannon, Andrei Swidinsky, and Zhonghua Xu (Eds.)*
245 **Shale: Subsurface Science and Engineering** *Thomas Dewers, Jason Heath, and Marcelo Sánchez (Eds.)*
246 **Submarine Landslides: Subaqueous Mass Transport Deposits From Outcrops to Seismic Profiles** *Kei Ogata, Andrea Festa, and Gian Andrea Pini (Eds.)*
247 **Iceland: Tectonics, Volcanics, and Glacial Features** *Tamie J. Jovanelly*
248 **Dayside Magnetosphere Interactions** *Qiugang Zong, Philippe Escoubet, David Sibeck, Guan Le, and Hui Zhang (Eds.)*
249 **Carbon in Earth's Interior** *Craig E. Manning, Jung-Fu Lin, and Wendy L. Mao (Eds.)*
250 **Nitrogen Overload: Environmental Degradation, Ramifications, and Economic Costs** *Brian G. Katz*
251 **Biogeochemical Cycles: Ecological Drivers and Environmental Impact** *Katerina Dontsova, Zsuzsanna Balogh-Brunstad, and Gaël Le Roux (Eds.)*
252 **Seismoelectric Exploration: Theory, Experiments, and Applications** *Niels Grobbe, André Revil, Zhenya Zhu, and Evert Slob (Eds.)*
253 **El Niño Southern Oscillation in a Changing Climate** *Michael J. McPhaden, Agus Santoso, and Wenju Cai (Eds.)*
254 **Dynamic Magma Evolution** *Francesco Vetere (Ed.)*
255 **Large Igneous Provinces: A Driver of Global Environmental and Biotic Changes** *Richard. E. Ernst, Alexander J. Dickson, and Andrey Bekker (Eds.)*
256 **Coastal Ecosystems in Transition: A Comparative Analysis of the Northern Adriatic and Chesapeake Bay** *Thomas C. Malone, Alenka Malej, and Jadran Faganeli (Eds.)*
257 **Hydrogeology, Chemical Weathering, and Soil Formation** *Allen Hunt, Markus Egli, and Boris Faybishenko (Eds.)*
258 **Solar Physics and Solar Wind** *Nour E. Raouafi and Angelos Vourlidas (Eds.)*
259 **Magnetospheres in the Solar System** *Romain Maggiolo, Nicolas André, Hiroshi Hasegawa, and Daniel T. Welling (Eds.)*
260 **Ionosphere Dynamics and Applications** *Chaosong Huang and Gang Lu (Eds.)*
261 **Upper Atmosphere Dynamics and Energetics** *Wenbin Wang and Yongliang Zhang (Eds.)*
262 **Space Weather Effects and Applications** *Anthea J. Coster, Philip J. Erickson, and Louis J. Lanzerotti (Eds.)*
263 **Mantle Convection and Surface Expressions** *Hauke Marquardt, Maxim Ballmer, Sanne Cottaar, and Jasper Konter (Eds.)*
264 **Crustal Magmatic System Evolution: Anatomy, Architecture, and Physico-Chemical Processes** *Matteo Masotta, Christoph Beier, and Silvio Mollo (Eds.)*
265 **Global Drought and Flood: Observation, Modeling, and Prediction** *Huan Wu, Dennis P. Lettenmaier, Qiuhong Tang, and Philip J. Ward (Eds.)*
266 **Magma Redox Geochemistry** *Roberto Moretti and Daniel R. Neuville (Eds.)*
267 **Wetland Carbon and Environmental Management** *Ken W. Krauss, Zhiliang Zhu, and Camille L. Stagg (Eds.)*
268 **Distributed Acoustic Sensing in Geophysics: Methods and Applications** *Yingping Li, Martin Karrenbach, and Jonathan B. Ajo-Franklin (Eds.)*
269 **Congo Basin Hydrology, Climate, and Biogeochemistry: A Foundation for the Future (English version)** *Raphael M. Tshimanga, Guy D. Moukandi N'kaya, and Douglas Alsdorf (Eds.)*
269 **Hydrologie, climat et biogéochimie du bassin du Congo: une base pour l'avenir (version française)** *Raphael M. Tshimanga, Guy D. Moukandi N'kaya, et Douglas Alsdorf (Éditeurs)*
270 **Muography: Exploring Earth's Subsurface with Elementary Particles** *László Oláh, Hiroyuki K. M. Tanaka, and Dezsö Varga (Eds.)*
271 **Remote Sensing of Water-Related Hazards** *Ke Zhang, Yang Hong, and Amir AghaKouchak (Eds.)*
272 **Geophysical Monitoring for Geologic Carbon Storage** *Lianjie Huang (Ed.)*
273 **Isotopic Constraints on Earth System Processes** *Kenneth W. W. Sims, Kate Maher, and Daniel P. Schrag (Eds.)*
274 **Earth Observation Applications and Global Policy Frameworks** *Argyro Kavvada, Douglas Cripe, and Lawrence Friedl (Eds.)*
275 **Threats to Springs in a Changing World: Science and Policies for Protection** *Matthew J. Currell and Brian G. Katz (Eds.)*
276 **Core-Mantle Co-Evolution: An Interdisciplinary Approach** *Takashi Nakagawa, Madhusoodhan Satish-Kumar, Taku Tsuchiya, and George Helffrich (Eds.)*
277 **Compressional Tectonics: Plate Convergence to Mountain Building (Tectonic Processes: A Global View, Volume 1)** *Elizabeth J. Catlos and Ibrahim Çemen (Eds.)*
278 **Extensional Tectonics: Continental Breakup to Formation of Oceanic Basins (Tectonic Processes: A Global View, Volume 2)** *Ibrahim Çemen and Elizabeth J. Catlos (Eds.)*
279 **Strike-Slip Tectonics: Oceanic Transform Faults to Continental Plate Boundaries (Tectonic Processes: A Global View, Volume 3)** *Ibrahim Çemen and Elizabeth J. Catlos (Eds.)*
280 **Landscape Fire, Smoke, and Health: Linking Biomass Burning Emissions to Human Well-Being** *Tatiana V. Loboda, Nancy H. F. French, and Robin C. Puett (Eds.)*
281 **Clouds and Their Climatic Impacts: Radiation, Circulation, and Precipitation** *Sylvia Sullivan and Corinna Hoose (Eds.)*
282 **Fast Processes in Large-Scale Atmospheric Models: Progress, Challenges, and Opportunities** *Yangang Liu and Pavlos Kollias (Eds.)*
283 **Helicities in Geophysics, Astrophysics, and Beyond** *Kirill Kuzanyan, Nobumitsu Yokoi, Manolis K. Georgoulis, and Rodion Stepanov (Eds.)*
284 **Noisy Oceans: Monitoring Seismic and Acoustic Signals in the Marine Environment** *Gaye Bayrakci and Frauke Klingelhoefer (Eds.)*
285 **Alfvén Waves Across Heliophysics: Progress, Challenges, and Opportunities** *Andreas Keiling (Ed.)*

Geophysical Monograph 285

# Alfvén Waves Across Heliophysics

## *Progress, Challenges, and Opportunities*

**Andreas Keiling**
***Editor***

This Work is a co-publication of the American Geophysical Union and John Wiley and Sons, Inc.

AGU
ADVANCING EARTH AND SPACE SCIENCES

WILEY

This edition first published 2024


**Published under the aegis of the AGU Publications Committee**

Matthew Giampoala, Vice President, Publications
Steven A. Hauck, II, Chair, Publications Committee
For details about the American Geophysical Union, visit us at www.agu.org.

The right of Andreas Keiling to be identified as the editor of this work has been asserted in accordance with law.

*Registered Office*
John Wiley & Sons, Inc., 111 River Street, Hoboken, NJ 07030, USA

*Editorial Office*
111 River Street, Hoboken, NJ 07030, USA

For details of our global editorial offices, customer services, and more information about Wiley products, visit us at www.wiley.com.

Wiley also publishes its books in a variety of electronic formats and by print-on-demand. Some content that appears in standard print versions of this book may not be available in other formats.



*Library of Congress Cataloging-in-Publication Data Applied for*

Hardback: 9781394195954

Cover design: Wiley
Cover image: Courtesy of Tom Elsden

Set in 10/12pt TimesNewRomanMTStd by Straive, Chennai, India

# CONTENTS

# LIST OF CONTRIBUTORS

**Bertrand Bonfond**
STAR Institute
University of Liège
Liège, Belgium

**Peter A. Damiano**
Geophysical Institute
University of Alaska Fairbanks
Fairbanks, AK, USA

**Gina A. DiBraccio**
Planetary Magnetospheres Laboratory
NASA Goddard Space Flight Center
Greenbelt, MD, USA

**Tom Elsden**
Department of Mathematics and Statistics
University of St Andrews
St Andrews, UK

**Jared R. Espley**
Planetary Magnetospheres Laboratory
NASA Goddard Space Flight Center
Greenbelt, MD, USA

**Christopher M. Fowler**
Department of Physics and Astronomy
West Virginia University
Morgantown, WV, USA

**Gerhard Haerendel**
Max Planck Institute for Extraterrestrial Physics
Garching near Munich, Germany

**Jasper S. Halekas**
Department of Physics and Astronomy
University of Iowa
Iowa City, IA, USA

**Michael D. Hartinger**
Department of Earth, Planetary, and Space Sciences
University of California Los Angeles
Los Angeles, CA, USA

*and*

Space Science Institute
Boulder, CO, USA

**Jay R. Johnson**
School of Engineering
Andrews University
Berrien Springs, MI, USA

**Andreas Keiling**
Space Sciences Laboratory
University of California Berkeley
Berkeley, CA, USA

**Eun-Hwa Kim**
Princeton Plasma Physics Laboratory
Princeton University
Princeton, NJ, USA

**Robert L. Lysak**
School of Physics and Astronomy
University of Minnesota
Minneapolis, MN, USA

**Norbert Magyar**
Centre for Mathematical Plasma-Astrophysics
Department of Mathematics
Katholieke Universiteit Leuven
Leuven, Belgium

**Ian R. Mann**
Department of Physics
University of Alberta
Edmonton, Canada

**Ivan P. Pakhotin**
Department of Physics and Astronomy
University of Calgary
Calgary, Canada

**Norberto Romanelli**
Department of Astronomy
University of Maryland
College Park, MD, USA

*and*

Planetary Magnetospheres Laboratory
NASA Goddard Space Flight Center
Greenbelt, MD, USA

**Alexander J. B. Russell**
School of Mathematics and Statistics
University of St Andrews
St Andrews, UK

**James W. R. Schroeder**
Department of Physics and Engineering
Wheaton College
Wheaton, IL, USA

**David Southwood**
Department of Physics
Imperial College
London, UK

**Ali Sulaiman**
School of Physics and Astronomy
University of Minnesota
Minneapolis, MN, USA

**Kazue Takahashi**
The Johns Hopkins University Applied Physics Laboratory
Laurel, MD, USA

**Andrew N. Wright**
Department of Mathematics and Statistics
University of St Andrews
St Andrews, UK

# PREFACE

Heliophysics encompasses all kinds of physics with a focus on the interaction between the Sun and other bodies in the solar system. Given the enormous space, space scientists specialize in phenomena associated with a certain region in space, each region being associated with a specific plasma regime. Accordingly, they typically share and discuss their results in separate conferences and workshops. However, phenomena can occur across different plasma regimes, which provides opportunities to extend our understanding of the influence of different plasma parameters. There has been a growing trend to acknowledge this interdisciplinary connection in conferences and workshops.

Personally, I started 13 years ago to bring different space science communities together. The chosen format was the Chapman conference series, which are small, topically focused meetings under the umbrella of the AGU. Not initially planned, it turned into a pursuit for more. Between 2011 and 2023, together with colleagues, we organized five Chapman conferences, all of which aimed to promote interdisciplinary scientific discussions among solar/heliospheric scientists and magnetospheric/ionospheric scientists. Each conference touched upon a different phenomenon or region in space, but all can be seen as part of the bigger picture: solar–terrestrial–planetary coupling and its consequences on space weather.

In connection with each Chapman conference, we published books in AGU's Geophysical Monograph Series: *Auroral Phenomenology and Magnetospheric Processes: Earth and Other Planets* (2012, vol. 197), *Magnetotails in the Solar System* (2015, vol. 207), *Low-Frequency Waves in Space Plasmas* (2016, vol. 216), and *Electric Currents in Geospace and Beyond* (2018, vol. 235). This book, *Alfvén Waves Across Heliophysics: Progress, Challenges, and Opportunities*, is the fifth addition to this miniseries, reviewing recent advances in Alfvén wave research. It is a topic dearest to my heart, being the foundation of my research career, starting with my Ph.D. It is still a very active research field, whose beginning can easily be traced to Hannes Alfvén's pioneering research almost a century ago. Starting with observations in the laboratory, observations in geospace, other planets, and the Sun followed, which took many decades to accomplish. It has unequivocally been proven that the Alfvén wave is a universal phenomenon, and its physics and applications continue to be intensely investigated across the heliosphere.

Perhaps, it is the solar and planetary science communities that have recently produced the most significant advances in our knowledge thanks to new space missions, but it is the geospace community that has the longest history of direct observations of Alfvén waves. Thus, solar and planetary wave researchers may be able to lean on the geospace community in understanding how these waves operate and under what conditions to learn how processes may scale with environmental conditions. One way to facilitate this is by bringing them all together in one meeting. We did this in my hometown Berlin in May 2023, as another Chapman conference.

In preparation for this meeting, 11 reviews were written for this book and presented as tutorials at the conference. The reviews cover three domains, corresponding to three parts in the book: Alfvén waves in the Sun; Alfvén waves at planets (excluding Earth); and Alfvén waves in geospace (Earth). In addition, some reviews take comparative approaches across different domains, and one review looks at Alfvén waves in the laboratory setting. While the solar wind is an invaluable part of Alfvén wave research, and it was initially planned as a contribution to this book, due to unpreventable and unfortunate circumstances, the contributions on this subject were not able to be included.

I trust the readers of this book will find it well worth the read, and I hope it can serve as a reference book for today's and future generations of space scientists. Because of its interdisciplinary nature, it should appeal to a broad community of space scientists and should also be of interest to astronomers and astrophysicists studying space plasmas beyond our solar system.

This latest Chapman conference and book—as well as the four preceding—have been a great journey. Beverly, my wife with a doctorate in microbiology (an expertise that came in handy, unexpectedly, when we had to deal with an Ebola wave during one conference, and a COVID-19 wave during the planning of the most recent conference), helped with all the conferences with regard to logistics, local matters, and cultural engagement. I am grateful to her and so many others (too many to list individually) who helped in organizing the conferences. Similarly, I am grateful to my earlier book co-editors, the chapter authors, and reviewers. Importantly, I heartily thank AGU Meetings staff, AGU Publications staff, and the team at Wiley for providing support throughout. Lastly, my editorial work on this book was supported by the National Science Foundation (AGS-2016788, AGS-2116971).

**Andreas Keiling**
*University of California at Berkeley, USA*

# 1

# Introduction to Alfvén Waves Across Heliophysics: Progress, Challenges, and Opportunities

**Andreas Keiling**

## ABSTRACT

*Alfvén Waves Across Heliophysics: Progress, Challenges, and Opportunities* is an interdisciplinary collaboration among different space science communities. In the 12 chapters of this book, advances of our knowledge about Alfvén waves across heliophysics are reviewed. Here, a small selection of examples is given to illustrate the range of Alfvén waves in various space plasmas of our solar system, including a selection of relevant space missions launched in the 21st century. Because of its interdisciplinary nature, this book should appeal to a broad community of space scientists, and it should also be of interest to astronomers/astrophysicists who are studying space plasmas beyond our solar system.

Alfvén waves are fundamental to the dynamics of space plasmas. This recognition is one of the great achievements in space research, going back to the last century, when Hannes Alfvén published his groundbreaking paper in 1942, which ultimately earned him the Nobel Prize in 1970. Since entering the 21st century, great advances in Alfvén wave research have been made *across* the entire field of heliophysics, coming from several directions: new space missions (Table 1.1) to unexplored heliospheric regions, sophisticated rocket campaigns in the auroral zone, enlarged magnetometer arrays and radar networks, and significant advances in the capabilities of simulations to allow modeling many of the relevant processes. This book compiles this new understanding of Alfvén waves with contributions from different space science communities. Overall, the book is divided into three main parts, dealing separately with Alfvén waves at the Sun (Part I), the inner and outer planets (Part II, which excludes Earth), and Earth (Part III). In addition, they include chapters that take a comparative approach across plasma regimes. To whet the appetite, a small selection of examples from each part is given next. While the solar wind is an invaluable part of Alfvén wave research, and it was initially planned as a contribution to this book, due to unpreventable and unfortunate circumstances, we were not able to include the contribution on this subject.

In geospace, spacecraft missions, such as THEMIS, Reimi, and Van Allen Probes, have provided unprecedented observational links between Alfvén waves in space and auroral fine structure in the ionosphere. Recent studies found that Alfvén waves are not just a marker of dynamic processes in space plasma but also an active factor. For example, Alfvén wave activity is responsible, at least partly, for the acceleration of relativistic electrons of the radiation belt. Furthermore, Alfvén waves are important in the energization of ionospheric plasma. There is also a growing recognition of the global, hemispheric energy transport by Alfvén waves in the magnetosphere–ionosphere system. It is also important to understand the limitations of current Alfvén wave theory in explaining related phenomena in order to identify missing pieces of physics.

In addition to advances in understanding Alfvén waves in our own geospace, much knowledge has been gained about other planetary magnetospheres in the last decade.

*Space Sciences Laboratory, University of California Berkeley, Berkeley, CA, USA*

*Alfvén Waves Across Heliophysics: Progress, Challenges, and Opportunities, Geophysical Monograph 285*, First Edition.
Edited by Andreas Keiling.

DOI: 10.1002/9781394195985.ch01

**Table 1.1** A selection of spacecraft missions (launched since 2000) relevant to *current* Alfvén wave research.

| Launch | Mission |
|---|---|
| **2000** | Cluster |
| **2000** | IMAGE |
| **2000** | CHAMP |
| **2004** | MESSENGER |
| **2005** | Venus Express |
| **2005** | Reimi |
| **2007** | THEMIS |
| **2010** | Solar Dynamics Observatory |
| **2011** | Juno |
| **2012** | Van Allen Probes |
| **2013** | MAVEN |
| **2013** | Swarm |
| **2015** | MMS |
| **2016** | Arase |
| **2018** | Parker Solar Probe |
| **2018** | BepiColombo |
| **2020** | Solar Orbiter |

For example, the current Juno mission provides ample evidence of Alfvénic activity at Jupiter's polar regions, which led to the conclusion that Alfvénic activity plays a major role in auroral acceleration processes for Jupiter's aurora. In fact, Alfvén waves have been thought to play a key role in many facets, including but not limited to moon–magnetosphere interactions, auroral acceleration processes, turbulent heating in the middle magnetosphere, and more. Prior to the Juno mission, evidences for Alfvénic interactions were limited to the equatorial magnetosphere (e.g., from Galileo), but now we are starting to appreciate how the Alfvén waves are transported and dissipated at the high magnetic latitudes, so we can probe their role in magnetosphere–ionosphere coupling at Jupiter. Although Juno is revealing new and exciting results on the importance of Alfvén waves, there are many unresolved questions regarding how and where they are dissipated, the role of whistler-mode versus Alfvén-mode waves in auroral processes, and where Alfvén waves are dominant in the auroral region. The structure and dynamics of Jupiter's magnetosphere are qualitatively different than those of Earth's due to Jupiter's strong magnetic field, its rapid rotation, and internal plasma source from Io. Nonetheless, the Jovian community may be able to lean on the terrestrial community in understanding how and under what conditions these processes operate to learn how processes may scale with environmental conditions.

New solar missions, such as Solar Orbiter and Parker Solar Probe, have provided new insights into Alfvén wave research in the solar atmosphere and the nascent solar wind. In the solar corona, Alfvén waves are important, because they are now being observed regularly as decayless waves in many coronal structures. In the last decade, observational evidence for Alfvén waves in the photosphere and chromosphere has also been reported. During solar flares, Alfvén waves launched by reconnection may play a significant role in transporting flare energy through the solar atmosphere and dissipating it through heating and particle acceleration. However, there are many challenges in understanding theoretically how this would work, particularly in complex coronal magnetic fields, and in diagnosing it using only remote-sensing flare observations. Currently, new simulation results are being obtained on how the waves potentially play a role in heating the corona. Further, Alfvén waves are being utilized for seismology of the solar atmosphere, yielding magnetic field maps. It is our opinion that Alfvén wave research related to the Sun provides some of the most exciting new results.

In the remainder of this book, many more examples with greater details are given to build up a comprehensive picture, which allows contrasting similarities and differences of Alfvén waves in different plasma regions/regimes. It is hoped that this book will serve as a reference book for today's and future generations of space scientists. Because of its interdisciplinary nature, this book should appeal to a broad community of space scientists, and it should also be of interest to astronomers/astrophysicists who are studying space plasmas beyond our solar system.

## ACKNOWLEDGMENTS

A.K. acknowledges the National Science Foundation for their support (grants AGS-2016788, AGS-2116971) throughout the entire editorial process of this book. The author thanks the science program committee in helping formulate the proposal for the AGU Chapman Conference (Berlin, Germany, 2023), from which some of the text was used for this introduction chapter.

# Part I
# Sun

# 2

# Alfvén Waves in the Solar Atmosphere

**Norbert Magyar**

**ABSTRACT**

Alfvén waves are propagating disturbances of magnetic tension and are one of the fundamental waves in plasma at large scales. The study of Alfvén waves in the solar corona has a long and continuing tradition, started way before Alfvén waves were found to be ever-present in the solar atmosphere. Their study is important for mainly two reasons. First, Alfvén waves might play a key role in the long-standing chromospheric and coronal heating problems, as well as the solar-wind acceleration problem, both for transmitting the energy in photospheric convection higher up in the atmosphere and more directly as a means of cascading and dissipating this energy through various mechanisms. Second, observed properties of Alfvén waves can be used to infer properties of the plasma through seismology. In this review, an overview of the basic theoretical understanding of magnetohydrodynamic (MHD) waves is presented, in an attempt to clarify the role that Alfvén waves play in the greater class of MHD waves encountered in the solar atmosphere. Furthermore, observational and theoretical findings on Alfvén waves in the solar atmosphere of the previous decade are reviewed, followed by a short summary of these results and still open questions.

## 2.1. INTRODUCTION

First derived by Alfvén (1942) as electromagnetic-hydrodynamic waves, confirmed experimentally in liquid mercury by Lundquist (1949), and needing the approval of such authorities in physics as Fermi to be taken seriously after years of disapproval by other world-renowned scientists (Dessler, 1970; Russell, 2018), the waves now bearing Hannes Alfvén's name were from their conception central to a better understanding of solar atmospheric processes. Alfvén himself first realized that such waves should in principle exist, through deep physical intuition, while thinking about the nature of sunspots (Alfvén & Lindblad, 1945), known to be strongly magnetized since Hale (1908). Other phenomena in which Alfvén waves might play a substantial role have been proposed shortly after, among the firsts being the then recently discovered coronal heating (Alfvén & Lindblad, 1947) and the existence of cosmic rays (Fermi, 1949). This interdisciplinary monograph, dedicated to reviewing recent advances in understanding Alfvén waves in the wide field of space physics, of which this chapter is part of, stands as proof of the everlasting impact that H. Alfvén made to the field, even 80 years on.

Although in his pioneering work Alfvén only derived a wave driven purely by the magnetic tension force, it had become increasingly crystallized in the 1950s that electroconducting fluids or plasmas whose dynamics can be described through the magnetohydrodynamic (MHD) equations admit various wave modes. This is caused by the interplay between various restoring forces, mostly magnetic and gas pressure forces. The different wave modes are clearly distinguishable under ideal and homogeneous conditions, leading to their usual identification as

*Centre for Mathematical Plasma-Astrophysics, Department of Mathematics, Katholieke Universiteit Leuven, Leuven, Belgium*

*Alfvén Waves Across Heliophysics: Progress, Challenges, and Opportunities, Geophysical Monograph 285*, First Edition.
Edited by Andreas Keiling.

DOI: 10.1002/9781394195985.ch02

Alfvén, fast, and slow waves. Without entering too much into detail here, and instead allowing for an extended discussion on this subject in a dedicated section of this chapter, it is already important to point out the usually blurred lines between these adjectives, the properties assigned to these adjectives based on linear analysis under homogeneous conditions, and the actual wave properties, especially when accounting for inhomogeneities or other nonideal conditions present. It is important to stress that waves identified as "Alfvén" waves, either in laboratory experiments, observations, or numerical studies, would seldom share only the properties of a pure linear Alfvén wave encountered in an ideal and homogeneous plasma. Therefore, while going along with the usual broad interpretation of what an Alfvén wave is when selecting the studies to be included in this review, let us always keep in mind that pure Alfvén waves are a very special case of a plasma wave, probably seldom encountered in their pure form in nature, as will be argued later on.

Alfvén waves play multiple important roles in our quest for a better understanding of the physics of the solar atmosphere. First, they potentially act as means of transmitting the energy available in the convective buffeting of the photosphere, thought as the ultimate energy source for heating the corona and accelerating the solar wind, along magnetic field lines reaching into the solar atmosphere. Once inside the chromosphere or corona, Alfvén wave energy could be dissipated by various proposed mechanisms. Dissipation in the almost ideal conditions of the solar plasma presupposes that small enough scales are generated (large spatial gradients in Alfvén wave perturbation field), at which thermalization is supposedly taken over by kinetic processes. These small-scale generation mechanisms range from linear processes such as phase mixing and resonant absorption to nonlinear shock dissipation and generation of Alfvén wave turbulence. Second, Alfvén waves could be the causing factor of various phenomena observed in the solar atmosphere, such as in increasing height from the photosphere, photospheric vortices, umbral and penumbral shocks, spicules, coronal jets, switchbacks, and so on, some of which phenomena will be discussed in detail later on. Last but not least, through the seismology of the observed waves, where the correct identification of the waves as outlined earlier is particularly important, physical properties of the local plasma through which the Alfvén waves are propagating could be inferred by comparing measured oscillation properties with the ones derived from MHD wave theory. These oscillation properties often depend on various properties of the plasma, such as magnetic field intensity, mass density, gravitational scale height, polytropic index, to name a few.

This chapter is dedicated to reviewing advances in the research and application of Alfvén waves in the solar atmosphere published in the last decade. According to the Smithsonian Astrophysical Observatory/National Aeronautics and Space Administration (SAO/NASA) Astrophysics Data System, there have been roughly 500 peer-reviewed publications since 2012, uniformly distributed through the years, on the subject. The search criteria were based on words found in the abstract, specifically the combination of "Alfvén," "wave," "solar," and "photosphere" or "chromosphere" or "corona." The study of Alfvén waves in the extended solar atmosphere and inner heliosphere, such as the outer corona and solar wind, has a separate chapter dedicated to it within this monograph. It is also important to note that this chapter only considers plasma for which the fluid approximation is valid; that is, for which MHD is a good approximation, and does not touch upon kinetic dynamics, such as kinetic Alfvén waves (e.g., Chen et al., 2021). At the spatial and temporal scales that waves are observed in the solar atmosphere, the fluid approximation holds to a very good degree. Besides giving my perspective on the current state of research on Alfvén waves in the solar atmosphere and guiding the reader through a summary of these advances, the task is to present the most important findings in a hopefully comprehensive manner in the following text.

Many helpful reviews of Alfvén waves in the solar atmosphere have been published in the preceding years. These discuss in detail some of the milestone observational, numerical, and theoretical findings of the previous decades, such as the theoretical breakthroughs of phase mixing, mode conversion, and resonant absorption (Goossens et al., 2011; Roberts, 2000), Alfvén wave turbulence (Schekochihin, 2022), the observational firsts of directly imaged large-amplitude transverse waves in coronal loops (Aschwanden et al., 1999; Nakariakov et al., 1999), the ubiquitous propagating waves first observed using the Coronal Multi-Channel Polarimeter (CoMP, Mathioudakis et al., 2012; Tomczyk et al., 2007), torsional Alfvén waves in the chromosphere (Jess et al., 2009, 2012), and advances in the field of coronal seismology (De Moortel & Nakariakov, 2012), among other related areas, such as decayless kink waves and the effects of thermal misbalance (Nakariakov et al., 2016; Nakariakov & Kolotkov, 2020). Additionally, the dynamics of Alfvén waves have been a part of several textbooks on MHD and plasma dynamics (Aschwanden, 2005; Goedbloed & Poedts, 2004; Priest, 2014; Somov, 2006), with even a dedicated book on the subject published decades ago but still remaining relevant (Hasegawa & Uberoi, 1982).

The first section of this review deals with the often overlooked but particularly important aspects of nomenclature and semantics of MHD waves, a problem already mentioned above, with some theoretical background to support the narrative. The second section is dedicated to recent observational identifications of Alfvén waves in the solar atmosphere, categorized according to the different atmospheric layers. In the third section, recent theoretical and numerical advances, particularly focusing on the ones applied to solar atmospheric physics, are reviewed. Finally, I conclude this review by summarizing some of the key recent findings and offering some future perspectives on further advancing our understanding of Alfvén waves and implications for solar atmospheric dynamics.

## 2.2. ALFVÉN, ALFVÉNIC, TRANSVERSE, KINK, FAST KINK? NOMEN EST OMEN

There is a great deal of disagreement within the solar physics community about the usage of specific adjectives to designate waves, with multiple adjectives in use to refer even to the same observation or numerical result, with examples in the title of this section. This is particularly true for waves that manifest as transverse oscillations or displacements of magnetic field lines, with their group speed vector aligned to the local mean magnetic field. As Alfvén waves are in fact transverse displacements of magnetic field lines (see Figure 2.1), it is important to dive deeper into this problem.

Let us recollect here the timeline of a still ongoing debate on the nature of the CoMP-like omnipresent propagating transverse waves in the solar corona. "Transverse" is the least controversial, if not descriptive, and most widely accepted way to refer to these waves. First reported by Tomczyk et al. (2007), these waves were referred to by the authors as Alfvén waves. This was quickly dismissed by Van Doorsselaere et al. (2008), stating that the waves seen by CoMP were instead fast kink waves of flux tubes, the propagating counterpart of standing waves in coronal loops first observed and identified as fast kink waves by Aschwanden et al. (1999) and Nakariakov et al. (1999). Under their (critical) assumption that the observed waves are propagating on flux tubes, Van Doorsselaere et al. (2008) stated that Alfvén waves are unable to transversely displace these flux tubes. Waves in a flux tube or magnetic cylinder were first investigated theoretically by Zaitsev and Stepanov (1975) and Edwin and Roberts (1983). Based on their analysis, fast kink waves are the only wave solutions that, in a low beta plasma as the solar corona, have group speeds close to the average Alfvén speed and could cause transverse displacements of the tube axis. However, the story does not stop there, as Goossens et al. (2009) first pointed out that "fast" is not the correct adjective to characterize kink waves and suggested using "Alfvénic" instead. Nevertheless, the authors still cautioned that waves propagating on an inhomogeneity or nonuniformity, such as kink waves on a flux tube, should not be characterized based on the distinct categories found in a homogeneous plasma, as waves in inhomogeneous plasma usually have mixed properties. That is, waves could display Alfvén, fast, or slow properties to varying degrees, depending on the

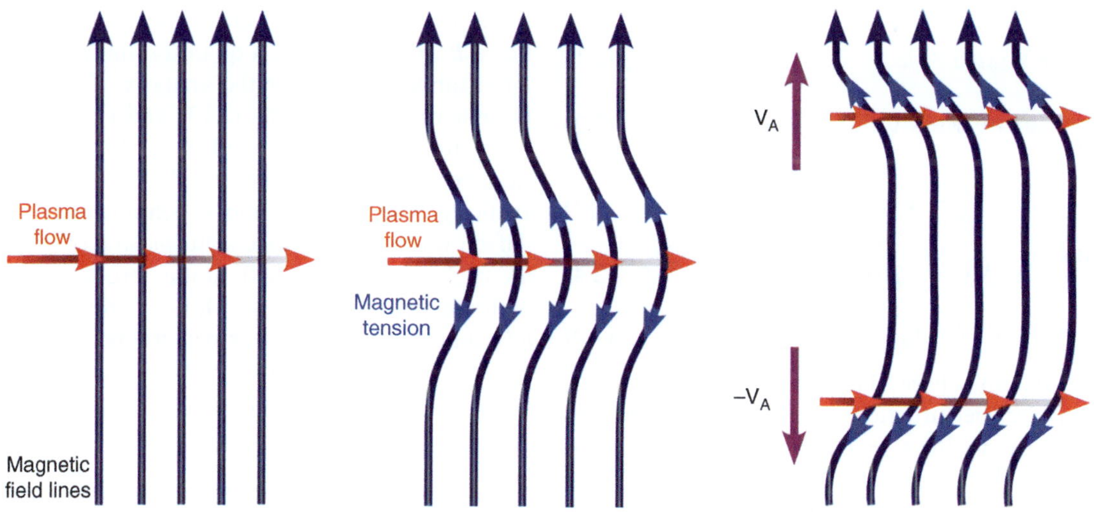


**Figure 2.1** Schematic representation of the perturbation of straight and uniform magnetic field lines, represented by parallel arrows, by a plasma flow transverse to the magnetic field lines, with this initial condition shown in the image on the left. Considering an ideal plasma, magnetic field lines are frozen-in; therefore, the plasma flow drags the magnetic field lines, building up magnetic tension through the bending of the field lines as in the image in the center. This combined velocity and magnetic field perturbation splits into two propagating perturbations of magnetic tension and associated velocity perturbations, representing parallel and antiparallel propagating Alfvén waves propagating at the Alfvén speed $V_A$.

local inhomogeneity through which they propagate. This important fact will be expanded upon later in this section. Goossens et al. (2012) doubled down on the problem of characterizing the transverse displacement of flux tubes. They suggested that kink waves, specifically the fundamental radial mode that survives even in an incompressible plasma under the conditions in which the fast waves vanish, behave like surface Alfvén waves, restricting the usage of the "fast" adjective to radial overtones, which require the plasma to be compressible. The term surface Alfvén waves has previously been used to refer to incompressible waves existing on a Cartesian plasma discontinuity, whose amplitude decreases exponentially with distance from the discontinuity (e.g., Roberts, 1981). Spurred by the diversity of proposed terms, it is not surprising that there are various preferred ways to refer to observed transverse waves in the solar atmosphere. Still, it is important to correctly characterize these waves, as terms used to describe them imply different properties, with crucial implications for their dynamics, as we will see in the following text, where the theoretical background of the aforementioned concepts will be gradually built up.

### 2.2.1. Waves in a Homogeneous Plasma of Infinite Extent: Alfvén, Fast, and Slow

Let us briefly present the clear distinction that can be made between different linear wave modes in a homogeneous plasma. For this, I largely follow the analysis presented in Goossens et al. (2012, 2019). We start by listing the equations of ideal MHD:

$$\frac{\partial \rho}{\partial t} + \nabla \cdot (\rho \boldsymbol{v}) = 0, \tag{2.1}$$

$$\rho \frac{\partial \boldsymbol{v}}{\partial t} + \rho(\boldsymbol{v} \cdot \nabla)\boldsymbol{v} = -\nabla p + \frac{1}{\mu_0}(\nabla \times \boldsymbol{B}) \times \boldsymbol{B}, \tag{2.2}$$

$$\frac{\partial \boldsymbol{B}}{\partial t} = \nabla \times (\boldsymbol{v} \times \boldsymbol{B}), \tag{2.3}$$

$$\frac{\partial p}{\partial t} + (\boldsymbol{v} \cdot \nabla)p = -\gamma p \nabla \cdot \boldsymbol{v}, \tag{2.4}$$

$$\nabla \cdot \boldsymbol{B} = 0, \tag{2.5}$$

where the symbols have their usual meanings. Next, let us consider perturbations around some homogeneous equilibrium with $\boldsymbol{v}_0 = 0; \boldsymbol{B}_0 = B_0\hat{z}; \rho_0; p_0$. Motions atop this equilibrium displace plasma elements from $\boldsymbol{r} \rightarrow \boldsymbol{r} + \boldsymbol{\xi}$, where $\boldsymbol{\xi}$ is the Lagrangian displacement, $\boldsymbol{v} = d\boldsymbol{\xi}/dt$. Linear perturbations, denoted with a prime, of Equations (2.1)–(2.5), expressed in terms of the linearized Lagrangian displacement $\boldsymbol{v} = \partial\boldsymbol{\xi}/\partial t$, read:

$$\rho' = -\rho_0 \nabla \cdot \boldsymbol{\xi}, \tag{2.6}$$

$$\rho_0 \frac{\partial^2 \boldsymbol{\xi}}{\partial t^2} = -\nabla p' + \frac{1}{\mu_0}(\nabla \times \boldsymbol{B}') \times \boldsymbol{B}_0, \tag{2.7}$$

$$\boldsymbol{B}' = \nabla \times (\boldsymbol{\xi} \times \boldsymbol{B}_0), \tag{2.8}$$

$$p' = -\gamma p_0 \nabla \cdot \boldsymbol{\xi}. \tag{2.9}$$

It is clear that the equation of motion can be written only in terms of $\boldsymbol{\xi}$. Looking for wavelike solutions, we assume that the displacement has the form $\boldsymbol{\xi} = \hat{\boldsymbol{\xi}} \exp(i(\boldsymbol{k} \cdot \boldsymbol{r} - \omega t))$. Next, a new set of variables is defined for a clear demonstration of distinct wave properties:

$$X \equiv k_z \hat{\xi}_z \text{ (parallel displacement)}, \tag{2.10}$$

$$iY \equiv \nabla \cdot \boldsymbol{\xi} = i\,\boldsymbol{k} \cdot \hat{\boldsymbol{\xi}} \text{ (compression)}, \tag{2.11}$$

$$iZ \equiv (\nabla \times \boldsymbol{\xi})_z = i(\boldsymbol{k} \times \hat{\boldsymbol{\xi}})_z \text{ (parallel vorticity)}. \tag{2.12}$$

Then, the equations governing wavelike solutions admitted by Equations (2.6)–(2.9) expressed in terms of the new variables are:

$$\omega^2 X - k_z^2 v_S^2 Y = 0, \tag{2.13}$$

$$k^2 v_A^2 X + \left(\omega^2 - k^2\left(v_A^2 + v_S^2\right)\right) Y = 0, \tag{2.14}$$

$$\left(\omega^2 - \omega_A^2\right) Z = 0, \tag{2.15}$$

where $v_s = \sqrt{\gamma p_0/\rho_0}$ is the sound speed, $v_A = B_0/\sqrt{\mu_0 \rho_0}$ is the Alfvén speed, and $\omega_A = k_z v_A$ is the Alfvén frequency. We obtained two sets of uncoupled equations. Equation (2.15) describes pure Alfvén waves, which are exclusively and completely characterized by parallel vorticity[1]. They cause no compression ($Y = 0$), and do not perturb the parallel component of velocity ($X = 0$), or equivalently, they do not cause total pressure variations, $P' = p' + B'^2/2\mu_0$, a quantity that will be of crucial importance later on for the analysis of inhomogeneous plasma waves. Equations (2.13) and (2.14) describe the magnetosonic waves, fast and slow, which cause compression, perturb total pressure, but do not propagate parallel vorticity. The analysis above demonstrates that linear Alfvén waves in a homogeneous plasma of infinite extent are in their own class, with unique properties not shared with magnetosonic waves. It follows from this that given any superposition of linear perturbations atop a homogeneous plasma, the subset of Alfvén waves is always completely identifiable and quantifiable. The appearance

1 There are some special cases of Alfvén waves carrying no parallel vorticity. For example, $\boldsymbol{k} = k_z\hat{z}$, that is, Alfvén waves with no amplitude modulation ($k_x = 0$, $k_y = 0$) perpendicular to the energy propagation direction or group speed vector ($\hat{z}$ direction), which is the case first described by Alfvén (1942). Some other special cases of pure Alfvén waves carrying no parallel vorticity are also possible; for example, in cylindrical geometry when the radial dependence of $m = 0$ Alfvén wave amplitudes on separate magnetic isosurfaces is such that it results in an irrotational velocity perturbation in the plane perpendicular to the magnetic field along $\hat{z}$.

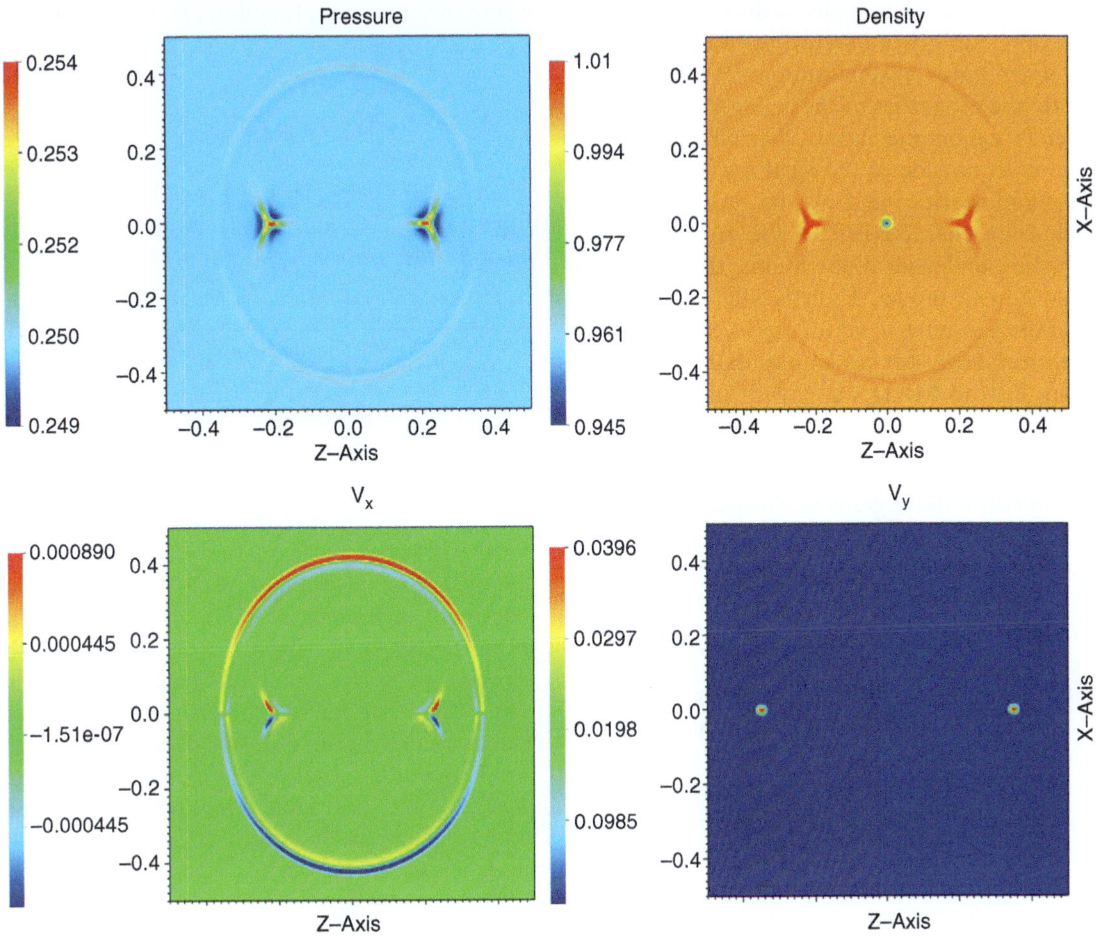


**Figure 2.2** Results of a 2.5D (two spatial dimensions, *x* and *z*, and three vector components) MHD numerical simulation with a straight and uniform magnetic field along the *z* axis, showing pressure, density, and two components of plasma velocity. The initial condition is a Gaussian perturbation in pressure and out-of-plane velocity ($V_y$) at $(x,z) = (0,0)$. The Alfvén waves propagating parallel and antiparallel to the magnetic field appear only in out-of-plane velocity. The isotropic waves are the fast waves, while the three-lobed perturbations lagging behind are the slow waves. The remnant density perturbation at the position of the initial pulse is the entropy wave, of zero frequency and phase speed. The value of plasma-β is 0.5. Units are arbitrary.

of fast, slow, and Alfvén waves as a result of a localized perturbation in a homogeneous plasma with a straight and uniform magnetic field can be seen in Figure 2.2.

It must be emphasized again that the result above is valid only for linear perturbations. Once nonlinear terms of these perturbations are taken into account, couplings between the $X$, $Y$, and $Z$ variables appear, allowing for wavelike solutions without linear counterparts, and, more importantly, for various couplings between Alfvén, fast, and slow modes. A relevant example of such a coupling in this case is the nonlinear Alfvén wave. While in an incompressible plasma ($Y = 0$) nonlinear Alfvén waves are exact solutions propagating without suffering distortion, in compressible plasma these drive, through ponderomotive forces, parallel displacement, and pressure and density perturbations. The coupling strength is proportional to the magnetic pressure gradient, $F_p \sim B_0^2 k_z^2 \nabla \xi^2$. The finite amplitude Alfvén wave induces flows along the magnetic field direction only for noncircular polarization, that is, $z$-dependent total magnetic field perturbation. However, not even circularly polarized Alfvén waves in a homogeneous compressible plasma are stable: these can get distorted and decay over time due to the parametric decay instability (PDI, Galeev & Oraevskii, 1963; Goldstein, 1978), generating reflection and slow wave perturbations. PDI appears due to any preexisting perturbation or noise of the background plasma, already hinting at the impact of plasma nonuniformity on the coupling of wave modes and spoiling of distinct wave types, the subject of the next subsection.

Given the observations above, let us draw some conclusions on the observability of pure Alfvén waves in nature. Obviously, if by pure Alfvén waves one means waves having strictly the properties of the linear Alfvén wave in a homogeneous and infinite plasma, then almost certainly these do not exist outside of the idealized conditions of our model. Nonlinearities and plasma noise couple the dynamics that are characteristic of the separate classes of waves existing under ideal conditions, that of Alfvén and magnetoacoustic modes. Therefore the term "Alfvén wave," when used to refer to waves observed in the solar atmosphere, necessarily carries an approximative meaning, denoting instead MHD waves with properties that are dominantly those of pure Alfvén waves.

### 2.2.2. Waves in an Inhomogeneous Plasma

***Inhomogeneity Perpendicular to the Magnetic Field***

The solar atmosphere appears inhomogeneous and highly structured. This structuring, either of the chromosphere or corona, referred to as fibrils, filaments, strands, or threads, largely outlines the local mean magnetic field, and is identifiable down to the resolution limit of observing instruments (Aschwanden & Peter, 2017; Williams et al., 2020). The highly filamentary appearance of the corona revealed by the Sun-grazing comet Lovejoy (Raymond et al., 2014) within hundreds of Megameters from the solar surface is yet another proof of the permeating inhomogeneity. The outer corona is also observed to be highly nonuniform, with a "wood-grain" structure as revealed through processed coronagraph images (DeForest et al., 2018).

Motivated by these observations, let us continue with the linear harmonic analysis of inhomogeneous plasma. We switch to cylindrical coordinates, motivated by the popularity of the flux tube model, and allow for inhomogeneity only in the radial direction, $\boldsymbol{B}_0 = B_0(r)\hat{z}; p_0(r); \rho_0(r)$, with the condition that total pressure is constant throughout the domain, $p_0(r) + \frac{B_0^2(r)}{2\mu_0} = const.$ We seek solutions of Equations (2.1)–(2.5) that satisfy harmonic Lagrangian displacements of the form $\xi = \hat{\xi}(r)\exp(i(m\varphi + k_z z - \omega t))$. The azimuthal wavenumber is an integer $m \geq 0$, such that the solution is periodic in $\hat{\varphi}$. As shown by Appert et al. (1974), Sakurai et al. (1991), and Goossens et al. (1992, 1995), the resulting linear MHD waves can be described by equations for Eulerian total pressure perturbations $P'$, and the radial component of displacement $\hat{\xi}_r$, with the hat dropped:

$$D\frac{d(r\xi_r)}{dr} = -C_2 r P', \tag{2.16}$$

$$\frac{dP'}{dr} = \rho_0\left(\omega^2 - \omega_A^2\right)\xi_r, \tag{2.17}$$

where the coefficient functions $D$, $C_2$ are given by

$$D = \rho_0\left(v_S^2 + v_A^2\right)\left(\omega^2 - \omega_A^2\right)\left(\omega^2 - \omega_C^2\right), \tag{2.18}$$

$$C_2 = \omega^4 - \left(v_S^2 + v_A^2\right)\left(\omega^2 - \omega_C^2\right)\left(\frac{m^2}{r^2} + k_z^2\right), \tag{2.19}$$

where $\omega_C^2 = v_S^2\omega_A^2/\left(v_S^2 + v_A^2\right)$ is called the cusp frequency. Equations (2.16) and (2.17) can also be rewritten as a second-order ordinary differential equation for $P'$ as:

$$\frac{\rho\left(\omega^2 - \omega_A^2\right)}{r}\frac{d}{dr}\left[\frac{r}{\rho\left(\omega^2 - \omega_A^2\right)}\frac{dP'}{dr}\right] = \left[\frac{m^2}{r^2} - \Gamma(\omega^2)\right]P', \tag{2.20}$$

where $\Gamma(\omega^2)$ is given by

$$\Gamma(\omega^2) = \frac{\left(\omega^2 - k_z^2 v_S^2\right)\left(\omega^2 - \omega_A^2\right)}{\left(v_S^2 + v_A^2\right)\left(\omega^2 - \omega_C^2\right)}. \tag{2.21}$$

As we will see more clearly in the following text, the linear coupling of total pressure perturbations and radial displacement in Equations (2.16) and (2.17) is a harbinger of the mixed Alfvén–magnetosonic properties that waves generally possess in an inhomogeneous plasma, as already noted by Hasegawa and Uberoi (1982). To better illustrate why that is the case, let us investigate the connection between the total pressure, radial displacement, and the $X$, $Y$, and $Z$ variables employed in the homogeneous analysis. These were derived by Sakurai et al. (1991) and Goossens et al. (2019):

$$\rho_0\left(\omega^2 - \omega_A^2\right)\xi_\varphi = i\frac{m}{r}P', \tag{2.22}$$

$$\rho_0\left(\omega^2 - \omega_A^2\right)X = i\,k_z^2\frac{v_S^2}{v_S^2 + v_A^2}P', \tag{2.23}$$

$$\rho_0\left(\omega^2 - \omega_C^2\right)Y = -i\frac{\omega^2}{v_S^2 + v_A^2}P', \tag{2.24}$$

$$Z = i\left(\frac{\xi_\varphi}{r} - \frac{d\xi_\varphi}{dr}\right) - \frac{m}{r}\xi_r. \tag{2.25}$$

The system above shows that, in general, all wave variables that for a homogeneous plasma represented the separate classes of Alfvén and magnetosonic waves are coupled by the total pressure and radial displacement, which in turn are coupled on their own. This is the theoretical basis for the statement made by Goossens et al. (2019) that, in general, MHD waves in an inhomogeneous plasma present mixed properties, and cannot be clearly divided into Alfvén, fast, and slow modes any longer, already in the linear regime. In fact, the coupling between compression and parallel vorticity, the signatures of magnetosonic and Alfvén waves, respectively, can be written in an elegant way, for $m \neq 0$ or non-axisymmetric modes:

$$Y \equiv -i\nabla\cdot\xi = N_C P'; \quad Z \equiv -i(\nabla\times\xi)_z = mN_V P', \tag{2.26}$$

with

$$N_C = \frac{i\omega^2}{\rho_0\left(v_S^2 + v_A^2\right)\left(\omega^2 - \omega_C^2\right)};$$
$$N_V = -\frac{1}{r\left\{\rho_0\left(\omega^2 - \omega_A^2\right)\right\}^2}\frac{d}{dr}\left\{\rho_0\left(\omega^2 - \omega_A^2\right)\right\}. \quad (2.27)$$

These relations demonstrate that vortical motions of waves generate compressional motions and vice versa, and that a wave can induce varying amounts of vorticity and compression, depending on the local degree of inhomogeneity. Thus, waves not only have mixed Alfvén–magnetosonic properties in an inhomogeneous plasma, but these properties vary spatially and temporally with the local inhomogeneity, even within a single wave packet. An elementary demonstration of mixed properties is illustrated in Figure 2.3, adapted from Magyar et al. (2019a).

Exceptions to the full coupling of wave variables do exist, and at least in the linear regime pure Alfvén waves can still be found in the equilibrium above with radial inhomogeneity. For example, when $m=0$, denoting axisymmetric perturbations, Equation (2.22) decouples from the rest of the equations and yields as solutions Alfvén waves that represent torsional oscillations on separate Alfvén speed isosurfaces. Moreover, Ruderman and Petrukhin (2022) have shown recently that pure Alfvén waves can exist even in generally inhomogeneous plasma configurations that obey some properties. They consider a straight but perpendicularly varying magnetic field $B_0(x,y)\hat{z}$. Another criterion is that the Alfvén speed has to be separable into a product of functions for perpendicular and $\hat{z}$ variations: $V_A = V_\perp(x,y)V_\parallel(z)$. They show that pure Alfvén waves exist in this equilibrium, with polarization tangent everywhere to the level lines of $V_\perp(x,y)$, if the magnetic field magnitude is a function of $V_\perp(x,y)$. Indications that Alfvén waves exist in arbitrary inhomogeneous plasma configurations can be found in numerical simulations of resonant absorption. For example, Terradas et al. (2008b) and Pascoe et al. (2011) have shown that Alfvén waves still appear in resonant layers of transversely oscillating magnetic flux tubes with an inhomogeneous internal structure consisting of strands of different widths and densities. Transverse waves that displace magnetic flux tubes as a whole are the focus of Section "Surface and Body Waves of Magnetic Flux Tubes".

### *Inhomogeneity along the Magnetic Field*

In the analysis above, we have considered a plasma with arbitrary inhomogeneity perpendicular to the straight background magnetic field, while keeping the plasma homogeneous along the magnetic field. In the converse case, a plasma homogeneous perpendicularly to the magnetic field but inhomogeneous along it is no less important. Inhomogeneity along the magnetic field is generally present in the solar atmosphere, due to gravitational stratification of the plasma and magnetic field

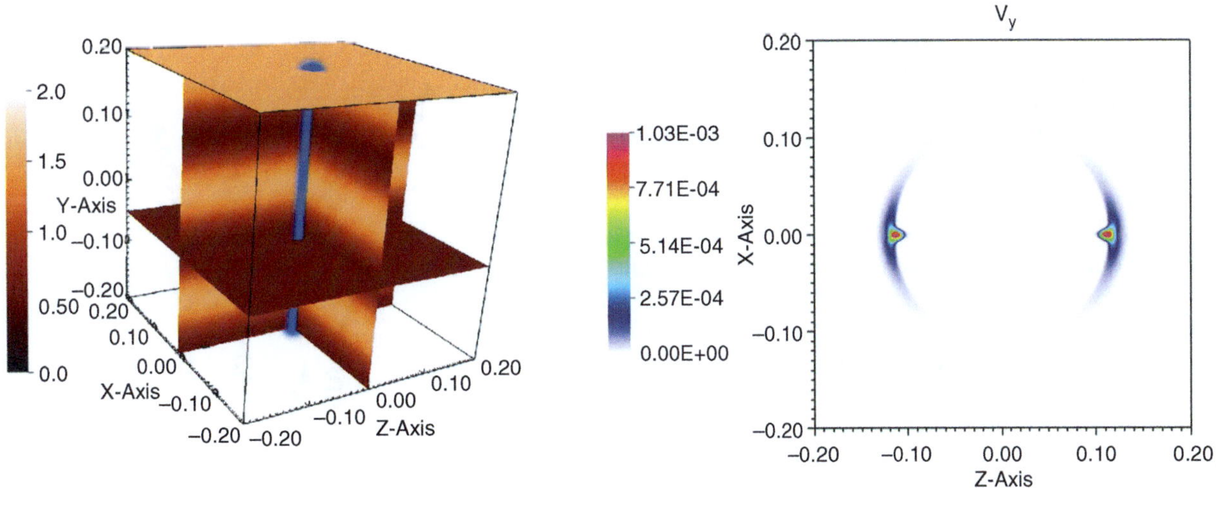


**Figure 2.3** Replicating the 2.5D MHD simulation shown in Figure 2.2 in a full 3D setup, obtained by extruding the 2.5D setup along the out-of-plane direction, along which a sinusoidal variation of background density is induced. The obtained initial condition is shown in the image on the left, which shows density in various slices along the three directions and the perturbation in $V_y$ centered around $(x,z) = (0,0)$ and constant along the $y$ direction. The image on the right shows the out-of-plane component of the velocity of a slice at some $y$ value, after some time. Comparing this picture to the $V_y$ in Figure 2.2, where it uniquely identified Alfvén waves, it appears that $V_y$ in the inhomogeneous 3D setup displays mixed Alfvén–fast charactestics, with isotropic "wings" appearing around a stronger "Alfvén" core perturbation. Units are arbitrary. Magyar et al. (2019a) / IOP Publishing.

expansion, among other sources. As we will see shortly, pure Alfvén waves do exist in this case, albeit Alfvén waves propagating parallel and antiparallel to the magnetic field become linearly coupled due to the inhomogeneity, that is, they undergo continuous reflection (Heinemann & Olbert, 1980; Hollweg, 1990; Hollweg & Isenberg, 2007). The reflection of Alfvén waves and their subsequent nonlinear interaction through counterpropagating wave collisions are central to models of turbulent coronal heating and solar-wind acceleration (e.g., Chandran & Perez, 2019; Dmitruk et al., 2001, 2002; Matthaeus et al., 1999; van Ballegooijen et al., 2011; Perez & Chandran, 2013; van Ballegooijen & Asgari-Targhi, 2016, 2017; Woolsey & Cranmer, 2014). To clearly illustrate the linear coupling of Alfvén waves due to the inhomogeneity, there is a better choice of variables than the Lagrangian displacement used previously. These are the Elsässer (1950) variables, defined as $z^{\pm} = \boldsymbol{v} \pm \boldsymbol{B}/\sqrt{\mu_0 \rho}$. As we are limiting this presentation to linear Alfvén waves, it is sufficient to consider the incompressible limit of Equations (2.1)–(2.5), which is obtained by taking $\nabla \cdot \boldsymbol{v} = 0$ and dropping Equation (2.4), as the gas pressure is now given by the divergence of Equation (2.2). Then one can rewrite the full incompressible MHD system using the Elsässer variables (Magyar et al., 2019b):

$$\frac{\partial z^{\pm}}{\partial t} + z^{\mp} \cdot \nabla z^{\pm} = -\frac{1}{\rho}\nabla P - \frac{1}{4}(z^{+} - z^{-})\nabla \cdot (z^{+} - z^{-}), \tag{2.28}$$

where $P$ is the total pressure. Note that there are two equations, for $z^{+}$ and $z^{-}$. For a homogeneous plasma, taking either $z^{+} = 0$; $z^{-} \neq 0$ or $z^{-} = 0$; $z^{+} \neq 0$ in the abovementioned equation, we find that Elsässer variables represent exact nonlinear Alfvén wave solutions of the incompressible MHD equations, propagating either parallel or antiparallel to the background magnetic field, respectively.

Proceeding to analyze the inhomogeneous setup, we linearize Equation (2.28) by considering small Eulerian perturbations of the Elsässer variables $z^{\pm} = z_0^{\pm} + \delta z^{\pm}$, and take a straight and uniform magnetic field $\boldsymbol{B}_0 = B_0 \hat{z}$ and $\rho_0(z)$, thus a variable Alfvén speed $V_{A0}(z)$ along the magnetic field. The resulting linearized equations for perturbations with dropped $\delta$ read:

$$\frac{\partial z^{\pm}}{\partial t} \mp V_{A0}(z)\frac{\partial z^{\pm}}{\partial z} \pm z_{\parallel}^{\mp}\frac{\partial V_{A0}(z)}{\partial z}\hat{z} = -\frac{1}{\rho_0(z)}\nabla P - \frac{1}{2}(z^{+} - z^{-})\frac{\partial V_{A0}(z)}{\partial z} - \frac{1}{2}V_{A0}(z)\frac{\partial (z_{\parallel}^{+} - z_{\parallel}^{-})}{\partial z}\hat{z}, \tag{2.29}$$

where $z_{\parallel}^{\pm}$ denotes the component along $\hat{z}$ of the perturbed Elsässer variable. Note that the parallel component is uncoupled, that is, it has no source terms driven by the perpendicular component. If one considers only pure Alfvén waves initially, these only perturb the perpendicular component and cause no gas or magnetic pressure perturbations. The equations for the perpendicular component are then:

$$\frac{\partial z_{\perp}^{\pm}}{\partial t} \mp V_{A0}(z)\frac{\partial z_{\perp}^{\pm}}{\partial z} = -\frac{1}{2}\left(z_{\perp}^{+} - z_{\perp}^{-}\right)\frac{\partial V_{A0}(z)}{\partial z}, \tag{2.30}$$

where the linear coupling of parallel and antiparallel propagating Alfvén waves to the magnetic field is proportional to the gradient of the Alfvén speed along the magnetic field. Thus, regions with large such gradient or small scale height compared to the wavelength of the Alfvén waves, as in the transition region, are particularly reflective to Alfvén waves. The transmissivity of Alfvén waves through the solar atmosphere has been studied in detail recently, in even more complex equilibria, and including various partial ionization effects of the chromosphere, something that we will focus on in Section 2.3. An illustration of Alfvén waves reflecting during their propagation through a coronal hole is shown in Figure 2.4.

***Surface and Body Waves of Magnetic Flux Tubes***

In the cylindrically symmetric model with a general inhomogeneity along the $r$ direction, we have earlier established the existence of waves with mixed properties, which in general propagate both parallel vorticity

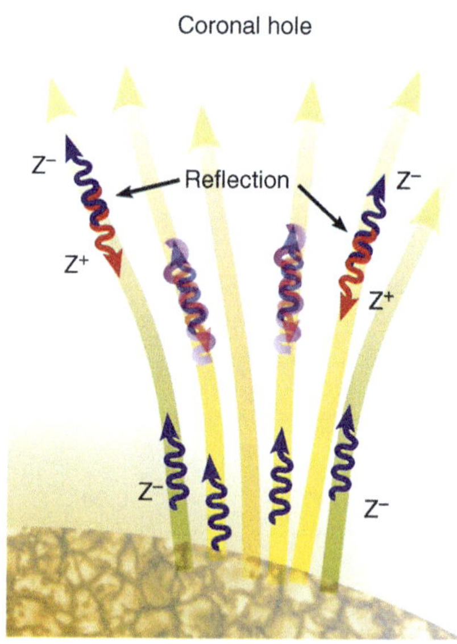


**Figure 2.4** Illustration representing a coronal hole with Alfvén waves driven at the base of the coronal hole, described by $Z^{-}$, propagating along magnetic field lines. Due to gravitational stratification and magnetic field line expansion, the Alfvén speed along these magnetic field lines is varying, leading to linear coupling as per Equation (2.30), generating reflected Alfvén waves, described by $Z^{+}$.

and compression, and perturb total pressure and radial displacement linearly. However, we have not yet investigated the types of wave modes that can exist in specific equilibria. The simplest choice is an interface at some radius $r = R$, at which the equilibrium quantities vary in a piecewise manner, defining a flux tube. Let us define the flux tube by a jump in background density:

$$\rho_0(r) = \begin{cases} \rho_{0i}, & \text{if } r \leq R, \\ \rho_{0e}, & \text{if } r > R, \end{cases} \tag{2.31}$$

where $R$ is the radius of the flux tube and the subscripts $i$, $e$ denote the interior and exterior of the flux tube, respectively.

The flux tube model is central to modeling waves observed in solar atmospheric structures such as coronal loops, plumes, arcades, etc., and has been studied by Zaitsev and Stepanov (1975), Edwin and Roberts (1983), and Goossens et al. (2012). Coronal structures, appearing as closed arcs in the case of coronal loops and elongated open-ended rays in the case of coronal plumes, are thought to outline the local magnetic field and appear brighter than the ambient plasma. Rewriting Equation (2.20) for this specific equilibrium, we obtain

$$\frac{1}{r}\frac{d}{dr}\left(r\frac{dP'}{dr}\right) = \left[\frac{m^2}{r^2} - \Gamma(\omega^2)\right] P', \tag{2.32}$$

with similar homogeneous equations satisfied both outside and inside $R$. Depending on the sign of the constant $\Gamma(\omega^2)$ outside and inside the flux tube, the solutions to the Bessel equation are either evanescent ($\sim e^{-kr}$, $\Gamma(\omega^2) < 0$) or wavelike ($\sim e^{-ikr}$, $\Gamma(\omega^2) > 0$). For $\Gamma_e(\omega^2) < 0$, wave solutions with discrete $\omega$ exist that are bound to the flux tube (Goossens et al., 2009). These are coherent or global wave motions of the flux tube, in which individual magnetic field lines with different Alfvén and sound speeds inside and outside of the flux tube oscillate collectively at the same frequency. Depending on the sign of $\Gamma_i(\omega^2)$, these are referred to as surface ($\Gamma_i(\omega^2) < 0$), or body modes ($\Gamma_i(\omega^2) > 0$). When $\rho_i > \rho_e$, as usually assumed in the case of coronal loops, only body waves exist, while for $\rho_i < \rho_e$, more representative of a photospheric flux tube above a magnetic bright point, both surface and body waves exist (Priest, 2014). Here we continue the discussion for an overdense flux tube, representative of coronal conditions. The quantity $\Gamma(\omega^2)$ can be interpreted as the radial wavenumber, and is usually denoted as $k_i^2 = \Gamma_i(\omega^2), k_e^2 = -\Gamma_e(\omega^2)$. The solutions to Equation (2.32) are given in terms of Bessel and modified Bessel functions of first and second kind, respectively:

$$P_i'(r) = A_i J_m(k_i r), \tag{2.33}$$

$$P_e'(r) = A_e K_m(k_e r), \tag{2.34}$$

where $A_{i,e}$ are constants. The expressions for the radial displacement can be found through Equations (2.16) and (2.17). The dispersion relation of the waves can be found by establishing the continuity of the total pressure and radial displacement across the interface at $R$:

$$\begin{aligned} &k_i \rho_e \left(\omega^2 - \omega_{Ae}^2\right) J_m'(k_i R) K_m(k_e R) \\ &= k_e \rho_i \left(\omega^2 - \omega_{Ai}^2\right) J_m(k_i R) K_m'(k_e R), \end{aligned} \tag{2.35}$$

where the primed Bessel functions denote their derivatives with respect to the argument. The dispersion relation is usually simplified by assuming that the wavelength along the tube is much larger than the tube radius ($k_z R \ll 1$), referred to as the thin tube limit, leading to the well-known phase speed for modes with $m \geq 1$ referred to as the kink speed:

$$v_K = \sqrt{\frac{\rho_i v_{Ai}^2 + \rho_e v_{Ae}^2}{\rho_i + \rho_e}}. \tag{2.36}$$

In fact, the term "kink" denotes only axisymmetric waves with $m = 1$, which are the only waves that displace the central axis of the magnetic flux tube. The kink wave with phase speed $v_{Ae} > v_k > v_{Ai}$ is generally referred to as the fast kink body MHD wave in the literature. This nomenclature originating from Edwin and Roberts (1983) is challenged by Goossens et al. (2009, 2012, 2014), who suggest the adjectives Alfvénic or surface Alfvén instead, based on several observations that are summarized here. First, they argue that kink waves in thin tubes are nearly incompressible, and their main restoring force is magnetic tension, which are dominantly Alfvén-like characteristics rather than fast-like. Second, they note that the fundamental radial mode (the smallest solution to Equation (2.21)) of the kink wave, which has no nodes in the solution inside the tube, survives in an incompressible plasma that admits no fast waves. Kink solutions for overtones of the radial wavenumber present nodes inside the tube, disappear under incompressible conditions, and have predominantly fast-like properties. Overall, it is demonstrated that the properties of the fundamental radial mode of kink waves in piecewise flux tubes in a compressible plasma are the same as those of surface Alfvén waves that exist on a planar discontinuity in incompressible plasma. Third, it is argued that, as the fundamental radial mode is the only trapped kink mode for thin tubes, it is probably the most relevant mode, as for observed kink waves in solar plasma structures the thin tube limit is a good approximation. Illustrations of surface Alfvén waves on a planar discontinuity and kink waves on a flux tube are presented in Figure 2.5.

What happens if the density varies continuously instead of a piecewise variation between the inside and outside

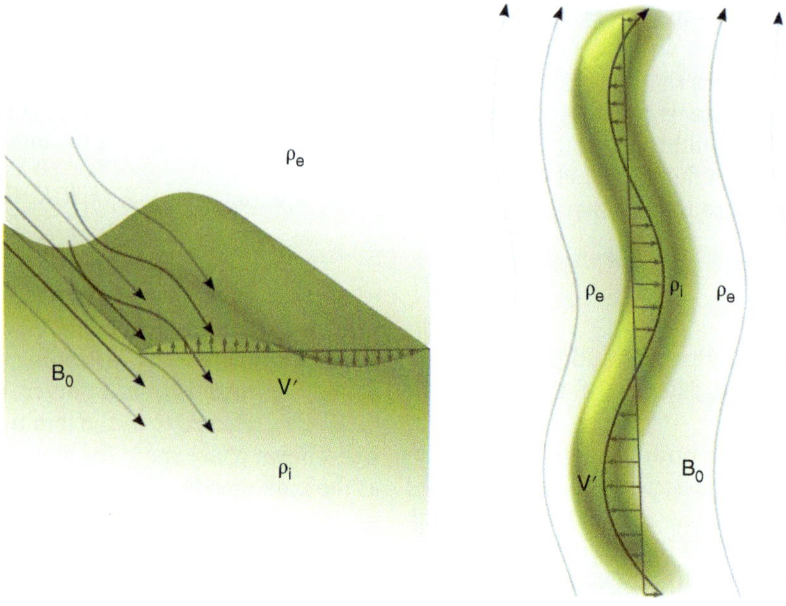


**Figure 2.5** Schematic representation of a surface Alfvén wave on a planar interface (left), and of a kink wave in a magnetic cylinder (right). The interface and the cylinder are defined by a jump in background density.

of the flux tube? This case has been studied in detail by several authors, and it leads to the phenomenon of quasi-modes and resonant absorption (Goedbloed & Poedts, 2004). Without entering into mathematical details here, which are quite involved for resonant absorption, it is important to note in the context of this paper that kink waves propagating in flux tubes with continuous variation of the Alfvén speed at their boundary undergo damping. This damping due to resonant absorption can be understood as an energy transfer between the global kink wave and local waves that are excited at the resonant surface inside the boundary where $\omega_k = \omega_A(r)$. It is particularly relevant to note here that the resonantly excited waves are almost pure Alfvén waves around the resonant layer. These are only "almost pure" because, albeit very small, pressure perturbations and compression are still present (Goossens et al., 2020). This implies that, even if only kink waves are excited initially in a flux tube, almost pure Alfvén waves are generated in resonant layers. Note that for the kink mode, the resonant Alfvén waves also have an $m = 1$ geometry, unlike the torsional Alfvén waves described by the uncoupled Equation (2.22), which correspond to $m = 0$. It should also be added that Goossens et al. (2012) established that kink waves of nonuniform flux tubes are still remarkably similar to surface Alfvén waves on a true discontinuity, therefore the discussion above on the validity of adjectives such as Alfvénic or surface Alfvén to refer to the kink wave still holds, according to their analysis.

### *Waves in Twisted Flux Tubes*

In our discussion of general radially inhomogeneous MHD in Section 2.2.1, we have only considered a straight background magnetic field instead of a general magnetic configuration. This was motivated by the fact that the linear coupling of variables corresponding to the separate wave modes existing in homogeneous MHD (Alfvén and magnetosonic), that is, the phenomenon of mixed properties, is already present in this relatively simple equilibria. However, a background angular component of the magnetic field, or in other words a twisted magnetic field, is thought to be relevant in many solar atmospheric applications, from flares to flux ropes in prominences and to coronal loops (e.g., Aschwanden, 2019; Fleishman & Toptygin, 2013a,b; Guo et al., 2021). In order to analyze radially inhomogeneous equilibria with twisted magnetic fields, we choose the following background quantities: $\boldsymbol{B}_0 = B_{z0}(r)\hat{z} + B_{\varphi 0}(r)\hat{\varphi}; p_0(r); \rho_0(r)$. Equilibrium requires the balance of total pressure and magnetic tension:

$$\frac{d}{dr}\left(p_0 + \frac{B_0^2}{2\mu_0}\right) = -\frac{B_{\varphi 0}^2}{\mu_0 r}. \tag{2.37}$$

Linear MHD waves atop this equilibrium, with similar notation as in the case of the straight magnetic field in Section 2.2.1, can still be described by two coupled ordinary differential equations for total pressure and radial displacement (Goossens et al., 2019):

$$D\frac{d(r\xi_r)}{dr} = C_1 r\xi_r - C_2 rP', \tag{2.38}$$

$$D\frac{dP'}{dr} = C_3\xi_r - C_1 P', \tag{2.39}$$

where $D$ and $C_2$ are given in Equations (2.18) and (2.19), respectively, and the coefficient functions $C_1$ and $C_3$ are given by:

$$C_1 = \frac{2}{\mu_0 r}B_{\varphi 0}^2\omega^4 - \left(v_S^2 + v_A^2\right)\left(\omega^2 - \omega_C^2\right)\frac{2mf_B}{\mu_0 r^2}B_{\varphi 0}, \tag{2.40}$$

$$C_3 = D\left[\rho_0\left(\omega^2 - \omega_A^2\right) + \frac{2B_{\varphi 0}}{\mu_0}\frac{d}{dr}\left(\frac{B_{\varphi 0}}{r}\right)\right] + \left(\frac{2\omega B_{\varphi 0}^2}{\mu_0 r}\right)^2 - 4\rho_0\left(v_S^2 + v_A^2\right)\left(\omega^2 - \omega_C^2\right)\omega_A^2\frac{B_{\varphi 0}^2}{\mu r^2}, \tag{2.41}$$

and finally, the quantity $f_B = \boldsymbol{k}\cdot\boldsymbol{B}_0 = k_z B_{z0} + mB_{\varphi 0}/r$. It is clear that the presence of twist yields a significantly more complicated set of equations than Equations (2.16) and (2.17) for a straight field. Nevertheless, these can still be combined into a single equation for either $P'$ or $\xi_r$ (Sakurai et al., 1991). In order to clearly illustrate the mixed properties of the waves and to study the effect of twist, let us write the Lagrangian displacement perpendicular and parallel to the equilibrium magnetic field lines in the plane of the magnetic surfaces, $\xi_\perp = (\xi_\varphi B_{z0} - \xi_z B_{\varphi 0})$ and $\xi_\parallel = \boldsymbol{\xi}\cdot\boldsymbol{B}_0/B_0$, respectively (Goossens et al., 2019):

$$\rho_0\left(\omega^2 - \omega_A^2\right)\xi_\perp = \frac{i}{B_0}\left[\left(\frac{m\,B_{z0}}{r} - k_z B_{\varphi 0}\right)P' - \frac{2f_B B_{\varphi 0}B_{z0}}{\mu_0 r}\xi_r\right], \tag{2.42}$$

$$\rho_0\left(\omega^2 - \omega_A^2\right)\xi_\parallel = \frac{if_B}{B_0}\frac{v_S^2}{v_S^2 + v_A^2}\left(P' - \frac{2B_{\varphi 0}^2}{\mu_0 r}\xi_r\right), \tag{2.43}$$

$$\nabla\cdot\boldsymbol{\xi} = -\frac{\omega^2}{\rho_0\left(v_S^2 + v_A^2\right)\left(\omega^2 - \omega_C^2\right)}\left(P' - \frac{2B_{\varphi 0}^2}{\mu_0 r}\xi_r\right). \tag{2.44}$$

As established previously, waves generally perturb displacement components both parallel and perpendicular to the magnetic field and cause compression, a signature of mixed properties. While in a straight magnetic field pure Alfvén waves were still present in radially inhomogeneous equilibria, manifesting as torsional $m = 0$ Alfvén waves on magnetic surfaces, due to the presence of the twist no pure Alfvén waves are present in this setup. This can be seen in the RHS of Equation (2.42), which is nonzero even for $m = 0$, coupling total pressure and radial displacement to motions perpendicular to magnetic field lines, which would necessarily be perturbed by Alfvén waves. This shows again that rather ideal conditions are required in order for pure Alfvén waves to exist, and it reinforces the previous narrative on the likely scarcity of pure Alfvén waves under realistic conditions. There have been a number of studies on the properties of waves that exist in twisted flux tubes, which will be touched upon in Section 2.4.

### 2.2.3. Concluding Remarks to this Section

Now that we have seen theoretical considerations, let us return to the discussion on the proper appellation or identification for transverse waves observed in the solar atmosphere. Transverse waves are observed through various methods, including direct imaging, spectroscopic measurement of Doppler shifts and nonthermal line broadenings, and radio emission, among others. In order to identify a wave as transverse, some assumption of the local background magnetic field direction is necessary. This is based on the observed anisotropic structuring of the plasma (e.g., spicules, coronal loops, and plumes), in that the direction presenting the least variation is assumed to outline the magnetic field direction as a result of the frozen-in condition and smaller transport coefficients (e.g., thermal conduction) across the magnetic field. The collinearity of the magnetic field and of the observed structuring can also be verified based on photospheric magnetic field extrapolations (Wiegelmann et al., 2005; Wiegelmann & Sakurai, 2021). The direction of propagation of the observed transverse waves is another important parameter. Through time series of imaging observations, the alignment of phase cross-correlations with the inferred background magnetic field direction can be studied. Only waves that propagate along the magnetic field can be candidates for identification as Alfvén waves, based on their dispersion relation in Equation (2.15). Here by propagation direction we mean the direction of the group speed vector or in other words the direction of energy propagation. The phase of Alfvén waves can propagate obliquely, with speed $v_A\cos(\theta)$, where $\theta$ is the angle between the wavevector and the magnetic field. Wave phase speeds can also be determined from phase cross-correlation, which are compared to estimates for local Alfvén and sound speeds. Intensity variations of spectral lines, if observed in the plasma undergoing the transverse oscillation, can be a signature of temperature or density perturbations as both affect the emissivity of the plasma. These perturbations are typical to magnetosonic waves.

Now, if an observation is made of a wave in the solar atmosphere that fits the properties of pure Alfvén wave known from the linear analysis of homogeneous and infinite plasma (transversality, along the field propagation, no intensity variations), it is tempting to identify these waves as being Alfvén waves. However, the main message of the analysis presented above is that one has to be cautious when labeling waves observed in realistic, nonideal conditions (plasma nonhomogeneity, finite amplitudes, etc.) using the fast, slow, Alfvén adjectives. Incidentally, the structuring of the plasma invoked in estimating the magnetic field orientation is proof that the plasma in which the waves are observed is inhomogeneous, leading to waves with mixed properties as discussed earlier. The main issue with mislabeling is that, together with their names, the properties of linear, ideal waves are also conferred upon the observed waves identified as such. This can lead to erroneous energy flux, seismological, and nonlinear behavior estimates. For example, Goossens et al. (2013) have shown that estimates of the energy content strongly depend on whether waves with similar transverse displacement are identified as kink waves of flux tubes or Alfvén waves. In coronal seismology, misidentification of the observed wave (e.g., Alfvén instead of kink) can lead to wrong estimates for the inferred plasma parameter, for example, magnetic field intensity. Nonlinearly, and specifically related to turbulence generation, Alfvén waves behave very differently compared to kink modes and all other wave modes in MHD for that matter. Alfvén waves are the only modes that are exact solutions of the incompressible MHD equations, meaning that their nonlinear evolution is only taking place through collisions between counterpropagating Alfvén waves. On the other hand, for example, kink waves of flux tubes can nonlinearly self-cascade for unidirectional propagation, a property referred to as uniturbulence (Magyar et al., 2019b). For example, identifying the omnipresent propagating waves seen in CoMP as Alfvén or Alfvénic (yes, the suffix makes a huge difference) may have a strong impact upon models relying on their turbulent evolution, for example, for coronal heating or space weather applications.

Let us now reflect on the various ways that transverse waves observed in the solar atmosphere are referred to. We reiterate that the solar atmosphere is a generally inhomogeneous plasma structured across the magnetic field, as shown by a number of works cited above. The first adjective is "Alfvén," most famously used by Tomczyk et al. (2007) to refer to the observed transverse waves in the solar corona. While we have argued that pure Alfvén waves can exist in inhomogeneous plasma under some conditions, these conditions are rather special, such as Alfvén waves existing on magnetic isosurfaces, with their polarization everywhere parallel to these. If we imagine the corona and chromosphere as plasmas displaying general, even turbulent inhomogeneities across the magnetic field (e.g., the coronal veil, Malanushenko et al., 2022), these isosurfaces would have to be precisely and selectively driven in order to excite only Alfvén waves. This is unlikely having in mind the random nature of the possible driver for these waves, the convective flows in the photosphere. Still, Alfvén waves could be driven on such isosurfaces as resonances of global kink waves. However, the small damping rate of kink waves in the solar atmosphere (Morton et al., 2021), and their volume-filling quality, would still mean that most of the energy would be in kink waves, as also observed in simulations (Magyar & Van Doorsselaere, 2022).

Next, we investigate the adjectives "kink" and "fast kink," used to refer to the standing transverse oscillations of coronal loops (Aschwanden et al., 1999; Nakariakov et al., 1999), or to CoMP waves by Van Doorsselaere et al. (2008). First of all, the term "kink" assumes that the oscillating structure is well approximated by the flux tube model or displays some kind of cylindrical symmetry. Otherwise, azimuthal wavenumbers such as $m = 1$ are not appropriate to describe waves in these structures. The flux tube model approximates spicules, loops, or plumes that are observed in relative isolation probably well enough for the purpose of theoretical wave mode analysis. Studies have shown that even if tubular plasma structures are not exactly cylindrically symmetric and have internal structure, the kink wave solution of cylindrical flux tubes derived above is well approximating the wave mode that transversely displaces the structure as a whole (Magyar & Van Doorsselaere, 2016; Pascoe et al., 2011). However, for generally inhomogeneous or turbulent cross sections with structures and transverse waves of various polarizations on many scales, as the solar corona is believed to be, the term "kink" loses its precise meaning in the lack of an $m = 1$ dependence of the perturbation field. The term "fast" best describes kink waves with axial or parallel wavelengths on the order or shorter than the tube radius and radial overtones. Most of the transverse waves observed in the solar atmosphere have much longer parallel wavelengths than the structures they propagate on (thin tube limit), most probably being the fundamental radial modes. Kink waves in thin tubes were shown to have dominant Alfvén-like rather than fast-like characteristics, producing negligible compression and driven mostly by magnetic tension.

Finally, the adjectives "Alfvénic" and "surface Alfvén" are discussed. Both were used to describe CoMP waves (e.g., Morton et al., 2015), or transverse waves, in generally inhomogeneous plasma across the magnetic field (e.g., Magyar & Van Doorsselaere, 2022). The term "surface Alfvén" intends to express the localization of the transverse waves to inhomogeneities (i.e., trapped waves), and their similarity to surface Alfvén waves on a planar discontinuity, following Goossens et al. (2012). It also highlights the fact that waves are modified by the presence of inhomogeneity, have mixed Alfvén–magnetoacoustic properties, or properties that are only encountered for waves in inhomogeneous plasma. As an example of the latter, in an incompressible and homogeneous plasma, linear wave solutions (Alfvén and pseudo-Alfvén, the incompressible vestige of slow waves) do not perturb total pressure, yet incompressible linear surface Alfvén waves on a discontinuity do perturb it. The term "Alfvénic" reflects the fact that transverse waves in a generally inhomogeneous plasma of parallel wavelengths longer than the typical perpendicular scale of inhomogeneity have dominantly Alfvén-like properties. In this sense, it is the term that can most broadly describe transverse waves in an inhomogeneous plasma, without making implicit assumptions about the geometry of the inhomogeneity as "kink" does, while still highlighting the dominant character of the mixed properties transverse wave.

Having clarified the meaning behind various terms used to describe transverse waves, in this chapter of a monograph dedicated to Alfvén waves, the adjective "Alfvén" will be used in the broadest sense to refer to waves with dominant Alfvén-like characteristics, while always keeping in mind that pure Alfvén waves are in fact restricted to an ideal, infinite, and homogeneous plasma with linear dynamics.

## 2.3. OBSERVATIONS OF ALFVÉN WAVES IN THE SOLAR ATMOSPHERE

While observations of waves in the solar atmosphere have been reported since the sighting of global perturbations by Moreton and Ramsey (1960), and subsequently Extreme ultraviolet Imaging Telescope (EIT) waves (Thompson et al., 1999), these waves propagate across the magnetic field, and are dominantly fast-like. The first observations of transverse waves that could be categorized as Alfvénic were those by Nakariakov et al. (1999) and Aschwanden et al. (1999), manifesting as large amplitude, impulsively driven standing kink waves of coronal loops. In subsequent years, many similar events have been spotted, with hundreds of analyzed events (Nechaeva et al., 2019), making standing kink waves the most thoroughly analyzed and most frequently observed transverse waves with Alfvénic properties in the solar atmosphere. Theory and observations of standing kink waves have been recently reviewed by Nakariakov and Kolotkov (2020), and will not be discussed here in detail, but only in some aspects. Readers interested in standing kink waves are kindly directed to these articles. The focus of this section will rather be on observations of transverse waves in the solar atmosphere that are referred to as Alfvén or Alfvénic by the authors in the respective publications. With the advent of new telescopes observing solar dynamics in a wide range of wavelengths, many waves with Alfvén-like characteristics were identified and analyzed in various regions of the solar atmosphere. In the following text, I will briefly review observations of Alfvén waves in various settings, in order of increasing distance from the solar surface, published over the last decade.

### 2.3.1. Alfvén Waves in the Solar Photosphere

The photosphere is in general dominated by gas pressure, with its main dynamics given by the convective turbulence appearing as granules. Under these conditions, the remote detection of transverse oscillations is hardly possible. However, intergranular lanes are known to be hotspots of magnetic activity, and often harbor intense flux tubes knows as magnetic bright points that are constantly shuffled by the surrounding flows (Vargas Domínguez & Utz, 2022). These could act as ideal waveguides for transverse waves. Swirls in intergranular lanes have been detected by numerous studies (e.g., Liu et al., 2019; Wedemeyer-Böhm et al., 2012). Larger magnetic structures, such as magnetic pores and sunspots, are also likely candidates for the detection of transverse waves. Despite the potential of photospheric magnetic elements for supporting transverse waves, torsional Alfvén waves were directly detected only recently by Stangalini et al. (2021, see Figure 2.6), in a magnetic pore, although indirect evidence of photospheric vortex flows driving chromospheric torsional motions already existed (Morton et al., 2013). The oscillations were detected by analyzing time series of instantaneous circular polarization (CP) maps of the Interferometric BIdimensional Spectropolarimeter (IBIS) at the Dunn Solar Telescope. Torsional Alfvén waves appeared as angular rotations that are in antiphase at the two opposite sides of the magnetic tube structure of the pore, determined by cross-correlating the CP maps. The search and study of Alfvén waves in photospheric structures will probably accelerate in the coming years thanks to the commissioning of observatories with unprecedented spatial and temporal resolution, such as the Daniel K. Inouye Solar Telescope (DKIST) (Jaeggli et al., 2022).

### 2.3.2. Alfvén Waves in the Solar Chromosphere

The solar chromosphere is arguably the most physically complex region of the solar atmosphere, due to effects related to partial ionization, nonlocal thermodynamic equilibrium radiative transfer, nonequilibrium ionization,

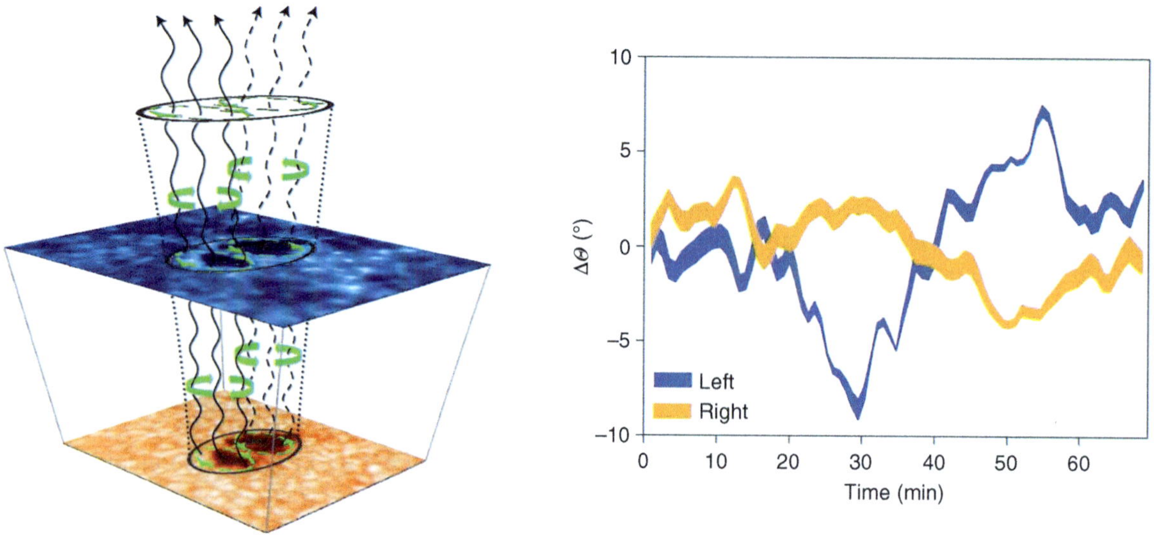


**Figure 2.6** Left: Schematic representation of the $m = 1$ antisymmetric torsional Alfvén oscillations in the magnetic pore observed by Stangalini et al. (2021) / Springer Nature. Right: Measured angular rotation oscillations of the two lobes of the magnetic pore, as obtained from a cross-correlation tracking analysis at the edges of the flux tube. Stangalini et al. (2021) / Springer Nature.

and so on (Carlsson et al., 2019). In this sense, the interpretation of chromospheric dynamics purely in terms of MHD waves requires a cautionary note. Waves in chromospheric magnetic structures have been studied by several authors (Verth & Jess, 2016) since the discovery of omnipresent transverse swaying of spicules by De Pontieu et al. (2007, see Figure 2.7) using data from the Hinode satellite, and also omnipresent torsional motions (De Pontieu et al., 2012) seen using the Swedish 1-m Solar Telescope (SST).

Spicules, also known as fibrils or mottle depending on their morphology, are dynamic jets that originate in the lower chromosphere with typical diameter on the order of hundreds of kilometers, and are generally short lived, with a lifetime of less than 10 min. As spicules are relatively easy to track due to their inherent brightness, most of the observed Alfvén waves in the solar chromosphere propagate on spicules. Spicules can support various wave types, and Sekse et al. (2013) interpreted the complex dynamics of spicules in terms of field aligned flows, swaying or displacements, and torsional motions in observations with the CRisp Imaging Spectro-Polarimeter (CRISP) at the SST. In their attempt to disentangle the various motion types, Sharma et al. (2018) analyzed the complex transverse motions of spicules in a three-dimensional manner.

As in the case of coronal oscillations, properties of the medium in which waves are seen propagating, in this case of the spicules, can be inferred by combining wave theory, for the purpose of which spicules are often modeled as flux tubes, with measured wave properties. Kuridze et al. (2012, 2013) characterized transverse oscillations in chromospheric mottles near the solar disk center, and used seismology to infer local magnetic field and plasma density scale heights. They found a strong decrease in magnetic field intensity with height, and a density scale height of around 280 km. Besides other wave characteristics, the damping of transverse waves in spicules has also been documented. Morton (2014) measured transverse wave motions interpreted as kink waves in spicule-like jets in the atmosphere above an off-limb sunspot, in which spatial damping of these waves was reported for the first time.

Doppler shifts of spectral lines emitted near the boundaries of spicules are an excellent way to detect torsional oscillations, for which the Doppler shifts are in antiphase across the spicules. Tavabi et al. (2014) observed and analyzed Alfvénic waves in polar spicules using the Hinode SOT Ca II H emission line, while Tavabi and Koutchmy (2019) used the Mg II line of the Interface Region Imaging Spectrograph (IRIS) to detect large-amplitude Doppler shifts corresponding to 20–30 km/s amplitude oscillations of periods of approximately 100 s that are consistent with propagating Alfvén waves in off-limb observations of spicules. A transverse wave of a similar period was observed also by Cho et al. (2019). They detected transverse oscillations of an H-alpha surge in the chromosphere corresponding to a polar coronal hole, with a period of approximately 2 min. Surges are ejections of chromospheric plasma into coronal heights, up to 200 Mm, representing upward rising motions or a "jet" along relatively straight trajectories that ultimately diffuse and fall back. An observation that

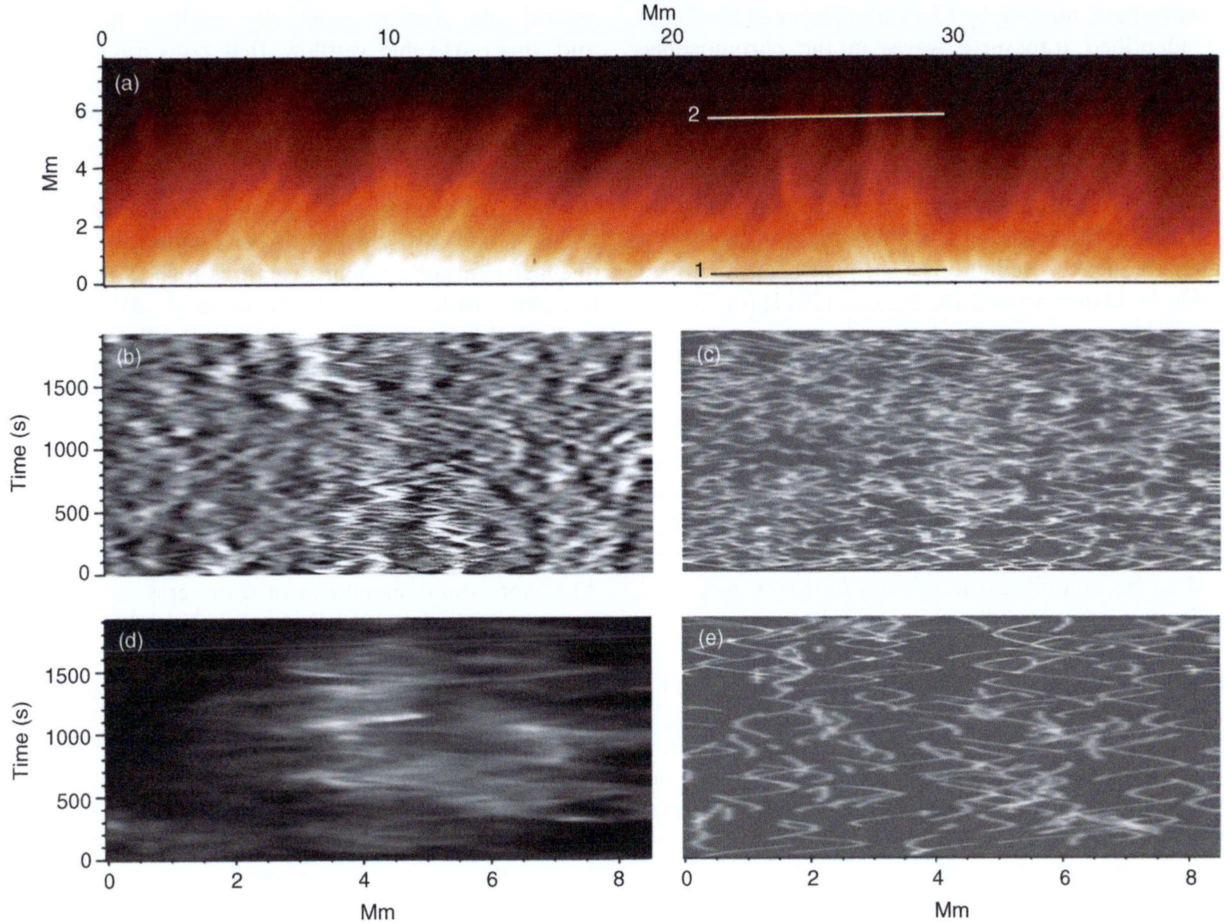


**Figure 2.7** Images illustrating the omnipresence of transverse swaying of spicules in the chromosphere. (a) Hinode Solar Optical Telescope (SOT) image in the Ca II H 3968 Å line showing the abundance of spicules in the chromosphere. (b) and (d) Spacetime diagrams for the lines 1 and 2 outlined in (a), respectively, showing clearly the swaying of the spicules over time. (c) and (e) Monte Carlo simulations of transverse motions in simulated spicules approximate well the observed dynamics. Reproduced with permission from De Pontieu et al. (2007).

is relevant for the estimation of transverse energy able to propagate into the solar corona was made by Liu et al. (2014), who investigated the characteristics of Alfvénic fluctuations of chromospheric spicules, and found a balanced flux of outward- and inward-propagating waves, which could indicate a strong reflectivity of the transition region for these waves. Although in a different setting, a strong flux of downward-propagating Alfvén waves has also been observed by Matthews et al. (2015) through broad spectral lines with strong red Doppler shifts in flare-related sunquake events. Besides waves in well-defined line intensity structures such as jets and spicules, oscillatory motions have also been observed in vortex flows. Tziotziou et al. (2019) found oscillatory motions in a long-lasting chromospheric vortex flow, with rotational components consistent with the presence of torsional Alfvén waves. Such vortex motions in the solar atmosphere have recently been reviewed by Tziotziou et al. (2023). Besides the direct imaging of spicules and Doppler shift analysis, other methods have been employed for the detection of Alfvénic waves, such as variations in the nonthermal width of spectral lines emitted by the spicule, which could be a signature of unresolved torsional motions. Zubkova et al. (2014) detected such variations of the nonthermal widths of spectral lines in spicules and did interpret these as propagating torsional Alfvén waves. Still, Chelpanov and Kobanov (2022) caution that periodic variations of nonthermal spectral line broadening should not be automatically interpreted as indicators of the presence of Alfvén waves, as these are often accompanied by intensity variations of similar periods. Chae et al. (2021, 2022) reported the first spectroscopic detection of Alfvénic waves around a sunspot region at chromospheric heights in the Hα line and Ca II 854.2 nm line. Radio observations have also been used recently to detect transverse waves in the chromosphere, notably by Guevara Gómez et al. (2022), who identified transverse waves in the chromosphere using Atacama Large Millimeter/submillimeter Array (ALMA) observations. The study also finds an almost equal power in waves propagating toward and away from the Sun.

Transverse oscillations in spicules typically have frequencies on the order of 80–300 s (Jess et al., 2015). Higher-frequency oscillations have also been identified, notably by Okamoto and De Pontieu (2011) for the first time, with average periods of 45 s. Since then, numerous instances of high-frequency transverse oscillations have been reported. Median Alfvénic wave periods of 83 s were reported by Jafarzadeh et al. (2017) spotted through Ca II H observations of lower solar chromospheric fibrils, and they estimate that the associated energy flux is on the order of 15 kW $m^{-2}$. Torsional Alfvén waves of high frequency (approximately 12–42 mHz or 24–83 s) were found to be ubiquitous in spicules by Srivastava et al. (2017). According to their analysis, these waves could be relevant for coronal heating and mass transfer for the solar wind. Yoshida et al. (2019) detected and analyzed high-frequency transverse oscillations in Doppler shifts using data from the Chromospheric Lyman-Alpha Spectro-Polarimeter (CLASP) sounding rocket experiment, the observed oscillations having periods of 20–50 s. Based on CRISP observations, Shetye et al. (2021) report on 30 examples of transverse oscillations in spicules with periods below 100 s. The authors study both plane-of-sky and line-of-sight motions, and find that most transverse oscillations are helical in nature. They additionally find that the oscillations are either standing or transversely propagating, due to a lack of apparent propagation along the structures. Automatic detection of waves in spicules allowed for much larger statistical studies to be carried out, as shown by Bate et al. (2022). They studied the properties of high-frequency transverse oscillations in spicules, and found mean periods of 54 s and mean velocity amplitudes of 21 km/s. In agreement with previous findings, an approximately equal amount of energy is observed to be contained in upward- and downward-propagating waves.

Jess et al. (2023) offer a comprehensive review of waves in the lower solar atmosphere, including of transverse waves in the chromosphere, and the advances that are to be expected in their study in terms of future capabilities of next-generation solar telescopes.

### 2.3.3. Alfvén Waves in the Solar Corona

The solar corona is the above-million Kelvin temperature, fully ionized (i.e., it does not contain neutrals), tenuous and optically thin plasma that sits above the lower solar atmosphere consisting of the photosphere, chromosphere, and transition region, and extends out to several solar radii to eventually result in the supersonic and super-Alfvénic outflow that constitutes the solar wind. The corona is highly dynamic and inhomogeneous, with omnipresent propagating transverse wave activity first detected through CoMP dopplergrams, as reported by Tomczyk et al. (2007).

These Alfvénic waves are studied intensely (Morton et al., 2016) also by the direct imaging of transverse displacements of coronal structures such as polar plumes (e.g., McIntosh et al., 2011; Morton et al., 2015; Thurgood et al., 2014; see Figure 2.8), using the Solar Dynamics Observatory, onboard the Atmospheric Imaging Assembly satellite (SDO/AIA), including through automatic transverse wave detection algorithms (e.g., Weberg et al., 2018). They were also observed in spectroscopic data from the Hinode Extreme-ultraviolet Imaging Spectrometer (EIS) instrument (e.g., Tian et al., 2012). Threlfall et al. (2013) compared observations from CoMP and AIA, and found signatures of transverse motions along the same magnetic structure through both instruments. Some of the properties of the Alfvénic waves detected through direct imaging are average velocity amplitudes of 20 km/s and average periods of hundreds of seconds (although see also Morton & McLaughlin, 2013, who find velocity amplitudes less than 3 km/s through direct imaging with Hi-C). The average velocity amplitudes previously found by Tomczyk et al. (2007) were only 0.5 km/s, deemed far too low to contribute to the energization of the solar corona. These low values were probably due to the coarse spatial and temporal resolution of the CoMP instrument, as well as the optically thin emission from the corona, resulting in the line-of-sight superposition of multiple oscillating structures within a single pixel and the averaging out of Doppler shifts (Antolin et al., 2017; McIntosh & De Pontieu, 2012; Pant et al., 2019; Pant & Van Doorsselaere, 2020). From the frequency spectrum of propagating Alfvénic waves, already investigated by Tomczyk and McIntosh (2009), an enhancement in power is found at frequencies corresponding to periods of around 5 min, which could indicate that the omnipresent propagating Alfvénic waves might be excited by p-modes, or that both might have in common as driver the convective flows in the photosphere. P-modes are standing global sound waves of the Sun, which acts as a resonant cavity. The driving mechanism behind CoMP waves was investigated in detail by Aschwanden et al. (2018), who found that these waves are probably convection driven, through the observed vortex motions in the photosphere. Other avenues toward Alfvénic wave generation were proposed by Karampelas and Van Doorsselaere (2020), who show that transverse waves can also be driven by steady-flow drivers through vortex shedding, which is the result of an oscillating acceleration transverse to the flow direction induced by the presence of an obstacle,

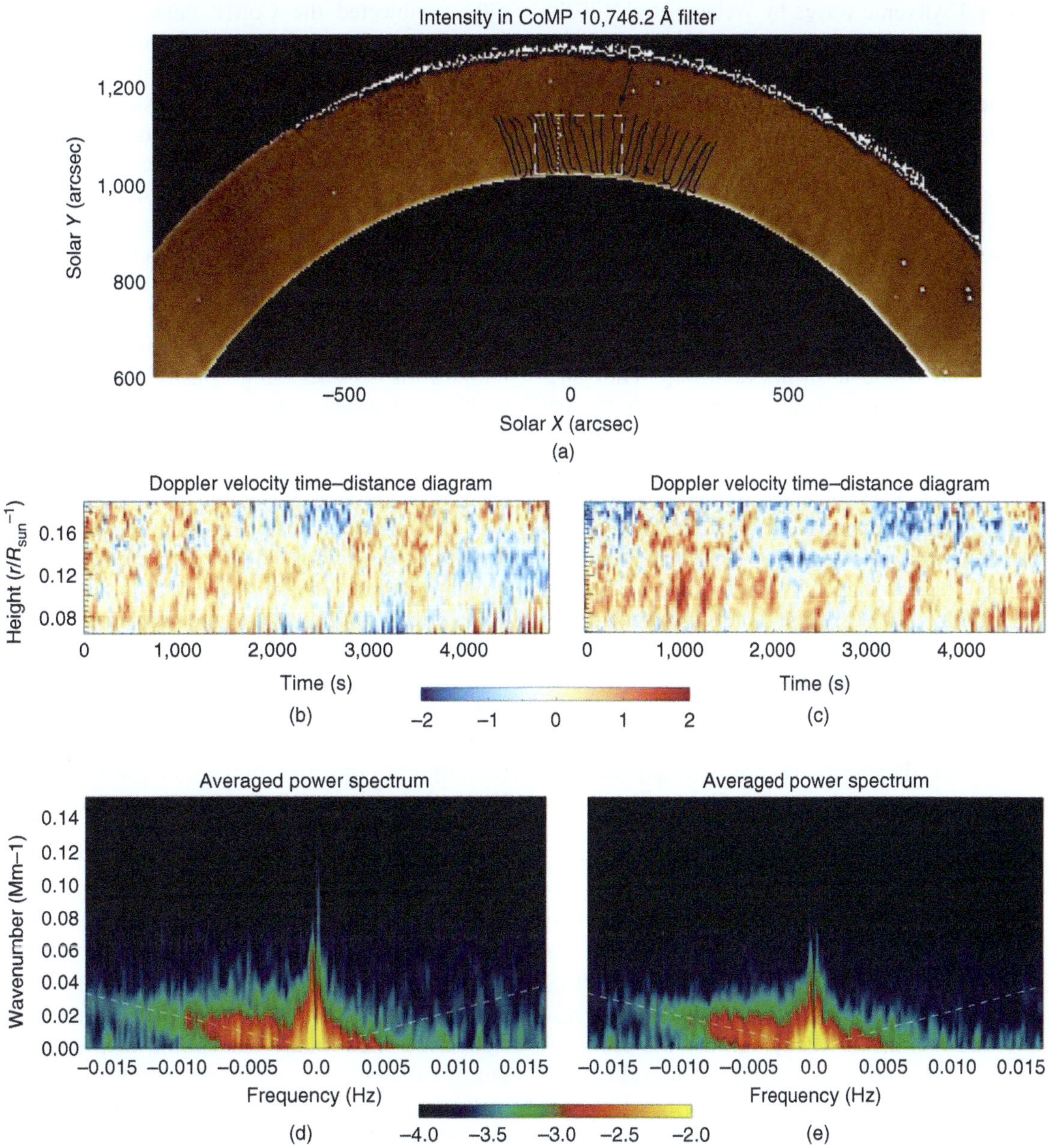


**Figure 2.8** Alfvénic waves in the solar atmosphere as seen by CoMP. (a) CoMP intensity image of the northern polar region of the Sun. (b) and (c) The predominantly upward-propagating waves can be seen in time–distance diagrams of the Doppler shifts within the box and tracks in image (a), respectively. (d) and (e) The Fourier-analyzed time–distance diagrams show the dominance of outward-propagating waves, but also reveal the presence of Sunward-propagating waves. Morton et al. (2015) / Springer Nature / CC BY 4.0.

here the coronal loop. Cranmer (2018) discusses the generation of low-frequency waves, with periods longer than 30 min, by magnetic reconnection in the Sun's "magnetic carpet" in the lower solar corona. Additionally, falling blobs of coronal rain were also shown to be able to induce transverse oscillations of the magnetic fields along which these fall by Verwichte and Kohutova (2017). Somewhat related to the previous work, Antolin et al. (2018) argue that transverse waves can be generated by colliding flows or clumps of plasma in the solar corona. The seismological capabilities of Alfvénic waves in the solar corona have been used, as suggested by Magyar and Van Doorsselaere (2018), to map the magnetic field in full-disk images of the corona (Yang et al., 2020). The density profile of a localized region in the corona was determined seismologically based on the properties of the observed Alfvénic waves by Weberg et al. (2020). Their results indicate that the relative density is larger than expected, signaling a more extended transition region than in current models (Avrett and Loeser, 2008). This could imply a more efficient wave propagation and transmission from the lower solar corona than current models would indicate.

The spatial damping of propagating Alfvénic waves has been investigated in a number of works. Hahn et al. (2012 and Hahn and Savin (2013, 2014) investigated the nonthermal width of spectral lines in a polar coronal hole, assumed to be the signature of propagating Alfvén waves and found a strong damping. They suggested that the lost energy could amount for up to 70% of that required to heat the plasma. Significant energy loss is also found through a similar analysis by Bemporad and Abbo (2012). Gupta (2017) also looked at the nonthermal broadening in the off-limb corona up to a distance of 140 Mm, and found a significant decrease in wave energy flux with height, with damping lengths in the range of 25–170 Mm. The damping length shows strong dependence on the peak temperature of the analyzed spectral line, suggesting temperature-dependent damping of Alfvénic waves through the role of thermal conduction. Although Alfvénic waves are only weakly compressible and therefore are not strongly affected by thermal conduction, Gupta (2017) speculates that (nonlinear) conversion of Alfvénic waves to slow waves, with their subsequent damping by thermal conduction, could explain the findings, without presenting a theoretical model to support this scenario. Gupta et al. (2019) studied the damping of Alfvénic waves through the nonthermal line widths as measured by Hinode/EIS, in a long off-limb coronal loop up to 1.4 Solar radii, and also found significant damping. On the contrary, Morton et al. (2021) find a weak damping of propagating Alfvénic waves as seen through their Doppler shift signature by CoMP, with the source of inconsistency with the damping measured through nonthermal line widths being unclear. Del Zanna et al. (2019) do not find evidence of Alfvén wave damping through nonthermal line widths acquisitioned through Hinode/EIS, in the quiet Sun up to 1.5 solar radii, and caution that due to uncertainties in estimating the instrumental width, a firm conclusion on whether damping is observed through nonthermal line width measurements using Hinode/EIS is not possible. Tiwari et al. (2019) investigated the damping of propagating Alfvénic waves through CoMP Doppler shift measurements and found that the damping is dependent on frequency, suspecting resonant absorption as the main damping mechanism. Additionally, they found that the damping decreases with the length of the loops in which the waves are propagating. Subsequently, Tiwari et al. (2021) subjected the CoMP data archive to an extensive statistical study of Alfvénic waves, and provided a catalog of 120 observed loops that sustain oscillations. They have confirmed previous results of damping rate dependency on frequency and coronal loop length. MHD waves, including transverse Alfvénic waves, have also been reviewed recently in open coronal structures by Banerjee et al. (2021).

Signatures of the nonlinear evolution of Alfvén waves in the transition region and lower solar corona have also been observed. Hahn et al. (2022) report the first observational evidence of parametric decay instability, manifesting as coexisting density and nonthermal line width oscillations, with a frequency relationship that is consistent with PDI. Earlier, De Moortel et al. (2014) and Liu et al. (2014) presented evidence for Alfvén wave turbulence developing in solar coronal loops, based on the observed height-dependent spectrum of Alfvénic waves observed with CoMP, with excess high-frequency oscillations being present toward the apex of the loops. Theoretical advances on the nonlinear evolution of Alfvénic waves in the solar corona will be discussed in the next chapter.

Besides the well-known transverse waves of coronal loops and plumes, torsional Alfvén waves have also recently been directly observed at coronal heights, using IRIS (Kohutova et al., 2020). The supporting structure on which the torsional Alfvén waves were spotted was a surge resulting from magnetic reconnection between prominence threads and ambient magnetic field. The observations have shown a clear antiphase oscillation of Doppler shifts at opposite edges of the flux tube. The authors concluded that magnetic reconnection is in general a likely source of torsional Alfvén waves. Earlier, signatures of resonant absorption, in the same form of antiphase oscillation of Doppler shifts, have been spotted by Okamoto et al. (2015), for the first time. They observed a transversely oscillating prominence, which at the same time also displayed the azimuthal motions, interpreted as the result of resonantly absorbed Alfvén waves. Transverse waves in prominences have been detected earlier by Okamoto et al. (2007) and Lin et al. (2009). A statistical study of these waves was performed by Hillier et al. (2013) using Hinode data, with observed periods ranging between 50 and 6000 s and amplitudes between 0.2 and 23 km/s. They highlight the omnipresence of transverse waves in prominences, much like the observed omnipresence of CoMP waves. Torsional Alfvén waves have also been observed in a solar flare, or more specifically in flare loops with enhanced density (Aschwanden & Wang, 2020). Reflected Alfvén waves propagating toward the Sun were first detected by DeForest et al. (2014) beyond 6 solar radii, using the Coronagraph 2

(COR2) observations of the Solar TErrestrial RElations Observatory (STEREO).

Alfvén waves may also be present as large-amplitude wave packets in the solar corona. Switchbacks are localized deflections in the magnetic field that were detected in large numbers by the Parker Solar Probe (PSP) satellite (e.g., Dudok de Wit et al., 2020; Raouafi et al., 2023). Although their origin remains unclear, some authors suggest that these could be large-amplitude propagating Alfvén waves that originate in the lower corona in reconnection events such as interchange reconnection (e.g., Fisk & Kasper, 2020; Sterling & Moore, 2020), or that they form in situ through the developing turbulence in the outer solar corona, at several solar radii (Johnston et al., 2022; Shoda et al., 2021; Squire et al., 2020). Recently, switchbacks have been spotted propagating in the solar corona using the Metis coronagraph on board Solar Orbiter (Telloni et al., 2022), where they manifest as propagating S-shaped vortices in white light images.

## 2.4. ADVANCES IN THE THEORY OF ALFVÉN WAVES

Although the linear theory of Alfvén waves in a homogeneous and infinite plasma has long been established, their existence in inhomogeneous plasma, as we have seen in Section 2.2, and their nonlinear evolution are still not entirely understood. Specifically, the effects of finite amplitude, the generally inhomogeneous background quantities both along and across the mean magnetic field, partial ionization in the lower solar atmosphere, among others, all have to be accounted for fully in order to have a complete description of Alfvén waves propagating through the solar atmosphere. Here, I will review some of most relevant analytical and numerical works that advanced our understanding of the dynamics of Alfvén waves in the solar atmosphere, which were published over the last decade.

### 2.4.1. Theory of Alfvén Waves in the Lower Solar Atmosphere

Let us begin by reviewing the behavior of Alfvén waves in the photosphere and chromosphere, where mode conversion induced by the presence of equipartition layers defined as regions where plasma $\beta = 1$, partial ionization effects, and abrupt changes in plasma parameters in the transition region have a strong influence on the dynamics. The problem of Alfvén wave propagation under homogeneous chromosperic conditions was studied by Soler et al. (2013a), who looked at pure Alfvén wave propagation in a partially ionized and unstructured plasma, and found analytical formulas for temporal and spatial attenuation of the Alfvén waves and appearing cutoff frequencies due to ion-neutral collisions. The cutoff frequency is the lowest frequency at which propagating modes can still exist, and it was also discussed in detail for Alfvén waves by Vranjes and Kono (2014). Alfvén wave propagation in the partially ionized chromosphere was studied numerically by Tu and Song (2013), using a 1D model of the solar atmosphere, in which they find that the average energy of Alfvén waves that is transmitted into the corona is 10% of the driving source energy in the lower chromosphere. Soler et al. (2015) studied analytically the wave heating resulting from Alfvén wave damping in the solar chromosphere. They found that the damping is dominated by Ohmic diffusion for low heights in the chromosphere and weak magnetic field, and by ambipolar diffusion in strong magnetic fields or higher in the chromosphere, concluding that the scales at which heating occurs in the chromosphere are currently unresolved by observations. Transverse wave propagation in lower atmospheric flux tubes was studied by a number of authors. Kink waves were studied by Soler et al. (2012, 2013b), who investigate analytically, using multifluid equations in the cold plasma approximation, the propagation of MHD modes in partially ionized flux tubes. They find that the resonant absorption of kink waves is only weakly affected by partial ionization effects in the range of wave frequencies that are observed, and that it produces a stronger damping of the kink waves than ion-neutral collisions for typical chromospheric conditions. The problem was also investigated by Yu et al. (2017), without including the effects of partial ionization. They show that kink waves can be efficiently damped by resonant absorption under photospheric conditions, with a similar efficiency as under coronal conditions. Torsional Alfvén waves in such flux tubes were investigated analytically by Soler et al. (2017). They calculated reflection and transmission rates to the corona, in an expanding flux tube model, finding that the frequency of maximum transmissivity is between 0.04 and 0.3 Hz, but that it depends on the degree of expansion, which also influences the heating rate. A similar study was carried out numerically by Soler et al. (2021), focusing on the transmission of wave energy into the corona. It is found that large amounts of energy, orders of magnitude higher than required to balance coronal radiative losses, escape into the corona, if the frequency of the waves from the photosphere matches the natural resonances of the coronal loop. However, most of this energy ends up again in the chromosphere, being unable to dissipate in the corona. On the other hand, the chromosphere appears to be heated much more efficiently by partial ionization effects, where this is able to balance the local radiative losses. Ballai (2019) studied the effects of nonequilibrium ionization on linear Alfvén waves in partially ionized plasmas, finding that it provides additional coupling between ions and neutrals, which enhances damping and renders the waves dispersive. Alfvén wave propagation in the solar chromosphere was studied numerically by solving the 1D two-fluid equations by Popescu Braileanu et al. (2019), who find, in agreement with previous works, that ion-neutral collisions cause alterations to Alfvén wave behavior. Russell and Fletcher (2013) studied numerically the one-dimensional propagation of Alfvénic waves from the corona toward the chromosphere, using the two-fluid model of plasma and neutrals. Such studies are relevant for large fluxes of Alfvénic waves triggered by flares. They concluded that high-frequency waves above 0.1 Hz are transmitted 20% into the chromosphere, and that their damping through ion-neutral collisions also depends strongly on the frequency, with the strongest damping for waves with frequency of 1 Hz or less.

Mode conversion to and from Alfvén waves, a feature of 2.5D equilibria (two spatial dimensions and three vector components) in light of the discussion in Section 2.2 or a feature of fully 3D equilibria if through mode conversion we understand the phenomenon of mixed properties, is an important topic that received considerable attention from the solar physics community. Cally (2017), Cally and Khomenko (2015, 2018, 2019), Khomenko et al. (2014, 2018), Khomenko and Cally (2019), among other works, studied fast to Alfvén mode conversion in the chromosphere taking into account various effects, such as ambipolar diffusion and Hall currents, and various magnetic field geometries, such as expanding flux tubes and bundles of flux tubes, through both analytical and numerical approaches. Strong mode coupling is found by Cally (2022) between Alfvén and slow waves in the lower solar atmosphere in three dimensions, analytically. It is found that 50% of Alfvén waves generated at the bottom of the chromosphere reach the top boundary, and that slow waves can be 40% effective in converting to Alfvén waves over the same distance, with these values strongly depending on the inclination of the magnetic field.

The dissipation of torsional Alfvén waves is being explored as a significant contributor to upper chromospheric heating. Kuźma et al. (2021), building on their previous 2.5D simulations (Kuźma et al., 2020), ran 3D two-fluid numerical simulations of continuously driven torsional Alfvén waves in the solar chromosphere. They show that Alfvén waves damping due to ion-neutral collisions results in significant heating. A similar 2.5D two-fluid numerical study, considering instead impulsively generated Alfvén waves, was carried out by Pelekhata et al. (2021). The study confirms previous findings of significant heating as a result of Alfvén wave damping through ion-neutral collisions. Kraskiewicz et al. (2023) arrive to the same conclusion of significant Alfvén wave heating in the solar chromosphere using a 1.5D two-fluid numerical model. Torsional Alfvén waves in solar photospheric vortices were studied numerically by Shelyag et al. (2013) including non-gray radiative transfer and a nonideal equation of state. They find that structures referred to as "solar tornadoes" in the photosphere are in fact oscillating structures displaying propagating Alfvén waves and not steady vortices. Torsional waves in solar tornadoes were studied analytically by Vasheghani Farahani et al. (2017). Their model consists of a rotating twisted magnetic cylinder embedded in a plasma with a straight magnetic field. It is shown that the torsional Alfvén waves propagating in such a tube have their properties modified, with their phase speed increasing with the amount of background twist and decreasing for increasing rotation rate of the tube. They also note that torsional waves become collective modes, meaning that torsional Alfvén waves are not independent on separate magnetic isosurfaces, and that the phase speed depends on the Alfvén speed both inside and outside of the tube. Murawski et al. (2015) also studied the properties of torsional Alfvén waves in a solar atmosphere with a realistic temperature profile, and found a weak reflection of these on the transition region, and an excess of magnetic energy in nonlinearly generated magnetoacoustic waves. Observed properties of torsional Alfvén waves in the solar chromosphere can also be exploited seismologically for diagnostics of plasma ionization, as shown by Ballai (2020). The nonlinear evolution of torsional Alfvén waves in a partially ionized plasma was studied by Ballester et al. (2020). They show that when ambipolar diffusion is taken into account, Alfvén waves are quickly damped by it and by the nonlinear coupling to slow modes. Alfvén wave propagation in radiative MHD models was presented by Chatterjee (2020), through a model of the solar atmosphere, from the photosphere into the corona. They find that wave propagation and dissipation depend strongly on the magnetic Prandtl number, the ratio between viscous and magnetic diffusion rate. Strong shock heating is found in the coronal part of the simulation only for Prandtl number values around unity, while the classical Prandtl number estimates for the corona are on the order of $10^{10}$ (e.g., Rempel, 2017).

Wójcik et al. (2017) investigated numerically Alfvén wave propagation along an axisymmetric flux tube embedded into the solar atmosphere. They found deviations from energy equipartition, that is, equipartition between kinetic and magnetic energy, a property of Alfvén waves in homogeneous media, for Alfvén waves propagating along expanding flux tubes. They used the model to determine conditions for which strong wave reflection occurs in the chromosphere. The effects of thermal misbalance, that is, plasma heating and cooling depending on background parameters at different powers, were studied for nonlinear Alfvén waves propagating in upper chromospheric conditions by Belov et al. (2020).

Density perturbations induced ponderomotively by nonlinear Alfvén waves cause a local misbalance between heating and cooling, and the authors have shown that for some temperature ranges this could lead to enhanced steepening and damping of the waves. For more details on the importance of thermal misbalance in compressive wave processes and seismology, see Kolotkov et al. (2021).

The generation of spicules could be linked to large-amplitude Alfvén waves, as shown by Martínez-Sykora et al. (2017), who established a connection between the generation of spicules and Alfvénic waves in the chromosphere through ion-neutral collisions or ambipolar diffusion in realistic, radiative numerical simulations of photospheric convection. The effect of Alfvén waves on spicule dynamics and heating was also studied by Sakaue and Shibata (2020) who simulated nonlinear Alfvén wave propagation through the chromosphere. The main finding of this one-dimensional MHD numerical study is that chromospheric dynamics induced by Alfvén waves saturate at certain values of nonlinearity, that is, the dynamics become independent of the amplitude after some value. A related study was performed by Wang and Yokoyama (2020) including the effects of radiative losses. They show that when driven by observation-inspired wave driver in the photosphere, the chromosphere displays a radiative loss profile resembling the one derived from observations. Additionally, the wave energy flux entering the coronal part of the simulations appears to be sufficient to maintain the quiet region corona.

Nutto et al. (2012) ran numerical simulations of a solar network region and found strong mode conversion on equipartition layer between acoustic and magnetic modes. They also find strong reflections above the equipartition. Khomenko and Cally (2012) studied fast to Alfvén conversion in 2.5D simulations in a sunspot-like magnetic configuration with an equipartition layer. They show that fast waves are strongly reflected at the layer, and some of their energy is transferred to Alfvén waves above the layer. The conversion to Alfvén waves is found to be particularly important for a high degree of inclination of the magnetic field with respect to the incident fast wavefront. Note that pure fast and Alfvén waves do exist in general 2.5D configurations, as Alfvén waves are polarized in the third, ignorable direction. Scalisi et al. (2021) investigated analytically the propagation of torsional Alfvén waves originating in magnetic bright point flux tubes, focusing on the mass flux that is generated through the ponderomotive force.

### 2.4.2. Theory of Alfvén Waves in the Solar Corona

The solar corona is arguably a much friendlier environment for theoretical inquiry into the properties of waves than the lower solar atmosphere, being fully ionized and optically thin, thus large-scale phenomena are well described by the ideal MHD equations. In this sense, the theoretical considerations presented in Section 2.2 apply well to describing the propagation of Alfvén waves in the solar corona. However, as mentioned earlier, Alfvén wave propagation in a generally inhomogeneous corona, as the corona is known to be, and the nonlinear evolution of Alfvén waves, either through self-interaction or collisions, are still not entirely understood. The omnipresent propagating Alfvénic waves in the solar corona were modeled in various plasma configurations, starting from the well-known piecewise magnetic flux tube with no azimuthal or along-the-field variation, to fully inhomogeneous and expanding setups. Pascoe et al. (2012, 2013) looked into Alfvénic waves propagating in flux tubes with smooth inhomogeneity using 3D MHD numerical simulations and discussed the mode coupling that is present between the global kink wave that displaces the tube and local azimuthal Alfvén waves. They found a Gaussian damping profile of the kink waves with distance rather than exponential, and suggested that observed damping profiles could be exploited seismologically. This was put in practice by Pascoe et al. (2022) who proposed seismologically inferring the transverse structuring of the corona through the observed damping of propagating Alfvénic waves. Soler et al. (2014) investigated the details of seismology applied to transverse oscillations of flux tubes and found that the assumed inhomogeneity profiles (sinusoidal, linear, parabolic variation) have a strong influence on the inferred plasma parameters, with errors as large as 50% for some scenarios. Antolin et al. (2017) focused on the observational signatures of transverse kink waves, allowing for nonlinear evolution leading to the Kelvin–Helmholtz instability (KHI) for standing waves (e.g., Antolin et al., 2014; Magyar et al., 2015; Terradas et al., 2008b). They note that while the observed kinetic energy inferred from Doppler motions decreases with height due to resonant absorption, a corresponding increase in line width from phase mixing and KHI is not observed, concluding that most of the wave energy of transverse waves is not observed or hidden. Observational signatures of multiple flux tubes undergoing kink oscillations were studied by De Moortel and Pascoe (2012), who also conclude that most of the wave energy is unobservable through optically thin Doppler shifts. Magyar and Van Doorsselaere (2016) derived synthetic Extreme Ultraviolet (EUV) emission images from a 3D numerical simulation of a bundle of strands forming a flux tube that is driven transversely. They show that the strand-like appearance of the flux tube through EUV emission is maintained while the bundles are deformed and mixed due to the ensuing nonlinear dynamics. Kuźma and Murawski (2018) performed 3D numerical simulations of multistranded flux tubes undergoing propagating transverse oscillations driven by an azimuthal driver at the base of the corona. They find that the oscillation of the central strand is torsional Alfvén-like, while the off-central strands undergo Alfvénic swaying. Torsional Alfvén waves driven at the base of the corona were shown to be able to generate density structuring in an initially homogeneous atmosphere by Díaz-Suárez and Soler (2021a), a relevant finding in the quest to understand the origins of the observed coronal structuring. Díaz-Suárez and Soler (2021b) studied the nonlinear evolution of torsional Alfvén waves in flux tubes through 3D numerical simulations and found that standing waves in coronal loops can generate turbulence through the development of KHI. They redid their analysis in Díaz-Suárez and Soler (2022) taking into account the presence of twist in flux tubes, which was found to delay the onset of the KHI. Pascoe et al. (2020) investigated the observational signatures of coronal loop transverse oscillations with implications for seismology based on the evolution of the inhomogeneous layer of the loops. It is found that the disparity between estimates for loop inhomogeneity by various methods (seismological, forward modeling) could be a telltale sign of the nonlinear evolution of coronal loops as a result of wave dynamics. Forward modeling of the line-width broadening induced by large-amplitude propagating Alfvén waves in the solar corona was investigated by Chmielewski et al. (2013), who conclude that it is plausible that observed line-width broadenings in the corona as described in the previous section are the result of large-amplitude Alfvén waves.

The energy content in Alfvénic waves propagating along a bundle of magnetic strands in the linear regime was investigated by Van Doorsselaere et al. (2014), who conclude that the energy content is overestimated by a factor of at least the filling factor of strands in the cross-sectional area, when compared to the energy in pure Alfvén waves propagating in the same volume. Interestingly, resonant absorption of kink waves, a process which is thought to be present whenever Alfvénic waves propagate along structures, can occur even in the absence of denser flux tube structures, as shown by Howson et al. (2019). They considered a setup with an expanding magnetic field but no density variations, with the flux tube being defined only through the magnetic field variation. This highlights that energy content estimations are problematic even for a seemingly uniform plasma, as judged only through emission.

The presence of a twisted magnetic field in flux tubes can introduce various effects on propagating kink waves. Terradas and Goossens (2012) studied the problem analytically, and concluded that the polarization of kink waves (in case of linear polarization) is changing along the flux tube due to the effect of the twist, suggesting that it could be exploited seismologically to measure the amount of twist in observed oscillations. On the other hand, they did not find an impact on kink frequency of standing modes, for the limit of weak twists and the thin tube approximation. Terradas et al. (2018) repeated the twisted loop study numerically, and in the nonlinear regime. They show that twist reduces the growth rate of KHI-like instabilities in the loop, and that after some value of the twist instabilities do not take place within the time it takes the kink waves to damp, a result also found analytically by Barbulescu et al. (2019). Alfvénic waves in prominences were modeled in a number of studies. As a numerical counterpart to the first claimed observation (Okamoto et al., 2015) of resonant absorption in prominence threads, Antolin et al. (2015) simulated transverse oscillations of a prominence modeled as a flux tube. They showed that the observational signatures of the transverse oscillations of the model, such as a thread-like structure due to the development of KHI, and typical signatures of resonant absorption, such as in-plane motions and line-of-sight velocities, match well with the observations. As plasma in prominences is partially ionized, effects of partial ionization on transverse wave propagation and dissipation were studied by Melis et al. (2021). They find that heating due to ambipolar diffusion is sufficient to balance radiative losses in the prominence, while in the coronal part of the model heating (due to other effects) is negligible. The seismological potential of transverse prominence oscillations has also been exploited, specifically for estimating the density profile along the prominence axis based on the ratio of the first two standing harmonics. Soler et al. (2015) found that this method is only sensitive to the average density profile and cannot distinguish between specific density profiles that would lead the same average profile. This conclusion was strengthened using the model of Martínez-Gómez et al. (2022), who additionally point out that the damping time due to resonant absorption also depends on the density profile, which dependence could also be exploited seismologically.

The nonlinear behavior of propagating Alfvénic waves has been investigated by a number of authors. Magyar et al. (2017) show that finite amplitude transverse waves suffer deformations or self-deform as a result of unidirectional propagation, and that a bundle of flux tubes driven by transverse waves transitions into a turbulent-like cross section. This phenomena, termed "uniturbulence," has been investigated in a number of other works, both analytical and numerical (Ismayilli et al., 2022; Magyar et al., 2019, 2021; Van Doorsselaere et al., 2020a, 2020b, 2021). These findings have implications for coronal heating. Models of the solar corona and solar wind heated through Alfvén wave turbulence require a source of incoming waves or reflections for turbulence to be generated through wave collisions, while Alfvénic waves propagating on transverse inhomogeneities can initiate a turbulent cascade through pure outward propagation. It is unclear whether incoming waves have enough energy to power the Alfvén wave cascade, especially in open field regions (van Ballegooijen & Asgari-Targhi, 2016, 2017; Verdini et al., 2019). Guo et al. (2019a) looked into the nonlinear interaction of kink and torsional Alfvén waves propagating within the same flux tube through 3D numerical simulations. They find that both types of standing waves can lead to KHI-like instability in the flux tube. They also find more heating when both kink and Alfvén waves are driven. Vasheghani Farahani et al. (2012) investigated analytically the nonlinear evolution of torsional Alfvén waves propagating along open magnetic field lines in the solar corona. The result of this study is that torsional Alfvén waves in flux tubes have nonlinear properties that are different from those of plane Alfvén waves, the evolution of torsional Alfvén waves being slower and independent of plasma-$\beta$, unlike that of plane waves. The same problem was addressed numerically by Shestov et al. (2017), who found that the nonlinear torsional Alfvén wave induces, besides field-aligned flows, tube waves and transverse flows in the radial direction associated with the fast sausage wave. They also find what they term "nonlinear phase mixing," which arises as a consequence of the specific radial profile of the torsional Alfvén wave perturbation, with larger amplitude shells propagating faster, therefore bending the wavefront. Phase mixing of finite-amplitude Alfvén waves has been studied in various magnetic field configurations, including magnetic X-type null points by Prokopyszyn et al. (2019), showing that nonlinearities result in a reduced amount of energy that is transferred to the plasma, as the nonlinearly induced density perturbations shift the positions where resonances can appear. Waves in such an X-type configuration were also studied before by Pucci et al. (2014), where an analysis of energy dissipation and heating by Alfvén waves is presented. Alfvén wave propagation around three-dimensional X-type null points was studied numerically by Thurgood and McLaughlin (2013), who point out that Alfvén waves accumulate near the fan-plane, where they cause a buildup in currents. They find that the fast and Alfvén waves are still decoupled linearly in such configurations, meaning that wave behavior around 2D null points extends naturally to the full 3D case. Ruderman and Petrukhin (2017) and Petrukhin et al. (2018) investigated the phase mixing of Alfvén waves in an expanding equilibrium magnetic field configuration, and found an enhanced damping compared to planar setups. However, in spite of this enhanced damping, Alfvén waves are still insignificantly damped for realistic values of the viscosity for coronal plume conditions. The authors suggest that increased viscosity, for example, by turbulence, by at least six orders of magnitude, could lead to efficient damping and heating. Tsiklauri (2016) included the effects of plasma flow on the phase mixing of Alfvén waves and explored observational implications. They show that, if open field region structures or plumes are heated by phase mixing, the induced plasma flow as a result cools these structures, rendering them to appear cooler than the background plasma. Boocock and Tsiklauri (2022a,b) investigated both analytically and numerically the phase mixing of torsional Alfvén waves propagating on diverging magnetic field lines in a stratified solar corona. It is found that nonlinear effects associated with the torsional waves, such as mode coupling to compressive modes and their wave steepening, exceed the heating that would be expected from phase mixing on its own. The phase mixing of Alfvénic waves in the generally inhomogeneous solar corona in the linear regime was studied numerically by Magyar and Van Doorsselaere (2022) with implications for the perpendicular spectrum that such inhomogeneities imprint upon the velocity and magnetic field perturbations. They found that the waves acquire the perpendicular spectrum of the structuring. Novel damping mechanisms for propagating Alfvén waves, other than phase mixing, resonant absorption, or turbulent cascade, have also been proposed. In order to explain the strong damping of Alfvén waves in the corona as observed in the measured nonthermal line widths in some studies, Zhao et al. (2015) propose a nonlinear damping mechanism termed "vector Alfvén wave decay," which can be thought of as a vector-type parametric decay instability. This vector decay is shown to be efficient even for the low frequencies of $<0.01$ *Hz* that are usually observed, and could account for the observed damping.

The heating resulting from the dissipation of Alfvénic or Alfvén waves in the solar corona is an important area of research, and one of the main areas of application for Alfvén wave theory (e.g., Arregui, 2015; Van Doorsselaere et al., 2020b). Pagano and De Moortel (2017) investigated the contribution of mode coupling and phase mixing of Alfvén waves in heating the corona, which are linear wave processes. The result of this study is that phase mixing is unlikely to have a significant contribution to heating the corona on a global scale. Pagano et al. (2018) strengthened these conclusions by including also higher harmonic standing waves in their numerical model. Additionally, Pagano et al. (2020) looked into the effect of coronal loops structure on wave heating through phase mixing, and found a significant impact of the initial density structure of the loop. Still, they conclude that heating through phase mixing is insufficient to balance for radiative losses in the corona. Damping of Alfvén waves by phase mixing was revisited analytically in 2.5D model by Prokopyszyn and Hood (2019). The heating rate that they find is three orders of magnitude lower than required to maintain the corona against radiative losses. Van Damme et al. (2020) study the effect of phase mixing on chromospheric evaporation through the heating it produces, in a 2D numerical setup consisting of a coronal loop with the relevant thermodynamic terms present. Although they find that phase mixing does result in a modest heating of the plasma, the resulting chromospheric evaporation is very weak, unable to change the density profile of the loop. Ebrahimi et al. (2020) studied analytically the phase mixing of kink waves in inhomogeneous flux tubes in the presence of viscosity. In this scenario, resonant Alfvén waves get phase-mixed and generate smaller and smaller scales. For viscosities corresponding to realistic Reynolds numbers, no significant heating from phase mixing results during observable transverse motions of impulsively triggered oscillations. Contrary to these setups finding insignificant wave heating, Shi et al. (2021) report on the first 3D MHD simulation of a flux tube driven by transverse waves of observed amplitudes that undergoes enough heating in the process to balance radiative losses. The heating is produced through the dissipation of the induced KHI around the oscillating loop. These results were constrained by De Moortel and Howson (2022), who showed through numerical simulations that balancing radiative losses through transverse wave heating in flux tubes is only possible for resonant driving and loops with densities more appropriate for the quiet sun than active regions, and only in the loop shell region. Heating in multistranded loop configurations as a consequence of Alfvénic wave damping was investigated by Guo et al. (2019b), who found that multistranded loops are heated faster than monolithic ones given the same wave driver, that is, compared to Guo et al. (2019a). Based on this result, they suggest that the generally inhomogeneous nature of the solar corona across the magnetic field needs to be taken into account in future wave-based heating models. Heating in complex coronal plasma configurations in closed-field regions was investigated by Howson et al. (2020) through 3D MHD numerical simulations, building upon their previous work investigating MHD waves in braided magnetic fields (Howson et al., 2019). They note that phase mixing has a volume-filling character due to the general inhomogeneity and that counterpropagating wave collisions induce a cascade of energy to small scales. Driven waves with frequencies lower than the fundamental standing kink mode of the system are not able to power coronal heating, while for higher-frequency waves the rate of dissipation is likely not enough in realistic settings to balance radiative losses. They found that sufficient heating occurs only for driven wave amplitudes that are not observed in the solar atmosphere. Velocity shears can induce mode coupling of MHD waves in the solar atmosphere, as studied analytically in the linear regime by Hollweg et al. (2013). It is shown that Alfvén waves propagating in such a shear perturb other wave modes and also result in their reflection and refraction, depending on the propagation direction. Some of the driven waves are compressive, and can damp through Landau resonances for high enough frequencies. The authors also conjecture the possibility of coherent interactions between the different forward-propagating signals that could lead to turbulence, hinting at dynamics resembling at least in concept the ones first demonstrated numerically by Magyar et al. (2017) for Alfvénic waves propagating on perpendicular density inhomogeneities.

Effects of thermal misbalance on propagating Alfvén waves have also been investigated for coronal conditions. Belov et al. (2021) note how longitudinal plasma motions generated by Alfvén waves through ponderomotive forces can trigger thermal misbalance, resulting in a complex interaction of induced bulk flows and the initial Alfvén wave. Belov et al. (2022) find a difference in the evolution of plane Alfvén waves and torsional Alfvén waves in a thermally active plasma, with stronger effects for plane waves.

The propagation and dynamics of large-amplitude spherically polarized Alfvén waves, that is, not perturbing the magnetic field magnitude, a candidate description for switchbacks observed in the solar wind by Parker Solar Probe, were investigated numerically by Tenerani et al. (2020). They find that such switchbacks are surprisingly stable when propagating through a homogeneous plasma with a straight magnetic field, but eventually get distorted by the parametric decay instability. Based on their survivability up to hundreds of Alfvén crossing times relative to their length, the authors suggest that switchbacks originating in the lower solar corona might reach heights scanned by the Parker Solar Probe. It is also suggested that this might only be the case under homogeneous conditions, and background gradients might lead to faster deformation. This idea was tested by Magyar et al. (2021), who have additionally considered a gravitationally stratified plasma in an expanding magnetic field, with the same switchback model. They note that for weak expansions, switchbacks unfold much faster than in a homogeneous setup due to the gravitational stratification, which acts through multiple mechanisms to destabilize these. However, for certain super-expansions of the magnetic field, that is, more than spherical expansion, switchbacks are able to retain some of their deflections, albeit they still suffer heavy deformations. Raouafi et al. (2023) discuss in detail the dynamics of propagating switchbacks in their recent review paper. It is interesting to note here that the switchback models in the abovementioned studies were 2.5D, not because the simulation of 3D switchback propagation would not be feasible computationally, but because there are no known analytical switchback solutions in three dimensions that are localized. That is, there are solutions for which the total magnetic field magnitude is constant everywhere and there is a localized wrap of any magnitude that represents a switchback, in an otherwise straight and uniform magnetic field. In other words, there are no known solutions of Alfvén wave packages localized in three-dimensional space of finite amplitude and spherical polarization.

## 2.5. SUMMARY AND OUTLOOK

Within this concluding section, let us reflect and put into context the fundamental MHD theory presented in Section 2.2 and the rather descriptively enlisted advances in the theory and observation of Alfvén waves in the solar atmosphere in Sections 2.3 and 2.4. Alfvén waves are the most researched and invoked waves among large-scale plasma oscillations, probably stemming from their deceivingly simple appearance in homogeneous and linear theory, and the associated intuitive understanding. However, we have seen that Alfvén waves are really only simple to understand under idealized conditions, and in such form they most probably do not exist in the solar atmosphere. Any finite amplitude or nonlinear effects, and the nonuniformity or inhomogeneity of the plasma, render Alfvén wave perturbations coupled to perturbations of the plasma that are characteristic of other wave modes, and to total pressure perturbations in general. Special cases exist in which pure Alfvén waves still survive under realistic conditions, but these appear to be restricted to specific magnetic isosurfaces and not volume-filling, leading to the conclusion that they probably amount to a small percentage of total wave energy in the solar atmosphere. On the other hand, we have seen that the adjective "Alfvén" and the properties it carries still survive to a good approximation in transverse waves that are mostly driven by magnetic tension, the hallmark of Alfvénic behavior. These Alfvénic waves come in many forms, from torsional oscillations in cylindrical geometry to the transverse swaying of various plasma structures in the lower atmosphere and corona.

We now know that transverse oscillatory behavior is all-pervading in the solar atmosphere, with a multitude of observational evidence both through direct imaging and spectroscopic signatures such as Doppler shifts and nonthermal line broadening. Some of the presented case studies find clear examples of single events in which the properties of the waves can be measured to a very good approximation, with examples now found in all regions of the solar atmosphere. Observations of omnipresent waves are important for estimating the energy flux that is entering the corona, in the hope of constraining and providing input parameters for coronal heating and solar-wind acceleration models. The measured frequencies, amplitudes, damping times or lengths can provide a valuable tool for inferring properties of the medium in which the waves propagate, through MHD seismology. While in the case of single events the application of seismology is necessarily limited to the environment of the event, in the case of ever-present waves plasma properties could be obtained for extended regions and durations, as already demonstrated in some works. Taking a leap forward and imagining what seismology could offer in the future, maps of various plasma parameters could be obtained in real time from the automated analysis of omnipresent waves and application of the corresponding seismological tools.

In what concerns the theoretical understanding of Alfvén wave dynamics, topics related to the inhomogeneity of the background plasma and nonlinear behavior are still current and forecasted to remain relevant in the future. It is clear that wave behavior is much richer and can lead to surprising effects in the presence of the aforementioned conditions and additional effects related to partial ionization and nonequilibrium ionization, as encountered in the chromosphere, and thermodynamic source and sink terms, and their synthesis in thermal misbalance. While current numerical simulations are capable of capturing most of the large-scale physics of the solar atmosphere and their effects on wave propagation, these can often get too complex for a clear understanding of the underlying mechanisms at play in shaping the dynamics. For example, effects of phase mixing, resonant absorption, ponderomotive forces, mode coupling and conversion, parametric decay, nonlinear deformations, instabilities, etc., can all be present at the same time. Combined with generally inhomogeneous conditions and a broadband driver, the effects of distinct mechanisms on the wave dynamics are hard to distinguish. That is the reason why analytical and numerical studies restricted to the effects of specific mechanisms are relevant, as they facilitate a more complete understanding of these. Advances in both studying the impact of specific mechanisms on waves and in the general simulation of wave-driven regions of the solar atmosphere have been made over the last decade. One of the outcomes of these efforts is that Alfvénic wave heating is still a very real candidate in at least partly explaining coronal heating and solar wind acceleration, and an almost established contributor to chromospheric heating. Equally relevant are the new ways that were explored for applying seismology, in order to infer plasma parameters that are otherwise hard to measure, such as magnetic field magnitude and density scale height.

Exciting observational findings on Alfvén waves in the solar atmosphere are expected to be delivered within the next few years. The full commissioning and science phase of DKIST has already started to offer great new opportunities for spotting waves in various atmospheric regions (e.g., Fehlmann et al., 2023). One of the critical science objectives of DKIST is to establish the origin and role played by torsional Alfvén waves in chromospheric heating, to be accomplished by its spectropolarimetric observing capabilities of unprecedented resolution and cadence (Rast et al., 2021). The European Solar Telescope (EST) will become the most powerful ground-based European telescope to study the Sun in the coming decades in visible and near infrared bands (Quintero Noda et al., 2022). The study of wave coupling and transmission throughout the lower solar atmosphere, in which Alfvén waves are expected to play a prominent role, is one of the principal science objectives of EST. The high-cadence simultaneous polarimetry of multiple photospheric and chromospheric spectral lines is expected to greatly expand our knowledge of Alfvén wave evolution and transmission in the lower solar atmosphere. Observations from Solar Orbiter with the Extreme Ultraviolet Imager (EUI) instrument have already delivered some previously unobserved wave activity, and will offer unique opportunities for imaging waves at the solar poles (e.g., Rochus et al., 2020). Parker Solar Probe is expected to plunge deeper into the solar corona, reaching just below 10 solar radii at perihelion, providing some great opportunities to study Alfvén waves in situ in the solar corona (Kasper et al., 2021). The Multi-Slit Solar Explorer (MUSE), a recently selected space mission, will offer unprecedented spectral and imaging diagnostics of the solar corona (e.g., De Pontieu et al., 2022). Thanks to its multislit capabilities, it will be possible to follow individual wave packet propagation to accurately determine the group speed and wave propagation direction, and thus better discriminate between Alfvénic and other wave modes. Among other possibilities, MUSE observations will have the required cadence and resolution to better observe the interplay between Alfvén waves and field-aligned motions as in spicules. The soon to be launched Aditya-L1 mission will allow us to observe the solar atmosphere in various new emission lines in the near-ultraviolet (e.g., Tripathi et al., 2017). With launch targeted toward 2025, SOLAR-C will represent the next-generation EUV spectrometer with currently unparalleled sensitivity and resolution (e.g., Shimizu et al., 2019).

Some of the still unsolved, overarching questions related to Alfvén waves in the solar atmosphere concern their generation, propagation, and eventual cascade and thermalization, which still lack clear description. More specifically, how exactly are the Alfvénic waves observed in the solar corona driven by photospheric dynamics, and what significance do peak energies around 5 min have as measured by CoMP. Do p-modes have an influence in this process as commonly suggested? Additionally, the lack of standing waves in CoMP-like observations, even along closed field lines, remains puzzling. Quantifying the energy available in Alfvén waves in the different layers of the solar atmosphere remains a challenge and warrants a continuing investigation. Transmission and reflection of Alfvén waves through the lower atmosphere with its complex physical processes is intensely researched and most probably a key factor in determining the wave energy flux and mass loss rate into the corona. Many linear and nonlinear channels have been proposed for the cascade of Alfvén wave energy to small scales, which is necessary for their thermalization due to the very small dissipation coefficients in the solar atmosphere. Works on MHD wave heating of the solar corona usually assume viscous or resistive damping with either classical coefficients or turbulently enhanced values, which can be orders of magnitude different, leaving a lot of room for speculation and uncertainty. Moreover, numerical simulations cannot achieve the extremely low dissipation coefficient values that are thought to be present in the corona, therefore simulations have orders of magnitude lower (magnetic) Reynolds numbers than expected. Although this problem concerns more the kinetic part of the coronal heating mechanism, it is important to remember that the bridging of the two approaches is seldom looked into in detail. A notable example is the recent attempt to unite large, MHD-scale, and kinetic turbulence under a single model by Meyrand et al. (2021) and Squire et al. (2022). The incorporation of inhomogeneities across the magnetic field in a consistent way in theories and models of Alfvén-wave-induced solar atmospheric turbulence is another important step that has to be taken in the future, as there are indications that energy cascade rates might be affected by inhomogeneity (Magyar and Nakariakov, 2021).

In conclusion, Alfvén waves, the legacy of Hannes Alfvén, remain a topic of active research for the solar atmosphere, being one of the fundamental plasma processes that might ultimately represent the solution to many of the Sun's mysteries that we are trying to unravel.

## ACKNOWLEDGMENTS

N.M. acknowledges Research Foundation – Flanders (Fonds voor Wetenschappelijk Onderzoek – Vlaanderen) for their support through a postdoctoral fellowship. This research was supported by the International Space Science Institute (ISSI) in Bern, through ISSI International Team project #560: "Turbulence at the Edge of the Solar Corona: Constraining Available Theories Using the Latest Parker Solar Probe Measurements."

## REFERENCES

Alfvén, H. (1942). Existence of electromagnetic-hydrodynamic waves. *Nature*, October, *150*, 405–406.

Alfvén, H., & Lindblad, B. (1945). Magneto-hydrodynamic waves and sunspots. *Monthly Notices of the Royal Astronomical Society*, *105*(1), 3–16.

Alfvén, H., & Lindblad, B. (1947). Granulation, magneto-hydrodynamic waves, and the heating of the solar corona. *Monthly Notices of the Royal Astronomical Society*, June, *107*, 211–219.

Antolin, P., De Moortel, I., Van Doorsselaere, T., & Yokoyama, T. (2017). Observational signatures of transverse magnetohydrodynamic waves and associated dynamic instabilities in coronal flux tubes. *The Astrophysical Journal*, February, *836*, 219.

Antolin, P., Okamoto, T. J., De Pontieu, B., Uitenbroek, H., Van Doorsselaere, T., & Yokoyama, T. (2015). Resonant absorption of transverse oscillations and associated heating in a solar prominence. II. Numerical aspects. *The Astrophysical Journal*, August, *809*, 72.

Antolin, P., Pagano, P., De Moortel, I., & Nakariakov, V. M. (2018). In Situ generation of transverse magnetohydrodynamic waves from colliding flows in the solar corona. *The Astrophysical Journal Letters*, *861*(2), 6.

Antolin, P., Yokoyama, T., & Van Doorsselaere, T. (2014). Fine strand-like structure in the solar corona from magnetohydrodynamic transverse oscillations. *The Astrophysical Journal Letters*, June, *787*, L22.

Appert, K., Gruber, R., & Vaclavik, J. (1974). Continuous spectra of a cylindrical magnetohydrodynamic equilibrium. *Physics of Fluids*, July, *17*, 1471–1472.

Arregui, I. (2015). Wave heating of the solar atmosphere. *Philosophical Transactions of the Royal Society of London Series A*, April, *373*, 20140261.

Aschwanden, M. J. (2005). *Physics of the solar corona. An introduction with problems and solutions*, 2nd ed. Chichester, UK: Springer, New York, Berlin.

Aschwanden, M. J. (2019). Helical twisting number and braiding linkage number of Solar Coronal Loops. *The Astrophysical Journal*, April, *874*, 131.

Aschwanden, M. J., Fletcher, L., Schrijver, C. J., & Alexander, D. (1999). Coronal loop oscillations observed with the transition region and coronal explorer. *The Astrophysical Journal*, August, *520*, 880–894.

Aschwanden, M. J., Gošic, M., Hurlburt, N. E., & Scullion, E. (2018). Convection-driven generation of ubiquitous coronal waves. *The Astrophysical Journal*, October, *866*, 73.

Aschwanden, M. J., & Peter, H. (2017). The Width distribution of loops and strands in the solar corona—Are we hitting rock bottom? *The Astrophysical Journal*, May, *840*, 4.

Aschwanden, M. J., & Wang, T. (2020). Torsional Alfvénic oscillations discovered in the magnetic free energy during solar flares. *The Astrophysical Journal*, March, *891*, 99.

Avrett, E. H., & Loeser, R. (2008). Models of the solar chromosphere and transition region from SUMER and HRTS observations: Formation of the extreme-ultraviolet spectrum of hydrogen, carbon, and oxygen. *The Astrophysical Journal Supplement Series*, *175*(1), 229–276.

Ballai, I. (2019). Linear waves in partially ionized plasmas in ionization non-equilibrium. *Frontiers in Astronomy and Space Sciences*, May, *6*, 39.

Ballai, I. (2020). Diagnostics of plasma ionisation using torsional Alfén waves. *Astronomy & Astrophysics*, March, *635*, L2.

Ballester, J. L., Soler, R., Terradas, J., & Carbonell, M. (2020). Nonlinear coupling of Alfvén and slow magnetoacoustic waves in partially ionized solar plasmas. *Astronomy & Astrophysics*, September, *641*, A48.

Banerjee, D., Krishna Prasad, S., Pant, V., McLaughlin, J. A., Antolin, P., Magyar, N., et al. (2021). Magnetohydrodynamic waves in open coronal structures. *Space Science Reviews*, September, *217*, 76.

Barbulescu, M., Ruderman, M. S., Van Doorsselaere, T., & Erdélyi, R. (2019). An analytical model of the Kelvin-Helmholtz instability of transverse coronal loop oscillations. *The Astrophysical Journal*, January, *870*, 108.

Bate, W., Jess, D. B., Nakariakov, V. M., Grant, S. D. T., Jafarzadeh, S., Stangalini, M., et al. (2022). High-frequency waves in chromospheric spicules. *The Astrophysical Journal*, May, *930*, 129.

Belov, S. A., Vasheghani Farahani, S., & Molevich, N. E. (2022). Propagating torsional Alfvén waves in thermally active solar plasma. *Monthly Notices of the Royal Astronomical Society*, *515*(4), 5151–5158.

Belov, S., Molevich, N., & Zavershinskii, D. (2020). Thermal Misbalance influence on the nonlinear shear Alfvén waves under solar atmosphere conditions. *Solar Physics*, *295*(11), 160.

Belov, S., Vasheghani Farahani, S., Molevich, N., & Zavershinskii, D. (2021). Longitudinal plasma motions generated by shear Alfvén waves in plasma with thermal misbalance. *Solar Physics*, *296*(6), 122.

Bemporad, A., & Abbo, L. (2012). Spectroscopic signature of Alfvén waves damping in a polar coronal hole up to 0.4 solar radii. *The Astrophysical Journal*, June, *751*, 110.

Boocock, C., & Tsiklauri, D. (2022a). Enhanced phase mixing of torsional Alfvén waves in stratified and divergent solar coronal structures—II. Non-linear simulations. *Monthly Notices of the Royal Astronomical Society*, *510*(2), 2618–2627.

Boocock, C., & Tsiklauri, D. (2022b). Enhanced phase mixing of torsional Alfvén waves in stratified and divergent solar coronal structures—Paper I. Linear solutions. *Monthly Notices of the Royal Astronomical Society*, *510*(2), 1910–1922.

Braileanu, B. P., Lukin, V. S., Khomenko, E., & de Vicente, Á. (2019). Two-fluid simulations of waves in the solar chromosphere. *Astronomy & Astrophysics*, June, *627*, A25.

Cally, P. S. (2017). Alfvén waves in the structured solar corona. *Monthly Notices of the Royal Astronomical Society*, *466*(1), 413–424.

Cally, P. S. (2022). On the fragility of Alfvén waves in a stratified atmosphere. *Monthly Notices of the Royal Astronomical Society*, February, *510*, 1093.

Cally, P. S., & Khomenko, E. (2015). Fast-to-alfvén mode conversion mediated by the hall current. I. Cold plasma model. *The Astrophysical Journal*, November, *814*, 106.

Cally, P. S., & Khomenko, E. (2018). Fast-to-Alfvén mode conversion in the presence of ambipolar diffusion. *The Astrophysical Journal*, March, *856*, 20.

Cally, P. S., & Khomenko, E. (2019). Fast-to-Alfvén mode conversion and ambipolar heating in structured media. I.

Simplified cold plasma model. *The Astrophysical Journal*, October, *885*, 58.

Carlsson, M., Pontieu, B. D., & Hansteen, V. H. (2019). New view of the solar chromosphere. *Annual Review of Astronomy and Astrophysics*, August, *57*, 189–226.

Chae, J., Cho, K., Lim, E.-K., & Kang, J. (2022). Propagating Alfvénic waves observed in the chromosphere around a small sunspot: Tales of 3-minute waves and 10-minute waves. *The Astrophysical Journal*, July, *933*, 108.

Chae, J., Cho, K., Nakariakov, V. M., Cho, K.-S., & Kwon, R.-Y.(2021). Spectroscopic detection of Alfvénic waves in the chromosphere of sunspot regions. *The Astrophysical Journal Letters*, June, *914*, L16.

Chandran, B. D. G., & Perez, J. C. (2019). Reflection-driven magnetohydrodynamic turbulence in the solar atmosphere and solar wind. *Journal of Plasma Physics*, August, *85*, 905850409.

Chatterjee, P. (2020). Testing Alfvén wave propagation in a "realistic" set-up of the solar atmosphere. *Geophysical & Astrophysical Fluid Dynamics*, March, *114*, 213.

Chelpanov, A., & Kobanov, N. (2022). Problems in observation and identification of torsional waves in the lower solar atmosphere. *Solar Physics*, December, *297*, 154.

Chen, L., Zonca, F., & Lin, Y. (2021). Physics of kinetic Alfvén waves: A gyrokinetic theory approach. *Reviews of Modern Plasma Physics*, February, *5*, 1.

Chmielewski, P., Srivastava, A. K., Murawski, K., & Musielak, Z. E. (2013). Pulse-driven non-linear Alfvén waves and their role in the spectral line broadening. *Monthly Notices of the Royal Astronomical Society*, *428*(1), 40–49.

Cho, I-H., Moon, Y.-J., Cho, K-S., Nakariakov, V. M., Lee, J-Y., & Kim, Y-H. (2019). A new type of jet in a polar limb of the solar coronal hole. *The Astrophysical Journal Letters*, October, *884*, L38.

Cranmer, S. R. (2018). Low-frequency Alfvén waves produced by magnetic reconnection in the Sun's magnetic carpet. *The Astrophysical Journal*, *862*(1), 6.

De Moortel, I., & Howson, T. A. (2022). (When) Can wave heating balance optically thin radiative losses in the corona? *The Astrophysical Journal*, December, *941*, 85.

De Moortel, I., McIntosh, S. W., Threlfall, J., Bethge, C., & Liu, J. (2014). Potential evidence for the onset of alfvénic turbulence in trans-equatorial coronal loops. *The Astrophysical Journal*, February, *782*, L34.

De Moortel, I., & Nakariakov, V. M. (2012). Magnetohydrodynamic waves and coronal seismology: an overview of recent results. *Royal Society of London Philosophical Transactions Series A*, July, *370*, 3193–3216.

De Moortel, I., & Pascoe, D. J. (2012). The effects of line-of-sight integration on multistrand coronal loop oscillations. *The Astrophysical Journal*, February, *746*, 31.

De Pontieu, B., Carlsson, M., Rouppe van der Voort, L. H. M., Rutten, R. J., Hansteen, V. H., & Watanabe, H. (2012). Ubiquitous torsional motions in Type II spicules. *The Astrophysical Journal*, May, *752*, L12.

De Pontieu, B., Mcintosh, S., Carlsson, M., Hansteen, V. H., Tarbell, T. D., Schrijver, C. J., et al. (2007). Chromospheric Alfvénic waves strong enough to power the solar wind. *Science*, December, *318*, 1574–1577.

De Pontieu, B., Testa, P., Martínez-Sykora, J., Antolin, P., Karampelas, K., Hansteen, V., et al. (2022). Probing the physics of the solar atmosphere with the Multi-slit Solar Explorer (MUSE). I. Coronal heating. *The Astrophysical Journal*, *926*(1), 36.

DeForest, C. E., Howard, R. A., Velli, M., Viall, N., & Vourlidas, A. (2018). The highly structured outer solar corona. *The Astrophysical Journal*, *862*(1), 18.

DeForest, C. E., Howard, T. A., & McComas, D. J. (2014). Inbound waves in the solar corona: A direct indicator of Alfvén surface location. *The Astrophysical Journal*, *787*(2), 10.

Dessler, A. J. (1970). Nobel Prizes: 1970 Awards Honor Three in Physics and Chemistry. *Science*, November, *170*, 604–606.

Díaz-Suárez, S., & Soler, R. (2021a). Overdense threads in the solar corona induced by torsional Alfvén waves. *The Astrophysical Journal Letters*, *922*(2), 6.

Díaz-Suárez, S., & Soler, R. (2021b). Transition to turbulence in nonuniform coronal loops driven by torsional Alfvén waves. *Astronomy & Astrophysics*, *648*, 17.

Díaz-Suárez, S., & Soler, R. (2022). Transition to turbulence in nonuniform coronal loops driven by torsional Alfvén waves. II. Extended analysis and effect of magnetic twist. *Astronomy & Astrophysics*, *665*, 12.

Dmitruk, P., Matthaeus, W. H., Milano, L. J., & Oughton, S. (2001). Conditions for sustainment of magnetohydrodynamic turbulence driven by Alfvén waves. *Physics of Plasmas*, May, *8*, 2377–2384.

Dmitruk, P., Matthaeus, W. H., Milano, L. J., Oughton, S., Zank, G. P., & Mullan, D. J. (2002). Coronal heating distribution due to low-frequency, wave-driven turbulence. *The Astrophysical Journal*, August, *575*, 571–577.

Dudok de Wit, T., Krasnoselskikh, V. V., Bale, S. D., Bonnell, J. W., Bowen, T. A., Chen, C. H. K., et al. (2020). Switchbacks in the Near-Sun Magnetic Field: Long memory and impact on the turbulence cascade. *The Astrophysical Journal Supplement*, February, *246*, 39.

Ebrahimi, Z., Soler, R., & Karami, K. (2020). Phase mixing of Kink MHD waves in the solar corona: Viscous dissipation and heating. *The Astrophysical Journal*, April, *893*, 157.

Edwin, P. M., & Roberts, B. (1983). Wave propagation in a magnetic cylinder. *Solar Physics*, October, *88*, 179–191.

Elsässer, W. M. (1950). The hydromagnetic equations. *Physical Review*, July, *79*, 183–183.

Fehlmann, A., Giebink, C., Kuhn, J., Mickey, D., & Scholl, I. (2023). The Daniel K. Inouye Solar Telescope (DKIST) Cryogenic near-infrared spectro-polarimeter. *Solar Physics*, *298*(1), 5.

Fermi, E. (1949). On the origin of the cosmic radiation. *Physical Review*, April, *75*, 1169–1174.

Fisk, L. A., & Kasper, J. C. (2020). Global circulation of the open magnetic flux of the Sun. *The Astrophysical Journal Letters*, May, *894*, L4.

Fleishman, G. D., & Toptygin, I. N. (2013a). Stochastic particle acceleration by helical turbulence in solar flares. *Monthly Notices of the Royal Astronomical Society*, March, *429*, 2515–2526.

Fleishman, G. D., & Toptygin, I. N. (2013b). *Cosmic electrodynamics: Electrodynamics and magnetic hydrodynamics of cosmic plasmas*. New York: Springer New York.

Galeev, A. A., & Oraevskii, V. N. (1963). The stability of Alfvén waves. *Soviet Physics Doklady*, May, *7*, 988.

Goedbloed, J. P. H., & Poedts, S. (2004). *Principles of magnetohydrodynamics*. UK: Cambridge University Press.

Goldstein, M. L. (1978). An instability of finite amplitude circularly polarized Afvén waves. *The Astrophysical Journal*, January, *219*, 700–704.

Gómez, J. C. G., Jafarzadeh, S., Wedemeyer, S., & Szydlarski, M. (2022). Propagation of transverse waves in the solar chromosphere probed at different heights with ALMA sub-bands. *Astronomy & Astrophysics*, September, *665*, L2.

Goossens, M., Andries, J., Soler, R., Van Doorsselaere, T., Arregui, I., & Terradas, J. (2012). Surface Alfvén waves in solar flux tubes. *The Astrophysical Journal*, July, *753*, 111.

Goossens, M., Arregui, I., Soler, R., & Van Doorsselaere, T. (2020). Resonant absorption: Transformation of compressive motions into vortical motions. *Astronomy & Astrophysics*, *641*, 8.

Goossens, M., Erdélyi, R., & Ruderman, M. S. (2011). Resonant MHD waves in the solar atmosphere. *Space Science Reviews*, July, *158*, 289–338.

Goossens, M., Hollweg, J. V., & Sakurai, T. (1992). Resonant behaviour of MHD waves on magnetic flux tubes. III—Effect of equilibrium flow. *Solar Physics*, April, *138*, 233–255.

Goossens, M. L., Arregui, I., & Van Doorsselaere, T. (2019). Mixed properties of MHD waves in non-uniform plasmas. *Frontiers in Astronomy and Space Sciences*, April, *6*, 20.

Goossens, M., Ruderman, M. S., & Hollweg, J. V. (1995). Dissipative MHD solutions for resonant Alfven waves in 1-dimensional magnetic flux tubes. *Solar Physics*, March, *157*, 75–102.

Goossens, M., Terradas, J., Andries, J., Arregui, I., & Ballester, J. L. (2009). On the nature of kink MHD waves in magnetic flux tubes. *Astronomy & Astrophysics*, August, *503*, 213–223.

Goossens, M., Van Doorsselaere, T., Soler, R., & Verth, G. (2013). Energy content and propagation in transverse solar atmospheric waves. *The Astrophysical Journal*, *768*(2), 12.

Guo, M., Van Doorsselaere, T., Karampelas, K., Li, B., Antolin, P., & De Moortel, I. (2019a). Heating effects from driven transverse and Alfvén waves in coronal loops. *The Astrophysical Journal*, *870*(2), 10.

Guo, M., Van Doorsselaere, T., Karampelas, K., & Li, B. (2019b). Wave heating in simulated multistranded coronal loops. *The Astrophysical Journal*, *883*(1), 9.

Guo, J. H., Ni, Y. W., Qiu, Y., Zhong, Z., Guo, Y., & Chen, P. F. (2021). Magnetic twists of solar filaments. *The Astrophysical Journal*, August, *917*, 81.

Gupta, G. R. (2017). Spectroscopic evidence of Alfvén wave damping in the off-limb solar corona. *The Astrophysical Journal*, February, *836*, 4.

Gupta, G. R., Zanna, G. D., & Mason, H. E. (2019). Exploring the damping of Alfvén waves along a long off-limb coronal loop, up to 1.4 solar radii. *Astronomy & Astrophysics*, July, *627*, A62.

Hahn, M., Fu, X., & Savin, D. W. (2022). Evidence for parametric decay instability in the lower solar atmosphere. *The Astrophysical Journal*, July, *933*, 52.

Hahn, M., Landi, E., & Savin, D. W. (2012). Evidence of wave damping at low heights in a polar coronal hole. *The Astrophysical Journal*, June, *753*, 36.

Hahn, M., & Savin, D. W. (2013). Observational quantification of the energy dissipated by alfvén waves in a polar coronal hole: evidence that waves drive the fast solar wind. *The Astrophysical Journal*, September, *776*, 78.

Hahn, M., & Savin, D. W. (2014). Evidence for wave heating of the quiet-sun corona. *The Astrophysical Journal*, October, *795*, 111.

Hale, G. E. (1908). On the probable existence of a magnetic field in sun-spots. *The Astrophysical Journal*, November, *28*, 315.

Hasegawa, A., & Uberoi, C. (1982). *The Alfvén wave*. Oak Ridge, TN (USA): U.S. Department of Energy Technical Information Center.

Heinemann, M., & Olbert, S. (1980). Non-WKB Alfven waves in the solar wind. *Journal of Geophysical Research*, March, *85*, 1311–1327.

Hillier, A., Morton, R. J., & Erdélyi, R. (2013). A statistical study of transverse oscillations in a quiescent prominence. *The Astrophysical Journal Letters, December*, *779*, L16.

Hollweg, J. V. (1990). On WKB expansions for Alfven waves in the solar wind. *Journal of Geophysical Research*, September, *95*, 14873–14879.

Hollweg, J. V., & Isenberg, P. A. (2007). Reflection of Alfvén waves in the corona and solar wind: An impulse function approach. *Journal of Geophysical Research (Space Physics)*, August, *112*, A08102.

Hollweg, J. V., Kaghashvili, E. K., & Chandran, B. D. G. (2013). Velocity-shear-induced Mode Coupling in the Solar Atmosphere and Solar Wind: Implications for Plasma Heating and MHD Turbulence. *The Astrophysical Journal*, June, *769*, 142.

Howson, T. A., De Moortel, I., Antolin, P., Van Doorsselaere, T., & Wright, A. N. (2019). Resonant absorption in expanding coronal magnetic flux tubes with uniform density. *Astronomy & Astrophysics*, *631*, 11.

Howson, T. A., De Moortel, I., & Reid, J. (2020). Phase mixing and wave heating in a complex coronal plasma. *Astronomy & Astrophysics*, *636*, 13.

Howson, T. A., De Moortel, I., Reid, J., & Hood, A. W. (2019). Magnetohydrodynamic waves in braided magnetic fields. *Astronomy & Astrophysics*, *629*, 12.

Ismayilli, R., Van Doorsselaere, T., Goossens, M., & Magyar, N. (2022). Non-linear damping of surface Alfvén waves due to Uniturbulence. *Frontiers in Astronomy and Space Sciences*, *8*, id. 241.

Jaeggli, S. A., Lin, H., Onaka, P., Yamada, H., Anan, T., Bonnet, M., et al. (2022). The Diffraction-Limited Near-Infrared Spectropolarimeter (DL-NIRSP) of the Daniel K. Inouye Solar Telescope (DKIST). *Solar Physics*, October, 297.

Jafarzadeh, S., Solanki, S. K., Gafeira, R., van Noort, M., Barthol, P., Blanco Rodríguez, J., et al. (2017). Transverse oscillations in slender Ca ii H fibrils observed with Sunrise/SuFI. *The Astrophysical Journal Supplement Series*, March, *229*, 9.

Jess, D. B., Jafarzadeh, S., Keys, P., Stangalini, M., Verth, G., & Grant, S. D. (2023). Waves in the lower solar atmosphere: The dawn of next-generation solar telescopes. *Living Reviews in Solar Physics*, January, *20*, 1.

Jess, D. B., Mathioudakis, M., Erdélyi, R., Crockett, P. J., Keenan, F. P., & Christian, D. J. (2009). Alfvén Waves in the Lower Solar Atmosphere. *Science*, March, *323*, 1582.

Jess, D. B., Morton, R. J., Verth, G., Fedun, V., Grant, S. D. T., & Giagkiozis, I. (2015). Multiwavelength studies of MHD waves in the solar chromosphere. An overview of recent results. *Space Science Reviews*, July, *190*, pp. 103–161.

Jess, D. B., Shelyag, S., Mathioudakis, M., Keys, P., Christian, D., & Keenan, F. (2012). Propagating wave phenomena detected in observations and simulations of the lower solar atmosphere. *The Astrophysical Journal*, February, *746*, 183.

Johnston, Z., Squire, J., Mallet, A., & Meyrand, R. (2022). On the properties of Alfvénic switchbacks in the expanding solar wind: Three-dimensional numerical simulations. *Physics of Plasmas*, July, *29*, 072902.

Karampelas, K., & Van Doorsselaere, T. (2020). Generating transverse loop oscillations through a steady-flow driver. *The Astrophysical Journal Letters*, *897*(2), L35.

Kasper, J. C., Klein, K. G., Lichko, E., Huang, J., Chen, C. H. K., Badman, S. T., et al. (2021). Parker solar probe enters the magnetically dominated solar corona. *Physical Review Letters*, December, *127*, 255101.

Khomenko, E., & Cally, P. S. (2012). Numerical simulations of conversion to alfvén waves in sunspots. *The Astrophysical Journal*, January, *746*, 68.

Khomenko, E., & Cally, P. S. (2019). Fast-to-Alfvén mode conversion and ambipolar heating in structured media. II. Numerical simulation. *The Astrophysical Journal,* October, *883*, 179.

Khomenko, E., Collados, M., Dı′ az, A., & Vitas, N. (2014). Fluid description of multi-component solar partially ionized plasma. *Physics of Plasmas*, September, *21*, 092901.

Khomenko, E., Vitas, N., Collados, M., & de Vicente, A. (2018). Three-dimensional simulations of solar magneto-convection including effects of partial ionization. *Astronomy & Astrophysics*, October, *618*, A87.

Kohutova, P., Verwichte, E., & Froment, C. (2020). First direct observation of a torsional Alfvén oscillation at coronal heights. *Astronomy & Astrophysics*, January, *633*, L6.

Kolotkov, D. Y., Zavershinskii, D. I., & Nakariakov, V. M. (2021). The solar corona as an active medium for magnetoacoustic waves. *Plasma Physics and Controlled Fusion*, December, *63*, 124008.

Kraskiewicz, J., Murawski, K., Zhang, F., & Poedts, S. (2023). Monochromatic two-fluid Alfvén waves in the partially ionised solar chromosphere. *Solar Physics*, January, *298*, 11.

Kuridze, D., Morton, R. J., Erdelyi, R., Dorrian, G. D., Mathioudakis, M., Jess, D. B., & Keenan, F. P. (2012). Transverse oscillations in chromospheric mottles. *The Astrophysical Journal*, April, *750*, 51.

Kuridze, D., Verth, G., Mathioudakis, M., Erdélyi, R., Jess, D. B., Morton, R. J., et al. (2013). Characteristics of transverse waves in chromospheric mottles. *The Astrophysical Journal*, November, *779*, 82.

Kuźma, B., & Murawski, K. (2018). Numerical simulations of transverse oscillations of a finely structured solar flux tube. *The Astrophysical Journal*, *866*(1), 8.

Kuźma, B., Murawski, K., & Poedts, S. (2021). 3D numerical simulations of propagating two-fluid, torsional Alfvén waves and heating of a partially ionized solar chromosphere. *Monthly Notices of the Royal Astronomical Society*, July, *506*, 989–996.

Kuźma, B., Wójcik, D., Murawski, K., Yuan, D., & Poedts, S. (2020). Numerical simulations of the lower solar atmosphere heating by two-fluid nonlinear Alfvén waves. *Astronomy & Astrophysics*, *639*, 10.

Lin, Y., Soler, R., Engvold, O., Ballester, J. L., Langangen, Ø., Oliver, R., et al. (2009). Swaying threads of a solar filament. *The Astrophysical Journal, October*, *704*, 870.

Liu, J., McIntosh, S. W., De Moortel, I., Threlfall, J., & Bethge, C. (2014). Statistical evidence for the existence of Alfvénic turbulence in solar coronal loops. *The Astrophysical Journal*, February, *797*(1), 10.

Liu, J., Nelson, C. J., Snow, B., Wang, Y., & Erdélyi, R. (2019). Evidence of ubiquitous Alfvén pulses transporting energy from the photosphere to the upper chromosphere. *Nature Communications*, August, *10*, 3504.

Lundquist, S. (1949). Experimental investigations of magneto-hydrodynamic waves. *Physical Review*, December, *76*, 1805–1809.

Magyar, N., & Van Doorsselaere, T. (2022). Phase mixing and the 1/f spectrum in the solar wind. *The Astrophysical Journal*, October, *938*, 98.

Magyar, N., & Nakariakov, V. M. (2021). Three-dimensional simulations of the inhomogeneous low solar wind. *The Astrophysical Journal*, January, *907*, 55.

Magyar, N., Utz, D., Erdélyi, R., & Nakariakov, V. M. (2021). Could switchbacks originate in the lower solar atmosphere? II. Propagation of switchbacks in the solar corona. *The Astrophysical Journal*, *914*(1), 13.

Magyar, N., & Van Doorsselaere, T. (2016). The instability and non-existence of multi-stranded loops when driven by transverse waves. *The Astrophysical Journal*, June, *823*, 82.

Magyar, N., & Van Doorsselaere, T. (2018). Assessing the capabilities of dynamic coronal seismology of Alfvénic waves through forward modeling. *The Astrophysical Journal*, April, *856*, 144.

Magyar, N., Van Doorsselaere, T., & Goossens, M. (2017). Generalized phase mixing: Turbulence-like behaviour from unidirectionally propagating MHD waves. *Scientific Reports*, November, *7*, 14820.

Magyar, N., Van Doorsselaere, T., & Goossens, M. (2019a). The nature of Elsässer variables in compressible MHD. *The Astrophysical Journal*, March, *873*, 56.

Magyar, N., Van Doorsselaere, T., & Goossens, M. (2019b). Understanding Uniturbulence: Self-cascade of MHD Waves in the Presence of Inhomogeneities. *The Astrophysical Journal*, September, *882*, 50.

Magyar, N., Van Doorsselaere, T., & Marcu, A. (2015). Numerical simulations of transverse oscillations in radiatively cooling coronal loops. *Astronomy & Astrophysics*, October, *582*, A117.

Malanushenko, A., Cheung, M. C. M., DeForest, C. E., Klimchuk, J. A., & Rempe, M. (2022). The coronal veil. *The Astrophysical Journal*, March, *927*, 1.

Martínez-Gómez, D., Soler, R., Terradas, J., & Khomenko, E. (2022). Transverse kink oscillations of inhomogeneous prominence threads: Numerical analysis and Hα forward modelling. *Astronomy & Astrophysics*, February, *658*, A106.

Martínez-Sykora, J., Pontieu, B., Hansteen, V., Voort, L. R., Carlsson, M., & Pereira, T. (2017). On the generation of solar spicules and Alfvénic waves. *Science*, June, *356*, 1269–1272.

Mathioudakis, M., Jess, D. B., & Erdélyi, R. (2012). Alfvén waves in the solar atmosphere. *Space Science Reviews*, November, *175*, 1–27.

Matthaeus, W. H., Zank, G. P., Oughton, S., Mullan, D. J., & Dmitruk, P. (1999). Coronal heating by magnetohydrodynamic turbulence driven by reflected low-frequency waves. *The Astrophysical Journal Letters*, September, *523*, L93–L96.

Matthews, S. A., Harra, L. K., Zharkov, S., & Green, L. M. (2015). Spectroscopic signatures related to a sunquake. *The Astrophysical Journal*, October, *812*, 35.

McIntosh, S. W., de Pontieu, B., Carlsson, M., Hansteen, V., Boerner, P., & Goossens, M. (2011). Alfvénic waves with sufficient energy to power the quiet solar corona and fast solar wind. *Nature*, July, *475*, 477–480.

McIntosh, S. W., & Pontieu, B. D. (2012). Estimating the dark energy content of the solar corona. *The Astrophysical Journal*, December, *761*, 138.

Melis, L., Soler, R., & Ballester, J. L. (2021). Alfvén wave heating in partially ionized thin threads of solar prominences. *Astronomy & Astrophysics, June*, *650*, A45.

Meyrand, R., Squire, J., Schekochihin, A. A., & Dorland, W. (2021). On the violation of the zeroth law of turbulence in space plasmas. *Journal of Plasma Physics*, *87*(3), 535870301.

Moreton, G. E., & Ramsey, H. E. (1960). Recent observations of dynamical phenomena associated with solar flares. *Publications of the Astronomical Society of the Pacific*, October, *72*, 357.

Morton, R. J. (2014). Magneto-seismological insights into the penumbral chromosphere and evidence for wave damping in spicules. *Astronomy& Astrophysics*, June, *566*, A90.

Morton, R. J., & McLaughlin, J. A. (2013). Hi-C and AIA observations of transverse magnetohydrodynamic waves in active regions. *Astronomy & Astrophysics*, May, *553*, L10.

Morton, R. J., Tiwari, A. K., Doorsselaere, T. V., & McLaughlin, J. A. (2021). Weak damping of propagating MHD kink waves in the quiescent corona. *The Astrophysical Journal*, December, *923*, 225.

Morton, R. J., Tomczyk, S., & Pinto, R. (2015). Investigating Alfvénic wave propagation in coronal open-field regions. *Nature Communications*, July, *6*, 7813.

Morton, R. J., Tomczyk, S., & Pinto, R. F. (2016). A global view of velocity fluctuations in the corona below 1.3 R⊙ with CoMP. *The Astrophysical Journal*, September, *828*, 89.

Morton, R. J., Verth, G., Fedun, V., Shelyag, S., & Erdélyi, R. (2013). Evidence for the photospheric excitation of incompressible chromospheric waves. *The Astrophysical Journal*, April, *768*, 17.

Murawski, K., Solov'ev, A., Musielak, Z. E., Srivastava, A. K., & Kraśkiewicz, J. (2015). Torsional Alfvén waves in solar magnetic flux tubes of axial symmetry. *Astronomy & Astrophysics*, *577*, 11.

Nakariakov, V. M., Anfinogentov, S. A., Antolin, P., Jain, R., Kolotkov, D. Y., Kupriyanova, E. G., et al. (2021). Kink oscillations of coronal loops. *Space Science Reviews*, September, *217*, 73.

Nakariakov, V. M., & Kolotkov, D. Y. (2020). Magnetohydrodynamic waves in the solar corona. *Annual Review of Astronomy and Astrophysics*, August, *58*, 441–481.

Nakariakov, V. M., Ofman, L., DeLuca, E. E., Roberts, B., & Davila, J. M. (1999). TRACE observation of damped coronal loop oscillations: Implications for coronal heating. *Science*, August, *285*, 862–864.

Nakariakov, V. M., Pilipenko, V., Heilig, B., Jelínek, P., Karlický, M., Klimushkin, D. Y., et al. (2016). Magnetohydrodynamic oscillations in the solar corona and Earth's magnetosphere: Towards consolidated understanding. *Space Science Reviews*, April, *200*, 75–203.

Nechaeva, A., Zimovets, I. V., Nakariakov, V. M., & Goddard, C. R. (2019). Catalog of decaying kink oscillations of coronal loops in the 24th solar cycle. *The Astrophysical Journal Supplement Series*, April, *241*, 31.

Nutto, C., Steiner, O., Schaffenberger, W., & Roth, M. (2012). Modification of wave propagation and wave travel-time by the presence of magnetic fields in the solar network atmosphere. *Astronomy & Astrophysics*, February, *538*, A79.

Okamoto, T. J., Antolin, P., De Pontieu, B., Uitenbroek, H., Van Doorsselaere, T., & Yokoyama, T. (2015). Resonant absorption of transverse oscillations and associated heating in a solar prominence. I. Observational aspects. *The Astrophysical Journal*, *809*(1), 12.

Okamoto, T. J., & Pontieu, B. D. (2011). Propagating waves along spicules. *The Astrophysical Journal*, July, *736*, L24.

Okamoto, T. J., Tsuneta, S., Berger, T. E., Ichimoto, K., Katsukawa, Y., Lites, B. W., et al. (2007). Coronal transverse magnetohydrodynamic waves in a solar prominence. *Science*, *318*, 1577.

Pagano, P., & De Moortel, I. (2017). Contribution of mode-coupling and phase-mixing of Alfvén waves to coronal heating. *Astronomy & Astrophysics*, *601*, 13.

Pagano, P., De Moortel, I., & Morton, R. J. (2020). Effect of coronal loop structure on wave heating through phase mixing. *Astronomy & Astrophysics*, *643*, 17.

Pagano, P., Pascoe, D. J., & De Moortel, I. (2018). Contribution of phase-mixing of Alfvén waves to coronal heating in multi-harmonic loop oscillations. *Astronomy & Astrophysics*, *616*, 12.

Pant, V., & Van Doorsselaere, T. V. (2020). Revisiting the relation between nonthermal line widths and transverse MHD wave amplitudes. *The Astrophysical Journal*, August, *899*, 1.

Pant, V., Magyar, N., Van Doorsselaere, T., & Morton, R. J. (2019). Investigating "Dark" energy in the solar corona using forward modeling of MHD waves. *The Astrophysical Journal*, *881*(2), 14.

Pascoe, D. J., Goddard, C. R., & Van Doorsselaere, T. (2020). Oscillation and evolution of coronal loops in a dynamical solar corona. *Frontiers in Astronomy and Space Sciences*, *7*, 61.

Pascoe, D. J., Hood, A. W., de Moortel, I., & Wright, A. N. (2012). Spatial damping of propagating kink waves due to mode coupling. *Astronomy & Astrophysics*, March, *539*, A37.

Pascoe, D. J., Hood, A. W., De Moortel, I., & Wright, A. N. (2013). Damping of kink waves by mode coupling. II. Parametric study and seismology. *Astronomy & Astrophysics*, March, *551*, A40.

Pascoe, D. J., Van Doorsselaere, T., & De Moortel, I. (2022). Coronal seismology using damping of propagating kink waves. *The Astrophysical Journal*, *929*(1), 10.

Pascoe, D. J., Wright, A. N., & De Moortel, I. (2011). Propagating coupled Alfvén and kink oscillations in an arbitrary inhomogeneous corona. *The Astrophysical Journal*, April, *731*, 73.

Pelekhata, M., Murawski, K., & Poedts, S. (2021). Solar chromosphere heating and generation of plasma outflows by impulsively generated two-fluid Alfvén waves. *Astronomy & Astrophysics*, *652*, 8.

Perez, J. C., & Chandran, B. D. G. (2013). Direct numerical simulations of reflection-driven, reduced magnetohydrodynamic turbulence from the Sun to the Alfvén critical point. *The Astrophysical Journal*, October, *776*, 124.

Petrukhin, N. S., Ruderman, M. S., & Shurgalina, E. G. (2018). Phase mixing of Alfvén waves in axisymmetric non-reflective magnetic plasma configurations. *Monthly Notices of the Royal Astronomical Society*, *474*(2), 2289–2301.

Priest, E. (2014). *Magnetohydrodynamics of the Sun*. Cambridge, UK: Cambridge University Press.

Prokopyszyn, A. P. K., & Hood, A. W. (2019). Investigating the damping rate of phase-mixed Alfvén waves. *Astronomy & Astrophysics*, *632*, 12.

Prokopyszyn, A. P. K., Hood, A. W., & De Moortel, I. (2019). Phase mixing of nonlinear Alfvén waves. *Astronomy & Astrophysics, April*, *624*, A90.

Pucci, F., Onofri, M., & Malara, F. (2014). Evolution of magnetohydrodynamic waves in low layers of a coronal hole. *The Astrophysical Journal*, *796*(1), 17.

Quintero Noda, C., Schlichenmaier, R., Bellot Rubio, L. R., Löfdahl, M. G., Khomenko, E., Jurčák, J., et al. (2022). The European Solar Telescope. *Astronomy & Astrophysics*, *666*, 36.

Raouafi, N. E., Matteini, L., Squire, J., Badman, S. T., Velli, M., Klein, K. G., et al. (2023). Parker solar probe: Four years of discoveries at solar cycle minimum. *Space Science Reviews*, February, *219*, 8.

Rast, M., Bello González, N., Bellot Rubio, L., Cao, W., Cauzzi, G., DeLuca, D., et al. (2021). Critical science plan for the Daniel K. Inouye Solar Telescope (DKIST), *Solar Physics*, April, *296*, 70.

Raymond, J. C., McCauley, P. I., Cranmer, S. R., & Downs, C. (2014). The solar corona as probed by Comet Lovejoy (C/2011 W3). *The Astrophysical Journal*, June, *788*, 152.

Rempel, M. (2017). Extension of the MURaM Radiative MHD Code for Coronal Simulations. *The Astrophysical Journal, January*, *834*, 10.

Roberts, B. (1981). Wave propagation in a magnetically structured atmosphere. I—Surface waves at a magnetic interface. *Solar Physics*, January, *69*, 27–38.

Roberts, B. (2000). Waves and oscillations in the corona—(Invited Review). *Solar Physics*, *193*, 139–152.

Rochus, P., Auchère, F., Berghmans, D., Harra, L., Schmutz, W., Schühle, U., et al. (2020). The solar orbiter EUI instrument: The extreme ultraviolet imager. *Astronomy & Astrophysics*, *642*, 21.

Ruderman, M. S., & Petrukhin, N. S. (2017). Phase mixing of Alfvén waves propagating in non-reflective magnetic plasma configurations. *Astronomy & Astrophysics*, *600*, 11.

Ruderman, M. S., & Petrukhin, N. S. (2022). Existence of purely Alvén waves in magnetic flux tubes with arbitrary cross-sections. *Physics*, July, *4*, 865–872.

Russell, A. J. B. (2018). 75th Anniversary of 'Existence of Electromagnetic-Hydrodynamic Waves'. *Solar Physics*, May, *293*(83).

Russell, A. J. B., & Fletcher, L. (2013). Propagation of alfvénic waves from corona to chromosphere and consequences for solar flares. *The Astrophysical Journal*, February, *765*, 81.

Sakaue, T., & Shibata, K. (2020). Energy transfer by nonlinear Alfvén waves in the solar chromosphere and its effect on spicule dynamics, coronal heating, and solar wind acceleration. *The Astrophysical Journal*, *900*(2), 19.

Sakurai, T., Goossens, M., & Hollweg, J. V. (1991). Resonant behaviour of MHD waves on magnetic flux tubes. I—Connection formulae at the resonant surfaces. *Solar Physics*, June, *133*, 227–245.

Scalisi, J., Oxley, W., Ruderman, M. S., & Erdélyi, R. (2021). Propagation of torsional Alfvén pulses in zero-beta flux tubes. *The Astrophysical Journal*, *911*(1), 8.

Schekochihin, A. A. (2022). MHD turbulence: A biased review. *Journal of Plasma Physics*, October, *88*, 155880501.

Sekse, D. H., van der Voort, L. R., Pontieu, B. D., & Scullion, E. (2013). Interplay of three kinds of motion in the disk counterpart of Type II spicules: upflow, transversal, and torsional motions. *The Astrophysical Journal*, May, *769*, 44.

Sharma, R., Verth, G., & Erdélyi, R. (2018). Evolution of complex 3D motions in spicules. *The Astrophysical Journal*, January, *853*, 61.

Shelyag, S., Cally, P. S., Reid, A., & Mathioudakis, M. (2013). Alfvén waves in simulations of solar photospheric vortices. *The Astrophysical Journal Letters*, October, *776*, L4.

Shestov, S. V., Nakariakov, V. M., Ulyanov, A. S., Reva, A. A., & Kuzin, S. V. (2017). Nonlinear evolution of short-wavelength torsional Alfvén waves. *The Astrophysical Journal*, May, *840*, 64.

Shetye, J., Verwichte, E., Stangalini, M., & Doyle, J. G. (2021). The nature of high-frequency oscillations associated with short-lived spicule-type events. *The Astrophysical Journal*, October, *921*, 30.

Shimizu, T., Shimizu, T., Imada, S., Kawate, T., Ichimoto, K., Suematsu, Y., Hara, H., et al. (2019). The Solar-C_EUVST mission. *Proceedings of the SPIE*, *11118*, 11.

Shi, M., Van Doorsselaere, T., Guo, M., Karampelas, K., Li, B., Antolin, P. (2021). The first 3D coronal loop model heated by MHD waves against radiative losses. *The Astrophysical Journal*, *908*(2), 9.

Shoda, M., Chandran, B. D. G., & Cranmer, S. R. (2021). Turbulent generation of magnetic switchbacks in the Alfvénic solar wind. *The Astrophysical Journal*, July, *915*, 52,.

Soler, R., Andries, J., & Goossens, M. (2012). Resonant Alfvén waves in partially ionized plasmas of the solar atmosphere. *Astronomy & Astrophysics*, January, *537*, A84.

Soler, R., Carbonell, M., & Ballester, J. L. (2013b). Magnetoacoustic waves in a partially ionized two-fluid plasma. *The Astrophysical Journal Supplement*, October, *209*(1), 19.

Soler, R., Carbonell, M., & Ballester, J. L. (2015). On the spatial scales of wave heating in the solar chromosphere. *The Astrophysical Journal*, September, *810*, 146.

Soler, R., Carbonell, M., Ballester, J. L., & Terradas, J. (2013a). Alfvén waves in a partially ionized two-fluid plasma. *The Astrophysical Journal*, April, *767*, 171.

Soler, R., Goossens, M., Terradas, J., & Oliver, R. (2014). The behavior of transverse waves in nonuniform solar flux tubes. II. Implications for coronal loop seismology. *The Astrophysical Journal*, February, *781*, 111.

Soler, R., Terradas, J., Oliver, R., & Ballester, J. L. (2017). Propagation of torsional Alfvén waves from the photosphere to the corona: Reflection, transmission, and heating in expanding flux tubes. *The Astrophysical Journal*, May, *840*, 20.

Soler, R., Terradas, J., Oliver, R., & Ballester, J. L. (2021). Resonances in a coronal loop driven by torsional alfvén waves propagating from the photosphere. *The Astrophysical Journal*, *909*(2), 16.

Somov, B. V. (2006). *Plasma Astrophysics, Part I: Fundamentals and practice*. 1st edn. Moscow, Russia: Springer New York.

Squire, J., Chandran, B. D. G., & Meyrand, R. (2020). In-situ switchback formation in the expanding solar wind. *The Astrophysical Journal Letters*, March, *891*, L2.

Squire, J., Meyrand, R., Kunz, M. W., Arzamasskiy, L., Schekochihin, A. A., & Quataert, E. (2022). High-frequency heating of the solar wind triggered by low-frequency turbulence. *Nature Astronomy*, *6*, 715–723.

Srivastava, A. K., Shetye, J., Murawski, K., Doyle, J. G., Stangalini, M., Scullion, E., et al. (2017). High-frequency torsional Alfvén waves as an energy source for coronal heating. *Scientific Reports*, March, *7*, 43147.

Stangalini, M., Stangalini, M., Erdélyi, R., Boocock, C., Tsiklauri, D., Nelson, C. J., Del Moro, D., et al. (2021). Torsional oscillations within a magnetic pore in the solar photosphere. *Nature Astronomy*, May, *5*, 691–696.

Sterling, A. C., & Moore, R. L. (2020). Coronal-jet-producing minifilament eruptions as a possible source of parker solar probe switchbacks. *The Astrophysical Journal*, June, *896*, L18.

Tavabi, E., & Koutchmy, S. (2019). Chromospheric peculiar off-limb dynamical events from *IRIS* observations. *The Astrophysical Journal*, September, *883*, 41.

Tavabi, E., Koutchmy, S., Ajabshirizadeh, A., Ahangarzadeh Maralani, A. R., & Zeighami, S. (2015). Alfvénic waves in polar spicules. *Astronomy & Astrophysics*, December, *573*, A4.

Telloni, D., Zank, G. P., Stangalini, M., Downs, C., Liang, H., Nakanotani, M., et al. (2022). Observation of a magnetic switchback in the solar corona. *The Astrophysical Journal Letters*, September, *936*, L25.

Tenerani, A., Velli, M., Matteini, L., Réville, V., Shi, C., Bale, S., et al. (2020). Magnetic field kinks and folds in the solar wind. *The Astrophysical Journal Supplement Series*, February, *246*, 32.

Terradas, J., Andries, J., Goossens, M., Arregui, I., Oliver, R., & Ballester, J. L. (2008b). Nonlinear instability of kink oscillations due to shear motions. *The Astrophysical Journal Letters*, November, *687*, L115–L118.

Terradas, J., Arregui, I., Oliver, R., Ballester, J. L., Andries, J., & Goossens, M. (2008a). Resonant absorption in complicated plasma configurations: Applications to multistranded coronal loop oscillations. *The Astrophysical Journal*, June, *679*, 1611–1620.

Terradas, J., & Goossens, M. (2012). Transverse kink oscillations in the presence of twist. *Astronomy & Astrophysics*, December, *548*, A112.

Terradas, J., Magyar, N., & Van Doorsselaere, T. (2018). Effect of magnetic twist on nonlinear transverse kink oscillations of line-tied magnetic flux tubes. *The Astrophysical Journal*, January, *853*, 35.

Thompson, B. J., Gurman, J. B., Neupert, W. M., Newmark, J.S., Delaboudinière, J.-P., Cyr, O.C.S., et al. (1999). *SOHO*/EIT observations of the 1997 April 7 coronal transient: Possible evidence of coronal Moreton waves. *The Astrophysical Journal*, June, *517*, L151–L154.

Threlfall, J., De Moortel, I., McIntosh, S. W., & Bethge, C. (2013). First comparison of wave observations from CoMP and AIA/SDO. *Astronomy & Astrophysics*, August, *556*, A124.

Thurgood, J. O., & McLaughlin, J. A. (2013). 3D Alfvén wave behaviour about proper and improper magnetic null points. *Astronomy & Astrophysics*, *558*, 10.

Thurgood, J. O., Morton, R. J., & McLaughlin, J. A. (2014). First direct measurements of transverse waves in solar polar plumes using SDO/AIA. *The Astrophysical Journal Letters*, *790*(1), 7.

Tian, H., McIntosh, S. W., Wang, T., Ofman, L., De Pontieu, B., Innes, D. E., Peter, H. (2012). Persistent Doppler shift oscillations observed with Hinode/EIS in the solar corona: Spectroscopic signatures of Alfvénic waves and recurring upflows. *The Astrophysical Journal*, *759*(2), 17.

Tiwari, A. K., Morton, R. J., & McLaughlin, J. A. (2021). A statistical study of propagating MHD kink waves in the quiescent corona. *The Astrophysical Journal*, September, *919*, 74.

Tiwari, A. K., Morton, R. J., Régnier, S., & McLaughlin, J. A. (2019). Damping of propagating kink waves in the solar corona. *The Astrophysical Journal*, May, *876*, 106.

Tomczyk, S., & McIntosh, S. W. (2009). Time-distance seismology of the solar corona with CoMP. *The Astrophysical Journal*, June, *697*, 1384–1391.

Tomczyk, S., McIntosh, S. W., Keil, S. L., Judge, P. G., Schad, T., Seeley, D. H., & Edmondson, J. (2007). Alfvén waves in the solar corona. *Science*, August, *317*, 1192–1196.

Tripathi, D., Ramaprakash, A. N., Khan, A., Ghosh, A., Chatterjee, S., Banerjee, D., et al. (2017). The solar ultraviolet imaging telescope on-board Aditya-L1. *Current Science*, *113*, 616–619.

Tsiklauri, D. (2016). Alfvén wave phase-mixing in flows. Why over-dense, solar coronal, open magnetic field structures are cool. *Astronomy & Astrophysics*, *586*, 10.

Tu, J., & Song, P. (2013). A study of Alfvén wave propagation and heating the chromosphere. *The Astrophysical Journal*, October, *777*, 53.

Tziotziou, K., Scullion, E., Shelyag, S., Steiner, O., Khomenko, E., Tsiropoula, G. et al. (2023). Vortex motions in the solar atmosphere. *Space Science Reviews*, *219*(1), 1.

Tziotziou, K., Tsiropoula, G., & Kontogiannis, I. (2019). A persistent quiet-Sun small-scale tornado. II. Oscillations. *Astronomy & Astrophysics*, March, *623*, A160.

van Ballegooijen, A. A., & Asgari-Targhi, M. (2016). Heating and acceleration of the fast solar wind by Alfvén wave turbulence. *The Astrophysical Journal*, April, *821*, 106.

van Ballegooijen, A. A., & Asgari-Targhi, M. (2017). Direct and inverse cascades in the acceleration region of the fast solar wind. *The Astrophysical Journal*, January, *835*, 10.

van Ballegooijen, A. A., Asgari-Targhi, M., Cranmer, S. R., & DeLuca, E. E. (2011). Heating of the solar chromosphere and corona by Alfvén wave turbulence. *The Astrophysical Journal*, July, *736*, 3.

Van Damme, H. J., De Moortel, I., Pagano, P., & Johnston, C. D. (2020). Chromospheric evaporation and phase mixing of Alfvén waves in coronal loops. *Astronomy & Astrophysics*, *635*, 11.

Van Doorsselaere, T., Gijsen, S. E., Andries, J., & Verth, G. (2014). Energy propagation by transverse waves in multiple flux tube systems using filling factors. *The Astrophysical Journal*, *795*(1), 7.

Van Doorsselaere, T., Goossens, M., Magyar, N., Ruderman, M. S., & Ismayilli, R. (2021). Nonlinear damping of standing kink waves computed with Elsässer variables. *The Astrophysical Journal*, *910*(1), 8.

Van Doorsselaere, T., Li, B., Goossens, M., Hnat, B., & Magyar, N. (2020a). Wave pressure and energy cascade rate of kink waves computed with Elsässer variables. *The Astrophysical Journal*, August, *899*, 100.

Van Doorsselaere, T., Nakariakov, V. M., & Verwichte, E. (2008). Detection of waves in the solar corona: Kink or Alfvén? *The Astrophysical Journal Letters*, March, *676*, L73–L75.

Van Doorsselaere, T., Srivastava, A. K., Antolin, P., Magyar, N., Vasheghani Farahani, S., Hui Tian, D., et al. (2020b). Coronal heating by MHD waves. *Space Science Reviews*, *216*(8).

Vasheghani Farahani, S., Ghanbari, E., Ghaffari, G., & Safari, H. (2017). Torsional wave propagation in solar tornadoes. *Astronomy & Astrophysics*, *599*, 6.

Vasheghani Farahani, S., Nakariakov, V. M., Verwichte, E., & Van Doorsselaere, T. (2012). Nonlinear evolution of torsional Alfvén waves. *Astronomy & Astrophysics*, *544*, 5.

Verdini, A., Grappin, R., & Montagud-Camps, V. (2019). Turbulent heating in the accelerating region using a multishell model. *Solar Physics*, May, *294*, 65.

Verth, G., & Jess, D. B. (2016). MHD wave modes resolved in fine-scale chromospheric magnetic structures. In: *Low-Frequency Waves in Space Plasmas* (pp. 431–448). Geophysical Monograph Series, Vol. 216. *Wiley.*

Verwichte, E., & Kohutova, P. (2017). Excitation and evolution of vertically polarised transverse loop oscillations by coronal rain. *Astronomy & Astrophysics*, *601*, 4.

Vranjes, J., & Kono, M. (2014). On the Alfvén wave cut-off in partly ionized collisional plasmas. *Physics of Plasmas*, January, *21*, 012110.

Wang, Y., & Yokoyama, T. (2020). Simulation of Alfvén wave propagation in the magnetic chromosphere with radiative loss: Effects of nonlinear mode coupling on chromospheric heating. *The Astrophysical Journal*, *891*(2), 9.

Weberg, M. J., Morton, R. J., & McLaughlin, J. A. (2018). An automated algorithm for identifying and tracking transverse waves in solar images. *The Astrophysical Journal*, January, *852*, 57.

Weberg, M. J., Morton, R. J., & McLaughlin, J. A. (2020). Using transverse waves to probe the plasma conditions at the base of the solar wind. *The Astrophysical Journal*, *894*(1), 10.

Wedemeyer-Böhm, S., Scullion, E., Steiner, O., Rouppe van der Voort, L., de La Cruz Rodriguez, J., Fedun, V., & Erdélyi, R. (2012). Magnetic tornadoes as energy channels into the solar corona. *Nature*, June, *486*, 505–508.

Wiegelmann, T., Lagg, A., Solanki, S. K., Inhester, B., & Woch, J. (2005). Comparing magnetic field extrapolations with measurements of magnetic loops. *Astronomy and Astrophysics*, *433*(2), 4.

Wiegelmann, T., & Sakurai, T. (2021). Solar force-free magnetic fields. *Living Reviews in Solar Physics*, *18*(1).

Williams, T., Walsh, R. W., Winebarger, A. R., Brooks, D. H., Cirtain, J. W., De Pontieu, B., et al. (2020). Is the high-resolution coronal imager resolving coronal strands? Results from AR 12712. *The Astrophysical Journal*, April, *892*, 134.

Wójcik, D., Murawski, K., Musielak, Z. E., Konkol, P., & Mignone, A. (2017). Numerical simulations of torsional Alfvén waves in axisymmetric solar magnetic flux tubes. *Solar Physics*, *292*(2), 13.

Woolsey, L. N., & Cranmer, S. R. (2014). Turbulence-driven coronal heating and improvements to empirical forecasting of the solar wind. *The Astrophysical Journal*, June, *787*, 160.

Yang, Z., Bethge, C., Tian, H., Tomczyk, S., Morton, R., Del Zanna, G., et al. (2020). Global maps of the magnetic field in the solar corona. *Science*, August, *369*, 694–697.

Yoshida, M., Suematsu, Y., Ishikawa, R., Okamoto, T. J., Kubo, M., Kano, R., et al. (2019). High-frequency wave propagation along a spicule observed by CLASP. *The Astrophysical Journal*, December, *887*, 2.

Yu, D. J., Van Doorsselaere, T., & Goossens, M. (2017). Resonant absorption of the slow sausage wave in the slow continuum. *Astronomy & Astrophysics*, June, *602*, A108.

Zaitsev, V. V., & Stepanov, A. V. (1975). *ssled. Geomagn. Aeron. Fiz. Solntsa*, *37*, 3.

Zanna, G. D., Gupta, G. R., & Mason, H. E. (2019). Hinode EIS line widths in the quiet corona up to 1.5 Rsun. *Astronomy & Astrophysics*, November, *631*, A163.

Zhao, J. S., Voitenko, Y., Guo, Y., Su, J. T., & Wu, D. J. (2015). Nonlinear damping of Alfvén waves in the solar corona below 1.5 solar radii. *The Astrophyiscal Journal*, October, *811*, 88.

Zubkova, A. V., Kobanov, N. I., Sklyar, A. A., Kostyk, R. I., & Shchukina, N. G. (2014). Periodic variations of the H alpha profile width in the chromosphere of coronal holes as a possible indicator of Alfvén waves. *Astronomy Letters*, April, *40*, 222–229.

# 3

## Alfvén Waves in Solar Flares

**Alexander J. B. Russell**

### ABSTRACT

Solar flares are dramatic events in which magnetic reconnection in the corona leads to heating of plasma to tens of megakelvins (MK) and acceleration of particles to high energies. They also centrally involve transport between the corona (where the magnetic reconnection occurs) and the lower solar atmosphere (where most energy is radiated from). There is substantial evidence for the presence of Alfvénic waves/turbulence in solar flares, for example, in the ubiquitous nonthermal broadening of flare spectral lines. The physical role that Alfvénic waves have in the flare has attracted considerable attention, especially since 2007–2010. This article reviews what spectroscopic observations reveal about the properties and importance of Alfvénic waves, turbulence, and transport in solar flares; mechanisms for wave excitation by magnetic reconnection at high Lundquist numbers and braking of the sunward reconnection jet; and models of wave energy transport to the lower atmosphere and the resulting heating and dynamics. The article finishes with discussion of the outlook for new progress.

### 3.1. INTRODUCTION

Solar flares are spectacular centerpieces of astrophysics that combine natural drama with the important impacts of space weather. For the scientifically curious, they possess a physical richness that interleaves diverse astrophysical plasma processes, span environments from the solar interior to the solar wind, and leave their mark across the full range of the electromagnetic spectrum. They are a prime example of a plasma phenomenon for which scientific advance requires not only greater understanding of components that are enigmatic and challenging of themselves, but also integration of components into the complete system.

This review examines the role of Alfvénic waves in solar flares. We hope to provide an overview of major points in this rapidly maturing field, including: observational constraints on the locations, timings, and importance of Alfvénic waves in solar flares; generation of waves/turbulence by magnetic reconnection at high Lundquist numbers and braking of the sunward reconnection jet; and downward transport to the lower solar atmosphere and the ensuing dynamics there. The article finishes with discussion of the outlook for new progress and a summary.

This article is part of an interdisciplinary collection on Alfvén waves in the heliosphere, and it is important to highlight at the outset that the investigation of Alfvén waves in solar flares differs from other plasma environments. In environments blessed with in situ observations, such as planetary magnetospheres and the solar wind, extensive datasets directly confirm the presence of Alfvén waves, establish key attributes such as wavelength and Poynting flux, document variations under different conditions, and connect these properties with phenomena such as particle energization and auroral emission. In contrast, detection of waves in solar environments relies on remote observation, which has been sufficiently challenging that imaging of waves in the solar corona was achieved only

*School of Mathematics and Statistics, University of St Andrews, St Andrews, UK*

*Alfvén Waves Across Heliophysics: Progress, Challenges, and Opportunities, Geophysical Monograph 285*, First Edition.
Edited by Andreas Keiling.

DOI: 10.1002/9781394195985.ch03

at the end of the 20th century. Continued advances have since established that magnetohydrodynamic (MHD) waves are ubiquitous in the Sun's corona and lower atmosphere, and sophisticated seismological techniques are used to infer local plasma properties from wave observations. There is also compelling evidence that upgoing MHD waves make important contributions to heating magnetically open parts of the corona and accelerating the fast solar wind. These topics are reviewed in other articles in this collection.

For flares, remote observations show beyond doubt that solar flares excite MHD waves in their surroundings. As early as 1960, Hα observations revealed the occurrence of Moreton waves in the chromosphere, which spread out from the flare site like a blast wave (Moreton & Ramsey, 1960). Later, coronal imaging found the coronal counterpart in the form of extreme ultraviolet (EUV) waves (see recent discussion about interpretation of these observations by Long et al. (2017)). Flare-induced shock waves are implicated in the generation of type II radio bursts (Uchida, 1960) and further evidence that flares launch disturbances into their surroundings is provided by the excitation of transverse kink oscillations of nearby coronal loops (Aschwanden et al., 1999; Nakariakov et al., 1999). Finally, wave oscillations are a leading candidate to explain quasi-periodic pulsations in flare light curves and radio emission, although slower compressible wave modes, such as slow modes and sausage modes, are typically considered responsible rather than Alfvén waves (see reviews by Nakariakov and Melnikov (2009) and McLaughlin et al. (2018)).

Solar flares also produce rapid, significant, and persistent changes of the photospheric magnetic field (Cameron & Sammis, 1999; Kosovichev & Zharkova, 2001; Wang et al., 1994). These photospheric field changes occur after the start of the flare, and they are believed to be a response to the restructuring of the coronal magnetic field. Sudol and Harvey (2005) documented many cases where components of the photospheric magnetic field changed by more than 0.01 T in less than 10 min (also see additional studies by Petrie and Sudol (2010) and Petrie (2013), and the review by Wang and Liu (2015)). Alfvénic wave fronts are a natural agent for communicating the magnetic field changes, motivating consideration of large-amplitude Alfvénic waves in flares (Fletcher & Hudson, 2008). In this context, resonant coupling between the downgoing Alfvén wave front and acoustic waves may excite seismic waves in the solar interior known as sunquakes (Fisher et al., 2012; Hudson et al., 2008; Russell et al., 2016; Wang & Liu, 2010). Furthermore, Johnstone et al. (2012) found an observational connection between photospheric magnetic field changes and enhanced chromospheric UV emissions, consistent with Alfvénic fronts heating the chromosphere on their way to the underlying photosphere.

Unfortunately, directly observing Alfvén waves at the heart of the flare is challenging for several fundamental reasons. One factor is time resolution: an upper limit on wave periods is the duration of the impulsive phase of a solar flare (defined in Section 3.2), which is typically of the order of several minutes. More challengingly, high-cadence radio and X-ray observations vary on timescales as short as tens of milliseconds (Benz, 1986; Kiplinger et al., 1983). Imaging high-frequency waves is clearly more challenging than resolving the 3–5-min periods typical of waves in the background Sun, as faster cadences starve instruments of photons. Another limitation is spatial resolution: as a benchmark on current capabilities, the *High-Resolution Coronal Imager* (Kobayashi et al., 2014; Rachmeler et al., 2019), which has an upcoming flare mission planned for 2024, has a pixel resolution of approximately 0.13″, or about 100 km on the Sun. Coronal waves with scales shorter than this cannot currently be spatially resolved. Finally, solar flares create challenging conditions for coronal observation such as bright backgrounds and large dynamic range. These issues can be mitigated by observing strategies such as restricting attention to low-energy flares, or limb flares that are partially occulted by the solar disk; but they nonetheless provide significant practical constraints.

For a long time, Alfvén waves and Poynting flux lay outside the main set of concepts used in flare models. Instead, debate centered on whether energy was delivered to the lower atmosphere by thermal conduction (the so-called "thermal" flare model) or electron beams (the "nonthermal" flare model). There were nonetheless thinkers willing to point out how strange it would be if solar flares were a rare exception to the universal importance of Poynting flux in astrophysical plasmas, and if the burstiness of the flare energy release did not excite waves. Further encouragement to consider Alfvén waves and Poynting flux could be taken from nonthermal broadening of flare spectral lines, magnetic field changes, the persistence of the electron number and supply problems in the collisional thick-target model (Brown et al., 2009; Fletcher & Hudson, 2008), and analogies to other systems such as Earth's magnetosphere, in which the importance of wave transport was becoming increasingly clear.

The intellectual conditions changed dramatically during 2007–2010, which we attribute to a confluence of events. First, 2007 was a breakthrough year for observations of transverse waves in the solar corona and the tipping point for the narrative that MHD waves are "ubiquitous" in the corona, which quickly became consensus in solar physics (De Pontieu et al., 2007; Lin et al., 2007; Okamoto et al., 2007; Tomczyk et al., 2007). Second, during 2008–2010, simulations of magnetic reconnection crossed the Biskamp criterion (Lundquist number $S_L > 10^4$) beyond which an initial Sweet-Parker reconnection layer is disrupted by plasmoid instability (Bhattacharjee et al., 2009; Daughton et al., 2011; Huang & Bhattacharjee, 2010; Lapenta, 2008). It rapidly became clear that magnetic reconnection at high Lundquist numbers is fast (normalized reconnection rates are 0.01–0.1 and independent of $S_L$) and highly dynamic. This development gave fresh impetus to the point long made by flare observations that the coronal energy release is highly variable, hence likely to launch waves. It also ushered in a new era in which some aspects of wave generation by magnetic reconnection can be probed using simulations. Third, increasingly sophisticated spaceborne spectrometers provided novel opportunities to identify where and when unresolved plasma motions occur in solar flares: the Hinode Extreme-Ultraviolet Imaging Spectrometer (Hinode/EIS, Culhane et al., 2007) launched on 22 September 2006 (UTC) and Interface Region Imaging Spectrograph (IRIS; De Pontieu et al., 2014) launched on 28 June 2013; together these missions have substantially advanced knowledge of waves/turbulence in solar flares.

We also highlight two major papers in this period. "Impulsive Phase Flare Energy Transport by Large-Scale Alfvén Waves and the Electron Acceleration Problem" by Fletcher and Hudson (2008) was especially influential and it ignited interest in wave energy transport among flare researchers. The paper "Chromospheric Evaporation by Alfvén Waves" by Haerendel (2009), while less frequently cited initially, also played an important role in launching this topic, introducing many of the ideas that will feature in this review.

The last 15 years have been an exciting period for the study of Alfvénic waves in solar flares, during which the topic has matured rapidly. Today, many of the central issues and tools have been clearly identified and new advances are being driven by advances of simulation and observing capabilities. As we hope the article will convey, this is an exciting time to study the role of Alfvénic waves in solar flares and we believe that many opportunities lie ahead.

### 3.2. SOLAR FLARES BACKGROUND

As a general definition, solar flares are events in which magnetic energy stored in the Sun's corona is converted to enhanced radiative emissions. The energies involved can be very large, in extreme events exceeding $10^{26}$ J ($10^{33}$ erg). The bulk of the radiated energy is normally emitted from the lower solar atmosphere in the white light and infrared parts of the spectrum; however, the pre-flare energy storage and release occur in the corona, so flare physics is closely associated with coronal processes and energy transport. Furthermore, flares can produce emissions across the electromagnetic spectrum, including radio, microwave, infrared, white light, EUV, X-rays, and gamma rays. An observational overview can be found in the article by Fletcher et al. (2011).

Since 1966, the International Astronomical Union has classified solar flares according to the peak X-ray flux measured by satellite observatories in Earth orbit (in the 1–8 Å range). The scale employed for this purpose employs scientific notation, but uses letters X, M, C, B, and A to denote the magnitude (respectively $10^{-4}$, $10^{-5}$, $10^{-6}$, $10^{-7}$, and $10^{-8}$ W m$^{-2}$). For example, an X8.2 flare is one for which the peak X-ray flux at Earth was $8.2 \times 10^{-4}$ W m$^{-2}$.

Flares are diverse phenomena: they exhibit power law distributions across broad ranges of energies; occur in active regions of different magnetic morphologies and complexities; and happen throughout the life cycle of individual active regions. Flare light curves exhibit several types of evolution, for example, impulsive, gradual, or long duration; and flares can be, but are not always, associated with coronal mass ejections (CMEs). To streamline the discussion, we focus on the canonical flare evolution of an impulsive phase followed by a gradual phase.

Observationally, the impulsive phase corresponds to a rapid rise in the 1–8 Å X-ray flux, during which impulsive phenomena are detected such as hard X-rays, microwave emission, and explosive upflows of plasma from the lower atmosphere. The impulsive phase is associated with the primary energy release, whereby magnetic reconnection liberates energy from the coronal magnetic field. A large portion of the released energy is ultimately transported to the lower solar atmosphere, forming flare ribbons and driving upflows known as "chromospheric evaporation" filling coronal loops with relatively dense plasma that radiates at high temperatures. The impulsive phase typically concludes within 10 min, and it is followed by a longer-lasting gradual phase, during which the X-ray emission decreases from its peak as the flare loops cool by radiation and thermal conduction. Timescales suggest that there is continued heating during the gradual phase, but it is less intense than during the impulsive phase.

The observational features of solar flares are intimately related to the topology of the coronal magnetic field. While different flares exhibit various morphologies, this article will consider the CSHKP "standard" model of an eruptive two-ribbon flare (Carmichael, 1964; Sturrock, 1966; Hirayama, 1974; Kopp & Pneuman, 1976). This phenomenology is especially important for space weather, since it is common among the largest X-class flares. It is also relevant to many of the flare observations, simulations, and theoretical works discussed in this article.

The CSHKP model proposes that magnetic reconnection occurs in an extended current sheet in the

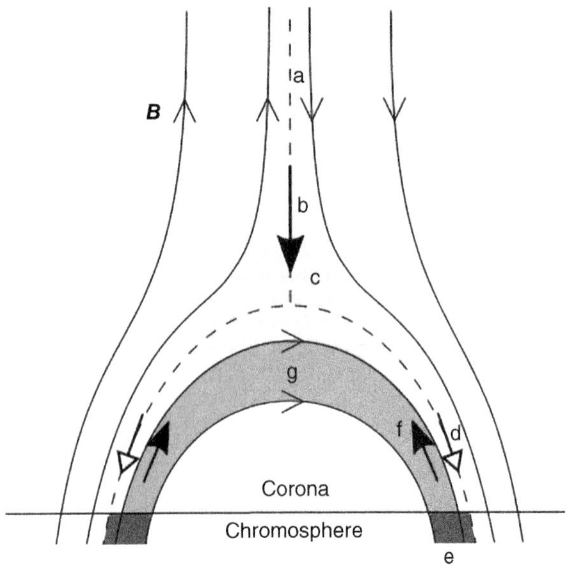


**Figure 3.1** CSHKP model of a two-ribbon flare. Magnetic field lines sketch the magnetic geometry, with dashed lines representing magnetic separatrices. Features include: **a:** reconnecting current sheet; **b:** sunward reconnection outflow (speed approximately $v_A$); **c:** above-the-loop region (light gray) associated with braking of the sunward reconnection outflow, particle trapping, and nonthermal electrons; **d:** field-aligned energy transport (unfilled arrow heads) to the chromosphere by electron beams, thermal conduction, Alfvénic waves, or a combination thereof; **e:** flare ribbons (bright emission from the heated chromosphere); **f:** chromospheric evaporation (dense upflows from the chromosphere to the corona); **g:** flare loops in X-rays and EUV, which are bright having been filled by evaporation.

corona, building on earlier ideas by Sweet (1958). This configuration is sketched in Figure 3.1. Energy released by magnetic reconnection is transported along the magnetic field, forming flare ribbons in the chromosphere that are magnetically connected to the coronal current sheet and/or reconnection outflow jet. This resembles auroral energy deposition near the magnetic footpoints of the reconnection site in a planetary magnetotail.

We also draw attention to the "above-the-loop" region that lies between the top of the flare loops and the bottom of the current sheet. The magnetic field line sketch in Figure 3.1 suggests the existence of a region of low magnetic field strength (field strength is inversely proportional to field line spacing), which coincides with where the sunward reconnection outflow jet is forced to brake. This region was not emphasized historically; however, given that flow braking typically redirects large-scale bulk motion into small-scale turbulence (cf. the pool below a waterfall, or the interface region around supernova remnants) and the pocket of weak magnetic field forms a magnetic trap for particles, we believe that the above-the-loop region should be of great physical interest.

The above-the-loop region has also become interesting for observational reasons, starting with the detection of a hard X-ray source at this location (Ishikawa et al., 2011; Krucker & Lin, 2008; Liu et al., 2008; Masuda et al., 1994; Petrosian et al., 2002; Sui & Holman, 2003). Microwave observations analyzed by Fleishman et al. (2022) found a magnetic field strength of 200–400 G in this region, compared to kilogauss field strengths elsewhere in the corona. Various investigations have found that almost all the electrons in this region are accelerated during flares, creating a nonthermal population with a number density of $10^{10}$–$10^{11}$ cm$^{-3}$ (Chen et al., 2020; Fleishman et al., 2022; Krucker et al., 2010; Krucker & Battaglia, 2014). This density is comparable to the electron beam density that would be required to power radiative emissions from flare ribbons in powerful events (Ishikawa et al., 2011; Krucker et al., 2011), although the particle flux to the lower atmosphere depends on what fraction of particles are in the magnetic trap's loss cone.

The 2D version of the CSHKP standard flare model shown in Figure 3.1 is a sufficient guide to follow the rest of this article. We emphasize, though, that the true dynamics occur in 3D, which has two major implications. First, even when the time-averaged dynamics are 2D, the 3D nature of the physical processes involved is crucial, as will be seen when discussing the generation of Alfvén waves in Section 3.6. Second, the global magnetic configuration is 3D, and readers interested in the 3D aspects of the global topology are directed toward Moore and Labonte (1980), Moore et al. (1997), Aulanier et al. (2012), Amari et al. (2014), Janvier et al. (2014), Janvier et al. (2015), and references therein.

## 3.3. TERMINOLOGY: ALFVÉN AND ALFVÉNIC

It is worthwhile to explain what we mean by Alfvén waves versus Alfvénic waves. As is well known, a homogeneous equilibrium supports three MHD wave modes: the Alfvén, fast, and slow modes. The Alfvén wave, discovered by Alfvén (1942), is characterized by field-aligned transport at the Alfvén speed,

$$v_A = \frac{B}{\sqrt{\mu_0 \rho}}. \tag{3.1}$$

It is a transverse wave, with magnetic field perturbations perpendicular to the background magnetic field. The restoring force is magnetic tension, and it is incompressible (unlike the magnetoacoustic modes). Finally, propagating Alfvén waves satisfy the Walén relation,

$$\frac{\delta \boldsymbol{B}}{B_0} = \mp \frac{\delta \boldsymbol{v}_\perp}{v_A}, \tag{3.2}$$

where the negative (positive) sign is for Alfvén waves propagating parallel (antiparallel) to the background magnetic field. In space physics, the Walén relation is often expressed as the ratio of the electric field perturbation to the magnetic field perturbation equaling the Alfvén speed. It is also a statement that the Alfvén wave's kinetic and magnetic energies are equipartitioned.

Alfvén waves also exist as nonlinear solutions for homogeneous equilibria. Finite-amplitude circularly polarized Alfvén waves, which have uniform magnetic pressure, retain $v_{\parallel} = 0$ and are strictly incompressible. On the other hand, finite-amplitude linearly polarized Alfvén waves perturb the magnetic pressure, which leads to "ponderomotive" flows of plasma, such that $v_{\parallel} \neq 0$. This nonlinear Alfvén wave is no longer strictly incompressible, and it produces density perturbations that travel with the wave. This outcome illustrates that one must beware of inadvertently introducing fragility into classification schemes for MHD waves by insisting too strongly on idealized properties such as $\nabla \cdot \boldsymbol{v} = 0$, which may not transfer to more realistic situations such as nonvanishing wave amplitudes.

In solar physics, a prominent debate around terminology (e.g., Goossens et al., 2009; Van Doorsselaere et al., 2008) has stemmed from the need to consider inhomogeneous equilibria, most pressingly dense structures aligned with the magnetic field known as coronal loops. Magnetic cylinder models of the type examined by Edwin and Roberts (1983) support linearized wave modes that are distinct to those of uniform medium, of which kink waves and azimuthal Alfvén waves are most pertinent to the present discussion. The azimuthal Alfvén waves are polarized such that velocity perturbations lie in flux surfaces of constant density, and they behave similarly to the Alfvén mode for a uniform medium. They include the "torsional Alfvén wave," which is strictly incompressible in these models and has an azimuthal wave number of $m = 0$, but this class can also be extended to include higher azimuthal wave numbers, for which the waves are weakly compressible.

The kink wave is a mode of the inhomogeneous structure, and it relies on small departures from $\nabla \cdot \boldsymbol{v} = 0$, which make it qualitatively distinct from a so-called "pure" Alfvén wave. Nonetheless, kink and Alfvén waves share many properties. Most importantly, in the slender tube limit (i.e., when parallel wavelengths are much greater than the tube radius), kink waves are transverse waves that transport energy along the magnetic field at the kink speed,

$$c_k = \sqrt{\frac{\rho_0 v_A^2 + \rho_e v_{Ae}^2}{\rho_0 + \rho_e}}, \tag{3.3}$$

where $\rho_0$ and $v_A$ are the mass density and Alfvén speed inside the cylinder, and $\rho_e$ and $v_{Ae}$ are their counterparts outside the cylinder. Inspecting the abovementioned formula, one sees that the kink speed is a mean of the Alfvén speeds in the cylinder and its environment, which bound its value. The wave is therefore said to be "Alfvénic" (Edwin & Roberts, 1983). Also in common with Alfvén waves, kink waves are only weakly compressible in the corona, and the dominant restoring force is magnetic tension (Goossens et al., 2009). We further add that (considering for simplicity the case of a homogeneous equilibrium magnetic field) the linearized induction equation implies the following generalization of the Walén relation for propagating waves:

$$\frac{\delta \boldsymbol{B}_\perp}{B_0} = -\frac{\delta \boldsymbol{v}_\perp}{c}, \tag{3.4}$$

where $c$ is the wave speed, set to $\pm c_k$ for kink waves compared to $\pm v_A$ for Alfvén waves. Hence, in both cases, the fluctuation ratio yields an Alfvénic speed.

It is helpful at this point to define a class of "Alfvénic waves" with similar properties, of which both "pure" Alfvén waves and kink waves are members. Properties of this class of MHD waves are:

- Field-aligned transport at Alfvénic speeds.
- Perturbations predominantly transverse to $\boldsymbol{B}_0$.
- Low compressibility.
- Magnetic tension the dominant restoring force.
- Propagating waves have $\delta \boldsymbol{B}_\perp / B_0 = -\delta \boldsymbol{v}_\perp / c$, where $c$ is the (Alfvénic) phase speed.

The semantics and history of "Alfvén" versus "Alfvénic" are discussed in depth in Section 2 of the article by Magyar in this collection (Chapter 2). In agreement with what we have written here, Magyar concludes: "The term Alfvénic reflects the fact that transverse waves in a generally inhomogeneous plasma of parallel wavelengths longer than the typical perpendicular scale of inhomogeneity have dominantly Alfvén-like properties. In this sense, it is the term that can most broadly describe transverse waves in an inhomogeneous plasma, without making implicit assumptions about the geometry of the inhomogeneity."

It is clearly desirable to make a precise distinction as to whether a transverse wave is a "pure" Alfvén wave or a kink wave, when there is sufficient information on the equilibrium density and the perturbations to do so. This is usually achievable for coronal loops that persist throughout the period of study. In other situations, remote observations are ambiguous as to the existence and extent of transverse density structuring. Furthermore, at the heart of a solar flare, the notion of a gently perturbed equilibrium is distinctly questionable, but MHD simulations of turbulence, reconnection, and turbulent magnetic relaxation show that Alfvénic fluctuations are ubiquitous even under such conditions. We therefore believe that it is often appropriate to identify a wave as Alfvénic without attempting finer distinctions.

Finally, one might ask why not simply name this broader class Alfvén waves? Doing so could, for example, better align with terminology such as "poloidal Alfvén waves" used in magnetospheric physics. Here, we can only answer that it is not the custom in solar physics to do so.

In this article, sections addressing observations will typically refer to Alfvénic waves, which are taken to be a class of MHD waves that have the properties itemized in the earlier text. Occasionally, we refer to this general class as Alfvén waves, as in the article's title; we do this sparingly and to provide consistency with other fields that do not use the "Alfvénic" terminology, acknowledging the interdisciplinary nature of the collection to which this article belongs. When using mathematical models to explore issues such as energy transport and wave dissipation, we usually consider "pure" Alfvén waves for concreteness, which is indicated in the text by naming them specifically as Alfvén waves.

## 3.4. NONTHERMAL LINE WIDTHS

The first question to ask about Alfvénic waves in solar flares is, "Do flare-generated Alfvénic waves leave a fingerprint in solar observations?" In answer, we spotlight nonthermal broadening of spectral lines in solar flares. If excess spectral line widths in flares can be attributed to transverse waves, one might reasonably expect these observations to advance diagnostics of Alfvén waves in solar flares, similarly to how hard X-ray diagnostics have advanced the study of flare nonthermal electrons.

### 3.4.1. Spectral Broadening by Unresolved Plasma Motions

A spectral line emitted by stationary plasma has a line width determined by Doppler broadening due to random thermal motions of the particles. Departures provide information about unresolved plasma motion—a principle that has long been exploited by astrophysicists and solar physics.

In the simplest models, nonthermal broadening is evaluated using the formula

$$(\Delta\lambda)_{nth} = \sqrt{(\Delta\lambda)^2_{obs} - (\Delta\lambda)^2_{inst} - (\Delta\lambda)^2_{th}}, \qquad (3.5)$$

where $(\Delta\lambda)_{obs}$ characterizes the observed line profile, $(\Delta\lambda)_{inst}$ is the instrumental broadening, and $(\Delta\lambda)_{th}$ is the line width in the absence of unresolved motions.

Estimating the thermal width $(\Delta\lambda)_{th}$ requires the ion temperature. One of the more precise methods for obtaining a temperature is to infer the electron temperature from intensity ratios of temperature-sensitive spectral lines, then assume thermal equilibrium between species such that $T_i \approx T_e$. Allowing for the possibility that $T_i \neq T_e$, some investigators have used the alternative simple heuristic of assuming that the ion temperature is close to the peak formation temperature of the line they are using, although this assumption carries significant uncertainty, since ionization states are populated for a broad range of temperatures around the peak formation temperature. A more sophisticated approach uses differential emission measure (DEM) analysis, which combines information from multiple ionization states.

Once obtained, nonthermal widths can be converted to velocity units using the Doppler formula,

$$v_{nth} = c\frac{(\Delta\lambda)_{nth}}{\lambda_0}, \qquad (3.6)$$

where $\lambda_0$ is the rest wavelength of the line center. The resulting $v_{nth}$ provides a representative magnitude for unresolved line-of-sight velocities.

Alternatively, the nonthermal velocity can be expressed as

$$v_{nth} = \sqrt{\frac{2k_B}{m_i}(T_D - T_e)}, \qquad (3.7)$$

where the Doppler temperature of the emitting ion is defined by

$$\sqrt{\frac{2k_B T_D}{m_i}} = \frac{c}{\lambda_0}\sqrt{(\Delta\lambda)^2_{obs} - (\Delta\lambda)^2_{inst}}. \qquad (3.8)$$

Finally, we point out that it is not always so simple. The "sum-of-squares" formulas used in the earlier text are rigorous for Gaussian line profiles, which are common in flare spectra, but not universal. Instrumental quirks (such as convolution with the instrumental spread function) or plasma conditions (such as non-Maxwellian ion populations) can produce non-Gaussian line profiles. In such cases, $v_{nth}$ typically becomes a curve-fitting parameter. It is also common that line widths are reported using the full width at half-maximum (FWHM) instead of fitting the Gaussian width, which introduces a factor $2\sqrt{\ln(2)} \approx 1.665$ in some papers.

### 3.4.2. Whole-Active-Region X-Ray Spectra

X-ray spectroscopy of the Sun developed rapidly after World War II, as solar scientists developed payloads for rocket, satellites, and space station programs, including OSO 6 (Doschek et al., 1971), Intercosmos-4 (Grineva et al., 1973), and Skylab (Doschek et al., 1975). These spectrometers did not spatially resolve features within flaring active regions, but they nonetheless provided considerable insights into the properties of flare plasma. The occurrence of nonthermal line broadening during impulsive flares was well established by the late 1970s, and it was a major scientific interest for the second-generation

X-ray spectrometers that flew aboard P78-1 (launched 24 Feb 1979), Solar Maximum Mission (SMM; launched 14 Feb 1980), and Hinotori (launched 21 Feb 1981). Doschek (1990) provides a general discussion of X-ray flare spectra results from this period. A decade later, Yohkoh (launched 30 Aug 1991) continued these investigations using an X-ray spectrometer roughly an order of magnitude more sensitive than previous instruments.

The main properties of nonthermal broadening were already known by the time of the systematic series of studies by Doschek et al. (1979), Doschek et al. (1980), and Feldman et al. (1980), which analyzed data from the SOLFLEX Bragg crystal spectrometers on the P78-1 satellite. These observations covered a sample of M and X class flares, with a focus on the Fe XXV and Ca XIX lines. The main conclusions about line widths were:

1. Line profiles produced in hot flare plasma (temperatures $\geq$ 10 MK) display nonthermal broadening of hundreds of kilometers per second, up to approximately 160 km s$^{-1}$ in the SOLFLEX data.

2. The broadening is greatest during the impulsive phase, which is when Fe XXV lines are usually first detected. Nonthermal widths decrease steadily from the initial peak.

3. Nonthermal velocities in the gradual phase are considerably lower than during the impulsive phase but remain elevated, with approximately 60 km s$^{-1}$ being common.

4. The magnitude of nonthermal motions is a weak function of temperature. Hotter lines have larger nonthermal velocities.

These conclusions were confirmed by SMM (Antonucci et al., 1982; Antonucci et al., 1984; Culhane et al., 1981; Gabriel et al., 1981) and Hinotori (Tanaka et al., 1982). In many observations, the peak $v_{nth}$ was recorded when counts first became sufficient to fit the line; thus, the true maximum almost certainly occurred when the Fe XXV line was too faint to be reliably fitted. Since greater sensitivity allows measurements earlier in the flare, the best estimates of the peak $v_{nth}$ in whole-active-region spectra came from the Yohkoh Bragg Crystal Spectrometer (BCS). We therefore adopt the peak $v_{nth}$ for looptop sources of 250 km s$^{-1}$ from Khan et al. (1995), consistent with Tanaka et al. (1982) and Mariska et al. (1996), rather than the SOLFLEX value quoted in Point 1. Figure 3.2 shows plots by Alexander et al. (1998) that illustrate Points 1 to 4 in two flares observed by Yohkoh.

Since nonthermal broadening is observed early in the flare, it is interesting to explore the timing of its peak relative to other impulsive phase phenomena. Timing studies by Antonucci et al. (1984), Alexander et al. (1998) (see Figure 3.2), Harra et al. (2001), and Ranns et al. (2001) found evidence that:

5. Nonthermal widths can increase a few minutes or more *before* the onset of impulsive phenomena such as the hard X-ray burst and appearance of high-speed upflows, although they are not always seen to do so.

The likely cause of nonthermal broadening is also informed by comparisons across different events. Using SMM data, Antonucci et al. (1982) and Antonucci (1984) found that blueshifts of 300–400 km s$^{-1}$ occurred for disk

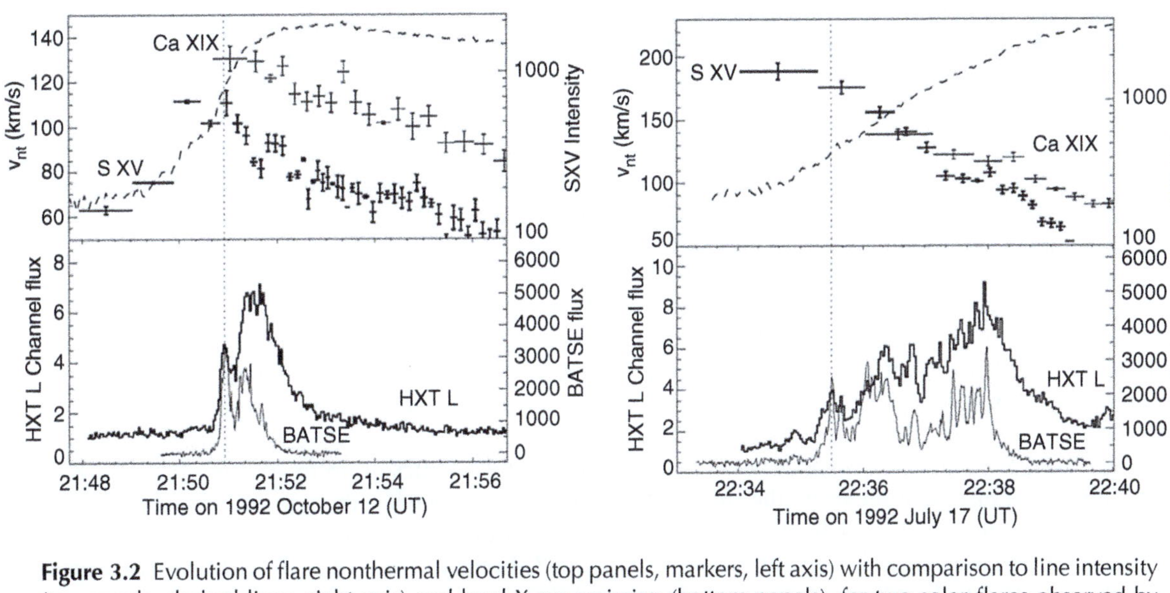


**Figure 3.2** Evolution of flare nonthermal velocities (top panels, markers, left axis) with comparison to line intensity (top panels, dashed lines, right axis) and hard X-ray emission (bottom panels), for two solar flares observed by Yohkoh. Nonthermal widths of hundreds of kilometers per second are seen minutes before the hard X-ray peak. Reproduced with permission from Alexander et al. (1998). © AAS.

flares but were absent for limb flares, consistent with bulk upward radial flows at a significant fraction of the sound speed. In contrast, it was found that:

6. The nonthermal velocity does not show significant center-to-limb variation.

These center-to-limb dependencies were confirmed by statistical surveys with Yohkoh, covering more than 100 events (Mariska et al., 1993; Mariska, 1994). Points 5 and 6 together provide evidence that a significant part of the nonthermal broadening is not due to superposition of emission from upflows at multiple speeds.

A final major result from whole-active-region studies is the finding by Fludra et al. (1989) that:

7. The nonthermal and upflow velocities are correlated in most flares.

Points 5 and 7 suggest that the nonthermal broadening is connected to the primary energy release.

### 3.4.3. Spatially Resolved EUV Observations—General Remarks

We now turn to observations that resolve specific locations within active regions, moving to lines in the EUV. There is considerable heritage to EUV spectroscopy, which includes instruments aboard SMM and the Solar and Heliospheric Observatory (SOHO). In the last decade, the instruments that have arguably contributed most to new knowledge are Hinode/EIS (Culhane et al., 2007) and IRIS (De Pontieu et al., 2014).

Spectral coverage for EIS includes lines of Fe XXIV and Fe XXIII, while IRIS coverage includes Fe XXI. Applying the trend that peak nonthermal velocities are lower in cooler lines, one should expect nonthermal velocities observed by EIS and IRIS to be lower than for Fe XXV lines in X-ray spectra. In other words, one should bear in mind that EIS and IRIS do not have access to the hottest flare plasma, which is where the strongest motions seem to occur.

Single-slit spectrographs can create raster "images" by scanning the slit position. It is important to be aware that pixel values in different stripes of a raster correspond to different times. The rastering process can also be slow, leading to long effective cadences. These considerations mean that rasters are better suited to studying the spatial distribution of $v_{nth}$ during the phase in which nonthermal velocities are decaying, than they would be for searching for the peak value of $v_{nth}$, which is short-lived. In contrast, the stare mode of operation keeps the slit pointed at one location. Sometimes, the slit is at an interesting location, in the best cases from before the start of the impulsive phase. IRIS has the advantage that as a slit-jaw spectrometer it captures contextual slit-jaw images that show exactly where the spectra correspond to.

Looking to the future, the Multi-slit Solar Explorer (MUSE; De Pontieu et al., 2020) is designed to simultaneously obtain EUV spectra along 37 slits, allowing observation of a 170″ × 170″ field of view with a cadence around 1 s and raster time of 12 s. Furthermore, its spectral coverage is well suited to flare studies (Cheung et al., 2022). We anticipate that this future mission will have a large scientific impact on the study of nonthermal broadening and Alfvén wave/turbulence in solar flares, a point we return to in Section 3.8.2.

Sections 3.4.4 to 3.4.6 summarize current knowledge from spatially resolved EUV spectra. They are organized according to spatial location, starting with observations of the flare plasma sheet, followed by above-the-loop and looptop sources, and finishing with the flare footpoints. Interpretation of the nonthermal broadening is discussed in Section 3.4.7.

### 3.4.4. Flare/CME Plasma Sheets

The reconnecting current sheet is a central feature of the standard flare model (Figure 3.1) and associated plasma sheets are sometimes imaged behind CMEs. Clear views are relatively rare, requiring viewing the plasma sheet end-on, but several cases have been documented. As elaborated in the following text, spectroscopic measurements reveal these sheets to be a site of significant nonthermal broadening, especially at early times in the eruption.

SOHO's Ultraviolet Coronagraph Spectrometer (UVCS) investigated several post-CME plasma sheets, typically for heliocentric distances between 1.5 and 2.0 $R_\odot$ and using Fe XVIII emission. The "Halloween Event" (4 November 2003) study by Ciaravella and Raymond (2008) has the distinction of measuring the line widths at relatively early times. Their results detected $v_{nth} \approx 380$ km s$^{-1}$ for their earliest data point, which was followed by rapid decay over 10 min. A few hours later, $v_{nth}$ around 50–100 km s$^{-1}$ is typical in post-CME plasma sheets (Bemporad, 2008; Ciaravella et al., 2002; Ciaravella & Raymond, 2008; Schettino et al., 2010; Susino et al., 2013).

More recently, Hinode/EIS has also investigated nonthermal broadening within post-CME plasma sheets. For the textbook flare on 10 September 2017, EIS collected raster spectroscopic images that extend from the solar limb to approximately 1.24 $R_\odot$ heliocentric altitude. Nonthermal velocities of up to 200 km s$^{-1}$ for Fe XXIV were detected, displaying the typical behavior of being strong early in the flare and decreasing thereafter (Cheng et al., 2018; French et al., 2020; Li et al., 2018; Warren et al., 2018).

The spatial distribution of $v_{nth}$ found by Warren et al. (2018) and Li et al. (2018) is interesting too. First, $v_{nth}$

increases with altitude in the plasma sheet. Second, the profile of $v_{nth}$ perpendicular to the plasma sheet appears single-peaked at higher altitudes; but at lower altitudes it appears double-peaked, with ridges of $v_{nth}$ coinciding with where one would expect magnetic separatrices, as the end of the current sheet opens in a cusp. These observations are consistent with reconnection models in which the strongest turbulence is generated in the vicinity of a dominant reconnection site above the maximum altitude of the EIS observations and turbulence/waves exiting from the end of the current sheet propagate near the magnetic separatrices. Alternatively, the double-ridge structure may indicate generation of small-scale motions by instabilities, for instance, Kelvin–Helmholtz instability fed by the velocity shear at the edge of the outflowing reconnection jet (Loureiro et al., 2013).

### 3.4.5. Above-the-Loop Region and Loops

We now move down to the above-the-loop regions associated with coronal hard X-ray and microwave emission. This region is interesting from the perspectives of wave/turbulence generation by braking of the reconnection outflow jet and particle acceleration, and it displays some of the largest nonthermal velocities outside the plasma sheet.

Figure 3.3 shows a limb flare analyzed by Shen et al. (2023). This region displays the canonical $v_{nth}$ evolution of steady decline from an initial peak greater than approximately 100 km s$^{-1}$. Similar $v_{nth}$ values for the above-the-loop region have been reported for different flares by Tian et al. (2014) and Polito et al. (2018).

Covering a larger field of view, Figure 3.4 shows a nonthermal velocity map from the 15 May 2013 X1.2 flare (Doschek et al., 2014; Kontar et al., 2017; Stores et al., 2021). Blue contours show Solar Dynamics Observatory (SDO) Atomospheric Imaging Assembly (AIA, Lemen et al., 2012) 94 Å intensity, indicating flare loops; white contours show SDO/AIA 304 Å intensity, indicating flare ribbons; and the orange pixels show the Fe XXIV nonthermal velocity obtained from an EIS raster. This map supports several important conclusions:

1. The greatest nonthermal velocity is at the apex of the above-the-loop region.
2. The value of $v_{nth}$ decreases by around 50%, as one descends toward the footpoints.
3. Ridges of $v_{nth}$ are located further out than the AIA 94 Å loops.

Point 3 is consistent with the broadening being associated with energy transport from the reconnection site in the corona. It might also indicate that $v_{nth}$ is greatest in the hottest loops, which occurs further out than the brightest loops (Tsuneta, 1996). All three points are further evidence that the nonthermal broadening is related to the primary energy release, not upflows from the chromosphere.

We also highlight a few earlier works of this type. In one of the first EIS studies, Hara et al. (2008) mapped line widths for a C2-class long-duration flare near the limb. Their results demonstrated that Ca XVII line widths are largest at the top of the above-the-loop region, and widths of the cooler Fe X and Fe XII lines are greatest at the top of the bright flare loops. The nonthermal broadening was consistent with motions at 50–100 km s$^{-1}$.

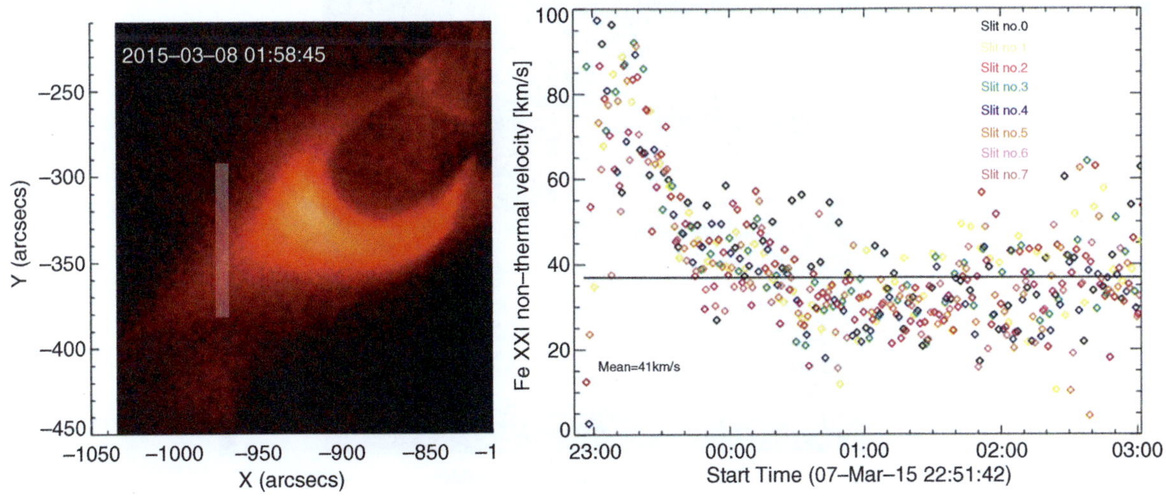


**Figure 3.3** Evolution of nonthermal velocities in the above-the-loop region of a long-duration M9.2 solar flare. The left panel is an Hinode X-ray Telescope (XRT) context image that shows the locations sampled by IRIS during the narrow raster. The right panel shows the time development of nonthermal velocities in this region. Reproduced with permission from Shen et al. (2023) / Frontiers Media S.A. / CC BY 4.0.

Hara et al. (2011) subsequently studied Fe XXIII and Fe XXIV during the impulsive phase of a B9.5 long-duration flare near disk center, finding nonthermal broadening of approximately 100 km $s^{-1}$ at the first timepoint, steadily decreasing thereafter. Harra et al. (2013) produced maps of $v_{nth}$ for four near-limb flares of M and X classifications. Three of the flares investigated by Harra et al. (2013) were eruptive and displayed pre-flare enhancements of $v_{nth}$ in the above-the-loop region. The largest eruptive flare (X2.1) was viewed close to the limb and values of $v_{nth}$ in the above-the-loop region displayed an initial peak between 100 and 150 km $s^{-1}$, decaying to less than 50 km $s^{-1}$. Doschek et al. (2014) analyzed four near-limb flares, including the 15 May 2013 X1.2 flare shown in Figure 3.4. They also analyzed the long-duration X1.7 flare on 27 January 2012, which was viewed at an orientation presenting an arcade of flare loops in the plane of sky. For this latter event, Doschek et al. (2014) measured plane-of-sky velocities by applying local correlation tracking (LCT) to AIA 131 Å images, which are dominated by Fe XXI emission. The LCT analysis revealed velocity variations by 100–200 km $s^{-1}$ over distances of just a few megameters (also see, McKenzie, 2013). This result reinforces the view that nonthermal broadening in the above-the-loop region is indeed due to plasma motions. Finally, we mention that flare nonthermal velocities around 100 km $s^{-1}$ have also been reported using SOHO's Solar Ultraviolet Measurements of Emitted Radiation (SUMER; Landi et al., 2003) and Coronal Diagnostic Spectrometer (CDS; Brosius, 2012) instruments.

### 3.4.6. Footpoints

Whole-active-region X-ray spectra for flares near disk center often have two main components: one stationary and the other blueshifted (Antonucci et al., 1982, 1984, 1990; Doschek et al., 1989; Fludra et al., 1989). It is reasonable to identify the stationary component with the looptop source discussed in the previous section. The blueshifted component is believed to be located at the magnetic footpoints and associated with chromospheric evaporation. The nonthermal widths of the blueshifted and stationary components can be significantly different (Doschek et al., 1989; Fludra et al., 1989).

Following earlier work using SOHO/CDS (Brosius, 2003), EIS made it possible to study footpoint sources across a broad temperature range with high spatial and spectral resolution. For example, Milligan and Dennis (2009) and Milligan (2011) studied EIS spectra for the footpoints of a C1.1 flare, reporting nonthermal broadening coincident with hard X-ray footpoint sources. Their results are summarized by Figure 3.5. Cooler lines (≤4 MK) exhibit modest blueshifts and redshifts (50 km $s^{-1}$ or less) and the nonthermal widths of 55–80 km $s^{-1}$, with no clear relation between their nonthermal widths and temperatures. The hottest lines, Fe XXIII and Fe XXIV (10–20 MK), had profiles consistent with a dominant stationary component plus a strongly blueshifted component. The blueshifted components exhibited upward velocities of around 250 km $s^{-1}$ and fitted nonthermal widths were reported to be around 100–120 km $s^{-1}$. Young et al. (2013) found similar results

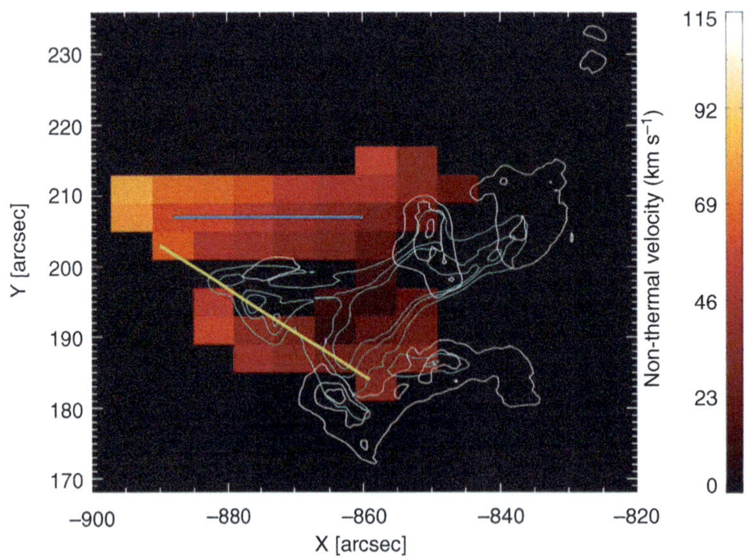


**Figure 3.4** Map of nonthermal velocity in the corona during an X1.2 flare. Orange pixels show nonthermal velocity measured from the EIS Fe XIV line. Cyan contours show AIA 94 Å intensity, outlining flare loops. White contours show AIA 304 Å intensity highlighting flare ribbons. Shen et al. (2023) / IOP Publishing / CC BY 4.0.

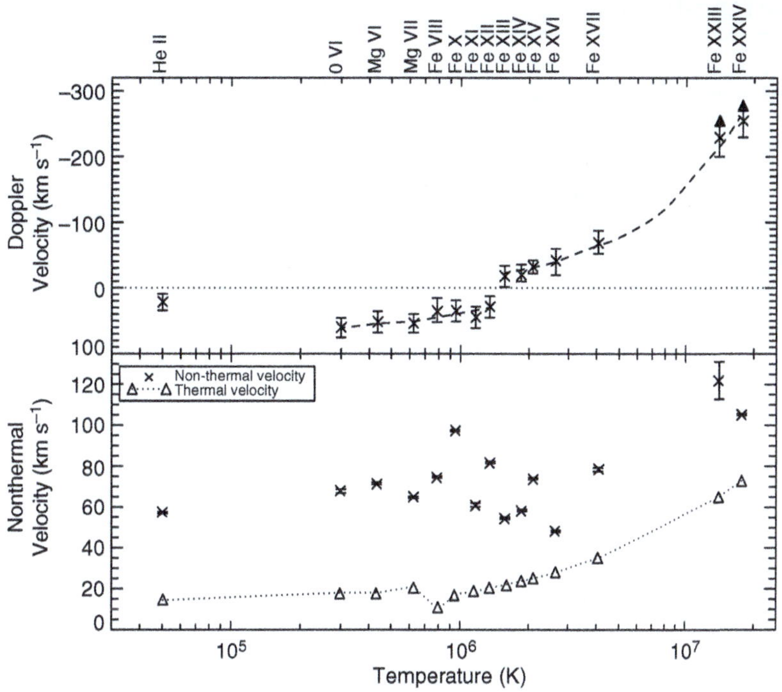


**Figure 3.5** Line-of-sight Doppler velocity (top panel) and nonthermal velocity (bottom panel) in footpoints of a C1.1 flare observed by Hinode/EIS. Hot lines (10–20 MK) display nonthermal widths up to 120 km $s^{-1}$, compared to 55–80 km $s^{-1}$ for cooler lines (≤4 MK). Reproduced with permission from Milligan (2011). © AAS.

for a flare kernel during the rise phase of a M1.1 class flare, confirming the occurrence of nonthermal broadenings of 100–120 km $s^{-1}$ in Fe XXIII and Fe XXIV, with significantly less broadening of cooler lines.

The temperature dependence of the Doppler velocity is an expected result, since 1D hydrodynamic simulations show that rapid deposition of energy in the chromosphere causes an explosive upflow, accompanied by a momentum-conserving cool downflow at lower altitudes (Fisher et al., 1985). The enhancement of nonthermal velocity for temperatures above 10 MK indicates that $v_{nth}$ is significantly larger for the explosively upflowing plasma than it is deeper in the chromosphere, which would be consistent with waves dissipating in the heated layer.

An important advance since Milligan (2011) has been the resolving of footpoint sources. Brosius (2013) reported a completely blueshifted Fe XXIII (approximately 14 MK) footpoint source for one C1 class flare with EIS. Soon after, IRIS made the significant breakthrough of regularly spatially resolving Fe XXI (approximately 10 MK) at the footpoints, obtaining totally blueshifted profiles (Graham & Cauzzi, 2015; Lee et al., 2017; Li et al., 2015; Polito et al., 2015, 2016; Tian et al., 2014, 2015; Young et al., 2015). When Fe XXI is first detected, the line is typically blueshifted by 100–400 km $s^{-1}$, sometimes taking it to the edge of the instrument's spectral read-out window, and it has a large nonthermal width, often around 100–120 km $s^{-1}$. Over the next several minutes, the blueshift steadily decreases to zero, and the nonthermal broadening decreases to around 50 km $s^{-1}$ or less.

If the footpoint broadening is interpreted as due to Alfvénic waves, then the hot flare lines provide the velocity amplitude above where the chromosphere is heated. A limitation is that sufficient Fe XXI intensity to fit the line is produced tens of seconds after the presumed energy deposition (see Figure 3 of Graham & Cauzzi (2015)). As for whole-active-region X-ray spectra, high-temperature EUV lines provide a lower bound for the maximum $v_{nth}$, but the true maximum of $v_{nth}$ almost certainly occurs before there is sufficient intensity to fit the lines.

To gain insight into times before and during the onset of flaring in individual pixels, one can turn to lines that exist at chromosphere and transition region temperatures. Jeffrey et al. (2018) investigated this question for a B class flare using IRIS Si IV (approximately 80,000 K). The evolution is plotted in Figure 3.6 (top panel) at 1.7 s cadence. The Si IV nonthermal velocity starts at a pre-flare value of about 9 km $s^{-1}$, increases over 10 s to a maximum of 30 km $s^{-1}$, and returns to pre-flare levels within a minute. The increase in $v_{nth}$ leads the increase in the IRIS Si IV and Reuven Ramaty High Energy Solar Spectroscopic Imager (RHESSI, Lin et al., 2002) 6–12 keV X-ray intensities by about 20 s. Also shown (bottom panel) is the position of the line center and the corresponding Doppler

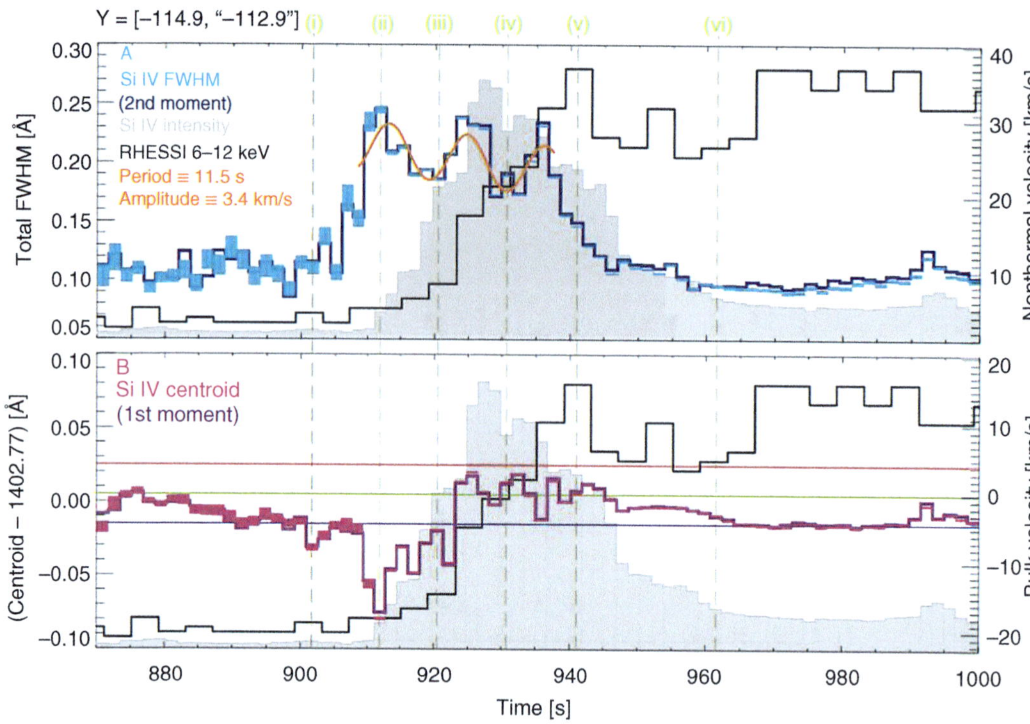


**Figure 3.6** Evolution of Si IV in the footpoint of a B-class flare. The top panel (a) shows the total line width (left axis) and the equivalent $v_{nth}$. The line width increases from its pre-flare value of 10 km s$^{-1}$ to a peak of 30 km s$^{-1}$. This increase in nonthermal broadening leads the increases of the Si IV intensity and 6–12 keV X-rays by 10–20 s. The bottom panel (b) shows the evolution of the line centroid. Reproduced from Jeffrey et al. (2018) under a CC BY 4.0 license.

velocity of the line shift (negative values are blueshifts). Jeffrey et al. (2018) suggested that their observations are consistent with unresolved plasma motions in the flare ribbons heating the plasma with a dissipation time of approximately 10 s.

The timings might tentatively connect the nonthermal broadening with footpoint heating before the impulsive phase (also see, Fletcher et al. (2013) and Simões et al. (2015)). Analysis of spatial scales along the ribbons for the same flare by French et al. (2021) have added to the timeline. Their work showed that an instability, hypothesized to be tearing instability of the coronal current sheet, first develops at a single "key scale," then spreads to smaller and larger scales, ending in fully developed turbulence with a power law slope of 2.3. The rise in nonthermal widths reported by Jeffrey et al. (2018) coincides with the stage in which the turbulence is developed across multiple scales. Additional examples of IRIS Si IV nonthermal broadening preceding flaring have been reported by Chitta and Lazarian (2020).

### 3.4.7. Interpretation

The consensus explanation for nonthermal broadening is unresolved mass motions. Various other line-broadening mechanisms exist, such as pressure broadening, opacity broadening, and the Stark effect, for example, discussion by Milligan (2011). The difficulty for many of the alternative hypotheses is the ubiquity of nonthermal broadening during flares, which occurs in the flare plasma sheet, above-the-loop region, loops and ribbons, and for a multitude of optically thin lines.

One simple alternative to unresolved motions would be that the line width reveals the true ion temperature, which differs from the electron temperature and the expected formation temperature of the line. This hypothesis has been previously rejected, with Antonucci (1984) writing that, "The observed profiles would imply a difference in ion and electron temperatures sometimes exceeding one order of magnitude. At the density inferred for the thermal plasma of flares, it would be difficult to explain the persistence of large differences in electron and ion temperatures over the observed flare rise times." The inferred approximately 100 MK ion temperatures would also be many times the peak formation temperature of the observed lines. This hypothesis therefore appears less plausible than unresolved mass motions.

Several studies have looked at the role of non-Maxwellian particle distributions, including Jeffrey et al. (2016), Dudík et al. (2017), and Polito et al. (2018).

Broad EUV spectral lines in flares often have Gaussian profiles, which are consistent with a Maxwellian ion distribution; however, a kappa profile provides a better fit in some cases. Dudík et al. (2017) concluded that kappa distributions do not remove the presence of nonthermal broadening, but on the contrary imply larger unresolved motions.

Supporting evidence that line broadening is due to unresolved motions can be found in direct imaging of plasma motions. LCT has been applied to SDO/AIA 131 Å images, finding variations in local velocities by hundreds of kilometers per second over distances of a few megameters (Doschek et al., 2014; McKenzie, 2013). Meanwhile, MHD simulations have demonstrated generation of small-scale plasma motions of 100 km $s^{-1}$ or greater in the braking or "interface" region, which propagate to the chromospheric footpoints as Alfvénic waves. When line profiles are forward modeled for these simulations, they produce broad Gaussian profiles that match observations (Ruan et al., 2023; Shen et al., 2023). These simulations are discussed further in Section 3.6.3.

A historical debate about the form of the unresolved motions has been between a superposition of field-aligned flows, or motions that are either isotropic or perpendicular to the magnetic field. Arguments against attributing broadening to superposition of field aligned flows have been made since at least the 1980s: a simple model involving vertical upflows is inconsistent with the center-to-limb behavior; upflows struggle to explain line widths in loop-top sources; and such models cannot account for events in which nonthermal broadening precedes occurrence of upflows. As spectral and spatial resolutions have improved, the symmetric broad profiles of resolved observations have added further weight against the idea that broadening is due to a mixture of upflows (Polito et al., 2019). Timings further make it likely that broadening is related to the energy release itself, rather than evaporative upflows.

Linear polarization (Stokes *P*) also provides information relevant to this debate. For a resolved magnetic field, *P* depends on the angle between the magnetic field and the radial direction, and the inferred magnetic field direction typically agrees with expectations from intensity images. Examining the 10 September 2017 textbook flare using the Coronal Multi-channel Polarimeter (CoMP) instrument, French et al. (2019) also found that *P* drops to low values in the plasma sheet and in the above-the-loop region, which is where large nonthermal broadening is detected in this flare (Li et al., 2018; Warren et al., 2018). The reduction in linear polarization signal was attributed to unresolved magnetic fields with mixed orientations. Putting together the bigger picture, the unresolved velocities (nonthermal broadening) are accompanied by unresolved magnetic field orientations (low values of *P*). This association would not occur for field-aligned flows, but it is consistent with large-amplitude Alfvénic waves.

This article takes the view that the inferred motions most likely have an Alfvénic nature. Maps of $v_{nth}$ in the corona (Figure 3.4) are consistent with field-aligned transport, as is the location of broadening at the flare footpoints. The transport is rapid and therefore better associated with the Alfvén speed than the sound speed. The unresolved motions are accompanied by unresolved variations in the direction of the magnetic field, indicating that they are not hydrodynamic in nature. If one ascribes unresolved motions to turbulence, fluctuations in strongly magnetized plasma typically have an Alfvénic character. At the same time, further work is required to conclusively establish whether unresolved motions are primarily perpendicular to the background magnetic field, for example, viewing angle studies at modern spatial resolution.

## 3.5. WAVE ENERGY DENSITY AND POYNTING FLUX

Combining the values of $v_{nth}$ in previous sections with a few assumptions and physical properties, one can estimate the energy density and Poynting flux of the inferred Alfvén waves. For these calculations, we assume that magnetic fluctuations are perpendicular to the background magnetic field, and there is equipartition of wave energy between kinetic and magnetic energies.

The averaged wave energy density can be expressed as

$$\langle E_{wave}\rangle = \frac{m_i n\langle v^2\rangle}{2} + \frac{\langle \delta B^2\rangle}{2\mu_0} = m_i n\langle v^2\rangle. \tag{3.9}$$

Coronal densities for flaring solar active regions are typically around $10^{17}$ $m^{-3}$, so approximating the coronal $m_i$ as the proton mass and considering $v_{nth} \approx 250$ km $s^{-1}$ for the above-the-loop region, the energy density of the waves is estimated as $\langle E_{wave}\rangle \approx 10$ J $m^{-3}$.

For comparison, the thermal energy density of the ions (or electrons) is

$$E_{th} = \frac{nk_B T}{\gamma - 1}. \tag{3.10}$$

Setting the adiabatic index as $\gamma = 5/3$ and considering the number density used in the earlier text, a temperature of 10 MK gives $E_{th} \approx 21$ J $m^{-3}$. In these estimates, the wave energy density is a significant fraction (about half) of the thermal energy, and high enough to play a significant role in flare energetics, for example, by accelerating particles in the above-the-loop region. This is especially true if the waves are acting as an intermediary via which energy is converted to other forms, as discussed by Kontar et al. (2017).

The comparison of energy densities can also be performed by expressing the ratio as

$$\frac{\langle E_{wave}\rangle}{E_{th}} = \frac{(\gamma - 1)m_i\langle v^2\rangle}{k_B T} = \gamma(\gamma - 1)\frac{\langle v^2\rangle}{c_s^2}, \tag{3.11}$$

which makes clear that the ratio is independent of the density. Here, $c_s = \sqrt{\gamma p/\rho} = \sqrt{\gamma k_B T/m_i}$ is the ion sound speed. From this formulation, we see that the wave energy density is a significant fraction of the thermal energy density because the $v_{nth}$ is a significant fraction of the ion sound speed.

We are also interested in the ability of waves to transport energy to the lower atmosphere. Poynting's theorem (with the energy density of the electric field neglected for nonrelativistic MHD) states that

$$\frac{\partial}{\partial t}\left(\frac{B^2}{2\mu_0}\right) + \nabla \cdot \boldsymbol{S} = -\boldsymbol{j} \cdot \boldsymbol{E}, \tag{3.12}$$

such that magnetic energy is transported by the Poynting vector,

$$\boldsymbol{S} = \frac{1}{\mu_0}\boldsymbol{E} \times \boldsymbol{B}. \tag{3.13}$$

We will estimate the Poynting flux for an Alfvén wave, assuming the ideal Ohm's law,

$$\boldsymbol{E} = -\boldsymbol{v} \times \boldsymbol{B}. \tag{3.14}$$

Assuming that magnetic and velocity perturbations are perpendicular to the background magnetic field, and focusing on transport parallel to the background magnetic field, we are interested in

$$\boldsymbol{S}_{\parallel} = \frac{1}{\mu_0}(\boldsymbol{B}_0 \times \delta\boldsymbol{v}) \times \delta\boldsymbol{B} = -\frac{1}{\mu_0}(\delta\boldsymbol{B} \cdot \delta\boldsymbol{v})\boldsymbol{B}_0. \tag{3.15}$$

Finally, eliminating $\delta\boldsymbol{B}$ using the Walén relation for propagating Alfvén waves,

$$\frac{\delta\boldsymbol{B}}{B_0} = \mp\frac{\delta\boldsymbol{v}}{v_A}, \tag{3.16}$$

the averaged parallel Poynting flux is

$$\begin{aligned}\langle S_{\parallel}\rangle &= \frac{B_0^2}{\mu_0 v_A}\langle v^2\rangle = v_A m_i n\langle v^2\rangle = v_A\langle E_{wave}\rangle \\ &= B_0\sqrt{\frac{m_i n}{\mu_0}}\langle v^2\rangle.\end{aligned} \tag{3.17}$$

This result can also be obtained heuristically using the principle that the wave energy density is transported at the Alfvén speed.

Magnetograms reveal active region magnetic field strengths of up to several kilogauss in the photosphere. In the active regions, microwave observations and magnetic field extrapolations also imply kilogauss field strengths in the corona (Brosius & White, 2006). For the above-the-loop region, field strengths are lower, for example, around 300–400 G (0.03–0.04 T) for the 10 September 2017 textbook flare (Fleishman et al., 2022) and the 15 May 2012 flare (Kontar et al., 2017).

We now estimate $\langle S_{\parallel}\rangle$ for the apex and footpoints, setting the number density of $10^{17}$ m$^{-3}$ for both cases. For the above-the-loop region, a magnetic field strength of 0.04 T and velocity amplitude of 250 km s$^{-1}$ yield a Poynting flux of $2.9 \times 10^7$ W m$^{-2}$ ($2.9 \times 10^{10}$ erg s$^{-1}$ cm$^{-2}$). In this case, the Alfvén speed is 2800 km s$^{-1}$, and $\langle E_{wave}\rangle \approx 10$ J m$^{-3}$ as calculated earlier in this section. Near the footpoint, a magnetic field strength of 0.1 T and velocity amplitude of 120 km s$^{-1}$ yield a Poynting flux of $1.7 \times 10^7$ W m$^{-2}$. For this case, the Alfvén speeds is 6900 km s$^{-1}$ and $\langle E_{wave}\rangle \approx 2.4$ J m$^{-3}$.

The Poynting fluxes calculated in the earlier text exceed $10^7$ W m$^{-2}$, and they are just slightly below the solar luminosity per unit area of $7 \times 10^7$ W m$^{-2}$. Both estimates are consistent with the normal range of energy deposition rates considered in flare modeling, which is $10^6$–$10^8$ W m$^{-2}$ (Fisher et al., 1985). The estimated Poynting fluxes are also comparable to electron energy deposition rates for specific events (Antonucci et al., 1990; Lee et al., 2017) and broadband white light and UV emissions (Fletcher & Hudson, 2008). Various 1D radiative-hydrodynamic simulations, including the wave-driven model of Reep and Russell (2016), have shown that energy fluxes of $10^7$ W m$^{-2}$ are capable of driving flare-like dynamics, including strong footpoint heating and explosive chromospheric evaporation. We therefore conclude that if nonthermal broadening is attributed to Alfvénic waves, the waves carry enough energy to contribute meaningfully to flare dynamics.

There are some events for which nonthermal electrons have been estimated to carry substantially larger energy fluxes into the flare footpoints than the values discussed in the earlier text. Specifically, Krucker et al. (2011) derived an energy deposition rate by electron beams in the region of $5 \times 10^9$ W m$^{-2}$ for the brightest footpoints during an X6.5 flare. While that estimate is sensitive to the low-energy cutoff assumed for the electron beam, and this flare appears to be an outlier compared to more common observational estimates of order $10^7$ W m$^{-2}$, it poses the question of whether Alfvén waves could plausibly deliver a similar Poynting flux in extreme circumstances.

The available parameters are the number density, magnetic field strength, and wave amplitude. As a reference point for number density, Milligan (2011) determined a footpoint density of $10^{17.5}$ m$^{-3}$ for a C1.1 class flare, which is slightly higher than the density value we used in the earlier text, for a relatively low energy flare. Meanwhile, 1D radiative hydrodynamics simulations show that large energy deposition rates produce dense chromospheric evaporations, with densities around $10^{19}$ m$^{-3}$ not uncommon. We might therefore be justified in considering a footpoint density of $10^{19}$ m$^{-3}$ for an X6.5 flare with energy deposition rate of $5 \times 10^9$ W m$^{-2}$. Increasing the assumed density by two orders of magnitude (e.g., from $10^{17}$ m$^{-3}$ to $10^{17}$ m$^{-3}$) increases the Poynting flux by one order of magnitude. Similarly, given this is an extreme event, the footpoint magnetic field strength might be raised to 2 kG and the peak value of $v_{nth}$ to 250 km s$^{-1}$, noting Point 7 in Section 3.4.2 that stronger upflows correlate with larger nonthermal widths. With these modifications, the Poynting flux would be $1.4 \times 10^9$ W m$^{-2}$, which is comparable to the energy deposition rate derived by Krucker et al. (2011). This exercise, while speculative, suggests that Alfvénic waves can deliver sufficient energy fluxes to contribute meaningfully to the dynamics of even the most extreme flares.

The main conclusion of this section is that observationally derived parameters for $v_{nth}$, density, and magnetic field strength imply that the wave energy density at the apex is comparable to the ion thermal energy, and Poynting fluxes are large enough to meaningfully affect the lower atmosphere (typically several times $10^7$ W m$^{-2}$). The effects of this energy transport are explored in Section 3.7.

## 3.6. GENERATION OF ALFVÉNIC WAVES/TURBULENCE

This article now moves to theory and simulations, starting with the generation of Alfvénic waves in solar flares. A major shift in flare theory during the last two decades has been the progression from laminar steady-state models to dynamic ones in which the magnetic field and flow may be nonlaminar. In this respect, theory has been catching up with the empirical fact that solar flares are highly dynamic with variability on sub-second timescales (Benz, 1986; Kiplinger et al., 1983). Here we focus on two promising wave sources: dynamic magnetic reconnection and braking of the sunward reconnection outflow. Our main goals are to understand qualitatively what processes are responsible for generating the waves/turbulence observed in solar flares, and to constrain the likely spectra of frequencies and wavenumbers.

As context, we note that many features of the Sweet–Parker model of reconnection (Parker, 1957a; Sweet, 1958) apply well to the CSHKP solar flare model shown in Figure 3.1, in particular the magnetic geometry, and Alfvénic outflows from the ends of the current sheet. Despite those successes, the resistive, steady-state, and laminar Sweet–Parker model faces several well-known problems. The best-known one concerns the reconnection rate, which Sweet–Parker theory gives as $v_{in}/v_A \sim S_L^{-1/2}$, where

$$S_L = \frac{v_A L}{\eta} \tag{3.18}$$

is the Lundquist number based on the half-length of the current sheet, $L$, and $\eta$ is the magnetic diffusivity. A typical solar Lundquist number is $S_L \sim 10^{12}$, for which classic Sweet–Parker reconnection predicts a reconnection rate of order $10^{-6}$. In contrast, observations of solar flares indicate that reconnection rates are of order 0.01–0.1. A second problem is that solar flares exhibit a "switch-on" effect, in which the impulsive phase starts suddenly following pre-flare evolution. Steady-state models such as classic Sweet–Parker or Petschek reconnection (Petschek, 1964) are, by definition, unable to address this triggering.

The dynamic reconnection models that we discuss in connection with wave generation (in Sections 3.6.1 and 3.6.2) are fast (their reconnection rates are 0.01–0.1 consistent with flare observations) and they switch on (via tearing instability). As of recently, there is no longer a rate or triggering problem, although major work is ongoing to understand 3D reconnection at high Lundquist numbers, in MHD and kinetic frameworks.

### 3.6.1. Plasmoid-Mediated Reconnection

It has been known for many decades that Sweet–Parker layers are unstable to tearing instability (Furth et al., 1963) above a critical Lundquist number $S_c$. When $S_L < S_c$, the Sweet–Parker flow stabilizes the tearing instability (Bulanov et al., 1978), whereas for $S_L > S_c$ tearing instability develops. The canonical instability threshold is $S_c \sim 10^4$ (Biskamp, 1986), although some researchers have suggested it might be as low as $10^3$.

While several of the key ideas in the theory of plasmoid-mediated reconnection were developed before 2008 (Biskamp, 1986; Loureiro et al., 2007; Shibata & Tanuma, 2001), a sea change occurred when computing power made it possible to simulate Sweet–Parker layers with $S_L > 10^4$ in 2.5D. This threshold was crossed when an MHD simulation by Lapenta (2008) demonstrated the switching on of a fast reconnection regime at the critical Lundquist number. The fast reconnection was subsequently confirmed by other investigators and theoretical understanding quickly developed (Bhattacharjee et al., 2009; Cassak et al., 2009; Huang & Bhattacharjee, 2010; Uzdensky et al., 2010). The papers by Huang and Bhattacharjee (2013) and Loureiro and Uzdensky (2016) offer detailed overviews.

Of interest to wave generation, the nonlinear evolution for $S_L > S_c$ produces a chain of plasmoids separated by reconnecting current sheets. This configuration is highly dynamic, with plasmoids constantly being created, growing, moving, merging, and being expelled from the reconnection layer. Current layers within the plasmoid chain are on average marginally stable, such that if they are Sweet–Parker layers then $S_l \sim S_c$ yields a reconnection rate of $S_c^{-1/2} \sim 0.01$, where the lower case subscript

$l$ is used to distinguish the half-length of the marginally stable Sweet–Parker layers from the half-length $L$ of the global reconnection layer. This is consistent with MHD simulation results. If the widths of the smallest current are thinner than the ion inertial length or gyroradius, they exhibit kinetic reconnection, yielding a faster global reconnection rate of order 0.1 (Birn et al., 2001; Daughton et al., 2009; Shepherd & Cassak, 2010).

From a wave-generation perspective, we take a special interest in the fact that the global outflows from plasmoid-mediated reconnection exhibit large time variability. The outflows are not continuous; they are instead "lumpy" because of the expulsion of discrete plasmoids. Distributions of plasmoid sizes have been investigated analytically and numerically, with Fermo et al. (2010), Uzdensky et al. (2010), Loureiro et al. (2012), and Huang and Bhattacharjee (2012) having developed the kinetic theory of plasmoids. The theoretical distribution function of plasmoids with respect to flux is a power law $f(\psi) \sim \psi^{-1}$, falling off more steeply for the largest plasmoids of which there is typically only one at any given time (Huang & Bhattacharjee, 2012, 2013).

2.5D simulations have been produced that explore plasmoid-mediated reconnection directly in a solar flare context. The simulation by Karpen et al. (2012) forms a CSHKP current sheet self-consistently via the magnetic breakout mechanism, which evolves gradually until it becomes tearing unstable, switching on fast plasmoid-mediated reconnection. During the fast reconnection phase, plasmoids are frequently ejected from the current sheet in the sunward outflow and collide with the underlying closed magnetic loops, exciting disturbances that propagate along the magnetic field toward the lower atmosphere (e.g., animated version of their Figure 13). Burstiness of outflows is also seen in flare observations, for example, Figure 3 of Cheng et al. (2018).

The frequent but irregular expulsion of flux ropes with a distribution of sizes leads us to expect the excitation of waves with a corresponding distribution of wavelengths and frequencies. For a solar flare, the distribution of plasmoid widths, and hence wavelengths, is expected to span from the effective thickness of the dynamic reconnection layer based on the fluctuation electric field (see, Beg et al., 2022; Shibata & Tanuma, 2001) down to the ion scales. For the textbook flare, this range of wavelengths corresponds to 10 Mm or greater at the outer scale (Li et al., 2018), whereas ion scales in the corona are of the order of meters. For timescales, we expect the largest flux ropes to be ejected at intervals comparable to the Alfvén travel time along the lower altitude half of the flare current sheet, setting an outer period $\tau_{max} \sim L/v_A$. We tentatively connect this outer period to the timescale of elementary flare bursts, which are typically several seconds to a few tens of seconds in duration (de Jager & de Jonge, 1978). Superimposed would be variability determined by the interval between smaller plasmoids, which occur more frequently according to the $f(\psi) \sim \psi^{-1}$ distribution. The shortest timescale that emerges is the length of a marginally stable reconnecting current sheet divided by the Alfvén speed, which yields $\tau_{min} \sim S_c\eta/v_A^2$. For coronal values, $\tau_{min}$ can be shorter than milliseconds. Spikes and other subsecond variability observed in flares (Benz, 1986; Kiplinger et al., 1983) might therefore be related to expulsion of flux ropes of a variety of sizes from the flare current sheet.

### 3.6.2. Reconnection Layer Turbulence

The real Sun is of course 3D, and the additional dimensional freedom qualitatively changes the nature of magnetic reconnection compared to 2.5D. Foremost, magnetic field lines are not constrained to lie on flux surfaces but can instead be stochastic. Field line wandering is the crucial ingredient in the turbulent reconnection paradigm advanced by Lazarian and Vishniac (1999), and it modifies particle acceleration because particles are no longer trapped in magnetic islands (Dahlin et al., 2015, 2017). Furthermore, stochasticity arises naturally in reconnection because flux ropes formed by tearing instability undergo secondary 3D instabilities such as kink instability, which break any initial 2.5D symmetry. Knowledge of wave generation during reconnection should therefore come from 3D models.

Early numerical tests of reconnection in the presence of turbulence relied on artificially driven turbulence (Kowal et al., 2009; Loureiro et al., 2009; Matthaeus & Lamkin, 1986). However, there is now a substantial collection of 3D simulations that show that when $S_L > S_c$, turbulence is self-generated and self-sustaining within the reconnection layer. For example, MHD simulations are shown by Oishi et al. (2015), Huang and Bhattacharjee (2016), Striani et al. (2016), Beresnyak (2017), Kowal et al. (2017, 2020), Yang et al. (2020), Beg et al. (2022), and Daldorff et al. (2022); and kinetic simulations by Daughton et al. (2011), Liu et al. (2013), Daughton et al. (2014), Stanier et al. (2019), Zhang et al. (2021), and Li et al. (2019).

The zoo of 3D self-generated turbulent reconnection (SGTR) simulations is still being systematically organized. Simulations that explicitly include global reconnection outflows and have a guide field comparable to the reconnecting magnetic field component, such as Huang and Bhattacharjee (2016) and Beg et al. (2022), typically form dynamic frayed flux ropes like those shown in Figure 3.7. These structures appear to make reconnection fast by a plasmoid-mediated mechanism. In such cases, generation of Alfvénic waves by the processes discussed in Section 3.6.1 is likely to carry over to 3D. Additionally, a new source of Alfvénic waves in 3D is that

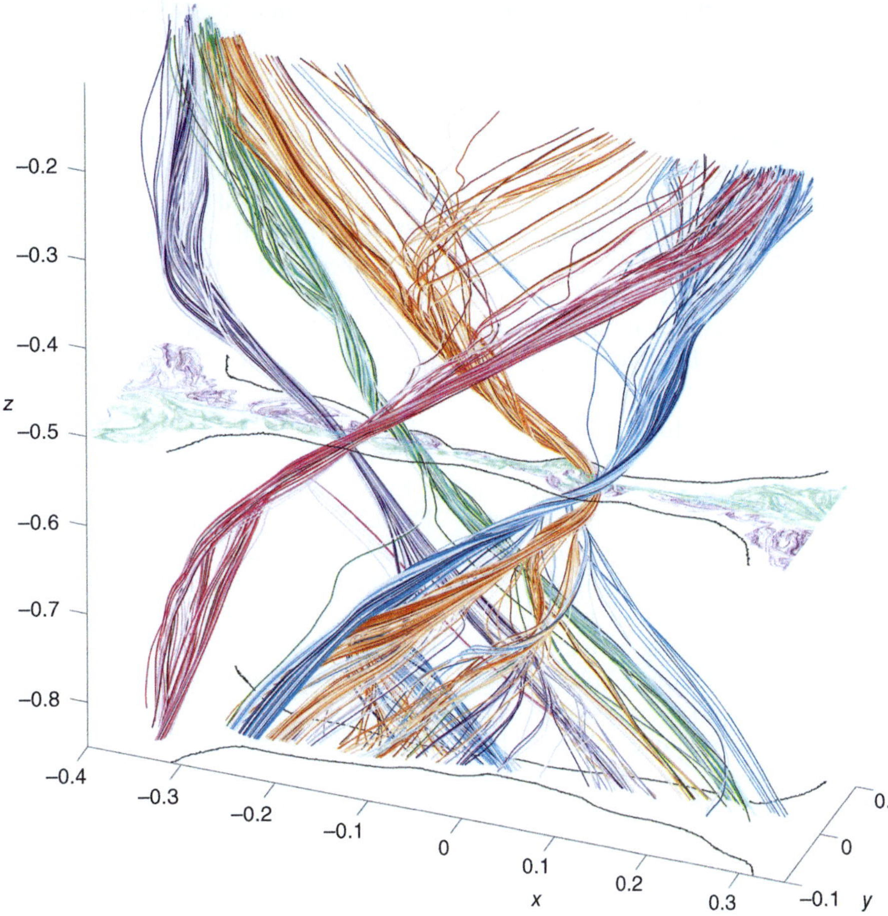


**Figure 3.7** Frayed flux ropes within 3D self-generated magnetic reconnection. Over short distances in $z$, magnetic field lines reveal dynamically-evolving twisted oblique flux ropes. Over longer distances, field lines become highly mixed. Flux rope dynamics, formation of localized twists, and the magnetic stochasticity all have implications for wave generation inside the reconnection layer. Reproduced from Beg et al. (2022) under a CC BY 4.0 license.

reconnection is expected to form magnetic twists locally, which propagate as Alfvénic waves. Since the magnetic field is stochastic over longer distances (notice the mixing of field lines toward the $z$ boundaries in Figure 3.7), the stochastic phase mixing (Similon & Sudan, 1989) described in Section 3.7.3 will transfer wave energy to shorter perpendicular scales within the reconnection layer.

Simulations with periodic boundaries along the current layer typically behave differently, with the turbulent layer thickening over time and a reconnection process that bears greater resemblance to Lazarian and Vishniac (1999). We suspect that those latter model setups are less applicable to CSHKP flares, and that they are affected by magnetic flux not being expelled from the current layer. However, it remains an ongoing task to comprehensively map out how 3D reconnection changes with parameters such as plasma beta and guide field strength, and it is possible that reconnection could change between 3D plasmoid-mediated and Lazarian–Vishniac regimes as a solar flare progresses.

The properties of fluctuations in the reconnection layer are of significant interest for generation of Alfvénic waves. Spectral and anisotropy properties have been investigated by several authors. Huang and Bhattacharjee (2016) found elongated eddies aligned with the local magnetic field, scale-independent anisotropy, and an energy spectrum displaying an inertial range with $\tilde{E}_k \sim k^{-\alpha}$, $2.3 < \alpha < 2.5$. In a different setup (with periodic boundaries) and using different analysis techniques, Kowal et al. (2017) found somewhat different properties consistent with Kolmogorov-like turbulence and the Goldreich and Sridhar (1995, 1997) model, also highlighting that turbulence properties vary with the plasma beta.

The present situation is therefore that we have high confidence that self-generated turbulence is an intrinsic feature of 3D reconnection at high Lundquist numbers. In the CSHKP flare model (Figure 3.1), the leading edges of the flare ribbons are connected to the flare current sheet, so waves launched within the reconnection layer would be the first to be incident on the lower atmosphere. Simulations show that fluctuations in the reconnection layer have power law spectra, and solar observations support this too (Chen et al., 2018; French et al., 2021). However, work is ongoing to understand the properties of the turbulence and how they depend on the guide field and plasma beta. The theory of MHD turbulence is similarly in a period of active development, with ongoing efforts to understand the role of reconnection within turbulence (e.g., Dong et al., 2022; Loureiro & Boldyrev, 2017).

### 3.6.3. Braking of Reconnection Outflows

The CSHKP flare model includes a sunward reconnection outflow (**b** in Figure 3.1) that impinges on the closed magnetic loops below. The flow is therefore forced to brake in the above-the-loop region (**c** in Figure 3.1). There are various physical possibilities for how this might occur, but everyday experience of filling a sink or bath from the tap instructs us that conversion of bulk-directed kinetic energy into turbulent (i.e., random small scale) motions provides a natural and effective solution. The ideas that converting kinetic energy of the reconnection outflow into small-scale vortical motions could be central to flare physics and explain that looptop nonthermal broadening has existed for at least 30 years (Haerendel, 2009; Larosa & Moore, 1993; Larosa et al., 1994; Tsuneta, 1995). However, simulations and observations have recently brought this mechanism to the fore.

We begin by deducing the properties of fluctuations produced by flow braking in magnetized plasma. Bending of magnetic field lines is opposed by the magnetic tension force, which is inversely proportional to the field line's radius of curvature. This force exerts a stabilizing influence on the development of small-scale motions in 2D planes containing the background magnetic field $\boldsymbol{B}$. However, the magnetic tension does not have this stabilizing effect on motions in planes perpendicular to $\boldsymbol{B}$. Applying these principles to the CSHKP model, turbulent braking requires dimensional freedom in the direction normal to the plane of Figure 3.1, hence we look for it only in flare simulations that have 3D freedom. We further anticipate that the small-scale motions produced by flow braking will have velocities virtually perpendicular to $\boldsymbol{B}$, and wavevectors satisfying $k_{\parallel} \ll k_{\perp}$ ($\boldsymbol{k} \cdot \boldsymbol{B} \approx 0$). Under these assumptions, magnetic perturbations are also perpendicular to $\boldsymbol{B}$, and evaluating the curls shows that vorticity $\boldsymbol{\omega}$ and current density $\boldsymbol{j}$ are virtually parallel (or antiparallel) to $\boldsymbol{B}$. Given the expected directions of velocity, vorticity, and current density, these small-scale motions take the form of Alfvénic waves. Finally, while the waves should be anisotropic with $k_{\parallel} \ll k_{\perp}$, the parallel wavelength $\lambda_{\parallel}$ cannot be infinite due to finite extent of the wave-braking region. The $\lambda_{\parallel}$ is likely to be determined by the width of the above-the-loop region, with the frequency determined by $f = v_A/\lambda_{\parallel}$.

The physical reasoning stated in the earlier text has very recently received the support of MHD simulations that study the CSHKP flare model with 3D freedom and high resolution (Chen et al., 2020; Ruan et al., 2023; Shen et al., 2022, 2023). These works have indeed found that flow braking produces small-scale vortical motions of an Alfvénic nature in the above-the-loop region. Furthermore, forward modeling these simulations produces symmetric Gaussian lines that are nonthermally broadened to widths consistent with EIS and IRIS observations.

Figure 3.8 shows an example from Ruan et al. (2023). The left panel displays the nonthermal width of a synthesized EIS Fe XXIV line, which reaches 200 km s$^{-1}$ near the apex in the above-the-loop region. Cyan contours represent AIA 131 Å intensity, showing the locations of flare loops. The phenomenology has much in common with the observation by Stores et al. (2021) shown in Figure 3.4.

The simulations also help to elaborate on the physics of how the small-scale motions are produced. Ruan et al. (2023) highlighted that collision of sunward reconnection flows with the underlying arcade forms reflected upflows that interleave with downflows in the above-the-loop region, as shown in the right panel of Figure 3.8. This interleaving relies on the vertical component of the plasma velocity being able to reverse along the direction perpendicular to the plane of Figure 3.1, hence it was not captured by earlier 2.5D simulations. Ruan et al. (2023) propose that the velocity shears between the downflows and reflected upflows feed a Kelvin–Helmholtz instability that generates small-scale vortical motions. This process appears to dominate in the above-the-loop region where nonthermal velocities are greatest. Shen et al. (2023) computed power spectra in the above-the-loop region for their simulation, obtaining a one-dimensional kinetic energy spectrum proportional to $k^{-2.0}$.

Globally, flow braking must be accomplished with force balance so that linear momentum is conserved. As pointed out by Haerendel (2009), the upward force that slows the bulk reconnection downflow at the looptops is ultimately balanced by an opposite force on the body of the Sun, requiring transmission of force between the loop apex and the lower atmosphere. Alfvénic waves are the natural agent. Thus, a comprehensive view of wave generation by flow braking encompasses not only looptop turbulence, but also waves that are transmitted to the lower atmosphere. These include quasi-stationary waves that transmit the slowly varying component of the force, in addition to which excitation of higher-frequency

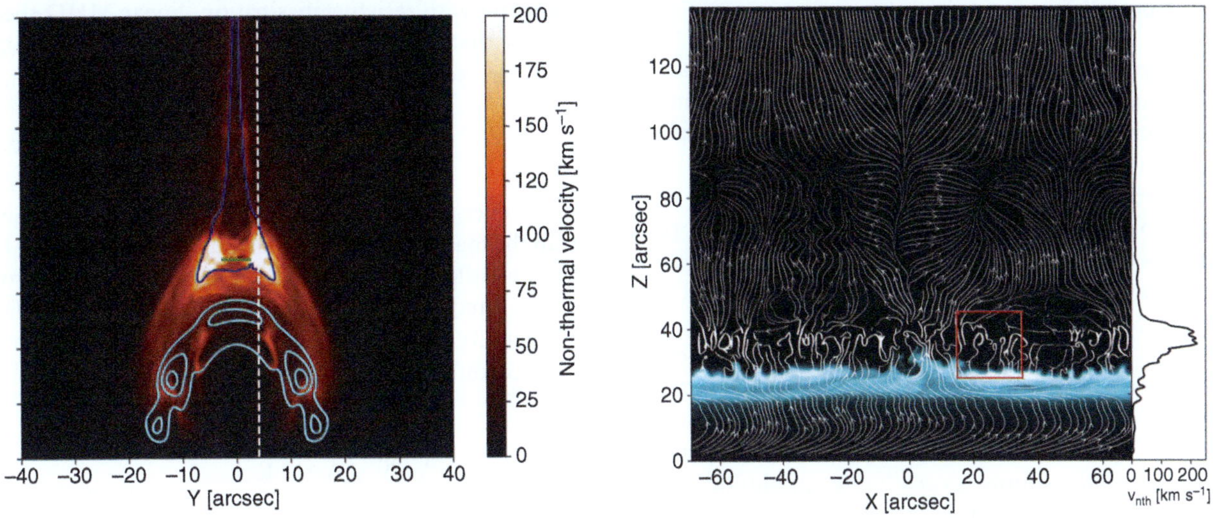


**Figure 3.8** MHD simulation demonstrating generation of small-scale motions by braking of the sunward reconnection flow. The left panel shows the nonthermal line widths of synthetic Fe XXIV profiles, with 131 Å intensity overlayed as cyan contours. The simulation produces nonthermal velocities of hundreds of kilometers per second that peak at the apex of the above-the-loop region, consistent with the observation shown in Figure 3.4. The right panel presents a side view, on a cross section at the location indicated by the white dashed line on the left panel. Streamlines (white) show small-scale motions above the flare loops (cyan). The line plot at the right-hand side shows $v_{nth}$, which peaks in the turbulent-braking region. Reproduced from Ruan et al. (2023) / IOP Publishing / CC BY 4.0.

propagating Alfvénic waves is also expected, given that reconnection outflows and their braking are highly variable in space and time. Ruan et al. (2023) confirmed this viewpoint, reporting that fluctuations propagate along the magnetic field to the footpoints as Alfvén waves.

Generation of Alfvénic waves by flow braking is largely agnostic about the details of the reconnection process so long as reconnection is fast, despite those details being important for wave generation and particle acceleration within the reconnection layer itself. The simulation by Ruan et al. (2023) used a nonuniform resistivity with a profile that typically produces Petschek reconnection, while Shen et al. (2022) considered two uniform resistivity cases on either side of the plasmoid instability threshold. For wave generation by flow braking, what matters is simply that there is a sunward reconnection flow to be stopped, and potentially the kinetic energy density of that flow.

## 3.7. TRANSPORT TO THE LOWER SOLAR ATMOSPHERE

### 3.7.1. General Remarks

We now examine the transport of energy to the lower solar atmosphere and the ensuing dynamics. Sections 3.7.2 and 3.7.3 describe the evolution of Alfvén waves in the corona, Section 3.7.4 addresses energy transmission across the transition region, and Section 3.7.5 considers the response of the lower atmosphere.

The earliest paper we are aware of that explored Alfvénic transport to the lower atmosphere in depth is by Emslie and Sturrock (1982). That pioneering work examined many of the points covered in this section, seeking a solution to the problem of temperature minimum region (TMR) heating during flares. The topic of flare-generated Alfvén waves was later explored by Melrose (1992) and Wheatland and Melrose (1994), with an emphasis on the propagation of Alfvénic fronts created by switching on of the energy release. The landmark paper of Fletcher and Hudson (2008) started a wider discussion of wave transport in solar flares, inspired in part by then recent microwave measurements of the coronal magnetic field strength and evidence that the photospheric magnetic field changes during flares. The paper is especially well known for highlighting the interesting possibility for Alfvénic waves/turbulence to accelerate particles locally in the loop legs and footpoints, which was further explored by Brown et al. (2009), McClements and Fletcher (2009), and Haerendel (2012). Important work by Haerendel (2009) also helped to define the modern version of the problem, proposing flow braking as a generator of Alfvénic waves, and raising the possibility that wave heating could be a driver of chromospheric evaporation.

### 3.7.2. Coronal Transport

We begin by examining how $v_{nth}$ is expected to vary in the corona, noting that observations such as the nonthermal velocity map in Figure 3.4 show a decrease by about 50% from the apex to the footpoints.

It is instructive to first neglect reflection and dissipation to obtain a simple reference case. Under these assumptions, using the principle that the energy density of the wave is transported at the Alfvén speed, $\langle S_{\|}\rangle A$ remains constant in a Lagrangian frame moving with the wave, where $A$ is the cross-sectional area of the flux tube. Then, since $A = \Phi/B$ where $\Phi$ is the (constant) magnetic flux, and $\langle S_{\|}\rangle = v_A m_i n \langle v^2\rangle$ as discussed in Section 3.5, the wave velocity amplitude is independent of $B$ and scales simply as $n^{-1/4}$ (the amplitude of the magnetic field perturbation meanwhile scales as $n^{1/4}$). This well-known scaling result can also be derived using WKB analysis.

In Section 3.5, we considered typical values for $v_{nth}$ of 250 km s$^{-1}$ in the looptop and 120 km s$^{-1}$ in the footpoints (based on the observations described in Section 3.4). The resulting velocity ratio would be accounted for by the density alone if the footpoints were denser than the looptop by a factor 19. In a hydrostatic equilibrium, such a density ratio corresponds to just under three density scale heights. For comparison, the scale height $H = k_B T/m_i g$ for a 10 MK plasma is 300 Mm (0.4 $R_\odot$), so $3H$ is considerably taller than normal flare arcades at flare temperature. It is possible that such a density contrast might be produced by chromospheric evaporation, which is a dynamic effect; however, since nonthermal velocity maps appear to show a gradual decrease in $v_{nth}$ along the length of the loops (e.g., Figure 3.4), it appears likely that reflection and/or dissipation are significant.

Alfvén waves partially reflect if they have parallel wavelengths longer than the local length scale of the equilibrium (parallel to the magnetic field); otherwise, they propagate without significant reflection, as in a WKB solution. In this context, we highlight the possibility of reflection being enhanced by small-scale density variations, which have previously been invoked for the Sun's open corona (van Ballegooijen & Asgari-Targhi, 2016). A hypothesis of reflection from small-scale density variations is difficult to falsify with current observations, so we leave this as an open possibility.

Waves propagating through a plasma can be dissipated by resistivity, viscosity, or particle acceleration. The corona is typically regarded as a low-dissipation plasma; however, the properties of flaring plasma are markedly different to normal coronal properties, and in flares there is the possibility for short wavelengths and high frequencies, so we will consider these mechanisms explicitly.

Starting with resistive dissipation, linear MHD theory predicts that the Alfvén waves damp resistively with a characteristic damping length

$$L_{res} = \frac{v_A}{\eta k^2} \tag{3.19}$$

for the energy density, where we have used $\eta = 1/\mu_0\sigma$ to mean the magnetic diffusivity and are using S.I. units. Different parallel and perpendicular diffusivities can also be incorporated, as shown in Section 3.7.5. Using the Spitzer conductivity (Spitzer, 1962), a convenient formula for the magnetic diffusivity for coronal applications is

$$\eta = 1.0\left(\frac{\ln\Lambda}{20}\right)\left(\frac{T}{10^6\ \mathrm{K}}\right)^{-3/2}\ \mathrm{m^2\ s^{-1}}. \tag{3.20}$$

Setting the Coulomb logarithm to 20 and the flare temperature to 10 MK yields a magnetic diffusivity of 0.032 m$^2$ s$^{-1}$. Subsequently, obtaining $L_{res}$ of 50 Mm or less for an Alfvén speed of 2800 km s$^{-1}$ requires wavelengths below 5 m. We conclude that resistivity does not significantly damp large-scale Alfvén waves in the corona, but it may provide dissipation for the shortest scale waves produced during flares, such as those reached by turbulent cascade.

For context, the ion kinetic scales (e.g., Shibata & Tanuma, 2001) are the ion inertial length,

$$\delta_i = 3.0\left(\frac{n}{10^{16}\ \mathrm{m^{-3}}}\right)^{-1/2}\ \mathrm{m}, \tag{3.21}$$

and the ion Larmor radius,

$$r_{L,i} = 0.1\left(\frac{B}{100\ \mathrm{G}}\right)^{-1}\left(\frac{T}{10^6\ \mathrm{K}}\right)^{1/2}\ \mathrm{m}. \tag{3.22}$$

For a number density of $10^{17}$ m$^{-3}$, $\delta_i$ is about 1 m, whereas for a flare magnetic field strength of 400 G and temperature of 10 MK, $r_{L,i}$ is about 8 cm. Comparing these values to $L_{res}$ confirms that resistive damping could be significant for the shortest wavelengths and therefore affect the dissipative range of any turbulent cascade, although collisionless effects due to $\delta_i$ also become significant at similar scales.

Viscous damping of Alfvén waves is more complicated than resistive damping because the Braginskii viscous stress tensor is highly anisotropic in magnetized plasma. In this context, we will use $\eta$ with a numerical subscript to represent the viscosity coefficients, which are given by

$$\eta_0 = 1.1\times 10^{-2}\left(\frac{\ln\Lambda}{20}\right)^{-1}\left(\frac{T}{10^6\ \mathrm{K}}\right)^{5/2}\ \mathrm{kg\ m^{-1}\ s^{-1}},$$
$$4\eta_1 = \eta_2 = 1.25(\Omega_i\tau_i)^{-2}\eta_0, \tag{3.23}$$

where $\Omega_i = eB/m_i$ is the proton gyrofrequency and

$$\tau_i = 7.5 \times 10^{-2}\left(\frac{T}{10^6\ \mathrm{K}}\right)^{3/2}\left(\frac{n}{10^{16}\ \mathrm{m}^{-3}}\right)^{-1}\ \mathrm{s} \quad (3.24)$$

is the proton collision time.

For our usual above-the-loop parameters, $\tau_i$ is 0.24 s and $\Omega_i$ is $3.8 \times 10^6$ Hz, so the plasma is strongly magnetized, with $\Omega_i\tau_i \sim 10^6 \gg 1$. Then, setting the Coulomb logarithm to 20, we estimate $\eta_0 = 3.5$ kg m$^{-1}$ s$^{-1}$, and obtain $\eta_1 = 1.3 \times 10^{-12}$ kg m$^{-1}$ s$^{-1}$ with $\eta_2 = 4\eta_1$.

In the linear theory, Equation (8.40) of Braginskii (1965),

$$L_{visc}^{linear} = \frac{v_A \rho_0}{\eta_1 k_\perp^2 + \eta_2 k_\parallel^2}. \quad (3.25)$$

Considering cases with $k_\perp^2 \gg k_\parallel^2$, obtaining $L_{visc}^{linear}$ of 50 Mm or less for an Alfvén speed of 2800 km s$^{-1}$ requires wavelengths below 3 m. We also remark that the temperature scaling of $\eta_1$ and $\eta_2$ is $\sim T^{-1/2}$ compared to $\sim T^{-3/2}$ for the resistivity, so viscous dissipation reduces more slowly with temperature than resistive dissipation does. These results show that linear viscous dissipation becomes effective in flares at similar scales to resistive dissipation and ion inertial effects.

The vast ratio of the parallel-to-perpendicular viscosity coefficients means that nonlinear viscous damping via the $\eta_0$ viscosity coefficient should also be considered for coronal Alfvén waves (Nocera et al., 1986; Russell, 2023). Nonlinear wave damping is potentially more important in flaring plasma than it is in the quiet corona, because the ratio of viscosity coefficients increases nonlinearly with temperature, as $\eta_0/\eta_1 \sim T^3$. Thus, the ratio of the viscosity coefficients is a factor 1000 greater for 10 MK plasma than it is for 1 MK plasma, significantly increasing the relative effectiveness of damping via the $\eta_0$ viscosity coefficient. The nonlinear damping also increases in absolute terms, since $\eta_0 \sim T^{5/2}$.

The nonlinear decay length for linearly polarized shear Alfvén waves due to parallel wavelengths and parallel viscosity (Russell, 2023) can be expressed as

$$L_{visc}^{nonlinear} = 1.2 \times 10^{12}\left(\frac{\ln\Lambda}{20}\right)\left(\frac{f}{1\ \mathrm{Hz}}\right)^{-2}\left(\frac{T}{10^6\ \mathrm{K}}\right)^{-5/2} \times \left(\frac{B}{100\ \mathrm{G}}\right)^5\left(\frac{n}{10^{16}\ \mathrm{m}^{-3}}\right)^{-3/2}\left(\frac{\langle v^2\rangle^{\frac{1}{2}}}{10^5\ \mathrm{m\ s}^{-1}}\right)^{-2}\ \mathrm{m}, \quad (3.26)$$

where we have dropped the finite-beta correction since a 10-MK plasma has a sound speed of 370 km s$^{-1}$ compared to an assumed Alfvén speed of 2800 km s$^{-1}$, giving $(c_s/v_A)^2 = 0.017$. Considering our usual above-the-loop values for $\ln\Lambda$, $B$, $T$, and $n$, and an rms velocity of 250 km s$^{-1}$, this expression reduces to

$$L_{visc}^{nonlinear} = 1.9 \times 10^{10}\left(\frac{f}{1\ \mathrm{Hz}}\right)^{-2}\ \mathrm{m}. \quad (3.27)$$

Obtaining $L_{visc}^{nonlinear}$ of 50 Mm or less from this result requires wave frequencies of 20 Hz or greater. We can therefore neglect nonlinear viscous damping for frequencies of around 1 Hz for our chosen parameters, but nonlinear viscous damping appears potentially significant for the higher-frequency part of the wave spectrum. We also remark that while nonlinear viscous damping is not significant at lower frequencies for our chosen parameters, a broader parameter survey may yet identify conditions for which it is.

The appearance of nonlinear viscous damping in hot tenuous plasma is associated with low collisionality, and collisionless effects may become important for higher-frequency waves. The traditional derivation of Braginskii MHD requires $k_\parallel \lambda_{mfp} < 1$, where $\lambda_{mfp} = v_{Ti}\tau_i$ is the ion mean-free path along the magnetic field with ion thermal speed $v_{Ti} = \sqrt{k_B T/m_i}$. This corresponds to the Reynolds number (based on $\eta_0$, $l = k_\parallel^{-1}$ and the ion sound speed) being greater than unity (Section 6.3 of Russell (2023)). Alternatively, it can be expressed as the following condition on the frequency,

$$f < f_c = \frac{v_A/v_{Ti}}{2\pi\tau_i} = 51\left(\frac{T}{10^6\ \mathrm{K}}\right)^{-2} \times \left(\frac{B}{100\ \mathrm{G}}\right)\left(\frac{n}{10^{16}\ \mathrm{m}^{-3}}\right)^{1/2}. \quad (3.28)$$

The parameter values we considered for the above-the-loop region yield $f_c = 7$ Hz. Considering frequencies lower than $f_c$ returns $L_{visc}^{nonlinear}$ of 390 Mm or greater, which is too large to matter for most flares. Damping of higher-frequency waves with $f > f_c$ may be collisionless in nature. For example, Squire et al. (2016) have shown that linearly polarized shear Alfvén waves with amplitudes $\delta B/B_0$ exceeding a threshold $\beta^{-1/2}$ are interrupted by parallel fire-hose instability.

The last coronal dissipation mechanism we consider is particle acceleration. This is a particularly promising avenue, considering the abundant evidence that solar flares are efficient particle accelerators. This is especially true in the above-the-loop region where nonthermal velocities are greatest and a very large fraction of the electrons can be nonthermal (see discussion and references in Section 3.2). Kontar et al. (2017) used joint RHESSI–EIS observations to infer that the timescale for Alfvénic energy density to energize electrons is 1–10 s, which is consistent with the turbulence timescale for Goldreich–Sridhar-type turbulence (see discussion in Kontar et al. (2017)). On the theory side, the idea that

Alfvénic waves and/or turbulence accelerate particles to high energies (Fermi, 1949; Parker, 1957b) is only slightly younger than the famously long-standing coronal heating problem, but uncertainty over the details of particle acceleration by waves has endured. This topic is rightly the subject of separate reviews, including Miller et al. (1997), Zharkova et al. (2011), and Petrosian (2012). We also highlight papers by Fletcher and Hudson (2008), Brown et al. (2009), McClements and Fletcher (2009), Bian, Kontar, and Brown (2010), and Haerendel (2012).

In summary, nonthermal broadening measurements indicate that the velocity amplitude of Alfvénic waves decreases by about 50% from the apex to the footpoint. This is unlikely to be accounted for by simple WKB density scaling and is likely to require either reflection (e.g., from small-scale inhomogeneities) or dissipation. Linear resistive and viscous dissipation both require short length scales (e.g., perpendicular wavelengths) of several meters. Thus, these processes do not damp large-scale Alfvénic waves, but they may be important at the conclusion of a turbulent cascade. Nonlinear viscous damping appears unlikely to damp Alfvén waves in flares generally, although it may damp the high frequency component of the wave spectrum. Collisionless processes may be important, and we note that when collisions are scarce Alfvén waves can develop significant pressure anisotropy, triggering instabilities such as fire-hose instability. Finally, given that solar flares are powerful particle accelerators, the most obvious answer to how the waves damp may be that their energy is transferred to nonthermal particles, although the details of particle acceleration by waves/turbulence remain an active research subject.

### 3.7.3. Magnetic Convergence and Phase Mixing

Having explored the processes governing wave amplitude, we now explore how transport affects the perpendicular wavelength. Here, we consider concentration of wave energy as the magnetic field converges toward the footpoints, as well as several types of phase mixing, which describes the shortening of perpendicular wavelengths, as wave fronts are distorted during propagation along field lines.

The simplest effect that modifies perpendicular wavelengths is the convergence (or divergence) of field lines: if field line spacing becomes closer, then maintaining the same number of phase cycles implies that the transverse wavelength is shorter. If convergence occurs in one transverse direction, $k_\perp \sim B$; if it occurs in both transverse directions, $k_\perp \sim B^{1/2}$.

The basic form of phase mixing (Heyvaerts & Priest, 1983) occurs due to transverse gradients of $v_A$, for example, due to density gradients. The corona has a fine-stranded structure, with the strands most frequently having full widths of 500–550 km (Aschwanden & Peter, 2017; Brooks et al., 2013; Williams et al., 2020). These values are comparable to 310 km elemental widths inferred from observations of coronal rain by Antolin and Rouppe van der Voort (2012); while widths for other types of fibrillar structure in the Sun are similar but sometimes narrower, often around 100 km. Accordingly, basic phase mixing implies that Alfvénic waves incident on the transition region should have perpendicular wavelengths not more than several hundred kilometers after propagating through coronal fine structure.

Phase mixing can also occur due to the magnetic field line geometry. Starting with the case of a smooth laminar field, neighboring field lines have different arc lengths from their apex to the lower boundary. Thus, if waves travel at the same speed along both field lines, then the part of the wave front that travels on the shorter field line reaches the boundary first. These time differences distort the wave front, shortening the perpendicular wavelength. This geometrical phase-mixing effect has been treated analytically by Weinberg (1962) and applied to solar flare energy transport by Emslie and Sturrock (1982). It has also been demonstrated in a simulation by Russell and Stackhouse (2013). However, this form of phase mixing is likely the weakest of the three.

The third type of phase mixing operates in nonlaminar magnetic fields. In stochastic magnetic fields, the separation distance between two field lines that are initially neighbors increases exponentially with respect to distance along the field. In this scenario, field-aligned transport produces an exponential shortening of perpendicular wavelengths, which accesses small scales at an exponential rate (Similon & Sudan, 1989). Considering the role of magnetic stochasticity in various models of fast magnetic reconnection (Section 3.6.2) and the possibility for reconnection in flow-braking turbulence (Section 3.6.3) to create magnetic stochasticity, it appears highly plausible that this type of phase mixing will apply to Alfvénic waves in solar flares. Future work with MHD flare simulations will be valuable for identifying the degree of stochasticity produced in flares and its impact on perpendicular wavelengths.

The effects described in the earlier text may provide a first stage in shortening the transverse wavelength. As waves develop larger $k_\perp$, the strengthened velocity and magnetic shears can trigger Kelvin–Helmholtz (KH) and tearing instabilities, which may accelerate the cross-scale transport. Heyvaerts and Priest (1983) included this possibility in their landmark paper on phase mixing, concluding that propagating waves are KH-stable but standing waves are KH-unstable. Applied to flare transport, one might therefore anticipate instabilities of phase-mixed waves playing a role where waves reflect from the transition region. Haerendel (2009) proposed that a two-stage mechanism of this type could operate in flares, and it is well publicized that KH instability of phase-mixed waves hastens dissipation in the boundaries of kink-oscillating coronal loops (Antolin et al., 2017; Terradas et al., 2008). Turbulent cascade may also play a role (e.g., Howes & Nielson, 2013, and references therein).

### 3.7.4. Energy Transmission to the Chromosphere

The transition region is normally marked by a change in temperature over a relatively short distance, from around 10,000 K in the chromosphere to about 1 MK in the corona (here we consider a pre-flare coronal temperature). Pressure balance implies that the Alfvén speed changes on the same scale. For example, assuming a neutral-dominated chromosphere, pressure balance yields $2n_{cor}T_{cor} = n_{chr}T_{chr}$, where $n_{cor}$ is the proton density and $n_{chr}$ is the total density of protons and hydrogen atoms, so that the total mass density increases by a factor of approximately 200. This in turn decreases the Alfvén speed by a factor 14, where the chromospheric $v_A$ is calculated using the total mass density because neutrals are collisionally coupled to the ions.

Alfvénic waves with parallel wavelengths much longer than the depth of the transition region see it as a step in the wave speed. The energy transmission coefficient (Emslie & Sturrock, 1982; Fletcher & Hudson, 2008; Russell & Fletcher, 2013) is

$$\mathcal{T} = \frac{4\theta^{1/2}}{(1+\theta^{1/2})^2}, \tag{3.29}$$

where $\theta = (v_{A,chr}/v_{A,cor})^2 \approx 0.5\ T_{chr}/T_{cor}$. Thus, $\mathcal{T} \approx 2\sqrt{2}\sqrt{T_{chr}/T_{cor}} \approx 0.28$ for the initial temperatures discussed in the earlier text, meaning that 28% of the wave energy crosses the transition region per incidence. This permits significant transmission of wave energy to the lower atmosphere, from the first times waves are generated in the corona. For shorter wavelengths, transmission exceeds this lower bound, approaching 100% transmission for the shortest wavelengths (Reep et al., 2018; Russell & Fletcher, 2013).

Alfvén waves have $k_{\|} = \omega/v_A$ so the parallel wavelength scales as $\lambda_{\|} \sim v_A$. The parallel wavelength is therefore shortened considerably (factor about 14) when the wave is transmitted to the chromosphere. If the new wavelength is shorter than the scale height for the Alfvén wave in the chromosphere, it will propagate there without further reflection. Otherwise, wave energy will continue to be reflected. Putting the whole system together, the percentage of incident energy retained in the chromosphere increases with frequency. Numerical studies by Russell and Fletcher (2013) (their Figure 5) and Reep et al. (2018) (their Figure 3) have investigated the functional dependence using semi-empirical models of the lower atmosphere. As a rule of thumb, energy transmission for frequencies above 1 Hz exceeds 20–50% per incidence, depending on the atmospheric model.

### 3.7.5. Chromospheric Heating

The chromospheric dissipation of downgoing waves was first investigated by Emslie and Sturrock (1982), using a WKB solution of the linearized single-fluid MHD equations. Their Figure 2 neatly summarizes how the heating rate peaks at different heights for different wave parameters. The heating rate for waves with 1 Hz frequencies and transverse wavelengths of 1–10 km or greater peaks in the TMR, and Emslie and Sturrock (1982) proposed that this energy deposition might account for the TMR warming by 100–200 K in flares (Machado et al., 1978). Figure 2 of Emslie and Sturrock (1982) also shows that waves with higher frequencies and/or shorter perpendicular wavelengths are stopped higher in the chromosphere. While this aspect was not developed further by Emslie and Sturrock (1982), wave heating at these higher altitudes causes a much greater temperature increase on account of the lower density, and it has subsequently been proposed as a driver of chromospheric evaporation, joining energy deposition by thermal conduction and particle beams (Haerendel, 2009; Reep & Russell, 2016).

One of the major dissipation processes in the chromosphere is ion-neutral friction. At chromospheric temperatures, protons and neutral hydrogen are coupled by resonant charge exchange, which operates with a momentum transfer collision frequency:

$$\upsilon_{H,p} = 1.18\times10^4\left(\frac{n_p}{10^{18}\ \mathrm{m^{-3}}}\right)\left(\frac{T}{10^4\ \mathrm{K}}\right)^{1/2} \times\left(1-0.125\log\left(\frac{T}{10^4\ \mathrm{K}}\right)\right)^2 \mathrm{s^{-1}}. \tag{3.30}$$

This formula is consistent with Equation (A10) of De Pontieu et al. (2001) and Equation (23) of Russell and Fletcher (2013) after correcting the typo pointed out in Reep and Russell (2016). At the TMR, the hydrogen ionization degree becomes so low that singly ionized metal ions dominate and the collision frequency is instead calculated from

$$\upsilon_{H,M^+} = 2.1\times10^3\left(1+\frac{1}{A_M}\right)^{-1/2}\left(\frac{n_{M^+}}{10^{18}\ \mathrm{m^{-3}}}\right)\mathrm{s^{-1}} \tag{3.31}$$

(De Pontieu et al., 2001; Russell & Fletcher, 2013). For chromospheric densities and temperatures, the ions and neutrals are strongly coupled with millisecond momentum transfer timescales. Thus, while neutral particles are not directly tied to the magnetic field by Lorentz force, they are tied to it indirectly by collisions with ions (this is a very different parameter regime to the Earth's ionosphere, in which the coupling timescale is closer to hours, allowing existence of Hall and Pedersen currents). While the collisional coupling in the chromosphere is very strong, it is imperfect, which leads to heating.

Resistive damping can also play a significant role in the chromosphere. Compared to the corona, the lower temperature enhances resistive damping by a factor greater than 1000. Resistive dissipation is also more effective in the chromosphere because $k_{\parallel} = \omega/v_A \sim n^{1/2}$ is larger due to the higher density, $k_{\perp}$ is likely to be larger due to additional phase mixing, and collisions with neutrals enhance the resistivity.

The effects of electron resistivity and ion-neutral friction can be combined using anisotropic Cowling resistivity. Here, we present a simplified version valid for low ionization degree and small $\omega/\nu_{ni}$, and refer readers to Reep and Russell (2016) for more general equations. The parallel diffusivity is

$$\eta_{\parallel} = \frac{m_e}{\mu_0 n_e e^2}(\upsilon_{ei} + \upsilon_{en}), \tag{3.32}$$

while the perpendicular diffusivity is

$$\eta_{\perp} = \eta_{\parallel} + \eta_C, \qquad \eta_C = \frac{B^2 \rho_n}{\mu_0 \upsilon_{ni} \rho_t^2}. \tag{3.33}$$

The wave energy damping length (Reep & Russell, 2016) is then

$$L_D = \frac{v_A}{\eta_{\parallel} k_{\perp}^2 + \eta_{\perp} k_{\parallel}^2} = \frac{v_A}{\eta_{\parallel} k^2 + \eta_C k_{\parallel}^2}. \tag{3.34}$$

This formula implies that damping by ion-neutral friction occurs through $k_{\parallel} = \omega/v_A$, whereas damping by electron resistivity is determined by $k$.

Russell and Fletcher (2013) modeled wave propagation from the corona to the chromosphere, and the ensuing damping, modeling the neutrals and plasma as separate fluids that are collisional-coupled. For simplicity, that work solved the linearized equations in the case $k_{\perp} = 0$. The results showed that a 1-s wave pulse, for example, deposits most of its energy at the TMR by ion-neutral friction, but the greatest temperature increases occur in the upper chromosphere.

The full problem includes nonlinear feedbacks, since heating the atmosphere causes it to evolve. The ensuing evolution has been investigated by adding wave heating models to 1D radiative hydrodynamic codes (Kerr et al., 2016; Reep & Russell, 2016; Reep et al., 2018). Figure 3.9 shows results from Reep and Russell (2016) that compare the atmospheric response for two cases of Alfvénic wave heating (top two rows) and one case of electron beam heating (bottom row), using the HYDrodynamics and RADiation Code (HYDRAD). These results demonstrate that wave heating can produce gentle or explosive evaporations with upflows at several hundred kilometers per second, resembling those produced by electron beam heating.

The significant caveat is that producing chromospheric evaporation requires relatively high frequencies and/or short perpendicular scales below 10 km, consistent with expectations from Emslie and Sturrock (1982), Haerendel (2009), and Russell and Fletcher (2013). It is currently unclear what fraction of wave energy produced by reconnection and flow braking enters the lower atmosphere with the properties needed to damp strongly in the upper atmosphere. The stochastic phase mixing (Similon & Sudan, 1989) discussed in Section 3.7.3 offers a powerful mechanism for transferring energy to short perpendicular wavelengths, as do the mechanisms of phase mixing combined with instabilities or turbulent cascade.

The general similarity of the upflows produced by short-wavelength Alfvénic waves and electron beams was further confirmed using the RAdiative hydroDYNamics code (RADYN) by Kerr et al. (2016), who also synthesized various chromospheric spectral lines. An intriguing finding was that Mg II profiles for their wave-heated simulation were different to those for the electron beam simulation, with the wave-heating simulation better matching IRIS observations. This finding suggests that careful study of lines formed in the mid to upper chromosphere may be capable of distinguishing between different heating mechanisms.

Further differences have been identified by Reep et al. (2018) who used a second-generation wave driver that replaced the assumption of instantaneous transport with a new method that captures the effects of finite wave travel time. In principle, the fact that Alfvénic waves take time to propagate (whereas electron beams travel at close to the speed of light) can lead to different dynamics and observational signatures. Reep et al. (2018) especially highlighted that as waves heat a particular layer in the chromosphere, the temperature and ionization fraction increase, decreasing the parallel and Cowling resistivities in that layer. Subsequent waves therefore penetrate deeper into the atmosphere before damping, such that waves "bore a hole" deeper into the chromosphere. This behavior is opposite to an electron beam, which heats at progressively greater heights as chromospheric evaporation raises the coronal density (assuming that the beam properties remain constant). As a result, differences develop in the ionization, velocity, and pressure profiles that might lead to further opportunities to observationally distinguish heating by Alfvénic waves from heating by electron beams or thermal conduction.

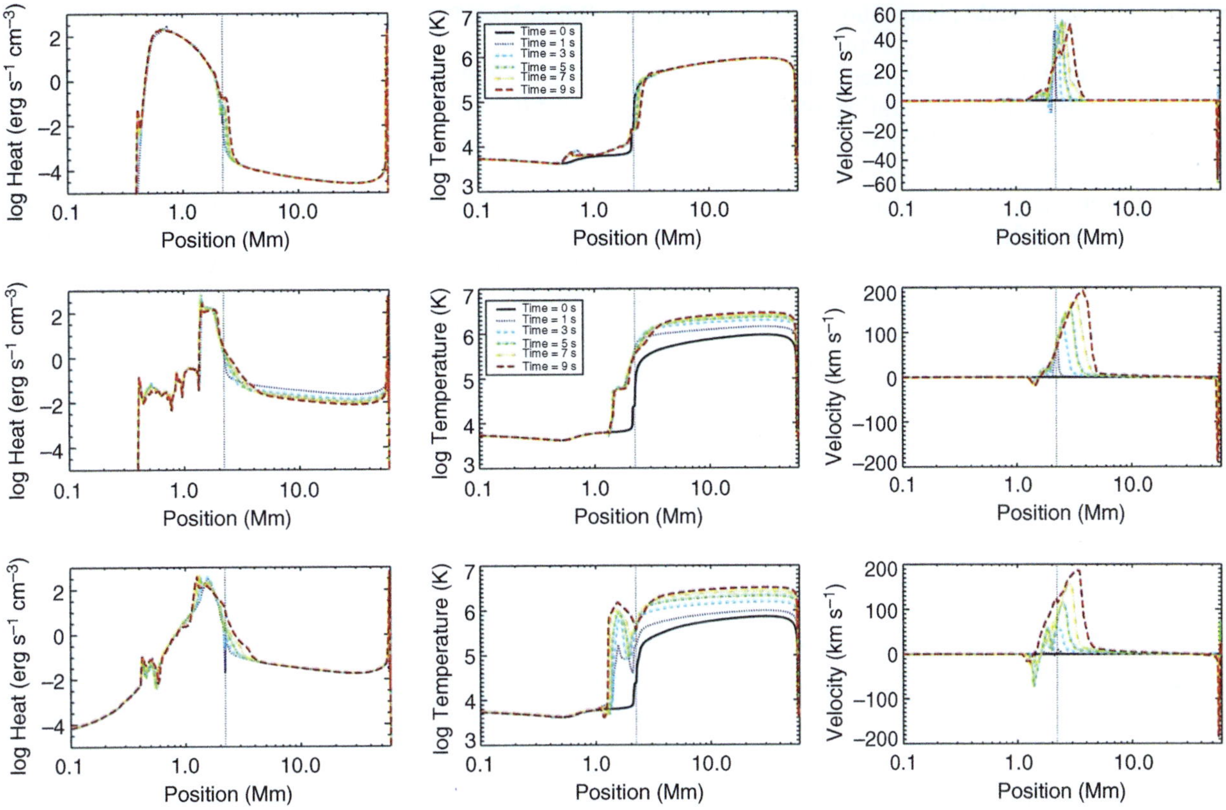


**Figure 3.9** Evolution of model solar atmospheres in three scenarios. The top row shows the response to waves injected with $f = 10\ Hz$ and $k_\perp = 10^{-3}\ m^{-1}$, the middle row to waves injected with $f = 10\ Hz$ and $k_\perp = 4 \times 10^{-2}\ m^{-1}$, and the bottom row to an electron beam with low-energy cutoff of 20 keV and spectral index $\delta = 5$. All simulations have an energy flux of $10^7$ W m$^{-2}$, consistent with estimates in Section 3.5. The upflows (right column) for the middle and bottom rows are virtually indistinguishable. Reproduced with permission from Reep and Russell (2016). © AAS.

## 3.8. FUTURE OUTLOOK

In the last 15 years, the topic of Alfvén waves in solar flares has matured significantly and several game-changing developments have given us high expectations for coming years. Here, we identify four major shifts that are anticipated or already happening, and the scientific questions they will help address.

### 3.8.1. High-Performance 3D Simulations

We have recently been fortunate to see 3D simulations achieve resolutions that transform the landscape for exploring the role of turbulence in solar flares. In particular, we highlighted in Section 3.6.2 how advanced simulations opened up the investigation of 3D magnetic reconnection at high Lundquist numbers, and in Section 3.6.3 how they have begun illuminating the generation of Alfvénic motions by flow braking in the above-the-loop region.

We anticipate that the new generation of 3D simulations will address the generation mechanisms for Alfvénic waves/turbulence in solar flares, establish the resulting wave properties, and connect to observations. Many things need to happen to fully grasp the opportunity; but much can be achieved by sustaining the development of 3D CSHKP flare simulations and allowing Moore's law to reveal a wider inertial range of the spectra. We are also hopeful that these topics will be approached not only with MHD, but also incorporating kinetic-type techniques to better model the reconnection process, and to self-consistently model particle acceleration. We believe this should be a high priority: done in coordination with theory and observations, the potential prizes include solving magnetic reconnection and the major workings of solar flares.

### 3.8.2. MUSE

Successive generations of spectroscopic observations have been the crucial source of knowledge about the amplitudes of small-scale plasma motions in solar flares, with hot lines especially important. The major axes of improvement for the instrument design have long been spatial resolution, spectral resolution, cadence, and instrument sensitivity. We have also seen, with Hinode/EIS and IRIS especially, the value of imaging, even with the low effective cadence that comes from rastering a single slit across a flare.

A wish list for a new instrument to study Alfvénic waves in solar flares would include: coverage of hot lines to capture the plasma with the largest nonthermal velocities; coverage of non-flare lines in the corona and transition region to investigate motions in these layers; spatial resolution comparable to or better than IRIS to spatially resolve footpoint emission; the best possible sensitivity to measure velocities as early as possible in the flare; short exposure times; and ability to produce images with rapid cadence.

The Multi-slit Solar Explorer (MUSE; De Pontieu et al., 2020) delivers in many of these respects. It will carry a multi-slit spectrograph (MUSE/SG) and a high-resolution EUV contextual imager (MUSE/CI). MUSE/SG builds on the heritage of single-slit EUV spectrographs such as EIS and IRIS, but it is designed to obtain spectra from 37 slits simultaneously for bright lines of Fe IX 171 Å, Fe XV 284 Å, Fe XIX 108 Å, and Fe XXI 108 Å, which are formed at temperatures around 0.7 MK, 2 MK, 10 MK, and 12 MK. Its design aims for spatial resolution of 0.4″ (about 300 km on the Sun), slit spacing of 4.45″, and a 170″ × 170″ field of view (FOV). Combined with a cadence of 1 s or less, it will be possible to either stare at many fixed slit locations simultaneously, providing comprehensively sampled measurements of $v_{nth}$ at timescales relevant to MHD turbulence and particle acceleration, or to raster the whole FOV in 12 s or less. The EUV imager MUSE/CI, meanwhile, will provide 304 Å and 195 Å images at 0.33″ resolution (for comparison the resolution of SDO/AIA is 1.5″; Lemen et al., 2012) with a 580″ × 290″ FOV and cadence of 4 s for single-channel operation (8 s for dual-channel operation).

Science objectives for MUSE related to flares and eruptions have been discussed by Cheung et al. (2022), and their Section 5.4 is especially relevant to the current article. We believe that MUSE has excellent potential to deliver a step change in observations of waves and turbulence in solar flares by measuring spectral properties with unrivaled spatial coverage and cadence. We also highlight that the combination of MUSE/SG and MUSE/CI is perfect for investigating motions in 3D. MUSE is currently estimated for launch in 2027, a time frame that gives impetus to advancing the theory and modeling in preparation.

### 3.8.3. DKIST

New observational opportunities also include the new generation of ground-based solar telescopes. At the current state of the art, the Daniel K. Inouye Solar Telescope (DKIST; Rimmele et al., 2020) is equipped with a 4 m aperture and a high-order adaptive optics system, which are designed to resolve scales as short as 0.02″, or 20 km on the Sun. Its initial suite of five instruments, which includes four polarimeters, allow observations at visible and near-infrared wavelengths, from 380 to 5000 nm. First light was achieved in December 2019, and it has more recently started collecting scientific data. Other 4 m class solar telescopes are in development, including the European Solar Telescope (EST), which in combination with DKIST would provide near-continuous solar observing.

Advancing understanding of solar flares and CMEs is a major science driver behind these telescopes, and Section 4 of the DKIST Critical Science Plan (Rast et al., 2021) details how DKIST is expected to contribute to flare science. Here, we briefly outline some of the major opportunities relevant to Alfvénic waves in solar flares.

A major objective for DKIST is to measure coronal magnetic fields, using principles demonstrated by Lin et al. (2000, 2004) using a 0.46-m aperture telescope. The much larger DKIST has targets of "spatial resolution of 1–2 arcsec, sensitivity of order 10 G or less ... with a temporal resolution of tens of seconds to minutes" (Rimmele et al., 2020). Such measurements would provide valuable contextual information for limb flares, complementing existing approaches of deducing magnetic field information from microwave observations (e.g., Chen et al., 2020; Fleishman et al., 2022) and intensity images. Routine measurements of coronal magnetic field strength would be a very welcome achievement for constraining the Poynting fluxes in flares (see Section 3.5).

Ground-based polarimetry can also identify regions of strong turbulence and/or wave activity. As discussed in Section 3.4.7, French et al. (2019) showed that the linear polarization (Stokes $P$) fell to low values in the flare plasma sheet and the above-the-loop region of the X8.2 flare on 10 September 2017. The reduction at these locations was attributed to the orientation of the magnetic field varying on unresolved scales and/or along the line of sight. Stokes $P$ is therefore complementary to the nonthermal broadening explored in Section 3.4: nonthermal broadening indicates unresolved motions, whereas low $P$ can identify unresolved variations in the magnetic field orientation. When the two occur together, the case for interpreting the unresolved motions as Alfvénic waves or turbulence is strengthened, since a model of field-aligned flows does not explain the reduction in $P$.

Another major objective of DKIST is to obtain magnetic field data at multiple heights in the Sun's lower atmosphere, allowing scientists to probe the 3D magnetic structure. As discussed in Section 3.1, solar flares and eruptions are powered by reconfiguration of the coronal magnetic field, which produces rapid changes to the magnetic field in the lower atmosphere that are likely communicated by large-amplitude Alfvénic wave fronts. DKIST offers a good prospect of probing these magnetic field changes, and their association with particle acceleration (Fletcher & Hudson, 2008), sunquakes (Hudson et al., 2008; Russell et al., 2016), and radiative emissions from the lower atmosphere (Johnstone et al., 2012; Russell & Fletcher, 2013).

Imaging and spectroscopy of flare ribbons at the high cadence and high resolution provided by DKIST will help to explore energy transport in flares. As discussed in Section 3.7.5, 1D radiative hydrodynamics simulations show that the lower atmosphere responds differently to heating by Alfvénic waves than it does to heating by electron beams, although there are also some major similarities (Kerr et al., 2016; Reep & Russell, 2016; Reep et al., 2018). DKIST data are especially important to constrain the size (Krucker et al., 2011) and height (Martínez Oliveros et al., 2012) of the white-light sources.

Finally, since flare ribbons are believed to be connected to the coronal reconnection layer (Figure 3.1), their morphology and dynamics shed light on many of the central issues concerning Alfvénic waves in flares. For example, French et al. (2021) used images from the IRIS Slit-Jaw Imager to track the exponential growth of power at different wavenumbers along the ribbons of a confined B-class flare. Instability began at a "key spatial scale" (1.75 Mm in the east ribbon), spread to smaller and larger scales and saturated with a $k^{-2.3}$ power law across all scales. Nonthermal broadening in the ribbons rose as the instability spread to all scales. The timeline and power law slope are consistent with the models of broadband Alfvénic wave generation within the reconnection layer discussed in Sections 3.6.1 and 3.6.2. Higher-resolution studies using DKIST, covering more events, are therefore an exciting prospect to advance understanding of the wave generation and the wave spectrum.

### 3.8.4. Integrated Modeling

There is valuable scientific work to be done advancing the subtopics covered in this article individually, and integrating the component parts into the whole system. As computing power advances further, we look forward to a combined treatment of the fast reconnection process and outflow braking in CHSKP flare geometries in the relatively near future. Thence, modeling approaches can be brought to bear that have already been developed through other work, including connections with the lower atmosphere, particle energization, radiative effects, and forward modeling to compare with observations (e.g., Cheung et al., 2019; Gordovskyy et al., 2020; Ruan et al., 2020). The end goal is to model flares from reconnection to radiation, capturing the energy pathways in between.

We also highlight the role for simulations tailored to specific events. 3D data-driven simulations (e.g., Amari et al., 2014) have greatly advanced understanding of the mechanisms responsible for triggering solar eruptions and flares. The necessity that such simulations model an entire active region and its environment means that the shortest resolved scale is significantly larger than for more idealized simulations. Hence, we must wait before wave excitation processes are captured in full data-driven simulations. Once this occurs, however, the ability to directly compare synthetic and real observations will be a great aid to refine our knowledge of Alfvénic waves in flares.

## 3.9. SUMMARY

There is abundant support from observations, simulations, and theory that solar flares generate Alfvénic waves with enough energy density and Poynting flux to play a significant role in the flare. The likely sources of these waves are the fast magnetic reconnection process required in solar flares and braking of the sunward reconnection flow. Additionally, the sudden onset of the flare impulsive phase may launch Alfvénic fronts, which propagate downward into the lower solar atmosphere. Once generated, Alfvénic waves evolve according to principles of field-aligned propagation, phase mixing, instability, and turbulent processes, with energy being lost to particle acceleration and/or heating the corona and lower atmosphere.

The last 15 years have brought the role of Alfvénic waves into considerably sharper focus. It is now widely acknowledged that they are likely to be a key component in the workings of solar flares, and nonthermal line widths have emerged as a key diagnostic that constrains wave energy density and Poynting flux. The above-the-loop region has been shown to be a key location for generating waves/turbulence, generation mechanisms have become addressable with high-performance 3D simulations, particle acceleration by Alfvénic waves/turbulence has gained greater observational support, and various investigations have shown the feasibility of waves directly heating in the lower atmosphere.

We finish with the message that key problems in flare physics are opening up, as modeling and theory embrace the inherent 3D and dynamic nature of real plasma and new observatories provide better windows on solar processes. These advances mean that there is a realistic prospect of solving the grand challenges of fast reconnection, flare particle acceleration, and flare energy transport within the span of a scientific career, and hence improving understanding and forecasting of space weather. Alfvénic waves are likely to be a central part of those solutions.

## ACKNOWLEDGMENTS

Alexander Russell thanks the many colleagues who have contributed to the development of the ideas presented in this article. The article benefited from discussions with participants of the 2023 AGU Chapman Conference on "Advances in Understanding Alfvén Waves in the Sun and the Heliosphere," and constructive comments and suggestions from two anonymous reviewers. I am grateful to creators and copyright holders for permitting the reproduction of figures in this work. Researching and writing the article made use of NASA's Astrophysics Data System Bibliographic Services. Most of the scholarship and writing was undertaken at the University of Dundee, UK, where Alexander Russell was employed before moving to the University of St Andrews in August 2023.

## REFERENCES

Alexander, D., Harra-Murnion, L. K., Khan, J. I., & Matthews, S. A. (1998). Relative timing of soft X-ray nonthermal line broadening and hard X-ray emission in solar flares. *The Astrophysical Journal*, *494*, L235–L238. doi: 10.1086/311175.

Alfvén, H. (1942). Existence of electromagnetic-hydrodynamic waves. *Nature*, *150*, 405–406.

Amari, T., Canou, A., & Aly, J.-J. (2014). Characterizing and predicting the magnetic environment leading to solar eruptions. *Nature*, *514*, 465–469. doi: 10.1038/nature13815.

Antolin, P., De Moortel, I., Van Doorsselaere, T., & Yokoyama, T. (2017). Observational signatures of transverse magnetohydrodynamic waves and associated dynamic instabilities in coronal flux tubes. *The Astrophysical Journal*, *836*, 219. doi: 10.3847/1538-4357/aa5eb2.

Antolin, P., & Rouppe van der Voort, L. (2012). Observing the fine structure of loops through high-resolution spectroscopic observations of coronal rain with the CRISP instrument at the Swedish solar telescope. *The Astrophysical Journal*, *745*, 152. doi: 10.1088/0004-637x/745/2/152.

Antonucci, E. (1984). Line broadenings and shifts during the impulsive phase of flares. *Memorie della Societa Astronomica Italiana*, *55*, 699–712.

Antonucci, E., Dodero, M. A., & Martin, R. (1990). Velocity-temperature distribution in the evaporating plasma during the impulsive phase of solar flares. *The Astrophysical Journal Supplement Series*, *73*, 147. doi: 10.1086/191446.

Antonucci, E., Gabriel, A. H., Acton, L. W., Culhane, J. L., Doyle, J. G., Leibacher, J. W., et al. (1982). Impulsive phase of flares in soft X-ray emission. *Solar Physics*, *78*, 107–123. doi: 10.1007/bf00151147.

Antonucci, E., Gabriel, A. H., & Dennis, B. R. (1984). The energetics of chromospheric evaporation in solar flares. *The Astrophysical Journal*, *287*, 917–925. doi: 10.1086/162749.

Aschwanden, M. J., Fletcher, L., Schrijver, C. J., & Alexander, D. (1999). Coronal loop oscillations observed with the transition region and coronal explorer. *The Astrophysical Journal*, *520*, 880–894. doi: 10.1086/307502.

Aschwanden, M. J., & Peter, H. (2017). The width distribution of loops and strands in the solar corona—Are we hitting rock bottom? *The Astrophysical Journal*, *840*, 4. doi: 10.3847/1538-4357/aa6b01.

Aulanier, G., Janvier, M., & Schmieder, B. (2012). The standard flare model in three dimensions. I. Strong-to-weak shear transition in post-flare loops. *Astronomy and Astrophysics*, *543*, A110. doi: 10.1051/0004-6361/201219311.

Beg, R., Russell, A. J. B., & Hornig, G. (2022). Evolution, structure, and topology of self-generated turbulent reconnection layers. *The Astrophysical Journal*, *940*, 94. doi: 10.3847/1538-4357/ac8eb6.

Bemporad, A. (2008). Spectroscopic detection of turbulence in post-CME current sheets. *The Astrophysical Journal*, *689*, 572–584. doi: 10.1086/592377.

Benz, A. O. (1986). Millisecond radio spikes. *Solar Physics*, *104*, 99–110. doi: 10.1007/bf00159950.

Beresnyak, A. (2017). Three-dimensional spontaneous magnetic reconnection. *The Astrophysical Journal*, *834*, 47. doi: 10.3847/1538-4357/834/1/47.

Bhattacharjee, A., Huang, Y.-M., Yang, H., & Rogers, B. (2009). Fast reconnection in high-Lundquist-number plasmas due to the plasmoid instability. *Physics of Plasmas, 16*, 112102. doi: 10.1063/1.3264103.

Bian, N. H., Kontar, E. P., and Brown, J. C. (2010). Parallel electric field generation by Alfvén wave turbulence. *Astronomy and Astrophysics*, *519*, A114. doi: 10.1051/0004-6361/201014048.

Birn, J., Drake, J. F., Shay, M. A., Rogers, B. N., Denton, R. E., Hesse, M., et al. (2001). Geospace Environmental Modeling (GEM) magnetic reconnection challenge. *Journal of Geophysical Research*, *106*, 3715–3720. doi: 10.1029/1999ja900449.

Biskamp, D. (1986). Magnetic reconnection via current sheets. *Physics of Fluids*, *29*, 1520–1531. doi: 10.1063/1.865670.

Braginskii, S. I. (1965). Transport processes in a plasma. *Reviews of Plasma Physics*, *1*, 205.

Brooks, D. H., Warren, H. P., Ugarte-Urra, I., & Winebarger, A. R. (2013). High spatial resolution observations of loops in the solar corona. *The Astrophysical Journal, 772*, L19. doi: 10.1088/2041-8205/772/2/l19.

Brosius, J. W. (2003). Chromospheric evaporation and warm rain during a solar flare observed in high time resolution with the coronal diagnostic spectrometer aboard the solar and heliospheric observatory. *The Astrophysical Journal*, *586*, 1417–1429. doi: 10.1086/367958.

Brosius, J. W. (2012). Extreme-ultraviolet spectroscopic observation of direct coronal heating during a C-class solar flare. *The Astrophysical Journal*, *754*, 54. doi: 10.1088/0004-637x/754/1/54.

Brosius, J. W. (2013). Chromospheric evaporation in solar flare loop strands observed with the extreme-ultraviolet imaging spectrometer on board Hinode. *The Astrophysical Journal, 762*, 133. doi: 10.1088/0004-637x/762/2/133.

Brosius, J. W., & White, S. M. (2006). Radio measurements of the height of strong coronal magnetic fields above sunspots at the solar limb. *The Astrophysical Journal*, *641*, L69–L72. doi: 10.1086/503774.

Brown, J. C., Turkmani, R., Kontar, E. P., MacKinnon, A. L., & Vlahos, L. (2009). Local re-acceleration and a modified thick target model of solar flare electrons. *Astronomy and Astrophysics*, *508*, 993–1000. doi: 10.1051/0004-6361/200913145.

Bulanov, S. V., Syrovatskiì, S. I., & Sakai, J. (1978). Stabilizing influence of plasma flow on dissipative tearing instability. *Soviet Journal of Experimental and Theoretical Physics Letters*, *28*, 177–179.

Cameron, R., & Sammis, I. (1999). Tangential field changes in the great flare of 1990 May 24. *The Astrophysical Journal*, *525*, L61–L64. doi: 10.1086/312328.

Carmichael, H. (1964). *A Process for Flares*. Proceedings of the AAS-NASA Symposium held 28–30 October, 1963 at the Goddard Space Flight Center, Greenbelt, MD. Edited by Wilmot N. Hess. Washington, DC: National Aeronautics and Space Administration, Science and Technical Information Division, 1964, 451–456.

Cassak, P. A., Shay, M. A., & Drake, J. F. (2009). Scaling of Sweet-Parker reconnection with secondary islands. *Physics of Plasmas*, *16*, 120702. doi: 10.1063/1.3274462.

Chen, B., Shen, C., Gary, D. E., Reeves, K. K., Fleishman, G. D., Yu, S., et al. (2020). Measurement of magnetic field and relativistic electrons along a solar flare current sheet. *Nature Astronomy*, *4*, 1140–1147. doi: 10.1038/s41550-020-1147-7.

Cheng, X., Li, Y., Wan, L. F., Ding, M. D., Chen, P. F., Zhang, J., & Liu, J. J. (2018). Observations of turbulent magnetic reconnection within a solar current sheet. *The Astrophysical Journal*, *866*, 64. doi: 10.3847/1538-4357/aadd16.

Cheung, M. C. M., Rempel, M., Chintzoglou, G., Chen, F., Testa, P., Martínez-Sykora, J., et al. (2019). A comprehensive three-dimensional radiative magnetohydrodynamic simulation of a solar flare. *Nature Astronomy*, *3*, 160–166. doi: 10.1038/s41550-018-0629-3.

Cheung, M. C. M., Martínez-Sykora, J., Testa, P., De Pontieu, B., Chintzoglou, G., Rempel, M., et al. (2022). Probing the physics of the solar atmosphere with the Multi-slit Solar Explorer (MUSE). II. Flares and eruptions. *The Astrophysical Journal*, *926*, 53. doi: 10.3847/1538-4357/ac4223.

Chitta, L. P., & Lazarian, A. (2020). Onset of turbulent fast magnetic reconnection observed in the solar atmosphere. *The Astrophysical Journal*, *890*, L2. doi: 10.3847/2041-8213/ab6f0a.

Ciaravella, A., & Raymond, J. C. (2008). The current sheet associated with the 2003 November 4 coronal mass ejection: Density, temperature, thickness, and line width. *The Astrophysical Journal*, *686*, 1372–1382. doi: 10.1086/590655.

Ciaravella, A., Raymond, J. C., Li, J., Reiser, P., Gardner, L. D., Ko, Y. -K., & Fineschi, S. (2002). Elemental abundances and post-coronal mass ejection current sheet in a very hot active region. *The Astrophysical Journal*, *575*, 1116–1130. doi: 10.1086/341473.

Culhane, J. L., Harra, L. K., James, A. M., Al-Janabi, K., Bradley, L. J., Chaudry, R. A., et al. (2007). The EUV imaging spectrometer for Hinode. *Solar Physics*, *243*, 19–61. doi: 10.1007/s01007-007-0293-1.

Culhane, J. L., Rapley, C. G., Bentley, R. D., Gabriel, A. H., Phillips, K. J., Acton, L. W., et al. (1981). X-ray spectra of solar flares obtained with a high-resolution bent crystal spectrometer. *The Astrophysical Journal*, *244*, L141–L145. doi: 10.1086/183499.

Dahlin, J. T., Drake, J. F., & Swisdak, M. (2015). Electron acceleration in three-dimensional magnetic reconnection with a guide field. *Physics of Plasmas*, *22*, 100704. doi: 10.1063/1.4933212.

Dahlin, J. T., Drake, J. F., & Swisdak, M. (2017). The role of three-dimensional transport in driving enhanced electron acceleration during magnetic reconnection. *Physics of Plasmas*, *24*, 092110. doi: 10.1063/1.4986211.

Daldorff, L. K. S., Leake, J. E., & Klimchuk, J. A. (2022). Impact of 3D structure on magnetic reconnection. *The Astrophysical Journal*, *927*, 196. doi: 10.3847/1538-4357/ac532d.

Daughton, W., Nakamura, T. K. M., Karimabadi, H., Roytershteyn, V., & Loring, B. (2014). Computing the reconnection rate in turbulent kinetic layers by using electron mixing to identify topology. *Physics of Plasmas*, *21*, 052307. doi: 10.1063/1.4875730.

Daughton, W., Roytershteyn, V., Albright, B. J., Karimabadi, H., Yin, L., & Bowers, K. J. (2009). Transition from collisional to kinetic regimes in large-scale reconnection layers. *Physical Review Letters*, *103*, 065004. doi: 10.1103/PhysRevLett.103.065004.

Daughton, W., Roytershteyn, V., Karimabadi, H., Yin, L., Albright, B. J., Bergen, B., & Bowers, K. J. (2011). Role of electron physics in the development of turbulent magnetic reconnection in collisionless plasmas. *Nature Physics*, *7*, 539–542. doi: 10.1038/nphys1965.

de Jager, C., & de Jonge, G. (1978). Properties of elementary flare bursts. *Solar Physics*, *58*, 127–137. doi: 10.1007/bf00152559.

De Pontieu, B., Martens, P. C. H., & Hudson, H. S. (2001). Chromospheric damping of Alfvén waves. *The Astrophysical Journal*, *558*, 859–871. doi: 10.1086/322408.

De Pontieu, B., McIntosh, S. W., Carlsson, M., Hansteen, V. H., Tarbell, T. D., Schrijver, C. J., et al. (2007). Chromospheric Alfvénic waves strong enough to power the solar wind. *Science*, *318*, 1574.

De Pontieu, B., Martínez-Sykora, J., Testa, P., Winebarger, A. R., Daw, A., Hansteen, V., et al. (2020). The Multi-slit Approach to Coronal Spectroscopy with the Multi-slit Solar Explorer (MUSE). *The Astrophysical Journal*, *888*, 3. doi: 10.3847/1538-4357/ab5b03.

De Pontieu, B., Title, A. M., Lemen, J. R., Kushner, G. D., Akin, D. J., Allard, B., et al. (2014). The Interface Region Imaging Spectrograph (IRIS). *Solar Physics*, *289*, 2733–2779. doi: 10.1007/s11207-014-0485-y.

Dong, C., Wang, L., Huang, Y.-M., Comisso, L., Sandstrom, T. A., & Bhattacharjee, A. (2022). Reconnection-driven energy cascade in magnetohydrodynamic turbulence. *Science Advances*, *8*, eabn7627. doi: 10.1126/sciadv.abn7627.

Doschek, G. A. (1990). Soft X-ray spectroscopy of solar flares: An overview. *The Astrophysical Journal Supplement Series*, *73*, 117. doi: 10.1086/191443.

Doschek, G. A., Feldman, U., Dere, K. P., Sandlin, G. D., Vanhoosier, M. E., Brueckner, G. E., et al. (1975). Forbidden lines of highly ionized iron in solar flare spectra. *The Astrophysical Journal*, *196*, L83-L86. doi: 10.1086/181749.

Doschek, G. A., Feldman, U., Kreplin, R. W., & Cohen, L. (1980). High-resolution X-ray spectra of solar flares. III—General spectral properties of X1-X5 type flares. *The Astrophysical Journal*, *239*, 725–737. doi: 10.1086/158158.

Doschek, G. A., Feldman, U., Seely, J. F., & McKenzie, D. L. (1989). High-resolution X-ray spectra of solar flares. IX. Mass upflow in the long-duration flare of 1979 June 5. *The Astrophysical Journal*, *345*, 1079. doi: 10.1086/167977.

Doschek, G. A., Kreplin, R. W., & Feldman, U. (1979). High-resolution solar flare X-ray spectra. *The Astrophysical Journal*, *233*, L157–L160. doi: 10.1086/183096.

Doschek, G. A., McKenzie, D. E., & Warren, H. P. (2014). Plasma dynamics above solar flare soft X-ray loop tops. *The Astrophysical Journal*, *788*, 26. doi: 10.1088/0004-637x/788/1/26.

Doschek, G. A., Meekins, J. F., Kreplin, R. W., Chubb, T. A., & Friedman, H. (1971). Iron-line emission during solar flares. *The Astrophysical Journal*, *170*, 573. doi: 10.1086/151243.

Dudík, J., Polito, V., Dzifčáková, E., Del Zanna, G., & Testa, P. (2017). Non-Maxwellian analysis of the transition-region line profiles observed by the interface region imaging spectrograph. *The Astrophysical Journal*, *842*, 19. doi: 10.3847/1538-4357/aa71a8.

Edwin, P. M., & Roberts, B. (1983). Wave propagation in a magnetic cylinder. *Solar Physics*, *88*, 179–191. doi: 10.1007/bf00196186.

Emslie, A. G., & Sturrock, P. A. (1982). Temperature minimum heating in solar flares by resistive dissipation of Alfvén waves. *Solar Physics*, *80*, 99–112. doi: 10.1007/bf00153426.

Feldman, U., Doschek, G. A., Kreplin, R. W., & Mariska, J. T. (1980). High-resolution X-ray spectra of solar flares. IV - General spectral properties of M type flares. *The Astrophysical Journal*, *241*, 1175–1185. doi: 10.1086/158434.

Fermi, E. (1949). On the origin of the cosmic radiation. *Physical Review*, *75*, 1169–1174. doi: 10.1103/PhysRev.75.1169.

Fermo, R. L., Drake, J. F., & Swisdak, M. (2010). A statistical model of magnetic islands in a current layer. *Physics of Plasmas*, *17*, 010702. doi: 10.1063/1.3286437.

Fisher, G. H., Bercik, D. J., Welsch, B. T., & Hudson, H. S. (2012). Global forces in eruptive solar flares: The Lorentz force acting on the solar atmosphere and the solar interior. *Solar Physics*, *277*, 59–76. doi: 10.1007/s11207-011-9907-2.

Fisher, G. H., Canfield, R. C., & McClymont, A. N. (1985). Flare loop radiative hydrodynamics. V—Response to thick-target heating. VI—Chromospheric evaporation due to heating by nonthermal electrons. VII—Dynamics of the thick-target heated chromosphere. *The Astrophysical Journal*, *289*, 414–441. doi: 10.1086/162901.

Fleishman, G. D., Nita, G. M., Chen, B., Yu, S., & Gary, D. E. (2022). Solar flare accelerates nearly all electrons in a large coronal volume. *Nature*, *606*, 674–677. doi: 10.1038/s41586-022-04728-8.

Fletcher, L., Dennis, B. R., Hudson, H. S., Krucker, S., Phillips, K., Veronig, A., et al. (2011). An observational overview of solar flares. *Space Science Reviews*, *159*, 19–106. doi: 10.1007/s11214-010-9701-8.

Fletcher, L., Hannah, I. G., Hudson, H. S., & Innes, D. E. (2013). Flare ribbon energetics in the early phase of an SDO flare. *The Astrophysical Journal*, *771*, 104. doi: 10.1088/0004-637x/771/2/104.

Fletcher, L., & Hudson, H. S. (2008). Impulsive phase flare energy transport by large-scale Alfvén waves and the electron acceleration problem. *The Astrophysical Journal*, *675*, 1645–1655. doi: 10.1086/527044.

Fludra, A., Lemen, J. R., Jakimiec, J., Bentley, R. D., & Sylwester, J. (1989). Turbulent and directed plasma motions in solar flares. *The Astrophysical Journal*, *344*, 991. doi: 10.1086/167866.

French, R. J., Judge, P. G., Matthews, S. A., & van Driel-Gesztelyi, L. (2019). Spectropolarimetric insight into plasma sheet dynamics of a solar flare. *The Astrophysical Journal*, *887*, L34. doi: 10.3847/2041-8213/ab5d34.

French, R. J., Matthews, S. A., Jonathan Rae, I., & Smith, A. W. (2021). Probing current sheet instabilities from flare ribbon dynamics. *The Astrophysical Journal*, *922*, 117. doi: 10.3847/1538-4357/ac256f.

French, R. J., Matthews, S. A., van Driel-Gesztelyi, L., Long, D. M., & Judge, P. G. (2020). Dynamics of late-stage reconnection in the 2017 September 10 solar flare. *The Astrophysical Journal*, *900*, 192. doi: 10.3847/1538-4357/aba94b.

Furth, H. P., Killeen, J., & Rosenbluth, M. N. (1963). Finite-resistivity instabilities of a sheet pinch. *Physics of Fluids*, *6*, 459–484. doi: 10.1063/1.1706761.

Gabriel, A. H., Phillips, K. J. H., Acton, L. W., Wolfson, C. J., Culhane, J. L., Rapley, C. G., et al. (1981). Observations of the limb solar flare on 1980 April 30 with the SMM X-ray polychromator. *The Astrophysical Journal*, *244*, L147–L151. doi: 10.1086/183500.

Goldreich, P., & Sridhar, S. (1995). Toward a theory of interstellar turbulence. II. Strong Alfvenic turbulence. *The Astrophysical Journal*, *438*, 763.

Goldreich, P., & Sridhar, S. (1997). Magnetohydrodynamic turbulence revisited. *The Astrophysical Journal*, *485*, 680–688. doi: 10.1086/304442.

Goossens, M., Terradas, J., Andries, J., Arregui, I., & Ballester, J. L. (2009). On the nature of kink MHD waves in magnetic flux tubes. *Astronomy and Astrophysics*, *503*, 213–223. doi: 10.1051/0004-6361/200912399.

Gordovskyy, M., Browning, P. K., Inoue, S., Kontar, E. P., Kusano, K., & Vekstein. G. E. (2020). Forward modeling of particle acceleration and transport in an individual solar flare. *The Astrophysical Journal*, *902*, 147. doi: 10.3847/1538-4357/abb60e.

Graham, D. R., & Cauzzi, G. (2015). Temporal evolution of multiple evaporating ribbon sources in a solar flare. *The Astrophysical Journal*, *807*, L22. doi: 10.1088/2041-8205/807/2/l22.

Grineva, Y. I., Karev, V. I., Korneev, V. V., Krutov, V. V., Mandelstam, S. L., Vainstein, L. A., et al. (1973). Solar X-ray spectra observed from the 'intercosmos-4' satellite and the 'vertical-2' rocket. *Solar Physics*, *29*, 441–446. doi: 10.1007/bf00150824.

Haerendel, G. (2009). Chromospheric evaporation via Alfvén waves. *The Astrophysical Journal*, *707*, 903–915. doi: 10.1088/0004-637x/707/2/903.

Haerendel, G. (2012). Solar Auroras. *The Astrophysical Journal*, *749*, 166. doi: 10.1088/0004-637x/749/2/166.

Hara, H, Watanabe, T., Harra, L. K., Leonard Culhane, J., & Young, P. R. (2011). Plasma motions and heating by magnetic reconnection in a 2007 May 19 flare. *The Astrophysical Journal*, *741*, 107. doi: 10.1088/0004-637x/741/2/107.

Hara, H., Watanabe, T., Matsuzaki, K., Harra, L. K., Leonard Culhane, J., Cargill, P., et al. (2008). 2006 December 17 long duration flare observed with the Hinode EUV Imaging spectrometer. *Publications of the Astronomical Society of Japan*, *60*, 275. doi: 10.1093/pasj/60.2.275.

Harra, L. K., Matthews, S. A., & Culhane, J. L. (2001). Nonthermal velocity evolution in the precursor phase of a solar flare. *The Astrophysical Journal*, *549*, L245–L248. doi: 10.1086/319163.

Harra, L. K., Matthews, S., Culhane, J. L., Cheung, M. C. M., Kontar, E. P., & Hara, H. (2013). The location of non-thermal velocity in the early phases of large flares—Revealing pre-eruption flux ropes. *The Astrophysical Journal*, *774*, 122. doi: 10.1088/0004-637x/774/2/122.

Heyvaerts, J., & Priest, E. R. (1983). Coronal heating by phase-mixed shear Alfven waves. *Astronomy and Astrophysics*, *117*, 220–234.

Hirayama, T. (1974). Theoretical model of flares and prominences. I: Evaporating flare model. *Solar Physics*, *34*, 323–338. doi: 10.1007/bf00153671.

Howes, G. G., & Nielson, K. D. (2013). Alfvén wave collisions, the fundamental building block of plasma turbulence. *I. Asymptotic solution. Physics of Plasmas*, *20*, 072304.

Huang, Y.-M., & Bhattacharjee, A. (2010). Scaling laws of resistive magnetohydrodynamic reconnection in the high-Lundquist-number, plasmoid-unstable regime. *Physics of Plasmas*, *17*, 062104–062104. doi: 10.1063/1.3420208.

Huang, Y.-M., & Bhattacharjee, A. (2012). Distribution of plasmoids in high-Lundquist-number magnetic reconnection. *Physical Review Letters*, *109*, 265002. doi: 10.1103/PhysRevLett.109.265002.

Huang, Y.-M., & Bhattacharjee, A. (2013). Plasmoid instability in high-Lundquist-number magnetic reconnection. *Physics of Plasmas*, *20*, 055702. doi: 10.1063/1.4802941.

Huang, Y.-M., & Bhattacharjee, A. (2016). Turbulent magnetohydrodynamic reconnection mediated by the plasmoid instability. *The Astrophysical Journal*, *818*, 20. doi: 10.3847/0004-637x/818/1/20.

Hudson, H. S., Fisher, G. H., & Welsch, B. T. (2008). Flare energy and magnetic field variations. *Subsurface and Atmospheric Influences on Solar Activity*, January 01, 2008.

Ishikawa, S., Krucker, S., Takahashi, T., & Lin, R. P. (2011). On the relation of above-the-loop and footpoint hard X-ray sources in solar flares. *The Astrophysical Journal*, *737*, 48. doi: 10.1088/0004-637x/737/2/48.

Janvier, M., Aulanier, G., Bommier, V., Schmieder, B., Démoulin, P., & Pariat, E. (2014). Electric currents in flare ribbons: Observations and three-dimensional standard model. *The Astrophysical Journal*, *788*, 60. doi: 10.1088/0004-637x/788/1/60.

Janvier, M., Aulanier, G., & Démoulin, P. (2015). From coronal observations to MHD simulations, the building blocks for 3D models of solar flares (invited review). *Solar Physics*, *290*, 3425–3456. doi: 10.1007/s11207-015-0710-3.

Jeffrey, N. L. S., Fletcher, L., & Labrosse, N. (2016). First evidence of non-Gaussian solar flare EUV spectral line profiles and accelerated non-thermal ion motion. *Astronomy and Astrophysics*, *590*, A99. doi: 10.1051/0004-6361/201527986.

Jeffrey, N. L. S., Fletcher, L., Labrosse, N., & Simões, P. J. A. (2018). The development of lower-atmosphere turbulence early in a solar flare. *Science Advances*, *4*, 2794. doi: 10.1126/sciadv.aav2794.

Johnstone, B. M., Petrie, G. J. D., & Sudol, J. J. (2012). Abrupt longitudinal magnetic field changes and ultraviolet emissions accompanying solar flares. *The Astrophysical Journal*, *760*, 29. doi: 10.1088/0004-637x/760/1/29.

Karpen, J. T., Antiochos, S. K., & DeVore, C. R. (2012). The mechanisms for the onset and explosive eruption of coronal mass ejections and eruptive flares. *The Astrophysical Journal*, *760*, 81. doi: 10.1088/0004-637x/760/1/81.

Kerr, G. S., Fletcher, L., Russell, A. J. B., & Allred, J. C. (2016). Simulations of the MG II K and CA II 8542 lines from an Alfven wave-heated flare chromosphere. *Astrophysical Journal*, *827*, 101. doi: 10.3847/0004-637X/827/2/101.

Khan, J. I., Harra-Murnion, L. K., Hudson, H. S., Lemen, J. R., & Sterling, A. C. (1995). Yohkho soft X-ray spectroscopic observations of the bright loop-top kernels of solar flares. *The Astrophysical Journal*, *452*, L153. doi: 10.1086/309724.

Kiplinger, A. L., Dennis, B. R., Frost, K. J., Orwig, L. E., & Emslie, A. G. (1983). Millisecond time variations in hard X-ray solar flares. *The Astrophysical Journal*, *265*, L99–L104. doi: 10.1086/183966.

Kobayashi, K., Cirtain, J., Winebarger, A. R., Korreck, K., Golub, L., Walsh, R. W., et al. (2014). The High-Resolution Coronal Imager (Hi-C). *Solar Physics*, *289*, 4393–4412. doi: 10.1007/s11207-014-0544-4.

Kontar, E. P., Perez, J. E., Harra, L. K., Kuznetsov, A. A., Emslie, A. G., Jeffrey, N. L. S., et al. (2017). Turbulent kinetic energy in the energy balance of a solar flare. *Physical Review Letters*, *118*, 155101. doi: 10.1103/PhysRevLett.118.155101.

Kopp, R. A., & Pneuman, G. W. (1976). Magnetic reconnection in the corona and the loop prominence phenomenon. *Solar Physics*, *50*, 85–98. doi: 10.1007/bf00206193.

Kosovichev, A. G., & Zharkova, V. V. (2001). Magnetic energy release and transients in the solar flare of 2000 July 14. *The Astrophysical Journal*, *550*, L105–L108. doi: 10.1086/319484.

Kowal, G., Falceta-Gonçalves, D. A., Lazarian, A., & Vishniac, E. T. (2017). Statistics of reconnection-driven turbulence. *The Astrophysical Journal*, *838*, 91. doi: 10.3847/1538-4357/aa6001.

Kowal, G., Falceta-Gonçalves, D. A., Lazarian, A., & Vishniac, E. T. (2020). Kelvin-Helmholtz versus tearing instability: What drives turbulence in stochastic reconnection? *The Astrophysical Journal*, *892*, 50. doi: 10.3847/1538-4357/ab7a13.

Kowal, G., Lazarian, A., Vishniac, E. T., & Otmianowska-Mazur, K. (2009). Numerical tests of fast reconnection in weakly stochastic magnetic fields. *The Astrophysical Journal*, *700*, 63–85. doi: 10.1088/0004-637x/700/1/63.

Krucker, S., & Battaglia, M. (2014). Particle densities within the acceleration region of a solar flare. *The Astrophysical Journal, 780*, 107. doi: 10.1088/0004-637x/780/1/107.

Krucker, S., Hudson, H. S., Glesener, L., White, S. M., Masuda, S., Wuelser, J. -P., & Lin, R. P. (2010). Measurements of the coronal acceleration region of a solar flare. *The Astrophysical Journal, 714*, 1108–1119. doi: 10.1088/0004-637x/714/2/1108.

Krucker, S., Hudson, H. S., Jeffrey, N. L. S., Battaglia, M., Kontar, E. P., Benz, A. O., et al. (2011). High-resolution imaging of solar flare ribbons and its implication on the thick-target beam model. *The Astrophysical Journal, 739*, 96. doi: 10.1088/0004-637x/739/2/96.

Krucker, S., & Lin, R. P. (2008). Hard X-ray emissions from partially occulted solar flares. *The Astrophysical Journal, 673*, 1181–1187. doi: 10.1086/524010.

Landi, E., Feldman, U., Innes, D. E., & Curdt, W. (2003). Mass motions and plasma properties in the $10^7$ K flare solar corona. *The Astrophysical Journal, 582*, 506–519. doi: 10.1086/344524.

Lapenta, G. (2008). Self-feeding turbulent magnetic reconnection on macroscopic scales. *Physical Review Letters, 100*, 235001. doi: 10.1103/PhysRevLett.100.235001.

Larosa, T. N., & Moore, R. L. (1993). A mechanism for bulk energization in the impulsive phase of solar flares: MHD turbulent cascade. *The Astrophysical Journal, 418*, 912. doi: 10.1086/173448.

Larosa, T. N., Moore, R. L., & Shore, S. N. (1994). A new path for the electron bulk energization in solar flares: Fermi acceleration by magnetohydrodynamic turbulence in reconnection outflows. *The Astrophysical Journal, 425*, 856. doi: 10.1086/174031.

Lazarian, A., & Vishniac, E. T. (1999). Reconnection in a weakly stochastic field. *The Astrophysical Journal, 517*, 700–718. doi: 10.1086/307233.

Lee, K.-S., Imada, S., Watanabe, K., Bamba, Y., & Brooks, D. H. (2017). IRIS, Hinode, SDO, and RHESSI observations of a white light flare produced directly by nonthermal electrons. *The Astrophysical Journal, 836*, 150. doi: 10.3847/1538-4357/aa5b8b.

Lemen, J. R., Title, A. M., Akin, D. J., Boerner, P. F., Chou, C., Drake, J. F., et al. (2012). The Atmospheric Imaging Assembly (AIA) on the Solar Dynamics Observatory (SDO). *Solar Physics, 275*, 17–40. doi: 10.1007/s11207-011-9776-8.

Li, X., Guo, F., Li, H., Stanier, A., & Kilian, P. (2019). Formation of power-law electron energy spectra in three-dimensional low-β magnetic reconnection. *The Astrophysical Journal, 884*, 118. doi: 10.3847/1538-4357/ab4268.

Li, Y., Ding, M. D., Qiu, J., & Cheng, J. X. (2015). Chromospheric evaporation in an X1.0 flare on 2014 March 29 observed with IRIS and EIS. *The Astrophysical Journal, 811*, 7. doi: 10.1088/0004-637x/811/1/7.

Li, Y., Xue, J. C., Ding, M. D., Cheng, X., Su, Y., Feng, L., et al. (2018). Spectroscopic observations of a current sheet in a solar flare. *The Astrophysical Journal, 853*, L15. doi: 10.3847/2041-8213/aaa6c0.

Lin, H., Kuhn, J. R., & Coulter, R. (2004). Coronal Magnetic Field Measurements. *The Astrophysical Journal, 613*, L177–L180. doi: 10.1086/425217.

Lin, H., Penn, M. J., & Tomczyk, S. (2000). A new precise measurement of the coronal magnetic field strength. *The Astrophysical Journal, 541*, L83–L86. doi: 10.1086/312900.

Lin, R. P., Dennis, B. R., Hurford, G. J. , Smith, D. M., Zehnder, A., Harvey, P. R., et al. (2002). The Reuven Ramaty High-Energy Solar Spectroscopic Imager (RHESSI). *Sol. Phys. 210*, 3–32. doi: 10.1023/A:1022428818870

Lin, Y., Engvold, O., Rouppe van der Voort, L. H. M., & van Noort, M. (2007). Evidence of traveling waves in filament threads. *Solar Physics, 246*, 65–72.

Liu, W., Petrosian, V., Dennis, B. R., & Jiang, Y. W. (2008). Double coronal hard and soft X-ray source observed by RHESSI: Evidence for magnetic reconnection and particle acceleration in solar flares. *The Astrophysical Journal, 676*, 704–716. doi: 10.1086/527538.

Liu, Y.-H., Daughton, W., Karimabadi, H., Li, H., & Roytershteyn, V. (2013). Bifurcated structure of the electron diffusion region in three-dimensional magnetic reconnection. *Physical Review Letters, 110*, 265004. doi: 10.1103/PhysRevLett.110.265004.

Long, D. M., Bloomfield, D. S., Chen, P. F., Downs, C., Gallagher, P. T., Kwon, R. -Y., et al. (2017). Understanding the physical nature of coronal "EIT waves". *Solar Physics, 292*, 7. doi: 10.1007/s11207-016-1030-y.

Loureiro, N. F., Samtaney, R., Schekochihin, A. A., & Uzdensky, D. A. (2012). Magnetic reconnection and stochastic plasmoid chains in high-Lundquist-number plasmas. *Physics of Plasmas, 19*, 042303 doi: 10.1063/1.3703318.

Loureiro, N. F., Schekochihin, A. A., & Cowley, S. C. (2007). Instability of current sheets and formation of plasmoid chains. *Physics of Plasmas, 14*, 100703 doi: 10.1063/1.2783986.

Loureiro, N. F., Schekochihin, A. A., & Uzdensky, D. A. (2013). Plasmoid and Kelvin-Helmholtz instabilities in Sweet-Parker current sheets. *Physical Review E, 87*, 013102. doi: 10.1103/PhysRevE.87.013102.

Loureiro, N. F., & Uzdensky, D. A. (2016). Magnetic reconnection: from the Sweet-Parker model to stochastic plasmoid chains. *Plasma Physics and Controlled Fusion, 58*, 014021. doi: 10.1088/0741-3335/58/1/014021.

Loureiro, N. F., Uzdensky, D. A., Schekochihin, A. A., Cowley, S. C., & Yousef, T. A. (2009). Turbulent magnetic reconnection in two dimensions. *Monthly Notices of the Royal Astronomical Society, 399*, L146–L150. doi: 10.1111/j.1745-3933.2009.00742.x.

Loureiro, N. F., & Boldyrev, S. (2017). Role of magnetic reconnection in magnetohydrodynamic turbulence. *Physical Review Letters, 118*, 245101. doi: 10.1103/PhysRevLett.118.245101.

Machado, M. E., Emslie, A. G., & Brown, J. C. (1978). The structure of the temperature minimum region in solar flares and its significance for flare heating mechanisms. *Solar Physics, 58*, 363–387. doi: 10.1007/bf00157282.

Mariska, J. T. (1994). Flare plasma dynamics observed with the YOHKOH Bragg Crystal Spectrometer. II. Properties of the Fe xxv, CA xix, and S XV resonance lines. *The Astrophysical Journal, 434*, 756. doi: 10.1086/174778.

Mariska, J. T., Doschek, G. A., & Bentley, R. D. (1993). Flare plasma dynamics observed with the YOHKOH Bragg Crystal Spectrometer. I. Properties of the CA XIX resonance line. *The Astrophysical Journal, 419*, 418. doi: 10.1086/173494.

Mariska, J. T., Sakao, T., & Bentley, R. D. (1996). Hard and soft X-ray observations of solar limb flares. *The Astrophysical Journal*, *459*, 815. doi: 10.1086/176945.

Martínez Oliveros, J.-C., Hudson, H. S., Hurford, G. J., Krucker, S., Lin, R. P., Lindsey, C., et al. (2012). The height of a white-light flare and its hard X-Ray sources. *The Astrophysical Journal*, *753*, L26. doi: 10.1088/2041-8205/753/2/l26.

Masuda, S., Kosugi, T., Hara, H., Tsuneta, S., & Ogawara, Y. (1994). A loop-top hard X-ray source in a compact solar flare as evidence for magnetic reconnection. *Nature*, *371*, 495–497. doi: 10.1038/371495a0.

Matthaeus, W. H., & Lamkin, S. L. (1986). Turbulent magnetic reconnection. *Physics of Fluids 29*, 2513–2534. doi: 10.1063/1.866004.

McClements, K. G., & Fletcher, L. (2009). Inertial Alfvén wave acceleration of solar flare electrons. *The Astrophysical Journal*, *693*, 1494–1499. doi: 10.1088/0004-637x/693/2/1494.

McKenzie, D. E. (2013). Turbulent dynamics in solar flare sheet structures measured with local correlation tracking. *The Astrophysical Journal*, *766*, 39. doi: 10.1088/0004-637x/766/1/39.

McLaughlin, J. A., Nakariakov, V. M., Dominique, M., Jelínek, P., & Takasao, S. (2018). Modelling quasi-periodic pulsations in solar and stellar flares. *Space Science Reviews*, *214*, 45. doi: 10.1007/s11214-018-0478-5.

Melrose, D. B. (1992). Energy propagation into a flare kernel during a solar flare. *The Astrophysical Journal*, *387*, 403. doi: 10.1086/171092.

Miller, J. A., Cargill, P. J., Gordon Emslie, A., Holman, G. D., Dennis, B. R., LaRosa, T. N., et al. (1997). Critical issues for understanding particle acceleration in impulsive solar flares. *Journal of Geophysical Research*, *102*, 14631–14660. doi: 10.1029/97ja00976.

Milligan, R. O. (2011). Spatially resolved nonthermal line broadening during the impulsive phase of a solar flare. *The Astrophysical Journal*, *740*, 70. doi: 10.1088/0004-637x/740/2/70.

Milligan, R. O., & Dennis, B. R. (2009). Velocity characteristics of evaporated plasma using Hinode/EUV imaging spectrometer. *The Astrophysical Journal*, *699*, 968–975. doi: 10.1088/0004-637x/699/2/968.

Moore, R. L., & Labonte, B. J. (1980). The filament eruption in the 3B flare of July 29, 1973 - Onset and magnetic field configuration. *Solar and Interplanetary Dynamics, January 01*, 1980.

Moore, R. L., Schmieder, B., Hathaway, D. H., & Tarbell, T. D. (1997). 3-D magnetic field configuration late in a large two-ribbon flare. *Solar Physics*, *176*, 153–169. doi: 10.1023/a:1004990817154.

Moreton, G. E., & Ramsey, H. E. (1960). Recent observations of dynamical phenomena associated with solar flares. *Publications of the Astronomical Society of the Pacific*, *72*, 357. doi: 10.1086/127549.

Nakariakov, V. M., & Melnikov, V. F. (2009). Quasi-periodic pulsations in solar flares. *Space Science Reviews*, *149*, 119–151. doi: 10.1007/s11214-009-9536-3.

Nakariakov, V. M., Ofman, L., Deluca, E. E., Roberts, B., & Davila, J. M. (1999). TRACE observation of damped coronal loop oscillations: Implications for coronal heating. *Science*, *285*, 862–864.

Nocera, L., Priest, E. R., & Hollweg, J. V. (1986). Nonlinear development of phase-mixed Alfvén waves. *Geophysical and Astrophysical Fluid Dynamics*, *35*, 111–129. doi: 10.1080/03091928608245889.

Oishi, J. S., Mac Low, M.-M., Collins, D. C., & Tamura, M. (2015). Self-generated turbulence in magnetic reconnection. *The Astrophysical Journal*, *806*, L12. doi: 10.1088/2041-8205/806/1/l12.

Okamoto, T. J., Tsuneta, S., Berger, T. E., Ichimoto, K., Katsukawa, Y., Lites, B. W., et al. (2007). Coronal transverse magnetohydrodynamic waves in a solar prominence. *Science*, *318*, 1577.

Parker, E. N. (1957a). Sweet's mechanism for merging magnetic fields in conducting fluids. *Journal of Geophysical Research*, *62*, 509–520. doi: 10.1029/JZ062i004p00509.

Parker, E. N. (1957b). Acceleration of cosmic rays in solar flares. *Physical Review*, *107*, 830–836. doi: 10.1103/PhysRev.107.830.

Petrie, G. J. D. (2013). A spatio-temporal description of the abrupt changes in the photospheric magnetic and Lorentz-force vectors during the 15 February 2011 X2.2 flare. *Solar Physics*, *287*, 415–440. doi: 10.1007/s11207-012-0071-0.

Petrie, G. J. D., & Sudol, J. J. (2010). Abrupt longitudinal magnetic field changes in flaring active regions. *The Astrophysical Journal*, *724*, 1218-1237. doi: 10.1088/0004-637x/724/2/1218.

Petrosian, V. (2012). Stochastic acceleration by turbulence. *Space Science Reviews*, *173*, 535–556. doi: 10.1007/s11214-012-9900-6.

Petrosian, V., Donaghy, T. Q., & McTiernan, J. M. (2002). Loop top hard X-ray emission in solar flares: Images and statistics. *The Astrophysical Journal*, *569*, 459-473. doi: 10.1086/339240.

Petschek, H. E. (1964). Magnetic field annihilation. In *The Physics of Solar Flares*, Proceedings of the AAS-NASA Symposium held 28–30 October 1963 at the Goddard Space Flight Center, Greenbelt, MD. Edited by Wilmot N. Hess. Washington, DC: National Aeronautics and Space Administration, Science and Technical Information Division, 1964, p. 425.

Polito, V., Dudík, J., Kašparová, J., Dzifčáková, E., Reeves, K. K., Testa, P., & Chen, B. (2018). Broad non-Gaussian Fe XXIV line profiles in the impulsive phase of the 2017 September 10 X8.3-class flare observed by Hinode/EIS." *The Astrophysical Journal, 864*, 63. doi: 10.3847/1538-4357/aad62d.

Polito, V., Reep, J. W., Reeves, K. K., Simões, P. J. A., Dudík, J., Del Zanna, G., et al. (2016). Simultaneous IRIS and Hinode/EIS observations and modelling of the 2014 October 27 X2.0 class flare. *The Astrophysical Journal*, *816*, 89. doi: 10.3847/0004-637x/816/2/89.

Polito, V., Reeves, K. K., Del Zanna, G., Golub, L., & Mason, H. E. (2015). Joint high temperature observation of a small C6.5 solar flare with Iris/Eis/Aia. *The Astrophysical Journal*, *803*, 84. doi: 10.1088/0004-637x/803/2/84.

Polito, V., Testa, P., & De Pontieu, B. (2019). Can the superposition of evaporative flows explain broad Fe XXI profiles during solar flares? *The Astrophysical Journal*, *879*, L17. doi: 10.3847/2041-8213/ab290b.

Rachmeler, L. A., Winebarger, A. R., Savage, S. L., Golub, L., Kobayashi, K., Vigil, G. D., et al. (2019). The High-Resolution Coronal Imager, Flight 2.1. *Solar Physics*, *294*, 174. doi: 10.1007/s11207-019-1551-2.

Ranns, N. D. R., Harra, L. K., Matthews, S. A., & Culhane, J. L. (2001). The timing of non-thermal soft X-ray emission line broadenings in solar flares. *Astronomy and Astrophysics, 379*, 616–621. doi: 10.1051/0004-6361:20011342.

Rast, M. P., González, N. B., Rubio, L. B., Cao, W., Cauzzi, G., Deluca, E., et al. (2021). Critical science plan for the Daniel K. Inouye Solar Telescope (DKIST). *Solar Physics, 296*, 70. doi: 10.1007/s11207-021-01789-2.

Reep, J. W., & Russell, A. J. B. (2016). Alfvénic wave heating of the upper chromosphere in flares. *The Astrophysical Journal, 818*, L20. doi: 10.3847/2041-8205/818/1/l20.

Reep, J. W., Russell, A. J. B., Tarr, L. A., & Leake, J. E. (2018). A hydrodynamic model of alfvenic wave heating in a coronal loop and its chromospheric footpoints. *Astrophysical Journal, 853*, 101 doi: 10.3847/1538-4357/aaa2fe.

Rimmele, T. R., Warner, M., Keil, S. L., Goode, P. R., Knölker, M., Kuhn, J. R., et al. (2020). The Daniel K. Inouye Solar Telescope—Observatory overview. *Solar Physics, 295*, 172. doi: 10.1007/s11207-020-01736-7.

Ruan, W., Xia, C., & Keppens, R. (2020). A fully self-consistent model for solar flares. *The Astrophysical Journal, 896*, 97. doi: 10.3847/1538-4357/ab93db.

Ruan, W., Yan, L., & Keppens, R. (2023). Magnetohydrodynamic turbulence formation in solar flares: 3D simulation and synthetic observations. *The Astrophysical Journal, 947*, 67. doi: 10.3847/1538-4357/ac9b4e.

Russell, A. J. B. (2023). Nonlinear damping and field-aligned flows of propagating shear Alfvén waves with Braginskii viscosity. arXiv:2303.11128. Accessed March 01, 2023. doi:10.48550/arXiv.2303.11128.

Russell, A. J. B., & Fletcher, L. (2013). Propagation of Alfvenic waves from corona to chromosphere and consequences for solar flares. *Astrophysical Journal, 765*, 81 doi: 10.1088/0004-637X/765/2/81.

Russell, A. J. B., Mooney, M. K., Leake, J. E., & Hudson, H. S. (2016). Sunquake generation by coronal magnetic restructuring. *The Astrophysical Journal, 831*, 42. doi: 10.3847/0004-637x/831/1/42.

Russell, A. J. B., & Stackhouse, D. J. (2013). Solar flares and focused energy transport by MHD waves. *Astronomy and Astrophysics, 558*, A76. doi: 10.1051/0004-6361/201321916.

Schettino, G., Poletto, G., & Romoli, M. (2010). CMEs from AR 10365: Morphology and physical parameters of the ejections and of the associated current sheet. *The Astrophysical Journal, 708*, 1135–1144. doi: 10.1088/0004-637x/708/2/1135.

Shen, C., Chen, B., Reeves, K. K., Yu, S., Polito, V., & Xie, X. (2022). The origin of underdense plasma downflows associated with magnetic reconnection in solar flares. *Nature Astronomy, 6*, 317–324. doi: 10.1038/s41550-021-01570-2.

Shen, C., Polito, V., Reeves, K. K., Chen, B., Yu, S., & Xie, X. (2023). Non-thermal broadening of IRIS Fe XXI line caused by turbulent plasma flows in the magnetic reconnection region during solar eruptions. *Frontiers in Astronomy and Space Sciences, 10*, 19. doi: 10.3389/fspas.2023.1096133.

Shepherd, L. S., & Cassak, P. A. (2010). Comparison of secondary islands in collisional reconnection to hall reconnection. *Physical Review Letters, 105*, 015004. doi: 10.1103/PhysRevLett.105.015004.

Shibata, K., & Tanuma, S. (2001). Plasmoid-induced-reconnection and fractal reconnection. *Earth, Planets and Space, 53*, 473-482. doi: 10.1186/bf03353258.

Similon, P. L., & Sudan, R. N. (1989). Energy dissipation of Alfven wave packets deformed by irregular magnetic fields in solar-coronal arches. *The Astrophysical Journal, 336*, 442. doi: 10.1086/167023.

Simões, P. J. A., Graham, D. R., & Fletcher, L. (2015). Impulsive heating of solar flare ribbons above 10 MK. *Solar Physics, 290*, 3573–3591. doi: 10.1007/s11207-015-0709-9.

Spitzer, L. (1962). *Physics of fully ionized gases*, 2nd ed. New York; London: Interscience.

Squire, J., Quataert, E., & Schekochihin, A. A. (2016). A stringent limit on the amplitude of alfvénic perturbations in high-beta low-collisionality plasmas. *The Astrophysical Journal, 830*, L25. doi: 10.3847/2041-8205/830/2/l25.

Stanier, A., Daughton, W., Le, A., Li, X., & Bird, R. (2019). Influence of 3D plasmoid dynamics on the transition from collisional to kinetic reconnection. *Physics of Plasmas, 26*, 072121. doi: 10.1063/1.5100737.

Stores, M., Jeffrey, N. L. S., & Kontar, E. P. (2021). The spatial and temporal variations of turbulence in a solar flare. *The Astrophysical Journal, 923*, 40. doi: 10.3847/1538-4357/ac2c65.

Striani, E., Mignone, A., Vaidya, B., Bodo, G., & Ferrari, A. (2016). MHD simulations of three-dimensional resistive reconnection in a cylindrical plasma column. *Monthly Notices of the Royal Astronomical Society, 462*, 2970–2979. doi: 10.1093/mnras/stw1848.

Sturrock, P. A. (1966). Model of the high-energy phase of solar flares. *Nature, 211*, 695–697. doi: 10.1038/211695a0.

Sudol, J. J., & Harvey, J. W. (2005). Longitudinal magnetic field changes accompanying solar flares. *The Astrophysical Journal, 635*, 647–658. doi: 10.1086/497361.

Sui, L., & Holman, G. D. (2003). Evidence for the formation of a large-scale current sheet in a solar flare. *The Astrophysical Journal, 596*, L251–L254. doi: 10.1086/379343.

Susino, R., Bemporad, A., & Krucker, S. (2013). Plasma heating in a post eruption current sheet: a case study based on ultraviolet, soft, and hard X-ray data. *The Astrophysical Journal, 777*, 93. doi: 10.1088/0004-637x/777/2/93.

Sweet, P. A. (1958). The neutral point theory of solar flares. *Electromagnetic Phenomena in Cosmical Physics, January 01*, 1958.

Tanaka, K., Watanabe, T., Nishi, K., & Akita, K. (1982). High-resolution solar flare X-ray spectra obtained with rotating spectrometers on the HINOTORI satellite. *The Astrophysical Journal, 254*, L59–L63. doi: 10.1086/183756.

Terradas, J., Andries, J., Goossens, M., Arregui, I., Oliver, R., & Ballester, J. L. (2008). Nonlinear instability of kink oscillations due to shear motions. *The Astrophysical Journal, 687*, L115. doi: 10.1086/593203.

Tian, H., Li, G., Reeves, K. K., Raymond, J. C., Guo, F., Liu, W., et al. (2014). Imaging and spectroscopic observations of magnetic reconnection and chromospheric evaporation in a solar flare. *The Astrophysical Journal, 797*, L14. doi: 10.1088/2041-8205/797/2/l14.

Tian, H., Young, P. R., Reeves, K. K., Chen, B., Liu, W., & McKillop, S. (2015). Temporal evolution of chromospheric evaporation: Case Studies of the M1.1 Flare on 2014 September 6 and X1.6 Flare on 2014 September 10. *The Astrophysical Journal, 811*, 139. doi: 10.1088/0004-637x/811/2/139.

Tomczyk, S., McIntosh, S. W., Keil, S. L., Judge, P. G., Schad, T., Seeley, D. H., & Edmondson, J. (2007). Alfvén waves in the solar corona. *Science, 317*, 1192.

Tsuneta, S. (1995). Particle acceleration and magnetic reconnection in solar flares. *Publications of the Astronomical Society of Japan, 47*, 691-697.

Tsuneta, S. (1996). Structure and dynamics of magnetic reconnection in a solar flare. *The Astrophysical Journal, 456*, 840. doi: 10.1086/176701.

Uchida, Y. (1960). On the exciters of type II and type III solar radio bursts. *Publications of the Astronomical Society of Japan, 12*, 376.

Uzdensky, D. A., Loureiro, N. F., & Schekochihin, A. A. (2010). Fast magnetic reconnection in the plasmoid-dominated regime. *Physical Review Letters, 105*, 235002. doi: 10.1103/PhysRevLett.105.235002.

van Ballegooijen, A. A., & Asgari-Targhi. M. (2016). Heating and acceleration of the fast solar wind by Alfvén wave turbulence. *The Astrophysical Journal, 821*, 106.

Van Doorsselaere, T., Nakariakov, V. M., & Verwichte, E. (2008). Detection of waves in the solar corona: Kink or Alfvén? *The Astrophysical Journal, 676*, L73. doi: 10.1086/587029.

Wang, H., Ewell, M. W., Jr., Zirin, H., & Ai, G. (1994). Vector magnetic field changes associated with X-class flares. *The Astrophysical Journal, 424*, 436. doi: 10.1086/173901.

Wang, H., & Liu, C. (2010). Observational evidence of back reaction on the solar surface associated with coronal magnetic restructuring in solar eruptions. *The Astrophysical Journal, 716*, L195–L199. doi: 10.1088/2041-8205/716/2/l195.

Wang, H., & Liu, C. (2015). Structure and evolution of magnetic fields associated with solar eruptions. *Research in Astronomy and Astrophysics, 15*, 145–174. doi: 10.1088/1674-4527/15/2/001.

Warren, H. P., Brooks, D. H., Ugarte-Urra, I., Reep, J. W., Crump, N. A., & Doschek, G. A. (2018). Spectroscopic observations of current sheet formation and evolution. *The Astrophysical Journal, 854*, 122. doi: 10.3847/1538-4357/aaa9b8.

Weinberg, S. (1962). Eikonal method in magnetohydrodynamics. *Physical Review, 126*, 1899–1909. doi: 10.1103/PhysRev.126.1899.

Wheatland, M. S., & Melrose, D. B. (1994). Alfvenic fronts and the turning-off of the energy release in solar flares. *Publications of the Astronomical Society of Australia, 11*, 25–27. doi: 10.1017/s1323358000019615.

Williams, T., Walsh, R. W., Winebarger, A. R., Brooks, D. H., Cirtain, J. W., De Pontieu, B., et al. (2020). Is the high-resolution coronal imager resolving coronal strands? Results from AR 12712. *The Astrophysical Journal, 892*, 134. doi: 10.3847/1538-4357/ab6dcf.

Yang, L., Li, H., Guo, F., Li, X., Li, S., He, J., et al. (2020). Fast magnetic reconnection with turbulence in high lundquist number limit. *The Astrophysical Journal, 901*, L22. doi: 10.3847/2041-8213/abb76b.

Young, P. R., Doschek, G. A., Warren, H. P., & Hara, H. (2013). Properties of a solar flare kernel observed by Hinode and SDO. *The Astrophysical Journal, 766*, 127. doi: 10.1088/0004-637x/766/2/127.

Young, P. R., Tian, H., & Jaeggli, S. (2015). The 2014 March 29 X-flare: Subarcsecond resolution observations of Fe XXI λ1354.1. *The Astrophysical Journal, 799*, 218. doi: 10.1088/0004-637x/799/2/218.

Zhang, Q., Guo, F., Daughton, W., Li, H., & Li, X. (2021). Efficient nonthermal ion and electron acceleration enabled by the flux-rope kink instability in 3D nonrelativistic magnetic reconnection. *Physical Review Letters, 127*, 185101. doi: 10.1103/PhysRevLett.127.185101.

Zharkova, V. V., Arzner, K., Benz, A. O., Browning, P., Dauphin, C., Emslie, A. G., et al. (2011). Recent advances in understanding particle acceleration processes in solar flares. *Space Science Reviews, 159*, 357–420. doi: 10.1007/s11214-011-9803-y.

# 4

# Excitation and Absorption of Alfvén Waves on the Sun and at Earth: A Comparison

**Gerhard Haerendel**

**ABSTRACT**

The comparison of Alfvén waves in magnetosphere and solar atmosphere is carried out for three comparable situations: excitation by external flows, by the reconnection exhaust, and supporting the erosion of stored shear stresses. The three situations are discussed separately for Earth and Sun based on selected recent literature. The subsequent comparisons reveal that there is little direct comparability, neither for the excitation processes, nor for the absorption processes. This is owed to several important differences in density, magnetic mirror ratio, Pedersen conductivity, and uncertainties about the real quantities. Furthermore, the Sun possesses some configurational and dynamical properties without counterpart at the Earth. Nevertheless, the comparisons help identifying some of the shortcomings of present days' concepts and may encourage new routes to pursue. The latter applies in particular to the coupling of spicules and jets to Alfvén waves and to the effects of unbalanced currents along coronal loops.

## 4.1. INTRODUCTION

This chapter is not a review. It is an attempt to compare excitation and absorption of Alfvén waves (AWs) on the Sun and in the magnetosphere. The hope is that this comparison may lead to an appreciation of the peculiarities on either side, raise questions about common solutions, and even stimulate new interpretations.

The plasma environments of Sun and Earth are grossly different, in size, energy input, temperature, pressure, and there are similarities. Both are filled with a collision-free magnetized plasma. Only in the transition regions toward the surfaces, the chromosphere and ionosphere, collisions with neutral particles play a role. The main difference lies in the direction from which the energy is supplied, on the Sun from below, the photosphere, in the magnetosphere from above, the solar wind. However, large-scale instabilities and magnetic eruptions can temporarily store energy high in the corona. Then energy is also supplied from above. This paper is concentrating on Alfvén waves. They are present everywhere in both realms and of great importance in transporting forces and energy across large distances. The most important aspects are, therefore, the beginning and end of this transport, that is, their excitation and absorption. Losses may occur already on their way. The main losses, however, occur, in case of the magnetosphere, in the ionosphere, whereas, in the solar case, Alfvén waves, being excited from above and below, deposit their energy in the chromosphere and corona, respectively. They also are likely to power the solar wind. Another commonality exists in that many energy deposition processes have luminous or radio-frequency signatures. They are a key means for exploring appearance and nature of the Alfvén waves and on the Sun the only ones.

AWs are ubiquitous and the literature about them is rich, for both realms. Therefore, I narrow down the subject by considering just three types of occurrences, small-scale kinetic Alfvén waves (KAWs), large-scale AWs

*Max Planck Institute for Extraterrestrial Physics, Garching, Germany*

*Alfvén Waves Across Heliophysics: Progress, Challenges, and Opportunities, Geophysical Monograph 285*, First Edition.
Edited by Andreas Keiling.

DOI: 10.1002/9781394195985.ch04

or field-aligned currents generated by impulsive events, and AWs spontaneously extracting stored magnetic energy. In the magnetosphere, the first two classes essentially result from external motions, directly or indirectly driven by the solar wind. On the Sun, "external" means an origin either from chromospheric flows or jets, the energy coming from the photosphere, or from a preceding powerful configurational change, a magnetohydrodynamic (MHD) instability. In the magnetosphere, energy is stored by internal convection in the evening auroral oval. As line-tying in chromosphere and photosphere prevents such type of convection on the sun, I choose the shear field of erupting pre-flare loops as counterpart for stored magnetic energy. In the following two sections, I discuss the state of knowledge and understanding of such selected classes of AWs on Earth and on Sun, the excitation processes, and, directly following, the ways of energy conversion and absorption. Section 4.4 is devoted to the comparison.

## 4.2. ALFVÉN WAVES IN THE MAGNETOSPHERE

### 4.2.1. Kinetic Alfvén Waves Excited by External Flows

#### *Excitation Processes*

"The kinetic Alfvén wave was introduced into space plasma physics by Hasegawa (1976) ... to include the effects of finite electron pressure and finite ion gyroradius on the ideal MHD shear Alfvén wave, $\omega = kV_A$. He suggested that mode conversion from a surface wave on the plasma sheet boundary could excite the kinetic Alfvén wave and lead to particle acceleration on auroral field lines." With these sentences, Lysak and Lotko (1996) begin their paper on the dispersion relation for shear Alfvén waves and continue citing Goertz and Boswell (1979) who realized that just above the auroral ionosphere the plasma was cold enough so that $\beta \ll m_e/m_i$, with $\beta = 8\pi n T_e/B^2$. In this case, the dispersion relation assumed a form different from that of a hot plasma and was called "inertial" Alfvén wave. While electron pressure and finite ion gyroradius increase the parallel phase velocity of the KAW, electron inertia leads to a decrease. Lysak and Lotko (1996) expanded the dispersion relation to include the full kinetic effects of both electrons and ions. These theoretical works laid the foundation for the investigation of the roles of KAWs in the magnetosphere, the attraction being the possibility of parallel electron and transverse ion acceleration.

Groundbreaking observations on these effects came from the Freja mission (André et al., 1998; Erlandson et al., 1994; Lundin et al., 1994; Wahlund et al., 1994). They revealed the connection of these waves with density cavities with scales as low as the electron skin depth, transverse ion acceleration, strongly field-parallel electron distributions with diffuse energy spectra below 1 keV, and strong low-frequency electric and magnetic wave fields. The latter ranged between 1 and 20 Hz and between 100 and 256 Hz, that is, below the O+ and between the He+ and proton gyrofrequencies, respectively. As already proposed by Hasegawa (1976), the KAWs appear to be excited mainly at boundaries, both at nightside and dayside. This was confirmed by measurements with the FAST and Polar missions (Chaston et al., 2003; Keiling et al., 2000; Wygant et al., 2000). Stasiewicz et al. (2000) and Keiling (2009) presented two excellent reviews. Recent three-dimensional simulations of KAWs by Cheng et al. (2020) trace the waves from their generation in the tail to the ionosphere (see the next subsection). Mende et al. (2003) found strong fluxes of field-parallel electron fluxes immediately poleward of a breakup arc of a substorm onset, which clearly suggested, by flux level and energy distribution, that they originated from accelerated cold topside ionospheric plasma. Chaston et al. (2006, 2007) extended the investigations on the relation between KAWs and density cavities by proposing that downward directed waves were focused toward the boundaries of the density depletions (Figure 4.1). Haerendel and Frey (2014) studied the auroral products of these KAWs, the Alfvénic arcs, at the poleward border of the expanding auroral oval during a substorm. They related them to flows entering from the polar cap, which preconditions the entry of plasma into and agglomeration of magnetic flux onto the magnetosphere. With Cluster and IMAGE measurements, Hull et al. (2016) found what they called an Alfvénic surge and interpreted it as a cascade process of large-scale AWs.

An important property of the KAWs emerging from the above is that they are excited at boundaries, not only along the nighttime magnetospheric border during a substorm, but also along the dayside magnetopause. The best statistically convincing support comes from satellite measurements of the Defense Meteorological Satellite Program (DMSP) (Newell, 2000). What he names broadband aurora energy and number fluxes match best with the signature of Alfvénic aurora. His Figures 5 and 6, showing the spatial distributions, separately for summer and winter and separately for energy and number flux, clearly confirm this association with midday and midnight boundaries. Indeed, Chaston et al. (2005) observed kinetic Alfvén waves propagating out of a reconnection site at the magnetopause and carrying a Poynting flux directed toward the dayside aurora. However, the abovementioned statistics mean that excitation of KAWs results predominantly from normal incidence of particle

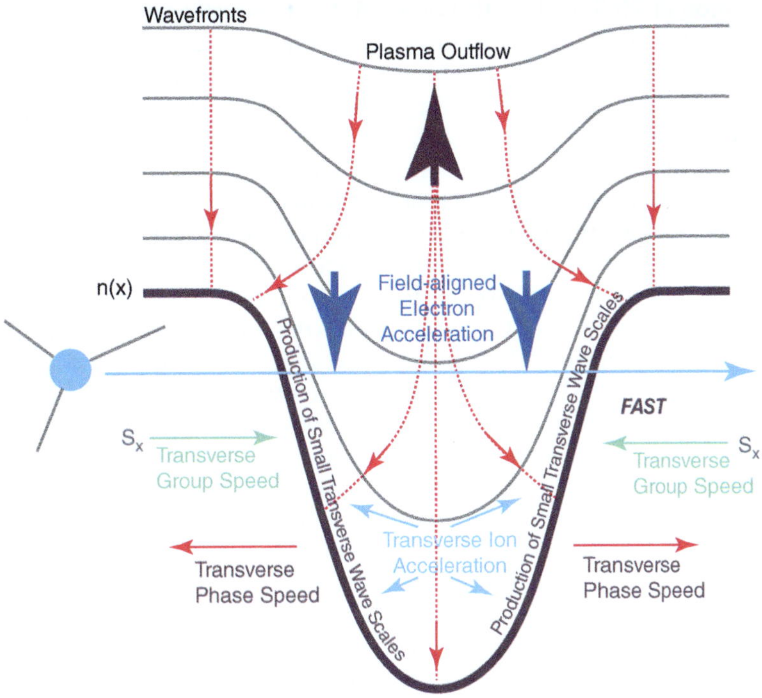


**Figure 4.1** Focusing of a shear Alfvén wave on a density cavity. Refraction on the cavity walls leads to phase mixing and production of smaller scales, ultimately to dissipation with transverse ion and parallel electron acceleration, thus steepening the walls. Source: Adapted from Chaston et al. (2006).

fluxes on some kind of magnetic boundary. Haerendel (2015a) presented a mechanism fitting with this conclusion. From the appearance of thin auroral arcs with erratic small-scale structures leading the arrival of flow bursts at the outer magnetospheric boundary during a substorm, he deduced the formation of a "stop layer" by the impact of high-beta plasma on a magnetic obstacle (Figure 4.2). A thickness of the order of $c/\omega_{pi}$ or proton gyroradius leads to separate entries of ions and electrons into the layer; the ions directly and the electrons by a self-excited diffusion process. Charge separation maintains an electric field that is retarding and stopping the ions and driving a Hall current. Balancing the densities of the electrons drifting along the layer with the normally entering ions leads to azimuthal structures also of the order of $c/\omega_{pi}$. These structures are the sources of KAWs, which generate the said auroral arcs of about 1-min duration.

At this point, I must clarify the origin of the high-beta plasma. Haerendel (2015b) has identified it with the leading part of the so-called dipolarization events. This name characterizes the rear part of the flow bursts, which, according to the studies with the THEMIS spacecraft by Runov et al. (2009), are dominated by a hair-needle-like magnetic field with a strong vertical component. What has received little attention so far is the high-beta leading edge of the flow bursts. Here, flow velocity

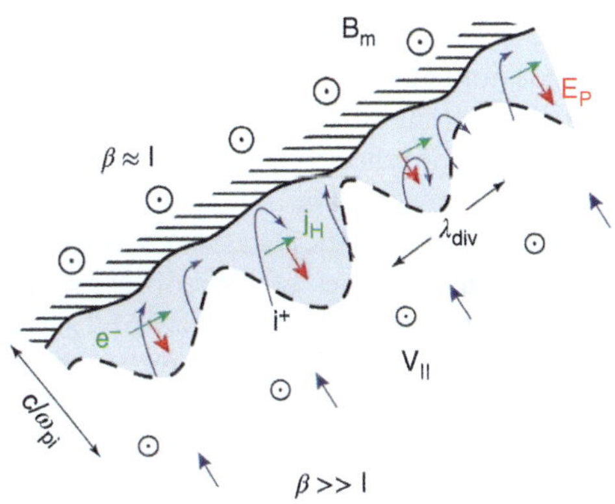


**Figure 4.2** High-beta plasma impinging on a magnetic boundary proceeds by direct entry of the ions and diffusive entry of the electrons. A normal electric field stops the ion, whereas the electrons perform a Hall drift. The matching of normal ion inflow and longitudinal electron drift enforces longitudinal structuring of order $c/\omega_{pi}$. Owing to the divergence of the Hall current, the structures are sources of kinetic Alfvén waves propagating the deposited kinetic energy along the magnetic field. Source: Adapted from Haerendel (2015a).

dominates over thermal motions of the ions (cf. Figures 5 and 6 of Runov et al. (2011)). When impinging on the magnetospheric border field, a "stop layer" can form. An alternate explanation, though not theoretically underpinned, is the suggestion of Hull et al. (2016) of a cascade process of a large-scale AW leading to a kinetic Alfvénic surge.

### *Absorption or Energy Deposition*

The question of where and how these waves are absorbed, finds an obvious answer in their ability to erode topside ionospheric plasma by transverse ion heating (André et al., 1998; Chaston et al., 2006; Erlandson et al., 1994; Lundin et al., 1994; Wahlund et al., 1994) and in the accompanying auroral signatures. Most of the energy transported by the KAWs is dumped into the topside ionosphere. This is clearly demonstrated by the observation of Mende et al. (2003) that the suprathermal electron flux within the Alfvénic arc exceeds by far the primary electron flux, which carries the high-altitude field-aligned current.

While the evidence for energy dumping by conversion into kinetic energy of accelerated ions and electrons is strong, there is also the possibility of electron acceleration during the propagation of the wave from the high-altitude generation site to the topside ionosphere owing to the parallel electric field component, as already proposed by Hasegawa (1976). Many observations from sounding rocket experiments and satellites have confirmed the working of this "Alfvénic acceleration" process, which is also capable of producing auroral structures (references in (Lysak et al., 2020)). According to Kletzing (1994), the acceleration is limited to twice the Alfvén speed. However, depending on the magnitude of the Alfvén speed, it can reach higher value, if reflections from the converging field are included (Artemyev et al., 2015).

An interesting discrepancy has been found to exist between the Poynting fluxes of AWs observed at 4–7 $R_E$ and the fluxes just above or below the auroral acceleration region, the first exceeding the latter by about an order of magnitude (Keiling et al., 2002; Wygant et al., 2000). An obvious solution of this discrepancy may be the spreading of the cross section of the wave by braking into smaller-scale waves, that is, KAWs. Haerendel and Frey (2014) discuss such a process and relate the scale-braking to trapping and multiple reflections in the ionospheric Alfvén resonator. This idea is close to the Alfvénic surge of Hull et al. (2016).

In conclusion, kinetic Alfvén waves are preferentially generated at magnetic boundaries, either directly by the stop-layer process or by a scale-braking process of larger-scale waves on their way toward the topside ionosphere, where most of their energy flux is converted into kinetic energy of ions and electrons. Precipitation of the electrons on the lower ionosphere generates various auroral signatures, whereas energized ions propagate upward and modify the composition of the Earth's plasma environment.

## 4.2.2. Generation of Alfvén Waves and Current Systems by Reconnection

### *Excitation Processes*

The most powerful liberation process of stored magnetic energy is the reconnection process. In the magnetosphere, there are two preferred locations, the dayside magnetopause and the inner edge of the tail on the nightside. While the main effect on the dayside is the interconnection with the Earth's magnetic field and the subsequent buildup of the magnetic tail and accompanying storage of magnetic energy, there is also a substantial amount of direct mass and energy inflow, as manifested by dayside and polar cap aurora. As discussed in the previous section, the impact of the solar wind plasma on the dayside magnetopause also generates kinetic Alfvén waves (Chaston et al., 2003; Keiling et al., 2000; Newell et al., 2010; Wygant et al., 2002). In this section, we will concentrate on the nightside, where the main input of energy into the magnetosphere takes place during substorms and magnetic storms. How this occurs has been a subject of intense investigations and controversies from the early 1990s until today. However, the initial uncertainty about where the liberation of energy stored in the tail is taking place has essentially given way to the common understanding that reconnection in the near-Earth tail around about 20 $R_E$ is the main process. Still existing divergences in understanding pertain to the way how the energy is entering the magnetosphere.

The kinetic Alfvén waves generated by the leading edge of the arriving flow bursts, as discussed in the previous section, deposit only a minor part of the energy carried by the flow burst. However, there is another, more direct source of kinetic Alfvén waves, the diffusion region of the reconnection process. This was discovered by the Cluster mission (Chaston et al., 2009). These waves radiate from the ion diffusion region in a cone-like pattern along its edges. Inside the diffusion region, the fluctuating fields constitute a wave resistivity against the inflow of the magnetic field. The related dissipation is of the same order as the energy flux of the ion jets. Cheng et al. (2020) have used a three-dimensional code to calculate the generation, dynamics, and global structure of KAWs in the reconnection region in the magnetotail and their transport along the magnetic field lines toward the ionosphere. They showed that, because of reflection at the inner boundary of the tail, only 1/14 of the Alfvénic wave energy from the diffusion regions penetrates into the dipolar magnetic field. In addition, the authors find a generation of KAWs at the boundary by "ion kinetic effects (Hall effects)," which is nothing else than the stop-layer formation discussed in the previous section.

The major part of the energy flux from reconnection in the tail resides in the ion flux and the dipolarized magnetic field of the flow bursts (Runov et al., 2009). How this energy is converted and delivered to the magnetosphere is the essence of a theory of the substorm. Braking and stopping the flow burst at the outer border of the near-dipolar magnetosphere involve generation and absorption of larger-scale Alfvén waves. Although this phase lasts only for a few minutes, it is conveniently described by the generation of a pair of field-aligned currents closing through the ionosphere. McPherron et al. (1973) had presented observations suggesting the existence of such a large-scale current system feeding the auroral electrojet (AEJ) during a substorm, the so-called substorm current wedge (SCW), with downward field-aligned currents in the morning sector and upward currents in the evening sector. It matched the characteristics of the large-scale Type I current system of Boström (1964).

The question, where the generator of the SCW was located, has occupied the attention of many researchers and led to various speculative concepts since 1990. Theories ranged from current diversion or disruption (Lui, 1990), ballooning instability (Roux et al., 1991) at the inner edge of the tail, to the braking flow bursts launched by reconnection in the near-Earth tail near 20 $R_E$ (Birn & Hesse, 1991; Birn et al., 1999; Haerendel, 1992; Shiokawa et al., 1997). However, it turned out that the flow braking could amount to at best 20% of the SCW current. Nevertheless, subsequent numerical modeling by Birn et al. (2004) and Birn and Hesse (2013) of the diversion and reflection of flow bursts, understood as low-entropy content magnetic bubbles, appeared to be the best available generation process of the field-aligned currents, downward in the morning sector and upward in the evening sector. Figure 4.3 shows, on the right, the diversion and reflection of the flow and, on the left, the field-aligned currents due to the vortexes generated. Haerendel (2009, 2015b) proposed that magnetic squeezing of hot plasma after arrival at the inner edge of the tail, "dying flow bursts," was the dominant force and energy source. Lately, the author addressed again the topic of the SCW (Haerendel, 2022). For instance, the theoretically derived energy fluxes from this process (Haerendel, 2015b) suffice to cover the observed total energy by auroral electrons in the night sector (Newell et al., 2010). He identified the generator current as a magnetic gradient current at the interface of the agglomerating magnetic field with the compressed plasma. Figure 4.4 shows a cartoon of this concept. The field-aligned currents emerging from the eastward flowing current are an expression of the magnetic shear stresses transporting forces and energy toward the ionosphere.

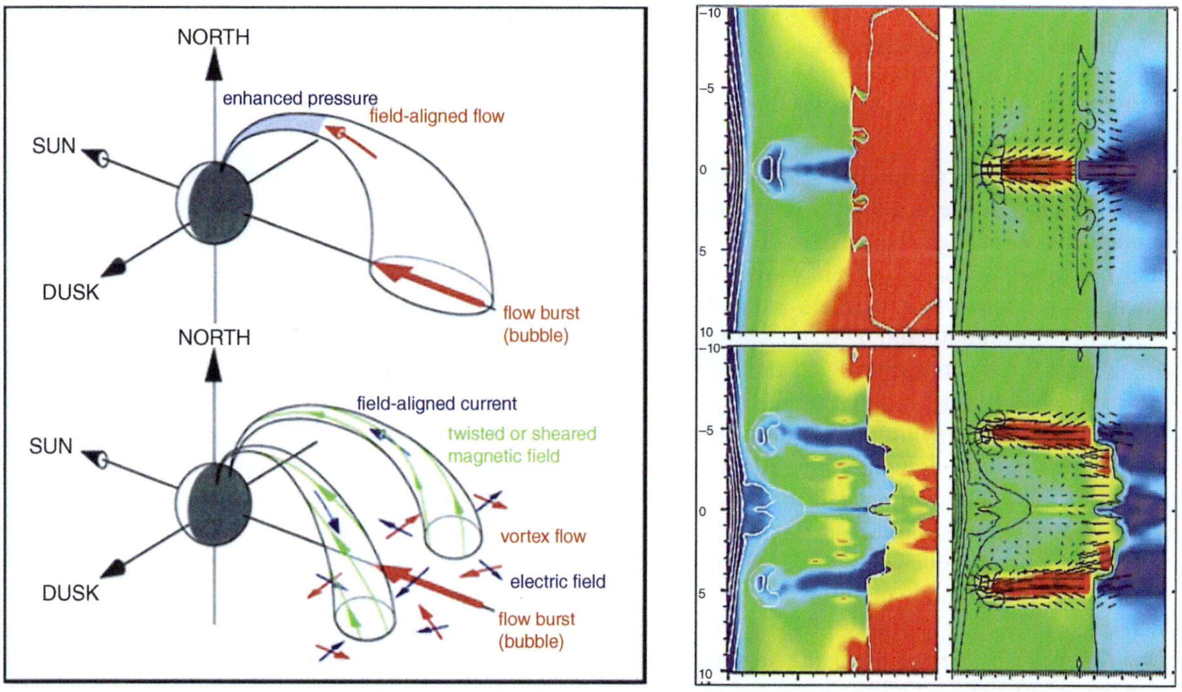


**Figure 4.3** (Right) A flow burst simulated as low-entropy flux tube arriving at the outer boundary of the near-dipolar magnetosphere is diverted by undergoing a vortex flow twisting the magnetic field. Source: From Birn et al. (2011). (Left) It generates a pair of opposing field-aligned currents, which close in the ionosphere. Source: Adapted from Birn et al. (2004).

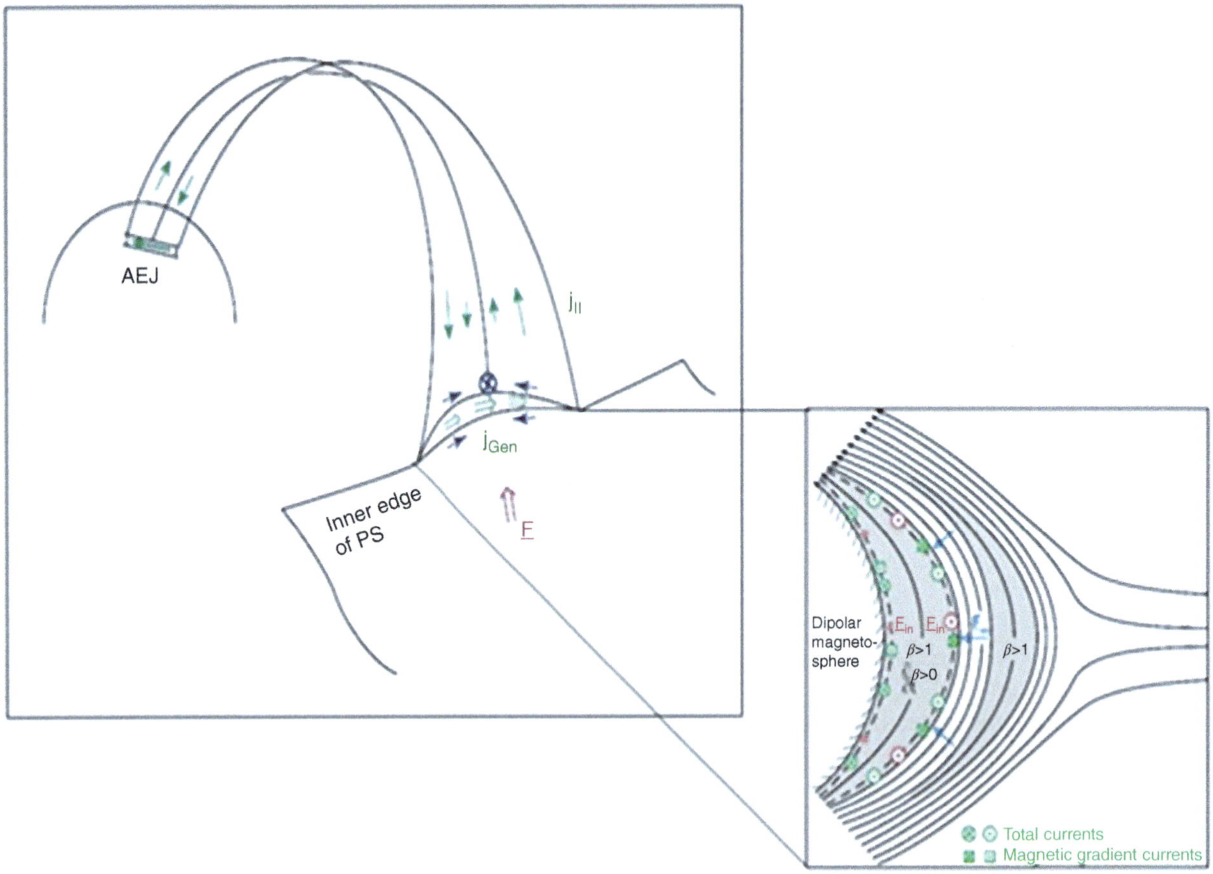


**Figure 4.4** "Dying" flow bursts stopped at the outer boundary of the near-dipolar magnetosphere. They exert a force on this boundary pushing it inward and drive an eastward magnetic gradient current, which diverges into a pair of field-aligned currents closing through the auroral electrojet (AEJ). Source: From Haerendel (2015a, 2022).

Now we have arrived at our key subject, the Alfvén waves. The normal force exerted by the dying flow bursts on the magnetospheric field has a short-term history of several propagation times of Alfvén waves toward the ionosphere. This means that the field-parallel currents form a sheet of downgoing and reflected waves. The fact that the driving force acts normally to the Earth's field leads to azimuthal structuring of the closing electrojet current and the accompanying auroral arcs. In total, we have a complex and strongly varying system of waves, the description of which by quasi-stationary field-aligned currents grossly simplifies the real situation. As the substorm is caused by a sequence of successively arriving flow bursts of a few minutes duration, all acting in the same sense as emitters of Alfvén waves, layer by layer is added to the outer border of the magnetosphere. Thus, closure currents, the AEJ, and their auroral manifestations expand poleward. This is traditionally designated as auroral breakup or poleward expansion. In short, one can say: The energy, liberated by reconnection in the tail, is carried by flow bursts to the outer border of the magnetosphere and there converted into magnetic energy and transferred to the ionosphere by Alfvén waves or magnetic shear stresses. The energy contained in the shear stresses can be released in two ways, by magnetic fractures and by driving dissipative ionospheric currents.

### *Absorption Processes*

The excited Alfvén waves and established field-aligned currents transport force and energy by the accompanying magnetic shear stresses. The latter are pointing eastward on the upward current leg and westward on the eastward leg of the SCW. Owing to the mirror force of the strongly converging magnetic field, high current densities can become sufficiently dense to suffer from a dearth of current carriers (Knight, 1973). The continuity of the current is maintained by electron acceleration in field-parallel potential drops. At the same time, the connectivity of the field lines is broken, and the magnetic shear stresses are released. It is the release of the shear stresses what provides the energy to be converted into kinetic energy of auroral electrons and ions.

This proceeds by Alfvén waves transporting the released magnetic energy as Poynting flux along the field lines into the auroral acceleration region, thousands of kilometers above the ionosphere. Precipitation of the accelerated particles enhances the ionospheric conductivity and creates the auroral light. While this energy influx originates from the released shear stresses, there are also normal forces acting. They drive the longitudinal AEJ, and the associated Poynting flux contributes to the total energy dumping via the resistive losses of the auroral electrojet. This involves the generation of a westward electric field and poleward Hall current. Balancing the normal Hall and Pedersen currents by the Cowling conductivity controls the westward electric field at a low value. This is equivalent to a blockage of the equatorward drift. This blockage acts back on the primary driving force in augmenting the strength of the current circuit. In all this complex interplay of forces and currents, Alfvén waves play the mediating role. Unfortunately, a detailed description of this interplay appears to escape present days' theoretical and numerical tools. So far, we have to cope with the simplified concepts described in the earlier text.

### *Wave Excitation and Erosion of Stored Magnetic Energy*

Flows or convection of magnetospheric plasma, driven by pressure forces against the friction with the collisional ionosphere, store energy in the sheared magnetic field which transmits the driving forces from above. For instance, the evening auroral oval during substorms is essentially a large magnetic energy storage in form of a westward magnetospheric convection. It consists of a broad current system with upward currents extending over a few degrees in latitude and several hours in local time, balanced by downward currents located in the equatorward part of the auroral oval. Longitudinal pressure forces exerted by the hot plasma intruded during a substorm are the driver of this huge current system of Type II (Boström, 1964). The generator currents are those, which constitute the Lorentz force balancing the pressure gradient.

In this system, auroral arcs appear spontaneously, for instance, from the poleward edge of the auroral oval (Haerendel et al., 2012), and propagate slowly equatorward. They extract energy out of the gross current circuit in addition to the energy deposited by ohmic losses of the ionospheric closure (Pedersen) currents (Haerendel, 1994, 2007). This takes place along the upward current sheet intersecting the generated auroral arc by releasing the corresponding shear stresses. What initiates the process is the appearance of a field-parallel potential drop, which decouples the upper part of the flux tube from the lower part, that is, it breaks the magnetic connection to the ionosphere. Because of the resemblance with the fracture of an elastic medium, the author named this process "magnetic fracture" (Haerendel, 1980). The thus generated perturbation travels as Alfvén wave up and down the current sheet triggering the transformation of shear stresses into tangential plasma motions. The energy contained therein is flowing as Poynting flux into the fracture zone/acceleration region, while the accelerated electrons image them onto the auroral rays or folds. The whole energy extraction and conversion process requires a few Alfvénic transit times between generator and ionosphere. Instead of dying out, the process propagates into the interior of the gross current circuit, where the magnetic shear stresses are stored. The current sheet thus consists of two oblique quasi-stationary Alfvén waves attached to the fracture zone propagating slowly into the higher sheared field (Figure 4.5). The viability of this concept has been verified quantitatively by a combined set of space and ground observations (Haerendel et al., 2012). The author presented a detailed analysis of the complex pattern of Alfvén waves emitted and reflected waves above a propagating arc (Haerendel, 2021).

Alfvén waves dominate the here described process in three ways. First, energy input into the magnetosphere sets up pressure forces and motions in the hot high-altitude plasma. As described in the earlier text, they are propagated toward the ionosphere by Alfvén waves. After settlement of the initial perturbations, which can also trigger global magnetic oscillations, a system of quasi-stationary field-aligned currents is established closing through the conducting lower ionosphere. This is the setup of the broad current system or energy storage described in the earlier text. Second, Alfvén waves carry the energy flux from the erosion front into the acceleration region of the auroral particles, the fracture zone. Third, the forces driving the magnetospheric convection are by no means constant. Short-term changes initiate Alfvén waves adjusting the enhanced energy input. This appears as sudden brightenings and unfoldings of the auroral arcs.

### *Absorption Processes*

The intimate connection of triggering Alfvén waves and energy conversion does not allow a separate treatment of the absorption of the waves. Indeed, the waves emitted from the propagating front of the current sheet shed their energy content into the fracture zone/auroral acceleration region. However, the attached field-aligned currents are not interrupted. Where they meet the ionosphere, the accelerated electrons enhance the ionization and electrical conductivity and thereby dump the major part of the released energy. In addition, local heating of ions and production of secondary electrons take place

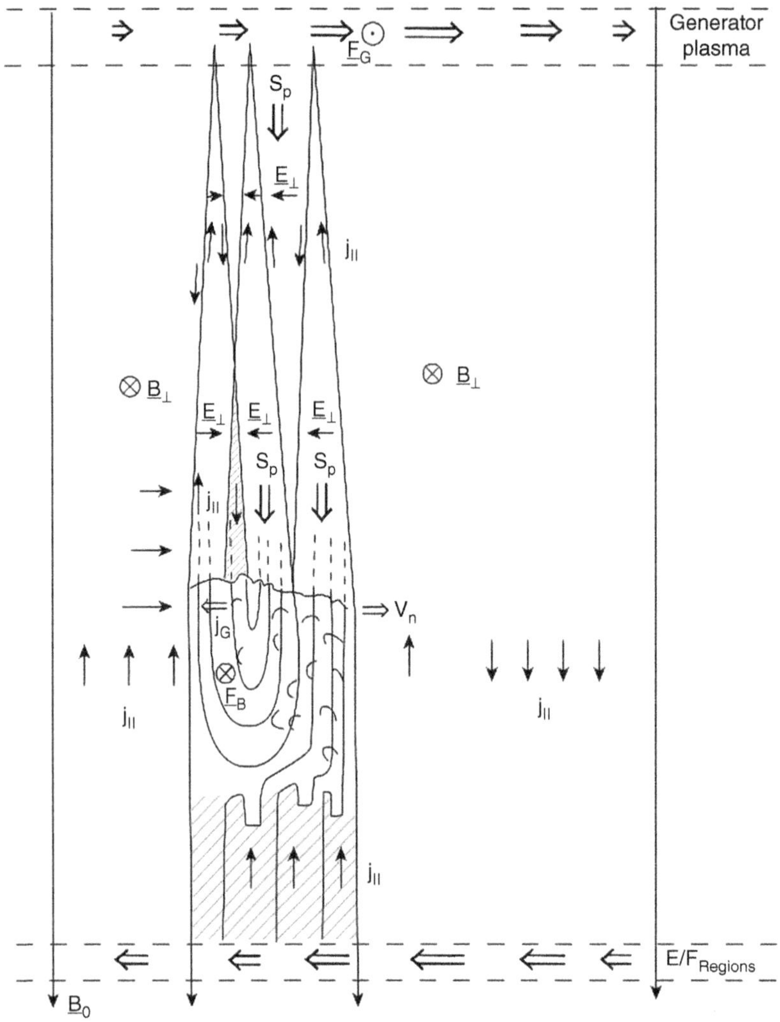


Figure 4.5 Slightly oblique Alfvén wave fronts attached to the auroral acceleration region or "fracture zone" in a straight field lines geometry. The upward currents are embedded in a large current circuit between generator region (outer magnetosphere) and the E and F regions of the ionosphere. The U-shaped potential of the acceleration region is structured by a 2D turbulence. Below it, potential fingers extend toward the ionosphere. The wave field contains downward Poynting flux ($S_p$) from the magnetic stress release in the front part, feedback from the generator in the center, and a stress restoring wave with balanced currents and upward Poynting flux in the rear. Source: Adapted from Haerendel (2021).

in the acceleration region. A third contribution comes from the closure current through the generated auroral arc and the related ohmic dissipation. These processes are well known and need no elaboration here.

The advantage of the magnetospheric arena is the fact that absorption processes of Alfvén waves manifest themselves largely by luminous auroral structures/arcs. This is not so easy on the Sun, where the contribution of Alfvén waves to the observable emissions is often hard to distinguish from other local energy conversion processes or is altogether invisible, as in the solar wind acceleration. Looking at the solar phenomena from the magnetosphere may bring in new questions to ask.

## 4.3. ALFVÉN WAVES IN THE SOLAR ATMOSPHERE

The solar atmosphere is steadily filled with various kinds of dynamic magnetic interaction and energy exchange processes leading to thermal emissions or high-energy radiation processes. Alfvén waves, although undoubtedly abundant in transporting energy and momentum, have optical signatures that are essentially hidden behind the dominating radiation. Only with the advance of spectral, angular, and temporal resolution techniques by modern ground-based and space-borne observatories has one begun to filter out signatures potentially indicative of Alfvén waves. Thus, solar Alfvén waves are very much a subject of the 21st century. However, already much earlier have theorists taken note of the role of Alfvén waves in the solar atmosphere and solar wind (Alfvén, 1947; Biermann, 1948; Schwarzschild, 1948). An overview of the state of art, at the beginning of the present century, was given by Cranmer and van Ballegooijen (2005). I will rather concentrate on the presently available indications of the excitation and absorption of AWs under the selection criteria defined in Section 4.1.

### *Excitation by External Flows or Jets*

All energy of the solar atmosphere is supplied through the photosphere, mostly via the magnetic field, but also by flows or jets excited by the convection and acoustic oscillations of the photosphere. Tracing the transport from the photosphere into the corona has been made possible by high spatially and temporally resolved extreme ultraviolet (EUV) observations of the intermediately positioned transition region (TR). This is owed to, first, the Transition Region and Coronal Explorer (TRACE) (De Pontieu et al., 2003) and subsequently to the Interface Region Imaging Spectrograph (IRIS) (De Pontieu et al., 2021), often supported by advanced instrumentation on ground optical telescopes. The transition region, by temperature between chromosphere and corona, manifests itself through so-called moss, highly structured and dynamic emissions formed above active regions and flares. Being formed at about 1 MK in a height range of 3–5 Mm, the mottled emissions with dark inclusions are structured by scales of 1–3 Mm. They are conductively heated in the legs of hot coronal loops (Fletcher & De Pontieu, 1999). In between appear cool ($10^4$ K) chromospheric plasma jets, which absorb EUV emissions of the moss causing rapid changes of its brightness (De Pontieu & Erdélyi, 2006). These jets appear as fibrils, mottles, or spicules in high-resolution chromospheric images taken from the ground in the wings of Hα. It is these cool plasma jets that I will focus on in the context of excitation of Alfvén waves by external flows.

A most important discovery by De Pontieu et al. (2004) was that chromospheric spicules result from the leakage of energy from photospheric oscillations and flows through the temperature minimum. This insight was gained by comparing observations with modeling, showing that p-mode acoustic oscillations with periods of 200–450 s can overcome the evanescence in the temperature minimum by tunneling along inclined magnetic flux tubes. The cool jets attain speeds of 20 km/s and carry a mass flux 100 times what is needed for feeding the solar wind. This path of energy transport can lead to shock formation or otherwise drive the upward mass flows as spicules or jets. Observations of the spicules reveal substantial transverse motions of the order of 10–25 km/s. Based on model calculations, the authors interpreted these motions as signs for the ubiquitous presence of Alfvén waves (De Pontieu et al., 2007).

The search for signatures of Alfvén waves is a common topic of many publications during the last two decades, the reason being that they must be the carrier of energy needed to heat the corona and drive the solar wind (Mathioudakis et al., 2013). The essential question for our context is whether and how the thus transported energy and momentum can couple to Alfvén waves and thereby carry part of the energy into corona and possibly the solar wind. While in the magnetosphere high-beta ion flows couple their energy to kinetic Alfvén waves by being stopped by transverse magnetic fields, the solar situation does not offer such a direct coupling process. The great question is then how much of the energy flux carried by the spicules is actually transported further into the corona.

From ground observations, Tomczyk et al. (2007) detected upward propagating waves by line-of-sight and intensity variations indicating phase velocities of 1–4 Mm/s in the corona, well above the transition region. Interestingly, the Fourier spectrum of the Doppler velocities suggested a substantial contribution from the p-mode oscillations (Figure 4.6). The amplitudes were, however, too low for carrying energy fluxes sufficient for coronal heating. A few years later, better-resolved observations with the Atmospheric Imaging Assembly (AIA) of the Solar Dynamics Observatory (SDO) by McIntosh et al (2011) revealed the ubiquitous presence of Alfvénic motions in the transition region and corona with much higher amplitudes, sufficient to accelerate the fast solar wind and heat the quiet corona (Figure 4.7). The periods ranging between 100 and 500 s again confirm an important role of the p-mode oscillations.

Three phenomena characterize the transition region: The 1Mio degree moss, dark absorption features of spicules or jets, and the presence of Alfvénic motions.

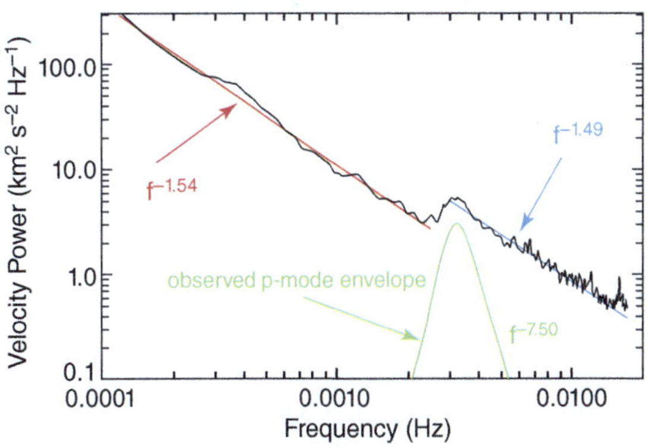


**Figure 4.6** Fourier spectrum of the Doppler velocities in the corona fitted by power laws and the p-mode spectrum (green line). Source: Tomcyk et al. (2007) / with permission of AAAS.

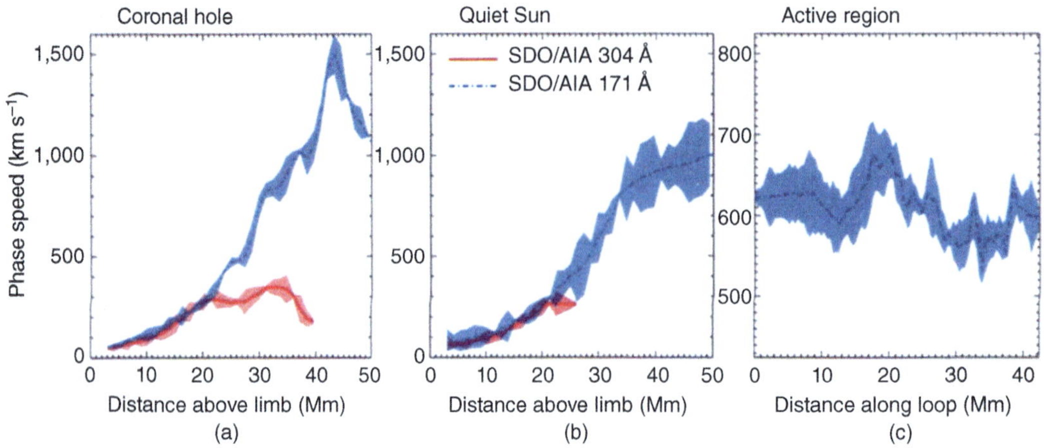


**Figure 4.7** Phase speeds Alfvénic motions derived from cross-correlation of space–time plots for (a) coronal holes, (b) the quiet Sun, and (c) a coronal loop in an active region measured with the SDO/AIA in two channels. Source: Adapted from McIntosh et al. (2011).

There must be an intimate relation. Both spicules and Alfvén waves are energy carriers. The question is whether the moss is steadily heated from above by heat conduction or impulsively by partial conversion of kinetic energy arriving from below (Tripathi et al., 2012). The rapid changes in the moss brightness on 1-min scales, caused by the dark inclusions from the jets or spicules (De Pontieu et al., 2003), may be taken as an indication for the latter. The fact that the periods of coronal Alfvén waves lie in the range of p-mode oscillations suggests a strong contribution of the spicules and jets to their excitation. Other than in the magnetosphere, where high-beta flows interact with transverse magnetic fields, the rising plasma in chromosphere and transition region needs an intermediate process for coupling energy to Alfvén waves. This process is likely to involve also some degree of dissipation. One way may be through resonant absorption of transverse oscillations in combination with Kelvin–Helmholtz instabilities (Figure 4.8) (Antolin et al., 2015, 2018; Cranmer & van Ballegooijen, 2005; Goossens et al., 2020; Van Doorsselaere et al., 2020). A related process is mode conversion involving a change from a specific wave mode into another, for instance, from the fast mode to transverse Alfvén waves (De Pontieu et al., 2021). All of these processes are rather speculative at the present state of research. Furthermore, although they may carry enough energy for heating the corona, they are likely to have too large scales and too low frequencies for being absorbed in the corona.

### *Dissipation of Flow-Excited Alfvén Waves*

While coronal heating by nanoflares, that is, stress-induced reconnection by braiding of coronal field lines, was for a long time the dominating concept since Parker's seminal paper (1972), the more recent literature addresses concepts that at least the quiet corona be heated by the dissipation of Alfvén waves emerging from below through the transition region. This conviction rests on the finding that the related energy flux clearly suffices to cover the radiative losses (approximately 100 W/m$^2$) (De Pontieu et al., 2007; McIntosh et al., 2011). Van Doorsselaere et al. (2020) presented an extensive review of proposed coronal heating processes. However, no viable process has yet been identified that could convert the energy of waves with periods in the range of 100–500 s and transverse scales of the order of 1 Mm effectively into heat, in spite of the effort of generations of researchers (see references in McIntosh et al., 2011). Looking at the magnetospheric situation, one is not too surprised, since there this role is taken by kinetic Alfvén waves. On the Sun, this would imply transverse scales of the order of 1 m, far outside of present days' observations. Nevertheless, one can speculate by postulating the generation of such waves in conjunction with the still unidentified heating of the moss in the transition region. The presence of such waves would appear as a contribution to the nonthermal line broadening of the observed UV spectra (Mathioudakis et al., 2013). In this context, it is tempting to refer to the theory by Shukla et al. (1999) of a highly effective electron heating in the corona by the electric field of high-frequency dispersive kinetic Alfvén waves. Their frequencies were chosen to range below the ion cyclotron frequency in a plasma with $m_e/m_i \ll \beta$, with a parallel wavelength of the order of 1% of the ion skin depth, $c/\omega_{pi}$, that is, approximately 1 cm. No excitation process is considered and many other aspects, like the ion heating, left aside.

Little more can be said at this place about the dissipation processes of the waves, which are commonly held responsible for the coronal heating. However, I will return to this topic in Section 4.4.1.

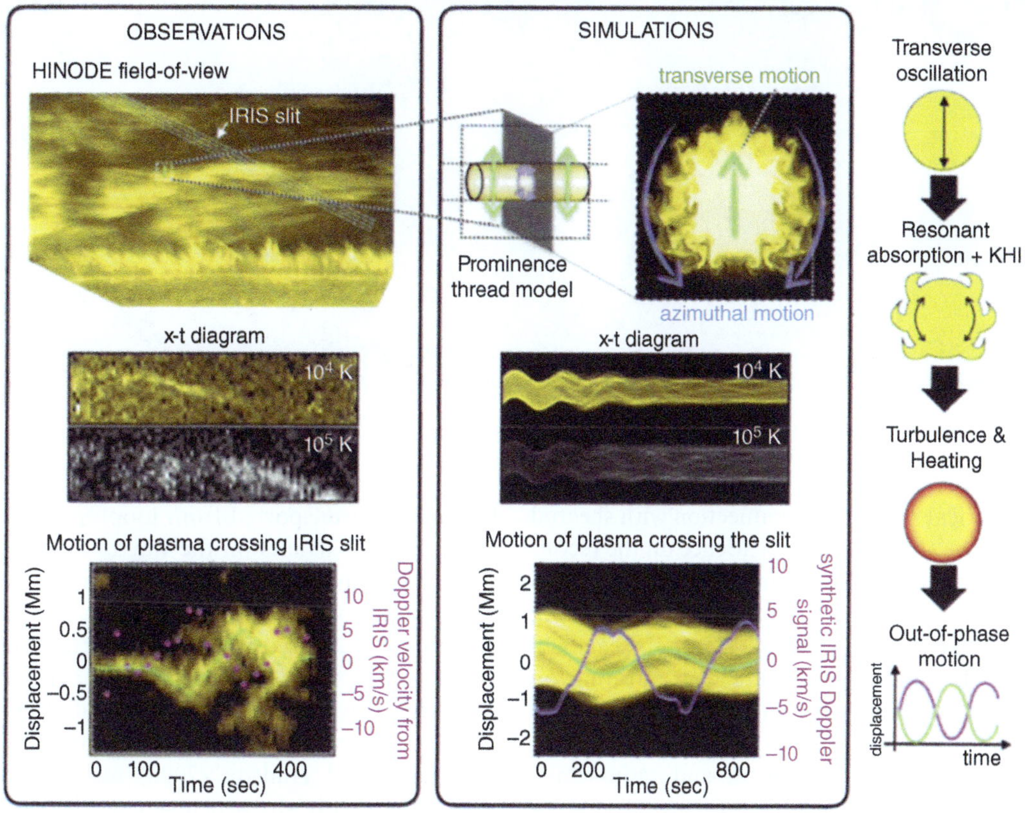


**Figure 4.8** Coordinated IRIS and Hinode/SOT observations of an active prominence at the limb, reported by Okamoto et al. (2015). Heating signatures were captured, seen by fading of the prominence threads in the SOT Ca II H line (yellow) accompanied by intensity increase in the hotter Si IV 1400 passband (gray). The Doppler velocity captured by IRIS in the Mg II k line (purple dots in bottom-left time–distance panel) shows an out-of-phase motion with respect to the plane-of-the-sky motion captured by SOT. 3D MHD simulations of an oscillating prominence thread with a kink mode performed by Antolin et al. (2015a) reproduce the heating signatures and phase relations, based on the combined effect from resonant absorption and the Kelvin-Helmholtz instability (KHI), as shown in the sketch. Source: Adapted from Van Doorsselaere et al. (2020).

### *Excitation of Alfvén Waves by Solar Flares*

The literature on solar flares is extremely rich. Fletcher et al. (2011) presented a very valuable overview of the observations in terms of the different radiative components during the evolution of a flare, with particular emphasis on the hard X-rays (HXR). The latter, dominating the impulsive phase, are by timing and chromospheric localization the best indicators of the key processes playing invisibly in the corona. It is generally agreed that the timing signals the onset of reconnection and the footprints (ribbons) track the magnetic field lines involved in that process. If the footprints are located on either side of the polarity inversion line (PIL, separating positive and negative vertical fields), it is a clear indication that the main source of energy is the conversion of antiparallel magnetic fields into the kinetic energy of plasma particles. The tendency of flare ribbons in Hα and UV/EUV to move away from the polarity inversion line (Fletcher et al., 2011) reveals the pileup of reconnected field, quite comparable to the poleward expansion in substorms (see Subsection "Excitation Processes") (Figure 4.9). These properties led to the two-dimensional reconnection model of Hirayama (1974), Sturrock (1966), Colgate (1978), and Kopp and Pneuman (1976), the so-called CSHKP flare. The establishment of a magnetic configuration prone to undergo reconnection involves frequently the eruption of magnetic flux by various MHD-type instabilities. They range between simple erupting loops (Hirayama, 1974; Török & Kliem, 2005), three-dimensional reconnection quasi-separatrix layers instead of a simple X-line (Aulanier et al., 2006; Démoulin et al., 1996), and erupting sigmoids (Aulanier et al., 2010). Schrijver (2009) claims that M- and X-class flares are due to the emergence of flux ropes along a strong-gradient polarity-inversion line. Even in the simpler first configuration, one often observes displacements of the HXR footpoints along the ribbons, indicating that reconnection proceeds with strongly sheared magnetic fields (Fletcher et al., 2011).

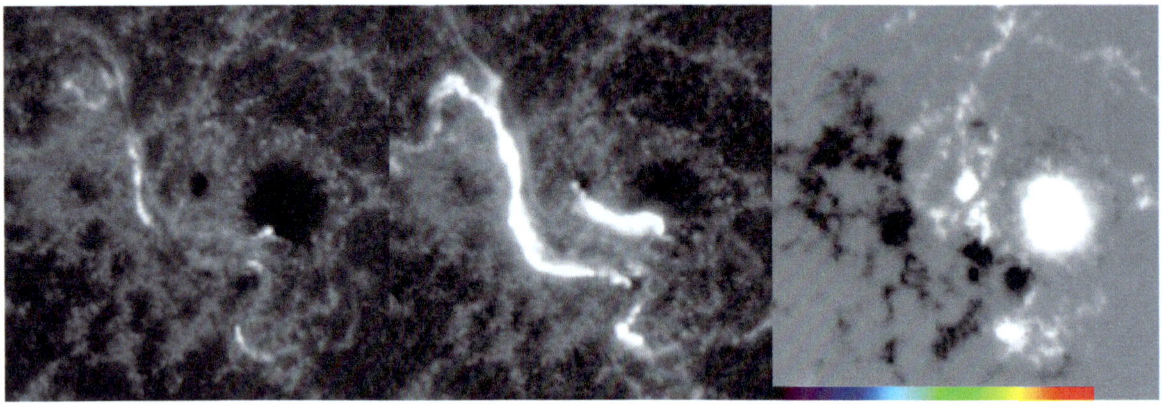

**Figure 4.9** Ribbons displaced along the PIL indicating sheared arcade fields and asymmetric emissions and increasing separation. Source: Qiu et.al., 2017 / IOP Publishing.

However, theoretical studies of reconnection with sheared or guide fields have shown little dependence on the reconnection rate (Cassak et al., 2007). Since the reconstruction of the flare geometry from the footprints is far from being straightforward (Schrijver, 2009), an existing shear field is not easy to determine. As we are concentrating on Alfvén waves, we will in the following stick to the CSHKP configuration.

The prime concept for the optical and radio emissions in the chromosphere is that they are excited by precipitating particles (mostly electrons) energized in the corona, preferentially in the plasma appearing in soft X-rays (SXR) and EUV on the looptops of arcades or post-flare cusp structures (see Fletcher et al., 2011) for references. There are basically three acceleration processes under consideration (Aschwanden, 2006): second-order Fermi acceleration, shock or first-order acceleration, and acceleration by parallel electric fields. The alternative that energy, transported from looptop as Poynting flux, can energize particles on the way or by interaction with the chromospheric plasma has found only relatively recent attention (Fletcher & Hudson, 2008). Emslie and Sturrock (1982) discussed the energy transport by Alfvén waves in the context of heating the temperature minimum. Similarly, Haerendel (2009) investigated chromospheric heating during flares by the dissipation of Alfvén waves.

Fletcher and Hudson (2008) regard the restructuring of the highly stressed magnetic field resulting from reconnection into time-varying structures as source of various types of MHD waves, among them the Alfvén mode. This is at variance from the quasi-steady Petschek reconnection or the CSHKP flare model. Without further specification, the authors assume that 3D perturbations of the post-reconnection flare loop result in the excitation of fast mode and Alfvén waves (Figure 4.10). More precisely, they consider inertial

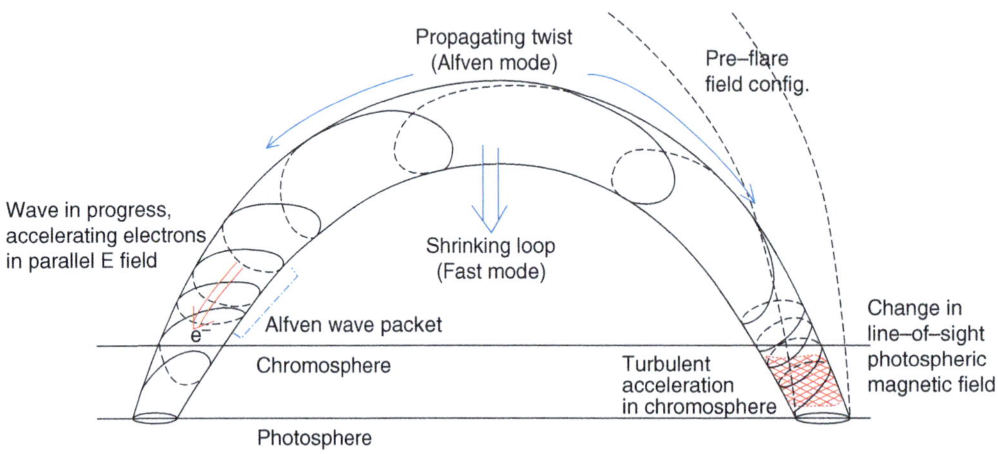


**Figure 4.10** A reconfiguring coronal field launches a torsional Alfvén wave pulse through the corona into the chromosphere. The inertial Alfvén wave can accelerate electrons by its $E_{||}$. The residual energy is damped in a cascade process heating the chromosphere and possibly leading to stochastic electron acceleration. Source: Fletcher and Hudson (2008) / Institute of Physics.

Alfvén waves by choosing a very low-beta plasma environment ($\beta \ll me/mi$) with a field strength of 500 G, a density of $10^9\,cm^{-3}$, and a temperature of $10^6$ K. The Alfvén waves thus reach speeds of the order of 0.1 c. Owing to their finite parallel electric field components, the waves can accelerate electrons with high thermal speed up to twice the Alfvén speed (Kletzing, 1994), that is, up to 10–30 keV. However, there are stringent conditions placed on the existence of sufficiently strong $E_{\parallel}$ components. It must exceed the so-called Dreicer field (Dreicer, 1959) in order to accelerate electrons freely above their thermal energy. The authors calculate some quantitative solutions but not in any self-consistent way. Electron energies gained by wave acceleration can be multiplied by reflections between the wave front and the mirror effect of a converging magnetic field, that is, by a first-order Fermi process (see also Artemyev et al. (2015)). These processes, playing during the propagation of the wave, will not dampen the energy flux significantly. When hitting the transition region and chromosphere, dissipation may lead to heating and turbulent acceleration. This part of the theory of Fletcher and Hudson (2008) will be considered in the following subsection.

At variance to the abovementioned highly cited paper, Haerendel (2009) investigated the evaporation of chromospheric plasma by Alfvén waves generated at the looptop during and after braking of the reconnection flow (Figure 4.11). He tailored the generation process after the corresponding process in the magnetosphere at substorm onset, but refrained from addressing the acceleration of high-energy electrons. Adopting much higher temperatures and densities and lower magnetic field strengths, all supported by observations (Krucker et al., 2008; Shibasaki, 2001), the looptop plasma is assumed to be of high beta. It is the pressure force of the magnetic field acting on the hot plasma that generates a transverse current across the looptops, which diverges into field-aligned currents set up by Alfvén waves. Speeds and scales differ completely from the inertial waves of Fletcher and Hudson (2008), but carry similar energy fluxes. Here, the emphasis is on the dissipation processes that I will discuss in the following text.

The two abovementioned contributions to the role played by Alfvén waves in transporting energy liberated by reconnection in the corona do not solve the major problem of which process accelerates the electrons responsible for the hard X-ray emissions and type III radio bursts during the impulsive phase. As for instance Melrose and Wheatland (2014) state, "no model has emerged that can account in a self-consistent way for important features from observations." I will not dwell here on the difficulties but conclude with these authors that Alfvén waves must play an essential role in the liberation of free energy stored in the coronal magnetic field

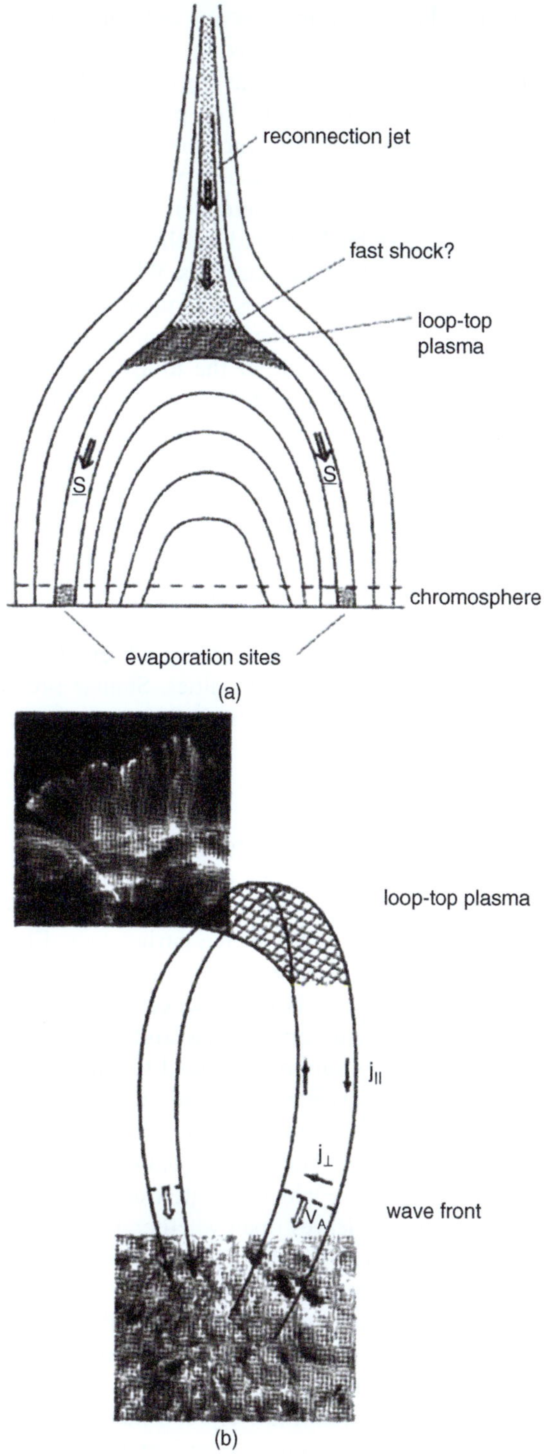


**Figure 4.11** Chromospheric heating by Alfvén waves. (a) Loop-top plasma layer arising from braking of the reconnection jet, possibly by a fast shock formation. (b) One flare loop out of an arcade (top left) is shown guiding energy in the Alfvén mode toward the highly structured chromosphere, where wave damping is aided by Kelvin–Helmholtz instability and phase mixing. Source: Adapted from Haerendel (2009).

prior to the flare. This involves not only the conversion of antiparallel magnetic field components but also the liberation of shear stresses built up in the MHD instability preceding the flare. The key process must be the generation of strong electric currents, both balanced as in the two abovementioned models or unbalanced as for instance in twisted loops (Haerendel, 2018). An indicator of acceleration by $E_{\parallel}$ would be the connection with upward flowing field-aligned currents. Where currents are generated by mechanical forces, as for instance by pressure forces at the looptop, they are balanced, and upward currents exist on either side of the loop. Therefore, the brightness of the ribbons on either side of the PIL should not be too different.

I will return to this in Section 4.3.1, and refer here only to the most problematic implication, the need for transverse scales of the order of 1 m for the currents to become dissipative and sustain parallel electric fields (Haerendel, 1994, 2012; Holman, 1985). The breaking of the observable scales of at best tens of kilometers to such scales does not only escape observations but presents major conceptual difficulties. Similar problems exist with respect to the dissipation of waves in the denser and collisional plasma of the chromosphere. This is addressed next.

#### *Dissipation of Alfvén Waves Excited by Solar Flares*

A detailed discussion of the dissipation processes has been provided in the two papers about the excitation of Alfvén waves referenced in the earlier text. Fletcher and Hudson (2008) excluding collisional damping take refuge to the development of turbulence owed to multiple reflections and interactions of fast and Alfvén modes. A cascade process is supposed to lead to chromospheric heating and Fermi acceleration. The latter may involve "transit-time" acceleration or gyroresonant interactions with whistler modes. The authors calculate a critical energy of 15 keV above which the latter energization may proceed. The paper leaves open whether $E_{\parallel}$ acceleration by the excited inertial Alfvén waves exceeds the turbulent acceleration in the chromosphere.

The paper by Haerendel (2009) considers only the damping of Alfvén waves in the chromosphere and the ensuing heating and evaporation. Five possible processes are analyzed quantitatively: ion-neutral damping, Coulomb collisions, anomalous resistivity, critical ionization velocity damping, and turbulent wave damping. The first two processes are discarded because they die out due to the quenching by the resulting heating and increasing ionization or decreasing resistance, respectively. An important consideration, also for the third process, is the impedance matching of the wave in the chromosphere with the magnetic diffusion timescale of the Coulomb or anomalous collisions. In the latter case, not only the impedance matching matters but also the critical current limit for triggering the relevant instability. Because of the low coronal density compared with that in the chromosphere, the conversion from wave to the energy of runaway electrons will take place above the chromosphere. Surprising is not the narrow width of the current sheet or tube of 3–10 m, required for the anomalous process, but the length of that region, ranging from 30 km to only 3 km, depending on the temperature. The quantities required by the process yield voltages of the order of 100 kV. Although this may appear promising, an interesting conflict arises. The energy flux implies draining the whole flux tube of its electron content in a transit time of an Alfvén wave (approximately 3 s). So, this process also drops out. The critical ionization velocity process would require very-high-frequency waves (2–10 Hz), which are unlikely to emerge from the looptop plasma. Finally, not unlike Fletcher and Hudson (2008), Haerendel (2009) is led to consider turbulence arising from multiple reflections of the waves and searches for the most relevant spatial scales allowing wave damping. Strong differences in phase velocity will imply phase mixing, and transverse shear flows will excite the Kelvin–Helmholtz instability. Requesting a damping time equal to the Alfvénic transit time from looptop leads to transverse scales well below 100 km, but well above the any direct damping scale. However, some still not fully explored steps are involved in the latter chain of arguments.

### 4.3.1. Alfvén Waves Extracting Stored Magnetic Energy

The solar atmosphere differs from the magnetosphere by the lack of a layer insulating it from the star. In the absence of local reconnection processes, the magnetic field is frozen in the photosphere. Therefore, gross internal flows like in the magnetosphere, as for instance driven by pressure forces, do not exist. Instead, there are dramatic reconfiguration processes initiated by MHD instabilities. They expand and move large volumes of magnetic flux and filling it with new free energy. If by stretching or twisting antiparallel fields are brought together, reconnection will be the mechanism for energy extraction. However, a non-negligible fraction of the stored energy may reside in shear stresses. Its extraction must involve field-aligned electric currents, as happens in the magnetosphere with embedded arcs (see Section "Wave Excitation and Erosion of Stored Magnetic Energy"). Both reconnection and release of shear stresses involve excitation of Alfvén waves, although it is often more convenient to describe them by quasi-static models. This is the approach used in the following subsection.

### *Extracting Energy out of Sheared Loops*

Shear stresses can be generated in two ways, by mechanical forces and by inductive processes. In the first case, the forces are propagated by Alfvén waves and balanced field-aligned current systems are established. In the second case, an unbalanced current, generated by the changing magnetic field, is induced in a magnetic flux loop. Melrose (2017) distinguishes between neutralized and unneutralized currents. The first have balanced field-aligned currents. The second have no return and close through the deep photosphere. There are various types of MHD instabilities creating such situations: torus instability (Kliem & Török, 2006), kink instability (Hood & Priest, 1981; Török & Kliem, 2005), and tether-cutting instability (Moore et al., 2001). Figure 4.12 shows a case of an erupting loop or arcade, which produces a reconnection situation of strongly sheared magnetic fields pervaded by opposing field-aligned currents as consequence of unbalanced currents induced by the eruption, for instance, by a changing field penetrating the loop. However, Dalmasse et al. (2015) have shown that photospheric shearing motions through the polarity inversion line can generate net currents along a loop.

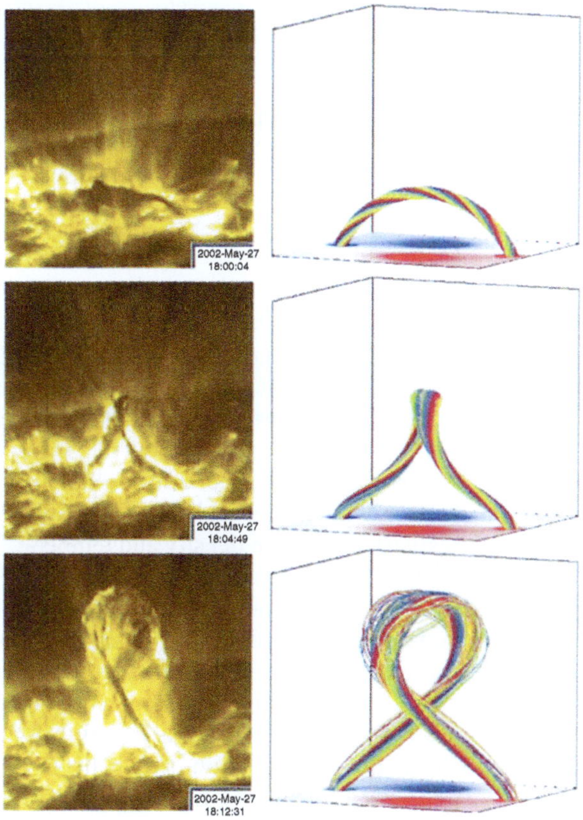


**Figure 4.12** (Left) TRACE 195 Å images of the confined filament eruption on 27 May 2002. (Right) Magnetic field lines outlining the core of the kink-unstable flux rope at $t = 0$, 24, and 37. The central part of the simulation box is shown and the underlying magnetogram. Source: Török and Kliem 2005 / Institute of Physics.

The configuration shown in Figure 4.12 (Haerendel, 2018) is of particular interest, because the common reconnection process, extracting free energy from antiparallel field components, may need support from another process extracting energy out of a strongly sheared field. On the other hand, theoretical studies of reconnection with "guide fields" reveal that the reconnection rate is not much affected (Cassak et al., 2007). Birn and Hesse (2010) found by MHD as well as PIC simulations that the magnetic energy of the guide field is carried from inflow to outflow without much change and is then converted into Poynting flux. Obviously, here no additional conversion process is needed. However, these simulations are only treating the local reconnection configuration, for instance, as appearing at the magnetopause with an oblique external magnetic field. The energy comes from merging of the local magnetic field. By contrast, the shear field in the coronal loop shown in Figure 4.12 stems from a net current, which is the storage of the energy. Release of this energy needs an unshearing process. The role of energy storage in a current circuit and its conversion by the release of shear stresses has been discussed in Section "Wave Excitation and Erosion of Stored Magnetic Energy" for the corresponding situation in the magnetosphere.

The possibility to measure vertical currents out of the photosphere, for instance with the Helioseismic and Magnetic Imager (HMI) magnetograph on the Solar Dynamics Explorer (SDI), is of great value in identifying unbalanced currents. This is particularly revealing, if it is coupled with an asymmetry of the energy deposition. Preference of the latter for upward currents would be a strong hint for acceleration by parallel electric fields in dissipative field-aligned currents, the auroral process. Yang et al. (2012) have investigated 172 time intervals in 22 flares for asymmetries of HXR footpoints and their relation to the respective magnetic field strengths, guided by the assumption that the mirror force will lead to lower brightness at the stronger field (Figure 4.12). However, only 75% were in support of that. Therefore, the reason for the 25% contradicting cases must have had another origin. Contributions from unbalanced current could be one. Unfortunately, no vertical current measurements were available for that study.

One case with a nearly 100% asymmetry in the Reuven Ramaty High Energy Solar Spectroscopic Imager (RHESSI) HXR brightness between the upward and downward current regions appeared in the flare of 15 February 2011 (Janvier et al., 2014; Musset et al., 2015) (Figure 4.13). This was discovered by the author (Haerendel, 2017). Another case, covered by several authors,

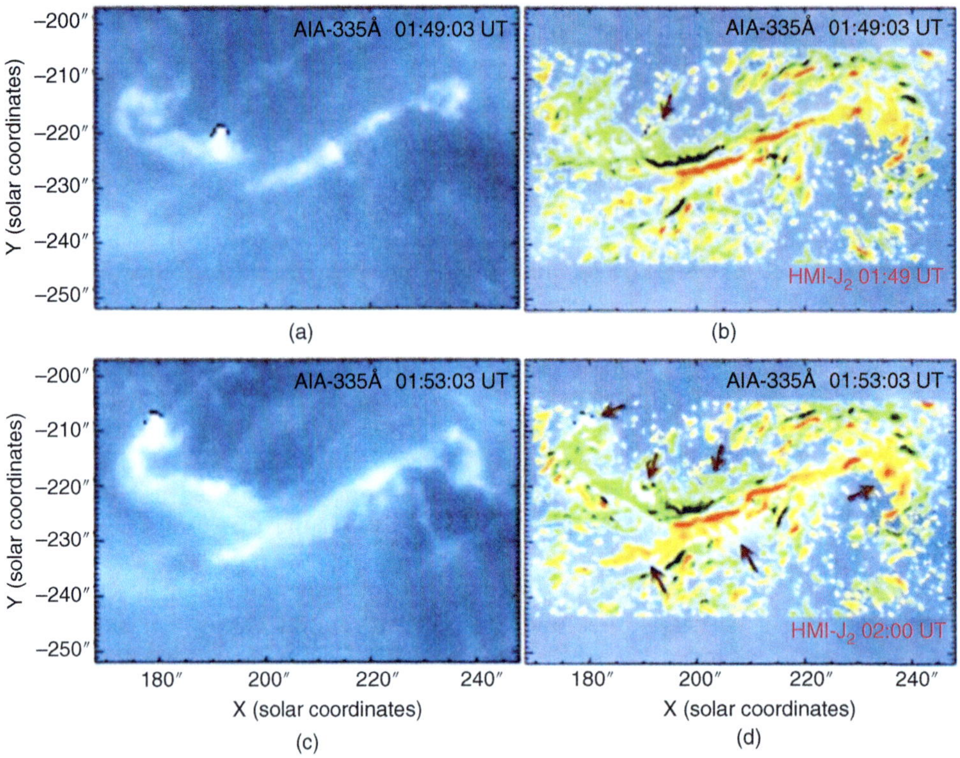


**Figure 4.13** Brightness maps in SDO/AIA 335 Å filter (left) and maps of vertical current up to 80 mA/m$^2$ from SDO/HMI vector magnetic field in the 6173 Å Fe I line during the flare on 15 February 2011. Source: From Janvier et al. (2014). Comparison of the two maps reveals a highly asymmetric brightness in favor of the upward current (Haerendel, 2017).

might be the flare of 22 June 2015 (Artemyev et al., 2021; Jing et al., 2016; Kang et al., 2019; Wheatland et al., 2018). However, this case is not clear. The superposed images from SDO/HMI and RHESSI are not sufficiently well resolved to ascertain that the impulsive phase showed asymmetric HXR brightness in the upward current footprint. In both cases, the HXR ribbons were very close to the PIL. This proximity solved a troublesome problem. In previous theoretical studies (Haerendel, 1994, 2012), the author was unable to suggest a convincing process generating highly concentrated currents with widths of the order of 1 m, needed for an anomalous resistivity to appear. The obvious proximity to a reconnection site with a current sheet of the order of c/ωpi suggested a symbiosis between reconnection and release of shear stresses (Figure 4.14). In (Haerendel, 2018) the author proposed that the transverse field components were reduced by the merging field traversing a dissipative field-aligned current layer, before undergoing reconnection of the antiparallel fields. This had a consequence. The conversion of this field component into heat and the energy of runaway electrons are enabled by an anomalous resistivity and a field-parallel electric field set up by a current-driven microturbulence. Associated with it is a magnetic diffusivity, $D_m$, while the current width, $\lambda_{perp}$, follows from the critical threshold for the onset of the anomalous resistivity. Combined, the two critical properties yield the diffusion time, $\tau_0$, and with that the penetration speed, $v_n$, of the magnetic field. The consequence is that the access to the reconnection site is controlled by the traversal through the dissipative current sheet and the concurrent destruction of the shear field. Critical current density and anomalous resistivity depend on density and temperature with some uncertainty parameters. In (Haerendel, 2018), it turns out that $v_n$ is substantially slower than the traditional reconnection speed of order 0.1 $v_A$. That leads to a lengthening of the diffusion region. There are other consequences, potentially observable, discussed in that paper.

How does the Alfvén speed enter this process? A flux tube approaching the reconnection site with normal velocity, $v_n$, finds itself subject to anomalous diffusion and reduction of the transverse field component, first in the neighborhood of the X-line. While this perturbation propagates with Alfvén speed along B, the approaching field with the external field-aligned currents meets the current sheet obliquely with an inclination $v_n/v_A$. Extending itself along the reconnection outflow channel, the dissipative current sheet of about 1 m width maps into

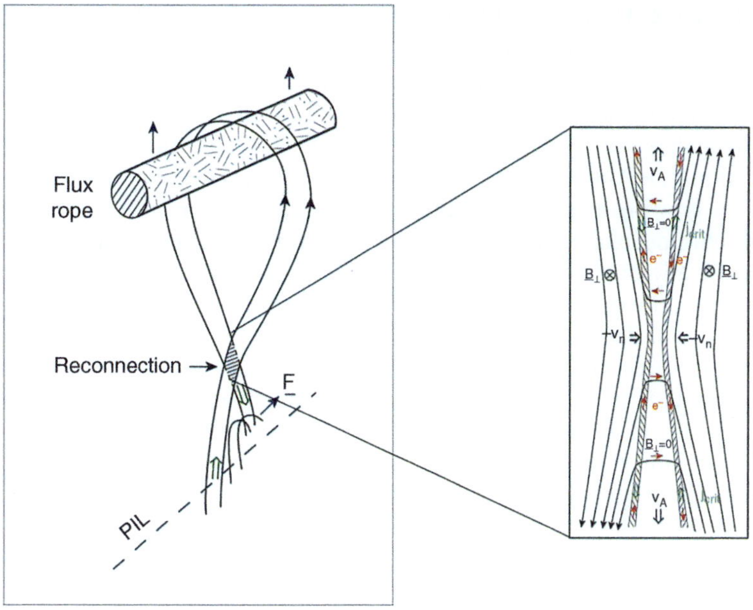


**Figure 4.14** Reconnection site generated by a sheared erupting flux rope. The reconnection inflow is moderated by traversal through a highly dissipative current layer framing the outflow channel. The magnetic energy content of the transverse field component is converted into heat and runaway electrons inside the oblique current layer. Source: Adapted from Haerendel (2018).

the photosphere with a width of the order of 10 km and the runaway electrons produce a hard X-rays emitting ribbon. This scenario produces realistic field-aligned voltages and energy fluxes. However, there are many questions raised. For instance, why is there not only guide-field reconnection? Or, why do the parallel electric fields not appear inside the reconnection outflow? The theory is an attempt to inject a new idea as to how the HXR electrons of the impulsive phase may be generated and to offer a reason for the striking current-related asymmetry. As the underlying event (Haerendel, 2017) is the only one observed so far, it needs confirmation or falsification by other asymmetric events.

I regard the process extensively discussed here as correspondent to the magnetospheric process treated in Subsection "Wave Excitation and Erosion of Stored Magnetic Energy," the release of shear stresses stored in a large-scale current system forming embedded auroral arcs. In the solar case, the storage consists of the strongly sheared flux tube created by a kink instability. The correspondent to the propagating oblique current sheet above the auroral arc is, on the Sun, the oblique dissipative current sheet moderating the inflow into the reconnection site. While in the auroral case liberated energy is flowing with Alfvén speed along the front into the acceleration region, it is in the solar case that the dissipative current sheet propagates with Alfvén speed along reconnection outflow channel.

In the scenarios considered here, the magnetosphere as well as the corona, the Alfvén waves are not excited somewhere and carry the energy toward another region. They only transport energy or the conversion process inside the same current sheet. Thus, excitation and dissipation proceed along the same current sheet.

The way these electrons deposit their energy in the collisional atmosphere and the luminous and radiofrequency emissions thereof are not different from those in other flare events. A separate treatment of dissipation is superfluous in this context.

## 4.4. A COMPARISON

### 4.4.1. Excitation by External Flows (Comparing Subsections "Excitation Processes" and "Excitation by External Flows or Jets")

The commonality lies in the origin of the wave energy from the kinetic energy of ion or plasma flows. In the magnetosphere, the conversion happens at magnetic boundaries, the sunward magnetopause, and the nightside boundary of the near-dipolar magnetosphere. Stoppage at perpendicular magnetic fields enables the conversion of kinetic into electromagnetic energy and by concomitant structuring the excitation of kinetic Alfvén waves. In the solar case, we lack an understanding of how the kinetic energy of the field-parallel flows of spicules and jets can be coupled to Alfvén waves. One takes refuge to resonant absorption (Chen & Hasegawa, 1974) and phase mixing (Heyvaerts & Priest, 1983). In spite of still ongoing research (Cally & Goossens, 2008: Goossens et al., 2002; Goosens et al., 2020), one may say that both processes are neither really understood, nor is there direct evidence for their operation (DePontieu et al., 2021). On the other hand, there is growing evidence for Alfvén waves emerging out of chromosphere and transition region. Furthermore, hardly any other mechanism than dissipation of Alfvén waves can explain the heating of the corona and solar wind. The latter aspect gives an important hint. The relevant waves must be small-scale, that is, KAWs. Typical coronal parameters are consistent with $me/mi \ll \beta \ll 1$, which differs from the conditions assumed by Fletcher and Hudson (2008). The observed oscillations of the spicules are probably not capable to excite transverse scales of the order of meters, as are required for generating KAWs. Indeed, the dispersion relation of KAWs implies also high frequencies. If present, they would hide in nonthermal line broadening. So, we arrive at the question: Is there a way that field-parallel flows can break into small scales?

I am not aware of any proposed mechanism. Therefore, I dare the following conjecture. The low-beta spicular flows are confined to field-parallel motions. What happens when the upflow speed falls to zero and reverses? The excitation time of the flow extends over at least 1.5 min. In other words, the downfall of plasma from the tip of the spicule must collide with the still rising matter. This must lead to a shock wave. Bart De Pontieu (1999) has demonstrated this case by numerical simulation in his thesis. The shocked plasma will be heated and break into smaller turbulent scales, exciting magneto-acoustic waves and interchange motions. Van Doorsselaere et al. (2020) have presented observations of prominence threads by Hinode/SOT (Solar Optical Telescope) and IRIS interpreted as evidence for absorption and heating by resonant absorption. Forward modeling led to a view as shown on the top of the right-hand column in Figure 4.8. This gives an idea of the turbulent motions after the collapse of a spicular flow although on a much larger transverse scale.

### 4.4.2. Absorption and Dissipation (Comparing Subsections "Absorption or Energy Deposition" and "Dissipation of Flow-Excited Alfvén Waves")

In the magnetosphere, kinetic or inertial Alfvén waves couple their energy to electrons by acceleration on the way to the ionosphere or in the topside ionosphere at steep density ledges (Chaston et al., 2006, 2007). The electrons are identified by their diffuse spectra (Newell, 2000) and their exciting erratically structured Alfvénic auroras (Haerendel & Frey, 2014). All of this is established by observations. On the Sun, the situation is less favorable. The waves responsible for heating the corona have not really been identified. The conviction that Alfvén waves take this role is owed to the lack of any other conceivable mechanism supplying from the chromosphere the energy richly available there. An additional piece of evidence may be local heating of the high-temperature moss in the transition region potentially shed by the waves on their way up. Observed large-scale and low-frequency waves are unlikely to achieve the heating. The responsible type of waves, smaller scale and higher frequency, may manifest itself by nonthermal line broadening. In conclusion, the media in which the waves deposit their energy, the ionosphere and the corona, are so different that little can be learned from the comparison. This is not the case with respect to their excitation.

### 4.4.3. Excitation by Reconnection (Comparing Subsections "Excitation Processes" and "Excitation of Alfvén Waves by Solar Flares")

Substorms in the magnetosphere and flares on the Sun have in common the origin by reconnection of antiparallel fields. The respective situations are generated by the solar wind in the Earth's realm and by loops erupting out of the photosphere or produced by MHD instabilities. This commonality has been frequently exploited in theories of solar flares. Again, as with the waves generated from the conversion of kinetic energy, the observational basis is much less favorable for the solar case than for the magnetosphere, where the flow of energy is pretty well understood. An important uncertainty of the solar case concerns the beta value of the medium where the reconnection outflow comes to a halt. Some configurations are so complex, that is, 3D reconnection, that comparison with the Earth's case appears far-fetched. Even concentrating on the CSHKP model can lead to very different conclusions. The two models of Fletcher and Hudson (2008) and of the author (Haerendel, 2009) are representative for the consequences of the assumed beta value. In the first case, it is the reconfiguration and shrinking of the loops forming out of the reconnected fields that transform into torsional Alfvén waves carrying the released energy. The second case is strongly modeled after the magnetospheric substorm. The exhaust flow from the reconnection process is stopped at the top of the underlying loops. The contracting magnetic field exerts a force on the high-beta plasma driving a transverse current from which Alfvén waves and balanced field-aligned currents emerge. Unfortunately, both cases, low and high beta, have in common that the mechanism converting the wave energy into kinetic energy of the electrons cannot be pinpointed. Therefore, systematic investigations of differences in energy spectra and morphology depending on the best-assessed beta value may be a promising direction to go. In the magnetosphere, shear stresses generated in the involved loops, the strong convergence of the field, and the low density of current carriers cooperate in producing field-parallel electric potential differences accelerating the electrons. In the corona, this is unlikely to happen, since the covergence of the field is weak. Haerendel (2009) has refrained from proposing an equivalent process for the flare and concentrated on the dissipation in the chromosphere. Fletcher and Hudson (2008) find it in multiple reflections of electrons first energized by an inertial Alfvén wave, a first-order Fermi process. Other authors have recruited turbulence. The truth is that, although many processes have been explored, we still lack a consistent scenario for the generation of the high-energy electrons of the impulsive phase. The existence of parallel electric fields, suggested by the magnetospheric counterpart, could not be underpinned by a process leading to the concentration of the currents in narrow transverse scales (Haerendel, 2012; Melrose & Wheatland, 2014). In this context, the ability to measure vertical currents in the photosphere, as with the HMI/AIA magnetograph, promises progress in the diagnostics. While brightness symmetries are not astounding, since most currents will be balanced, asymmetries in favor of upward currents will be clear indications of the presence of parallel electric fields. A first case is treated in the third comparison.

#### 4.4.4. Absorption and Dissipation (Comparing Subsections "Absorption Processes" and "Dissipation of Alfvén Waves Excited by Solar Flares")

The commonality of magnetosphere and Sun with respect to the dissipation of Alfvén waves stimulated by large reconnection events is not as great as one would expect. Although both ionosphere and chromosphere are filled with collision-dominated partially ionized plasmas, the responses to impinging waves are quite different. While both media cause a high reflectivity, collisional damping in the ionosphere allows fast buildup of quasi-stationary current systems, the auroral electrojet being the most striking example. By contrast, energy transported by Alfvén waves toward the chromosphere will be subject to three different processes, partial energy conversion into kinetic energy of electrons before arrival, multiple reflections, and excitation of a turbulent cascade providing partial damping. The least explored is the first path of the energy. Also the third path is far from being well understood. Most of these uncertainties are owed to the lack of knowledge about the prevailing classes of energy-transporting waves. Often preference is given to kinetic or inertial Alfvén waves because of their dissipative nature, although direct support by observations is practically nonexistent. Instead support must be sought from theory and indirect evidence such as asymmetries and correlations between vertical currents and HXR brightness.

#### 4.4.5. Extraction of Energy Stored in Sheared Fields (Comparing Subsections "Wave Excitation and Erosion of Stored Magnetic Energy" and "Extracting Energy out of Sheared Loops")

On the first sight, comparing the release of shear stresses stored, on the one hand, in the evening auroral oval and, on the other hand, stored in sheared coronal loops, appears a bit far-fetched. However, one important commonalty is the presence of an unbalanced current. Of course, currents must close somehow. In the magnetospheric case, the current is driven and closes over a wide range of longitudes. At conjugate magnetic latitudes, however, the currents have the same sense, either upward or downward. In the solar counterpart, the current is unidirectional along a loop. Therefore, while auroral arcs appear in the northern as well in the southern conjugate area, on the Sun, high-energy X-rays will be generated on only one side of the loop. The role of Alfvén waves is also opposite. In the auroral current sheet, the released energy is transported by the Alfvén mode into the acceleration region. In the configuration of reconnection moderated by the release of shear stresses, the acceleration region is transported by the Alfvén mode along the exhaust flow channel. Although the acceleration processes differ, owed to the mirror effect in the magnetosphere and attributed to anomalous resistivity in the corona, there is indeed much commonality. The only problem is that the solar case rests on only one theoretical paper (Haerendel, 2018). This means that a well-proven concept is compared with a speculation. Whether right or wrong, its treatment in this comparison is meant to stimulate the search for unbalanced currents with asymmetric hard X-ray ribbons.

The dissipation in both cases is largely due to the precipitation of the energized electrons.

### 4.5. CONCLUSIONS

This treatment of the three corresponding situations on Earth and Sun points out similarities, differences, and lack of knowledge, mostly on the solar side, because of the restricted observabilities. Few concrete hints are being offered for future directions of research. However, in the case of excitation of Alfvén waves by external flows, a promising direction may be to explore the generation of small transverse scales by collapsing spicules and coupling to kinetic Alfvén waves. In the case of solar flares, a major unknown is the beta value of the looptop plasma. In the earlier text, a recommendation is given for systematic investigations of differences in energy spectra and morphology depending on the best-assessed beta value. For flares in general and, in particular, for the situation of sheared erupting loops, searches for asymmetric impulsive energy depositions and concurrent determinations of vertical currents promise new insights into the acceleration processes. Beyond these few concrete hints, there remains some hope that by reading through these comparisons unexpected ideas may arise.

## REFERENCES


Alfvén, H. (1947). Magneto hydrodynamic waves, and the heating of the solar corona. *Monthly Notices of the Royal Astronomical Society*, *107*, 211. https://doi.org/10.1093/mnras/107.2.211

André, M., Norqvist, P., Andersson, L., Eliasson, L., Eriksson, A. I., Blomberg, L., et al. (1998). Ion energization mechanisms at 1700 km in the auroral region. *Journal of Geophysical Research*, *103*(A3), 4199–4222. doi:10.1029/97JA00855

Antolin, P., Okamoto, T. J., De Pontieu, B., Uitenbroek, H., Van Doorsselaere, T., & Yokoyama, T. (2015). Resonant absorption of transverse oscillations and associated heating in a solar prominence. II. Numerical aspects. *Astrophysical Journal*, *809*, 72. DOI. ADS.

Antolin, P., Schmit, D., Pereira, T. M. D., De Pontieu, B., & De Moortel, I. (2018). Transverse wave induced Kelvin-Helmholtz rolls in spicules. *Astrophysical Journal*, *856*, 44. DOI. ADS.

Artemyev, A. V., Rankin, R., & Blanco, M. (2015). Electron trapping and acceleration by kinetic Alfven waves in the inner magnetosphere. *Journal of Geophysical Research: Space Physics*, *120*, 10,305–10, 316. doi:10.1002/JA021781

Artemyev, A. V., Zimovets, I., Sharykin, I., Nishimura, Y., Downs, C., Weygand, J., et al. (2021). Comparative study of electric currents and energetic particle fluxes in a solar flare and earth magnetospheric substorm. *Astrophysical Journal*, *923*, 151. https://doi.org/10.3847/1538-4357/ac2df.

Aulanier, G., Paria, E., & Démoulin, P. (2006). Slip-running reconnection in quasi-separatrix layers. *Solar Physics*, *238*, 347–376. doi:10.1007/s11207-006-0230-2

Aulanier, G., Török, T., Démoulin, P., & DeLuca, E. E. (2010). Formation of torus-unstable flux ropes and electric currents in erupting sigmoids. *Astrophysical Journal*, *708*, 314–333. doi:10.1088/0004-637X/708/1/314

Aschwanden, M. J. (2006). *Physics of the Solar Corona*. Berlin, Heidelberg: Springer.

Biermann, L. (1948). Über die Ursache der chromosphärischen Turbulenz und des UV-Exzesses der Sonnenstrahlung. *Zeitschrift für Astrophysik*, *25*, 161.

Birn, J., & Hesse, M. (1991). Magnetotail equilibrium theory: The general three-dimensional solution. *Journal of Geophysical Research*, *92*, 11,101.

Birn, J., Hesse, M., Haerendel, G., Baumjohann, W., & Shiokawa, K. (1999). Flow braking and the substorm current wedge. *Journal of Geophysical Research*, *104*(A9), 19,895–19,904. doi: 10.1029/1999JA900173

Birn, J., Raeder, J., Wang, Y. L., Wolf, R. A., & Hesse, M. (2004). On the propagation of bubbles in the geomagnetic tail. *Annales Geophysicae*, *22*, 1773–1786.

Birn, J., & Hesse, M. (2010). Energy release and transfer in guide field reconnection. *Physics of Plasmas*, *17*, 012109. doi:10.1063/1.3299388

Birn, J., Nakamura, R., Panov, E. V., & Hesse, M. (2011). Bursty bulk flows and dipolarization in MHD simulations of magnetotail reconnection. *Journal of Geophysical Research*, *116*, A01210. doi:10.1029/2010JA016083

Birn, J., & Hesse, M. (2013). The substorm current wedge in MHD simulations. *Journal of Geophysical Research: Space Physics*, *118*, 3364–3376. doi:10.1002/jgra.555018

Boström, R. (1964). A model of the auroral electrojets. *Journal of Geophysical Research*, *69*, 4983–4999. doi:10.1029/JZ069i023p04983

Cally, P. S., & Goossens, M. (2008). Three-dimensional MHD wave propagation and conversion to Alfvén waves near the solar surface. I. Direct numerical solution. *Solar Physics*, *251*, 251. doi: 10.1007/s11207-007-9086-3

Cassak, P. A., Drake, J. F., & Shay, M. A. (2007). Catastrophic onset of fast magnetic reconnection with a guide field. *Physics of Plasmas*, *14*, 054502. doi:10.1063/1.2734948

Chaston, C. C., Carlson, C. W., Ergun, R. E., McFadden, J. P., & Strangeway, R. J. (2003). Properties of small-scale Alfvén waves and accelerated electrons from FAST. *Journal of Geophysical Research*, *108*(A4), 8003. doi:10.1029/2002JA009420

Chaston, C. C., Phan, T. D., Bonnell, J. W., Mozer, F. S., Acuña, M., Goldstein, M. L., et al. (2005). Drift-kinetic Alfvén waves observed near a reconnection X line in the Earth's magnetopause. *Physical Review Letters*, *95*, 065002.

Chaston, C. C., Genot, V., Bonnell, J. W., Carlson, C. W., McFadden, J. P., Ergun, R. E., et al. (2006). Ionospheric erosion by Alfvén waves. *Journal of Geophysical Research*, *111*, A03206. doi:10.1029/2005JA011367

Chaston, C. C., Hull, A. J., Bonnell, J. W., Carlson, C. W., Ergun, R. E., Strangeway, R. J., & McFadden, J. P. (2007). Large parallel electric fields, currents, and density cavities in dispersive Alfvén waves above the aurora. *Journal of Geophysical Research*, *112*, A05215. doi:10.1029/2006JA012007

Chaston, C. C., Johnson, J. R., Wilber, M., Acuña, M., Goldstein, M. L., & Reme, H. (2009). Kinetic Alfvén wave turbulence and transport through a reconnection diffusion region. *Physical Review Letters*, *102*, 015001. doi:10.1103/Phys.Rev. Lett.102.015001

Chen, L., & Hasegawa, A. (1974). A theory of long-period magnetic pulsations: 1. Steady state excitation of field line resonance. *Journal of Geophysical Research*, *79*(7), 1024–1032. https://doi.org/10.1029/JA079i007p01024.

Cheng, L., Lin, Y., Perez, J. D., Johnson, J. R., & Wang, X. (2020). Kinetic Alfvén waves from magnetotail to the ionosphere in global hybrid simulation associated with fast flows. *Journal of Geophysical Research: Space Physics*, *125*, e2019JA027062. https://doi.org/10.1029/2019JA027062

Colgate, S. A. (1978). A phenomenological model of solar flares. *Astrophysical Journal*, *221*, 1068–1083.

Cranmer, S. R., & van Ballegooijen, A. A. (2005). On the generation, propagation, and reflection of Alfven waves from the solar photosphere to the distant heliosphere. *Astrophysical Journal*, Suppl. S. 156, 265–293.

Dalmasse, K., Aulanier, G., Démoulin, P., Kliem, B., Török, T., & Pariat, E. (2015). The origin of net electric currents INin active regions. *Astrophysical Journal*, *810*, 17. doi: 10.1088/0004-637X/810/1/17

Démoulin, P., Priest, E. R., & Lonie, D. P. (1996). Three-dimensional magnetic reconnection without null points: 2. Application to twisted flux tubes. *Journal of Geophysical Research*, *101*, 7631–7646. https://doi.org/10.1029/95JA03558.

De Pontieu, B. (1999). Numerical simulations of spicules driven by weakly-damped Alfven waves, *Astronomy & Astrophysics*, *347*, 696–710.

De Pontieu, B., Tarbell, T., & Erdélyi, R. (2003). Correlations on arcsecond scales between chromospheric and transition region emission in active regions. *Astrophysical Journal*, *590*, 502–518.

De Pontieu, B., Erdélyi, R., & James, S. P. (2004). Solar chromospheric spicules from the leakage of photospheric oscillations and flows. *Nature*, *430*, 536. DOI. ADS.

De Pontieu, B., & Erdélyi, R. (2006). The nature of moss and lower atmospheric seismology. *Philosophical Transactions of the Royal Society A*, *364*, 383–394.

De Pontieu, B., McIntosh, S. W., Carlsson, M., Hansteen, V. H., Tarbell, T. D., Schrijver, C. J., et al. (2007). Chromospheric Alfvénic waves strong enough to power the solar wind. *Science*, *318*, 1574.

De Pontieu, B., Polito, V., Hansteen, V., Testa, P., Reeves, K. K., Antolin, P., et al. (2021). A new view of the solar interface region from the Interface Region Imaging Spectrograph (IRIS). *Solar Physics*, *296*, 84.

Dreicer, H. (1959). Electron and ion runaway in a fully ionized gas. *I. Physical Review*, *115*(2), 238–249. doi:10.1103/PhysRev.115.238.

Emslie, A. G., & Sturrock, P. A. (1982). Temperature minimum heating in solar flares by resistive dissipation of Alfvén waves. *Solar Physics*, *80*, 99. doi: 10.1007/BF00153426

Erlandson, R. E., Zanetti, L. J., Acuña, M. H., Eriksson, A. I., Eliasson, L., Boehm, M. H., & Blomberg, L. G. (1994). Freja observations of electromagnetic ion cyclotron ELF waves and transverse oxygen ion acceleration on auroral field lines. *Geophysical Research Letters*, *21*(17), 1855–1858.

Fletcher, L., & De Pontieu, B. (1999). Plasma diagnostics of transition region "moss" using SOHO/CDS and Trace. *Astrophysical Journal*, *520*, L135–138. doi:10.1086/312517

Fletcher, L., & Hudson, H. S. (2008). Impulsive phase flare energy transport by large-scale Alfvén waves and the electron acceleration problem. *Astrophysical Journal*, *675*, 1645. DOI. ADS.

Fletcher, L., Dennis, B. R., Hudson, H. S., Krucker, S., Phillips, K., Veronig, A., et al. (2011). An observational overview of solar flares. *Space Science Reviews, 159, 19*. DOI. ADS.

Goertz, C. K., & Boswell, R. W. (1979). Magnetosphere-ionosphere coupling. *Journal of Geophysical Research*, *84*, 7239.

Goossens, M., Andries, J., & Aschwanden, M. J. (2002). Coronal loop oscillations. An interpretation in terms of resonant absorption of quasi-mode kink oscillations. *Astronomy & Astrophysics*, *394*, L39. doi.org/10.1051/0004-6361:20021378

Goossens, M., Arregui, I., Soler, R., & Van Doorsselaere, T. (2020). Resonant absorption: Transformation of compressive motions into vortical motions. *Astronomy & Astrophysics*, *641*, A106. DOI. ADS.

Hasegawa, A. (1976). Particle acceleration by MHD surface wave and formation of aurora. *Journal of Geophysical Research*, *81*, 5083.

Keiling, A. (2009). Alfven waves and their roles in the dynamics of the Earth's magnetotail: A review. *Space Science Reviews*, *142*(1–4), 73–156. doi:10.1007/s11214-008-9463-8

Haerendel, G. (1980). Auroral particle acceleration—An example of a universal plasma process. *ESA Journal*, *4*, 197–210.

Haerendel, G. (1992). Disruption, ballooning or auroral avalanche—On the cause of substorms. *In Proceedings of the First International Conference on Substorms*, Kiruna, Sweden, 23–27 March 1992, European Space Agency Spec. Publ., ESA SP-337, 417–420.

Haerendel, G. (1994). Acceleration from field-aligned potential drops. *Astrophysical Journal*, Suppl. *90*, 765–774.

Haerendel, G. (2007). Auroral arcs as sites of magnetic stress release. *Journal of Geophysical Research*, *112*, A09214. doi: 10.1029/2007JA012378

Haerendel, G. (2009). Poleward arcs of the auroral oval during substorms and the inner edge of the plasma sheet. *Journal of Geophysical Research*, *107*, A06214. doi:10.1029/2009JA014138

Haerendel, G., Frey, H. U., Chaston, C. C., Amm, O., Juusola, L., Nakamura, R., et al. (2012). Birth and life of auroral arcs embedded in the evening auroral oval convection: A critical comparison of observations with theory. *Journal of Geophysical Research*, *117*, A12220. doi: 10.1029/2012JA018128.

Haerendel, G. (2012). Solar auroras. *The Astrophysical Journal*, *749*, 166. doi:10.1088/0004-637X/749/2/166.

Haerendel, G., & Frey, H. U. (2014). Role and origin of the poleward Alfvénic arc. *Journal of Geophysical Research*, *119*, 2945–2962. doi:10.1002/2014JA019786

Haerendel, G. (2015a). Substorm onset: Current sheet avalanche and stop layer. *Journal of Geophysical Research: Space Physics*, *120*, 1697–1714. doi:10.1002/2014JA020571

Haerendel, G. (2015b). Flow bursts, breakup arc, and substorm current wedge. *Journal of Geophysical Research: Space Physics*, *120*, 2796–2807. https://doi:org/10.1002/2014JA02054

Haerendel, G. (2017). Evidence for field-parallel electron acceleration in solar flares. *Astrophysical Journal*, *847*, 113. doi.org/10.3847/1538-4357/aa8995

Haerendel, G. (2018), Reconnection mediated by magnetic fractures and the solar flare. *Astrophysical Journal*, *855*, 95. https://doi.org/10.3847/1538-4357/aab0a9

Haerendel, G. (2021). Auroral arcs: The fracture theory revisited. *Journal of Geophysical Research: Space Physics*, *126*, e2020JA028194. doi:10.1029/2020JA028194.

Haerendel, G. (2022). How energy enters the magnetosphere during a substorm: A perspective after 60 years of working in the field. *Perspectives of Earth and Space Scientists, 3*, e2022CN000180. https://doi.org/10.1029/2022CN000180

Heyvaerts, J., & Priest, E. R. (1983). Coronal heating by phase-mixed shear Alfven waves. *Astronomy & Astrophysics, 117*, 220–234.

Hirayama, T. (1974). Theoretical model of flares and prominences. I: Evaporating flare model, *Solar Physics, 34*, 323–338.

Holman, G. D. (1985). Acceleration of runaway electrons and Joule heating in solar flares. *Astrophysical Journal, 293*, 584-594.

Hood, A. W., & Priest, E. R. (1981). Critical conditions for magnetic instabilities in force-free coronal loops. *Geophysical & Astrophysical Fluid Dynamics, 17*, 297. doi.org/10.1080/03091928108243687

Hull, A. J., Chaston, C. C., Frey, H. U., Fillingim, M. O., Goldstein, M. L., Bonnell, J. W., & Mozer, F. S. (2016), The "Alfvénic surge" at substorm onset/expansion and the formation of "Inverted Vs": Cluster and IMAGE observations. *Journal of Geophysical Research: Space Physics, 121*, 3978–4004. doi:10.1002/2015JA022000.

Janvier, M., Aulanier, G., Bommier, V., Schmieder, B., Démoulin, P., & Pariat, E. (2014). Electric currents in flare ribbons: Observations and three-dimensional standard model. *Astrophysical Journal, 788*, 60. doi: 10.1088/0004-637X/788/1/60

Jing, J., Xu, Y., Cao, W., Liu, Ch., Gary, D., & Wang, Ha. (2016). Unprecedented fine structure of a solar flare revealed by the 1.6 m New Solar Telescope. *Scientific Reports, 6*, 24319. doi: 10.1038/srep24319

Kang, J., Inoue, S., Kusano, K., Park, S.-H., & Moon, Y.-J. (2019). Onset mechanism of M6.5 solar flare observed in active region 1237. *Astrophysical Journal, 887*, 263. doi: 10.3847/1538-4357/ab5582

Keiling, A., Wygant, J. R., Cattell, C., Temerin, M., Mozer, F. S., Kletzing, C. A., et al. (2000). Large Alfvén wave power in the plasma sheet boundary layer during the expansion phase of substorms. *Geophysical Research Letters, 27*, 3169–3172. doi:10.1029/2000GL000127

Keiling, A., Wygant, J. R., Cattell, C., Peria, W., Parks, G., Temerin, M., et al. (2002). Correlation of Alfvén wave Poynting flux in the plasma sheet at 4–7 RE with ionospheric electron energy flux. *Journal of Geophysical Research, 107*(A7), 1132. doi:10.1029/2001JA900140

Keiling, A. (2009). Alfven waves and their roles in the dynamics of the Earth's magnetotail: A review. *Space Science Reviews, 142*(1–4), 73–156. doi:10.1007/s11214-008-9463-8

Kletzing, C. A. (1994). Electron acceleration by kinetic Alfven waves. *Journal of Geophysical Research, 99*(A6), 11095–11104.

Kliem, B., & Török, T. (2006). Torus instability. *Physical Review Letters, 96*, 255002. doi:10.1103/PhysRevLett.96.255002

Knight, S. (1973). Parallel electric fields. *Planetary and Space Science, 21*, 741–750.

Kopp, R. A., & Pneuman, G. W. (1976). Magnetic reconnection in the corona and the loop prominence phenomenon. *Solar Physics, 50*, 85.

Krucker, S., Battaglia, M., Cargill, P. J., Fletcher, L., Hudson, H. S., MacKinnon, A. L., et al. (2008). Hard X-ray emission from the solar corona. *Astronomy and Astrophysics Review, 16*, 155-208.

Lui, A. T. Y. (1991). A synthesis of magnetospheric substorm models. *Journal of Geophysical Research, 96*(A2), 1849–1856. doi:10.1029/90JA0243

Lundin, R., Eliasson, L., Haerendel, G., Boehm, M., & Holback, B. (1994). Large-scale auroral plasma density cavities observed by Freja. *Geophysical Research Letters, 21*(17), 1903– 1906. doi:10.1029/94GL00888

Lysak, R. L., & Lotko, W. (1996). On the kinetic dispersion relation for shear Alfvén waves. *Journal of Geophysical Research, 101*, 5085–5094.

Lysak, R. L., Echim, M., Karlsson, T., Marghitu, O., Rankin, R., Song, Y., et al. (2020). Quiet, discrete auroral arcs: Acceleration mechanisms. *Space Science Reviews, 216*, 92. doi:10.1007/s11214-020-00715-5

Mathioudakis, M., Jess, D. B., & Erdélyi, R. (2013). Alfvén waves in the solar atmosphere. *Space Science Reviews, 175*, 1–27. doi:10.1007/s11214-012-9944-7

McIntosh, S. W., de Pontieu, B., Carlsson, M., Hansteen, V., Boerner, P., & Goossens, M. (2011). Alfvénic waves with sufficient energy to power the quiet solar corona and fast solar wind. *Nature, 475*, 477. doi:10.1038/nature10235

McPherron, R. L., Russell, C. T., & Aubry, M. P. (1973). Satellite studies of magnetospheric substorms on August 15, 1968, 9, Phenomenological model for substorms. *Journal of Geophysical Research, 78*, 3131–3149.

Melrose, D. B., & Wheatland, M. S. (2014). Bulk energization of electrons in solar flares by Alfvén waves. *Solar Physics, 289*, 881–897. doi:10.1007/s11207-013-0376-7

Melrose, D. B. (2017). Current-driven flare and CME models. *Journal of Geophysical Research, 122*, 7963–7978. doi:10.1002/2017JA024035

Mende, S. B., Carlson, C. W., Frey, H. U., Peticolas, L. M., & Østgaard, N. (2003). FAST and IMAGE-FUV observations of a substorm onset. *Journal of Geophysical Research, 108*(A9), 1344. doi:10.1029/2002JA009787

Moore, R. L., Sterling, A. C., Hudson, H. S., & Lemen, J. R. (2001). Onset of the magnetic explosion in solar flares and coronal mass ejections. *Astrophysical Journal, 552*, 833. doi: 10.1086/320559

Musset, S., Vilmer, N., & Bommier, V. (2015). Hard X-ray emitting energetic electrons and photospheric electric currents. *Astronomy & Astrophysics, 580*, 106. doi: 10.1051/0004-6361/201424378

Newell, P. T. (2000). Reconsidering the inverted-V particle signature: Relative frequency of large-scale electron acceleration events. *Journal of Geophysical Research, 105*(7), 15,779–15, 794.

Newell, P. T., Sotirelis, T., & Wing, S. (2010). Seasonal variations in diffuse, monoenergetic, and broadband aurora. *Journal of Geophysical Research, 115*, A03216. doi: 10.1029/2009JA014805

Okamoto, T. J., Antolin, P., De Pontieu, B., Uitenbroek, H., Van Doorsselaere, T., & Yokoyama, T. (2015). Resonant absorption of transverse oscillations and associated heating in a solar prominence. *I. Observational aspects. Astrophysical Journal, 809*, 71, https://doi.org/10.1088/0004-637X/809/1/71. 1506.08965

Parker, E. N. (1972). Topological dissipation and the small-scale fields in turbulent gases. *Astrophysical Journal*, *74*, 499–510.

Qiu, J., Longcope, D. W., Cassak, P. A., & Priest, E. R. (2017). Elongation of flare ribbons. *Astrophysical Journal*, *838*, 17. doi: 10.3847/1538-4357/aa6341

Roux, A., Perraut, S., Robert, P., Morane, A., Pedersen, A., Korth, A., et al. (1991). Plasma sheet instability related to the westward traveling surge. *Journal of Geophysical Research*, *96*(A10), 17,697–17,714. doi:10.1029/91JA01106

Runov, A., Angelopoulos, V., Sitnov, M. I., Sergeev, V. A., Bonnell, J., McFadden, J. P., et al. (2009). THEMIS observations on an earthward-propagating dipolarization front. *Geophysical Research Letters*, *36*, L14106. doi:10.1029/2009GL038980.

Runov, A., Angelopoulos, V., Zhou, X.-Z., Zhang, X.-J., Li, S., Plaschke, F., & Bonnell, J. (2011). A THEMIS multiscale study of dipolarization fronts in the magnetotail plasma sheet. *Journal of Geophysical Research*, *116*, A05216. doi:10.1029/2010JA016316

Schrijver, C. J. (2009). Driving major solar flares and eruptions: A review. *Advances in Space Research*, *43*, 739. doi:10.1016/j.asr.2008.11.004

Schwarzschild, M. (1948). On noise arising from the solar granulation. *Astrophysical Journal*, *107*, 1.

Shibasaki, K. (2001). High-beta disruption in the solar atmosphere. *Astrophysical Journal*, *557*, 326–331.

Shiokawa, K., Baumjohann, W., & Haerendel, G. (1997). Braking of high-speed flows in the near-Earth tail. *Geophysical Research Letters*, *24*, 1179–1182. doi:10.1029/97GL01062

Shukla, P. K., Bingham, R., McKenzie, J. F., & Axford, W. I. (1999). Solar coronal heating by high-frequency dispersive Alfvén waves. *Solar Physics*, *186*, 61–66.

Stasiewicz, K., Bellan, P., Chaston, C., Kletzing, C., Lysak, R., Maggs, J., et al. (2000), Small scale Alfvénic structure in the aurora. *Space Science Reviews*, *92*, 432–533.

Sturrock, P. A. (1966). Model of the high-energy phase of solar flares. *Nature*, *211*, 695–697.

Török, T., & Kliem, B. (2005). Confined and ejective eruptions of kink-unstable flux ropes. *Astrophysical Journal*, *630*, L97. doi: 10.1086/462412.

Tomczyk, S., McIntosh, S. W., Keil, S. L., Judge, P. G., & Edmondson, J. (2007). Alfven waves in the solar corona. *Science*, *317*(5842), 1192–1196.

Tripathi, D., Mason, H. E., & Klimchuk, J. A. (2012). Active region moss: Doppler shifts from Hinode/Extreme ultraviolet imaging spectrometer observations. *Astrophysical Journal*, *753*, 37. doi:10.1088/0004-637X/753/1/37

Van Doorsselaere, T., Srivastava, A. K., Antolin, P., Magyar, N., Vasheghani Farahani, S., Tian, H., et al. (2020). Coronal heating by MHD waves. *Space Science Reviews*, *216*, 140. DOI. ADS.

Wahlund, J.-E., Louarn, P., Chust, T., de Feraudy, H., Roux, A., Holback, B., et al. (1994). On ion acoustic turbulence and the nonlinear evolution of kinetic Alfvén waves in aurora. *Geophysical Research Letters*, *21*(17), 1831–1834. doi:10.1029/94GL01289

Wheatland, M. S., Melrose, D. B., & Mastrano, A. (2018). Photospheric response to a Flare, *Astrophysical Journal*, *864*, 159. https://doi.org/10.3847/1538-4357/aad8ae

Wygant, J. R., Keiling, A., Cattell, C. A., Johnson, M., Lysak, R. L., Temerin, M., et al. (2000). Polar spacecraft based comparisons of intense electric fields and Poynting flux near and within the plasma sheet tail lobe boundary to UVI images: An energy source for the aurora. *Journal of Geophysical Research*, *105*, 18,675–18,692. doi:10.1029/1999JA900500

Wygant, J. R., Keiling, A., Cattell, C. A., Lysak, R. L., Temerin, M., Mozer, F. S., et al. (2002). Evidence for kinetic Alfvén waves and parallel electron energization at 4–6 RE altitudes in the plasma sheet boundary layer. *Journal of Geophysical Research*, *107*, A8. doi: 10.1029/2001JA900113

Yang, Y.-H., Cheng, C. Z., Krucker, S., Hsieh, M.-S., & Chen, N.-H. (2012). Asymmetry of hard X-ray emissions at conjugate footpoints in solar flares. *Astrophysical Journal*, *756*, 42. doi:10.1088/0004-637X/756/1/42

# Part II
# Planets

# 5

# Alfvén Waves at Mars

**Norberto Romanelli[1,2], Christopher M. Fowler[3], Gina A. DiBraccio[2], Jared R. Espley[2], and Jasper S. Halekas[4]**


## ABSTRACT

The solar wind upstream of Mars's bow shock can be described in terms of Alfvénic turbulence, with an incompressible energy cascade rate of $10^{-17}$ J m$^{-3}$ s$^{-1}$ at magnetohydrodynamics (MHD) scales. The solar wind has more Alfvén waves propagating outwards from the Sun (than inwards) and a median Alfvén ratio of ~0.33. Newly ionized planetary protons associated with the extended hydrogen corona generate waves at the local proton cyclotron frequency. These 'proton cyclotron waves' (PCW) mostly correspond to fast magnetosonic waves, although the ion cyclotron (Alfvénic) wave mode is possible for large Interplanetary Magnetic Field cone angles. PCW do not show significant effects on the solar wind energy cascade rates at MHD scales but could affect smaller scales. The magnetosheath displays high amplitude wave activity, with high occurrence rate of Alfvén waves. Turbulence appears not fully developed in the magnetosheath, suggesting fluctuations do not have enough time to interact in this small-size region. Some studies suggest PCW affect turbulence in the magnetosheath. Overall, wave activity is reduced inside the magnetic pile-up region and the Martian ionosphere. However, under certain conditions, upstream waves can reach the upper ionosphere. So far, there have not been conclusive observations of Alfvén waves in the ionosphere or along crustal magnetic fields, which could be due to the lack of adequate observations.


## 5.1. INTRODUCTION

More than 80 years ago, Hannes Alfvén proposed the existence of a new type of electromagnetic-hydrodynamic wave capable of propagating in a conductive fluid medium (Alfvén, 1942). Currently known as Alfvén waves, they are the result of the coupling between the plasma motion and magnetic field line stresses. Plasma flow across magnetic field lines bends them, perturbing the current density field, modifying the total magnetic field force, and ultimately affecting plasma motion itself. Alfvén waves, fast and slow magnetosonic waves, are the three low-frequency normal modes present in a magnetized ideal conductive fluid medium, whose evolution over long time scales is described by means of magnetohydrodynamics or MHD theory (e.g., Cramer, 2001). In the MHD regime, the Alfvén wave mode is incompressible, not presenting perturbations in the plasma density and particle pressure. The magnetic field strength is also unaffected by this normal wave mode. Identifying wave modes is important as, under many conditions, they constitute


*[1]Department of Astronomy, University of Maryland, College Park, MD, USA*

*[2]Planetary Magnetospheres Laboratory, NASA Goddard Space Flight Center, Greenbelt, MD, USA*

*[3]Department of Physics and Astronomy, West Virginia University, Morgantown, WV, USA*

*[4]Department of Physics and Astronomy, University of Iowa, Iowa City, IA, USA*


*Alfvén Waves Across Heliophysics: Progress, Challenges, and Opportunities, Geophysical Monograph 285*, First Edition.
Edited by Andreas Keiling.

DOI: 10.1002/9781394195985.ch05

a non-collisional coupling in space and astrophysical plasmas. Moreover, this identification also allows us to indirectly infer other physical processes occurring in a variety of environments, such as the solar wind and the magnetospheres of planets (e.g., Kivelson & Russell, 1995).

The pristine solar wind upstream from Mars has been studied in terms of incompressible Alfvénic turbulence (Parker, 1958; Pope, 2001). Nonlinear wave activity is modeled in terms of the interaction of counter-streaming Alfvén waves and by the presence of an inertial range, where energy is transferred without dissipation throughout several spatial and temporal scales (e.g., Andrés et al., 2020; Bruno & Carbone, 2013; Frisch, 1995). In the MHD scales, the magnetic field power spectral density displays a Kolmogorov-like form. In this regime, the power of magnetic field fluctuations follows a power law with the wave frequency, and the spectral index is equal to −5/3 (e.g., Alexandrova, 2008; Andrés et al., 2020; Ruhunusiri et al., 2017).

The interaction between the magnetized solar wind and the Martian ionosphere contributes to a planetary magnetosphere that bears similarities with those of Venus and comets (e.g., Acuña et al., 1998, 1999; Bertucci et al., 2011; Cravens & Gombosi, 2004; Dubinin et al., 2023; Halekas et al., 2021; Mazelle et al., 2004; Nagy et al., 2004). On the other hand, the presence of crustal magnetic fields and the interaction with the Interplanetary Magnetic Field (IMF) is responsible for features also present in intrinsically magnetized planetary magnetospheres. In particular, crustal magnetic fields create mini-magnetospheres, with characteristics that resemble those of magnetized planets but at localized scales; they can magnetically reconnect with the solar wind IMF, providing the solar wind access to localized regions of the ionosphere (in analogy to the cusp regions at Earth); and contribute to the dynamics of the environment. Indeed, crustal magnetic fields rotate with the planet, adding to the proper solar wind variability (e.g., Acuña et al., 1998, 1999; Brain et al., 2003; Bowers et al., 2023; DiBraccio et al., 2022; Harada et al., 2018). As a result, the interaction between the solar wind/IMF, the Martian ionosphere, and the remanent crustal magnetic fields results in a hybrid planetary magnetosphere (e.g., DiBraccio et al., 2022; Dubinin et al., 2023).

The first boundary the solar wind encounters when interacting with the Martian magnetosphere is the bow shock, as shown in Figure 5.1. Electromagnetic forces in this collisionless boundary slow down and heat the incoming supermagnetosonic solar wind. As the upstream solar wind Alfvén Mach number is high at Mars heliocentric distances ($M_A \sim 11$), backstreaming

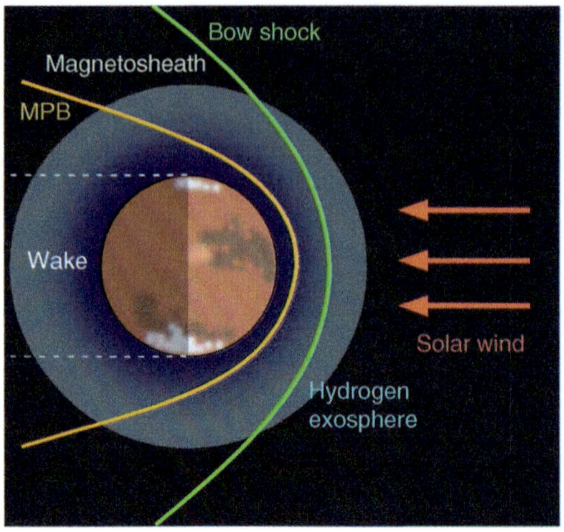


**Figure 5.1** Schematic representation of the Martian magnetosphere. The bow shock is the boundary that separates the upstream solar wind from the magnetosheath region. It contributes to solar wind heating and deceleration. The magnetic pile-up boundary (or MPB) is the inner boundary that separates the magnetosheath from the magnetic pile-up region. Downstream of the MPB, the plasma is mostly composed of planetary heavy ions. Due to Mars' relatively weak gravity field and the lack of an intrinsic global magnetic field, the hydrogen corona extends beyond Mars's bow shock. Source: Taken from Ruhunusiri et al. (2017).

protons are generated, which contribute to diverting the solar wind around the planet (Biskamp, 1973; Edmiston & Kennel, 1984; Gosling & Robson, 1985; Halekas et al., 2017; Phillips & Robson, 1972; Paschmann et al., 1980; Slavin & Holzer, 1981; Sonnerup, 1969). Thus, a foreshock is present at Mars under nominal solar wind conditions (Bale et al., 2005; Burgess et al., 2005; Eastwood et al., 2005; Jarvinen et al., 2022). The backstreaming protons interact with the incoming solar wind and give rise to low-frequency waves immersed in the Martian foreshock (e.g., Jarvinen et al., 2022; Meziane et al., 2017; Romanelli, Mazelle, & Meziane, 2018). In addition, the shock generates electromagnetic plasma waves capable of overcoming the solar wind velocity and traveling upstream (e.g., Mazelle et al., 2004). The average bow shock stand-off distance is $\sim 1.6 R_M$, where $1 R_M \sim 3390$ km, significantly smaller relative to those associated with intrinsically magnetized planets (e.g., Eastwood et al., 2005; Gruesbeck et al., 2018; Mazelle et al., 2004; Turc et al., 2023). For instance, the average Earth's bow shock stand-off distance is $\sim 14 R_E$, where $R_E$ stands for Earth's radius ($1 R_E \sim 6378$ km) (e.g., Fairfield, 1971; Formisano, 1979).

The region downstream of the bow shock, the magnetosheath, is characterized by compressed, heated, and slower solar wind plasma and high-amplitude wave activity (e.g., Fowler et al., 2017; Halekas, 2017; Halekas et al., 2017). The sources of these waves are numerous, from transmission through the bow shock, to local generation to energy transfer between modes and scales (e.g., Dubinin & Fraenz, 2016; Ruhunusiri et al., 2017). The inner boundary of the magnetosheath is the magnetic pile-up boundary or MPB, whose average stand-off distance is $\sim 1.2\,R_M$. Dowstream from the MPB, the background magnetic field drapes and piles up around Mars (e.g., Bertucci et al., 2011; Espley, 2018; Romanelli et al., 2014; Trotignon et al., 2006). Moreover, the plasma is mostly composed of planetary heavy ions and the magnetic field strength is generally larger than in the magnetosheath, resulting in a region of relatively dense and low plasma beta and generally less intense wave activity (Fowler et al., 2017; Nagy et al., 2004). Additionally, Mars also has a magnetotail consisting of magnetic lobes of opposite polarity separated by a current sheet whose location depends on the IMF (e.g., Crider et al., 2004; Curry et al., 2022; DiBraccio et al., 2017, 2018, 2022; Harada et al., 2015; Ramstad et al., 2020; Romanelli et al., 2015; Romanelli, Modolo, Leblanc, Chaufray, Hess, et al., 2018b). A significant fraction of the total planetary ion escape occurs down the magnetotail region, with particle fluxes dependent on position, particle energy, and external conditions (such as the orientation of the IMF), among other parameters (e.g., Curry et al., 2022; Dubinin et al., 2011).

Mars also has a hydrogen (H) exosphere that extends beyond the bow shock (Bhattacharyya et al., 2015; Chaffin et al., 2015). This is due to Mars's relatively weak gravity and small size magnetosphere due, in turn, to the lack of a global intrinsic magnetic field (Acuña et al., 1998). The neutral exospheric atoms are ionized by photoionization, charge exchange, and electron impact, and give rise to a newborn planetary proton population, initially nearly at rest with respect to Mars (Rahmati et al., 2015, 2017; Yamauchi et al., 2015). Seasonal variability of the H corona results in seasonal variability of the newborn planetary proton population, also creating seasonal variability in physical processes occurring inside and upstream of the magnetosphere of Mars. In particular, previous studies have reported seasonal and long-term variability of low-frequency wave activity upstream of Mars and effects inside the magnetosphere (e.g., Bhattacharyya et al., 2015; Halekas et al., 2017, 2020; Hughes et al., 2019; Jiang et al., 2023; Romanelli et al., 2016; Romeo et al., 2021; Ruhunusiri et al., 2017; Yamauchi et al., 2015).

In addition, the relatively small scale size of the Martian magnetosphere means that characteristic length scales of the plasma (such as the proton gyro radius) are comparable to the length scales of the system (e.g., bow shock stand-off distance), implying that the shocked solar wind does not have space to fully thermalize before encountering the planet. Kinetic effects are thus likely to play a key role in the Mars–solar wind interaction (e.g., Dong et al., 2015; Jarvinen et al., 2022; Kallio et al., 2011; Moses et al., 1988; Romanelli et al., 2019). In particular, although some of the observed wave properties may be understood in terms of MHD normal modes, previous analyses have shown statistical and case studies where wave polarization and frequency and the local ion and electron velocity distribution functions necessarily demand a kinetic description. This is the case, for instance, for most ultra-low frequency (ULF) waves observed upstream from the Martian bow shock and in the foreshock. The observed polarization is typically elliptical or circular, while MHD predicts linearly polarized modes. Moreover, particle populations in these regions are far from Maxwellians (e.g., Jarvinen et al., 2022; Mazelle et al., 2004; Meziane et al., 2017; Romanelli et al., 2016; Romeo et al., 2021; Russell, 1994). It is also worth mentioning that the presence of boundaries (bow shock and MPB) affects the possible wave modes excited in the Martian magnetosphere. Examples of associated processes are mode conversion, wave reflection as well as surface waves, among others.

The relatively small size of the hybrid Martian magnetosphere also suggests that waves resulting from the interaction with the solar wind should be present downstream of the bow shock and even in the ionosphere. Subsequent observations have confirmed that this is the case and the magnetosphere of Mars is now known to be a highly dynamic environment, where processes taking place upstream from the bow shock can significantly impact the underlying ionosphere (e.g., Fowler et al., 2017; Halekas et al., 2017; Jiang et al., 2023; Ruhunusiri et al., 2017). Studying the electromagnetic wave environment at Mars is thus very important as waves can facilitate energy propagation and transfer in the collisionless environment, heat the ionosphere and potentially modify planetary ion escape (e.g., Collinson et al., 2018; Ergun et al., 2006; Fowler, Andersson, Ergun, et al., 2018; Jarvinen et al., 2022). Furthermore, waves can lead or modify magnetic field reconnection conditions in the Martian magnetosphere (Bowers et al., 2023; Chen et al., 2022). In other words, the presence of an extended H exosphere, the relatively small size of the Martian magnetosphere, and the remanent crustal magnetic fields, among other parameters, provide a compelling plasma environment in the solar system to study electromagnetic plasma wave generation and evolution.

Different planetary missions to Mars have allowed the identification of several low-frequency waves. The first observations of wave activity at Mars were obtained by the Phobos-2 mission (e.g., Delva & Dubinin, 1998; Grard et al., 1989; Russell et al., 1990; Skalsky et al.,

1998). Since then, analysis of magnetic, electric, and plasma observations by Phobos-2, Mars Global Surveyor (MGS), Mars EXpress (MEX), and Mars Atmosphere and Volatile EvolutioN (MAVEN) missions to Mars presented us with a picture where certain types of waves are more likely to be observed in different regions of the Martian magnetosphere and upstream, with variability in their properties and occurrence rate. Despite this progress, comprehensive analysis of low-frequency waves in the Martian magnetosphere has been limited for several reasons. The Phobos-2 mission had a relatively high periapsis (~900 km) and lasted ~8.5 months, resulting in a relatively small amount of data. MGS lacked an ion instrument capable of measuring the bulk properties of local plasma. Additionally, most of its orbits were Sun-synchronous at ~400 km, affecting the spatial coverage of the Martian magnetosphere throughout this mission (September 11, 1997 to November 2, 2006). Since December 25, 2003, MEX has explored the Martian magnetosphere, providing valuable measurements that continue to improve our understanding of plasma processes. However, the lack of a magnetometer onboard MEX puts constraints on the characterization of wave modes around Mars. Inserted into orbit around Mars on September 21, 2014, MAVEN is the first and only spacecraft so far to observe all main regions of the magnetosphere carrying a magnetometer, an ion plasma analyzer, and an electric field instrument. Moreover, its precessing orbit also allows sampling of the entire magnetosphere under different solar wind and solar cycle conditions (Jakosky et al., 2015).

In this chapter, we provide a review focused on Alfvén waves, taking into account the presence of other wave modes. Several excellent reviews on low-frequency waves at Mars provide a comprehensive context for this work (Delva et al., 2011; Dubinin & Fraenz, 2016; Glassmeier & Espley, 2006; Mazelle et al., 2004; Russell, 1994). This chapter is centered on and highlights some of the many recent observations and conclusions provided by the ongoing MAVEN mission (Jakosky et al., 2015).

## 5.2. ALFVÉN WAVES UPSTREAM FROM THE MARTIAN BOW SHOCK

Alfvén waves of several wavelengths are present in the upstream solar wind. These waves are responsible for correlated fluctuations in the magnetic and velocity fields, which can affect the Martian plasma environment (Andrés et al., 2020; Belcher & Davis, 1971; Halekas et al., 2017). In addition, the nonlinear interaction between counter-streaming Alfvén waves is considered to be responsible for the energy cascade, by which energy is transferred throughout several scales of the system. Under many conditions, the sense of the nonlinear energy transfer is directed from the larger wavelengths (or timescales) to the smaller ones. This is the so-called direct incompressible energy cascade of the solar wind (Alexandrova, 2008; Andrés et al., 2020; Bruno & Carbone, 2013; Maron & Goldreich, 2001; Romanelli et al., 2022; Ruhunusiri et al., 2017).

A study conducted by Halekas et al. (2017) made use of 45 s average MAVEN Magnetometer (MAG) and Solar Wind Ion Analyzer (SWIA) observations to study the Alfvénic content in the Martian magnetosphere and upstream of the bow shock (Connerney et al., 2015; Halekas et al., 2015). Among the explored properties, the authors computed the normalized cross-helicity $\sigma_C$, the normalized residual energy $\sigma_r$, and the Alfvén ratio, defined as follows:

$$\sigma_C = \frac{2\langle \delta\mathbf{u} \cdot \delta\mathbf{u}_A \rangle}{\langle \delta\mathbf{u}^2 + \delta\mathbf{u}_A^2 \rangle} \tag{5.1}$$

$$\sigma_r = \frac{\langle \delta\mathbf{u}^2 - \delta\mathbf{u}_A^2 \rangle}{\langle \delta\mathbf{u}^2 + \delta\mathbf{u}_A^2 \rangle} \tag{5.2}$$

$$r_A = \frac{\langle \delta\mathbf{u}^2 \rangle}{\langle \delta\mathbf{u}_A^2 \rangle} \tag{5.3}$$

where $\delta\mathbf{u}$ and $\delta\mathbf{u}_A$ are the solar wind bulk and Alfvén incompressible velocity fluctuations, respectively. The latter is defined as $\mathbf{u}_A \equiv \mathbf{B}/\sqrt{\mu_0\rho_0}$, where $\rho_0$ is the mean mass plasma density and $\mu_0$ is the vacuum magnetic permeability. The angular bracket $\langle\rangle$ indicates a time average over each 30-min analyzed interval. Note that, by definition, $-1 \leq \sigma_C \leq 1$, $-1 \leq \sigma_r \leq 1$, and $r_A \geq 0$.

The normalized cross-helicity $\sigma_C$ is a measure of the linear correlation between velocity and magnetic field fluctuations. A value of $\sigma_C = \pm 1$ is consistent with an Alfvén wave propagating antiparallel or parallel to the background magnetic field direction. Usually, fluctuations with $|\sigma_C| \sim 1$ are described as Alfvénic. The normalized residual energy and the Alfvén ratio quantify the energy balance between the kinetic and magnetic field fluctuations. In particular, a value of $\sigma_r = 0$ is consistent with energy equipartition, and values of $\sigma_r < 0$ show a given event has more energy in the magnetic field fluctuations. Consistently, a value of $r_A = 0$ implies all energy is present in the magnetic field fluctuations and $r_A = 1$ is associated with energy equipartition.

Halekas et al. (2017) showed that the solar wind upstream from the Martian bow shock has a majority of Alfvén waves that propagate outward from the Sun. Figure 5.2a–c displays a larger amount of waves with negative (positive) normalized cross-helicities for +By/−Bx (−By/+Bx) Mars Solar Orbital (MSO) IMF configuration (Roberts et al., 1987). The MSO coordinate system is centered on Mars with the X-axis pointing toward the Sun, and the Z-axis perpendicular to Mars's orbital plane and positive toward the ecliptic north. The Y-axis completes

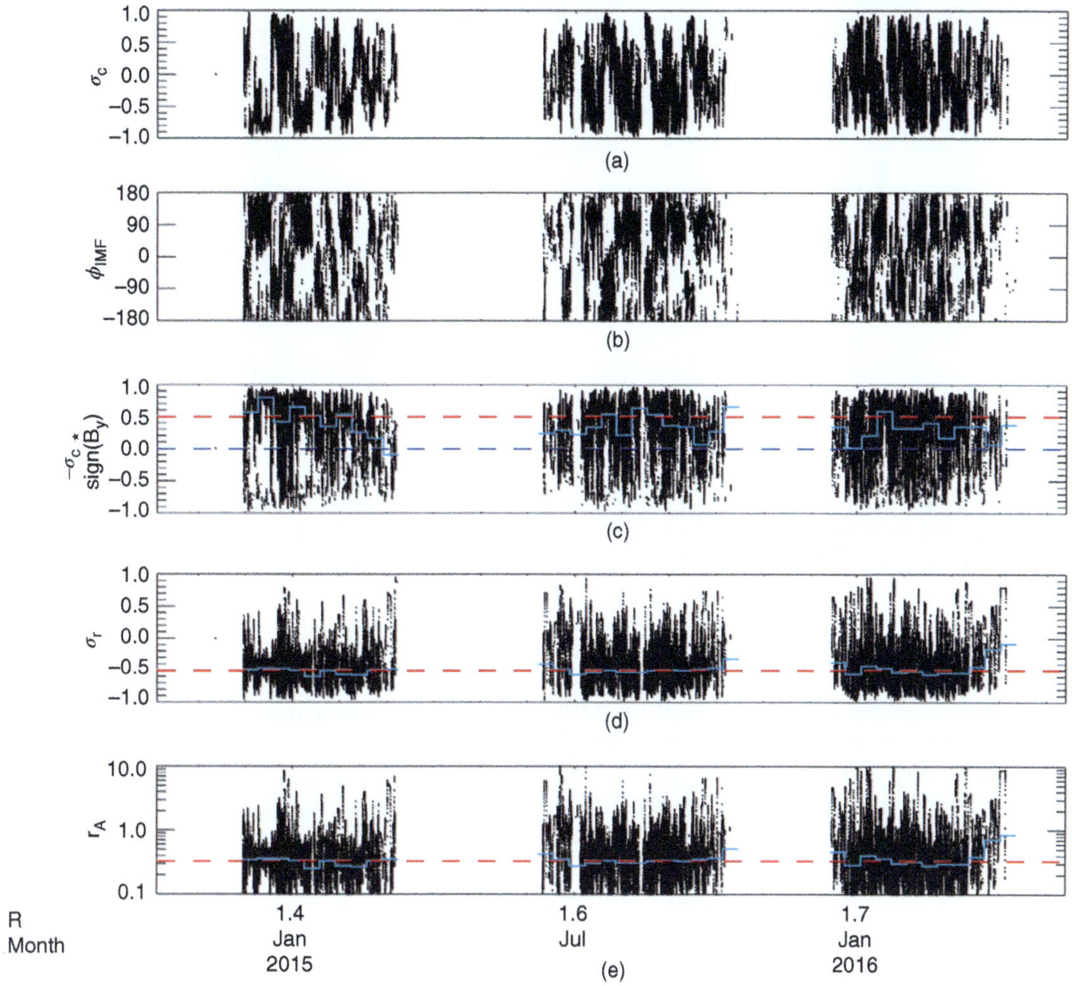


**Figure 5.2** Solar wind properties computed from 30-min time intervals upstream of Mars' bow shock as a function of time, between October 2014 and May 2016 (black points). (a) The normalized cross helicity ($\sigma_C$), (b) azimuthal angle of the IMF in the X-Y MSO plane $\phi_{IMF} = tan^{-1}(By/Bx)$, (c) $-\sigma_C$ multiplied by the sign of the IMF dawn-dusk (By) MSO component, (d) normalized residual energy ($\sigma_r$), (e) and Alfvén ratio ($r_A$). The light blue lines in panels (c-e) show the corresponding median values over 10-day intervals. The dashed red lines display the corresponding median values from March to May 2014 (MAVEN's cruise phase) for comparison. Note the dashed blue line in panel (c) is equal to zero and is shown for reference. Text labels indicate Mars's heliocentric distance (R) in astronomical units. MSO: Mars Solar Orbital. IMF: Interplanetary Magnetic Field. Source: Adapted from Halekas et al. (2017).

the right-handed system. Moreover, Figure 5.2d also shows that solar wind magnetic field fluctuations have more energy than velocity fluctuations, with normalized residual energies on the order of $\sim -0.5$. These results are consistent with previous observations of the solar wind. They provide additional information for the characterization and modeling of the occurrence of Alfvén waves with heliocentric distance, latitude, solar cycle, and solar wind properties (Bavassano et al., 1998; Breech et al., 2005; Bruno & Carbone, 2013; Tu & Marsch, 1991).

Interestingly, $\sigma_C$ was found to decrease in the quasi-parallel foreshock, possibly associated with effects from nonlinear compressive waves present in this region (see the top and middle row panels, Figure 13 in Halekas et al., 2017). On the other hand, the Alfvén ratio was found to increase as the solar wind approaches the Martian bow shock, as shown in Figure 5.3, making use of cylindrical Mars Solar Electric (MSE) coordinates. The MSE coordinate system is centered at Mars and its axes are defined as follows. The X-axis points toward the Sun, including correction for solar wind aberration due to Mars' orbital motion, and the Z-axis points along the convective electric field seen in the planet's rest frame ($\mathbf{E}_{SW} = -\mathbf{U}_{SW} \times \mathbf{B}_{SW}$), where $\mathbf{U}_{SW}$ and $\mathbf{B}_{SW}$ are the solar wind velocity and IMF, respectively. The Y-axis completes the right-handed system. These observations

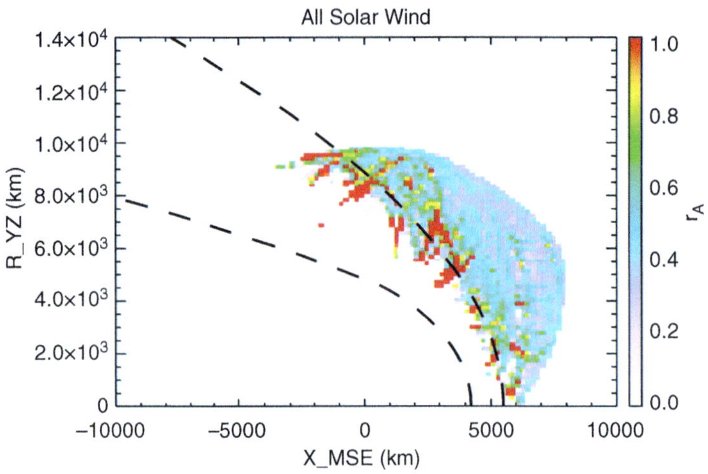


**Figure 5.3** The normalized Alfvén ratio for solar wind measurements between October 2014 and May 2016, in cylindrical MSE coordinates. The dashed black curves display the bow shock and magnetic pile-up fits from Trotignon et al. (2006). MSE: Mars Solar Electric. Source: Adapted from Halekas et al. (2017).

constitute another example of Mars affecting the solar wind, even upstream from the bow shock. The computed Alfvén ratio increase, from ~0.2 to ~1, is similar for both the quasi-parallel and quasi-perpendicular bow shock regions (not shown), suggesting it is likely not due to foreshock effects. As discussed in Halekas et al. (2017), the observed energy transfer from magnetic to kinetic fluctuations could be associated with solar wind mass loading by planetary pick-up ions, especially protons resulting from Mars's extended H exosphere (Halekas, 2017; Rahmati et al., 2017; Yamauchi et al., 2015).

Mass-loading is only one of several macroscopic effects associated with cumulative planetary ion pick-up. Analogous to comet pick-up ion physics, the newly ionized planetary proton population represents an additional, non-thermal component, of the total proton velocity distribution function (e.g., Delva et al., 2011, 2015; Gary, 1993; Mazelle et al., 2004). This distribution is capable of generating electromagnetic plasma waves upstream from the Martian bow shock, in association with different plasma instabilities (Gary, 1993). These waves affect the pristine solar wind before it encounters the bow shock and can be transmitted into the magnetosheath, having also effects over processes in the Martian magnetosphere (Andrés et al., 2020; Jiang et al., 2023; Romanelli et al., 2022; Romeo et al., 2021; Ruhunusiri et al., 2017, 2015).

Several parameters determine which plasma instability has the largest linear wave growth rate. Among them is the IMF cone angle, that is, the angle between the IMF and the solar wind velocity at the time of pick-up. As shown in Brinca and Tsurutani (1989), the ion–ion right-hand (RH) resonant instability is predominant when the IMF cone angle is smaller or equal to 75°. The ion–ion left-hand (LH) resonant instability is the most unstable for IMF cone angles larger than ~75° (Brinca and Tsurutani, 1989). Both ion–ion resonant instabilities have maximum linear wave growth rates for propagation parallel to the background magnetic field (Brinca, 1991; Gary, 1993). Therefore, the presence of pick-up ions is also responsible for the generation of ultra-low frequency waves, which, in the Hall-MHD formalism, correspond to the fast magnetosonic (right-hand polarization) and the ion cyclotron waves (left-hand polarization), respectively. Such wave modes tend, in turn, to the MHD fast magnetosonic and the Alfvén linearly polarized wave modes for very low frequencies, under parallel propagation conditions (Cramer, 2001). Interestingly, the Doppler shift is responsible for the observation of these waves (RH and LH) at a frequency near the local proton cyclotron frequency in the spacecraft frame, with a left-handed elliptical polarization (Brinca, 1991; Gary, 1993; Gary et al., 1989). It is because of this particular property that these waves are sometimes referred to as proton cyclotron waves (or PCW). However, as mentioned before, only under very specific conditions PCW do they actually correspond to the ion cyclotron (Alfvénic) wave branch. Note that the nomenclature PCW, is based on wave properties seen in the spacecraft reference frame, while the ion cyclotron wave mode is defined in the plasma rest frame.

The first report of PCW upstream from Mars' bow shock was made by Russell et al. (1990). These ultra-low frequency waves have also been detected and analyzed in greater detail thanks to magnetic field observations provided by MGS and MAVEN missions (Bertucci et al., 2013; Brain et al., 2002; Liu et al., 2020; Mazelle et al., 2004; Romanelli et al., 2013, 2016; Romeo et al., 2021; Wei & Russell, 2006; Wei et al., 2011, 2014). These waves were consistently found left-handed elliptically polarized in the spacecraft reference frame and propagating quasi-parallel to the mean IMF (with propagation angles on the order of 20°), which has been associated with nonlinear effects (e.g., Romanelli et al., 2013).

MGS and MAVEN allowed the identification of a long-term trend, with higher PCW occurrence rate near the Martian perihelion (Bertucci et al., 2013; Romanelli et al., 2013, 2016; Romeo et al. 2021). Such an annual trend is likely associated with seasonal changes in the exospheric hydrogen density and resulting newborn planetary proton density (Halekas, 2017; Rahmati et al., 2017, 2018; Yamauchi et al., 2015). Figure 5.4 shows the PCW occurrence rate as a function of time, as seen by MAVEN between October 2014 and February 2020. Romeo et al. (2021) reported that this rate displays an increase up to ~30–35% near (slightly after) the Martian perihelion and southern summer solstice. These values are an order of magnitude larger than the average value near the Martian aphelion (~2%), in agreement with previous studies (Bertucci et al., 2013; Romanelli et al., 2013, 2016). It is also worth noticing that the PCW occurrence rate increase takes place during part of the Martian dust storm season, marked by gray regions in Figure 5.4b (Romeo et al., 2021). In this context, Chaffin et al. (2021) reported that a regional dust storm observed well after the perihelion increased planetary H escape by a factor of five to ten, suggesting that dust dynamics affect exospheric H densities more than seasonal variations.

Recent studies have focused on the effects that PCWs have on Alfvénic turbulence in the pristine solar wind (Andrés et al., 2020; Romanelli et al., 2022; Ruhunusiri et al., 2017). In particular, Andrés et al. (2020) analyzed 4 months of MAVEN magnetic field and plasma data, to determine if the magnetic field power spectra and/or the energy cascade rates are affected by PCW. Figure 5.5 displays the magnetic field power spectral density (PSD) associated with 30-min intervals in the pristine solar wind (Gruesbeck et al., 2018) when the IMF cone angle was approximately constant. All these events (sets A and B) have a normalized fluctuation plasma density smaller than 20%, thus corresponding to nearly incompressible solar wind conditions. Set A contains cases with PCW, while these waves are absent in the events contained in Set B (Andrés et al., 2020). As can be seen, Figure 5.5, left panel, shows a clear peak in the PSD at the local proton cyclotron frequency for all events. These results also show a power law decay consistent

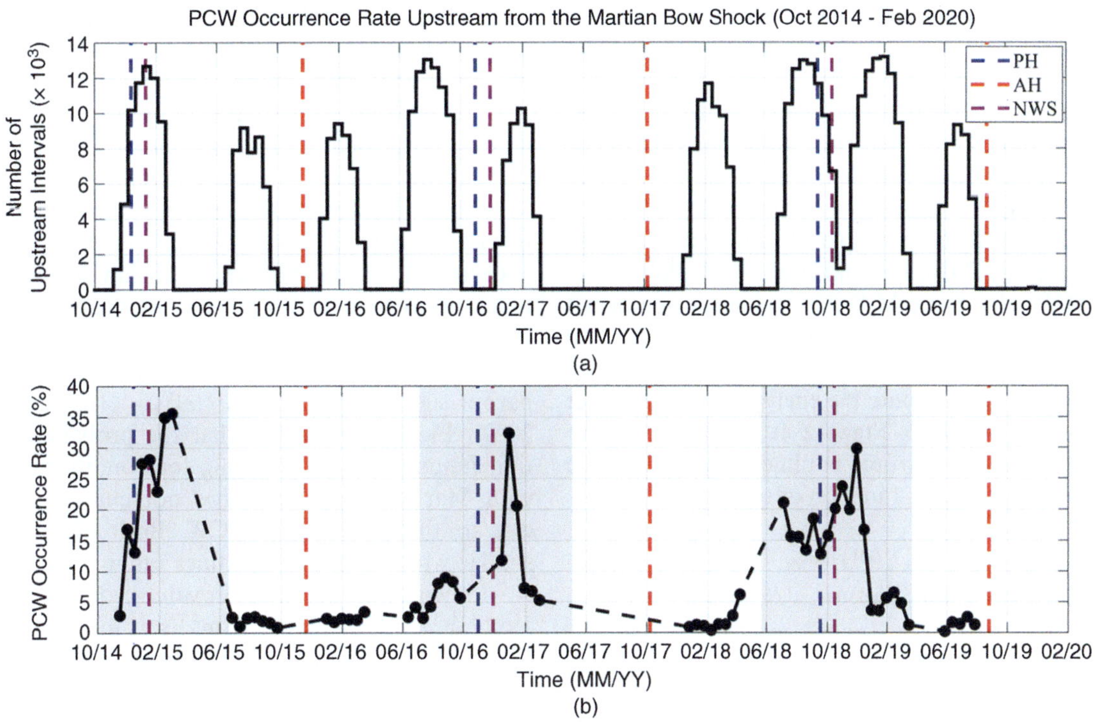


**Figure 5.4** (a) Histogram of the number of analyzed 512 s time intervals ($\times 10^3$) upstream from Mars's bow shock as a function of time (15-day bins), between October 2014 and February 2020. (b) Derived PCW occurrence rate (%) as a function of time. The dashed black line is associated with data gaps in the upstream region (see upper panel). Gray-shaded areas correspond to periods of dust storm seasonal activity. (a-b) The blue, purple, and red vertical dashed lines correspond to the Martian perihelion (PH), the northern winter solstice (NWS), and the aphelion (AH), respectively. PCW: Proton Cyclotron Waves. Source: Romeo et al. (2021) / John Wiley & Sons.

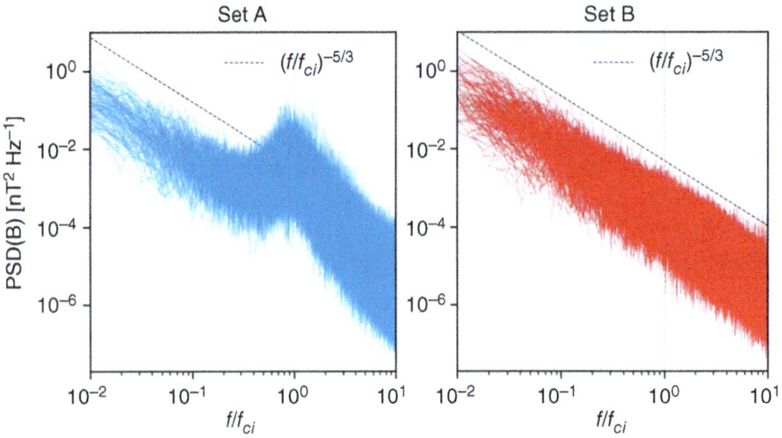


**Figure 5.5** Magnetic field power spectra density as a function of the observed normalized frequency for 30-min events, seen by MAVEN MAG when is upstream and magnetically disconnected from the bow shock of Mars. Set A (left) and Set B (right) consist of 184 events with PCW and 208 events without PCW, respectively. MAVEN: Mars Atmosphere and Volatile EvolutioN. MAG: Magnetometer. PCW: Proton Cyclotron Waves. Source: Taken from Andrés et al. (2020).

with a Kolmogorov spectrum. Indeed, wave power can be modeled as $P \propto P_0 f^{-\gamma}$, where $f$ is the frequency and the spectral index, $\gamma = -5/3$, is constant in the MHD scales for all cases in both sets. The observed decay is consistent with Alfvénic turbulence and does not appear to be affected by PCW at the MHD scales (Andrés et al., 2020).

Andrés et al. (2020) also estimated, for the first time, the absolute value of the incompressible energy cascade rate of the solar wind upstream from Mars bow shock at MHD scales. For this, the authors made use of the exact relation for incompressible MHD turbulence (see, e.g., Andrés et al., 2016; Ferrand et al., 2021; Politano & Pouquet, 1998a,b). Since MAVEN is a single spacecraft mission, Andrés et al. (2020) assumed full isotropy and made use of the Taylor hypothesis to integrate Equation 4 in their work and compute the energy cascade rate per unit volume, $\varepsilon$ (see, also, Stawarz et al., 2011). Under these conditions, the isotropic nonlinear energy cascade rate can be expressed as a function of time lags $\tau$ as:

$$\varepsilon = \rho_0 \langle [(\Delta \mathbf{u} \cdot \Delta \mathbf{u} + \Delta \mathbf{u}_A \cdot \Delta \mathbf{u}_A)\Delta u_\ell - (\Delta \mathbf{u} \cdot \Delta \mathbf{u}_A + \Delta \mathbf{u}_A \cdot \Delta \mathbf{u})\Delta u_{A\ell}]/(-4\tau U_0/3)\rangle \quad (5.4)$$

where $u_\ell = \mathbf{u} \cdot \hat{U}_0$, $u_{A\ell} = \mathbf{u}_A \cdot \hat{U}_0$, and $U_0$ is the mean plasma flow speed. The angular bracket $\langle \cdot \rangle$ denotes an ensemble average (Pope, 2001), which was taken as the time average assuming ergodicity. Note that Equation 5.4 makes use of the increments definition, $\Delta\alpha \equiv \alpha' - \alpha$, where fields are evaluated at position $\mathbf{x}$ or $\mathbf{x}' = \mathbf{x} + \boldsymbol{\ell}$. A prime is added to the corresponding field in the latter case. Equation 5.4 shows that under the considered assumptions, $\varepsilon$ is fully defined by velocity and magnetic field time increments, that Andrés et al. (2020) computed based on MAVEN MAG and SWIA data.

Andrés et al. (2020) reported an increase in the absolute value of the nonlinear solar wind incompressible energy cascade rate at MHD scales ($\langle|\varepsilon|\rangle_{MHD}$), when PCW are detected in the pristine solar wind. However, the known variability of PCW with the Martian heliocentric distance and the relatively small data set analyzed prevented the authors from determining the factor(s) responsible for such an increase. Making use of more than 5 years of MAVEN measurements and a similar methodology, Romanelli et al. (2022) concluded that $\langle|\varepsilon|\rangle_{MHD}$ displays a decreasing trend with the Martian heliocentric distance, in agreement with other studies (e.g., Andrés et al., 2021; Bandyopadhyay et al., 2020; Hadid et al., 2017). Moreover, the presented results suggest that PCW do not have a significant effect on $\langle|\varepsilon|\rangle_{MHD}$ (Romanelli et al., 2022). Figure 5.6a shows that the probability distribution function of log $\langle|\varepsilon|\rangle_{MHD}$ for conditions associated with Martian perihelion does not change significantly due to the presence of PCW. Both distributions are similar with log $\langle|\varepsilon|\rangle_{MHD}$ values ranging from $\sim -19$ to $\sim -15$, with $\langle|\varepsilon|\rangle_{MHD}$ median values equal to $1.4 \times 10^{-17}$ J m$^{-3}$ s$^{-1}$ and $1.5 \times 10^{-17}$ J m$^{-3}$ s$^{-1}$, for cases at perihelion with and without PCW, respectively. In contrast, Figure 5.6b displays a clear shift of the distribution with larger $\langle|\varepsilon|\rangle_{MHD}$ values around the Martian perihelion. The $\langle|\varepsilon|\rangle_{MHD}$ median value is equal to $1.5 \times 10^{-17}$ J m$^{-3}$ s$^{-1}$ and $4.6 \times 10^{-18}$ J m$^{-3}$ s$^{-1}$, for perihelion and aphelion conditions, respectively (Romanelli et al., 2022).

It is worth noticing that these results do not rule out the possibility of PCW effects on Alfvénic turbulence at kinetic scales. Complementary studies are needed to

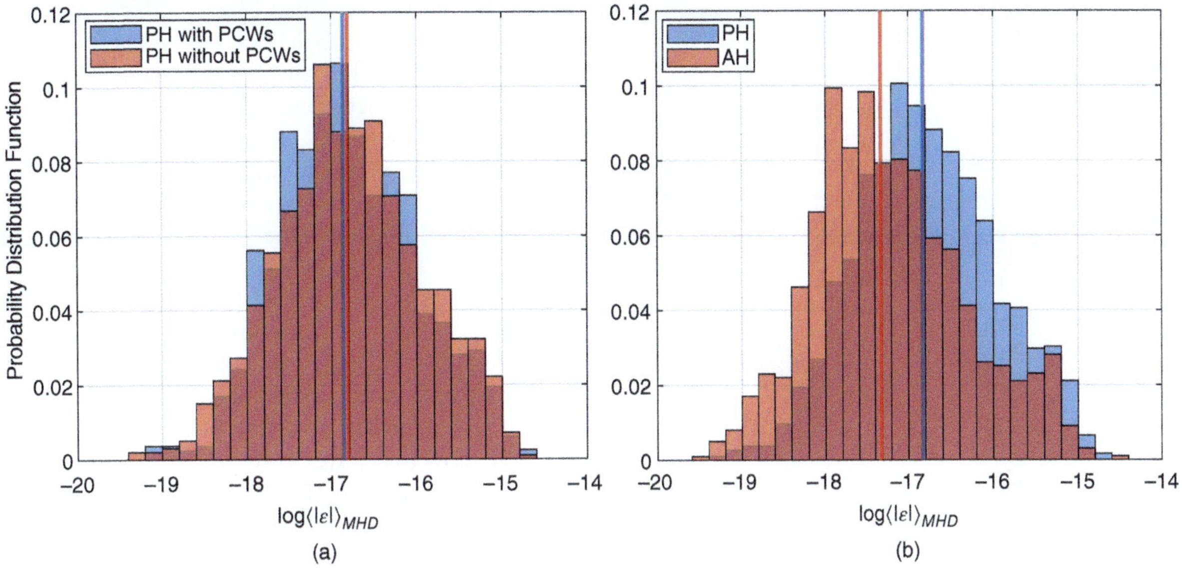


**Figure 5.6** Probability distribution function of $\log\langle|\varepsilon|\rangle_{MHD}$ for different Martian orbital locations and PCW activity. (a) Martian perihelion with PCW (blue) and without PCW (red) (b) Martian perihelion conditions (with and without PCW, blue) and aphelion conditions (with and without PCW, red). (a-b) The vertical solid lines correspond to the median of the respective distributions. PCW: Proton Cyclotron Waves. Source: Romanelli et al. (2022) / IOP Publishing / CC BY 4.0.

potentially identify these effects, using plasma observations with a higher sampling frequency and theoretical models of solar wind turbulence in this regime (Andrés et al., 2018; Galtier, 2008). Moreover, future studies could also focus on the effects that planetary foreshocks and associated wave activity have on solar wind turbulence. Planetary foreshocks are filled with electromagnetic plasma waves in different frequency ranges and can affect the nonlinear energy cascade rate in the MHD and kinetic scales (Burgess et al., 2012; Eastwood et al., 2005). The variability of solar wind properties and IMF strength and direction with heliocentric distance also offers the possibility to perform comparative analyses (e.g., Andrés et al., 2015; Glass et al., 2023; Jarvinen et al., 2022; Meziane et al., 2017; Romanelli & DiBraccio, 2021; Romanelli, Mazelle, & Meziane, 2018; Romanelli et al., 2020a). In this regard, numerical simulations also constitute a particularly useful approach to understand better these environments, improve data interpretation, and point toward fundamental physical processes taking place (e.g., Jarvinen et al., 2022).

## 5.3. ALFVÉN WAVES IN THE MARTIAN MAGNETOSHEATH

Identifying wave modes in planetary magnetosheaths is particularly challenging for several reasons. Waves in the magnetosheath can be generated upstream and convected by the solar wind, can be generated at the bow shock, and/or be associated with local plasma instabilities (e.g., Dubinin & Fraenz, 2016; Harada et al., 2019). In the case of Mars, the latter also includes effects derived from the presence of planetary protons and oxygen ions (e.g., Barabash et al., 1991; Chaffin et al., 2015; Clarke et al., 2017; Curry et al., 2015; Deighan et al., 2015; Dong et al., 2015; Feldman et al., 2011; Halekas, 2017; Rahmati et al., 2015, 2017; Yamauchi et al., 2015). Furthermore, normal wave modes can be coupled in a nonuniform medium such as the magnetosheath (Cramer, 2001).

It is also important to remark that conditions in the magnetosheath present asymmetries related to the local IMF and due to the deflection of the solar wind around Mars (Dubinin et al., 2018; Romanelli et al., 2020b). The former gives rise to temperature anisotropies, particularly downstream of the quasi-perpendicular bow shock (Espley et al., 2004; Halekas et al., 2020; Jin et al., 2022; Simon-Wedlund et al., 2022a, 2022b). The latter is responsible, for example, for superAlfvénic conditions in the Martian magnetosheath flanks, while the shocked solar wind is significantly decelerated near the subsolar MPB (Halekas et al., 2020).

A detailed study on the electromagnetic wave power spatial distribution around Mars at different frequency ranges has been reported by Fowler et al. (2017). The magnetosheath was found to be the region with the strongest magnetic field wave power in the ULF range (Dubinin & Fraenz, 2016; Fowler et al., 2017). This is associated with processes contributing to dissipating the bulk kinetic energy of the pristine solar wind (e.g., Auer et al., 1971;

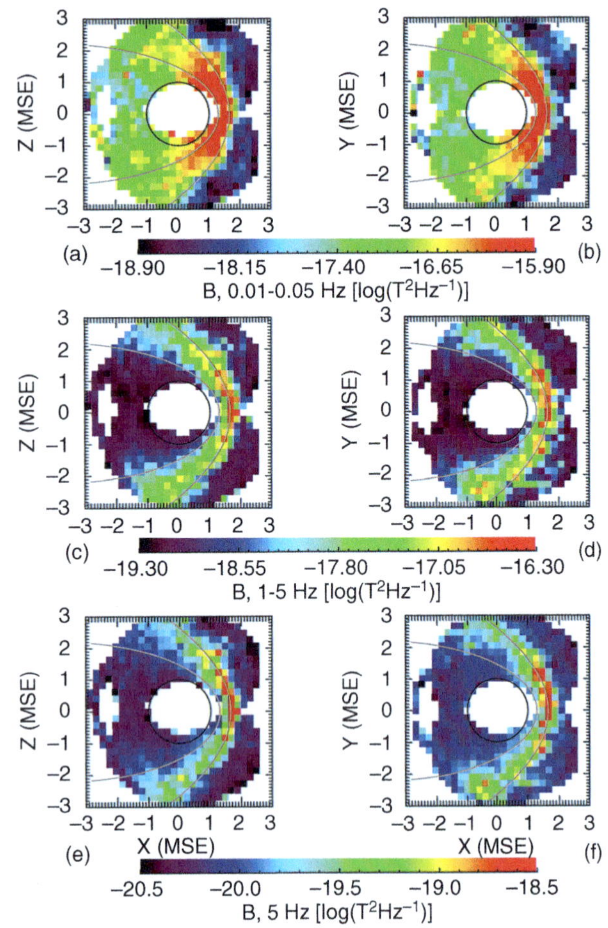


**Figure 5.7** Statistical magnetic field strength wave power for several frequency ranges, for altitudes greater than 600 km. (a and b) 0.01–0.05 Hz, (c and d) 1–5 Hz, and (e and f) 5 Hz. The left and right columns correspond to projections onto the X-Z and X-Y MSE planes, respectively. MSE: Mars Solar Electric. Source: Fowler et al. (2017) / John Wiley & Sons.

Burgess et al., 1989; Papadopoulos, 1971). Figure 5.7 left (right) column displays the magnetic field power distribution in the $X_{MSE} - Z_{MSE}$ ($X_{MSE} - Y_{MSE}$) plane.

Figures 5.7a and 5.7b show that most of the magnetic field wave power for frequencies between 0.01 Hz and 0.05 Hz lies in the magnetosheath, upstream of the terminator plane. Magnetic field wave power is also significant in the magnetotail, although it is approximately one order of magnitude weaker throughout this region. The absence of a significant decrease in magnetic field ULF wave power between the magnetosheath and the ionosphere, shown in Figures 5.7a and 5.7b, is consistent with the expected lack of complete ion thermalization in the magnetosheath. Indeed, the relatively small size of the Martian magnetosheath (its thickness along the Mars–Sun axis is ~1200 km) compared to ion spatial scales (e.g., the convected solar wind proton gyroradius is ~1000 km) implies that kinetic effects are relevant and suggests that thermalization is likely incomplete before plasma reaches the MPB and inner regions of the Martian magnetosphere (Jiang et al., 2023; Kallio et al., 2011; Moses et al., 1988; Ruhunusiri et al., 2017). Thermalization processes involve space and time scales related to local ion gyrofrequencies, which in the Martian magnetosheath are near the ULF range (Fowler et al., 2017).

Figures 5.7c and 5.7d show that magnetic wave power in the 1–5 Hz range is maximum near the subsolar region and along the Martian bow shock. Wave power takes smaller values in the Martian magnetosheath and is three orders of magnitude smaller in the tail. Figures 5.7e and 5.7f show that wave power at 5 Hz presents a similar spatial distribution to that of between 1 and 5 Hz, but is two orders of magnitude smaller. Local maxima in the wave power at magnetospheric boundaries make evident that the solar wind kinetic energy is transferred to particle heating in these very thin regions. Moreover, the associated particle velocity distribution function has potentially sufficient free energy for the local generation of electromagnetic waves. Magnetic field wave power is also significant slightly upstream of the average bow shock location. This is due to several factors: the non-stationarity of this boundary, the presence of reflected ions at the shock, over and undershoots, and electromagnetic plasma waves. Among the latter, whistler waves, likely generated at the shock, are capable of propagating into and affecting the upstream region (Eastwood et al., 2005; Edberg, Auster, et al., 2009; Edberg, Brain, et al., 2009; Gosling et al., 1982; Halekas, 2017; Livesey et al., 1982; Mazelle et al., 2004; Paschmann et al., 1981; Trotignon et al., 2006).

So far, the identification of wave modes in the Martian magnetosheath has been performed by computing transport ratios, interpreted in terms of MHD linear theory. Following the method first developed by Gary (1993) and extended by Song et al. (1994), Ruhunusiri et al. (2015) identified low-frequency wave modes as Alfvén and quasi-parallel slow waves (indistinguishable by the employed methodology), fast, quasi-perpendicular slow, and mirror mode waves. This was done based on four transport ratios: the transverse ratio ($T_R$), the compressional ratio ($C_R$), the phase ratio ($P_R$), and the Doppler ratio ($D_R$). These are defined in terms of velocity, magnetic field, and density fluctuations as follows:

$$T_R = \frac{\delta \mathbf{B} \cdot \delta \mathbf{B} - \delta B_{\parallel}^2}{\delta B_{\parallel}^2} \tag{5.5}$$

$$C_R = \frac{\delta n_i^2}{n_{i0}^2} \Big/ \frac{\delta \mathbf{B} \cdot \delta \mathbf{B}}{B_0^2} \tag{5.6}$$

$$P_R = \frac{\delta n_i}{n_{i0}} \Big/ \frac{\delta B_\parallel}{B_0} \tag{5.7}$$

$$D_R = \frac{\delta \mathbf{u}_i \cdot \delta \mathbf{u}_i}{u_{i0}^2} \Big/ \frac{\delta \mathbf{B} \cdot \delta \mathbf{B}}{B_0^2} \tag{5.8}$$

where $\delta B_\parallel$ is the component of the magnetic field fluctuation ($\delta\mathbf{B}$) parallel to the background magnetic field $\mathbf{B_0}$. In addition, $\delta n_i$, $n_{i0}$, $\delta\mathbf{u}_i$, and $u_{i0}$ are the ion number density fluctuation, mean value, the fluctuation in the velocity field, and the mean ion velocity, respectively. The wave mode identification method developed by Song et al. (1994) assumes that a single wave mode is dominant at a single frequency with a single wavenumber. In addition, it is applicable in a high beta plasma, suggesting the results reported by Ruhunusiri et al. (2015) are most accurate in the Martian magnetosheath. Table 1 in Ruhunusiri et al. (2015) reports the range of values for the transport ratios (Equations 5.5–5.8) used to identify low-frequency wave modes in the Martian magnetosphere.

Figure 5.8 displays maps of the wave mode occurrence ratio, based on MAVEN MAG, SWIA, and Suprathermal and Thermal Ion Composition (STATIC) (McFadden et al., 2015) data between October 7, 2014 and April 28, 2015, in cylindrical MSO coordinates. Ruhunusiri et al. (2015) found a very high wave occurrence rate of Alfvén and quasi-parallel slow waves in the pristine solar wind and the magnetosheath (panel a). Several of the wave events upstream from the bow shock are likely Alfvén waves, in agreement with previous studies of upstream ULF waves and solar wind turbulence (e.g., Andrés et al., 2020; Halekas et al., 2020). Moreover, the observed occurrence rate decrease from the upstream region into the magnetosheath may be indicative of upstream Alfvén wave convection. Ruhunusiri et al. (2015) also reported a significant amount of fast magnetosonic waves in the Martian magnetosheath that increases closer to the MPB (panel b). The source of these magnetosonic waves could be mode conversion at the bow shock or PCW and foreshock wave transmission from upstream to the magnetosheath. In addition, the authors also concluded there is a minor mirror mode wave population that is observed more frequently near the dayside magnetosheath region (panel d), in agreement with other magnetic field observations (Espley et al., 2004; Simon Wedlund et al., 2022a,b). Given that the wave identification method by Song et al. (1994) may not be applied downstream of the MPB (low beta plasma region) the authors did not provide final conclusions about waves in this region (Ruhunusiri et al., 2015).

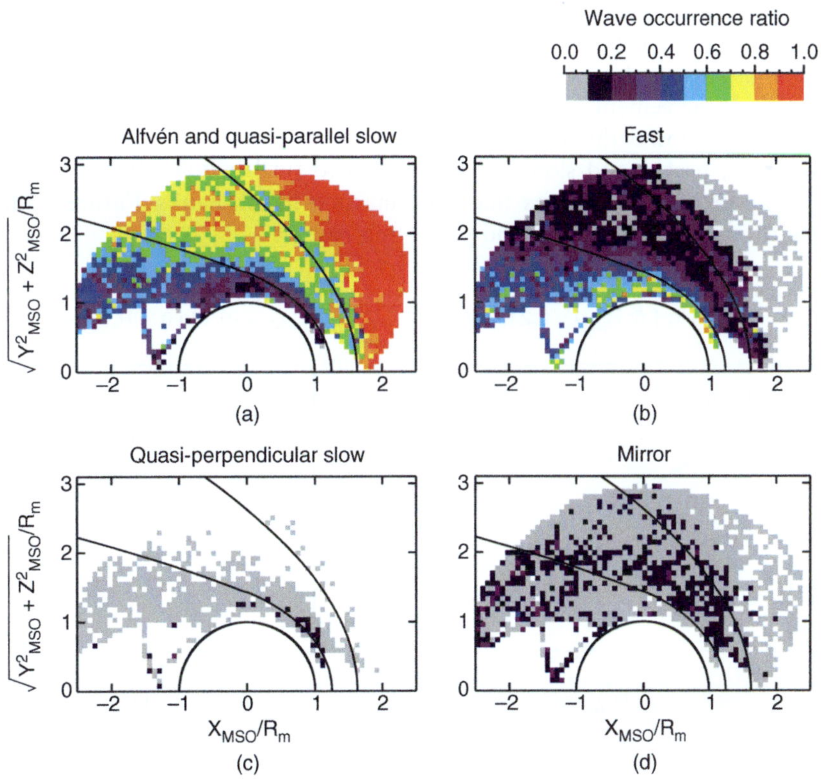


**Figure 5.8** Occurrence rate of several low-frequency wave modes. (a) Alfvén and quasi-parallel slow, (b) fast magnetosonic, (c) quasi-perpendicular slow, (d) mirror. Source: Taken from Ruhunusiri et al. (2015) based on the identification technique developed in Song et al. (1994) for high beta plasmas.

The previous analysis was not concerned with the nonlinear interaction between fluctuations of different temporal and spatial scales. Ruhunusiri et al. (2017) performed the first global characterization of turbulence in the Martian magnetosphere, based on the computation of spectral indices for perturbations observed in the magnetic field. As mentioned before, these indices are the slopes associated with the magnetic field power spectra distribution as a function of the observed frequency (in logarithmic scale). They provide information about the physical processes at play within a given frequency range (energy injection, transfer or cascade, and dissipation). In contrast with observations in the pristine solar wind, Ruhunusiri et al. (2017) concluded there is not an inertial range in the Martian magnetosheath following Kolmogorov's scaling (where $\gamma = -5/3$) expected when energy is transferred between scales.

Figure 5.9 shows the median magnetic field power spectral density as a function of the normalized frequency for several Martian solar longitude ($L_s$) values. $L_s$ is the angle between Mars and the Sun, measured from the northern hemisphere spring equinox where $L_s = 0°$. The Martian perihelion and aphelion correspond to $L_s = 251°$ and $L_s = 71°$, respectively. The northern hemisphere summer (winter) solstice occurs at $L_s = 90°$ ($L_s = 270°$). Ruhunusiri et al. (2017) found that the spectral indices in the magnetosheath (panel b) take larger values ($\gamma \sim -0.5$) for frequencies smaller than the local proton gyrofrequency and much more negative values ($\gamma \sim -2.7$) at higher frequencies. This is in agreement with observations at other planetary magnetosheaths and suggests that the shocked solar wind, heated and slowed down at the bow shock, is composed of fluctuations that do not have sufficient time to interact nonlinearly and give rise to a fully developed energy cascade rate (e.g., Hadid et al., 2015; Tao et al., 2015; Zimbardo et al., 2010). Although there seem to be some effects of PCW in the Martian magnetosheath spectra, they do not appear as pronounced as in the region upstream of the bow shock (panel a). Note the clear peak at $f/f_{H^+} \sim 1$, for $L_s \sim 270°$ when MAVEN is in the upstream region. In other words, the Martian magnetosheath appears to be filled with high amplitude Alfvén waves that have not reached a fully developed turbulent regime, at least near the terminator plane (Fowler et al., 2017; Ruhunusiri et al., 2015, 2017).

Similar conclusions have also been reached by Jiang et al. (2023). However, the authors also reported the presence of a Kolmogorov-like spectrum in the Martian magnetosheath for very low wave frequencies. The authors found magnetic field spectra in the Martian magnetosheath consisting of triple power laws, where the intermediate regime displays a plateau. An example of such events is shown in Figure 5.10. Note that their breaking frequency points are typically between $\sim 10^{-2}$ and $\sim 10^{-1}$ and $\sim 1$ and $\sim 10$ times the local gyrofrequency, respectively, depending on different parameters, such as the plasma beta (see Figure 7c and 7d in that work). Their statistical analysis allows the identification of a significant correlation between the occurrence rate of plateau power spectra and planetary proton pick-up ion parameters. Specifically, the authors suggest the formation of plateau-like spectra may be due to energy injection associated with PCW (Jiang et al., 2023). It is possible, however, that a fully developed energy cascade rate is reached further downstream from the planet, along the magnetosheath flanks. Future studies may shed light on this matter.

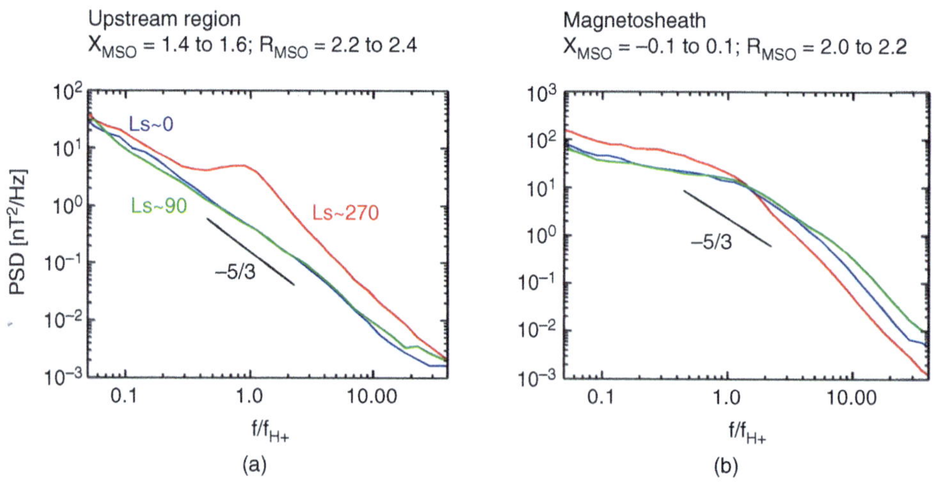


**Figure 5.9** Median magnetic field power spectra density as a function of the normalized frequency for three seasons in (a) the region upstream of the Martian bow shock and (b) the Martian magnetosheath. $f_{H^+}$ refers to the local proton cyclotron frequency. $L_s$ is the Martian solar longitude. Source: Ruhunusiri et al. (2017) / John Wiley & Sons.

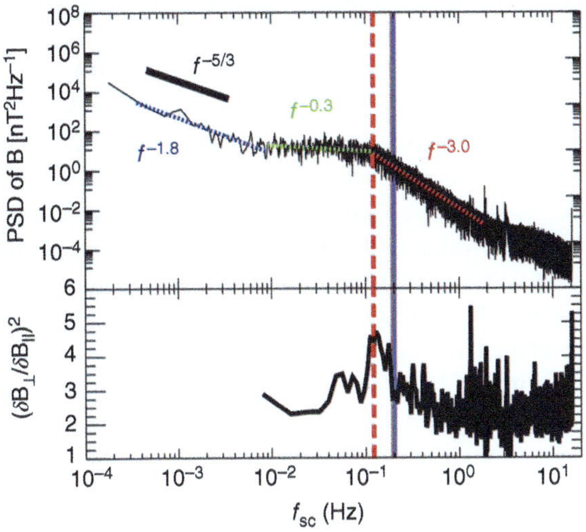


**Figure 5.10** (Upper) Magnetic field power spectral density for a Martian magnetosheath interval on June 4, 2016, between 21:10 UT and 22:43 UT. Vertical lines correspond to relevant ion-scale frequencies. The red dashed line shows the highest frequency breakpoint $f_{bk2}$= 0.12 Hz. The light red and light blue lines display $f_{\rho i}$ and $f_{di}$, respectively, with $f_{\rho i} = V_i/2\pi\rho_i$, $f_{di} = V_i/2\pi d_i$, where $V_i$ is the local plasma bulk velocity, $\rho_i$ is the local thermal proton gyro-radius, and $d_i$ is the proton inertial length. (Lower) Square of the magnetic field perpendicular fluctuation normalized by the magnetic field parallel fluctuation, as a function of the observed frequency in the spacecraft reference frame. Source: Jiang et al. (2023) / John Wiley & Sons.

## 5.4. ALFVÉN WAVES INSIDE THE MPB AND THE IONOSPHERE

While there have been studies of Alfvén waves in several regions of the Martian magnetosphere, there are far fewer studies focused on this phenomenon within the Martian MPB and ionosphere. This is likely due to a combination of factors such as a relatively low level of wave activity and the lack of adequate instrumentation and spacecraft sampling until more recent missions.

Figure 5.11 shows the compressional ratio $C_R$ is larger than 1 inside the Martian MPB. This result is observed both using MAVEN SWIA (panel a) and STATIC (panel b) data and suggests Alfvén waves are not commonly present in the magnetic pile-up region and magnetotail (Ruhunusiri et al., 2015). In addition, as can be seen in Figure 5.7, the magnetic field wave power decreases downstream of the MPB, particularly in the magnetotail. This could be due to relatively weak energy dissipation processes in the explored frequency ranges, as well as the much lower plasma beta values in this region (Fowler et al., 2017). The spectral indices in the magnetic pile-up region and the Martian wake are near −2 or smaller in both the MHD and kinetic ranges (Ruhunusiri et al., 2017). The lack of a spectral break at the local proton cyclotron frequency may result from the plasma containing primarily heavy planetary ions instead of protons (Ruhunusiri et al., 2017). Previous reports have found similar results for the Venusian wake at frequencies between 0.01 and 0.5 Hz (Vörös, Zhang, Leaner, Volwerk, Delva, & Baumjohann, 2008; Vörös, Zhang, Leaner, Volwerk, Delva, & Baumjohann, et al. 2008).

The absence of Alfvén wave observations in the Martian ionosphere and/or along crustal magnetic fields may also be the result of limited instrumentation and spacecraft orbital geometry. Early Mars orbiters, such as Phobos-2, carried magnetometers but did not sample the ionosphere. More recent missions such as MGS and MEX periodically sampled the ionosphere, but their ionospheric coverage was limited. While MGS MAG/Electron Reflectometer provided observations of the electromagnetic wave environment at Mars, the ~400 km circular orbit precluded any comprehensive study within the ionosphere. In contrast, MEX periapsis altitudes reach

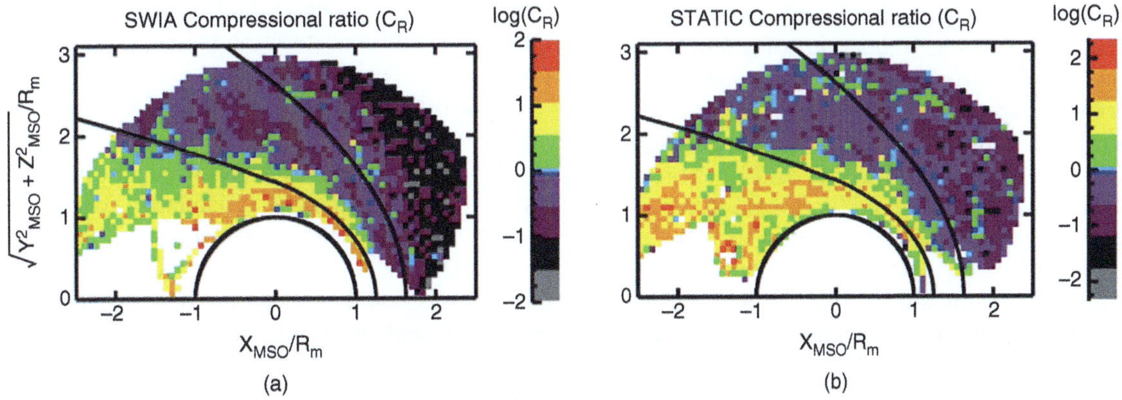


**Figure 5.11** Compressional ratio $C_R$ based on MAVEN MAG and SWIA (panel a) and STATIC (panel b) observations as a function of cylindrical MSO coordinates. MAVEN: Mars Atmosphere and Volatile EvolutioN. MAG: Magnetometer. SWIA: Solar Wind Ion Analyzer. STATIC: Suprathermal and Thermal Ion Composition. MSO: Mars Solar Orbital. Source: Ruhunusiri et al. (2015) / John Wiley & Sons.

down to ~275 km, providing in situ samples of the topside upper Martian ionosphere but it is unable to reach the main ionospheric peak, located at ~125 km altitude at the sub-solar point. Indeed, the main peak has been observed via remote sensing using the Mars Advanced Radar for Subsurface and Ionospheric Sounding radar on MEX (e.g., Dubinin et al., 2006; Gurnett et al., 2005; Orosei et al., 2015; Picardi et al., 2004). Moreover, the lack of a magnetometer onboard MEX precludes any detailed study of the electromagnetic wave environment. In the case of MAVEN, periapsis typically lies between 150 and 200 km, determined by targeting a specific neutral density corridor. MAVEN completed nine deep dip campaigns during the first few years of its mission, where periapsis altitude was lowered to ~125 km for roughly week-long periods. This enabled MAVEN to sample down to the main ionospheric peak at mid to high solar zenith angles (~45° – 90°). Also, MAVEN's precessing orbit provides comprehensive coverage throughout the magnetosphere and the Martian ionosphere.

The disparity in the study of Alfvén waves within the Martian ionosphere, compared to the rich and diverse collection of work related to their analysis within the terrestrial ionosphere, also likely lies in the underlying limitations of in situ plasma measurements made at bodies other than Earth. In spite of particle and field measurements becoming more commonplace on interplanetary missions, their sampling frequency is typically much lower than their terrestrial counterparts. For example, the ion (electron) distribution functions measured by the MEX Analyser of Space Plasma and Energetic Atoms 3 - Ion Mass Analyzer - (Electron Mass Spectrogram) are accumulated over 192 (4) s, and MAVEN STATIC and Solar Wind Electron Analyzer ion and electron distribution functions are accumulated over 4 and 2 s, respectively (Dubinin et al., 2006; McFadden et al., 2015; Mitchell et al., 2016). Such measurements typically resolve the macro-scale parameters of the system, however, they are usually limited in resolving kinetic scale physics based on electromagnetic waves at frequencies of a few Hz up to 1000s Hz in the Martian plasma environment. In contrast, terrestrial space physics missions such as Cluster, Fast Auroral SnapshoT (FAST), and Magnetospheric Multiscale (MMS) measure the ion and electron distribution functions at tens of milliseconds (Burch et al., 2016; Carlson et al., 1998; Escoubet et al., 1997). The fields measurements (i.e., magnetic and electric fields) on interplanetary missions suffer from similar limitations: magnetic field data are typically limited to DC measurements at 32 Hz (compared to AC measurements of kHz at Earth), while only a single year of 1D electric field wave power spectra is available for Mars, made by the MAVEN Langmuir Probe and Waves (LPW) instrument (Andersson et al., 2015; Fowler et al., 2017).

Recent studies have shown electromagnetic waves inside the MPB (e.g., Harada et al., 2016), in the ionosphere (e.g., Esman et al., 2022; Gurnett et al., 2010), and the Martian surface (e.g., Johnson et al., 2020; Mittelholz et al., 2021, 2023); some of them in connection with upstream processes leading to their generation (e.g., Fowler et al., 2020; Shane et al., 2019; Shane & Liemohn, 2021, 2022; Wang et al., 2023). These have been primarily whistler waves observed at the interaction region between the shocked solar wind and the upper ionosphere. A comprehensive study on the occurrence rate and spatial distribution of these waves has not been undertaken yet. Next, we briefly review these works, which illustrate some of the potential effects Alfvén waves generated upstream may have in the Martian ionosphere.

Fowler et al. (2020) showed that whistler waves are produced in the upper ionosphere via a process known as "magnetic pumping." A summary of their event is shown in Figure 2 of their work. The free energy to drive the generation of the whistler waves is sourced from ULF compressive magnetosonic waves that are generated due to the Mars–solar wind interaction and that propagate across the draped magnetic field into the dayside ionosphere. Adiabatic compression of the plasma along wave fronts in the magnetic field acts to drive the electron distribution function unstable to the generation of whistler waves by creating an anisotropic distribution (e.g., Kennel & Petschek, 1966). The whistler waves were shown to break the reversibility of the underlying ULF magnetic pumping via efficient wave-particle interactions with the anisotropic electrons, leading to localized heating of this population. Such heating may play important roles in, for example, driving ionospheric photochemistry (Fox & Bakalian, 2001), the dissociative recombination of $O_2^+$ (Lillis et al., 2017), and the creation of ambipolar electric fields ($-\nabla p_e$) that can drive ions to escape to space (Ergun et al., 2016). Further study is needed to fully evaluate the impact on the energy and dynamics of the Martian ionosphere and to determine the occurrence rate of this process.

On the other hand, Wang et al. (2023) observed whistler waves within localized "magnetic dips" at the interaction region between the solar wind and upper dayside ionosphere, in a manner postulated to be similar to whistler wave generation in the equatorial magnetosphere (e.g., LeDocq et al., 1998). The exact whistler generation mechanisms could not be determined, but temperature anisotropy and electron beam instabilities were proposed as likely candidates. The background plasma environment at Mars is such that these whistler waves propagate at much slower velocities than at Earth and so localized heating rather than propagation of the whistlers may be more important at Mars. Additional analysis is required to better understand the effects these waves have on the local plasma dynamics.

In addition, a comprehensive series of papers by Shane et al. (2019); Shane and Liemohn (2021); Shane and Liemohn (2022) used numerical approaches to demonstrate that whistler wave–particle interactions with suprathermal electrons can explain observed pitch angle distributions (PADs) on closed crustal field lines which intersect the Martian ionosphere. Many previous studies have shown that >100 eV PADs on dayside closed field lines are isotropic or trapped, however, these PADs are not expected based on collisional scattering and conservation of adiabatic invariants alone (which are important as the electrons mirror within the closed crustal field loops) (e.g., Brain et al., 2007; Liemohn et al., 2003). Numerical calculations of the bounce-averaged electron diffusion coefficients show that wave–particle interactions are more efficient than Coulomb collisions above the exobase region and can play an important role in shaping the PADs in a manner consistent with observations (Shane & Liemohn, 2021). Numerical simulation of the bounce-averaged electron diffusion equation, including whistler wave resonance, demonstrated that such whistler wave–particle interactions alter the PADs and are consistent with observations (Shane & Liemohn, 2022). The authors noted that future work should determine whether the frequency of whistler waves is high enough to explain the average behavior in observed PADs, and determine how such whistler waves would be generated.

Several studies have also shown that solar wind pressure pulses can be convected into the bow shock and "ring" the planetary magnetosphere (e.g., Collinson et al., 2018). Pressure pulses propagate through the magnetosphere in the form of fast magnetosonic waves (traveling across the draped magnetic field), where they can reach both the day and nightside ionosphere and drive heating of the light and heavy ion species through a variety of wave–particle interaction processes (Fowler, Andersson, Peterson, et al., 2018, 2021).

A final note should be made with respect to the localized crustal magnetic field regions at Mars. These regions act as "mini-magnetospheres" that rotate with the planet, and can sporadically magnetically reconnect with the solar wind IMF (e.g., Bowers et al., 2023; Weber et al., 2020) to create "mini cusp-like" regions. Given the similarity of these cusp regions with the auroral zones at Earth, one may expect electromagnetic waves to play similar crucial roles in the physical processes present in these regions (primarily precipitation and auroral physics (Andre & Yau, 1997)). In particular, Ergun et al. (2006) performed numerical calculations to show that electromagnetic waves associated with the Mars-solar wind interaction and field-aligned currents in regions of crustal magnetic fields could heat ionospheric $O^+$ and $O_2^+$ to escape energy and drive escape rates of $\sim 10^{25}$ $s^{-1}$, via cyclotron damping. Their predictions closely match total ion escape rates that have been calculated based on recent observations in the magnetotail region of Mars (e.g., Brain et al., 2015; Dong et al., 2015; Ramstad et al., 2015). However, the full range of ion energization mechanisms has not been comprehensively identified or quantified. There are still many open questions about the initial acceleration of ions in the Mars ionosphere, and the role that electromagnetic waves play in it (Hanley et al., 2022). In addition, a full 3D electric field in situ instrumentation at low and medium frequencies, together with magnetic field and particle measurements, are needed in order to identify the exact wave modes taking place in these regions.

## 5.5. CONCLUSIONS

Over the last few years, there has been significant progress in characterizing the Alfvénic input and content of the Martian magnetosphere. The pristine solar wind, upstream of Mars's bow shock, has been characterized in terms of Alfvénic turbulence, where the absolute value of the incompressible nonlinear energy cascade rate is on the order of $10^{-17}$ J $m^{-3}$ $s^{-1}$ at MHD scales. Studies have also shown that the cascade rate varies with Mars' heliocentric distance, in agreement with investigations focused on the solar wind closer to the Sun and upstream of other planetary magnetospheres (Andrés et al., 2020, 2021; Bandyopadhyay et al., 2020; Hadid et al., 2017; Romanelli et al., 2022). Moreover, values of normalized cross-helicity and residual energy show that the solar wind presents more Alfvén waves propagating outward from the Sun (than inward) and that the magnetic field fluctuations have more energy than the kinetic field counterpart (Andrés et al., 2020; Halekas et al., 2017). PCWs associated with the extended hydrogen corona do not appear to modify the solar wind energy cascade rate at MHD scales, although their energy could have potential effects on the kinetic regime (Romanelli et al., 2022). The magnetosheath is characterized by high amplitude wave activity and a high occurrence rate of Alfvén waves, although other wave modes are also present (Fowler et al., 2017; Ruhunusiri et al., 2015). The large amplitude waves are likely the result of (incomplete) thermalization processes of the solar wind. The relatively small size of the magnetosheath, together with potential effects from the bow shock, PCW, and other wave modes, is also responsible for a magnetic field turbulent spectrum that is not fully developed (Fowler et al., 2017; Jiang et al., 2023). Among the secondary wave populations in the magnetosheath, there exists a significant component of fast magnetosonic waves and a minor contribution of mirror mode waves, particularly in the dayside (Espley et al., 2004; Ruhunusiri et al., 2015; Simon Wedlund, Volwerk, Mazelle, Rojas Mata, et al. 2022). Downstream of the magnetic pile-up boundary, waves generally display less magnetic field power (Fowler et al., 2017). The PSD of the magnetic field does not show a Kolmogorov-like spectrum but rather is characterized by a spectral index more negative both below and above the local proton gyrofrequency (Ruhunusiri et al., 2017).

Early studies of the solar wind interaction with Mars predicted that significant electromagnetic wave activity would be present at the interface and within the Martian ionosphere, as a direct result of the relatively small scale size of the magnetospheric system (Ergun et al., 2006; Kallio et al., 2011). Subsequent studies have confirmed this and hint that a plethora of electromagnetic wave modes exist and interact with the ionospheric plasma to drive energization and dynamics (e.g., Collinson et al., 2018; Fowler, Andersson, Ergun, et al., 2018, 2021). Several of these studies have identified the presence of whistler mode waves and subsequent wave–particle interactions in this Mars–solar wind interaction region. However, Alfvén waves have not been conclusively detected in the ionosphere or along remanent crustal magnetic fields yet. This could be due to the instrumental cadence or orbital geometry associated with previous missions to Mars as well as the lack of consistent conditions leading to their excitation.

Despite this progress, there is still much to be done to definitively identify wave modes and the processes giving rise to them at Mars. The lack of sufficient studies is partly due to the relatively small amount of simultaneous ion plasma and magnetic field data before the current MAVEN mission and the dynamics behavior of the Martian environment, among other factors. Mars possesses a hybrid magnetosphere with elements of induced and intrinsic magnetospheres (Brain et al., 2003; DiBraccio et al., 2022; Dubinin et al., 2023; Mazelle et al., 2004). The extended hydrogen corona is also responsible for additional processes and their variability over long (seasonal) timescales (Halekas, 2017; Romanelli et al., 2016; Romeo et al., 2021). Also, the use of single spacecraft observations does not allow computing the Doppler shift for low-frequency waves, due to the relative motion between the plasma rest frame and the spacecraft reference frame. Given that the solar wind velocity is close to or larger than the phase speed of many waves inside and upstream of the Martian magnetosphere, the waves are observed at different speeds and, in some cases, with different polarization than that seen in the plasma reference frame (Brinca, 1991). As a result, so far many studies on waves have investigated the conditions that favor their excitation and basic properties, such as the observed frequency and polarization, as opposed to looking at the intrinsic wave frequency and wavelength and determining the corresponding wave mode.

Future studies could be focused on the determination of the direction of the energy cascade rate upstream of Mars and the effects that PCW and foreshock waves have on the MHD and kinetic regimes. Comparison between studies at different planetary foreshocks is also an interesting venue for future work. Moreover, an analysis of the processes occurring in the Martian bow shock and magnetic pile-up boundary would be beneficial to better understand the nature of the wave activity present inside the Martian magnetosphere, in particular the magnetosheath. Additional unknowns remain including wave mode identification downstream of the MPB and Alfvén wave detection near the ionosphere or along the crustal magnetic fields. If the latter are present, additional analyses should be focused on determining their occurrence rates (in time and space), their production mechanisms, their impacts on the local plasma, and their impact on more global parameters.

A combination of current in situ measurements and global simulations can address some of these unknowns. Hybrid numerical simulations are particularly useful to improve data analysis and point out physical processes potentially taking place in the Martian magnetosphere (Jarvinen et al., 2022, Modolo et al., 2016; Romanelli, Modolo, Leblanc, Chaufray, Martinez, et al., 2018). They are also important as wave amplitudes are often large, suggesting linear wave theory is limited (if not invalid) in interpreting spacecraft observations. However, answers to some of the previous questions will likely remain elusive until measurements equivalent in capability to those made by terrestrial space physics missions can be made at Mars. In particular, simultaneous multi-spacecraft and multi-probe electric and magnetic field measurements with high-time resolution particle instruments are needed to determine the exact wave modes, polarization, and Poynting flux. These observations would allow, in turn, the determination of the growth and dissipation of waves in various regions in the Martian magnetosphere. In this regard, multi-spacecraft missions such as the planned Escape and Plasma Acceleration and Dynamics Explorers (ESCAPADE) will certainly contribute to improving our current understanding of waves in this magnetosphere (Lillis et al., 2022).

## ACKNOWLEDGMENTS

The MAVEN project is supported by NASA through the Mars Exploration Program. The material is based upon work supported by NASA under award number 80GSFC21M0002. Support for this research was also provided by GSFC/EIMM within NASA's Planetary Science Division Research Program.

## REFERENCES


Acuña, M. H., Connerney, J. E. P., Ness, N. F., Lin, R. P., Mitchell, D., Carlson, C. W., et al. (1999). Global distribution of crustal magnetization discovered by the Mars Global Surveyor MAG/ER experiment. *Science*, *284*, 790.

Acuña, M. H., Connerney, J. E. P., Wasilewski, P., Lin, R. P., Anderson, K. A., Carlson, C. W., et al. (1998). Magnetic field and plasma observations at Mars: Initial results of the Mars Global Surveyor Mission. *Science*, *279*(5357), 1676–1680.

Alexandrova, O. (2008). Solar wind vs magnetosheath turbulence and Alfvén vortices. *Nonlinear Processes in Geophysics*, *15*(1), 95–108.

Alfvén, H. (1942). Existence of electromagnetic-hydrodynamic waves. *Nature*, *150*(3805), 405–406.

Andersson, L., Ergun, R. E., Delory, G. T., Eriksson, A., Westfall, J., Reed, H., et al. (2015). The Langmuir Probe and Waves (LPW) Instrument for MAVEN. *Space Science Reviews*, *195*(1–4), 173–198.

Andre, M., & Yau, A. (1997). Theories and observations of ion energization and outflow in the high latitude magnetosphere. *Space Science Reviews*, *80*, 27–48.

Andrés, N., Galtier, S., & Sahraoui, F. (2018). Exact law for homogeneous compressible hall magnetohydrodynamics turbulence. *Physical Review E*, *97*(1), 013204.

Andrés, N., Meziane, K., Mazelle, C., Bertucci, C., & Gómez, D. (2015). The ulf wave foreshock boundary: Cluster observations. *Journal of Geophysical Research: Space Physics*, *120*(6), 4181–4193.

Andrés, N., Mininni, P. D., Dmitruk, P., & Gomez, D. O. (2016). von Kármán–Howarth equation for three-dimensional two-fluid plasmas. *Physical Review E*, *93*(6), 063202.

Andrés, N., Romanelli, N., Hadid, L. Z., Sahraoui, F., DiBraccio, G., & Halekas, J. (2020). Solar wind turbulence around Mars: Relation between the energy cascade rate and the proton cyclotron waves activity. *The Astrophysical Journal*, *902*(2), 134.

Andrés, N., Sahraoui, F., Hadid, L. Z., Huang, S. Y., Romanelli, N., Galtier, S., et al. (2021). The evolution of compressible solar wind turbulence in the inner heliosphere: PSP, THEMIS, and MAVEN observations. *The Astrophysical Journal*, *919*(1), 19.

Auer, P. L., Kilb, R. W., & Crevier, W. F. (1971). Thermalization in the Earth's bow shock. *Journal of Geophysical Research*, *76*(13), 2927.

Bale, S. D., Balikhin, M. A., Horbury, T. S., Krasnoselskikh, V. V., Kucharek, H., Möbius, E., et al. (2005). Quasi-perpendicular shock structure and processes. *Space Science Reviews*, *118*(1–4), 161–203.

Bandyopadhyay, R., Goldstein, M., Maruca, B., Matthaeus, W., Parashar, T., Ruffolo, D., et al. (2020). Enhanced energy transfer rate in solar wind turbulence observed near the sun from parker solar probe. *The Astrophysical Journal Supplement Series*, *246*(2), 48.

Barabash, S., Dubinin, E., Pissarenko, N., Lundin, R., & Russell, C. T. (1991). Picked-up protons near Mars: Phobos observations. *Geophysical Research Letters*, *18*(10), 1805–1808.

Bavassano, B., Pietropaolo, E., & Bruno, R. (1998). Cross-helicity and residual energy in solar wind turbulence: Radial evolution and latitudinal dependence in the region from 1 to 5 AU. *Journal of Geophysical Research*, *103*(A4), 6521–6530.

Belcher, J. W., & Davis, L. J. (1971). Large-amplitude Alfvén waves in the interplanetary medium, 2. *Journal of Geophysical Research*, *76*(16), 3534.

Bertucci, C., Duru, F., Edberg, N., Fraenz, M., Martinecz, C., Szego, K., & Vaisberg, O. (2011). The induced magnetospheres of Mars, Venus, and Titan. *Space Science Reviews*, *162*(1–4), 113–171.

Bertucci, C., Romanelli, N., Chaufray, J.-Y., Gomez, D., Mazelle, C., Delva, M., et al. (2013). Temporal variability of waves at the proton cyclotron frequency upstream from Mars: Implications for Mars distant hydrogen exosphere. *Geophysical Research Letters*, *40*(15), 3809–3813.

Bhattacharyya, D., Clarke, J. T., Bertaux, J.-L., Chaufray, J.-Y., & Mayyasi, M. (2015). A strong seasonal dependence in the martian hydrogen exosphere. *Geophysical Research Letters*, *42*(20), 8678–8685.

Biskamp, D. (1973). Collisionless shock waves in plasmas. *Nuclear Fusion*, *13*(5), 719.

Bowers, C. F., DiBraccio, G. A., Slavin, J. A., Gruesbeck, J. R., Weber, T., Xu, S., et al. (2023). Exploring the solar wind-planetary interaction at mars: Implication for magnetic reconnection. *Journal of Geophysical Research: Space Physics*, *128*(2), e2022JA030989. e2022JA030989 2022JA030989.

Brain, D. A., Bagenal, F., Acuña, M. H., & Connerney, J. E. P. (2003). Martian magnetic morphology: Contributions from the solar wind and crust. *Journal of Geophysical Research: Space Physics*, *108*(A12).

Brain, D. A., Bagenal, F., Acuña, M. H., Connerney, J. E. P., Crider, D. H., Mazelle, C., et al. (2002). Observations of low-frequency electromagnetic plasma waves upstream from the Martian shock. *Journal of Geophysical Research: Space Physics*, *107*(A6).

Brain, D. A., Lillis, R. J., Mitchell, D. L., Halekas, J. S., & Lin, R. P. (2007). Electron pitch angle distributions as indicators of magnetic field topology near mars. *Journal of Geophysical Research: Space Physics*, *112*(A9).

Brain, D. A., McFadden, J. P., Halekas, J. S., Connerney, J. E. P., Bougher, S. W., Curry, S., et al. (2015). The spatial distribution of planetary ion fluxes near mars observed by maven. *Geophysical Research Letters*, *42*(21), 9142–9148.

Breech, B., Matthaeus, W. H., Minnie, J., Oughton, S., Parhi, S., Bieber, J. W., & Bavassano, B. (2005). Radial evolution of cross helicity in high-latitude solar wind. *Geophysical Research Letters*, *32*(6), L06103.

Brinca, A. (1991). Cometary linear instabilities: From profusion to perspective. *Geophysical Monograph Series*, *61*, 211–221.

Brinca, A. & Tsurutani, B. T. (1989). Influence of multiple ion species on low-frequency electromagnetic wave instabilities. *Journal of Geophysical Research: Space Physics*, *94*(A10), 13565–13569.

Bruno, R., & Carbone, V. (2013). The solar wind as a turbulence laboratory. *Living Reviews in Solar Physics*, *10*(1), 2.

Burch, J. L., Moore, T. E., Torbert, R. B., & Giles, B. L. (2016). Magnetospheric multiscale overview and science objectives. *Space Science Reviews*, *199*(1–4), 5–21.

Burgess, D., Lucek, E. A., Scholer, M., Bale, S. D., Balikhin, M. A., Balogh, A., et al. (2005). Quasi-parallel Shock Structure and Processes. *Space Science Reviews*, *118*(1–4), 205–222.

Burgess, D., Möbius, E., & Scholer, M. (2012). Ion acceleration at the Earth's bow shock. *Space Science Reviews*, *173*(1–4), 5–47.

Burgess, D., Wilkinson, W. P., & Schwartz, S. J. (1989). Ion distributions and thermalization at perpendicular and quasi-perpendicular supercritical collisionless shocks. *Journal of Geophysical Research*, *94*(A7), 8783–8792.

Carlson, C. W., Pfaff, R. F., & Watzin, J. G. (1998). The Fast Auroral SnapshoT (FAST) mission. *Geophysical Research Letters*, *25*(12), 2013–2016.

Chaffin, M. S., Chaufray, J. Y., Deighan, J., Schneider, N. M., McClintock, W. E., Stewart, A. I. F., et al. (2015). Three-dimensional structure in the mars h corona revealed by iuvs on maven. *Geophysical Research Letters*, *42*(21), 9001–9008.

Chaffin, M. S., Kass, D. M., Aoki, S., Fedorova, A. A., Deighan, J., Connour, K., et al. (2021). Martian water loss to space enhanced by regional dust storms. *Nature Astronomy*, *5*, 1036–1042.

Chen, L.-J., Halekas, J., Wang, S., DiBraccio, G. A., Romanelli, N., Ng, J., et al. (2022). Solitary magnetic structures developed from gyro-resonance with solar wind ions at mars and earth. *Geophysical Research Letters*, *49*(3), e2021GL097600. e2021GL097600 2021GL097600.

Clarke, J. T., Mayyasi, M., Bhattacharyya, D., Schneider, N. M., Mcclintock, W. E., Deighan, J. I., et al. (2017). Variability of D and H in the Martian upper atmosphere observed with the MAVEN IUVS echelle channel. *Journal of Geophysical Research: Space Physics*, *122*(2), 2336–2344.

Collinson, G., Wilson III, L. B., Omidi, N., Sibeck, D., Espley, J., Fowler, et al. (2018). Solar wind induced waves in the skies of mars: Ionospheric compression, energization, and escape resulting from the impact of ultralow frequency magnetosonic waves generated upstream of the martian bow shock. *Journal of Geophysical Research: Space Physics*, *123*(9), 7241–7256.

Connerney, J. E. P., Espley, J., Lawton, P., Murphy, S., Odom, J., Oliversen, R., & Sheppard, D. (2015). The MAVEN magnetic field investigation. *Space Science Reviews*, *195*(1–4), 257–291.

Cramer, N. (2001). *The Physics of Alfvén Waves*. John Wiley & Sons, Ltd.

Cravens, T., & Gombosi, T. (2004). Cometary magnetospheres: A tutorial. *Advances in Space Research*, *33*(11), 1968–1976. Comparative Magnetospheres.

Crider, D. H., Brain, D. A., Acuña, M. H., Vignes, D., Mazelle, C., & Bertucci, C. (2004). Mars global surveyor observations of solar wind magnetic field draping around Mars. *Space Science Reviews*, *111*(1), 203–221.

Curry, S. M., Luhmann, J. G., Ma, Y. J., Dong, C. F., Brain, D., Leblanc, F., et al. (2015). Response of Mars O+ pickup ions to the 8 March 2015 ICME: Inferences from MAVEN data-based models. *Geophysical Research Letters*, *42*(21), 9095–9102.

Curry, S. M., Tatum, P., Mitchell, D., Luhmann, J. G., McFadden, J., Ruhunusiri, S., et al. (2022). Ion acceleration in Mars' twisted magnetotail. *Monthly Notices of the Royal Astronomical Society: Letters*, *517*(1), L121–L125.

Deighan, J., Chaffin, M. S., Chaufray, J.-Y., Stewart, A. I. F., Schneider, N. M., Jain, S. K., et al. (2015). MAVEN IUVS observation of the hot oxygen corona at Mars. *Geophysical Research Letters*, *42*(21), 9009–9014.

Delva, M., Bertucci, C., Volwerk, M., Lundin, R., Mazelle, C., & Romanelli, N. (2015). Upstream proton cyclotron waves at venus near solar maximum. *Journal of Geophysical Research: Space Physics*, *120*(1), 344–354.

Delva, M., & Dubinin, E. (1998). Upstream ULF fluctuations near Mars. *Journal of Geophysical Research: Space Physics*, *103*(A1), 317–326.

Delva, M., Mazelle, C., & Bertucci, C. (2011). Upstream ion cyclotron waves at Venus and Mars. *Space Science Reviews*, *162*(1–4), 5–24.

DiBraccio, G. A., Dann, J., Espley, J. R., Gruesbeck, J. R., Soobiah, Y., Connerney, J. E. P., et al. (2017). Maven observations of tail current sheet flapping at Mars. *Journal of Geophysical Research: Space Physics*, *122*(4), 4308–4324.

DiBraccio, G. A., Luhmann, J. G., Curry, S. M., Espley, J. R., Xu, S., Mitchell, D. L., et al. (2018). The twisted configuration of the martian magnetotail: Maven observations. *Geophysical Research Letters*, *45*(10), 4559–4568.

DiBraccio, G. A., Romanelli, N., Bowers, C. F., Gruesbeck, J. R., Halekas, J. S., Ruhunusiri, S., et al. (2022). A statistical investigation of factors influencing the magnetotail twist at mars. *Geophysical Research Letters*, *49*(12), e2022GL098007. e2022GL098007 2022GL098007.

Dong, Y., Fang, X., Brain, D. A., McFadden, J. P., Halekas, J. S., Connerney, J. E., Curry, S. M., Harada, Y., Luhmann, J. G., & Jakosky, B. M. (2015). Strong plume fluxes at Mars observed by MAVEN: An important planetary ion escape channel. *Geophysical Research Letters*, *42*(21), 8942–8950.

Dubinin, E., & Fraenz, M. (2016). *Ultra-Low-Frequency Waves at Venus and Mars* (chapter 20, pp. 343–364). American Geophysical Union (AGU).

Dubinin, E., Fraenz, M., Fedorov, A., Lundin, R., Edberg, N., Duru, F., & Vaisberg, O. (2011). Ion energization and escape on Mars and Venus. *Space Science Reviews*, *162*(1–4), 173–211.

Dubinin, E., Fraenz, M., Pätzold, M., Halekas, J. S., Mcfadden, J., Connerney, J. E. P., et al. (2018). Solar wind deflection by mass loading in the martian magnetosheath based on maven observations. *Geophysical Research Letters*, *45*(6), 2574–2579.

Dubinin, E., Fraenz, M., Pätzold, M., Tellmann, S., Modolo, R., DiBraccio, G., et al. (2023). Magnetic fields and plasma motions in a hybrid martian magnetosphere. *Journal of Geophysical Research: Space Physics*, *128*(1), e2022JA030575. e2022JA030575 2022JA030575.

Dubinin, E., Fränz, M., Woch, J., Roussos, E., Barabash, S., Lundin, R., et al. (2006). Plasma Morphology at Mars. Aspera-3 Observations. *Space Science Reviews*, *126*(1–4), 209–238.

Eastwood, J. P., Lucek, E. A., Mazelle, C., Meziane, K., Narita, Y., Pickett, J., & Treumann, R. A. (2005). The Foreshock. *Space Science Reviews*, *118*(1–4), 41–94.

Edberg, N. J. T., Auster, U., Barabash, S., Bößwetter, A., Brain, D. A., Burch, J. L., et al. (2009a). Rosetta and Mars Express observations of the influence of high solar wind pressure on the Martian plasma environment. *Annales Geophysicae*, *27*(12), 4533–4545.

Edberg, N. J. T., Brain, D. A., Lester, M., Cowley, S. W. H., Modolo, R., Fränz, M., & Barabash, S. (2009b). Plasma boundary variability at Mars as observed by Mars Global Surveyor and Mars Express. *Annales Geophysicae*, *27*(9), 3537–3550.

Edmiston, J. P., & Kennel, C. F. (1984). A parametric survey of the first critical mach number for a fast mhd shock. *Journal of Plasma Physics*, *32*(3), 429–441.

Ergun, R. E., Andersson, L., Peterson, W. K., Brain, D., Delory, G. T., Mitchell, D. L., et al. (2006). Role of plasma waves in Mars' atmospheric loss. *Geophysical Research Letters*, *33*(14).

Ergun, R. E., Andersson, L. A., Fowler, C. M., Woodson, A. K., Weber, T. D., Delory, G. T., et al. (2016). Enhanced o2+ loss at mars due to an ambipolar electric field from electron heating. *Journal of Geophysical Research: Space Physics*, *121*(5), 4668–4678.

Escoubet, C. P., Schmidt, R., & Goldstein, M. L. (1997). Cluster – Science and Mission Overview. *Space Science Reviews*, *79*, 11–32.

Esman, T. M., Espley, J., Gruesbeck, J., Fowler, C. M., Xu, S., Elrod, M., et al. (2022). Martian ionospheric magnetic fluctuations below 200 km. *Journal of Geophysical Research: Space Physics*, *127*(9), e2022JA030470. e2022JA030470 2022JA030470.

Espley, J. R. (2018). The martian magnetosphere: Areas of unsettled terminology. *Journal of Geophysical Research: Space Physics*, *123*(6), 4521–4525.

Espley, J. R., Cloutier, P. A., Brain, D. A., Crider, D. H., & Acuña, M. H. (2004). Observations of low-frequency magnetic oscillations in the Martian magnetosheath, magnetic pileup region, and tail. *Journal of Geophysical Research: Space Physics*, *109*(A7).

Fairfield, D. H. (1971). Average and unusual locations of the earth's magnetopause and bow shock. *Journal of Geophysical Research (1896–1977)*, *76*(28), 6700–6716.

Feldman, P. D., Steffl, A. J., Parker, J. W., A'Hearn, M. F., Bertaux, J.-L., Alan Stern, S., et al. (2011). Rosetta-Alice observations of exospheric hydrogen and oxygen on Mars. *Icarus*, *214*(2), 394–399.

Ferrand, R., Galtier, S., & Sahraoui, F. (2021). A compact exact law for compressible isothermal hall magnetohydrodynamic turbulence. *Journal of Plasma Physics*, *87*(2).

Formisano, V. (1979). Orientation and shape of the Earth's bow shock in three dimensions. *Planetary and Space Science*, *27*(9), 1151–1161.

Fowler, C. M., Agapitov, O. V., Xu, S., Mitchell, D. L., Andersson, L., Artemyev, A., et al. (2020). Localized heating of the martian topside ionosphere through the combined effects of magnetic pumping by large-scale magnetosonic waves and pitch angle diffusion by whistler waves. *Geophysical Research Letters*, *47*(5), e2019GL086408. e2019GL086408 10.1029/2019GL086408.

Fowler, C. M., Andersson, L., Ergun, R. E., Harada, Y., Hara, T., Collinson, G., et al. (2018a). Maven observations of solar wind-driven magnetosonic waves heating the martian dayside ionosphere. *Journal of Geophysical Research: Space Physics*, *123*(5), 4129–4149.

Fowler, C. M., Andersson, L., Halekas, J., Espley, J. R., Mazelle, C., Coughlin, E. R., et al. (2017). Electric and magnetic variations in the near-mars environment. *Journal of Geophysical Research: Space Physics*, *122*(8), 8536–8559.

Fowler, C. M., Andersson, L., Peterson, W. K., Halekas, J., Nagy, A. F., Ergun, R. E., et al. (2018b). Correlations between enhanced electron temperatures and electric field wave power in the martian ionosphere. *Geophysical Research Letters*, *45*(2), 493–501.

Fowler, C. M., Hanley, K. G., McFadden, J. P., Chaston, C. C., Bonnell, J. W., Halekas, J. S., et al. (2021). Maven observations of low frequency steepened magnetosonic waves and associated heating of the martian nightside ionosphere. *Journal of Geophysical Research: Space Physics*, *126*(10), e2021JA029615. e2021JA029615 2021JA029615.

Fox, J. L., & Bakalian, F. M. (2001). Photochemical escape of atomic carbon from mars. *Journal of Geophysical Research: Space Physics*, *106*(A12), 28785–28795.

Frisch, U. (1995). *Turbulence. The legacy of A.N. Kolmogorov*.

Galtier, S. (2008). von Kármán–Howarth equations for hall magnetohydrodynamic flows. *Physical Reviw E*, *77*(1), 015302.

Gary, S. P. (1993). *Theory of space plasma microinstabilities*. Cambridge Atmospheric and Space Science Series. Cambridge Univ. Press.

Gary, S. P., Akimoto, K., & Winske, D. (1989). Computer simulations of cometary-ion/ion instabilities and wave growth. *Journal of Geophysical Research: Space Physics*, *94*(A4), 3513–3525.

Glass, A. N., Tracy, P. J., Raines, J. M., Jia, X., Romanelli, N., & DiBraccio, G. A. (2023). Characterization of foreshock plasma populations at mercury. *Journal of Geophysical Research: Space Physics*, *128*(2), e2022JA031111. e2022JA031111 2022JA031111.

Glassmeier, K.-H., & Espley, J. (2006). *ULF Waves in Planetary Magnetospheres* (pp. 341–359). American Geophysical Union (AGU).

Gosling, J. T., & Robson, A. E. (1985). *Ion reflection, gyration, and dissipation at supercritical shocks* (pp. 141–152). American Geophysical Union (AGU).

Gosling, J. T., Thomsen, M. F., Bame, S. J., Feldman, W. C., Paschmann, G., & Sckopke, N. (1982). Evidence for specularly reflected ions upstream from the quasi-parallel bow shock. *Geophysical Research Letters*, *9*(12), 1333–1336.

Grard, R., Pedersen, A., Klimov, S., Savin, S., Skalsky, A., Trotignon, J. G., & Kennel, C. (1989). First measurements of plasma waves near Mars. *Nature*, *341*(6243), 607–609.

Gruesbeck, J. R., Espley, J. R., Connerney, J. E. P., DiBraccio, G. A., Soobiah, Y. I., Brain, D., et al. (2018). The three-dimensional bow shock of mars as observed by Maven. *Journal of Geophysical Research: Space Physics*, *123*(6), 4542–4555.

Gurnett, D. A., Kirchner, D. L., Huff, R. L., Morgan, D. D., Persoon, A. M., Averkamp, T. F., et al. (2005). Radar soundings of the ionosphere of Mars. *Science*, *310*(5756), 1929–1933.

Gurnett, D. A., Morgan, D. D., Duru, F., Akalin, F., Winningham, J. D., Frahm, R. A., et al. (2010). Large density fluctuations in the Martian ionosphere as observed by the Mars Express radar sounder. *Icarus*, *206*(1), 83–94.

Hadid, L., Sahraoui, F., & Galtier, S. (2017). Energy cascade rate in compressible fast and slow solar wind turbulence. *The Astrophysical Journal*, *838*(1), 9.

Hadid, L. Z., Sahraoui, F., Kiyani, K. H., Retinò, A., Modolo, R., Canu, P., et al. (2015). Nature of the MHD and kinetic scale turbulence in the magnetosheath of Saturn: Cassini observations. *Astrophysical Journal Letters*, *813*(2), L29.

Halekas, J. S. (2017). Seasonal variability of the hydrogen exosphere of Mars. *Journal of Geophysical Research: Planets*, *122*(5), 901–911.

Halekas, J. S., Luhmann, J. G., Dubinin, E., & Ma, Y. (2021). *Induced magnetospheres*, chapter 25 (pp. 391–406). American Geophysical Union (AGU).

Halekas, J. S., Ruhunusiri, S., Harada, Y., Collinson, G., Mitchell, D. L., Mazelle, C., et al. (2017). Structure, dynamics, and seasonal variability of the Mars-Solar wind interaction: Maven solar wind ion analyzer in-flight performance and science results. *Journal of Geophysical Research: Space Physics*, *122*(1), 547–578.

Halekas, J. S., Ruhunusiri, S., Vaisberg, O. L., Harada, Y., Espley, J. R., Mitchell, D. L., et al. (2020). Properties of plasma waves observed upstream from mars. *Journal of Geophysical Research: Space Physics*, *125*(9), e2020JA028221. e2020JA028221 2020JA028221.

Halekas, J. S., Taylor, E. R., Dalton, G., Johnson, G., Curtis, D. W., McFadden, J. P., et al. (2015). The solar wind ion analyzer for MAVEN. *Space Science Reviews*, *195*(1–4), 125–151.

Hanley, K. G., Fowler, C. M., McFadden, J. P., Mitchell, D. L., & Curry, S. (2022). Maven-static observations of ion temperature and initial ion acceleration in the martian ionosphere. *Geophysical Research Letters*, *49*(18), e2022GL100182. e2022GL100182 2022GL100182.

Harada, Y., Andersson, L., Fowler, C. M., Mitchell, D. L., Halekas, J. S., Mazelle, C., et al. (2016). Maven observations of electron-induced whistler mode waves in the Martian magnetosphere. *Journal of Geophysical Research: Space Physics*, *121*(10), 9717–9731.

Harada, Y., Halekas, J. S., DiBraccio, G. A., Xu, S., Espley, J., Mcfadden, J. P., et al. (2018). Magnetic reconnection on dayside crustal magnetic fields at Mars: Maven observations. *Geophysical Research Letters*, *45*(10), 4550–4558.

Harada, Y., Halekas, J. S., McFadden, J. P., Mitchell, D. L., Mazelle, C., Connerney, J. E. P., et al. (2015). Magnetic reconnection in the near-mars magnetotail: Maven observations. *Geophysical Research Letters*, *42*(21), 8838–8845.

Harada, Y., Ruhunusiri, S., Halekas, J. S., Espley, J., DiBraccio, G. A., Mcfadden, J. P., et al. (2019). Locally generated ULF waves in the martian magnetosphere: Maven observations. *Journal of Geophysical Research: Space Physics*, *124*(11), 8707–8726.

Hughes, A., Chaffin, M., Mierkiewicz, E., Deighan, J., Jain, S., Schneider, N., et al. (2019). Proton aurora on Mars: A dayside phenomenon pervasive in southern summer. *Journal of Geophysical Research: Space Physics*, *124*(12), 10533–10548.

Jakosky, B. M., Lin, R., Grebowsky, J., Luhmann, J., Mitchell, D., Beutelschies, G., et al. (2015). The mars atmosphere and volatile evolution (maven) mission. *Space Science Reviews*, *195*(1–4), 3–48.

Jarvinen, R., Kallio, E., & Pulkkinen, T. I. (2022). Ultra-low frequency foreshock waves and ion dynamics at mars. *Journal of Geophysical Research: Space Physics*, *127*(5), e2021JA030078. e2021JA030078 2021JA030078.

Jiang, W., Li, H., Liu, X., Verscharen, D., & Wang, C. (2023). Statistical properties of plateau-like turbulence spectra in the martian magnetosheath: Maven observations. *Journal of Geophysical Research: Space Physics*, *128*(1), e2022JA030874. e2022JA030874 2022JA030874.

Jin, T., Lei, L., Yiteng, Z., Lianghai, X., & Fuhao, Q. (2022). Statistical analysis of the distribution and evolution of mirror structures in the Martian magnetosheath. *The Astrophysical Journal*, *929*(2), 165.

Johnson, C. L., Mittelholz, A., Langlais, B., Russell, C. T., Ansan, V., Banfield, D., et al. (2020). Crustal and time-varying magnetic fields at the InSight landing site on Mars. *Nature Geoscience*, *13*(3), 199–204.

Kallio, E., Chaufray, J.-Y., Modolo, R., Snowden, D., & Winglee, R. (2011). Modeling of Venus, Mars, and Titan. *Space Science Reviews*, *162*(1–4), 267–307.

Kennel, C. F., & Petschek, H. E. (1966). Limit on stably trapped particle fluxes. *Journal of Geophysical Research (1896–1977)*, *71*(1), 1–28.

Kivelson, M. G., & Russell, C. T. (1995). *Introduction to space physics*.

LeDocq, M. J., Gurnett, D. A., & Hospodarsky, G. B. (1998). Chorus source locations from VLF poynting flux measurements with the polar spacecraft. *Geophysical Research Letters*, *25*(21), 4063–4066.

Liemohn, M. W., Mitchell, D. L., Nagy, A. F., Fox, J. L., Reimer, T. W., & Ma, Y. (2003). Comparisons of electron fluxes measured in the crustal fields at mars by the mgs magnetometer/electron reflectometer instrument with a b field-dependent transport code. *Journal of Geophysical Research: Planets*, *108*(E12).

Lillis, R., Curry, S., Hara, T., Curtis, D., Taylor, E., Parker, J., et al. (2022). ESCAPADE: Mars' first smallsat science mission will unveil its unique hybrid magnetosphere. In *44th COSPAR Scientific Assembly. Held 16–24 July*, volume 44 (pp. 420).

Lillis, R. J., Deighan, J., Fox, J. L., Bougher, S. W., Lee, Y., Combi, M. R., et al. (2017). Photochemical escape of oxygen from mars: First results from maven in situ data. *Journal of Geophysical Research: Space Physics*, *122*(3), 3815–3836.

Liu, D., Yao, Z., Wei, Y., Rong, Z., Shan, L., Arnaud, S., et al. (2020). Upstream proton cyclotron waves: Occurrence and amplitude dependence on imf cone angle at mars-from maven observations. *Earth and Planetary Physics*, *4*(1), 51–61.

Livesey, W. A., Kennel, C. F., & Russell, C. T. (1982). ISEE-1 and -2 observations of magnetic field strength overshoots in quasi-perpendicular bow shocks. *Geophysical Research Letters*, *9*(9), 1037–1040.

Maron, J., & Goldreich, P. (2001). Simulations of Incompressible Magnetohydrodynamic Turbulence. *The Astrophysical Journal*, *554*(2), 1175–1196.

Mazelle, C., Winterhalter, D., Sauer, K., Trotignon, J. G., Acuna, M. H., Baumgärtel, K., et al. (2004). Bow shock and upstream phenomena at mars. *Space Science Reviews*, *111*, 115–181.

McFadden, J. P., Kortmann, O., Curtis, D., Dalton, G., Johnson, G., Abiad, R., et al. (2015). MAVEN SupraThermal and Thermal Ion Compostion (STATIC) Instrument. *Space Science Reviews*, *195*(1–4), 199–256.

Meziane, K., Mazelle, C. X., Romanelli, N., Mitchell, D. L., Espley, J. R., Connerney, J. E. P., et al. (2017). Martian electron foreshock from maven observations. *Journal of Geophysical Research: Space Physics*, *122*(2), 1531–1541.

Mitchell, D. L., Mazelle, C., Sauvaud, J. A., Thocaven, J. J., Rouzaud, J., Fedorov, A., et al. (2016). The MAVEN Solar Wind Electron Analyzer. *Space Science Reviews*, *200*(1–4), 495–528.

Mittelholz, A., Johnson, C. L., Fillingim, M., Grimm, R. E., Joy, S., Thorne, S. N., & Banerdt, W. B. (2023). Mars' external magnetic field as seen from the surface with insight. *Journal of Geophysical Research: Planets*, *128*(1), e2022JE007616. e2022JE007616 2022JE007616.

Mittelholz, A., Johnson, C. L., Fillingim, M., Joy, S. P., Espley, Halekas, J., et al. (2021). Space weather observations with insight. *Geophysical Research Letters*, *48*(22), e2021GL095432. e2021GL095432 2021GL095432.

Modolo, R., Hess, S., Mancini, M., Leblanc, F., Chaufray, J.-Y., Brain, D., et al. (2016). Mars-solar wind interaction: Lathys, an improved parallel 3-d multispecies hybrid model. *Journal of Geophysical Research: Space Physics*, *121*(7), 6378–6399.

Moses, S. L., Coroniti, F. V., & Scarf, F. L. (1988). Expectations for the microphysics of the Mars-solar wind interaction. *Geophysical Research Letters*, *15*(5), 429–432.

Nagy, A. F., Winterhalter, D., Sauer, K., Cravens, T. E., Brecht, S., Mazelle, C., et al. (2004). The plasma Environment of Mars. *Space Science Reviews*, *111*(1), 33–114.

Orosei, R., Jordan, R. L., Morgan, D. D., Cartacci, M., Cicchetti, A., Duru, F., et al. (2015). Mars Advanced Radar for Subsurface and Ionospheric Sounding (MARSIS) after nine years of operation: A summary. *Planetary and Space Science*, *112*, 98–114.

Papadopoulos, K. (1971). Ion thermalization in the Earth's bow shock. *Journal of Geophysical Research*, *76*(16), 3806.

Parker, E. N. (1958). Dynamics of the interplanetary gas and magnetic fields. *The Astrophysical Journal*, *128*, 664.

Paschmann, G., Sckopke, N., Asbridge, J., Bame, S., & Gosling, J. (1980). Energization of solar wind ions by reflection from the Earth's bow shock. *Journal of Geophysical Research: Space Physics*, *85*(A9), 4689–4693.

Paschmann, G., Sckopke, N., Papamastorakis, I., Asbridge, J. R., Bame, S. J., & Gosling, J. T. (1981). Characteristics of reflected and diffuse ions upstream from the earth's bow shock. *Journal of Geophysical Research*, *86*(A6), 4355–4364.

Phillips, P. E., & Robson, A. E. (1972). Influence of reflected ions on the magnetic structure of a collisionless shock front. *Physical Review Letters*, *29*(3), 154–157.

Picardi, G., Biccari, D., Seu, R., Marinangeli, L., Johnson, W. T. K., Jordan, R. L., et al. (2004). Performance and surface scattering models for the Mars Advanced Radar for Subsurface and Ionosphere Sounding (MARSIS). *Planetary and Space Science*, *52*, 149–156.

Politano, H., & Pouquet, A. (1998a). Dynamical length scales for turbulent magnetized flows. *Geophysical Research Letters*, *25*(3), 273–276.

Politano, H., & Pouquet, A. (1998b). von kármán–howarth equation for magnetohydrodynamics and its consequences on third-order longitudinal structure and correlation functions. *Physical Review E*, *57*(1), R21.

Pope, S. B. (2001). Turbulent flows.

Rahmati, A., Larson, D. E., Cravens, T. E., Lillis, R. J., Dunn, P. A., Halekas, J. S., et al. (2015). MAVEN insights into oxygen pickup ions at Mars. *Geophysical Research Letters*, *42*(21), 8870–8876.

Rahmati, A., Larson, D. E., Cravens, T. E., Lillis, R. J., Halekas, J. S., McFadden, J. P., et al. (2017). Maven measured oxygen and hydrogen pickup ions: Probing the martian exosphere and neutral escape. *Journal of Geophysical Research: Space Physics*, *122*(3), 3689–3706.

Rahmati, A., Larson, D. E., Cravens, T. E., Lillis, R. J., Halekas, J. S., McFadden, J. P., et al. (2018). Seasonal variability of neutral escape from mars as derived from maven pickup ion observations. *Journal of Geophysical Research: Planets*, *123*(5), 1192–1202.

Ramstad, R., Barabash, S., Futaana, Y., Nilsson, H., Wang, X.-D., & Holmström, M. (2015). The Martian atmospheric ion escape rate dependence on solar wind and solar euv conditions: 1. Seven years of mars express observations. *Journal of Geophysical Research: Planets*, *120*(7), 1298–1309.

Ramstad, R., Brain, D. A., Dong, Y., Espley, J., Halekas, J., & Jakosky, B. (2020). The global current systems of the Martian induced magnetosphere. *Nature Astronomy*, *4*, 979–985.

Roberts, D. A., Goldstein, M. L., Klein, L. W., & Matthaeus, W. H. (1987). Origin and evolution of fluctuations in the solar wind: Helios observations and helios-voyager comparisons. *Journal of Geophysical Research: Space Physics*, *92*(A11), 12023–12035.

Romanelli, N., Andrés, N., & DiBraccio, G. A. (2022). Variability of the incompressible energy cascade rate in solar wind turbulence around mars. *The Astrophysical Journal*, *929*(2), 145.

Romanelli, N., Bertucci, C., Gomez, D., Mazelle, C., & Delva, M. (2013). Proton cyclotron waves upstream from Mars: Observations from Mars global surveyor. *Planetary and Space Science*, *76*, 1–9.

Romanelli, N., Bertucci, C., Gómez, D., & Mazelle, C. (2015). Dependence of the location of the martian magnetic lobes on the interplanetary magnetic field direction: Observations from mars global surveyor. *Journal of Geophysical Research: Space Physics*, *120*(9), 7737–7747.

Romanelli, N., DiBraccio, G., Gershman, D., Le, G., Mazelle, C., Meziane, K., et al. (2020a). Upstream ultra-low frequency waves observed by messenger's magnetometer: Implications for particle acceleration at mercury's bow shock. *Geophysical Research Letters*, *47*(9), e2020GL087350. e2020GL087350 10.1029/2020GL087350.

Romanelli, N., DiBraccio, G., Halekas, J., Dubinin, E., Gruesbeck, J., Espley, J., et al. (2020b). Variability of the solar wind flow asymmetry in the martian magnetosheath observed by maven. *Geophysical Research Letters*, *47*(22), e2020GL090793. e2020GL090793 10.1029/2020GL090793.

Romanelli, N., DiBraccio, G., Modolo, R., Leblanc, F., Espley, J., Gruesbeck, J., et al. (2019). Recovery timescales of the dayside Martian magnetosphere to IMF variability. *Geophysical Research Letters*, *46*(20), 10977–10986.

Romanelli, N., & DiBraccio, G. A. (2021). Occurrence rate of ultra-low frequency waves in the foreshock of Mercury increases with heliocentric distance. *Nature Communications*, *12*, 6748.

Romanelli, N., Gómez, D., Bertucci, C., & Delva, M. (2014). Steady-state Magnetohydrodynamic Flow around an Unmagnetized Conducting Sphere. *The Astrophysical Journal*, *789*(1), 43.

Romanelli, N., Mazelle, C., Chaufray, J.-Y., Meziane, K., Shan, L., Ruhunusiri, S., et al. (2016). Proton cyclotron waves occurrence rate upstream from mars observed by maven: Associated variability of the martian upper atmosphere. *Journal of Geophysical Research: Space Physics*, *121*(11), 11–113.

Romanelli, N., Mazelle, C., & Meziane, K. (2018a). Nonlinear wave-particle interaction: Implications for newborn planetary and backstreaming proton velocity distribution functions. *Journal of Geophysical Research: Space Physics*, *123*, 1100–1117.

Romanelli, N., Modolo, R., Leblanc, F., Chaufray, J.-Y., Hess, S., Brain, D., et al. (2018b). Effects of the crustal magnetic fields and changes in the IMF orientation on the magnetosphere of Mars: Maven observations and lathys results. *Journal of Geophysical Research: Space Physics*, *123*(7), 5315–5333.

Romanelli, N., Modolo, R., Leblanc, F., Chaufray, J.-Y., Martinez, A., Ma, Y., et al. (2018c). Responses of the Martian magnetosphere to an interplanetary coronal mass ejection: MAVEN observations and LatHyS results. *Geophysical Research Letters*, *45*(16), 7891–7900.

Romeo, O. M., Romanelli, N., Espley, J. R., Mazelle, C., DiBraccio, G. A., Gruesbeck, J. R., & Halekas, J. S. (2021). Variability of upstream proton cyclotron wave properties and occurrence at mars observed by maven. *Journal of Geophysical Research: Space Physics*, *126*(2), e2020JA028616. e2020JA028616 2020JA028616.

Ruhunusiri, S., Halekas, J., Espley, J., Mazelle, C., Brain, D., Harada, Y., et al. (2017). Characterization of turbulence in the mars plasma environment with maven observations. *Journal of Geophysical Research: Space Physics*, *122*(1), 656–674.

Ruhunusiri, S., Halekas, J. S., Connerney, J. E. P., Espley, J. R., McFadden, J. P., Larson, D. E., et al. (2015). Low-frequency waves in the martian magnetosphere and their response to upstream solar wind driving conditions. *Geophysical Research Letters*, *42*(21), 8917–8924.

Russell, C. T. (1994). *Planetary upstream waves* (pp. 75–86). American Geophysical Union (AGU).

Russell, C. T., Luhmann, J. G., Schwingenschuh, K., Riedler, W., & Yeroshenko, Y. (1990). Upstream waves at mars: Phobos observations. *Geophysical Research Letters*, *17*(6), 897–900.

Shane, A., & Liemohn, M. (2021). Whistler wave interactions with superthermal electrons on martian crustal magnetic fields: Bounce-averaged diffusion coefficients and time scales. *Journal of Geophysical Research: Space Physics*, *126*(6), e2021JA029118. e2021JA029118 2021JA029118.

Shane, A., Liemohn, M., Florie, C., & Xu, S. (2019). Misbehaving high-energy electrons: Evidence in support of ubiquitous wave-particle interactions on dayside martian closed crustal magnetic fields. *Geophysical Research Letters*, *46*(21), 11689–11697.

Shane, A. D., & Liemohn, M. W. (2022). Modeling wave-particle interactions with photoelectrons on the dayside crustal fields of mars. *Geophysical Research Letters*, *49*(2), e2021GL096941. e2021GL096941 2021GL096941.

Simon-Wedlund, C., Volwerk, M., Mazelle, C., Halekas, J., Rojas-Castillo, D., Espley, J., & Möstl, C. (2022a). Making waves: Mirror mode structures around mars observed by the maven spacecraft. *Journal of Geophysical Research: Space Physics*, *127*(1), e2021JA029811. e2021JA029811 2021JA029811.

Simon-Wedlund, C., Volwerk, M., Mazelle, C., Rojas Mata, S., Stenberg Wieser, G., Futaana, Y., et al. (2022b). Statistical distribution of mirror mode-like structures in the magnetosheaths of unmagnetised planets: 1. mars as observed by the maven spacecraft. *EGUsphere*, 2022, 1–40.

Skalsky, A., Dubinin, E., Delva, M., Grard, R., Klimov, S., Sauer, K., & Trotignon, J. G. (1998). Wave observations at the foreshock boundary in the near-Mars space. *Earth, Planets and Space*, *50*, 439–444.

Slavin, J. A., & Holzer, R. E. (1981). Solar wind flow about the terrestrial planets 1. Modeling bow shock position and shape. *Journal of Geophysical Research*, *86*(A13), 11401.

Song, P., Russell, C. T., & Gary, S. P. (1994). Identification of low-frequency fluctuations in the terrestrial magnetosheath. *Journal of Geophysical Research: Space Physics*, *99*(A4), 6011–6025.

Sonnerup, B. U. O. (1969). Acceleration of particles reflected at a shock front. *Journal of Geophysical Research (1896–1977)*, *74*(5), 1301–1304.

Stawarz, J. E., Vasquez, B. J., Smith, C. W., Forman, M. A., & Klewicki, J. (2011). Third moments and the role of anisotropy from velocity shear in the solar wind. *The Astrophysical Journal*, *736*(1), 44.

Tao, C., Sahraoui, F., Fontaine, D., Patoul, J., Chust, T., Kasahara, S., & Retinò, A. (2015). Properties of Jupiter's magnetospheric turbulence observed by the Galileo spacecraft. *Journal of Geophysical Research (Space Physics)*, *120*(4), 2477–2493.

Trotignon, J. G., Mazelle, C., Bertucci, C., & Acuña, M. H. (2006). Martian shock and magnetic pile-up boundary positions and shapes determined from the Phobos 2 and Mars Global Surveyor data sets. *Planetary and Space Science*, *54*(4), 357–369.

Tu, C. Y., & Marsch, E. (1991). A case study of very low cross-helicity fluctuations in the solar wind. *Annales Geophysicae*, *9*, 319–332.

Turc, L., Roberts, O. W., Verscharen, D., Dimmock, A. P., Kajdič, P., Palmroth, M., et al. (2023). Transmission of foreshock waves through Earth's bow shock. *Nature Physics*, *19*(1), 78–86.

Vörös, Z., Zhang, T. L., Leaner, M. P., Volwerk, M., Delva, M., & Baumjohann, W. (2008a). Intermittent turbulence, noisy fluctuations, and wavy structures in the Venusian magnetosheath and wake. *Journal of Geophysical Research (Planets)*, *113*(E12), E00B21.

Vörös, Z., Zhang, T. L., Leubner, M. P., Volwerk, M., Delva, M., Baumjohann, W., & Kudela, K. (2008b). Magnetic fluctuations and turbulence in the Venus magnetosheath and wake. *Geophysical Research Letters*, *35*(11), L11102.

Wang, J., Yu, J., Chen, Z., Xu, X., Li, K., Cui, J., & Cao, J. (2023). MAVEN observations of whistler-mode waves within the magnetic dips in the Martian ionopause/ionosphere. *The Astrophysical Journal*, *165*(2), 56.

Weber, T., Brain, D., Xu, S., Mitchell, D., Espley, J., Halekas, J., et al. (2020). The influence of interplanetary magnetic field direction on martian crustal magnetic field topology. *Geophysical Research Letters*, *47*(19), e2020GL087757. e2020GL087757 10.1029/2020GL087757.

Wei, H. Y., Cowee, M. M., Russell, C. T., & Leinweber, H. K. (2014). Ion cyclotron waves at mars: Occurrence and wave properties. *Journal of Geophysical Research: Space Physics*, *119*(7), 5244–5258.

Wei, H. Y., & Russell, C. T. (2006). Proton cyclotron waves at mars: Exosphere structure and evidence for a fast neutral disk. *Geophysical Research Letters*, *33*(23), L23103.

Wei, H. Y., Russell, C. T., Zhang, T. L., & Blanco-Cano, X. (2011). Comparative study of ion cyclotron waves at mars, venus and earth. *Planetary and Space Science*, *59*(10), 1039–1047.

Yamauchi, M., Hara, T., Lundin, R., Dubinin, E., Fedorov, A., Sauvaud, J.-A., et al. (2015). Seasonal variation of martian pick-up ions: Evidence of breathing exosphere. *Planetary and Space Science*, *119*, 54–61.

Zimbardo, G., Greco, A., Sorriso-Valvo, L., Perri, S., Vörös, Z., Aburjania, G., Chargazia, K., & Alexandrova, O. (2010). Magnetic turbulence in the geospace environment. *Space Science Reviews*, *156*(1–4), 89–134.

# 6

# Alfvén Waves at the Giant Planets

**David Southwood**

**ABSTRACT**

The magnetospheres of the giant planets have two special features which impact Alfvén and other magnetohydrodynamic wave structure. There are sources of plasma deep within the equatorial magnetosphere where material that becomes ionized is emitted from moons and both the Jupiter and Saturn systems rotate rapidly. We look at how these features affect not only Alfvén wave structure but also pay attention to the function that Alfvén waves should be expected to fulfill in such systems. We extend the topic of the paper beyond the simple mode polarized purely with a transverse magnetic field to include effects that are inherent in nonuniform plasmas, such surface waves and field-line resonance where the field perturbation has compressional magnetic components. Field-aligned currents are fundamental in coupling the magnetosphere and ionosphere but also to maintaining rotation in the magnetosphere itself. The carriage of field-aligned current is a critical feature in all the circumstances we discuss. We concentrate on theory but review magnetic observations while finally noting that open questions remain.

## 6.1. INTRODUCTION

In a uniform plasma, the shear Alfvén mode is the one magnetohydrodynamic (MHD) mode in a uniform plasma that carries field-aligned current. This property is closely associated with its critical dynamical function. The mode is excited by departures of linear or angular momentum balance along the background magnetic field. In the presence of field-aligned current, the magnetic field is sheared or twisted, often meaning that the plasma flow is also sheared or vortical.

In any magnetosphere, the mode can be important in ionosphere–magnetosphere coupling. Although in most of the magnetospheric volume the plasma is collision-free, collisions between the ionized and the neutral in the ionosphere occur at the feet of the field lines. It is here that the field-guided Alfvén mode can experience collisional dissipation. The electrically resistive nature of the ionosphere is an important means of damping. Thus, the mode not only does transmit momentum, but it also can be important in transferring of energy from the magnetosphere to the ionosphere and upper atmosphere. In the rotating magnetospheres of the giant planets, Jupiter and Saturn, it is also expected to be important in transmitting changes in angular momentum from the atmosphere and ionosphere to the magnetosphere.

MHD does not specify how individual current carriers maintain the induced field-aligned current. Where MHD magnetospheric motions give rise to intense upward current, strong downward acceleration of magnetospheric electrons can result and aurorae are excited where the electrons encounter the atmosphere. Thus, Alfvén waves can also be expected to play an important part in auroral structures. Nonetheless, in this short review, we shall limit our attention to the wave behavior in context of the background MHD of the magnetospheres of the giant planets rather than in auroral processes themselves.

*Department of Physics, Imperial College, London, UK*

*Alfvén Waves Across Heliophysics: Progress, Challenges, and Opportunities, Geophysical Monograph 285*, First Edition.
Edited by Andreas Keiling.

DOI: 10.1002/9781394195985.ch06

The plan of this chapter is as follows. We first give a summary of the dynamical environment encountered in a fast-rotating giant magnetosphere. Then as an illustration before treating more generally the occurrence and roles of the waves at the giant planets, we start with a brief discussion of the iconic case of Alfvén wave generation at the giant planets by the planetary moons (e.g., Io at Jupiter and Enceladus at Saturn). The Io case is well studied. It illustrates some of the basic considerations needed for understanding Alfvén wave behavior at the giant planets. We then look more closely at how the background rotation impacts the morphology and the behavior of Alfvén waves due to the plasma largely being confined to an equatorial torus in both magnetospheres, and then examine some of the recent observational studies.

Less attention has been given to angular momentum transport by Alfvén waves, and yet transport of angular momentum from the neutral atmosphere and ionosphere to the magnetosphere is a basic feature of the giant planet magnetospheres. It would be difficult to believe that waves do not play some part. We start from the theoretical work of Lofting et al. (1997) with a derivation of Alfvén wave equations with the Coriolis effect included to illustrate how the Alfvén waves transport angular momentum and the relationship of their polarization to transport. Lastly, we give a brief discussion where future work might reasonably proceed and discuss possible links between large-scale Alfvén and related MHD modes and the steady diffusion outward of plasma from sources (moons) deep within the magnetosphere known to be present.

## 6.2. GIANT PLANET MAGNETOSPHERES: DYNAMICAL BACKGROUND

The magnetospheres of Saturn and Jupiter differ from the terrestrial magnetosphere in several ways. The rotation is fast (10-h period) and the spatial scales of the systems are considerably larger than at Earth; rotation cannot automatically be ignored. The primary source of plasma is from the moons, through volcanoes on Io at Jupiter and geysers on Enceladus at Saturn, which are in the near-equatorial inner magnetosphere. Material released from the moons becomes ionized and creates a plasma torus round the moon's orbit. A magnetospheric circulation system must then carry material outward to eventual loss in interplanetary space. The outward transport must be diffusive because, although mass moves outward, magnetic flux must be conserved overall. In the inner torus, the rotation is close to corotating with the planetary ionosphere and atmosphere at the feet of the field lines. Although the flow may be slower than corotation at larger distances, the plasma transport outward must also be accomplished with transfer of angular momentum from the ionosphere as long as the plasma remains on flux tubes connected to the planet. Similarly, the rotation results in a component of the centrifugal force parallel to the background field that keeps the torus plasma equatorially confined. During the outward motion of a flux tube, the planetary field will become increasingly distended due to the ionized material near the equator. At some point, the outward moving plasma will either encounter the magnetopause and be lost through the boundary or, more likely, as it rotates through the nighttime magnetotail sector, move so far from the planet that the stress in the distended planetary field can no longer counter the centrifugal force. The field will undergo reconnection and the material injected deep in the magnetosphere will escape down the planetary magnetotail, normally in a bubble shape known as a plasmoid. As noted earlier, magnetic flux is conserved overall, and although the net transport of ionized material is outward, there must be a mechanism for a return of magnetic flux inward. In the inner regions of the magnetosphere, this is normally envisaged as an interchange motion where flux tubes with higher mass move outward and underpopulated tubes move in (Southwood & Kivelson, 1989). At Jupiter, in the dayside morning sector of the magnetosphere, there is a distinct cushion region between the magnetopause and the outer edge of the region where the equatorial field is distorted by the torus/plasma sheet (see, e.g., Kivelson and Southwood, 2005). The field cushion is undistorted by any equatorial currents and so is ascribed to depleted flux tubes returning from the nightside after reconnection and release of plasma. Within the cushion region bubbles of plasma are sporadically seen where the field strength drops close to zero (field nulls) and the plasma density rises. The nulls have been attributed to material breaking off the outer edge of the plasma sheet and eventually escaping through the magnetopause. There is no equivalent region seen at Saturn. Reconnection events in the tail do inject high-energy plasma and those tubes must also contain depleted low-energy plasma after reconnection, and the flux must find its way back fairly directly to the diffusion system to return to the inner magnetosphere.

## 6.3. THE PLASMA SOURCES AND ALFVÉN WAVES

Near the moons that are the internal magnetospheric plasma sources, it has long been recognized that Alfvén waves are set up associated with the diversion of corotation flow about the conducting moon's body and by the mass pickup from the ongoing ionization of material injected into the surroundings by the moon. The internal sources of magnetospheric plasma have by chance given us something close to an MHD-wave point source. The Jovian moon, Io, has received most attention since the Voyager-1 flyby in 1979 (Ness et al., 1979) showed the first evidence of the wave. We shall concentrate on that in a brief review and use it to illustrate key distinctive features for Alfvén and other MHD waves in the giant planet context.

The electrically conducting interior of Io and its immediate surroundings where volcanic material emitted from the interior becomes ionized form an obstruction to the magnetospheric plasma flow which in the Io orbital region is close to corotating with the planet. The flux tubes that encounter Io and its close environment are slowed and thus locally a shear is created. Over the poles of Io, the flow appears to be brought almost to a halt (Frank & Patterson, 2001). The flow shear and its associated magnetic-field shear launch Alfvén waves traveling both north and south along the background field. The waves travel in the plasma frame of reference (see, e.g., Neubauer, 1980). It follows the waves reach the ionospheres at the north and south of the field somewhere downstream with respect to Io. They will be reflected there and return to the torus downstream of Io and then further bounces can be made further downstream.

There are lessons to learn from the example this provides and the well-documented observation and simulations of the structures that form. Although the simulations show that some excitation of the compressional MHD waves occurs, the primary excitation is a train of Alfvén waves behind the moon. The multiple bouncing structure was first deduced from the form of radio emissions excited by the electrons accelerated by upward field-aligned currents in the Alfvén waves (Gurnett & Goertz, 1981). The overall pattern of Alfvén waves was first analyzed by Bagenal (1983) using a planetary magnetic-field model and the plasma distributions deduced from the Voyager spacecraft flybys. The auroral pattern in the Jovian atmosphere associated with the accelerated electrons indicates reflection to the opposite hemisphere there, but is further complicated by the fact that the accelerated electrons contain bidirectional beams [see, e.g., Bonfond et al. (2017) and references therein]. Now it is known that the moons Europa and Ganymede also display auroral footprints in the Jovian atmosphere. The details of the Alfvén wave structures associated with moon interactions are to be addressed elsewhere in this volume and we will not pursue these aspects here. We shall rather use the well-known Alfvén wave structure that arises from an effective point source in a fast-rotating giant planet magnetosphere to explore implications for Alfvén and other wave excitations on larger scales.

The work of Bagenal (1983) and its subsequent development by Hinton et al. (2019) are illustrated in Figure 6.1, which is reproduced from Hinton et al. (2019). The sketch shows multiple bounces of an Alfvén structure that the authors envisage extends potentially right around Jupiter. The sketch serves to illustrate well how the Alfvén wave propagation varies inside and outside the torus that the source gives rise to. The slowing of the Alfvén wave as it transits the equatorial high-density torus is evident in the figure and it is clear that there is a large contrast in Alfvén speed inside the torus region with the exterior.

Hinton et al. (2019) examined the Alfvén wave propagation. They used an up-to-date magnetic field model

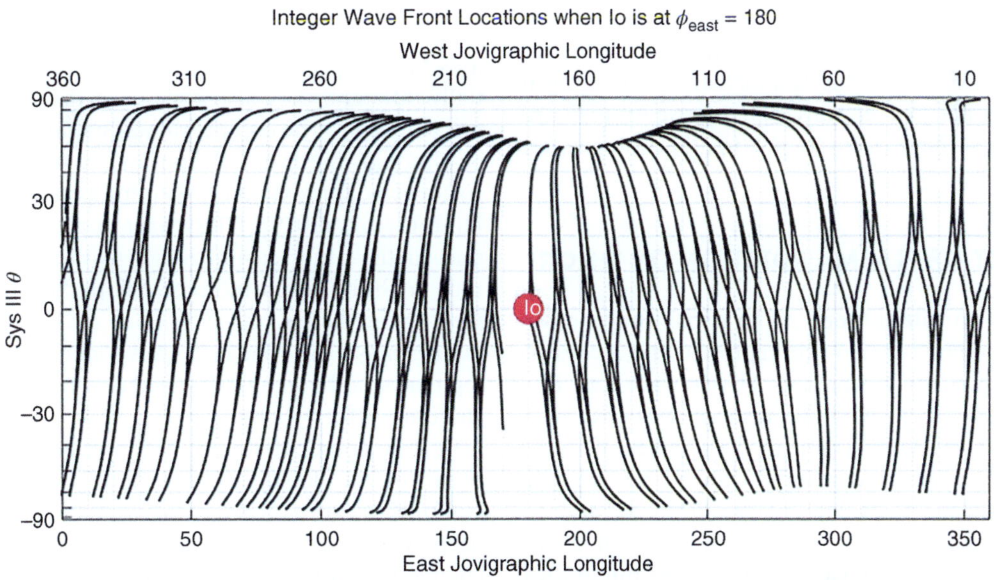


**Figure 6.1** The phase fronts of an Alfvén wave attached to the Jovian moon Io as derived by Hinton et al. (2019) / John Wiley & Sons. The vertical axis is the sine of the latitude. The horizontal axes show east (below) and west (above) System III longitudes. The sketch shows schematically how the wave is swept back from Io in the Jovian plasma flow and reflects off the ionosphere in each hemisphere. Io's notional location as the source of the pattern is 180°.

JRM09 based on the data obtained from the Juno mission. This was combined with their own 3D torus mass-density model to derive Alfvén transit times within the pattern in Figure 6.1. Round-trip travel times along the field and back to the original Io latitude vary between 25 and 30 min at different longitudes, but the travel times for the legs between the torus boundary and the ionosphere in the same hemisphere are much shorter, 2–3 min, making clear the effect of the deep trough in Alfvén speed in the torus.

## 6.4. LESSONS TO DRAW

A simple lesson to draw from the previous section is that interpretation of spacecraft observations needs to pay attention to the background plasma flow. The spacecraft will not be in the rotating plasma frame. For larger-scale sources much bigger than Io it means that there can be an angular Doppler shift in the frequency detected on a spacecraft. Khurana and Kivelson (1989) first noted that the Doppler shift could impact observations of MHD waves in analysis of disturbances with 5–30-min period during the Voyager-2 spacecraft flyby of Jupiter in 1979. The signals analyzed are compressional and thus not pure Alfvén waves, but nonetheless it is clear that Doppler shift needs be borne in mind in data interpretation in a rotating system. The angular wave number, $m$, is a useful guide to employ for assessing the impact of the Doppler effect if one has some estimate available of $m$. Taking the planetary rotation period approximately 600 min, then an observed period $T_{obs}$ anywhere comparable to the $600/m$ min means attention needs to be paid to possible Doppler effects. Bearing in mind that toroidal Alfvén wave signals have $m$ small and poloidal signals have $m$ large, polarization information can often provide a clue.

## 6.5. IMPACT OF PLASMA ROTATION ON THEORY

Were there no rotation and no Io orbital motion, the 25–30-min bounce times would represent the local fundamental period for a standing Alfvén wave and thus the lowest-order local field-line resonance (FLR). In a plasma stationary with respect to the sources of waves, standing Alfvén waves can form trapped between the reflecting ionospheres at each of the field line. Indeed, in the terrestrial system, energy from a global MHD fast compressional mode can be coupled to pump local FLRs where the frequency matches the eigenfrequency of a standing Alfvén wave. FLRs form a very important aspect of our understanding of the terrestrial MHD wave spectrum (e.g., Southwood, 1974; Wright & Elsden, 2016).

The background plasma rotation at Io precludes FLR happening with the Io waves. The reason is that after reflection in the ionosphere the Io wave returns along the field line to the Io latitude substantially displaced from the original source. FLR will only occur with a source with a frequency and azimuthal scale such that there is little phase change during the bounce time along the field. The plot in Figure 6.1 shows that in the Io case, although a bouncing standing Alfvén wave structure occurs, resonances could not form as when a reflected wave returns to near the equatorial location of the Io orbit the original signal has gone. The relative motion between Io and the wave embedded in the magnetospheric plasma flow precludes this happening in this case. For a coherent response between the transmitted and returning waves, the source scale size must be much larger than the angular distance moved by the plasma in the bounce, that is, if $\Delta\phi$ represents the angular size of the source, then $\Delta\phi \gg \Omega T_{bounce}$ where $\Omega$ is the plasma rotation rate with respect to the source. Thus, signals need to have a small angular wave number $m$.

Glassmeier et al. (1989) reported that during the Voyager-1 Jupiter flyby, field pulsations with periods of around 13.3 and 20 min were detected in the Io torus vicinity. They deduced that both signals had low $m$ number from the signal polarization. They further deduced that the higher frequency, which showed a compressional field component, was primarily poloidal and the 20-min signal was primarily polarized in the azimuthal direction and thus could be standing Alfvén wave. It was also proposed that the latter wave energy was largely trapped in the torus at low latitudes in the high-mass-density, low-Alfvén-speed region. As the periods were substantially different, they also concluded that the signals were not coupled to the compressional signals.

Although the bounce time for an Alfvén wave in the Io torus is approximately 30 min, as we have seen from Hinton et al. (2019), Glassmeier et al. (1989) points out that in the outer Jovian magnetosphere the equivalent time might be hours and even comparable to the rotation time. Nonetheless, observational evidence has become increasingly clear that in both the giant planets there are regular periodic signals detected that bear resemblance to standing Alfvén waves.

Early theoretical studies did look at how FLR might occur in the Jovian system (Kivelson & Southwood, 1986). Kivelson and Southwood (1986) realized that the deep Alfvén velocity trough created by the torus would change the nature of the coupling between the fast compressional mode and the formation of FLR by the Alfvén mode. Most importantly, they pointed out that the compressional signals would have a different variation along the field to the transverse mode and this would modify the manner in which the compressional signals would

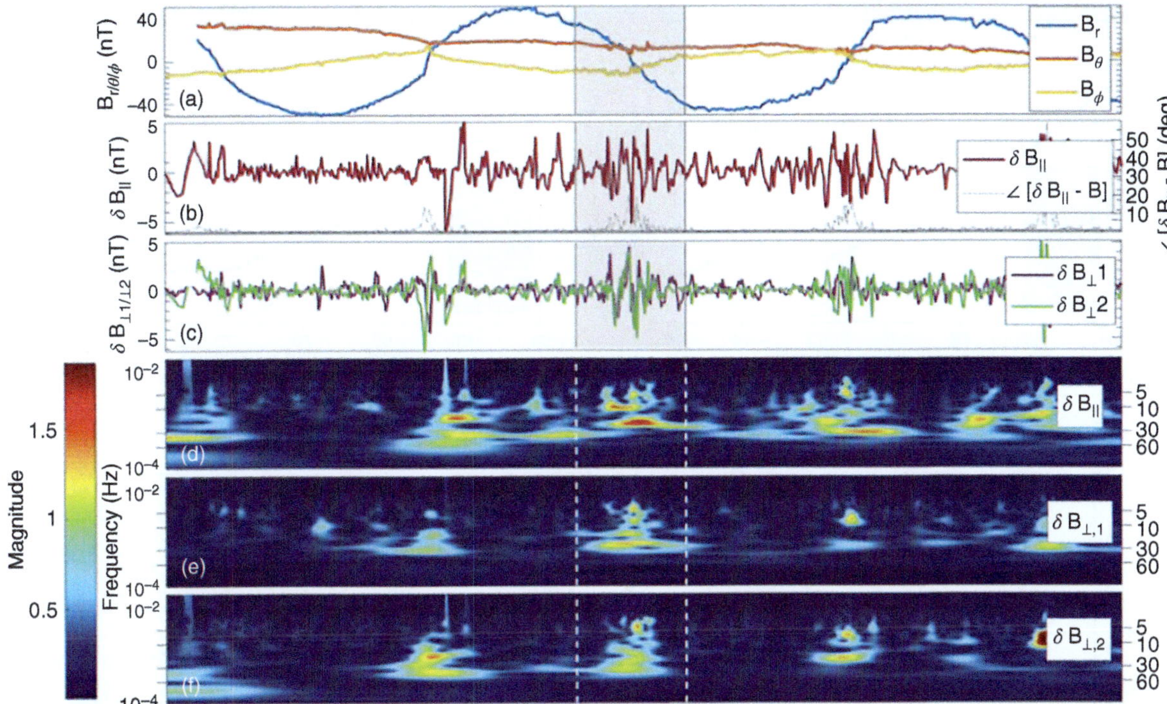


**Figure 6.2** From Manners and Masters (2019) / John Wiley & Sons, Galileo spacecraft magnetometer data are shown. A particular event seen in crossing the Io torus is highlighted by the shading and vertical markers. Panel (a) shows the three components of the planetary magnetic field in spherical polar system III coordinates. Panel (b) shows the compressional wave disturbance field in mean field-aligned (MFA) coordinates and panel (c) shows the two transverse components in MFA. Panels (d)–(f) show wavelet analysis for all three MFA components indicating the presence of harmonic structure.

pump the resonance. In particular, they showed that a compressional mode signal with fundamental standing structure along the field could excite higher harmonics of the Alfvén mode and might do so more readily than the fundamental.

Subsequent analysis of the magnetic field data from the Galileo Jupiter orbiter spacecraft and earlier flyby data suggests that indeed waves with periods 10–60 min are regularly seen at Jupiter and it does seem likely that low-frequency standing Alfvén mode signals do form FLRs in this band in the Jovian system (Lysak & Song, 2020; Manners & Masters, 2019; Manners & Masters, 2020; Manners et al., 2018; Wilson & Dougherty, 2000). Manners and Masters (2020) following previous authors call the transverse signals quasi-periodic oscillations (QPOs) with a period range spanning 5–60 min, and dominated by signals with periods 15, 30, and 40 min. The last feature suggests that they are due to a harmonic structure.

An example from Manners et al. (2018) which illustrates the occurrence of multiple harmonics analysis is shown in Figure 6.2. Consistent with Kivelson and Southwood's (1986) initial simple theory, the same from Manners et al. (2018) suggest that higher-order resonances are preferentially excited.

At Saturn, quasi-periodic signals which are largely transverse are also recorded in a similar range of periods. The terminology QP60 is often used; the period being around 60 min (Bunce et al., 2014; Mitchell et al., 2016; Yates et al., 2016). Once again, they are attributed to standing Alfvén waves. Rusaitis et al. (2021) usefully expanded the modeling of field and plasma for the standing Alfvén wave equations in the Saturn system by using the formalism for a more general background field first used in the terrestrial context by Singer et al. (1981). They use a modern model of the Saturn field (Khurana et al., 2006) and the plasma density model of Bagenal & Delamere (2011). Results are very different from what a dipole field would provide. In particular, the eigenmodes of the third and fourth harmonics for standing waves over a large range of magnetic shells have periods in the range 40–75 min. Figure 6.3 shows the results from Rusaitis et al. (2021).

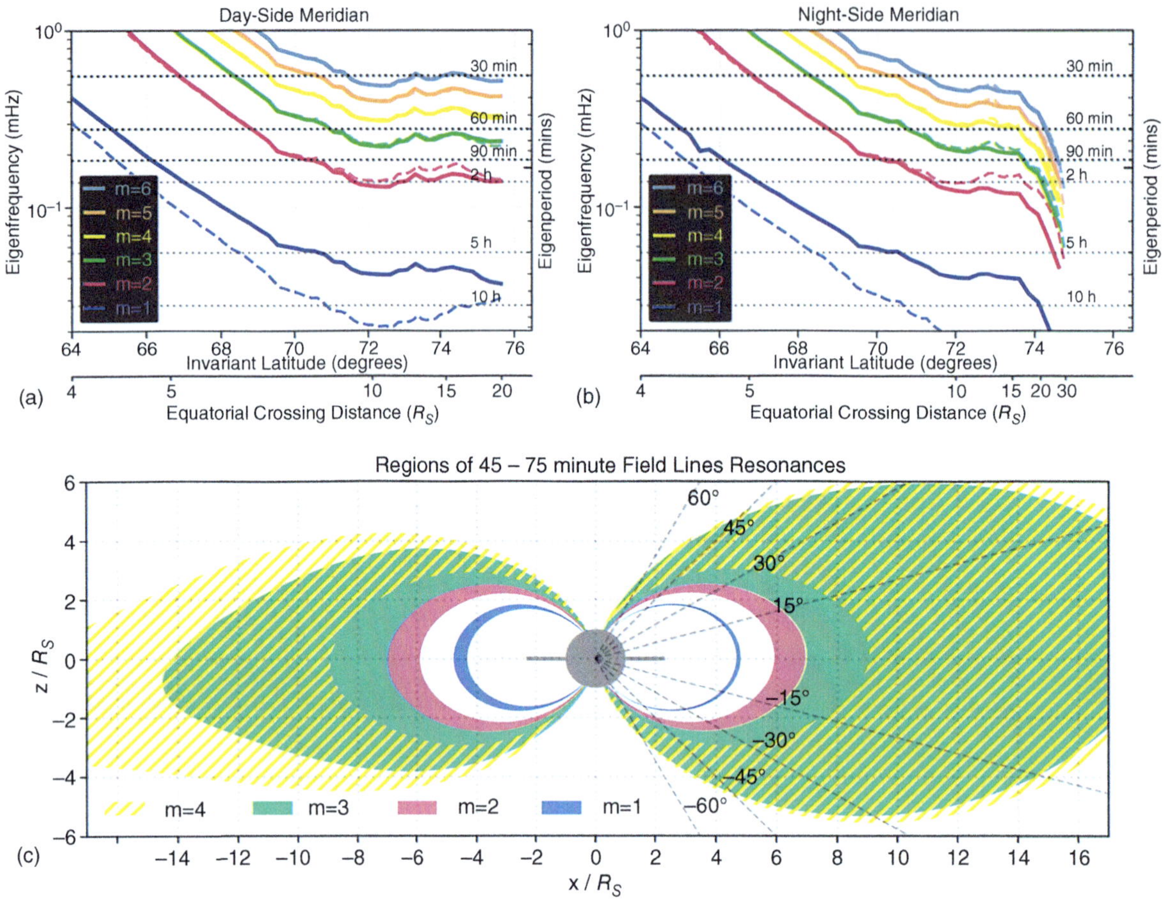


**Figure 6.3** From Rusaitis et al. (2021) / John Wiley & Sons, panels (a) and (b) show the computed standing Alfvén wave resonant frequencies for toroidal (solid) and poloidal (dashed) in Saturn's magnetosphere for the first six harmonics as a function of equatorial field-line crossing point for dayside and nightside. Panel (c) shows how the use of a sophisticated field model produces large regions of the outer magnetosphere where eigenperiods between 45 and 75 min vary little with radial distance. These periods correspond to harmonics between 4 and 6.

The QP60 signals appear in a variety of other data from Cassini (Palmaerts et al., 2016). A large review is given by Palmaerts (2017). It is clear that the 1-h period range represents an important aspect of short time-scale dynamics at Saturn. Moreover, as was clear in the earliest report by Kleindienst et al. (2009), the signals appear to show some modulation close to the planetary rotation period. We introduce next some theory to examine the effect of planetary rotation on Alfvén waves.

## 6.6. THE CORIOLIS EFFECT ON ALFVÉN WAVES IN A ROTATING MAGNETOSPHERE

Rotation is a fundamental feature of the giant planet magnetospheres. If there were not internal sources of plasma, steady rotation would mean that flux tubes would move as a whole at the same angular speed (Ferraro, 1937). However, as noted earlier, with continuous sources of plasma in the deep interior of the magnetosphere, there will be continuous plasma diffusion outward, and if rotation is maintained, there must be continuous provision of angular momentum from the ionosphere to the magnetosphere.

Any change in source rate must inevitably give rise to Alfvén mode waves to adjust the angular momentum distribution along the field. Little work has been published in a magnetospheric context attempting to directly address the effect of rotation on Alfvén waves. The first analysis of the effect of system rotation on the Alfvén wave mode was by Lehnert (1954) in the context of the Sun. Derivations in the context of planetary dynamo theory followed Hide (1966) and Moffat (1978). The study of Lofting et al. (1997) was the earliest with direct reference to the MHD waves in the magnetosphere of the outer planets.

In a rotating system, the first question is which frame of reference to use. For meteorology, there is normally no issue, as the atmosphere moves with the planet to a first approximation and so the rotating frame is a natural choice and we follow this here. Working in the rotating frame, we examine the linearized equations governing a small magnetic field perturbation, **b**, and an associated plasma displacement perturbation, $\xi$, in a cold plasma system with mass density, $\rho$, and the background magnetic field, **B**, where we assume the background system is under steady rotation where the (vector) angular velocity is $\mathbf{\Omega}$.

The field perturbation, **b**, and displacement perturbation, $\xi$, are related by the linearized MHD frozen field condition,

$$\boldsymbol{b} = \nabla \times (\xi \times \boldsymbol{B}) \tag{6.1}$$

The system rotation enters through the linearized equation of motion in the frame of reference of the rotating plasma,

$$\rho\frac{\partial^2 \xi}{\partial t^2} = +\frac{(\nabla \times \boldsymbol{b}) \times \boldsymbol{B}}{\mu_0} - \boldsymbol{\nabla}(\delta p) - \boldsymbol{\Omega} \times (\boldsymbol{\Omega} \times \boldsymbol{\delta r}) - 2\rho\left(\boldsymbol{\Omega} \times \frac{\partial \xi}{\partial t}\right), \tag{6.2}$$

where $\delta p$ is the perturbation in plasma pressure and $\boldsymbol{\delta r}$ is the disturbance perturbation in the radius vector from the rotation axis, **r**. The two terms in $\mathbf{\Omega}$ are the centrifugal force and the Coriolis force introduced by choosing the rotating reference frame. The centrifugal term can be absorbed into the gradient term as an effective centrifugal potential and the magnetic force is divided into terms associated with magnetic pressure and tension:

$$\rho\frac{\partial^2 \boldsymbol{\xi}}{\partial t^2} = -\boldsymbol{\nabla}\left(\delta p + \frac{\boldsymbol{B}.\boldsymbol{b}}{\mu_0} + \rho\Omega r\xi_r\right) + \frac{\boldsymbol{B}.\boldsymbol{\nabla}\boldsymbol{b}}{\mu_0} - 2\rho\left(\boldsymbol{\Omega} \times \frac{\partial \xi}{\partial t}\right). \tag{6.3}$$

$\delta p$ is a pressure perturbation assumed to be isotropic. In fact, at low frequencies in a collision-free plasma, gas pressure is likely to be anisotropic with respect to the background field direction with separate parallel and perpendicular components. The final term on the right is the Coriolis effect. It can have a major effect on Alfvén waves.

We look first at a very simple case based on the early work examined by Lehnert (1954), Hide (1966), Moffat (1978), and Lofting et al. (1997) who all looked at something similar. We shall assume the plasma is cold (i.e., plasma $\beta$ small), the rotation is sub-Alfvénic, the field and plasma density are uniform, and the background field $\boldsymbol{B}$ and the angular rotation velocity are in the $z$ direction. Although this case looks prima facie somewhat unrealistic, it is instructive. In a cold plasma, if the rotational flow is sub-Alfvénic, magnetic compressional effects can often reasonably be decoupled from the transverse shearing motions of the Alfvén mode with field and magnetic pressure governed by

$$\delta p + \frac{\boldsymbol{B}.\boldsymbol{b}}{\mu_0} + \rho\Omega r\xi_r \approx \text{constant}. \tag{6.4}$$

Nonetheless, what the equation shows is that, even if the plasma thermal pressure is negligible, field compression may occur to balance the changing centrifugal effect due to the wave.

Using Equation (6.1), after some manipulation, the transverse plasma displacement is found to obey the equation:

$$\mu_0\rho\frac{\partial^2 \xi}{\partial t^2} - (\boldsymbol{B}.\boldsymbol{\nabla})^2\xi = -2\mu_0\rho\left(\boldsymbol{\Omega} \times \frac{\partial \xi}{\partial t}\right). \tag{6.5}$$

Solutions are assumed to vary as $\exp i(k_x x + k_y y + k_z z \omega t)$. On resolving $x$ and $y$ components, one has a pair of equations for the transverse components:

$$\left[\omega^2 - (Ak_z)^2\right]\xi_x + 2i\omega\Omega\xi_y = 0, \tag{6.6}$$

$$\left[\omega^2 - (Ak_z)^2\right]\xi_y - 2i\omega\Omega\xi_x = 0, \tag{6.7}$$

where we have substituted for the Alfvén speed, $A = B/\sqrt{\mu_0\rho}$.

A quartic dispersion relation then is derived,

$$\left(\omega^2 - A^2k_z^2\right)^2 - 4(\omega\Omega)^2 = 0. \tag{6.8}$$

This can be solved to give

$$\omega^2 - A^2k_z^2 = \pm 2\omega\Omega, \tag{6.9}$$

see, for example, Lehnert (1954), Lofting et al. (1997), and Moffat (1978).

The four roots correspond to transverse waves moving in each direction along the background field where for each direction there are two polarizations possible, left-handed and right-handed.

Using Equation (6.9) in Equations (6.6) and (6.7), one sees that the roots $\omega_+$ corresponding to the upper sign in Equation (6.9) give

$$\xi_x = -i\xi_y, \tag{6.10}$$

and the root $\omega_-$ corresponding to the lower sign yields

$$\xi_x = i\xi_y. \tag{6.11}$$

Both modes are circularly polarized independent of frequency. A systematic sense of transverse polarization indicates that a mode is transmitting angular momentum along the background field. The mode with root $\omega_+$ is polarized in the opposite sense to the background rotation and has the phase speed above the Alfvén speed. It is carrying angular momentum opposed to the sense of the background rotation. The mode with root $\omega_-$ is polarized in the same sense as the background rotation. Its phase speed is below the Alfvén speed and is carrying angular momentum in the sense of the background rotation.

Lofting et al. (1997) had in mind the effect of introducing the Coriolis term on standing Alfvén waves or FLRs. When $|\omega| \gg |\Omega|$, Lofting et al. (1997) suggested that the effect might result in creation of a doublet structure for field-line resonances. However, this may be a red herring in the giant planet context where the mode is quite likely to be excited by an angular momentum imbalance along the background field. Only the resonant mode carrying the appropriate sense of angular momentum would be expected.

Glassmeier et al. (1989) use the magnetic Rossby number (Hide, 1966) ($k_z A/\Omega$) to judge the need for inclusion of Coriolis effects. As long as the wave frequency is of order $k_z A$ and $\omega$ is much larger than the rotation frequency, the effect is weak. Lofting et al. (1997) despite doing substantially more by way of derivation only look at the same limit. Taking the opposite extreme and looking at when the Coriolis effect is dominant yield another important special case. When the frequency is very small, $\delta\omega$, one can ignore the $\omega^2$ term in Equation (6.9), and so

$$\delta\omega = \pm\frac{B^2 k_z^2}{2\mu_0 \rho \Omega}. \tag{6.12}$$

As before, the two signs correspond to propagation in opposite senses along the field. These modes would both have circular polarization. Moreover, both modes would satisfy Equation (6.11), and so both modes carry angular momentum in the sense of background rotation.

Referring back to the equations shows that in this small frequency limit the Coriolis force is balancing the magnetic body force. This is what Hide (1966) identifies as magnetostrophic flow. The frequency given in Equation (6.12) is the frequency in the rotating plasma frame of reference. In an inertial frame, the wave would appear to have a frequency close to $\Omega$. One may well ask if waves are seen near the rotation period in the giant planet magnetospheres.

## 6.7. WAVES NEAR THE PLANETARY ROTATION PERIOD

At Jupiter, deciding if there are waves present in the system at frequencies near the planetary rotation period is not clear. The strong departure of the planetary field from axisymmetry potentially would hide any superposed wave near the rotation period. However, at Saturn, nature has been kind, as the background planetary field is remarkably axisymmetric. One can see that there are pervasive magnetic oscillations near the expected planetary rotation period detected throughout the magnetosphere that are driven by sources external to the planet (Espinosa et al., 2000, 2003a, 2003b). They are known as planetary period oscillations (PPOs), as the period is close to the planetary rotation period.

The PPOs are one of the most surprising features of the Saturn magnetospheric system. The earliest papers by Espinosa et al. (2000, 2003a, 2003b) called them cam disturbances because they broke the symmetry of the planetary field. The papers using the signal polarization showed that they could not be due to a rotating internal feature of the planetary field, but they must be due to a source external to the planet. The fact that the amplitude of the signals varied little with radial distance from Saturn also indicated that the source could not be internal to the planet (Southwood & Kivelson, 2007). The latter authors went on to show that the signals exhibited a single azimuthal angular wavelength ($m = 1$) and that the north–south symmetry indicated there was not an equatorial source.

Subsequently, Gurnett et al. (2009) showed that the Saturn kilometric radio emissions, which also appear to pulse at a period close to the planetary rotation period, had a remarkable property. Gurnett et al. (2009) found that the radio signals originating from the north and south appeared to pulse at a slightly different rate. Analysis of the magnetic PPO signals found that they were also composed of two periods differing by 1–2% (Andrews et al., 2010; Southwood, 2011). Both northern and southern signals had an angular wave number, $m = 1$, and, for both signals, the transverse magnetic field polarization corresponded systematically to rotation in the planetary rotation sense.

Jia et al. (2012a, 2012b) showed that the polarization of the signals detected in the magnetosphere could well be reproduced in a global simulation by imposing twin vortex motion in the ionosphere at the feet of magnetospheric field lines. Whether the vortices were created by similar motion in the neutral atmosphere was moot.

The first suggestion that the global signals might be associated with Alfvén waves was made by Southwood and Cowley (2014). Andrews et al. (2012) had done an exhaustive survey of the northern and southern signals by the Cassini magnetometer, and Southwood and Cowley (2014) proposed a global current system. Andrews et al. (2012) showed that in the polar caps of each hemisphere where the field can be expected to be open, that is, extending into interplanetary space, the northern signals only exhibit the northern period and in the southern polar cap only southern periods are seen. In contrast, at lower latitudes on the closed field lines, a mixed signal is detected.

Figure 6.4 shows two sketches from Southwood and Cowley (2014). The upper panel shows a sketch of the field-aligned current patterns for the signals oscillating with the northern period. A similar set of currents would drive the southern system. The lower panel shows a

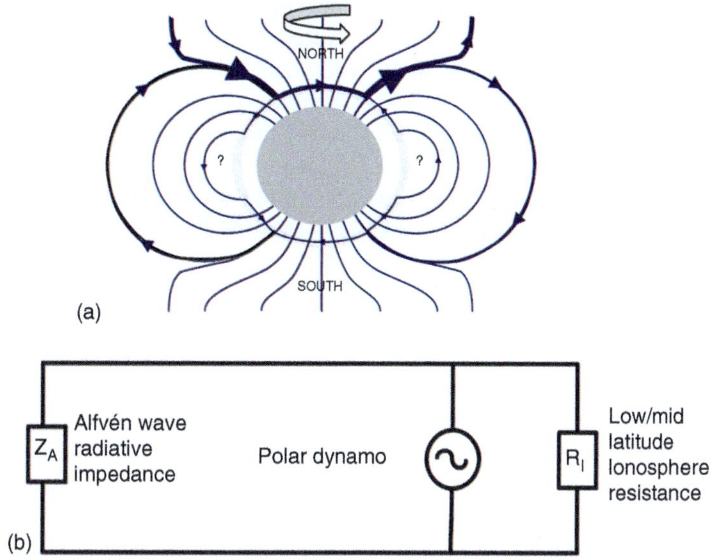


**Figure 6.4** A composite of two sketches from Southwood and Cowley (2014) / John Wiley & Sons. (a) The current configuration deduced for the northern-period PPO oscillations. The current system rotates about the planetary rotation/magnetic axis at the northern period. (b) A simple circuit analogy for the current system is shown in panel (a), in which the energy source is placed in the polar ionosphere.

notional electrical circuit analogy. The energy source is proposed to be the alternating current generated by rotation in the polar cap and an alternating voltage source is shown located in the polar cap. The current system closes partially through the closed field region where the ionospheric resistance absorbs energy. The second loop is where current flows into the open field in the polar cap. Here, the field-aligned currents drive Alfvén waves that propagate out into interplanetary space. Accordingly, the closure of the element of the circuit is represented by a radiative impedance associated with the outward propagation of energy from the polar cap. Of course, this model predicts that the magnetosphere is radiating Alfvén signals at the polar ionospheric rotation into interplanetary space.

Subsequently, Hunt et al. (2016) analyzed the distribution of field-aligned current both in the PPOs and the steady field-aligned currents. They showed indeed that the location of the PPO current peaks is close to the open and closed field-line boundary where the auroral-zone steady currents also peak as the sketch in Figure 6.4 indicates. The signals are almost certainly in the MHD regime, and in MHD, the magnetic field is a fundamental means for transmission of stress in the medium. However, identifying the global current structure, in particular the field-aligned current location and their closure, is a way to identify the transmission of energy.

The components transverse to the field have an elliptical signal polarization with the same sense as the planetary rotation. That rotation sense indicates that almost certainly the signals are associated with angular momentum transfer and that their dynamic function would be appropriate to transfer angular momentum from the ionosphere into the magnetosphere.

An evident discrepancy with the signals being purely Alfvénic is that they show no variation in frequency with magnetic shell. They are apparently a global oscillation. The hydromagnetic surface wave, a close relative of the Alfvén mode, provides a possible answer and we shall look at this further.

## 6.8. THE HYDROMAGNETIC SURFACE WAVE IN A ROTATING SYSTEM

Southwood (2011) first suggested that PPOs might be associated with a hydromagnetic surface wave. The existence of the mode requires there being a magnetic shell across which there is a large change in the plasma density; this is the surface where amplitudes peak. The boundary between the open and closed field lines provides such a natural surface where the density changes substantially. The mode is a hybrid of the transverse Alfvén mode and the compressional fast mode. On each side of the surface the mode resembles a compressional fast mode wave but one which is evanescent in both directions normal to the surface. Indeed, the PPOs exhibit a compressional field component, whereas a pure Alfvén mode would be transverse-polarized. In addition, we have already noted with Equation (6.4) that the modification produced by the wave effect on the centrifugal force requires such a compressional component. Indeed, in the PPOs, the field compression is in phase with the meridional motion just as Equation (6.4) suggests (Andrews et al., 2010, 2019; Provan et al., 2009).

Southwood (2011) noted that the polarization of the Saturn PPOs matched that of a hydromagnetic surface wave and the surface in question might be the steep boundary in plasma density reported by Gurnett et al. (2009), located close to the expected open-closed field-line boundary. Subsequently, Hunt et al. (2014) and Hunt et al. (2015) have identified the structure in invariant latitude of the PPO field-aligned currents at Saturn directly, and as noted earlier, the currents are indeed located close to the open-closed field-line boundary. Their amplitude and location of their peak are close to that of the non-oscillatory field-aligned currents classically identified with maintaining steady sub-corotation between the ionosphere and magnetosphere.

The simplest form of the surface wave occurs where there is a steep jump in plasma mass density perpendicular to the field (see, e.g., Lee & Roberts, 1986; Uberoi, 1982). In the simplest theory, there is an interface between two plasma populations where the mass density is uniform on each side but changes from $\rho_1$ to $\rho_2$ across it. Requiring dynamical continuity means that the normal displacement to the boundary and total pressure are continuous across the boundary. Provided the parallel wavelength much exceeds the perpendicular wavelength in the plane of the interface, the fast mode dispersion relation will require evanescence in the normal direction on both sides of the interface and thus the surface wave can exist. The phase speed is the Alfvén speed derived from the mean of the mass density on each side of the interface.

$$\omega^2 = \frac{2(Bk_z)^2}{\mu_0(\rho_1 + \rho_2)}. \tag{6.13}$$

As noted earlier, the hydromagnetic surface mode is a hybrid of the Alfvén mode and the fast mode. Critically, for our discussion here, it carries field-aligned currents whose amplitude peaks at the surface. Hunt et al.'s (2016) data analysis shows current peaking near the open-closed field-line boundary, itself likely to be the density gradient found by Gurnett et al. (2009) that led to Southwood's (2011) suggestion. If the waves are propagating transverse to the field with a definite sense, the transverse-field components are circularly polarized with the sense of rotation reversing across the surface. This latter feature is indeed detected in the PPOs when the Cassini spacecraft is crossing high-latitude field lines mapping to latitudes above the auroral zone (Provan et al., 2009).

Southwood (2011) did not give a theoretical derivation for the existence of the hydromagnetic surface wave in a rotating inhomogeneous system. Ideally, a derivation should be done in a spherical or cylindrical polar-based system where Lofting et al.'s (1997) work indicates how to go. This is beyond the remit of this review. Nevertheless, using the same approach as used by previous authors and using the assumptions of the various derivations of the hydromagnetic surface wave in the straight-field case, the mode appears to exist and should have a frequency in the rotating frame given by

$$\delta\omega = \pm\frac{B^2k_z^2}{\mu_0 2\Omega(\rho_1 + \rho_2)}. \tag{6.14}$$

Accordingly, Southwood and Cowley's (2014) suggestion of Alfvén waves radiating outward from each polar cap into the open field lines of the magnetotail might be slightly modified to refer to the hydromagnetic surface wave. Moreover, we have seen that introducing plasma rotation introduces a new mode near the planetary rotation period and that PPOs are likely to be associated with angular momentum transfer in the Saturn magnetosphere. Future theory work in more general field geometries should be useful in elucidating fully both the structure and function of PPOs.

## 6.9. BACKGROUND PLASMA DIFFUSION: SOME OPEN ISSUES

In both giant magnetospheres, there is a background outward transport of the material in the torus. This is believed to be diffusive, but any unsteady element requires equilibration along magnetic tubes and must involve field-aligned currents and thus Alfvén waves to transport angular momentum. Moreover, as material diffuses outward, not only is angular momentum transport required but a nonlinear effect of the waves themselves might provide diffusion.

The diffusion is believed to start in the inner torus through the interchange instability (see, e.g., Southwood & Kivelson, 1989). In this motion, the flux tubes that are denser than their surroundings move outward exchanging their position with less-dense tubes which move in. Field-aligned currents are fundamental to the process. There is evidence at Jupiter and Saturn that in the inner regions of the torus scales are relatively short. However, as material moves outward, denser material might be expected to rotate slower than less-dense tube and some self-organization might occur producing large sectors where outward motion dominates over inward and vice versa. Alfvén mode signals traveling between the equatorial torus plasma and the ionosphere would necessarily be part of the equilibration. The function of the signals in the 1-h period band described earlier may well represent part of the process.

One may also ask the degree to which at Saturn the PPOs actually could provide net diffusion to the plasma.

First, the PPOs are rotating at close to the planetary rate, while sub-rotation is reported in the plasma in the outer torus. Over multiple cycles, individual particles will experience net motion outward or inward. Could this effect be the basis for a net diffusion in the torus population as a whole?

Lastly, Jupiter, in particular, has a large highly energetic radiation-belt population within its magnetosphere. It has long been recognized that Alfvén and other MHD waves have a part to play in radiation-belt particle scattering in the terrestrial context (Southwood et al., 1969).

## 6.10. CONCLUDING REMARKS

In this review, we have illustrated how Alfvén wave behavior differs from the more familiar terrestrial environment. Not only are the effects of system rotation very different but also the dominance of the internal magnetospheric sources of plasma from the moons places very different constraints on the system.

We used a brief analysis of work on the well-documented Alfvén wave generation by Io as a way to illustrate how the giant planets differ from other environments due to plasma rotation and the internal sources.

We widened our study to include forms of the Alfvén wave where there is coupling to the compressional mode but where the carriage of field-aligned current is still fundamental to the structure. Following discussion of the effect of the Coriolis force in particular, we have proposed that the well-known PPOs that pervade the Saturn system are a modified form of the Alfvén wave where the Coriolis force is dominant, that is, the flow in the wave is magnetostrophic (Hide, 1966). We have speculated that theoretical analysis of the Coriolis effect on a hydromagnetic surface wave centered on the shells close to the open-closed field-line boundary should be productive if it can be done in a more appropriate geometry. We also pointed out that as the PPOs extend into the polar cap at Saturn, the system as a whole would be emitting Alfvén-like waves along open flux tubes into the interplanetary medium.

We closed by noting that diffusive processes are believed to dominate the circulation of plasma within the giant planet systems. Continuous transport of angular momentum from the ionosphere to the magnetosphere is a critical element of that circulation and Alfvén waves will likely play an important part, and here would be another potentially interesting area for theoretical analysis.

## AVAILABILITY STATEMENT

The realistic field model for Saturn referred to in the text by Krishan K. Khurana is available on the open-access repository Zenodo (Rusiatis et al., 2021).

## ACKNOWLEDGMENTS

The author acknowledges funding from UK Science and Technology Facilities Council grant: ST/W001071/1.

## REFERENCES

Andrews, D. J., Coates, A. J., Cowley, S. W. H., Dougherty, M. K., Lamy, L., Provan, G., & Zarka, P. (2010). Magnetospheric period oscillations at Saturn: Comparison of equatorial and high-latitude magnetic field periods with north and south Saturn kilometric radiation periods. *Journal of Geophysical Research*, *115*, A12252. https://doi.org/10.1029/2010JA015666

Andrews, D. J., Cowley, S. W. H., Dougherty, M. K., Lamy, L., Provan, G., & Southwood, D. J. (2012). Planetary period oscillations in Saturn's magnetosphere: Evolution of magnetic oscillation properties from southern summer to post-equinox. *Journal of Geophysical Research*, *117*, A4. https://doi.org/10.1029/2011JA017444

Andrews, D. J., Cowley, S. W. H., Provan, G., Hunt, G. J., Hadid, L. Z., Morooka, M. W., & Wahlund, J.-E. (2019). The structure of planetary period oscillations in Saturn's equatorial magnetosphere: Results from the Cassini mission. *Journal of Geophysical Research: Space Physics*, *124*, 8361–8395 https://doi.org/10.1029/2019JA026804

Bagenal, F. (1983). Alfvén wave propagation in the Io plasma torus. *Journal of Geophysical Research*, *88*(A4), 3013. https://doi.org/10.1029/ja088ia04p03013

Bagenal, F., & Delamere, P. A. (2011). Flow of mass and energy in the magnetospheres of Jupiter and Saturn. *Journal of Geophysical Research: Space Physics*, *116*(A5). A05029, https://doi.org/10.1029/2010ja016294

Bonfond, B., Grodent, D., Badman, S. V., Saur, J., Gérard, J.-C., & Radioti, A. (2017). Similarity of the Jovian satellite footprints: Spots multiplicity and dynamics. *Icarus*, *292*, 208–217. https://doi.org/10.1016/j.icarus.2017.01.009

Bunce, E. J., Grodent, D. C., Jinks, S. L., Andrews, D. J., Badman, S. V., Coates, A. J., et al. (2014). Cassini nightside observations of the oscillatory motion of Saturn's northern auroral oval. *Journal of Geophysical Research: Space Physics*, *119*(5): 3528–3543. https://doi.org/10.1002/2013ja019527

Espinosa, S. A., & Dougherty, M. K. (2000). Periodic perturbations in Saturn's magnetic field. *Geophysical Research Letters*, *27*(17), 2785–2788. https://doi.org/10.1029/2000gl000048

Espinosa, S. A., Southwood, D. J., & Dougherty, M. K. (2003a). Re-analysis of Saturn's magnetospheric field data in view of spin-periodic perturbations. *Journal of Geophysical Research*, *108*(A2), 1085. https://doi.org/10.1029/2001JA005083

Espinosa, S. A., Southwood, D. J., & Dougherty, M. K. (2003b). How can Saturn impose its rotation period in a non-corotating magnetosphere? *Journal of Geophysical Research*, *108*(A2), 1086. https://doi.org/10.1029/2001JA005084

Ferraro, V. C. A. (1937). The Non-uniform Rotation of the Sun and its Magnetic Field. *Monthly Notices of the Royal Astronomical Society*, *97*, 458. https://doi.org/10.1093/mnras/97.6.458

Frank, L. A., & Patterson, W. R. (2001). *Journal of Geophysical Research: Space Physics*, *106*(A4), 6131–6149. https://doi.org/10.1029/2000ja000159

Glassmeier, K.-H., Ness, N. F., Acuña, M. H., & Neubauer, F. M. (1989). Standing hydromagnetic waves in the Io plasma torus: Voyager 1 observations. *Journal of Geophysical Research*, *94*(A11), 15063. https://doi.org/10.1029/ja094ia11p15063

Gurnett, D. A., & Goertz, C. K. (1981). Multiple Alfven wave reflections excited by Io: Origin of the Jovian decametric arcs. *Journal of Geophysical Research*, *86*(A2), 717. https://doi.org/10.1029/ja086ia02p00717

Gurnett, D. A., Lecacheux, A., Kurth, W. S., Persoon, A. M., Groene, J. B., Lamy, L., et al. (2009). Discovery of a north-south asymmetry in Saturn's radio rotation period. *Geophysical Research Letters*, *36*, 16. https://doi.org/10.1029/2009GL039621

Hide, R. (1966). Free hydromagnetic oscillations of the Earth's core and the theory of the geomagnetic secular variation. *Philosophical Transactions of the Royal Society London*, *259*, 615–647. https://doi.org/10.1098/rsta.1966.0026

Hinton, P. C., Bagenal, F., & Bonfond, B. (2019). Alfvén wave propagation in the Io plasma torus. *Geophysical Research Letters*, *46*, 1242–1249. https://doi.org/10.1029/2018GL081472

Hunt, G. J., Cowley, S. W. H., Provan, G., Bunce, E. J., Alexeev, I. I., Belenkaya, E. S., et al. (2014). Field-aligned currents in Saturn's southern nightside magnetosphere: Subcorotation and planetary period oscillation components. *Journal of Geophysical Research: Space Physics*, *119*, 9847–9899. https://doi.org/10.1002/2014JA020506

Hunt, G. J., Cowley, S. W. H., Provan, G., Bunce, E. J., Alexeev, I. I., Belenkaya, E. S., et al. (2015). Field-aligned currents in Saturn's northern nightside magnetosphere: Evidence for interhemispheric current flow associated with planetary period oscillations. *Journal of Geophysical Research: Space Physics*, *120*, 7552–7584. https://doi.org/10.1002/2015JA021454

Hunt, G. J., Cowley, S. W. H., Provan, G., Bunce, E. J., Alexeev, I. I., Belenkaya, E. S., et al. (2016). Field-aligned currents in Saturn's magnetosphere: Local time dependence of southern summer currents in the dawn sector between midnight and noon. *Journal of Geophysical Research: Space Physics*, *121*, 7785–7804. https://doi.org/10.1002/2016JA022712

Jia, X., Hansen, K. C., Gombosi, T. I., Kivelson, M. G., Tóth, G., DeZeeuw, D. L., & Ridley, A. J. (2012a). Magnetospheric configuration and dynamics of Saturn's magnetosphere: A global MHD simulation. *Journal of Geophysical Research*, *117*, A05225. https://doi.org/10.1029/2012JA017575

Jia, X., Kivelson, M. G., & Gombosi, T. I. (2012b). Driving Saturn's magnetospheric periodicities from the upper atmosphere/ionosphere. *Journal of Geophysical Research*, *117*, A04215. https://doi.org/10.1029/2011JA017367

Khurana, K. K., Arridge, C. S., Schwarzl, H., & Dougherty, M. K. (2006). A model of Saturn's magnetospheric field based on latest Cassini observations. AGU Spring Meeting Abstracts, 2007 (P44A--1).

Khurana, K. K., & Kivelson, M. G. (1989). Ultralow frequency MHD waves in Jupiter's middle magnetosphere. *Journal of Geophysical Research*, *94*(A5), 5241. https://doi.org/10.1029/ja094ia05p05241

Kivelson, M. G., & Southwood, D. J. (1986). Coupling of global magnetospheric MHD eigenmodes to field line resonances. *Journal of Geophysical Research*, *91*, 4345. https://doi.org/10.1029/ja091ia04p04345

Kivelson, M. G., & Southwood, D. J. (2005). Dynamical consequences of two modes of centrifugal instability in Jupiter's outer magnetosphere. *Journal of Geophysical Research*, *110*, A12209. https://doi.org/10.1029/2005JA011176

Kleindienst, G., Glassmeier, K.-H., Simon, S., Dougherty, M. K., & Krupp, N. (2009). Quasiperiodic ULF-pulsations in Saturn's magnetosphere. *Annales Geophysicae*, *27*(2), 885–894. https://doi.org/10.5194/angeo-27-885-2009

Lee, M. A., & Roberts, B. (1986). On the behavior of hydromagnetic surface waves. *The Astrophysical Journal*, *301*, 430. https://doi.org/10.1086/163911

Lehnert, B. (1954). Magnetohydrodynamic waves under the action of the coriolis force. *The Astrophysical Journal*, *119*, 647. https://doi.org/10.1086/145869

Lofting, C. F. A., Dougherty, M. K., & Southwood, D. J. (1997). The Coriolis effect in a rapidly rotating magnetosphere. *Advances in Space Research*, *20*(2), 239–242. https://doi.org/10.1016/s0273-1177(97)00541-3

Lysak, R. L., & Song, Y. (2020). Field line resonances in Jupiter's magnetosphere. *Geophysical Research Letters*, *47*, e2020GL089473. https://doi.org/10.1029/2020GL089473

Manners, H., Masters, A., & Yates, J. N. (2018). Standing AlfvénWaves in Jupiter's magnetosphere as a source of 10 to 60-min quasiperiodic pulsations. *Geophysical Research Letters*, *45*, 8746–8754. https://doi.org/10.1029/2018GL078891

Manners, H., & Masters, A. (2019). First evidence for multiple-harmonic standing Alfvén waves in Jupiter's equatorial plasma sheet. *Geophysical Research Letters*, *46*, 9344–9351. https://doi.org/0.1029/2019GL083899

Manners, H., & Masters, A. (2020). The global distribution of ultralow-frequency waves in Jupiter's magnetosphere. *Journal of Geophysical Research: Space Physics*, *125*, e2020JA028345. https://doi.org/10.1029/2020JA028345

Mitchell, D. G., Carbary, J., Bunce, E., Radioti, A., Badman, S., Pryor, W., et al. (2016). Recurrent pulsations in Saturn's high latitude magnetosphere. *Icarus*, *263*, 94–100. https://doi.org/10.1016/j.icarus.2014.10.028

Moffat, H. K. (1978). *Magnetic field generation in electrically conducting fluids*. Cambridge Monographs on Mechanics and Applied Mathematics, Cambridge: Cambridge University Press.

Ness, N. F., Acuna, M. H., Lepping, R. P., Burlaga, L. F., Behannon, K. W., & Neubauer, F. M. (1979). Magnetic field studies at Jupiter by Voyager 2: Preliminary results. *Science*, *206*(4421), 966–972. https://doi.org/10.1126/science.206.4421.966

Neubauer, F. M. (1980). Nonlinear standing Alfvén wave current system at Io: Theory. *Journal of Geophysical Research: Space Physics*, *85*(A3), 1171–1178. https://doi.org/10.1029/ja085ia03p01171

Palmaerts, B. (2017). Multi-instrument study of the hourly pulsations in Saturn's magnetosphere. Thesis, Georg-August-Universität Göttingen.

Palmaerts, B., Roussos, E., Krupp, N., Kurth, W. S., Mitchell, D. G., & Yates, J. N. (2016). Statistical analysis and multi-instrument overview of the quasi-periodic 1-hour pulsations in Saturn's outer magnetosphere. *Icarus*, *271*, 1–18. https://doi.org/10.1016/j.icarus.2016.01.025

Provan, G., Andrews, D. J., Arridge, C. S., Coates, A. J., Cowley, S. W. H., Milan, S. E., et al. (2009). Polarization and phase of planetary-period magnetic field oscillations on high-latitude field lines in Saturn's magnetosphere. *Journal of Geophysical Research: Space Physics*, *114*, A2. https://doi.org/10.1029/2008ja013782

Rusaitis, L., Khurana, K. K., Kivelson, M. G., & Walker, R. J. (2021). Quasiperiodic 1-hour Alfven wave resonances in Saturn's magnetosphere: Theory for a realistic plasma/field model. *Geophysical Research Letters*, *48*, e2020GL090967. https://doi.org/10.1029/2020GL090967

Singer, H. J., Southwood, D. J., Walker, R. J., & Kivelson, M. G. (1981). Alfven wave resonance in a realistic magnetospheric magnetic field geometry. *Journal of Geophysical Research*, *86*, 4589. https://doi.org/10.1029/ja086ia06p04589

Southwood, D. J. (1974). Some features of field line resonances in the magnetosphere. *Planetary and Space Science*, *22*, 483. https://doi.org/10.1016/0032-0633(74)90078-6

Southwood, D. J., & Kivelson, M. G. (2007). Saturnian magnetospheric dynamics: Elucidation of a camshaft model. *Journal of Geophysical Research*, *112*, A12222, https://doi.org/10.1029/2007JA012254

Southwood, D. J. (2011). Direct evidence of differences in magnetic rotation rate between Saturn's northern and southern polar regions. *Journal of Geophysical Research*, *116*, A01201. https://doi.org/10.1029/2010JA016070

Southwood, D. J., & Cowley, S. W. H. (2014). The origin of Saturn's magnetic periodicities: Northern and southern current systems. *Journal of Geophysical Research: Space Physics*, *119*, 1563–1571. https://doi.org/10.1002/2013JA019632

Southwood, D. J., & Kivelson, M. G. (1989). Magnetospheric interchange motions. *Journal of Geophysical Research*, *94*, 299–308. https://doi.org/10.1029/ja094ia01p00299

Southwood, D. J., Dungey, J. W., & Etherington, R. J. (1969). Bounce resonant interaction between pulsations and trapped particles. *Planetary and Space Science*, *17*, 349. https://doi.org/10.1016/0032-0633(69)90068-3

Uberoi, C. (1982), A note on the existence of Alfven surface waves, *Solar Physics*, *78*, 351–354. https://doi.org/10.1007/BF00151614

Wilson, R. J., & Dougherty, M. K. (2000). Evidence provided by Galileo of ultra low frequency waves within Jupiter's middle magnetosphere. *Geophysical Research Letters*, *27*, 835–838. https://doi.org/10.1029/1999GL010750.

Wright, A. N., & Elsden, T. (2016). The theoretical foundation of 3D Alfvén resonances: Normal modes. *Astrophysical Journal*, *833*, 230. https://doi.org/10923.3847/1538-4357/833/2/230

Yates, J. N., Southwood, D. J., Dougherty, M. K., Sulaiman, A. H., Masters, A., Cowley, S. W. H., et al. (2016). Saturn's quasiperiodic magnetohydrodynamic waves. *Geophysical Research Letters*, *43*, 11,102–11,111. https://doi.org/10.1002/2016GL071069

# 7

# Alfvén Waves Related to Moon-Magnetosphere Interactions

**Bertrand Bonfond[1] and Ali Sulaiman[2]**

**ABSTRACT**

The electro–dynamic interactions between moons and the magnetosphere of their host planets have been investigated since the mid-20th century, and the implication of the Alfvén waves was recognized right away. However, in the first models, Alfvén waves were only considered as current carriers. It is only after the Voyager missions that the possibility of complex reflection patterns was considered and their ability to accelerate particles became fully appreciated only recently. In this chapter, we review the history of our understanding of the various cases of moon–magnetosphere interactions in our Solar System. The presence of the largest of these moons in the stream of the magnetospheric plasma generates powerful large-scale Alfvén waves, which can break up into smaller scales, reflect off density gradients and accelerate particles, which ultimately impact the atmosphere of the planet to generate auroras and trigger radio emissions. The best known case is the Io–Jupiter interaction, since its observational signatures are the richest and most obvious. As our means of investigation have improved, signatures of similar interactions have also been discovered for the other Galilean moons, as well as for moons orbiting Saturn. Interestingly, similar interactions can occur on rare occasions between the planets themselves and the solar wind, and most likely take place in exo-planetary systems as well.

## 7.1. HISTORY OF THE MOON-MAGNETOSPHERE INTERACTIONS

This chapter deals with a particular type of phenomenon involving Alfvén waves throughout the Solar System, the moon–magnetosphere interactions. As we will discuss further, the motion of a moon through the magnetospheric plasma can excite powerful Alfvén waves propagating in both directions along the magnetic field lines toward the host planet and generate spectacular auroral phenomena. Furthermore, the peculiar geometries of these systems sometimes allow clear disentanglement of the different steps of the process, from the local Alfvén wave generation at the moons, to their intense signature in the radio emissions, passing through their filamentation via a turbulent cascade, their reflection at density gradients, and their interaction with charged particles, making them fascinating plasma physics laboratories.

In most cases in our Solar System, the regime of the moon–magnetosphere interaction is sub-Alfvénic, i.e., the relative velocity between the incoming plasma and the obstacle $v_0$ is lower than the local Alfvén speed

$$v_A = \frac{B}{\sqrt{\mu_0 \rho}},$$

where $B$ is the magnetic field strength, $\mu_0$ is vacuum permeability, and $\rho$ is the plasma mass density; in other words, the Alfvén Mach number

$$M_A = \frac{v_0\sqrt{\mu_0 \rho}}{B},$$

[1] *STAR Institute, University of Liège, Liège, Belgium*
[2] *School of Physics and Astronomy, University of Minnesota, Minneapolis, MN, USA*

*Alfvén Waves Across Heliophysics: Progress, Challenges, and Opportunities, Geophysical Monograph 285*, First Edition.
Edited by Andreas Keiling.

DOI: 10.1002/9781394195985.ch07

where $v_0$ is the relative velocity between the moon and the magnetospheric plasma, is lower than one (with the exception of Titan, our Moon, and Callisto). The nature of the interaction would not change abruptly beyond this threshold, but the amplitude of the far-field magnetic disturbance decreases rapidly as the Alfvén Mach number increases (Ridley, 2007; Sarantos & Slavin, 2009). An illustration of this transition can be found in Figure 7.1, which shows results from a magneto-hydrodynamic simulation of the interaction between a magnetized planet and the solar wind for different Alfvén Mach numbers between 8.2 and 0.7. It can be seen that the magnetosphere transforms into Alfvén wings, carrying more and more energy along the Z-axis (i.e., along the magnetic field lines). Saur (2021) provided a thorough review of the moon–magnetosphere interaction, with an emphasis on the theory of the local interaction. Hence, the present review is rather focused on the observations and the way they advanced our understanding of the processes involved.

It is important to distinguish between two types of obstacles generating Alfvén waves: the mechanical obstacles and the electromagnetic obstacles. The former can be further separated into 1) inert obstacles, for which the moon's surface absorbs particles from the incoming plasma and generating a low density wake downstream, and 2) mass-loading obstacles, for which moons with a significant atmosphere modify the momentum of the incoming plasma as the newly ionized particles drag the field lines before being accelerated. On the other hand, electromagnetic obstacles can be formed by either an intrinsic or an induced magnetic field, which can itself originate either in the ionosphere, in a sub-surface ocean, or in an asthenosphere (a ductile and partially melted layer below a rocky crust) and in a metallic core. In practice, the effects are combined, as a moon can both have an atmosphere and a sub-surface conductive layer (e.g., Europa; see Section 7.2.1).

The case of Io and Jupiter is probably the most prominent example, at least historically speaking. The first hint of the peculiar relationship between Io and Jupiter came from Bigg (1964), who noted that the decametric radio emissions from Jupiter were modulated by Io's position both along its orbit and in the Jovian magnetic field. At the same time, but much closer to the Earth, the first experimental telecommunication satellite, a 30-m wide balloon named Echo 1, was losing altitude at a larger rate than expected. Drell et al. (1965) suggested an original theory to explain this strange behavior: since Echo 1 basically consisted of a giant sphere of electrically conductive metal, it was electro-magnetically interacting with the terrestrial magnetic field. As the flux tubes filled with magnetospheric plasma move relative to the spacecraft, an electric field forms across it perpendicular to both the motion direction and the magnetic field. Since the body of the spacecraft is conductive, an electric current would flow across the spacecraft. The Lorenz force, perpendicular to both the magnetic field and the current direction would be oriented in the direction opposite to Echo 1's velocity vector, would then slow the spacecraft down. At the same time, the magnetic field associated with this transverse electric current would shield the object from the external magnetic field. The interaction between the spacecraft and the surrounding plasma generates Alfvén waves, which propagate along the magnetic field lines (forming a structure called Alfvén wings) and which carry field-aligned currents, thus allowing for the current flowing across the spacecraft to close outside of it.

While the unexpected slowing down of Echo 1 turned out to be more easily explained by an underestimation of the radiation pressure and atmospheric drag at its orbiting altitude, the idea of an electric current associated with the motion of a conductive celestial body disturbing the ambient plasma flow proved to be relevant and attracted other researchers, who proposed that a similar system could be at play at Io. A first scenario was drafted by Piddington and Drake (1968), followed by a more detailed discussion by Goldreich and Lynden-Bell (1969). This first model is called the "unipolar inductor" model. It should be noted that, while this model was opposed to the "Alfvén wings" model later on, it was already clear in the authors' minds that the current was carried along the magnetic field lines via Alfvén waves. Furthermore, the concept of Alfvén wings was already defined by Drell et al. (1965). However, in the unipolar inductor model, as the Alfvén velocity was expected to be close to the speed of light because the magnetosphere plasma was expected to be extremely tenuous, the Alfvén waves reflecting on the Jovian ionosphere would directly fall back on Io before the satellite could move significantly along its orbit. Because feedback would get established very rapidly, the interaction could then be modeled as a direct current loop. The motion of Io, considered as an infinitely conductive body, across Jupiter's North-South aligned magnetic field lines, induces a motional electric field oriented radially in the anti-Jovian direction, and the current flows in the same direction. It would then connect to the Jovian ionosphere along the magnetic field lines (see Figure 7.2). Initial speculations suggested that the current would flow inside Io's interior, but later iterations (Webster et al., 1972) considered that the current would rather flow as Pedersen and Hall currents in Io's ionosphere. Indeed, the existence of such an ionosphere was later demonstrated through Pioneer 10 radio occultations (Kliore et al., 1975). One of the important characteristics of this family of models is that the current intensity is essentially limited by the Pedersen conductivity in Jupiter's ionosphere.

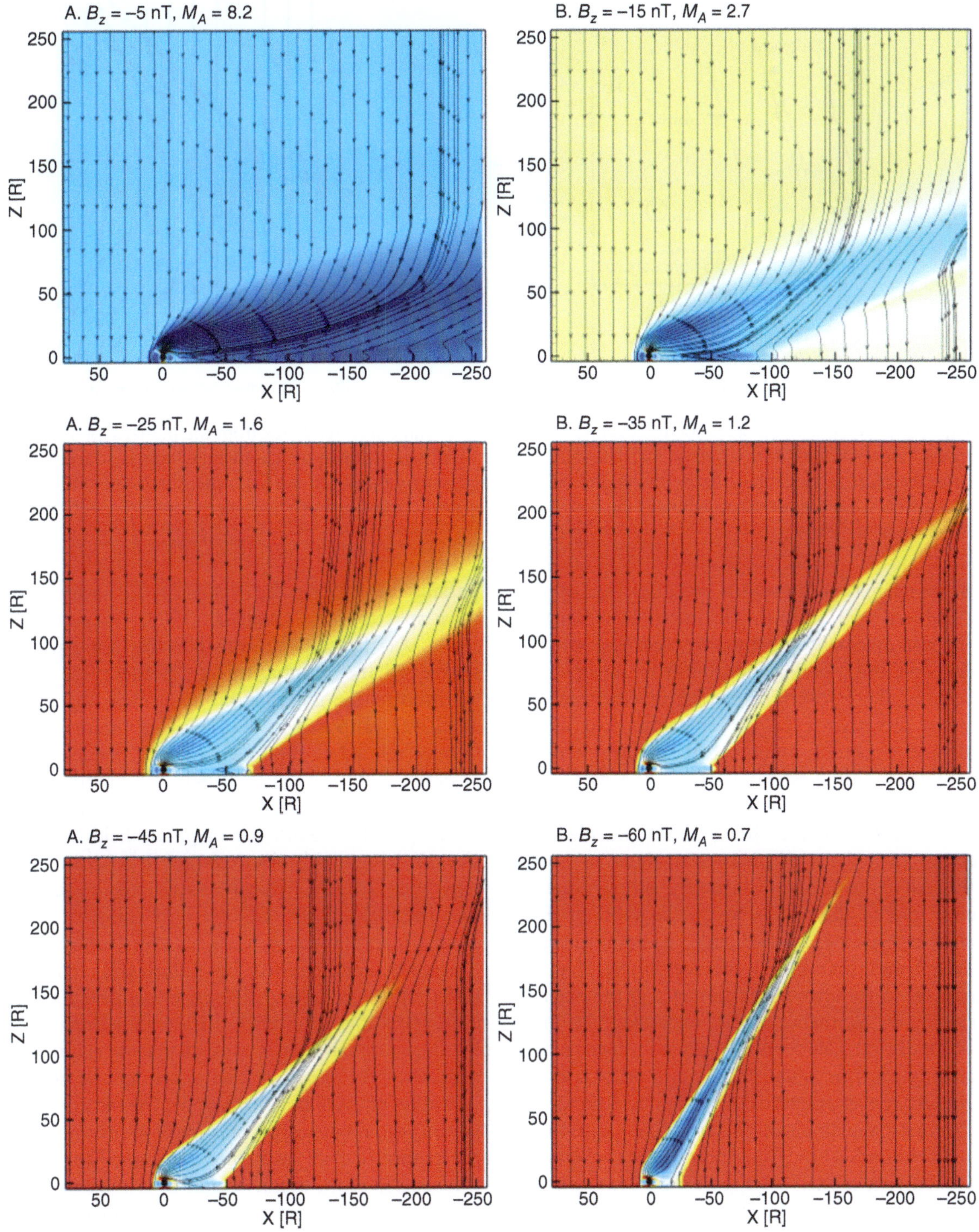


**Figure 7.1** Outputs from magneto-hydrodynamic simulations of the interaction between the solar wind and the Earth. The colors represent the electric field (dark blue is 0 mV/m and red is 10 mV/m) and the lines represent the magnetic field. Images of the velocity fields would display similar features. Starting from typical conditions and an Alfvén Mach number ($M_A$) of 8.2, we can see that the magnetotail transforms progressively into Alfvén wings as the $M_A$ decreases. When Alfvén wings form, energy, and currents can be efficiently carried away from the interaction site over long distances. Source: Reproduced Ridley (2007) / European Geosciences Union / CC BY 4.0.

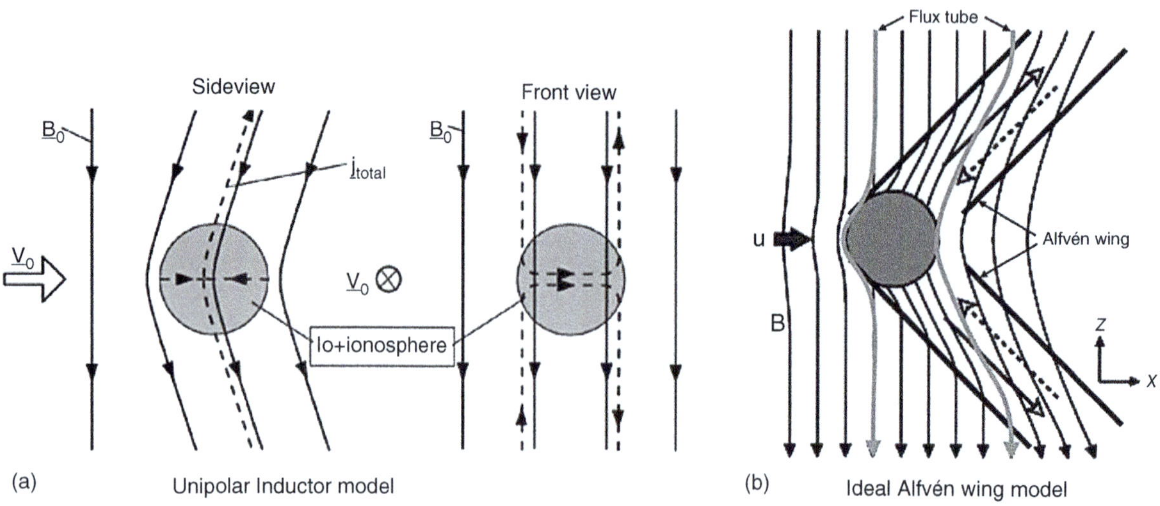


**Figure 7.2** (a) Schematic of the local interaction in the case of immediate feedback from the planet's ionosphere to the moon (the "unipolar inductor"). (b) Schematic of the local interaction in the case where there is no feedback from the planet's ionosphere to the moon (the "ideal Alfvén wing model"). In the first case, the Io flux tube and the Alfvén wing are identical, while in the second, they are decoupled. Source: Adapted from Saur (2004) and Kivelson et al. (2004).

When Voyager 1 performed a close encounter with Io on 5 March 1972 (closest distance: 20500 km), its trajectory was originally designed to cross the southern Io Alfvén wing, but it only skimmed it from the upstream side (Ness et al., 1979). The magnetometer, however, measured a disturbance of the magnetic field consistent with 5 MA current, a number of the same order of magnitude as the initial estimates of ~ 1 MA by Goldreich and Lynden-Bell (1969). Furthermore, the Voyager probes revealed evidence of Io's intense volcanism and the presence of a dense plasma torus along Io's orbit. Since the Alfvén velocity is inversely proportional to the square root of the mass density, the validity of the direct current hypothesis came under question, because the Alfvén travel time could become longer than the time taken by Io to move by its diameter in the plasma's reference frame. Neubauer (1980) expanded the linear model of Drell et al. (1965) into a fully non-linear model. He noted that, in this situation, the current would flow along the Alfvén characteristics, which are tilted, not only relative to the unperturbed field lines but to the perturbed field lines as well (contrary to the expectations of the DC current model, see Figure 7.2). Moreover, the non-linear Alfvén current tubes (the Alfvén wings) have a non-negligible resistance and, in this theoretical frame, since there is no feedback from Jupiter's ionosphere, it is this Alfvénic resistance that controls the amount of current, rather than the Pedersen conductance in Jupiter's ionosphere. With the Alfvén wings being distinct from the Io flux tube, the Alfvén waves' reflections on the density gradient outside of the plasma torus, or in Jupiter's ionosphere, would end up forming a complex pattern, possibly explaining the shape of the radio decametric wave spectrograms (Bagenal, 1983; Gurnett & Goertz, 1981). At the same time, since Io's atmosphere is the source of the plasma torus, another phenomenon in addition to the unipolar inductor/mass loading mechanism was proposed as a current generator: the pick-up current related to the newly ionized particles (Cheng & Paranicas, 1998; Goertz, 1980). As neutral atoms and molecules get ionized around Io, the electrons and ions start gyrating around the magnetic field lines in opposite directions with electrons going toward Jupiter and ions away from it. Vasyliūnas (2016), however, argued that the pick-up currents are only transient as they are almost instantaneously (i.e., on the electron-gyroperiod timescale) compensated by polarization currents. The main effect generating Alfvén waves and its associated currents is due to momentum exchange related to mass loading, since the newly ionized plasma is initially at rest relative to the moon, it exchanges momentum with the incoming plasma and slows down the resulting flow. This velocity gradient in the plasma flow results in a curl of the magnetic field and thus currents. The currents are transmitted along the field lines by Alfvén waves.

Another important discovery took place between the Voyagers flybys and the arrival of Galileo's finding of the Io auroral footprint, a set of spots magnetically connected to Io, which are followed by an extended tail along a contour mapping to Io's orbit (Figure 7.3a). Its first detection in the infrared domain (Connerney et al., 1993) was followed by its observation in the ultraviolet domain with the Hubble Space Telescope (Clarke et al., 1996; Prangé et al., 1996). On one hand, the existence

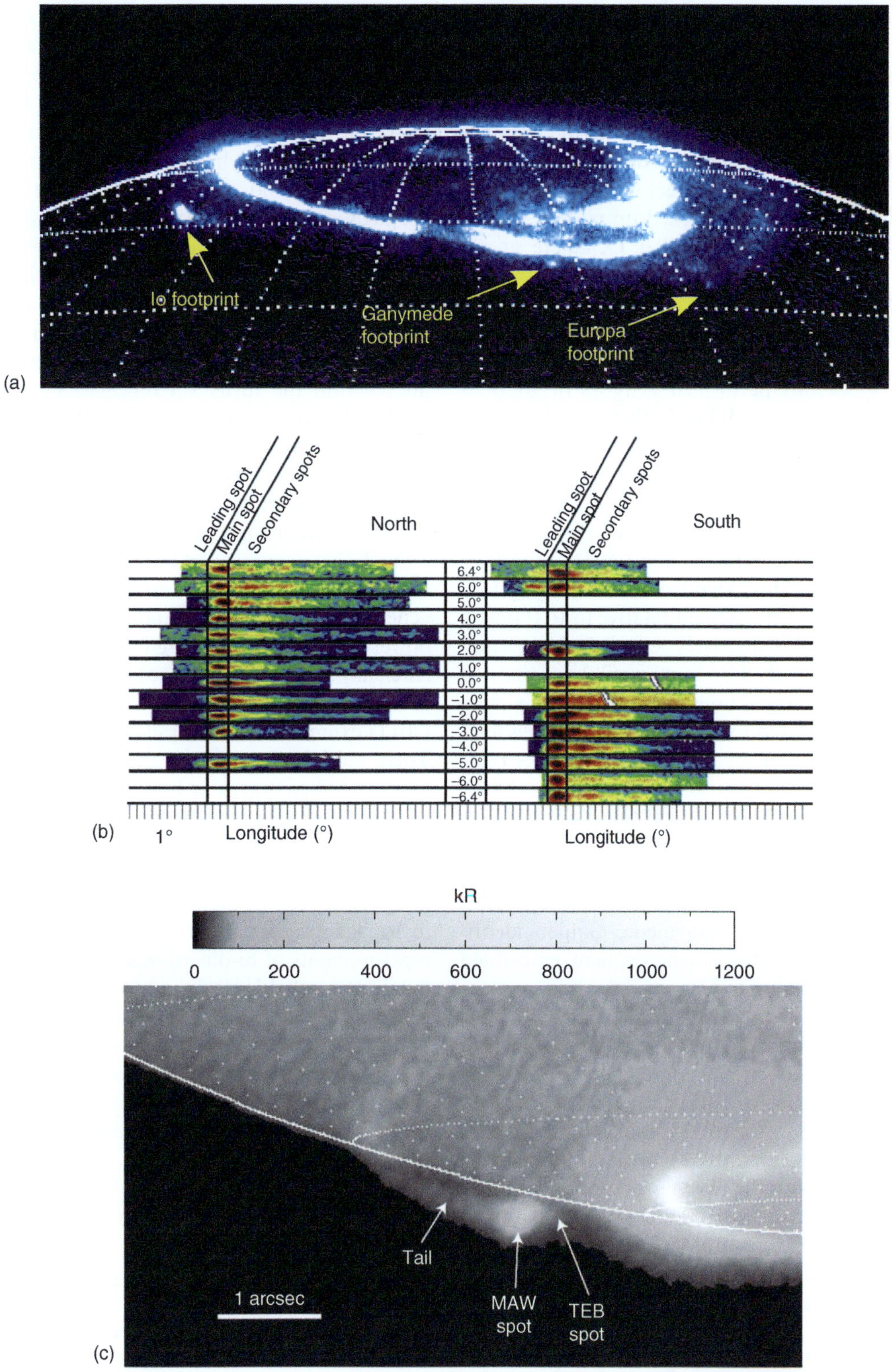


**Figure 7.3** (a) Image of the northern FUV aurora of Jupiter acquired with the Hubble Space Telescope's ACS camera on 7 February 2006. The Io, Europa, and Ganymede footprints are simultaneously visible. Source: Adapted from Bonfond., 2012 / American Geophysical Union. All Rights Reserved. (b) Scheme of the Io morphology as a function of the centrifugal latitude of Io in the torus. The longitudes are not measured on the planet, but mapped to the equatorial plane along the magnetic field lines according to the VIP4 magnetic field model for an easier comparison of both hemispheres. Source: Bonfond et al. (2008) / with permission of John Wiley & Sons. (c) Example of Io footprint tail and spots observed above the limb. Source: Adapted from Bonfond., 2010 / John Wiley Sons.

of several spots (Clarke et al., 2002; Gérard et al., 2006) clearly favored a de-coupling between Io and Jupiter's ionosphere, as the multiple spots were a prediction from the Alfvén wings reflection pattern (Gurnett & Goertz, 1981; Neubauer, 1980). On the other hand, the presence of a long, continuous tail left the door open for the establishment of a steady-state direct current loop associated with acceleration of the picked-up plasma toward corotation (Hill & Vasyliūnas, 2002; Delamere et al., 2003).

Galileo's exploration of the Jovian system from December 1995 to September 2003 marked another giant leap in our understanding of the moon–magnetosphere interactions. The spacecraft performed 7 flybys of Io, 9 of Europa, 6 of Ganymede, and 9 of Callisto, sending back a wealth of particle, magnetic field, and wave data documenting the interactions. For example, at Io, Galileo revealed a wake of stagnant plasma (Frank et al., 1996) and intense bi-directional electron beams (Frank et al., 1996; Frank & Paterson, 1999; Williams & Thorne, 2003; Williams et al., 1996). This extended wake of freshly ionized plasma remains quite stagnant relative to Io (Hinson et al., 1998). Hence, this discovery led to the suggestion that the field lines disturbed by Io may still be connected to it when the Alfvén waves bounce back, hence reviving the idea of a DC current loop. Several hybrid models have then been proposed to reconcile both approaches (e.g., Crary & Bagenal, 1997; Pontius, 2002); and Neubauer (1998) as well as Saur (2004) argued that both theories were extreme cases of a more general mixed Alfvén wings theory. At Europa, Galileo identified the induced magnetic field stemming from Europa's sub-surface ocean (Kivelson et al., 2000). At Ganymede, Galileo identified the presence of a mini-magnetosphere embedded within Jupiter's one, owing to Ganymede's intrinsic magnetic field (Kivelson et al., 1998). At Callisto, Galileo also found the signature of an induced magnetic field, even if its origin, a sub-surface conductive layer, or the ionosphere, remains unclear (Hartkorn & Saur, 2017; Khurana et al., 1998). Almost a decade later, Cassini would provide a similar wealth of information regarding the Kronian moons, with the most notable example being the discovery of the neutral clouds originating from geysers on the southern pole of Enceladus (Dougherty et al., 2006; Porco et al., 2006).

Before the discovery of footprint emissions, most theoretical efforts concentrated on the generation of the Alfvén waves and on their propagation (Wright, 1987; Wright & Southwood, 1987), but little consideration was given to their dissipation and their ability to interact with particles. Indeed, in ideal magneto-hydrodynamics (MHD), Alfvén waves do not develop a parallel electric field and are thus unable to accelerate particles along the field lines (Lysak, 2023). In the framework reducing the interaction to a direct current loop, should it concern the spots or the tail, the acceleration of the auroral electrons is attributed to quasi-static potential drops located above the Jovian ionosphere, described by the Knight relationship or a variation of it adapted to Jupiter (Ergun et al., 2009; Hill and Vasyliūnas, 2002; Matsuda et al., 2012). On the other hand, models focusing on the Alfvén waves, propagation started to account for the dispersive effects of the Alfvén waves (i.e., the kinetic and inertial regimes), for which an alternating parallel electric field could develop and accelerate the charged particles in a parallel direction. Crary (1997) proposed a model in which inertial Alfvén waves would accelerate electrons along the field lines within the torus via repeated Fermi acceleration. A decade later, Jones and Su (2008) considered both kinetic and inertial effects and concluded that inertial effects would be much more efficient at accelerating the electrons in both directions along the field lines outside the torus rather than inside it. Hess et al. (2010) further refined this scenario and assessed the energy balance for each step leading to auroral emissions: the power generation, the turbulent filamentation of the waves, the power transmission through the torus boundaries, the efficiency of the electron acceleration, and their ability to finally precipitate into the atmosphere, and Hess et al. (2011) extended this analysis to Europa and Enceladus. Their model showed that the electron acceleration by inertial Alfvén waves at high latitude forms broadband bi-directional electrons beams. The electrons accelerated toward Jupiter could precipitate into the atmosphere and generate auroral emissions. The electrons accelerated away from Jupiter then cross the equatorial plane close to Io (Jacobsen et al., 2010), which explains the Galileo measurements of bi-directional electron beams during Io flybys. Part of this electron beam would remain trapped on its field line, and part of it would precipitate into the atmosphere in the other end of the field line and also generate auroral emissions. Indeed, Bonfond et al. (2008) noted that the relative motion of the Io footprint spots and the occurrence of a leading spot upstream of the brightest one was not compatible with a scenario only relying on reflected Alfvén waves, but could be explained if electrons accelerated anti-planetward in one hemisphere could precipitate in the opposite hemisphere (see Figures 7.3b and 7.4). The brightest spot is generally associated with the main Alfvén wing (MAW), originating directly from Io. A second spot, always located downstream of the first one, is associated with the first reflection of the Alfvén wing (RAW). Its distance to the MAW spot is minimum in the northern hemisphere when Io is close to the southern edge of the torus (Hinton et al., 2019) (Figure 7.4). A third spot is associated with electrons accelerated away from Jupiter in one hemisphere and precipitating in the opposite hemisphere (Trans-hemispheric Electron Beam,

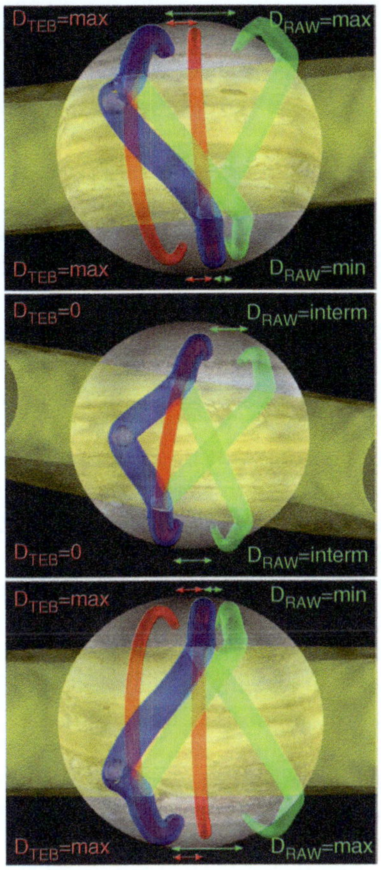


**Figure 7.4** Schematic presentation of the Alfvén wing reflection pattern and Trans-hemispheric Electron Beams (TEB, shown in red) when Io is in its northern-most (top), central (middle), or southern-most position relative to the plasma torus (shown in yellow). The main Alfvén wing (MAW) is shown in blue and the reflected Alfvén wing (RAW) is shown in green. On the top panel, in the North, the distance between the MAW spot and the TEB spot ($D_{TEB}$) is maximal as well as the distance between the MAW spot and the RAW spot ($D_{RAW}$). Note that in a linear case, $max(D_{RAW}) \sim 2max(D_{TEB})$. In the South, $D_{TEB}$ is also maximal, but the TEB spot is now upstream of the MAW spot. $D_{RAW}$ reaches its minimum. When Io is in the center of the torus (middle panel) the situation in the two hemispheres is symmetric, and the TEB spots are merged with the MAW spots. Source: Bonfond et al., 2017b / with permission of Elsevier.

TEB spot). Because electrons travel faster than Alfvén waves, the TEB spot is located upstream of the MAW spot in the northern hemisphere when Io is close to the southern edge of the torus (see Figure 7.4) (Bonfond et al., 2008). The altitude of the brightness peak is 850–900 km (Bonfond, 2010) (Figure 7.3c) and its vertical profile suggests a broadband energy distribution with a 1 keV mean energy. This observation constitutes another clue in favor of the Alfvénic acceleration process, since the broad vertical profile of the Io footprint main spot and tail is only compatible with a broadband, kappa-like electron energy distribution, and not with a narrow distribution as expected from a quasi-static potential drop (Bonfond et al., 2009; Bonfond, 2010; Bonfond, Saur, et al., 2017). The resulting anisotropic distributions of field-aligned electrons are unstable to the generation of plasma waves, which are commonly observed on the flux tubes connected to the Io footprint. A prominent one is the whistler-mode auroral hiss, which arises from a beam-plasma instability, undergoing wave growth via Landau resonance (e.g., Sulaiman et al., 2020). The resulting band of frequencies obeys a dispersion relation such that higher frequencies propagate at a larger angle to the background magnetic field, enabling detection of the interaction well before and after the satellite footprint flux tube is crossed (Figures 7.7 and 7.8c). Radio waves are also generated as a result of the moon–magnetosphere interaction, notably the Io Decametric Radio Emission. Here, the anisotropic electron distribution drives the cyclotron maser instability from which the growth of radio waves occurs (Zarka, 1998). The generation of a free-space mode as such enables moon–magnetosphere interactions to be detected from very remote distances (Bigg, 1964).

After Galileo's close encounters providing precious data on the local interaction, Juno's arrival at Jupiter in August 2016 offered a unique opportunity to explore the physical processes at the polar end of the field lines. While its observations confirmed the scenario inferred from remote sensing of the aurora, they also unveiled unexpected results, suggesting that the story told here is not the final word on the issue, as described in Section 7.2.1.

As discussed earlier, most of the ideas related to the generation of Alfvén waves by moon–magnetosphere interactions were motivated by observations of the Io-Jupiter system. However, the subsequent exploration of other systems revealed that the lessons learned at Io could essentially apply everywhere, even on Earth (see Section 7.2.4). It is also noteworthy that the modeling efforts reducing interactions to direct current systems are fruitful first steps, providing the correct orders of magnitude. However, at some point, they proved to be oversimplifications that obscured essential properties of the Alfvén waves, which turned out to be key to explaining the increasingly detailed observations, such as the decoupling between the Alfvén wing and the satellite flux tube, or the bi-directional broadband electron acceleration (Bonfond et al., 2008, 2009; Hess et al., 2010; Neubauer, 1980). Similar limitations have also been noted for the main auroral emissions at Jupiter (Bonfond et al., 2020).

## 7.2. MOON-MAGNETOSPHERE AND SIMILAR INTERACTIONS THROUGHOUT THE SOLAR SYSTEM

### 7.2.1. The Jovian Moons

#### *Io*

Because of the favorable combination of its intrinsic strength and relative proximity to the Earth, the Io–Jupiter electro–magnetic interaction is the easiest to scientifically investigate. Its signatures in the radio, infrared, and ultraviolet domains radiate enough energy to be observed from Earth, and Io has been visited by two spacecrafts (Voyager 1 and Galileo) and is currently explored at a close distance by Juno. The magnetic in-situ field and particle observations from Voyager 1 (Acuña et al., 1981; Belcher et al., 1981) and Galileo (Kivelson et al., 1996) provided signatures of the generation of large-scale Alfvén waves at Io. Furthermore, high-frequency/small scale fluctuations of the magnetic and electric fields (see Figure 7.5) suggested that the Alfvén waves were subject to turbulent filamentation into smaller scales, which would more easily allow the Alfvén waves to escape from the torus and trigger electron acceleration processes outside of it (Chust et al., 2005; Hess et al., 2010). Moreover, Galileo particle instruments showed the presence of bi-directional electron beams in Io's wake (Frank et al., 1996; Frank & Paterson, 1999; Williams & Thorne, 2003; Williams et al., 1996) and mono-directional above Io's poles (Mauk et al., 2001), indicative of an electron acceleration process taking place close to Jupiter. After their exit from the high-density torus, the Alfvén waves related either to the main Alfvén wing spot or to the following tail have been observed by Juno at high latitude (Gershman et al., 2019; Sulaiman et al., 2020) (Figure 7.7). Furthermore, Alfvénic Poynting fluxes in the highlatitudes were found to highly correlate with electron energy fluxes and are coupled by an efficiency (10%) that is fully consistent with acceleration from Alfvén wave filamentation via a turbulent cascade process (Hess et al., 2010; Sulaiman et al., 2023). Juno also observed the bidirectional broadband electron population (Szalay et al., 2018) expected from the models of acceleration via inertial Alfvén waves (Damiano et al., 2019; Hess et al., 2010; Jones & Su, 2008). The main difference between the electron populations related to the main Alfvén wing spot or the tail is the energy flux, rather than the energy or the pitch angle distributions (Szalay et al., 2020b). What was less expected from previous theoretical studies (e.g., Hess et al., 2013a) was the finding of proton acceleration along the Alfvén wing (Clark et al., 2020; Szalay et al., 2020c). Another surprise was the occasional identification of a double peak in the particle flux down the tail (Szalay et al., 2018) and of a split tail in the infrared images (Mura et al., 2018).

The Io auroral footprint is made up of at least 3 spots and an extended tail in the downstream direction. The spots are approximately 850 km long, which corresponds to 3–4 times the projected size of Io, <200 km wide (<2 times the projected width of Io) and 1200 km tall (full width at half maximum) (Bonfond, 2010). The relative distance between the spots evolves as Io oscillates north and south of the torus centrifugal equator. The UV-emitted power of the different spots varies with the System-III longitude of Io (System-III is a longitude system fixed with the magnetic field of Jupiter). The observed brightness results from a combination of processes (Bonfond et al., 2013b; Hess et al., 2013) and the footprint may even completely vanish when it crosses an auroral injection signature (Bonfond et al., 2012; Hess, Bonfond, & Delamere, 2013). On the other hand, Hue et al. (2019) have shown that the brightness of the main spot was not affected by the collapse of Io's atmosphere during the eclipse.

Moreover, high-resolution images of the $H_3^+$ auroral emissions of the footprints showed that the spots were formed of sub-structures, named sub-spots, which were fixed with the planet (contrary to the larger spots features that are moving with Io) (Moirano et al., 2021; Mura et al., 2018).

In addition to the infrared and ultraviolet domains, the morphology of the Io footprint has also been observed in the visible domain (Gladstone et al., 2007; Vasavada et al., 1999).

The precipitation of electrons into the polar atmosphere leads to a loss-cone pitch angle distribution, which is susceptible to the growth of the maser-cyclotron radio emission (Louis et al., 2017; Queinnec & Zarka, 1998). In addition to the Alfvénic acceleration, which is the main source of particle energization, evidence of upward migrating intermittent potential structures have also been highlighted by Hess et al. (2009). The recurrent formation and upward migration of these structures every 2–3 minutes may explain the short-timescale fluctuations of the Io footprint brightness (Bonfond et al., 2007; Bonfond, Hess, Gérard, et al., 2013). A visual summary of the chain of processes giving rise to the auroral footprints, together with the related plasma and radio waves, can be found on Figure 7.6.

#### *Europa*

Arising mostly from the sputtering of its icy surface by magnetospheric particles, the atmosphere and ionosphere of Europa are much less dense than Io's, leading to a weaker plasma interaction and 20 times smaller mass loss (~50kg $s^{-1}$) (Bagenal & Dols, 2020; Saur et al., 1999).

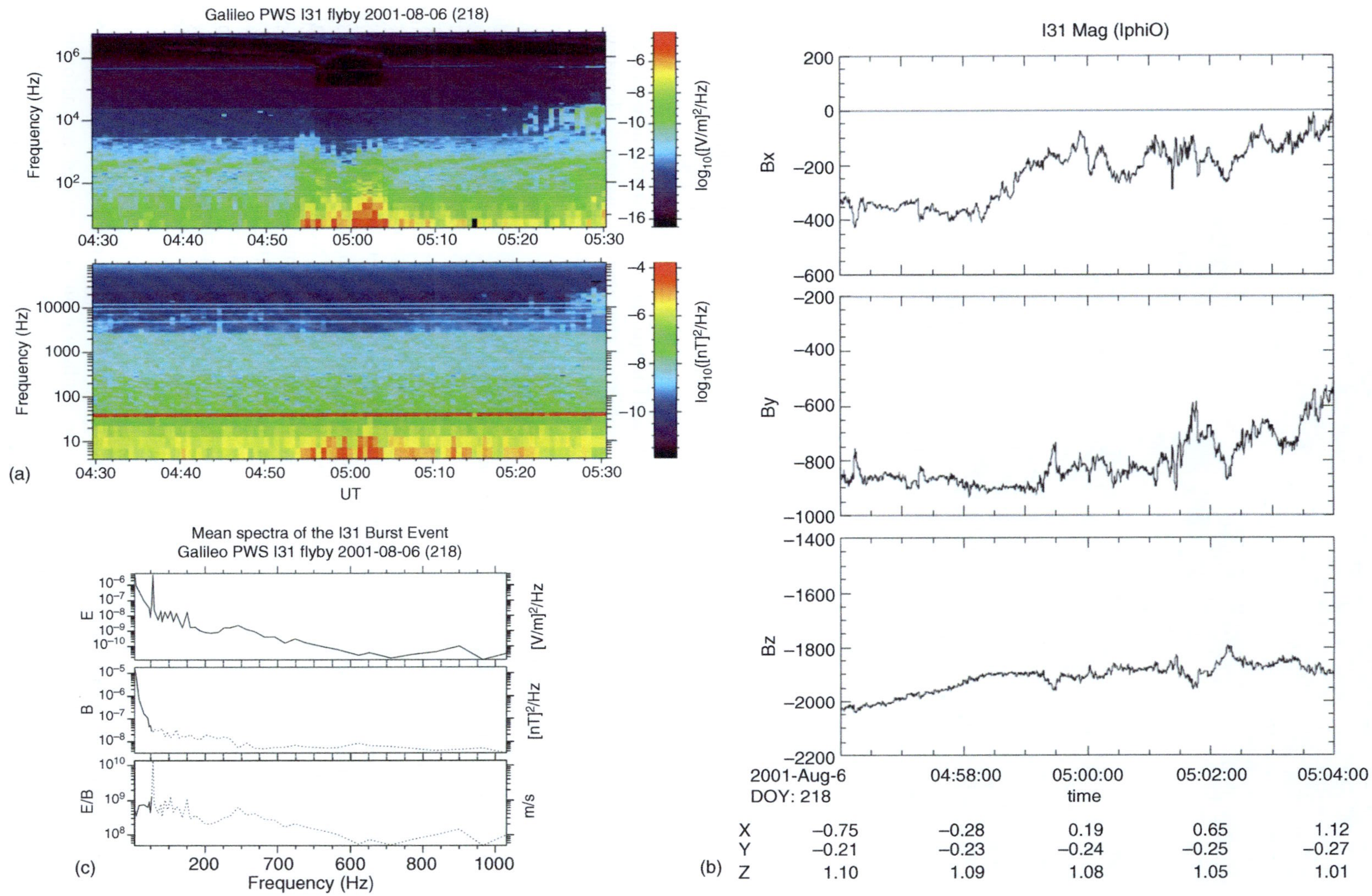


**Figure 7.5** (a) Calibrated dynamic spectra of the electric field (top) and magnetic field (bottom) acquired by the PWS (Plasma Wave Subsystem) instrument on board Galileo during the I31 flyby on 7 December 1995. The Alfvén wing crossing started about 5 minutes before the closest approach (04:59:20). (b) Magnetic field observations during the crossing of the Alfvén wings during the same event, but the time interval is narrowed down to the duration of the Alfvén wing crossing. The x-axis is parallel to the unperturbed flow, the y-axis is pointing toward Jupiter and the z-axis is parallel to Jupiter's rotation axis. (c) Electric and magnetic wave spectra, and the corresponding ratio of these two quantities. In this example, it can be seen that the crossing of the Alfvén wing is accompanied with the observation of intense higher frequency waves (with most of the energy remaining below 50 Hz), which have been interpreted as the signature of turbulent filamentation of the Alfvén waves originating from Io. Source: Chust et al. (2005) / with permission of Elsevier.

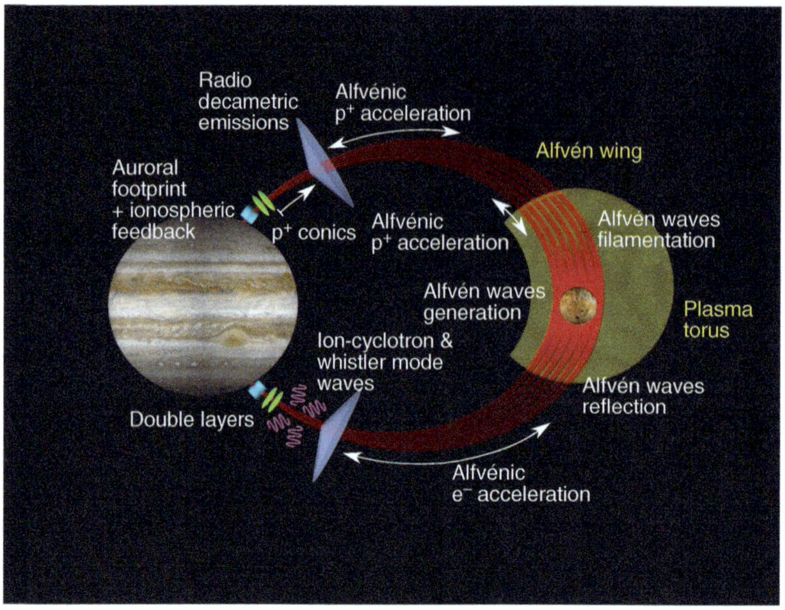


**Figure 7.6** Schematic of the chain of processes arising from the Io–Jupiter electro-magnetic interaction.

However, Europa is also the source of an induced magnetic field, which is one of the most compelling evidences for the existence of its subsurface ocean (Kivelson et al., 2000). Both effects contribute to making Europa an obstacle to the stream of magnetospheric plasma and to the generation of Alfvén wings (e.g., Schilling et al., 2008). Localized sources of neutrals related to cryovolcanoes (Arnold et al., 2019; Huybrighs et al., 2020; Jia et al., 2018; Roth et al., 2014) could even cause smaller structures, called Alfvén winglets, within the larger Alfvénic structure (Blöcker et al., 2016). Moreover, evidence for an extended plasma wake, or plume, were found in the Galileo data (Eviatar & Paranicas, 2005; Kivelson et al., 1999; Kurth et al., 2001).

Juno has also crossed the magnetic field lines connected to the Europa footprint tail and intriguing differences were found compared to the observations at Io (Allegrini et al., 2020). In addition to a broadband energy distribution with a mean energy of 3.6 keV, a peak at an energy of $\sim 2$ keV was also found, suggesting that the electron acceleration was at least partly electrostatic. At the Jovian end of the Alfvén wings, UV auroral spots (Bonfond, Grodent, et al., 2017; Clarke et al., 2002) and a downstream tails are also occasionally observed (Bonfond, Saur, et al., 2017; Grodent et al., 2006). High-resolution images of the infrared $H_3^+$ emissions also showed the occasional appearance of sub-dots, similarly to Io's (Moirano et al., 2021). As shown in Figure 7.3a, the UV Europa footprint is much weaker than the Io's, and it is often hard to identify among the surrounding auroral emissions. The identification of a second spot is even rarer, and, when it could be identified, its location and motion were found consistent with those of an reflected Alfvén wing (RAW) spot (Bonfond, Grodent, et al., 2017).

While the generated power is much weaker than for Io, decametric radio emissions related to the Europa footprint have also been observed by Voyager 1 and Cassini (Louis, Lamy, Zarka, Cecconi, & Hess, 2017). The interaction between Europa and the Jovian magnetosphere will become the object of great scrutiny in the next decade with the arrival of two dedicated spacecrafts, Europa Clipper, and JUICE.

### *Ganymede*

Ganymede is the largest moon of the Solar System (with a radius $R_G = 2634$ km) and is the only one with its own intrinsic magnetic field, which carves a magnetosphere within Jupiter's magnetosphere (Kivelson et al., 1998). It is this 2-3 $R_G$-wide mini-magnetosphere that forms an obstacle to the plasma flow (Neubauer, 1998; Jia et al., 2008). At the front of Ganymede's magnetosphere (i.e., the upstream side relative to the magnetospheric plasma flow), the magnetic field is anti-parallel to Jupiter's, a configuration favorable to magnetic reconnection. The transverse electric potential drop across Ganymede's magnetosphere is estimated at around 80 kV (Zhou et al., 2020), and the total current carried along the Alfvén waves lies around 0.5 MA (Lavrukhin and Alexeev, 2015). This unique configuration of a nested magnetosphere creates a variety of accelerated or trapped electron populations, either on closed field lines around Ganymede or between Ganymede and Jupiter (Frank et al., 1997; Williams, Mauk, & McEntire, 1997; Williams et al., 1997;

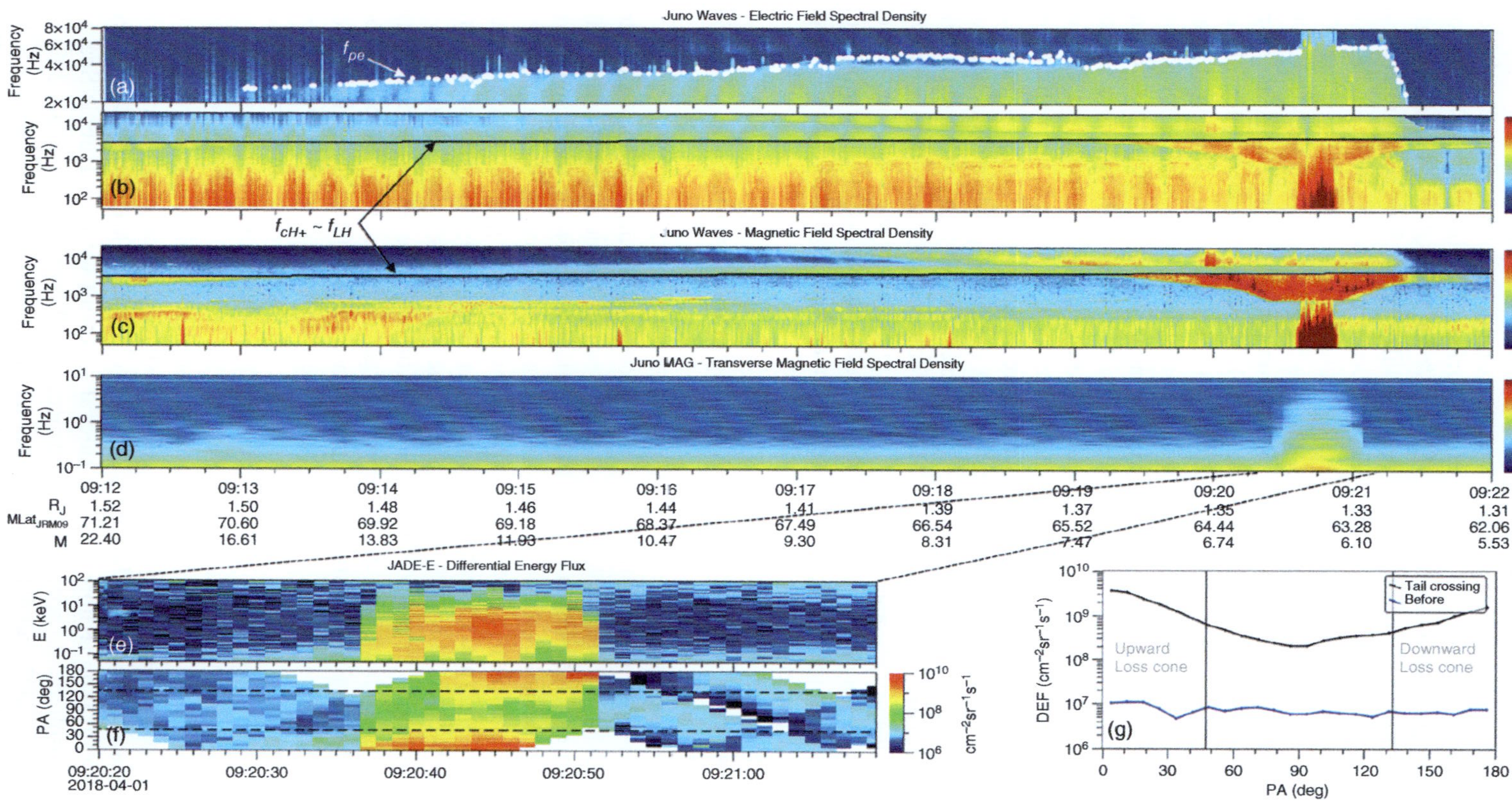


**Figure 7.7** (a-d) Electric and magnetic field frequency-time spectrograms measured in the MAW revealing MHD and plasma waves across multiple scales, namely, Alfvén, ion cyclotron, and whistler modes. (e-g) Electron data revealing broadband and bidirectional distributions of electrons, likely accelerated by inertial Alfvén waves at the high latitudes, and also the source of whistler-mode waves. Source: Sulaiman et al. (2020) / with permission of John Wiley & Sons.

Williams & Mauk, 1997; Eviatar et al., 2000). Furthermore, the particles energized by magnetic reconnection at the front and tail of the magnetopause are the sources of local electrostatic and electromagnetic emissions (Gurnett et al., 1996). Galileo also measured a deceleration of the convective magnetospheric plasma flow to less than one third of rigid corotation, which is compatible with a traversal of an Alfvén wing (Eviatar et al., 1998). At high latitudes, Ganymede's auroral footprint is formed of at least two spots (Bonfond, Hess, Bagenal, et al., 2013) and an elongated tail (Bonfond, Saur, et al., 2017). The presence of such a tail while Ganymede is not subject to any significant mass loading implies that this process isn't a necessary condition for the presence of a footprint tail. It favors interpretations of the auroral footprint tails implying repeated reflections of Alfvén waves, as opposed to explanations involving the progressive acceleration of plasma in the wake. The motion of the secondary spot suggested an origin related to trans-hemispheric electron beams; an interpretation confirmed by Juno's in situ measurements when crossing field lines connected to this spot (Hue et al., 2021). On field lines connected to the TEB spot, Juno only detected weak currents in addition to the bi-directional electron beams, contrary to measurements in the tail where significant currents and strong Alfvénic activity were also found (Szalay et al., 2020a). As Juno was crossing the Ganymede footprint tail, it also encountered the source region of decametric radio emissions (Louis et al., 2020). In the high-resolution infrared images, sub-dots, similar to those observed at Io, were also reported (Mura et al., 2018). The brightness of the main spot varies according to three separate timescales: several hours, tens of minutes, and 2-3 minutes (Grodent et al., 2009). The first one is related to the passage of Ganymede through the denser core of the plasma sheet, the second probably corresponds to interactions with local plasma inhomogeneities, such as plasma injections (see also Bonfond, Grodent, et al., 2017) and the shortest one could correspond either to the time between subsequent reconnections on Ganymede's magnetopause or to some ionospheric feedback mechanism, such as the recurrent apparition of double layer structures, as documented for Io (Hess et al., 2009).

#### *Callisto*

With a faint $O_2$, $CO_2$, and $H_2O$ atmosphere and an H corona dominated by sublimation rather than sputtering (Carlson, 1999; Cunningham et al., 2015; Roth et al., 2017), Callisto possesses a significant conductive ionosphere generated by photo-ionization (Kliore et al., 2002). A review of Callisto's atmosphere and space environment can be found in Galli et al. (2022). Signatures of an induced magnetic field, similar to Europa's, have been recorded by Galileo. It is, however, uncertain whether this induced field arises from currents in a sub-surface conductive layer (Khurana et al., 1998; Neubauer, 1998; Zimmer et al., 2000) or in the ionosphere (Hartkorn & Saur, 2017). Measurements acquired during some flybys by Galileo (namely C3 and C10) suggested the presence of a downstream plasma wake (or plume), while other (during C22) did not lead to the same conclusion (Eviatar & Paranicas, 2005; Gurnett et al., 2000). Moreover, Galileo may have encountered a field-aligned electron beam in Callisto's wake, similar to those observed at Io (Mauk & Saur, 2007). Of the four Galilean moons, Callisto is the one generating the weakest interaction, due to the lower plasma density, the weaker ambient magnetic field, and the lack of a large internal field (Saur et al., 2013). At the Jovian end of the field lines, two tentative detections of a Callisto auroral footprint were reported by Bhattacharyya et al. (2018), based on Hubble Space Telescope UV images. If the detected spots do indeed correspond to a Callisto footprint, it could be as bright as Ganymede's. However, the search for this footprint with more sensitive ultraviolet and infrared instruments on board the Juno spacecraft has been unfruitful so far (Moirano et al., 2024). In conclusion, while the signatures of the interaction similar to those observed for the other Galilean satellites are expected, their detection (of a dense wake, of electron beams, of an auroral footprint) remains uncertain, and the arrival of the JUICE spacecraft is thus much awaited.

### 7.2.2. The Kronian Moons

#### *Enceladus*

The moons of Saturn were not expected to trigger as strong electrodynamic interactions as those of Jupiter. Indeed, the alignment of the magnetic dipole with the rotation axis prevents oscillations of the local magnetic field at the moons to generate strong induced fields as on Jupiter's moons (only the orbital eccentricities and the local time variations would modify the local magnetic field). Moreover, the largest moon, Titan, orbits at a much larger distance from the planet, both in absolute and relative terms, than the Galilean moons. The other moons were deemed too small to hold a sub-surface ocean or even a significant ionosphere. However, the finding of a huge water ion cloud located south of Enceladus (radius: ~250 km) and originating from cryo-volcanoes led to the realization that Enceladus's environment formed a much larger obstacle than originally thought (Dougherty et al., 2006; Porco et al., 2006). Initial searches for the Enceladus auroral footprint with the Hubble Space Telescope were unsuccessful (Wannawichian et al., 2008), but later observations of the aurora from Cassini's UVIS (UltraViolet Imaging Spectrograph) demonstrated the presence of an auroral spot connected to Enceladus (Pryor et al., 2011)

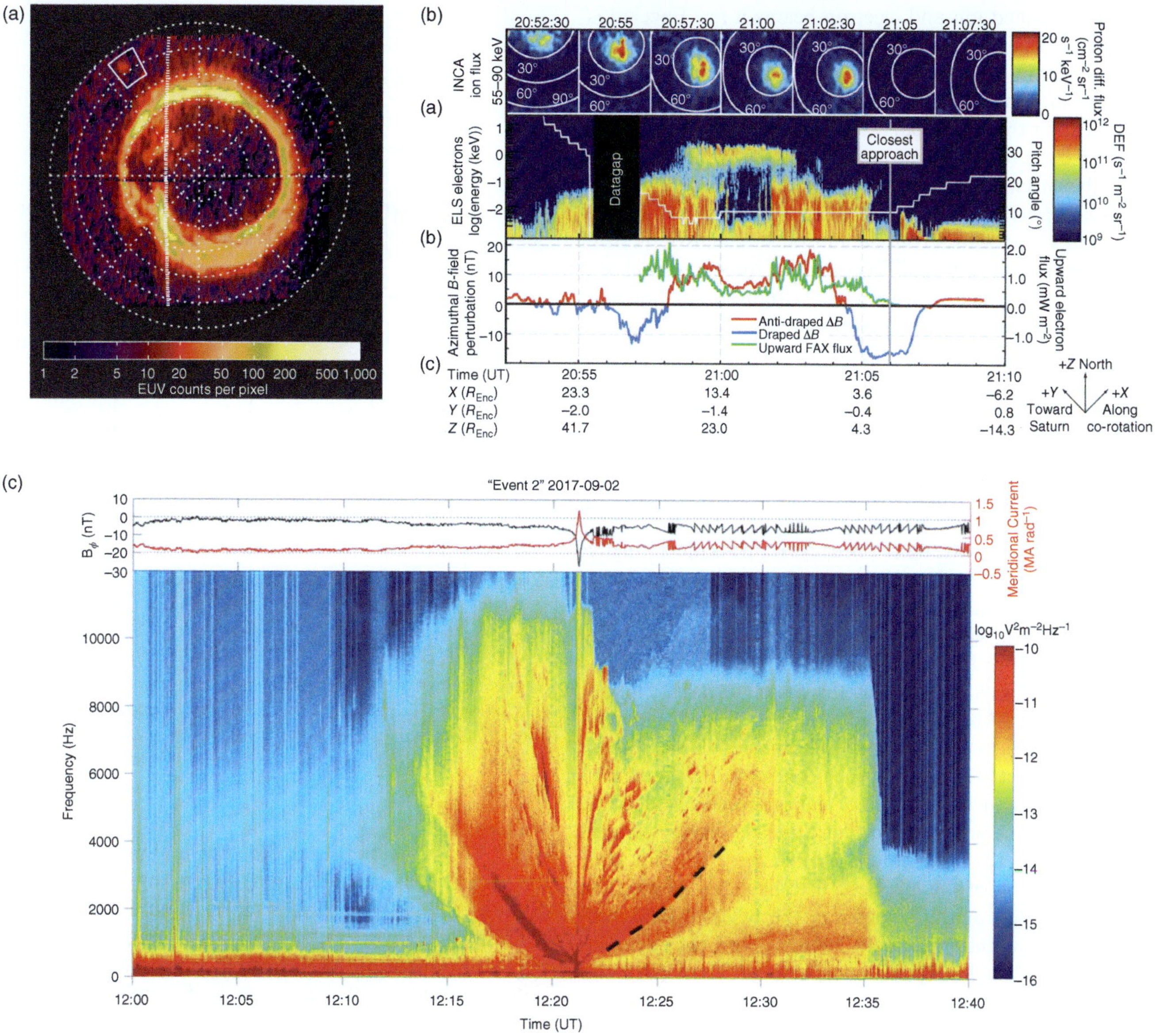


**Figure 7.8** (a) Image of the Extreme-UV northern aurora at Saturn acquired on 26 August 2008 by the UltraViolet Imaging Spectrograph on board Cassini. The spot highlighted on the top left side is the Enceladus footprint. (b) Particle and field measurements acquired around 11 August 2008 flyby of Enceladus by Cassini's Ion and Neutral Camera (INCA), Electron Spectrometer (ELS), and magnetometer. The observed ion and electron beams are originating from Saturn's north pole and were thus accelerated close to the planet and away from it, reminiscent of the electron beams creating the TEB spots at Jupiter. (c) High-resolution electric field dynamic spectrogram of the auroral hiss emissions as Cassini was crossing the Enceladus Alfvén wing at high latitude on 2 September 2017. A ray-tracing analysis indicates that the source region is co-located with the Enceladus footprint on Saturn. Source: (a) and (b) are from Pryor et al. 2011 / Springer Nature, and (c) is from Sulaiman et al. (2018).

(Figure 7.8a). The electro-dynamic interaction between Enceladus and Saturn similarly gives rise to Alfvén wings (Jia et al., 2011). The location of Enceladus' plume in its southern polar region leads to a north-south asymmetry in the orientation of the Alfvén wings. Furthermore, the absorption of electrons by the dust in the plume yields a reversal of the sign of the Hall conductivity (anti-Hall effect), as well as a reversal of the sign of the $B_y$ perturbation (i.e., the perturbation of the magnetic field along the anti-Saturn/Saturn direction) in the Alfvén wing (Kriegel et al., 2011; Simon et al., 2011). Moreover, Cassini's particle instruments detected the presence of a field aligned ion and electron beams a few Enceladus' radii downstream of the moon. While the energy flux of these beams was sufficient to generate observable auroral footprints on Saturn, they were oriented anti-planetward, indicating that these particles had been accelerated relatively close to Saturn and away from the planet, similarly to the mechanism explaining the TEBs and their associated spots at Jupiter (see Figure 7.8b). Field-aligned currents and whistler-mode auroral hiss emissions from the beam-plasma instability have been measured near and within the Enceladus Alfvén wing (Figure 7.8c), however, unlike Io, no radio emissions have been observed (Gurnett et al., 2011; Sulaiman et al., 2018).

### *Titan*

Titan is Saturn's largest moon (radius: 2575 km) and has a thick and hydrocarbon-rich $N_2$ atmosphere and a dense ionosphere, making it a significant source of plasma in Saturn's magnetosphere. However, Titan does not play the role Io plays on Jupiter, because it orbits much further away from Saturn, so that it even occasionally crosses the magnetopause. Hence, depending on its position in orbit around Saturn and on the solar wind conditions, Titan could be located either in the plasma sheet, in the magnetosphere lobes, or in the magnetosheath (Rymer et al., 2009). At such distances from the planet, the magnetic field is so weak that the plasma $\beta$ is larger than 1 (Neubauer et al., 2006). In such conditions, the dynamics of the plasma (or lack thereof) takes precedence over the magnetic tension and the interacting field lines get draped around the Kronian moon. Interestingly, the timescale of convection of the magnetic field becomes longer than the timescale of the changes in the external field, and the previous magnetic field configuration thus remains "fossilized" in the thick ionosphere of Titan (Bertucci et al., 2008).

Despite the complexity of the interaction between the ever-changing magnetospheric plasma (Simon et al., 2010) and Titan's thick exosphere and ionosphere (e.g., Sillanpää et al., 2007; Snowden et al., 2013; Wei et al., 2007), numerical simulations show the systematic development of Alfvén wings connected to Titan (Kallio et al., 2007; Sillanpää & Johnson, 2015). Whether these Alfvén wings trigger any significant signature on Saturn remains to be demonstrated.

### *Rhea*

Located in Saturn's E ring, Rhea does not have any significant magnetic field or any significant plasma source, nor is it a conducting object. Yet its surface absorbs plasma while the magnetic flux passes across the 1528 km wide moon. Close encounters with the Cassini spacecraft showed that this wake is subject to diamagnetic currents as the pressure gradients between the plasma

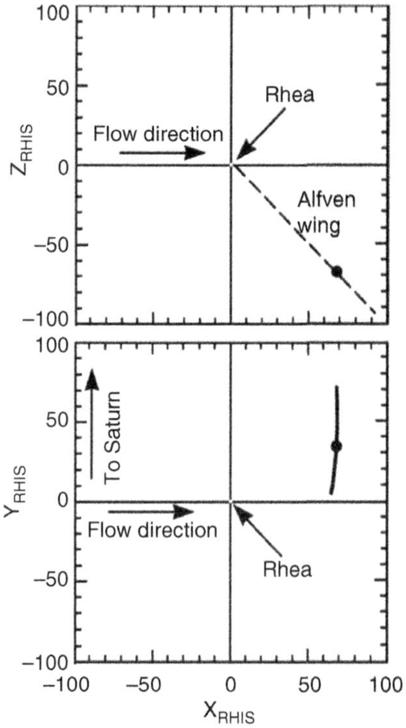


**Figure 7.9** Plot of the Cassini trajectory during the distant Rhea encounter of 3 June 2010. The black circle marks the passage through the Alfvén wing. Tracing the Alfvén waves back in time (dashed lines) shows that they originate from Rhea or its close wake. Source: Khurana et al., 2017 / John Wiley & Sons.

void (~4-6 $R_{Rh}$ long) and the surrounding plasma distorts and bends the field lines, generating Alfvén waves (Simon et al., 2012). Furthermore, a more distant flyby (102 $R_{Rh}$, see Figure 7.9) revealed that the Alfvén wings were still present far away from Rhea and that the associated currents probably close in Saturn's ionosphere (Khurana et al., 2017). Khurana et al. (2017) also argue that it is the deceleration of the plasma in the wake resulting from the pressure gradient rather than the diamagnetic currents themselves that generate most of the Alfvénic disturbance.

### 7.2.3. Triton

The time variability of Neptune's magnetic field that is intercepted by Triton is especially large due to the tilt between Neptune's magnetic and rotational axes, as well as Triton's large orbital obliquity. The resulting structure of the electro-dynamic interaction is highly asymmetric, with one of Triton's Alfvén wings shown to be oriented toward upstream, initiating the deflection of the incident flow farther than it would otherwise (Simon et al., 2022). It is, therefore, reasonable to expect a highly complex current system and propagation pattern of Alfvén waves, and, consequently, regions where dissipation, particle acceleration, and auroral generation would take place.

### 7.2.4. The Earth

At 1 astronomical unit (AU), the solar wind's Alfvén Mach number is generally on the order of 8, thus the conditions are not favorable to the formation of significant Alfvén wings, as they bend over to form the magnetotail lobes (see the top left panel in Figure 7.1). However, during the passage of an interplanetary coronal mass ejection (ICME) across the Earth, the density of the solar wind can occasionally become so low and the field strength so high that the interaction gets both sub-fast and sub-Alfvénic. Under such atypical conditions, magneto-hydrodynamic simulations show that the bow shock expands out and vanishes, and well-formed Alfvén wings develop (Ridley, 2007) (see Figure 7.1). Chané et al. (2012) found observational evidence that the Earth entered into such a regime for four hours between 24 and 25 May 2002. During this interval, the magnetic field wasn't unusually high (~9.8 nT), but the estimated density dropped down to 0.05 $cm^{-3}$ (compared to typical values between 0.6 and 20 $cm^{-3}$). The Geotail spacecraft was located on the dusk side of the Earth and its magnetic field and plasma velocity measurements suggest that the spacecraft entered and exited the Alfvén wing nine times. Figure 7.10 shows the expected shape and orientation of the Alfvén wings during this time interval compared to the typical shape of the terrestrial magnetosphere. Such exceptionally low density values were achieved because an ICME expanded right in the wake of another one (Chané et al., 2021).

#### *The Moon*

The Moon is also an inert object without any significant magnetic field or mass-loading source. Moreover, the Moon orbits so far away from the Earth that it spends most of its time in the supersonic and super Alfvénic interplanetary medium. And yet, the two Moon orbiting ARTEMIS spacecrafts found a bending of the magnetic field and a deceleration of the plasma flow in the Moon's wake, which are interpreted as characteristic signatures of Alfvén waves (Zhang et al., 2016) (see Figure 7.11). It is, however, unlikely that this configuration would allow a significant amount of energy and current to propagate along the magnetic field lines.

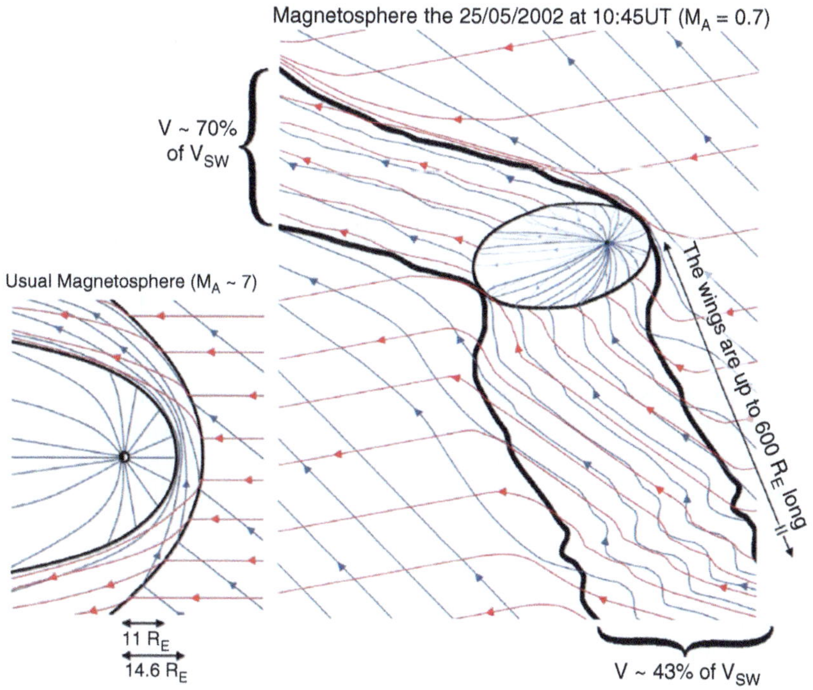


**Figure 7.10** Sketches of the interaction between the solar wind and the Earth's magnetosphere seen from the north. The magnetic field is shown in blue, the velocity in red. The left sketch shows the typical configuration of the magnetosphere while the right one shows the development of Alfvén wings at a time during which the interplanetary medium density at Earth dropped so much that the Alfvén Mach number was below one. The Geotail spacecraft was located on the dusk side of the Earth (i.e., toward the top of the image) and entered and exited from the Alfvén wing several times. Source: Chané et al. (2012) / John Wiley & Sons.

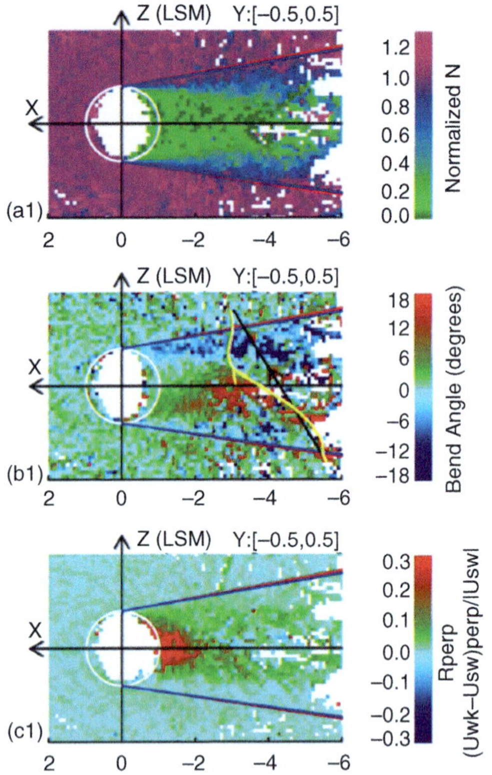


**Figure 7.11** Distribution of the average ARTEMIS measurements for the density normalized by the ion density in the solar wind (top), the bent angle of the magnetic field magnetic $\delta_{XZ}$ (middle), and the X component of the plasma flow deceleration rate in the Moon's wake (bottom). The density gradient stemming from the absorption of the plasma by th Moon's surface is the source of Alfvén wings, which manifest through a deceleration of the plasma flow and a bending of the magnetic field in the wake. Source: Zhang et al. (2016) / John Wiley & Sons.

### 7.2.5. Mercury

Since Mercury is closer to the Sun than the Earth (perihelion:0.31 AU and aphelion:0.47 AU), the typical magnetic field strength, and thus the typical Alfvén Mach number is also much lower than for the Earth (~2–3 compared to ~8 or above for the Earth). Hence, Alfvén Mach number excursions below 1 during ICME crossings are also expected to be more frequent, especially throughout the time intervals for which the field is strong, but the density is low after the initial interplanetary shock. Using a simple analytical models, Sarantos and Slavin (2009) explored the possibility for the formation of Alfvén wings at Mercury with a low Alfvén Mach number, since Mercury, under such configurations, is expected to be an intermediate state between Ganymede and the Earth. The BepiColombo mission may soon bring us further evidence of Alfvén wings originating from Mercury.

### 7.2.6. Exoplanets

As discussed before, the interaction between Solar System planets and the solar wind can occasionally become sub-Alfvénic, allowing for the propagation of energy at large distances. However, even at Mercury, the Alfvén speed remains lower than the solar wind speed, and these waves will never reach the solar atmosphere. Indeed, the Alfvén radius, i.e., the limit inside which the Alfvén speed is larger than the stellar wind speed, is $\lesssim$0.1 AU while Mercury orbits at least three times farther out. However, an increasingly large number of exoplanets are being discovered, and many of them orbit at close distances relative to their parent stars, including some orbiting inside the Alfvén radius, making it possible to trigger a planetary footprint on the star. Furthermore, observational evidences are accumulating in favor of such a Star Planet Magnetic Interaction (SPMI). These evidences can be sorted into two categories: one focuses on single targets and requires the monitoring of a star's luminosity through time, and the second consists of statistical investigations of anomalous stellar activity on large catalogs of stars with known close-in exoplanets. In the first category, the most compelling case probably concerns HD 179949, which shows a surplus of Ca II K chromospheric emissions which is synchronous with the orbital period of a known exoplanet (Shkolnik et al., 2003). Other tentative evidences were reported based on optical broadband photometry (Pagano et al., 2009; Walker et al., 2008) or X-ray emissions (Maggio et al., 2015). Unfortunately, the statistical approach has not provided conclusive evidence yet (e.g., Viswanath et al., 2020).

It should be noted that the electro–magnetic interaction between planets and stars may be much more varied and complex than those studied in our Solar System. For example, the relative motion of the planet in the plasma stream is most often dominated by the radial motion of the stellar wind rather than the orbital motion of the planet. Moreover, while the orientation of the stellar magnetic field may be mostly latitudinal for a planet orbiting very close; it is expected to become essentially longitudinal at larger distances. Then the planets and their magnetospheres may also display more varied configurations (presence of conductive layers, atmospheres, ionospheres and magnetospheres) compared to the moons of the Giant Planets. And finally, the star's intrinsic variability and activity certainly complicate the picture and probably explains why SPMI signatures are not systematically observed, even at HD 179949 (Shkolnik et al., 2008). But the most important argument to consider is the fact that a stellar atmosphere is quite different than a planetary one. Indeed some authors have noted that the power of the excess chromospheric emissions appeared three orders of magnitude larger than the expected energy arising from the Alfvénic interaction (Saur et al., 2013). This finding motivated the investigation of alternative scenarios, where the SPMI would modulate the star's emissions rather than directly providing the energy. The interested reader is referred to the recent reviews by Shkolnik and Llama (2018); Saur (2018); and Strugarek (2018).

## 7.3. SUMMARY

The electro–dynamic interaction between a plasma flow and large celestial bodies, especially at a low Alfvén Mach number, can generate powerful Alfvén waves propagating energy at large distances along the magnetic field lines. But the Alfvén waves generation is only the first step of a long chain of processes, including turbulence, complex reflection patterns, electric currents, energy transport and dissipation, momentum transfer, wave–particle interaction, aurora, joule heating of the ionosphere of the parent planet, and powerful radio emission. Examples of similar interactions abound in the Solar System, but the Jovian moons and the Io–Jupiter interaction in particular, are the best studied ones so far. Indeed, this case benefits both from a low Alfvén Mach number, strong energy generation, and relative stability of the incoming plasma characteristics, which all facilitate the measurements of the signatures of the Alfvén waves and their consequences, as well as their interpretation. For other systems, the puzzle is missing more pieces, as sometimes we only have access to hints of the local interaction. It is, however, remarkable that Alfvén wings have been observed for bodies as diverse as Rhea, Io, Europa, Ganymede, or the Earth. Electron beams, most probably accelerated by Alfvén waves, have been observed at Io, Ganymede, Callisto, and Enceladus, while auroral footprints have been reported for all four Galilean moons and for Enceladus. Despite the idiosyncrasies associated with each individual moon or planet, general patterns emerge and similar processes are observed, indicating that such processes are universal and probably very common in the universe. With missions specifically dedicated to the Galilean moons, such as JUICE and Europa Clipper, the BebiColombo mission arriving at Mercury or a future mission dedicated to the exploration of the Ice Giants, we can expect that the next decades will bring us a wealth of observational clues to further complete our understanding of the way Alfvén waves mediate the interaction between celestial objects and the plasma flows surrounding them. Even beyond the limits of our Solar System, evidences are starting to emerge, showing the importance of the Alfvén waves in the interaction between planets and their parent stars.

## REFERENCES


Acuña, M. H., Neubauer, F. M., & Ness, N. F. (1981). Standing Alfvén wave current system at Io: Voyager 1 observations. *Journal of Geophysical Research: Space Physics*, *86*(A10), 8513–8521. ISSN 2156-2202. doi: 10.1029/JA086iA10p08513.

Allegrini, F., Gladstone, G. R., Hue, V., Clark, G., Szalay, J. R., Kurth, W. S., et al. (2020). First report of electron measurements during a Europa Footprint Tail Crossing by Juno. *Geophysical Research Letters*, *47*(18), e2020GL089732. ISSN 1944-8007. doi: 10.1029/2020GL089732.

Arnold, H., Liuzzo, L., & Simon, S. (2019). Magnetic Signatures of a Plume at Europa During the Galileo E26 Flyby. *Geophysical Research Letters*, *46* (3), 1149–1157. ISSN 1944-8007. doi: 10.1029/2018GL081544.

Bagenal, F. (1983). Alfvén wave propagation in the Io plasma torus. *Journal of Geophysical Research: Space Physics*, *88*(A4), 3013–3025. ISSN 2156-2202. doi: 10.1029/JA088iA04p03013.

Bagenal, F., & Dols, V. (2020). The Space Environment of Io and Europa. *Journal of Geophysical Research: Space Physics*, *125*(5), e2019JA027485. ISSN 2169-9402. doi: 10.1029/2019JA027485.

Belcher, J. W., Goertz, C. K., Sullivan, J. D., & Acunã, M. H. (1981). Plasma observations of the Alfvén wave generated by Io. *Journal of Geophysical Research: Space Physics*, *86*(A10), 8508–8512. ISSN 2156-2202. doi: 10.1029/JA086iA10p08508.

Bertucci, C., Achilleos, N., Dougherty, M. K., Modolo, R., A. Coates, J., Szego, K., et al. (2008). The Magnetic Memory of Titan's Ionized Atmosphere. *Science*, *321*(5895), 1475–1478. doi: 10.1126/science.1159780. Publisher: American Association for the Advancement of Science.

Bhattacharyya, D., Clarke, J. T., Montgomery, J., Bonfond, B., Gérard, J.-C., & Grodent, D. (2018). Evidence for auroral emissions From Callisto's footprint in HST UV Images. *Journal of Geophysical Research: Space Physics*, *123*(1), 364–373. ISSN 2169-9402. doi: 10.1002/2017JA024791.

Bigg, E. K. (1964). Influence of the satellite Io on Jupiter's decametric emission. *Nature*, *203*(4949), 1008–1010. ISSN 1476-4687. doi: 10.1038/2031008a0.

Blöcker, A., Saur, J., & Roth, L. (2016). Europa's plasma interaction with an inhomogeneous atmosphere: Development of Alfvén winglets within the Alfvén wings. *Journal of Geophysical Research: Space Physics*, *121* (10), 9794–9828. ISSN 2169-9402. doi: 10.1002/2016JA022479.

Bonfond, B. (2010). The 3-D extent of the Io UV footprint on Jupiter. *Journal of Geophysical Research: Space Physics*, *115*(A9), A09217. ISSN 2156-2202. doi: 10.1029/2010JA015475.

Bonfond, B. (2012). When Moons Create Aurora: The Satellite Footprints on Giant Planets. In *Geophysical Monograph Series*. AGU. ISBN 978-0-87590-487-0. doi: 10.1029/2011GM001169.

Bonfond, B., Gérard, J.-C., Grodent, D., & Saur, J. (2007). Ultraviolet Io footprint short timescale dynamics. *Geophysical Research Letters*, *34* (6). ISSN 1944-8007. doi: 10.1029/2006GL028765.

Bonfond, B., Grodent, D., Gérard, J.-C., Radioti, A., Saur, J., & Jacobsen, S. (2008). UV Io footprint leading spot: A key feature for understanding the UV Io footprint multiplicity?

*Geophysical Research Letters*, *35*(5). ISSN 1944-8007. doi: 10.1029/2007GL032418.

Bonfond, B., Grodent, D., Gérard, J.-C., Radioti, A., Dols, V., Delamere, P. A., & Clarke, J. T. (2009). The Io UV footprint: Location, inter-spot distances and tail vertical extent. *Journal of Geophysical Research: Space Physics*, *114*(A7). ISSN 2156-2202. doi: 10.1029/2009JA014312.

Bonfond, B., Grodent, D., Gérard, J.-C., Stallard, T., Clarke, J. T., Yoneda, M., et al. (2012). Auroral evidence of Io's control over the magnetosphere of Jupiter. *Geophysical Research Letters*, *39*(1), L01105. ISSN 1944-8007. doi: 10.1029/2011GL050253.

Bonfond, B., Hess, S., Bagenal, F., Gérard, J.-C., Grodent, D., Radioti, A., et al. (2013). The multiple spots of the Ganymede auroral footprint. *Geophysical Research Letters*, *40*(19), 4977–4981. ISSN 1944-8007. doi: 10.1002/grl.50989.

Bonfond, B., Hess, S., Gérard, J. C., Grodent, D., Radioti, A., Chantry, V., et al. (2013). Evolution of the Io footprint brightness I: Far-UV observations. *Planetary and Space Science*, *88*, 64–75. ISSN 0032-0633. doi: 10.1016/j.pss.2013.05.023.

Bonfond, B., Grodent, D., Badman, S. V., Saur, J., Gérard, J. C., & Radioti, A. (2017). Similarity of the Jovian satellite footprints: Spots multiplicity and dynamics. *Icarus*, *292*, 208–217. ISSN 0019-1035. doi: 10.1016/j.icarus.2017.01.009.

Bonfond, B., Saur, J., Grodent, D., Badman, S. V., Bisikalo, D., Shematovich, V., et al. (2017). The tails of the satellite auroral footprints at Jupiter. *Journal of Geophysical Research: Space Physics*, *122*(8), 7985–7996. ISSN 2169-9402. doi: 10.1002/2017JA024370.

Bonfond, B., Yao, Z., & Grodent, D. (2020). Six observational pieces of evidence against corotation as the main cause for the aurora at Jupiter. Copernicus Meetings. doi: https://doi.org/10.5194/epsc2020-29.

Carlson, R. W. (1999). A tenuous carbon dioxide atmosphere on Jupiter's Moon Callisto. *Science*, *283*(5403), 820–821. doi: 10.1126/science.283.5403.820.

Chané, E., Saur, J., Neubauer, F. M., Raeder, J., & Poedts, S. (2012). Observational evidence of Alfvén wings at the Earth. *Journal of Geophysical Research: Space Physics*, *117*(A9). ISSN 2156-2202. doi: 10.1029/2012JA017628.

Chané, E., Schmieder, B., Dasso, S., Verbeke, C., Grison, B., Démoulin, P., & Poedts, S. (2021). Over-expansion of a coronal mass ejection generates sub-Alfvénic plasma conditions in the solar wind at Earth. *Astronomy & Astrophysics*, *647*, A149. ISSN 0004-6361, 1432-0746. doi: 10.1051/0004-6361/202039867. Publisher: EDP Sciences.

Cheng, A. F., & Paranicas, C. (1998). Model of field aligned potential drops near Io. *Geophysical Research Letters*, *25*(6), 833–836. ISSN 1944-8007. doi: 10.1029/98GL00407.

Chust, T., Roux, A., Kurth, W. S., Gurnett, D. A., Kivel son, M. G., & Khurana, K. K. (2005). Are Io's Alfvén wings filamented? Galileo observations. *Planetary and Space Science*, *53*(4), 395–412. ISSN 0032-0633. doi: 10.1016/j.pss.2004.09.021.

Clark, G., Mauk, B. H., Kollmann, P., Szalay, J. R., Sulaiman, A. H., Gershman, D. J., et al. (2020). Energetic proton acceleration associated with Io's footprint tail. *Geophysical Research Letters*, *47*(24), e2020GL090839. ISSN 1944-8007. doi: 10.1029/2020GL090839.

Clarke, J. T., Ballester, G. E., Trauger, J., Evans, R., Connerney, J. E. P., Stapelfeldt, K., et al. (1996). Far-Ultraviolet Imaging of Jupiter's Aurora and the Io "Footprint". *Science*, *274*(5286), 404–409, Oct. 1996. ISSN 0036-8075, 1095-9203. doi: 10.1126/science.274.5286.404.

Clarke, J. T., Ajello, J., Ballester, G., Jaffel, L. B., Connerney, J., Gérard, J.-C., et al. (2002). Ultraviolet emissions from the magnetic footprints of Io, Ganymede and Europa on Jupiter. *Nature*, *415* (6875), 997–1000. ISSN 1476-4687. doi: 10.1038/415997a.

Connerney, J. E. P., Baron, R., Satoh, T., & Owen, T. (1993). Images of Excited $H_3^+$ at the Foot of the lo Flux Tube in Jupiter's Atmosphere. *Science*, *262*(5136), 1035–1038. ISSN 0036-8075, 1095-9203. doi: 10.1126/science.262.5136.1035.

Crary, F. J. (1997). On the generation of an electron beam by Io. *Journal of Geophysical Research: Space Physics*, *102*(A1), 37–49. ISSN 2156-2202. doi: 10.1029/96JA02409.

Crary, F. J., & Bagenal, F. (1997). Coupling the plasma interaction at Io to Jupiter. *Geophysical Research Letters*, *24*(17), 2135–2138. ISSN 1944-8007. doi: 10.1029/97GL02248.

Cunningham, N. J., Spencer, J. R., Feldman, P. D., Strobel, D. F., France, K., & Osterman, S. N. (2015). Detection of Callisto's oxygen atmosphere with the Hubble Space Telescope. *Icarus*, *254*, 178–189. ISSN 0019-1035. doi: 10.1016/j.icarus.2015.03.021.

Damiano, P. A., Delamere, P. A., Stauffer, B., Ng, C.-S., & Johnson, J. R. (2019). Kinetic simulations of electron acceleration by dispersive scale Alfvén waves in Jupiter's magnetosphere. *Geophysical Research Letters*, *46* (6), 3043–3051. ISSN 1944-8007. doi: 10.1029/2018GL081219.

Delamere, P. A., Bagenal, F., Ergun, R., & Su, Y.-J. (2003). Momentum transfer between the Io plasma wake and Jupiter's ionosphere. *Journal of Geophysical Research: Space Physics*, *108*(A6), 1241. ISSN 2156-2202. doi: 10.1029/2002JA009530

Dougherty, M. K., Khurana, K. K., Neubauer, F. M., Russell, C. T., Saur, J., Leisner, J. S., & Burton, M. E. (2006). Identification of a dynamic atmosphere at Enceladus with the Cassini magnetometer. *Science*, *311*(5766), 1406–1409. doi: 10.1126/science.1120985.

Drell, S. D., Foley, H. M., & Ruderman, M. A. (1965). Drag and propulsion of large satellites in the ionosphere: An Alfvén propulsion engine in space. *Journal of Geophysical Research (1896-1977)*, *70*(13), 3131–3145. ISSN 2156-2202. doi: 10.1029/JZ070i013p03131.

Ergun, R. E., Ray, L., Delamere, P. A., Bagenal, F., Dols, V., & Su, Y.-J. (2009). Generation of parallel electric fields in the Jupiter–Io torus wake region. *J. Geophys. Res.*, *114*, A05201 doi: 10.1029/2008JA013968.

Eviatar, A., & Paranicas, C. (2005). The plasma plumes of Europa and Callisto. *Icarus*, *178*(2), 360–366. ISSN 0019-1035. doi: 10.1016/j.icarus.2005.06.007.

Eviatar, A., Williams, D. J., Paranicas, C., McEntire, R. W., Mauk, B. H., & Kivelson, M. G. (2000). Plasma flow in the magnetosphere of Ganymede. *Geophysical Research Letters*, *25*(8), 1257–1260, 1998. ISSN 1944-8007. doi: 10.1029/98GL50867.

Eviatar, A., Williams, D. J., Paranicas, C., McEntire, R. W., Mauk, B. H., & Kivelson, M. G. (2000). Trapped energetic electrons in the magnetosphere of Ganymede. *Journal of Geophysical Research: Space Physics*, *105*(A3), 5547–5553. ISSN 2156-2202. doi: 10.1029/1999JA900450.

Frank, L. A., & Paterson, W. R. (1999). Intense electron beams observed at Io with the Galileo spacecraft. *Journal of Geophysical Research: Space Physics*, *104*(A12), 28657–28669. ISSN 2156-2202. doi: 10.1029/1999JA900402.

Frank, L. A., Paterson, W. R., Ackerson, K. L., Vasyliunas, V. M., Coroniti, F. V., & Bolton, S. J. (1996). Plasma Observations at Io with the Galileo Spacecraft. *Science*, *274*(5286), 394–395. ISSN 0036-8075, 1095-9203. doi: 10.1126/science.274.5286.394.

Frank, L. A., Paterson, W. R., Ackerson, K. L., & Bolton, S. J. (1997). Low-energy electron measurements at Ganymede with the Galileo spacecraft: Probes of the magnetic topology. *Geophysical Research Letters*, *24*(17), 2159–2162. ISSN 1944-8007. doi: 10.1029/97GL01632.

Galli, A., Vorburger, A., Carberry Mogan, S. R., Roussos, E., Stenberg, G., Wieser, Wurz, P., et al. (2022). Callisto's atmosphere and its space environment: Prospects for the particle environment package on board JUICE. *Earth and Space Science*, *9*(5), e2021EA002172. ISSN 2333-5084. doi: 10.1029/2021EA002172.

Gershman, D. J., Connerney, J. E. P., Kotsiaros, S., DiBraccio, G. A., Martos, Y. M., -Vinãs, A. F., et al. (2019). Alfvénic fluctuations associated With Jupiter's auroral emissions. *Geophysical Research Letters*, *46*(13), 7157–7165. ISSN 1944-8007. doi: 10.1029/2019GL082951.

Gladstone, G. R., Stern, S. A., Slater, D. C., Versteeg, M., Davis, M. W., Retherford, K. D., et al. (2007). Jupiter's Nightside Airglow and Aurora. *Science*, *318*(5848), 229–231. ISSN 0036-8075, 1095-9203. doi: 10.1126/science.1147613.

Goertz, C. (1980). Io's interaction with the plasma torus. *Journal of Geophysical Research: Space Physics*, *85*(A6), 2949–2956. ISSN 2156-2202. doi: 10.1029/JA085iA06p02949.

Goldreich, P., & Lynden-Bell, D. (1969). Io, a jovian unipolar inductor. *Astrophysical Journal*, *156*, 59–78.

Grodent, D., Gérard, J.-C., Gustin, J., Mauk, B. H., Connerney, J. E. P., & Clarke, J. T. (2006). Europa's FUV auroral tail on Jupiter. *Geophysical Research Letters*, *33*, 6201–+. doi: 10.1029/2005GL025487.

Grodent, D., Bonfond, B., Radioti, A., Gérard, J.-C., Jia, X., Nichols, J. D., & Clarke, J. T. (2009). Auroral footprint of Ganymede. *Journal of Geophysical Research: Space Physics*, *114*(A7), A07212. ISSN 2156-2202. doi: 10.1029/2009JA014289.

Gurnett, D. A., & Goertz, C. K. (1981). Multiple Alfven wave reflections excited by Io: Origin of the Jovian decametric arcs. *Journal of Geophysical Research: Space Physics*, *86*(A2), 717–722. ISSN 2156-2202. doi: 10.1029/JA086iA02p00717.

Gurnett, D. A., Kurth, W. S., Roux, A., Bolton, S. J., & Kennel, C. F. (1996). Evidence for a magnetosphere at Ganymede from plasma-wave observations by the Galileo spacecraft. *Nature*, *384* (6609), 535–537. ISSN 1476-4687. doi: 10.1038/384535a0. Number: 6609 Publisher: Nature Publishing Group.

Gurnett, D. A., Persoon, A. M., Kurth, W. S., Roux, A., & Bolton, S. J. (2000). Plasma densities in the vicinity of Callisto from Galileo plasma wave observations. *Geophysical Research Letters*, *27*(13), 1867–1870. ISSN 1944-8007. doi: 10.1029/2000GL003751.

Gurnett, D. A., Averkamp, T. F., Schippers, P., Persoon, A. M., Hospodarsky, G. B., Leisner, J. S., et al. (2011). Auroral hiss, electron beams and standing Alfvén wave currents near Saturn's moon Enceladus. *Geophysical Research Letters*, *38*(6), L06102. ISSN 1944-8007. doi: 10.1029/2011GL046854.

Gérard, J.-C., Saglam, A., Grodent, D., & Clarke, J. T. (2006). Morphology of the ultraviolet Io footprint emission and its control by Io's location. *Journal of Geophysical Research*, *111*(A10), 4202–+. doi: 10.1029/2005JA011327.

Hartkorn, O., & Saur, J. (2017). Induction signals from Callisto's ionosphere and their implications on a possible subsurface ocean. *Journal of Geophysical Research: Space Physics*, *122*(11), 11,677–11,697. ISSN 2169-9402. doi: 10.1002/2017JA024269.

Hess, S., Zarka, P., Mottez, F., & Ryabov, V. B. (2009). Electric potential jumps in the Io-Jupiter flux tube. *Planetary and Space Science*, *57*(1), 23–33. ISSN 0032-0633. doi: 10.1016/j.pss.2008.10.006.

Hess, S. L. G., Delamere, P., Dols, V., Bonfond, B., & Swift, D. (2010). Power transmission and particle acceleration along the Io flux tube. *Journal of Geophysical Research: Space Physics*, *115*(A6), 23–33. ISSN 2156-2202. doi: 10.1029/2009JA014928.

Hess, S. L. G., Delamere, P., Dols, A. V., & Ray, L. C. (2011). Comparative study of the power transferred from satellite-magnetosphere interactions to auroral emissions. *Journal of Geophysical Research: Space Physics*, *116*(A1), A01202. ISSN 2156-2202. doi: 10.1029/2010JA015807.

Hess, S. L. G., Bonfond, B., Chantry, V., Gérard, J. C., Grodent, D., Jacobsen, S., & Radioti, A. (2013). Evolution of the Io footprint brightness II: Modeling. *Planetary and Space Science*, *88*, 76–85. ISSN 0032-0633. doi: 10.1016/j.pss.2013.08.005.

Hess, S. L. G., Bonfond, B., & Delamere, P. A. (2013). How could the Io footprint disappear? *Planetary and Space Science*, *89*, 102–110. ISSN 0032-0633. doi: 10.1016/j.pss.2013.08.014.

Hill, T. W., & Vasyliūnas, V. M. (2002). Jovian auroral signature of Io's corotational wake. *Journal of Geophysical Research: Space Physics*, *107*(A12), SMP 27–1–SMP 27–5. ISSN 2156-2202. doi: 10.1029/2002JA009514.

Hinson, D. P., Kliore, A. J., Flasar, F. M., Twicken, J. D., Schinder, P. J., & Herrera, R. G. (1998). Galileo radio occultation measurements of Io's ionosphere and plasma wake. *Journal of Geophysical Research*, *103*, 29343–29358. doi: 10.1029/98JA02659.

Hinton, P. C., Bagenal, F., & Bonfond, B. (2019). Alfvén wave propagation in the Io Plasma Torus. *Geophysical Research Letters*, *46*(3), 1242–1249. ISSN 1944-8007. doi: 10.1029/2018GL081472.

Hue, V., Greathouse, T. K., Bonfond, B., Saur, J., Gladstone, G. R., Roth, L. (2019). Juno-UVS observation of the Io footprint during solar eclipse. *Journal of Geophysical Research: Space Physics*, *124*(7), 5184–5199. ISSN 2169-9402. doi: 10.1029/2018JA026431.

Hue, V., Greathouse, T. K., Gladstone, G. R., Bonfond, B., Gérard, J.-C., Vogt, M. F., et al. (2021). Detection and characterization of circular expanding UV-emissions observed in Jupiter's Polar Auroral regions. *Journal of Geophysical Research: Space Physics*, *126*(3), e2020JA028971. ISSN 2169-9402. doi: 10.1029/2020JA028971.

Huybrighs, H. L. F., Roussos, E., Blöcker, A., Krupp, N., Futaana, Y., Barabash, S., et al. (2020). An active plume eruption on europa During Galileo Flyby E26 as indicated by energetic Proton depletions. *Geophysical Research Letters*, *47*(10), e2020GL087806. ISSN 1944-8007. doi: 10.1029/2020GL087806.

Jacobsen, S., Saur, J., Neubauer, F. M., Bonfond, B., Gérard, J.-C., & Grodent, D. (2010). Location and spatial shape of electron beams in Io's wake. *Journal of Geophysical Research*, *115*(A14), A04205. doi: 10.1029/2009JA014753.

Jia, X., Walker, R. J., Kivelson, M. G., Khurana, K. K., & Linker, J. A. (2008). Three-dimensional MHD simulations of Ganymede's magnetosphere. *Journal of Geophysical Research*, *113*(A12), 6212–+. doi: 10.1029/2007JA012748.

Jia, X., Kivelson, M. G., Khurana, K. K., & Kurth, W. S. (2018). Evidence of a plume on Europa from Galileo magnetic and plasma wave signatures. *Nature Astronomy*, *2*(6), 459–464. ISSN 2397-3366. doi: 10.1038/s41550-018-0450-z.

Jia, Y.-D., Russell, C. T., Khurana, K. K., Wei, H. Y., Ma, Y. J., Leisner, J. S., et al. (2011). Cassini magnetometer observations over the Enceladus poles. *Geophysical Research Letters*, *38*(19). ISSN 1944-8007. doi: 10.1029/2011GL049013.

Jones, S. T., & Su, Y.-J. (2008). Role of dispersive Alfvén waves in generating parallel electric fields along the Io-Jupiter fluxtube. *Journal of Geophysical Research*, *113*(A12), 12205–+. doi: 10.1029/2008JA013512.

Kallio, E., Sillanpää, I., Jarvinen, R., Janhunen, P., Dougherty, M., Bertucci, C., & Neubauer, F. (2007). Morphology of the magnetic field near Titan: Hybrid model study of the Cassini T9 flyby. *Geophysical Research Letters*, *34*(24). ISSN 1944-8007. doi: 10.1029/2007GL030827.

Khurana, K. K., Kivelson, M. G., Stevenson, D. J., Schubert, G., Russell, C. T., Walker, R. J., & Polanskey, C. (1998). Induced magnetic fields as evidence for subsurface oceans in Europa and Callisto. *Nature*, *395*(6704), 777–780. ISSN 1476-4687. doi: 10.1038/27394.

Khurana, K. K., Fatemi, S., Lindkvist, J., Roussos, E., Krupp, N., Holmström, M., et al. (2017). The role of plasma slowdown in the generation of Rhea's Alfvén wings. *Journal of Geophysical Research: Space Physics*, *122*(2), 1778–1788. ISSN 2169-9402. doi: 10.1002/2016JA023595.

Kivelson, M. G., Khurana, K. K., Walker, R. J., Russell, C. T., Linker, J. A., Southwood, D. J., & Polanskey, C. (1996). A Magnetic Signature at Io: Initial Report from the Galileo Magnetometer. *Science*, *273*(5273), 337–340. doi: 10.1126/science.273.5273.337. Publisher: American Association for the Advancement of Science.

Kivelson, M. G., Warnecke, J., Bennett, L., Joy, S., Khurana, K. K., Linker, J. A., et al. (1998). Ganymede's magnetosphere: Magnetometer overview. *Journal of Geophysical Research: Planets*, *103*(E9), 19963–19972. ISSN 2156-2202. doi: 10.1029/98JE00227.

Kivelson, M. G., Khurana, K. K., Stevenson, D. J., Bennett, L., Joy, S., Russell, C. T., et al. (1999). Europa and Callisto: Induced or intrinsic fields in a periodically varying plasma environment. *Journal of Geophysical Research*, *104*, 4609–4626. doi: 10.1029/1998JA900095.

Kivelson, M. G., Khurana, K. K., Russell, C. T., Volwerk, M., Walker, R. J., & Zimmer, C. (2000). Galileo Magnetometer Measurements: A Stronger Case for a Subsurface Ocean at Europa. *Science*, *289*(5483), 1340–1343. doi: 10.1126/science.289.5483.1340. Publisher: American Association for the Advancement of Science.

Kivelson, M. G., Bagenal, F., Kurth, W. S., Neubauer, F. M., Paranicas, C., & Saur, J. (2004). Magnetospheric interactions with satellites. Jupiter: The Planet, Satellites and Magnetosphere (pages 513–536).

Kliore, A. J., Fjeldbo, G., Seidel, B. L., Sweetnam, D. N., Sesplaukis, T. T., Woiceshyn, P. M., & Rasool, S. I. (1975). The atmosphere of Io from Pioneer 10 radio occultation measurements. *Icarus*, *24*(4), 407–410. ISSN 0019-1035. doi: 10.1016/0019-1035(75)90057-3.

Kliore, A. J., Anabtawi, A., Herrera, R. G., Asmar, S. W., Nagy, A. F., Hinson, D. P., & Flasar, F. M. (2002). Ionosphere of Callisto from Galileo radio occultation observations. *Journal of Geophysical Research: Space Physics*, *107*(A11), SIA 19–1–SIA 19–7. ISSN 2156-2202. doi: 10.1029/2002JA009365.

Kriegel, H., Simon, S., Motschmann, U., Saur, J., Neubauer, F. M., Persoon, A. M., et al. (2011). Influence of negatively charged plume grains on the structure of Enceladus' Alfvén wings: Hybrid simulations versus Cassini Magnetometer data. *Journal of Geophysical Research: Space Physics*, *116*(A10), A10223. ISSN 2156-2202. doi: 10.1029/2011JA016842.

Kurth, W. S., Gurnett, D. A., Persoon, A. M., Roux, A., Bolton, S. J., & Alexander, C. J. (2001). The plasma wave environment of Europa. *Planetary and Space Science*, *49*(3), 345–363. ISSN 0032-0633. doi: 10.1016/S0032-0633(00)00156-2.

Lavrukhin, A. S., & Alexeev, I. I. (2015). Aurora at high latitudes of Ganymede. *Astronomy Letters*, *41*(11), 687–692. ISSN 1562-6873. doi: 10.1134/S1063773715110043.

Louis, C. K., Lamy, L., Zarka, P., Cecconi, B., & Hess, S. L. G. (2017). Detection of Jupiter decametric emissions controlled by Europa and Ganymede with Voyager/PRA and Cassini/RPWS. *Journal of Geophysical Research: Space Physics*, *122*(9), 9228–9247. ISSN 2169-9402. doi: 10.1002/2016JA023779.

Louis, C. K., Lamy, L., Zarka, P., Cecconi, B., Imai, M., Kurth, W. S., et al. (2017). Io-Jupiter decametric arcs observed by Juno/Waves compared to ExPRES simulations. *Geophysical Research Letters*, *44*(18), 9225–9232. ISSN 1944-8007. doi: 10.1002/2017GL073036.

Louis, C. K., Louarn, P., Allegrini, F., Kurth, W. S., & Szalay, J. R. (2020). Ganymede-Induced Decametric Radio Emission: In Situ Observations and Measurements by Juno. *Geophysical Research Letters*, *47*(20), e2020GL090021. ISSN 1944-8007. doi: https://doi.org/10.1029/2020GL090021.

Lysak, R. L. (2023). Kinetic Alfvén waves and auroral particle acceleration: A review. *Reviews of Modern*

*Plasma Physics*, *7*(1), 6, Jan. 2023. ISSN 2367-3192. doi: 10.1007/s41614-022-00111-2.

Maggio, A., Pillitteri, I., Scandariato, G., Lanza, A. F., Sciortino, S., Borsa, F., et al. (2015). Coordinated X-ray and optical observations of Star–Planet interaction in HD 17156. *The Astrophysical Journal Letters*, *811*(1), L2. ISSN 2041-8205. doi: 10.1088/2041-8205/811/1/L2.

Matsuda, K., Terada, N., Katoh, Y., & Misawa, H. (2012). A simulation study of the current-voltage relationship of the Io tail aurora. *Journal of Geophysical Research (Space Physics)*, *117*(A16), 10214. doi: 10.1029/2012JA017790.

Mauk, B. H., & Saur, J. (2007). Equatorial electron beams and auroral structuring at Jupiter. *Journal of Geophysical Research*, *112*(A11), 10221–+. doi: 10.1029/2007JA012370.

Mauk, B. H., Williams, D. J., & Eviatar, A. (2001). Understanding Io's space environment interaction: Recent energetic electron measurements from Galileo. *Journal of Geophysical Research: Space Physics*, *106*(A11), 26195–26208. ISSN 2156-2202. doi: 10.1029/2000JA002508.

Moirano, A., Mura, A., Adriani, A., Dols, V., Bonfond, B., Waite, J. H., et al. (2021). Morphology of the Auroral Tail of Io, Europa, and Ganymede From JIRAM L-Band Imager. *Journal of Geophysical Research: Space Physics*, *126*(9), e2021JA029450. ISSN 2169-9402. doi: 10.1029/2021JA029450.

Moirano, A., Mura, A., Hue, V., et al. (2024). The Infrared Footprint Tracks of Io, Europa and Ganymede at Jupiter Observed by Juno-JIRAM. ESS Open Archive. doi: 10.22541/essoar.168394732.26574509/v1.

Mura, A., Adriani, A., Connerney, J. E. P., Bolton, S., Altieri, F., Bagenal, F., et al. (2018). Juno observations of spot structures and a split tail in Io-induced aurorae on Jupiter. *Science*, *361*(6404), 774–777. ISSN 0036-8075, 1095-9203. doi: 10.1126/science.aat1450.

Ness, N. F., Acunã, M. H., Lepping, R. P., Burlaga, L. F., Behannon, K. W., & Neubauer, F. M. (1979). Magnetic Field Studies at Jupiter by Voyager 1: Preliminary Results. *Science*, *204*(4396), 982–987. doi: 10.1126/science.204.4396.982.

Neubauer, F. (1980). Nonlinear standing Alfvén wave current system at Io: Theory. *Journal of Geophysical Research: Space Physics*, *85*(A3), 1171–1178. ISSN 2156-2202. doi: 10.1029/JA085iA03p01171.

Neubauer, F. M. (1998). The sub-Alfvénic interaction of the Galilean satellites with the Jovian magnetosphere. *Journal of Geophysical Research: Planets*, *103*(E9), 19843–19866. ISSN 2156-2202. doi: 10.1029/97JE03370.

Neubauer, F. M., Backes, H., Dougherty, M. K., Wennmacher, A., Russell, C. T., Coates, A., et al. (2006). Titan's near magnetotail from magnetic field and electron plasma observations and modeling: Cassini flybys TA, TB, and T3. *Journal of Geophysical Research: Space Physics*, *111*(A10). ISSN 2156-2202. doi: 10.1029/2006JA011676.

Pagano, I., Lanza, A. F., Leto, G., Messina, S., Barge, P., & Baglin, A. (2009). CoRoT-2a Magnetic Activity: Hints for Possible Star-Planet Interaction. *Earth, Moon, and Planets*, *105*(2), 373–378. ISSN 1573-0794. doi: 10.1007/s11038-009-9301-3.

Piddington, J. H., & Drake, J. F. (1968). Electrodynamic effects of Jupiter's Satellite Io. *Nature*, *217*:935–937. doi: 10.1038/217935a0.

Pontius, D. H. (2002). The Io current wing. *Journal of Geophysical Research: Space Physics*, *107*(A8), SMP 6–1–SMP 6–12. ISSN 2156-2202. doi: 10.1029/2001JA000331.

Porco, C. C., Helfenstein, P., Thomas, P. C., Ingersoll, A. P., Wisdom, J., West, R., et al. (2006). Cassini Observes the Active South Pole of Enceladus. *Science*, *311*(5766), 1393–1401. doi: 10.1126/science.1123013.

Prangé, R., Rego, D., Southwood, D., Zarka, P., Miller, S., & Ip, W. (1996). Rapid energy dissipation and variability of the Io-Jupiter electrodynamic circuit. *Nature*, *379*(6563), 323–325. ISSN 1476-4687. doi: 10.1038/379323a0.

Pryor, W. R., Rymer, A. M., Mitchell, D. G., Hill, T. W., Young, D. T., Saur, J., et al. (2011). The auroral footprint of Enceladus on Saturn. *Nature*, *472*(7343), 331–333. ISSN 1476-4687. doi: 10.1038/nature09928.

Queinnec, J., & Zarka, P. (1998). Io-controlled decameter arcs and Io-Jupiter interaction. *Journal of Geophysical Research: Space Physics*, *103*(A11), 26649–26666. ISSN 2156-2202. doi: 10.1029/98JA02435.

Ridley, A. J. (2007). Alfvén wings at Earth's magnetosphere under strong interplanetary magnetic fields. *Annales Geophysicae*, *25*(2), 533–542. ISSN 0992-7689. doi: 10.5194/angeo-25-533-2007.

Roth, L., Saur, J., Retherford, K. D., Strobel, D. F., Feldman, P. D., McGrath, M. A., & Nimmo, F. (2014). Transient water vapor at Europa's South Pole. *Science*, *343*(6167), 171–174. ISSN 0036-8075, 1095-9203. doi: 10.1126/science.1247051.

Roth, L., Alday, J., Becker, T. M., Ivchenko, N., & Retherford, K. D. (2017). Detection of a hydrogen corona at Callisto. *Journal of Geophysical Research: Planets*, *122*(5), 1046–1055. ISSN 2169-9100. doi: 10.1002/2017JE005294.

Rymer, A. M., Smith, H. T., Wellbrock, A., Coates, A. J., & Young, D. T. (2009). Discrete classification and electron energy spectra of Titan's varied magnetospheric environment. *Geophysical Research Letters*, *36*(15), L15109. ISSN 1944-8007. doi: 10.1029/2009GL039427.

Sarantos, M., & Slavin, J. A. (2009). On the possible formation of Alfvén wings at Mercury during encounters with coronal mass ejections. *Geophysical Research Letters*, *36*(4). ISSN 1944-8007. doi: 10.1029/2008GL036747.

Saur, J. (2004). A model of Io's local electric field for a combined Alfvénic and unipolar inductor far-field coupling. *Journal of Geophysical Research*, *109*(A18), 1210–+. doi: 10.1029/2002JA009354.

Saur, J. (2018). Electromagnetic Coupling in Star-Planet Systems. In H. J. Deeg and J. A. Belmonte (Eds.), *Handbook of exoplanets* (pp. 1877–1893). Springer International Publishing, Cham, 2018. ISBN 978-3-319-55333-7. doi: 10.1007/978-3-319-55333-7_27.

Saur, J. (2021). Overview of Moon-Magnetosphere interactions. In *Magnetospheres in the solar system* (pp. 575–593). American Geophysical Union (AGU). ISBN 978-1-119-81562-4. doi: 10.1002/9781119815624.ch36. Section: 36.

Saur, J., Neubauer, F. M., Strobel, D. F., & Summers, M. E. (1999). Three-dimensional plasma simulation of Io's interaction with the Io plasma torus: Asymmetric plasma flow.

*Journal of Geophysical Research: Space Physics*, *104*(A11), 25105–25126. ISSN 2156-2202. doi: 10.1029/1999JA900304.

Saur, J., Grambusch, T., Duling, S., Neubauer, F. M., & Simon, S. (2013). Magnetic energy fluxes in sub-Alfvénic planet star and moon planet interactions. *Astronomy & Astrophysics*, *552*, A119. ISSN 0004-6361, 1432-0746. doi: 10.1051/0004-6361/201118179.

Schilling, N., Neubauer, F. M., & Saur, J. (2008). Influence of the internally induced magnetic field on the plasma interaction of Europa. *Journal of Geophysical Research: Space Physics*, *113*(A3). ISSN 2156-2202. doi: 10.1029/2007JA012842.

Shkolnik, E., Walker, G. A. H., & Bohlender, D. A. (2003, November). Evidence for Planet-induced chromospheric activity on HD 179949. *The Astrophysical Journal*, *597*(2), 1092–1096. ISSN 0004-637X. doi: 10.1086/378583.

Shkolnik, E., Bohlender, D. A., Walker, G. A. H., & Cameron, A. C. (2008). The On/Off Nature of Star-Planet Interactions*. *The Astrophysical Journal*, *676*(1), 628. ISSN 0004-637X. doi: 10.1086/527351.

Shkolnik, E. L., & Llama, J. (2018). Signatures of Star-Planet interactions. In H. J. Deeg and J. A. Belmonte (Eds.), *Handbook of Exoplanets* (pp. 1737–1753). Springer International Publishing, Cham, 2018. ISBN 978-3-319-55333-7. doi: 10.1007/978-3-319-55333-7_20.

Sillanpää, I., & Johnson, R. E. (2015). The role of ion-neutral collisions in Titan's magnetospheric interaction. *Planetary and Space Science*, *108*, 73–86. ISSN 0032-0633. doi: 10.1016/j.pss.2015.01.007.

Sillanpää, I., Kallio, E., Jarvinen, R., & Janhunen, P. (2007). Oxygen ions at Titan's exobase in a Voyager 1-type interaction from a hybrid simulation. *Journal of Geophysical Research: Space Physics*, *112*(A12), A12205. ISSN 2156-2202. doi: 10.1029/2007JA012348.

Simon, S., Wennmacher, A., Neubauer, F. M., Bertucci, C. L., Kriegel, H., Saur, J., et al. (2010). Titan's highly dynamic magnetic environment: A systematic survey of Cassini magnetometer observations from flybys TA-T62. *Planetary and Space Science*, *58*(10), 1230–1251. ISSN 0032-0633. doi: 10.1016/j.pss.2010.04.021.

Simon, S., Saur, J., Kriegel, H., Neubauer, F. M., Motschmann, U., & Dougherty, M. K. (2011). Influence of negatively charged plume grains and hemisphere coupling currents on the structure of Enceladus' Alfvén wings: Analytical modeling of Cassini magnetometer observations. *Journal of Geophysical Research: Space Physics*, *116*(A4). ISSN 2156-2202. doi: 10.1029/2010JA016338.

Simon, S., Kriegel, H., Saur, J., Wennmacher, A., Neubauer, F. M., Roussos, E., et al. (2012). Analysis of Cassini magnetic field observations over the poles of Rhea. *Journal of Geophysical Research: Space Physics*, *117*(A7), A07211. ISSN 2156-2202. doi: 10.1029/2012JA017747.

Simon, S., Addison, P., & Liuzzo, L. (2022). Formation of a displaced plasma wake at Neptune's Moon Triton. *Journal of Geophysical Research: Space Physics*, *127*(1), e2021JA029958. ISSN 2169-9402. doi: 10.1029/2021JA029958.

D. Snowden, R. V. Yelle, M. Galand, A. J. Coates, A. Wellbrock, G. H. Jones, & Lavvas, P. (2013, September). Auroral electron precipitation and flux tube erosion in Titan's upper atmosphere. *Icarus*, *226*(1), 186–204. ISSN 0019-1035. doi: 10.1016/j.icarus.2013.05.021.

Strugarek, A. (2018). Models of Star-Planet magnetic interaction. In H. J. Deeg and J. A. Belmonte (Eds.), *Handbook of Exoplanets* (pp. 1833–1855). Springer International Publishing, Cham, 2018. ISBN 978-3-319-55333-7. doi: 10.1007/978-3-319-55333-7_25.

Sulaiman, A. H., Kurth, W. S., Hospodarsky, G. B., Averkamp, T. F., Ye, S.-Y., Menietti, J. D., et al. (2018). Enceladus auroral hiss emissions during Cassini's Grand Finale. *Geophysical Research Letters*, *45*(15), 7347–7353. ISSN 1944-8007. doi: 10.1029/2018GL078130.

Sulaiman, A. H., Hospodarsky, G. B., Elliott, S. S., Kurth, W. S., Gurnett, D. A., Imai, M., et al. (2020). Wave-particle interactions associated with Io's auroral footprint: Evidence of Alfvén, ion cyclotron, and whistler modes. *Geophysical Research Letters*, *47*(22), e2020GL088432. ISSN 1944-8007. doi: 10.1029/2020GL088432.

Sulaiman, A. H., Szalay, J. R., Clark, G., Allegrini, F., Bagenal, F., Brennan, M. J., et al. (2023). Poynting fluxes, field-aligned current densities, and the efficiency of the Io-Jupiter electrodynamic interaction. *Geophysical Research Letters*, *50*(10), e2023GL103456. ISSN 1944-8007. doi: 10.1029/2023GL103456.

Szalay, J. R., Bonfond, B., Allegrini, F., Bagenal, F., Bolton, S., Clark, G., et al. (2018). In situ observations connected to the Io footprint tail aurora. *Journal of Geophysical Research: Planets*, *123*(11), 3061–3077. ISSN 2169-9100. doi: 10.1029/2018JE005752.

Szalay, J. R., Allegrini, F., Bagenal, F., Bolton, S. J., Bonfond, B., Clark, G., et al. (2020a). Alfvénic acceleration sustains Ganymede's footprint tail aurora. *Geophysical Research Letters*, *47*(3), e2019GL086527. ISSN 1944-8007. doi: https://doi.org/10.1029/2019GL086527.

Szalay, J. R., Allegrini, F., Bagenal, F., Bolton, S. J., Bonfond, B., Clark, G., et al. (2020b). A new framework to explain changes in Io's footprint tail electron fluxes. *Geophysical Research Letters*, *47*(18), e2020GL089267. ISSN 1944-8007. doi: 10.1029/2020GL089267.

Szalay, J. R., Bagenal, F., Allegrini, F., Bonfond, B., Clark, G., Connerney, J. E. P., et al. (2020c). Proton acceleration by Io's Alfvénic interaction. *Journal of Geophysical Research: Space Physics*, *125*(1), e2019JA027314. ISSN 2169-9402. doi: 10.1029/2019JA027314.

Vasavada, A. R., Bouchez, A. H., Ingersoll, A. P., Little, B., Anger, C. D., & The Galileo SSI Team. (1999). Jupiter's visible aurora and Io footprint. *Journal of Geophysical Research*, *104*, 27133–+.

Vasyliūnas, V. M. (2016). Physical origin of pickup currents. *Annales Geophysicae*, *34* (1), 153–156. ISSN 0992-7689. doi: 10.5194/angeo-34-153-2016.

Viswanath, G., Narang, M., Manoj, P., Mathew, B., & Kartha, S. S. (2020, May). A statistical search for Star-Planet interaction in the ultraviolet using GALEX. *The Astronomical Journal*, *159*, 194. ISSN 0004-6256. doi: 10.3847/1538-3881/ab7d3b.

Walker, G. A. H., Croll, B., Matthews, J. M., Kuschnig, R., Huber, D., Weiss, W. W., et al. (2008). MOST detects variability on $\tau$ Bootis A possibly induced by its planetary companion. *Astronomy and Astrophysics*, *482*, 691–697. doi: 10.1051/0004-6361:20078952.

Wannawichian, S., Clarke, J. T., & Pontius, D. H. (2008). Interaction evidence between Enceladus' atmosphere and Saturn's magnetosphere. *Journal of Geophysical Research: Space Physics*, *113*(A7). ISSN 2156-2202. doi: 10.1029/2007JA012899.

Webster, D. L., Alksne, A. Y., & Whitten, R. C. (1972, June). Does Io's ionosphere influence Jupiter's Radio bursts? *The Astrophysical Journal*, *174*, 685. ISSN 0004-637X. doi: 10.1086/151530.

Wei, H. Y., Russell, C. T., Wahlund, J.-E., Dougherty, M. K., Bertucci, C., Modolo, R., et al. (2007). Cold ionospheric plasma in Titan's magnetotail. *Geophysical Research Letters*, *34*(24). ISSN 1944-8007. doi: 10.1029/2007GL030701.

Williams, D. J., & Mauk, B. (1997). Pitch angle diffusion at Jupiter's moon Ganymede. *Journal of Geophysical Research: Space Physics*, *102*(A11), 24283–24287. ISSN 2156-2202. doi: 10.1029/97JA02260.

Williams, D. J., & Thorne, R. M. (2003). Energetic particles over Io's polar caps. *Journal of Geophysical Research: Space Physics*, *108* (A11), 1397. ISSN 2156-2202. doi: 10.1029/2003JA009980.

Williams, D. J., Mauk, B. H., McEntire, R. E., Roelof, E. C., Armstrong, T. P., Wilken, B., et al. (1996). Electron beams and ion composition measured at lo and in its Torus. *Science*, *274*, 401–403.

Williams, D. J., Mauk, B., & McEntire, R. W. (1997). Trapped electrons in Ganymede's magnetic field. *Geophysical Research Letters*, *24*(23), 2953–2956. ISSN 1944-8007. doi: 10.1029/97GL03003.

Williams, D. J., Mauk, B. H., McEntire, R. W., Roelof, E. C., Armstrong, T. P., Wilken, B., et al. (1997). Energetic particle signatures at Ganymede: Implications for Ganymede's magnetic field. *Geophysical Research Letters*, *24* (17), 2163–2166. ISSN 1944-8007. doi: 10.1029/97GL01931.

Wright, A. N. (1987). The interaction of Io's Alfven waves with the Jovian magnetosphere. *Journal of Geophysical Research: Space Physics*, *92*(A9), 9963–9970. ISSN 2156-2202. doi: 10.1029/JA092iA09p09963.

Wright, A. N., & Southwood, D. J. (1987). Stationary Alfvenic structures. *Journal of Geophysical Research*, *92*, 1167–1175.

Zarka, P. (1998). Auroral radio emissions at the outer planets: Observations and theories. *Journal of Geophysical Research: Planets*, *103* (E9), 20159–20194. ISSN 0148-0227. doi: 10.1029/98JE01323.

Zhang, H., Khurana, K. K., Kivelson, M. G., Fatemi, S., Holmström, M., Angelopoulos, V., et al. (2016). Alfvén wings in the lunar wake: The role of pressure gradients. *Journal of Geophysical Research: Space Physics*, *121*(11), 10,698–10,711. ISSN 2169-9402. doi: 10.1002/2016JA022360.

Zhou, H., Tóth, G., Jia, X., & Chen, Y. (2020). Reconnection-driven dynamics at Ganymede's upstream magnetosphere: 3-D Global Hall MHD and MHD-EPIC simulations. *Journal of Geophysical Research: Space Physics*, *125*(8), e2020JA028162. ISSN 2169-9402. doi: https://doi.org/10.1029/2020JA028162.

Zimmer, C., Khurana, K. K., & Kivelson, M. G. (2000). Subsurface oceans on Europa and Callisto: Constraints from Galileo magnetometer observations. *Icarus*, *147*(2), 329–347. ISSN 0019-1035. doi: 10.1006/icar.2000.6456.

# 8

# Alfvén Waves and Aurora at Earth and Jupiter: Comparative Analysis

**Robert L. Lysak**

**ABSTRACT**

The formation of discrete auroral arcs is directly connected with the dynamics of Alfvén waves propagating along auroral field lines. The auroral acceleration process has been most directly studied at Earth and at Jupiter, due to multiple satellites such as S3-3, Polar, Viking, and FAST in polar orbit around Earth and the Juno satellite in polar orbit around Jupiter. These observations have indicated two primary mechanisms for the acceleration of auroral particles: a quasi-static acceleration due to parallel electric fields that produces a nearly monoenergetic distribution of electrons, and a broadband acceleration mainly due to acceleration in the parallel electric fields of kinetic Alfvén waves. Monoenergetic acceleration in upward current regions is more common at Earth, while at Jupiter, broadband acceleration appears to dominate and occurs in both upward and downward current regions. Parallel electric fields are favored when the ratio of the plasma density to magnetic-field strength is lowest due to the lack of available charge to carry the field-aligned current. Broadband acceleration is dominant when the parallel electric fields are fluctuating on rapid timescales and when the current systems are filamented, which can occur due to phase mixing or a turbulent cascade to smaller scales.

## 8.1. INTRODUCTION

The mechanism to accelerate and heat charged particles to high energy is one of the most important problems in cosmic plasma physics. One of the most visible manifestations of cosmic charged-particle acceleration is the formation of the aurora, not only at Earth but also on other planets, such as Jupiter. Planetary auroral zones are highly dynamic plasma regions characterized by fundamental plasma processes such as the generation of field-aligned currents, parallel electric fields, and the acceleration of both electrons and ions. The purpose of this review is to compare auroral acceleration processes at Earth and Jupiter, which have been the most extensively sampled with satellite measurements. We will focus on discrete auroral arcs, rather than the diffuse aurora that is associated with pitch angle scattering at both planets. In particular, this review will focus on the role of kinetic Alfvén waves at providing the energy for auroral acceleration, as well as giving a direct means for the acceleration of auroral particles.

Decades of observation of Earth's auroral zone by satellites, sounding rockets, radar measurements, ground magnetometers, and optical observations makes this region an ideal laboratory for the study of these fundamental processes. Observations indicate that there are at least two modes of particle acceleration operating in the auroral zone. In many cases, precipitating auroral electrons show a peak at a characteristic energy, implying that these electrons have fallen through a parallel potential drop (e.g., Evans, 1974; Frank & Ackerson, 1971; Mozer et al., 1980). Such observations suggest that a quasi-static parallel electric field has developed along auroral field lines. However, measurements from the

*School of Physics and Astronomy, University of Minnesota, Minneapolis, MN, USA*

*Alfvén Waves Across Heliophysics: Progress, Challenges, and Opportunities, Geophysical Monograph 285*, First Edition.
Edited by Andreas Keiling.

DOI: 10.1002/9781394195985.ch08

Fast Auroral SnapshoT (FAST) spacecraft (e.g., Chaston et al., 1999, 2000, 2006; Chaston, Bonnell, Carlson, et al., 2002; Chaston, Bonnell, Carlson, McFadden, Ergun, et al., 2003; Chaston, Bonnell, Peticolis, et al., 2002; Chaston, Peticolas, et al., 2003; Chaston, Bonnell, Carlson, McFadden, Strangeway, et al., 2003; Chaston, Carlson, et al., 2007; Chaston, Hull, et al., 2007; McFadden, Carlson & Ergun, 1999) have indicated that auroral electrons can have a broad energy distribution, consistent with studies from sounding rockets of field-aligned energetic electron bursts (e.g., Arnoldy et al., 1985; Clemmons et al., 1994; Johnstone & Winningham, 1982; Knudsen et al., 1998; Lynch et al., 1994, 1999; McFadden et al., 1986, 1987, 1990; Robinson et al., 1989; Semeter et al., 2001). These field-aligned bursts are consistent with models for the Alfvén wave acceleration of these electrons (e.g., Chaston et al., 1999, 2000; Chaston, Bonnell, Carlson, et al., 2002; Chaston, Bonnell, Peticolis, et al., 2002; Kletzing, 1994; Kletzing & Hu, 2001; Lysak & Song, 2003a,b, 2005; Thompson & Lysak, 1996; Watt et al., 2005, 2006), and thus the aurora produced by these electrons has been termed the "Alfvénic aurora."

The arrival of the Juno satellite at Jupiter in July 2016 provided detailed measurements of particle acceleration in a second planetary auroral zone. Juno's polar orbit has a perijove distance (i.e., closest approach to Jupiter) of $1.06\,R_J$ ($1\,R_J = 71{,}492$ km is the 1 bar radius of Jupiter) measured from the center of Jupiter (Connerney et al., 2017) and crosses the auroral zone at distances of less than $1.6\,R_J$ due to the difference between the rotational and magnetic axes of Jupiter (e.g., Bagenal et al., 2017). Juno also sees both broadband and quasi-static acceleration of electrons; however, surprisingly, the broadband acceleration seems to be dominant at Jupiter in contrast to the situation at Earth (Mauk et al., 2017; Salveter et al., 2022). Another major surprise is the lack of a strong field-aligned current signature, relative to Earth, on Jovian auroral field lines (Connerney et al., 2017; Kotsiaros et al., 2019). These surprising results suggest that Jupiter's auroral zone is more dynamic and time-dependent than that of Earth.

The fundamental physical processes occurring in these two very different regions of auroral particle acceleration are likely very similar; however, the environment in which they occur may be quite different. The most significant difference is that the terrestrial aurora is primarily driven by the solar wind–magnetosphere interaction, while at Jupiter, the planetary rotation dominates the magnetospheric dynamics, with the solar wind interaction playing a secondary role. This leads to a different morphology of the underlying field-aligned current system. At Earth, the solar wind interaction creates a two-cell convection pattern leading to the formation of so-called Region 1 and Region 2 currents (Iijima & Potemra, 1976). The Region 1 currents are more poleward and are upward on the dusk side of the magnetosphere and downward on the dawn side, while the more equatorward Region 2 currents have the opposite polarity. However, at Jupiter, the generation of auroral field-aligned currents lies in the region where the corotation breaks down, which leads to a current system that is downward on the poleward side of the auroral zone and upward on the equatorward side at all local times (Vasyliunas, 1983). The more equatorward, upward current region at Jupiter has been termed Zone I (Mauk et al., 2020), and the more poleward, downward current region has been termed Zone II. However, in contrast with Earth, where the downward current regions are generally lacking auroral emissions, at Jupiter the Zone II region has bright aurora associated with broadband, bidirectional acceleration. The current-generation region in the outer magnetosphere is thought to be a very turbulent region (e.g., Ng et al., 2018, 2022; Saur et al., 2003) that can lead to a structuring of the current system. At both planets, the energies to power the aurora are likely to be carried by the propagation of shear Alfvén waves along field lines and the interaction of these waves with the ionosphere and magnetosphere. Observations from the Polar satellite (e.g., Keiling et al., 2000, 2001, 2002; Wygant et al., 2000, 2002) at 4–5 $R_E$ indicated that strong Alfvén waves carrying significant Poynting flux are often present in the Earth's plasma-sheet boundary layer (PSBL) and are associated with field-aligned electron acceleration (Wygant et al., 2002). Although a comparable measure of the Poynting flux from Juno is not possible due to lack of a low-frequency electric field measurement, it is logical to assume that the kinetic Alfvén waves play a major role in the evolution of auroral arcs at Jupiter as well as at Earth.

A special case present at Jupiter but not at Earth is the presence of large moons in the inner magnetosphere. Aurora from Jupiter's moon Io have been observed with the Infrared Telescope Facility (Connerney et al., 1993) and the Hubble Space Telescope (Bonfond et al., 2008, 2013; Clarke et al., 1996), as well as from Ganymede and Europa (Clarke et al., 2002). Io and the other moons are known to produce Alfvén "wings" due to their motion with respect to the corotating magnetosphere of Jupiter (Goertz, 1980; Neubauer, 1980). Recently, Juno has made a number of high-latitude crossings of field lines magnetically connected to Io, indicating the presence of accelerated electrons and ions (Szalay et al., 2018, 2020) and Alfvén waves with a power-law spectrum extending up to the proton gyrofrequency (Gershman et al., 2019; Sulaiman et al., 2020). These observations support the suggestion that Alfvén waves are directly connected with auroral particle acceleration.

It has long been known (e.g., Chen & Hasegawa, 1974; Hasegawa & Chen, 1975, 1976; Hasegawa, 1976;

Goertz & Boswell, 1979; Stéfant, 1970) that Alfvén waves that have perpendicular wavelengths the order of the ion acoustic gyroradius ($\rho_s = c_s/\Omega_i$), where $c_s$ is the ion acoustic speed and $\Omega_i$ is the ion gyrofrequency) or the electron inertial length ($\lambda_e = c/\omega_{pe}$), where $c$ is the speed of light and $\omega_{pe}$ is the electron plasma frequency) develop parallel electric fields. Since plasma kinetic theory is necessary to accurately describe such structures, these waves are referred to as kinetic Alfvén waves. At Earth, the ion acoustic gyroradius and electron inertial length range from less than a kilometer to tens of kilometers in the auroral zone, consistent with the thicknesses of auroral arcs (e.g., Borovsky, 1993; Knudsen et al., 2001; Maggs & Davis, 1968; Partamies et al., 2010). In Jupiter's low-altitude auroral zone, observations indicate that the density can be as low as $10^{-3}$ cm$^{-3}$, which gives an electron inertial length of about 150 km (Sulaiman et al., 2022). At higher latitudes, Saur et al. (2018) estimate that the ion acoustic gyroradius is also the order of 100 km in the equatorial plane at distances where the currents are likely generated. Observations of the field-aligned currents on the Io flux tube show that these currents are filamented (Chust et al., 2005; Sulaiman, Szalay et al., 2023), suggesting that these kinetic effects are important in this region. Furthermore, theory and numerical modeling (e.g., Coffin et al., 2022; Damiano et al., 2019; Hess et al., 2010) also show how Alfvénic acceleration can produce broadband acceleration.

The remainder of this review is as follows: the next section will consider the dynamics of kinetic Alfvén waves and their role in producing broadband acceleration of electrons. The inhomogeneity of the plasma along auroral field lines will be seen to be essential in understanding these processes. Then we will describe the theory of quasi-static parallel electric fields and their effects on auroral particles, considering the aurora at both Earth and Jupiter. While these are often described by electrostatic models, the energy source in such models is provided by the Poynting flux associated with the propagation of Alfvén waves, and so a full understanding of these processes requires the dynamics of these waves to be considered. Finally, we will discuss some unresolved issues that remain in order to fully understand the dynamic processes that produce the discrete aurora.

## 8.2. BROADBAND ACCELERATION: THE "ALFVÉNIC AURORA"

The ideal MHD Alfvén wave has no parallel electric field, and so it cannot directly accelerate electrons along the magnetic field. However, as noted in the earlier text, for small perpendicular wavelengths, kinetic effects lead to the formation of such a field. A critical criterion in the development of parallel electric fields is whether the electron thermal speed is greater than or less than the Alfvén speed, or the condition that the electron plasma beta, $\beta = 2\mu_0 n T_e/B^2$, is greater than or less than the electron-to-ion mass ratio. For $\beta > m_e/m_i$, which may be called the warm plasma limit, the ion acoustic gyroradius defined in the earlier text is the critical scale length, while in the cold plasma limit ($\beta < m_e/m_i$), the electron inertial length is most important. It should be noted that this latter case is often called the "inertial Alfvén wave"; however, it is not a distinct wave mode and so it is more accurate to describe it as the inertial limit of the kinetic Alfvén wave. A description of the fully kinetic dispersion relation has been given by Lysak and Lotko (1996) and Lysak (2008), who found that the dispersion relation in the warm plasma limit can be simplified to read

$$\omega^2 = k_\parallel^2 V_A^2 \left(1 + k_\perp^2 \rho_s^2 + k_\perp^2 \rho_i^2\right). \tag{8.1}$$

In the cold plasma limit, the dispersion relation becomes

$$\omega^2 = k_\parallel^2 V_A^2 \frac{1 + k_\perp^2 \rho_i^2}{1 + k_\perp^2 \lambda_e^2}. \tag{8.2}$$

In these expressions, $\rho_i$ is the thermal ion gyroradius and a Padé approximation (e.g., Johnson & Cheng, 1997) has been used to approximate a Bessel function expression. A recent review of the theory of the kinetic Alfvén wave has been given by Lysak (2023). Here, we will concentrate on how this theory applies to the auroral zones of Earth and Jupiter.

In the warm plasma limit, it can be shown (e.g., Hasegawa, 1976) that the magnitude of the parallel electric field is given by

$$E_\parallel = \frac{k_\parallel k_\perp \rho_s^2}{1 + k_\perp^2 \rho_i^2} E_\perp. \tag{8.3}$$

On the other hand, in the inertial limit, the parallel electric field is (e.g., Goertz & Boswell, 1979)

$$E_\parallel = \frac{k_\parallel k_\perp \lambda_e^2}{1 + k_\perp^2 \lambda_e^2} E_\perp. \tag{8.4}$$

It is important to note that while the finite ion gyroradius affects the dispersion relation, it does not lead to a parallel electric field. In fact, in the warm plasma limit, the ion gyroradius reduces the parallel electric field. This effect has been seen in both fluid and kinetic simulations of kinetic Alfvén waves (Chaston, Bonnell, Carlson, McFadden, Strangeway, et al., 2003; Damiano et al., 2015, 2016).

The relative importance of these two limits depends on the assumed distribution of plasma along the auroral field line. At Earth, for typical auroral density profiles, the region below about 4–5 $R_E$ lies in the inertial regime, while the warm plasma limit applies at higher altitudes

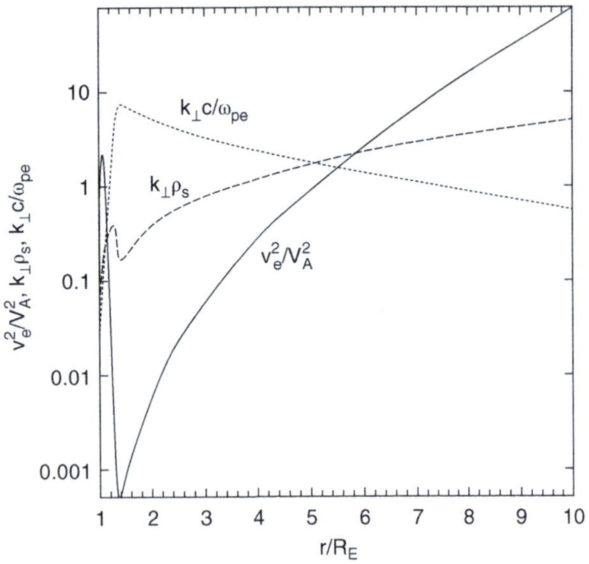


**Figure 8.1** Kinetic Alfvén wave parameters for a model auroral field line at Earth. The solid line gives the transition point between the inertial and kinetic limits, with the inertial limit being valid when $v_e^2/V_A^2 < 1$ for an assumed temperature of 1 keV. The dotted line gives the value of $k_\perp c/\omega_{pe}$ and the dashed line gives $k_\perp \rho_s$ assuming a perpendicular wavelength of 1 km Source: Lysak & Lotko, 1996 / John Wiley & Sons.

(e.g., Lysak & Carlson, 1981; Lysak & Lotko, 1996). Figure 8.1 shows the parameters along an auroral field line, with the electron temperature assumed to be 100 eV and the perpendicular wavelength assumed to be 1 km (Lysak & Lotko, 1996). However, at Jupiter, the magnetic field is much stronger and the outer magnetosphere has very low density, except in the plasma disk populated by heavy ions. In this case, the Alfvén speed is close to the speed of light except in the disk, and so the inertial limit is appropriate throughout the whole auroral field line except in the plasma disk. Figure 8.2 shows a contour plot of the electron temperature where the electron thermal speed equals the Alfvén speed for a model of Jupiter's magnetosphere based on the current sheet model of Connerney et al. (1981) and the plasma model of Bagenal and Delamere (2011). On this plot, the red colors indicate temperatures over 1 keV, while blue colors give lower temperatures, showing that for this temperature the inertial limit is valid on auroral field lines within 10–20 $R_J$ and the kinetic limit is relevant in the plasma sheet at larger distances. The innermost field line shown in the figure is approximately at the orbit of Io, and here the inertial limit holds until reaching within about 1 $R_J$ of Io.

The nature of Alfvénic acceleration is different in the two plasma regimes as well. In the warm plasma limit, the wave phase velocity is in the bulk of the electron distribution, and so electrons can be accelerated at the Landau resonance. The observations of Wygant et al. (2002) are consistent with this picture, showing a symmetric trapped population at lower velocity but with an extended, field-aligned tail in the direction of the downward wave phase velocity, as indicated by the net downward Poynting flux in this region. These results are consistent with numerical modeling by Watt and Rankin (2009).

On the other hand, in the cold plasma regime, there is a bulk acceleration of the electrons. In this limit, the Alfvén speed is much greater than the electron thermal speed and so the electron distribution essentially sees a time-dependent electric field. In a uniform, infinite plasma, a sinusoidal wave electric field would alternately accelerate and decelerate the electrons, leading to no net

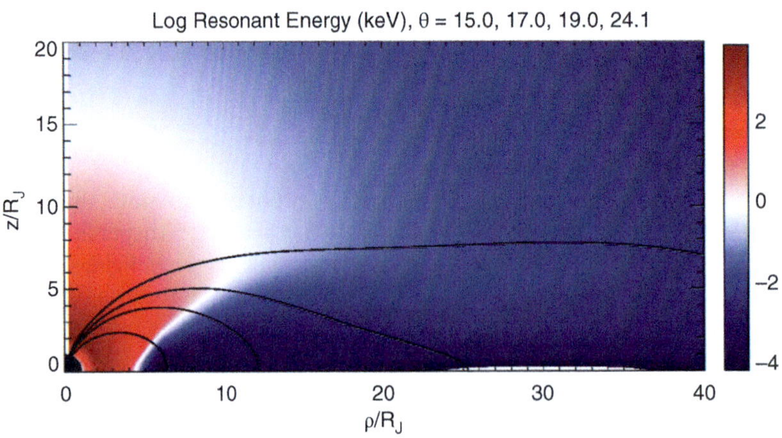


**Figure 8.2** The logarithm of the energy for which the electron thermal speed is equal to the Alfvén velocity for a model of Jupiter's magnetosphere. In the red regions, the inertial limit is appropriate, while the blue regions give the kinetic regime for a temperature of 1 keV. Source: Reproduced from Lysak & Lotko, 2020 / with permission of John Wiley & Sons.

acceleration. However, the auroral field lines at both Earth and Jupiter have a finite length, so that the electrons can be accelerated downward and leave the acceleration region before the electric field changes polarity to decelerate them. Numerical modeling of kinetic Alfvén waves with low frequency, including those associated with field-line resonances, shows that this mode of acceleration produces monoenergetic electrons (e.g., Tikhonchuk & Rankin, 2000). In this case, the relevant parameter is the ratio of the electron travel time through the acceleration region to the wave period. Electrons with energies of 100 eV travel almost 1 $R_E$/s (1 $R_E$ = 6380 km), while 10 keV electrons travel almost 1 $R_J$/s (1 $R_J$ = 71,492 km). Since the size of the acceleration region scales with the planetary radius, waves with periods longer than a few seconds, such as field-line resonances at Earth with periods of a few minutes, will tend to produce monoenergetic beams of electrons.

The production of broadband electrons in the inertial regime thus requires a more rapid variation in the parallel electric fields. One means of producing such fields is wave trapping in the ionospheric Alfvén resonator (IAR), a resonant cavity produced in the topside ionosphere due to the rapid increase in the Alfvén speed with increasing altitude due to the corresponding decrease in the plasma density. This structure was first noticed by Polyakov and Rapaport (1981) in the context of ionospheric sounding experiments. The importance of this structure for auroral acceleration was considered by Trakhtengertz and Feldstein (1984) and Lysak (1991, 1993), and its presence has been confirmed by sounding rockets (e.g., Cohen et al., 2013; Lynch et al., 2015), ground magnetometers (e.g., Hebden et al., 2005), and low-Earth-orbit satellites (e.g., Hirano et al., 2005; Miles et al., 2018; Pakhotin et al., 2018). At Jupiter, evidence for the IAR has been found in the modulation of radio waves in so-called S-bursts (Ergun et al., 2006; Su et al., 2006). Numerical modeling of an auroral flux tube at Jupiter indicates that the inertial electric field for waves in the IAR frequency range could lead to acceleration up to 100 kV (Lysak et al., 2021).

Observations from the FAST satellite have provided evidence that waves in the IAR at Earth are responsible for auroral electron acceleration (Chaston, Bonnell, Carlson et al., 2002; Chaston, Bonnell, Peticolas, et al., 2002; Chaston, Carlson, et al., 2007; Chaston, Hull, et al., 2007). They considered test particle simulations based on models of waves trapped in the IAR region and found good agreement with observed electron distributions. Observations show that such waves can drive bright, narrow auroral arcs (Chaston, Peticolas, et al., 2003) and can account for 25–39% of the total electron energy deposited in the ionosphere (Chaston, Carlson, et al., 2007). These events may represent a transitional phase in auroral dynamics, since they are often observed at the poleward edge of the auroral cavity (Chaston, Bonnell, Carlson, et al., 2002), as well as in the auroral arc that first brightens at substorm onset (Mende et al., 2003). More detailed analysis of FAST data indicates that broadband and quasi-static acceleration occur simultaneously in over 60% of events analyzed (Dombeck et al., 2018).

At Jupiter, estimates of the density profiles indicate that the IAR has resonant frequencies in the 0.1–20.0 Hz range (Lysak et al., 2021). Alfvén waves in the range up to 5.0 Hz have been observed by Juno in crossings of the Io flux tube at about 0.4 $R_J$ altitude (Gershman et al., 2019); however, higher-frequency waves are obscured by noise in the magnetometer measurements. Alfvén waves above 50 Hz are observed by the "Waves" instrument on Juno (Sulaiman et al., 2020) with a spectrum that matches up with the extrapolation of the magnetometer measurements. These waves show an apparent cutoff at the proton cyclotron frequency with an enhancement just below this frequency, which suggests that this is a true frequency spectrum and not the result of a Doppler shift. These observations suggest that an alternative explanation for these waves is an instability at the ion cyclotron resonance, a process that has also been suggested at Earth (Temerin et al., 1986). These higher-frequency waves would have rapidly fluctuating parallel electric fields that could produce broadband acceleration. Numerical simulations of the Io flux tube by Damiano et al. (2019) and Coffin et al. (2022) have been initiated with a short parallel wavelength perturbation, which produces waves with a period of about 4 s, according to Figure 2c of Damiano et al. (2019). Thus, the waves produced here are already in the range of IAR frequencies where broadband acceleration would be expected.

Another possibility is that waves are generated in the IAR due to the ionospheric feedback instability (e.g., Lysak, 1991; Miura & Sato, 1980). In this scenario, the electron precipitation in the presence of a background electric field can modulate the ionospheric conductivity, producing field-aligned currents at the gradients in the conductivity. The waves that are reflected back toward the ionosphere by the Alfvén speed gradients of the IAR can further enhance the conductivity, leading to a positive feedback. This instability tends to evolve to short perpendicular wavelengths, indicating that it can reach the kinetic regime where it can accelerate auroral particles (e.g., Lysak & Song, 2002). This instability can operate for lower-frequency waves associated with field-line resonances (e.g., Lu et al., 2007; Pokhotelov et al., 2002); however, in this case, the wave frequencies are too low to produce a broadband distribution. The feedback instability is also favored by low conductivity, as it is damped by recombination in higher-conductivity regions (e.g., Lysak & Song, 2002; Streltsov & Lotko, 2008). Thus, it might be relevant in the Zone II downward current region at Jupiter, which is characterized by bidirectional, broadband acceleration (Mauk et al., 2020).

However, it should be noted that most of the analysis of the feedback instability has assumed a height-integrated ionosphere. It has been suggested that the feedback instability is stabilized when the vertical structure of the ionosphere is resolved (Sydorenko & Rankin, 2017), although this suggestion has been disputed (Watanabe & Maeyama, 2018). However, the feedback interaction need not be unstable for the interplay between particle precipitation and modification of the ionospheric conductivity to be important. Sounding rocket observations at Earth (Cohen et al., 2013; Lynch et al., 2015) have shown that small-scale field-aligned currents are associated with conductivity gradients, consistent with a feedback interaction being present.

One possible indication that a feedback interaction is operating at Jupiter is the observation of structure in the footprint tails of Io, Europa, and Ganymede. In addition to the main spots associated with the Alfvén wings generated at Io, the auroral tails of these moons exhibit a detailed substructure that often shows an alternating position with respect to footprints of the moons (Mura et al., 2018). Detailed observations of this substructure indicate that these sub-dots corotate with Jupiter's magnetosphere rather than moving along with the motion of the moon (Moirano et al., 2021). In addition, the spacing of the sub-dots appears to be the same at all three moons, independent of their distance from Jupiter. This suggests that the sub-dots are a phenomenon characteristic of Jupiter's ionosphere, with the feedback interaction being a prime candidate.

Finally, it should be recognized that Alfvén waves are not the only mechanism that can produce broadband electron distributions in the auroral zones of Earth and Jupiter. In particular, Juno observations at Jupiter have shown examples in which an inverted-V distribution becomes broadband at the peak of the "V" (Mauk et al., 2018, 2023), suggesting that the broadband distribution was caused by an instability of the electron beam, likely due to whistler mode waves (Elliott et al., 2018). Another possibility is the formation of small-scale bidirectional electric fields, called "time domain structures" by Mozer et al. (2015), who observed such structures by the Van Allen Probes at Earth, where they were associated with bidirectional electron acceleration. Such structures have not been reported from Juno observations, possibly due to the limitations of the electric field observations. However, Elliott et al. (2018) and Sulaiman et al. (2022) have shown that the whistler mode turbulence has a spiky waveform, although it is not clear whether these measurements were parallel electric fields that could directly accelerate particles.

## 8.3. QUASI-STATIC AURORAL ACCELERATION

As noted in the earlier text, the observation of monoenergetic auroral electrons has led to the hypothesis that these electrons have been accelerated in a quasi-static potential drop. This conclusion has been strengthened by observations showing that positively charged ions can be accelerated upward to energies comparable to the energies of the downward accelerated electrons (e.g., Temerin et al., 1981). Models for the quasi-static acceleration can be classified in two categories, double-layer models that describe the localized breakdown of quasi-neutrality producing strong, localized parallel electric fields and global models that describe particle motion along the entire field line. While these models do not directly invoke the propagation of Alfvén waves, we will first briefly review them to point out how they interact with these waves.

The double-layer models were suggested by Block (1972), based on laboratory experiments, to account for the formation of parallel electric fields. The double layer is a localized region of charge separation (i.e., a layer of positive charge and a layer of negative charge) that can be self-consistently supported by the resulting particle distributions. The most basic double-layer models for upward parallel electric fields consist of an electron beam and a trapped population of ions originating in the magnetosphere and an ion beam and trapped electrons from the ionosphere. The existence of these solutions requires that the ion beams have a speed $u_0 > (T_e/m_i)^{1/2}$, the so-called Bohm criterion, and that the electron current through the structure satisfies $j_{\|} > ne(2T_e/m_e)^{1/2}$ (e.g., Goertz, 1979). Observations from the FAST satellite (e.g., Ergun et al., 2004) have provided experimental evidence for this type of structure.

Global models for quasi-static acceleration consider the motion of electrons along the auroral field line in the presence of parallel electric fields and the magnetic mirror force. The total energy of an (nonrelativistic) electron in a mirror magnetic field can be written as

$$W = \frac{1}{2}mv_{\|}^2 + \mu B(z) - e\Phi(z) + mg(z) = \frac{1}{2}mv_{\|}^2 + U(z). \tag{8.5}$$

In this equation, $\mu = mv_{\perp}^2/2B$ is the magnetic moment of the electron, which is conserved, and $z$ denotes the distance along the field line. For electrons, the gravitational potential can often be ignored. $U(z)$ defines an effective potential including the magnetic mirror force as well as the parallel potential $\Phi(z)$. While the magnetic mirror term increases monotonically as the electron proceeds down the field line, an upward parallel electric field corresponds to a potential that increases down the field line, so that the second term in the effective potential decreases, leading to a peak in the effective potential.

Knight (1973) noted that the parallel current consisted of only those electrons that can overcome this peak, leading to a current–voltage relation, later generalized to an anisotropic plasma by Fridman and Lemaire (1980) to read:

$$j_{\parallel} = \frac{B_I}{B_S} N_e \sqrt{\frac{T_{\parallel}}{2\pi m_e}} \left[1 - \frac{\exp(-xe\Phi/T_{\parallel})}{1+x}\right], \tag{8.6}$$

where $x = (T_{\parallel}/T_{\perp})/(B_I/B_S - 1)$, $B_I$ is the magnetic field at the ionosphere and $B_S$, $N_e$, and $T_{\parallel}$ and $T_{\perp}$ are the magnetic field, density, and parallel temperature and perpendicular temperature of the electron distribution at the source point. Remarkably, for typical auroral parameters, this relationship is approximately linear in the regime $1 \ll e\Phi/T_{\parallel} \ll x^{-1}$ which covers much of the relevant range for auroral acceleration. If the parallel electric field is localized to lower altitudes, the source point for this equation is located at the top of the parallel electric field region, since the mirror force will repel electrons with large magnetic moment causing them to mirror before reaching the parallel electric field. This accessibility condition is discussed in detail by Boström (2003), who noted that this was originally considered in the context of electrical discharges by Davisson (1925). Boström (2003) has also derived similar conditions for particle distributions other than bi-Maxwellians. It should also be noted that while the original Knight relation was formulated for steady-state conditions, numerical modeling of Alfvén waves with low frequency in a mirror geometry produces a similar current–voltage relation (e.g., Damiano & Johnson, 2012; Damiano & Wright, 2008).

It should be emphasized that the Knight relation assumes a potential drop and then determines the current that is produced by that potential drop. Thus, it does not provide a complete picture of the formation of these potentials. This relation must be combined with a model showing that the potential is self-consistent with the assumed particle distributions. An early model of this sort was developed by Chiu and Schulz (1978), who used a quasi-neutrality assumption, which was extended by Chiu and Cornwall (1980), who used Poisson's equation to find a self-consistent potential. Motivated by data from the FAST satellite, Ergun et al. (2000) introduced eight different particle populations and solved the Poisson equation using a Vlasov code. Their results indicated that one or two regions of localized parallel electric field could be formed, depending on the parameters assumed in the model.

Similar models have been proposed at Jupiter; however, in the rapidly rotating Jovian magnetosphere a centrifugal potential must be included in Equation (8.5). Ray et al. (2009) introduced such a term and noted that this formed a "choke point" of very low density where the downward gravitational force and the upward centrifugal force were balanced. In their model, this point occurred at about 2.4 $R_J$ on field lines connected to Jupiter's moon Io. The maximum current along the flux tube, according to Equation (8.6) in the case where $T_{\parallel} = T_{\perp}$, is $j_{\parallel} = ne(T_e/2\pi m_e)^{1/2} B_I/B_S$, where now the source point noted in the earlier discussion should be at a rather low altitude, keeping the magnetic field ratio small. They determined this limit to be about 10 μA/m² which corresponded to potentials of up to 100 kV.

While these models are informative, they suffer from two major drawbacks: first, they are steady-state models that show the self-consistent existence of parallel electric fields, but they neglect the question of how these structures evolve and form. In addition, the models are purely electrostatic and neglect the electromagnetic energy input needed to support these structures. These limitations were addressed by Song and Lysak (2001, 2006) who noted that the energy input for the formation of parallel electric fields was due to the propagation of Alfvén waves along the field line. They noted that the formation of parallel electric fields was fundamentally due to the displacement current term in Maxwell's equations:

$$\varepsilon_0 \frac{\partial E_{\parallel}}{\partial t} = \frac{1}{\mu_0}(\nabla \times \mathbf{B})_{\parallel} - J_{\parallel}. \tag{8.7}$$

If an Alfvén wave propagates along the magnetic field line, the curl of **B** in Equation (8.7) needs to be balanced by the field-aligned current. When the Alfvén wave propagates into a region where previously there was no current, the two terms on the right-hand side of this equation are not balanced and an electric field must develop to accelerate the electrons. In a cold plasma, the electron equation of motion becomes

$$\frac{\partial J_{\parallel}}{\partial t} = \frac{ne^2}{m_e} E_{\parallel}. \tag{8.8}$$

The combination of Equations (8.7) and (8.8) indicates that the parallel electric field develops on a timescale comparable to the inverse of the plasma frequency. This could lead to only a transitory electric field; however, if the resulting field-aligned current exceeds the threshold noted in the earlier text for double-layer formation, kinetic effects can set up a quasi-static parallel electric field.

The next question is where along the field line should the parallel electric field form? If we consider a flux tube carrying a constant field-aligned current, the current density $J_{\parallel}$, that is, the current per area, will increase due to the decreasing cross section of the magnetic flux tube, and so will be proportional to the background field strength. However, as noted in the earlier text, the threshold for double-layer formation occurs when the electron drift $v_d = J_{\parallel}/ne > (2T_e/m_e)^{1/2}$. Therefore, this quantity will be maximized at the point along the magnetic field line where the ratio $B/n$ is the largest. This implies that the parallel electric fields will form at low altitudes, where $B$ is large, but in density cavities where $n$ is low. Therefore, the altitude profile of the plasma density is a critical parameter for the formation of parallel electric fields.

It has long been recognized (e.g., Persoon et al., 1988; Strangeway et al., 1998) that auroral acceleration at Earth occurs in density cavities, in which the cold plasma from the ionosphere has been excluded. Furthermore, the auroral zone is also characterized by upgoing ion beams (e.g., Ghielmetti et al., 1978; McFadden et al., 1998) that have been associated with these density depletions (McFadden, Carlson, Ergun, Klumpar et al., 1999). In addition, the perpendicular heating of ions by wave–particle interactions can lead to upflow due to the enhanced magnetic mirror force on these particles (Sharp et al., 1979). The formation of these cavities has been associated with the ponderomotive force of Alfvén waves (Li & Temerin, 1993; Sydorenko et al., 2008), especially in the presence of field-line resonances (Rankin et al., 1995).

A detailed study of the relationship between kinetic Alfvén waves, ion beams, and density depletions based on FAST data was presented by Chaston et al. (2006). They considered a positive feedback scenario in which kinetic Alfvén waves incident on a density gradient can phase-mix to produce smaller perpendicular wavelengths. The small perpendicular wavelengths can accelerate ions perpendicular to the magnetic field (e.g., Chen et al., 2001; Johnson & Cheng, 2001), where they can be accelerated upward by the mirror force, deepening the density cavity. The parallel electric field of the kinetic Alfvén wave leads to an inward Poynting flux, which focuses the waves in the density depletion region and also serves to accelerate electrons downward to produce the aurora.

Juno observations give an indication that the outflow of ions is also associated with density cavities at Jupiter. Mauk et al. (2018, 2020) have shown that upflowing ions are often observed in conjunction with inverted-V electrons, suggesting that these ions are associated with density depletions as in the Chaston et al. (2006) observations at Earth. Juno observations during a crossing of the Io-generated Alfvén wings also show upward accelerated ions occurring over sizes comparable to the cross section of Io (Szalay et al., 2018) that are attributed to acceleration by Alfvén waves.

Density depletions are also associated with the auroral acceleration region at Jupiter. The density in Jupiter's polar regions can be measured by observing plasma-wave resonances and cutoffs (Elliott et al., 2021; Sulaiman et al., 2022). In a strongly magnetized plasma such as that in the low-altitude, high-latitude regions of Jupiter, the plasma frequency can be found either as the upper cutoff of the whistler-mode emissions, the lower cutoff of O-mode emissions, or by the frequency of plasma oscillations, while the lower hybrid frequency, which is the lower cutoff of the whistler mode, is close to the ion plasma frequency and so is also a measure of the density. Using this technique, Elliott et al. (2021) have noted that the density at low altitudes is in the 10–100 $cm^{-3}$ range in the region between the main auroral oval and the footprint of Io, while much lower densities are found in the main auroral region and equatorward of the Io flux tube. Sulaiman et al. (2022) further found that the density decreased rapidly in Zone I, reaching densities as low as 0.1 $cm^{-3}$. These observations have been confirmed by direct measurements of the low-energy plasma by the Jovian Auroral Distributions Experiment (JADE), suggesting that the density is primarily in the JADE energy range (Allegrini et al., 2021). More recently, Sulaiman, Santolik et al., (2023) have extended these measurements into Zone II and the polar cap by using the fact that the ratio of the electric to magnetic field magnitudes sharply increases at the O-mode cutoff, indicating that the densities in the more polar regions can be as low as $10^{-3}$ $cm^{-3}$. Recently, Damiano et al. (2023) have noted that electron energization up to megavolt energies can be produced in such low density cavities.

Therefore, even the quasi-static, monoenergetic electron acceleration at both Earth and Jupiter can be associated with the propagation of Alfvén waves along auroral field lines. These Alfvén waves serve to carry electromagnetic energy in the form of Poynting flux from generator regions in the outer magnetosphere toward the low-altitude region along auroral field lines where the $B/n$ ratio becomes the largest. In these regions, the displacement current must form in order to accelerate the electrons necessary to carry the field-aligned currents. Kinetic Alfvén waves also play a role in decreasing the density along the field lines, which creates favorable conditions for parallel electric fields to form and accelerate electrons to form the aurora.

## 8.4. DISCUSSION: SIMILARITIES AND DIFFERENCES BETWEEN EARTH AND JUPITER

As we have seen from the discussion in the earlier text, many of the same physical processes appear to be acting at both Earth and Jupiter. In both cases, both monoenergetic and broadband electron acceleration are operating. However, at Earth, monoenergetic acceleration appears to be dominant, while broadband acceleration is more common at Jupiter. In both cases, the auroral acceleration regions are mainly defined by the predominant sense of the field-aligned current. At Earth, the two-cell convection produced by the solar-wind interaction leads to a current system that shows a dawn–dusk asymmetry, with the upward current being on the poleward side at dusk and more earthward on the dawn side. At Jupiter, the upward current region is equatorward of the downward current at all local times due to the driving of the currents by the planetary rotation. Furthermore, while fine scale currents of the opposite sense are present at Earth, the field-aligned currents at Jupiter are often filamented (e.g., Chust et al., 2005; Sulaiman, Szalay et al., 2023), and can be taken to be upward or downward only in an average sense.

In both magnetospheres, the predominant auroral acceleration appears to occur close to the planet, rather than in the outer magnetosphere where the currents are generated. In both cases, the propagation of Alfvén waves is essential in carrying electromagnetic energy from the generator to the acceleration region. In both cases, the acceleration appears to occur where the $B/n$ ratio is high, since in such regions the magnetic shear requires more current to flow than the plasma can provide. However, at Jupiter, the magnetic field is much stronger (about 10 Gauss, 1 mT, in the polar ionosphere as opposed to 0.5 Gauss, 50 μT, at Earth). In addition, the densities appear to be much lower at Jupiter as well, with wave observations indicating density as low as $10^{-3}$ $cm^{-3}$, while at Earth, the density cavities have densities in the 0.1–1.0 $cm^{-3}$ range. Thus, it may be the case that the acceleration region is more extended at Jupiter, although this is hard to confirm with the limited number of perijove crossings of Jupiter's auroral zone.

One puzzling aspect of Jupiter's main auroral regions is the relative lack of Alfvén wave measurements at low altitudes, which are more common at Earth. Indeed, the strongest Alfvén-wave amplitudes appear in the diffuse auroral region and particularly on the flux tubes containing the Alfvén wings of Io as well as Europa and Ganymede. Despite the lack of wave activity, bidirectional, broadband electron distributions are seen at these altitudes, especially in the Zone II, primarily downward current region. The Alfvén wave activity may be stronger at higher altitudes along the field lines, which is a plausible source for the downward broadband distributions, but this cannot explain the broadband upward electron distributions in the Zone II region. Indeed, observations of Alfvén waves at higher altitudes have been seen by Juno (Lorch et al., 2022).

Indeed, while the Zone I upward current region at Jupiter has similarities to the upward current regions at Earth, the downward current regions seem to be quite different. At Earth, the downward current region is largely devoid of discrete auroral arcs, and sometimes exhibits the "black aurora," or dark arc-like regions within a band of diffuse aurora (e.g., Marklund et al., 1994, 1997). However, at Jupiter, the aurora in the downward current regions is often just as bright as in the upward current regions. This is due to the bidirectional broadband distributions in this region, which have upward electron fluxes that are stronger than the downward fluxes, leading to the net downward current, but in which the downward electron fluxes are sufficient to produce bright aurora. This would be consistent with a higher-altitude source of the Alfvén waves that accelerate these downward electrons.

One possible clue to the differences in the downward current region is the stronger gravitational pull of Jupiter. At both planets, an ambipolar field exists, since the heavy ions are confined closer to the planet than the electrons, leading to an upward ambipolar field that retards the upward motion of the electrons to maintain charge neutrality. The ambipolar potential, which is basically half the escape potential energy of the ions, is 9.25 eV at Jupiter, while it is about 0.3 eV at Earth. Since the downward current is generally carried by upward ionospheric electrons, they must escape this potential to carry current. Measurements of Jupiter's ionosphere are rather sparse, but modeling indicates that ionospheric temperatures are the order of 0.1 eV (e.g., Yelle & Miller, 2004), so that only a few electrons can escape to carry the current. This may imply that a low-altitude acceleration mechanism may be operative, perhaps from stochastic solitary structures that have been observed as well at Earth.

In summary, the basic physics of auroral acceleration is largely the same at Earth and Jupiter, but the vastly different environments in the two planets give significant differences, particularly in the downward current regions. Continued measurements from the Juno extended mission, which will sample the northern auroral zone and polar cap at lower altitudes, should shed more light on how these processes act at Jupiter. A mission involving multiple spacecraft passing through the auroral zone at Earth would also be beneficial at improving our understanding of these dynamic processes.

## ACKNOWLEDGMENTS

This work is the result of over 40 years of auroral studies and has benefited from the work of many of my colleagues, especially Yan Song, Ali Sulaiman, Fran Bagenal, Chris Chaston, and John Wygant. This work was supported by NASA grant 80NSSC20K1269.

## REFERENCES

Allegrini, F., Kurth, W. S., Elliott, S. S., Saur, J., Livadiotis, G., Nicolaou, G., et al. (2021). Electron partial density and temperature over Jupiter's main auroral emission using Juno observations. *Journal of Geophysical Research: Space Physics*, *126*, e2021JA029426. https://doi.org/10.1029/2021JA029426

Arnoldy, R. L., Moore, T. E., & Cahill, L. J. (1985). Low-altitude field-aligned electrons. *Journal of Geophysical Research*, *90*, 8445.

Bagenal, F., & Delamere, P. A. (2011). Flow of mass and energy in the magnetospheres of Jupiter and Saturn. *Journal of Geophysical Research*, *116*, A05209. doi:10.1029/2010JA016294

Bagenal, F., Adriani, A., Allegrini, F., Bolton, S. J., Bonfond, B., Bunce, E. J., et al. (2017). Magnetospheric science objectives of the Juno mission. *Space Science Reviews*, *213*, 219. doi: 10.1007/s11214-014-0036-8

Block, L. P. (1972). Potential double layers in the ionosphere. *Cosmic Electrodynamics*, *3*, 349.

Bonfond, B., Grodent, D., Gérard, J.-C., Radioti, A., Saur, J., & Jacobsen, S. (2008). UV Io footprint leading spot: A key feature for understanding the UV Io footprint multiplicity? *Geophysical Research Letters*, *35*, 5107.

Bonfond, B., Hess, S., Gérard, J.-C., Grodent, D., Radioti, A., Chantry, V., et al. (2013). Evolution of the Io footprint brightness I: Far-UV observations. *Planetary and Space Science*, *88*, 64.

Borovsky, J. E. (1993). Auroral arc thicknesses as predicted by various theories. *Journal of Geophysical Research*, *98*, 6101.

Boström, R. (2003). Kinetic and space charge control of current flow and voltage drops along magnetic flux tubes: Kinetic effects. *Journal of Geophysical Research*, *108*(A4), 8004. doi: 10.1029/2002JA009295

Chaston, C. C., Bonnell, J. W., Carlson, C. W., Berthomier, M., Peticolas, L. M., Roth, I., et al. (2002). Electron acceleration in the ionospheric Alfven resonator. *Journal of Geophysical Research*, *107*(A11), 1413. doi:10.1029/2002JA009272

Chaston, C. C., Bonnell, J. W., Carlson, C. W., McFadden, J. P., Ergun, R. E., & Strangeway, R. J. (2003). Properties of small-scale Alfvén waves and accelerated electrons from FAST. *Journal of Geophysical Research*, *108*(A4), 8003. doi:10.1029/2002JA009420

Chaston, C. C., Bonnell, J. W., Carlson, C. W., McFadden, J. P., Strangeway, R. J., & Ergun, R. E. (2003). Kinetic effects in the acceleration of auroral electrons in small scale Alfven waves: A FAST case study. *Geophysical Research Letters*, *30*(6), 1289. doi: 10.1029/2002GL015777

Chaston, C. C., Bonnell, J. W., Peticolas, L. M., Carlson, C. W., McFadden, J. P., & Ergun, R. E. (2002). Driven Alfven waves and electron acceleration: A FAST case study. *Geophysical Research Letters*, *29*(11), 10.1029/2001GL013842

Chaston, C. C., Carlson, C. W., Ergun, R. E., & McFadden, J. P. (2000). Alfven waves, density cavities and electron acceleration observed from the FAST spacecraft. *Physica Scripta*, *T84*, 64–68.

Chaston, C. C., Carlson, C. W., McFadden, J. P., Ergun, R. E., & Strangeway, R. J. (2007). How important are dispersive Alfvén waves for auroral particle acceleration? *Geophysical Research Letters*, *34*, L07101. doi:10.1029/2006GL029144

Chaston, C. C., Carlson, C. W., Peria, W. J., Ergun, R. E., & McFadden, J. P. (1999). FAST observations of inertial Alfvén waves in the dayside aurora. *Geophysical Research Letters*, *26*, 647.

Chaston, C. C., Genot, V., Bonnell, J. W., Carlson, C. W., McFadden, J. P., Ergun, R. E., et al. (2006). Ionospheric erosion by Alfvén waves, *Journal of Geophysical Research*, *111*, A03206. doi:10.1029/2005JA011367

Chaston, C. C., Hull, A. J., Bonnell, J. W., Carlson, C. W., Ergun, R. E., Strangeway, R. J., & McFadden, J. P. (2007). Large parallel electric fields, currents, and density cavities in dispersive Alfvén waves above the aurora. *Journal of Geophysical Research*, *112*, A05215. doi: 10.1029/2006JA012007

Chaston, C. C., Peticolas, L. M., Bonnell, J. W., Carlson, C. W., Ergun, R. E., McFadden, J. P. & Strangeway, R. J. (2003). Width and brightness of auroral arcs driven by inertial Alfven waves. *Journal of Geophysical Research*, *108*(A2), 1091. doi:10.1029/2001JA007537

Chen, L., & Hasegawa, A. (1974). Plasma heating by spatial resonance of Alfvén wave. *Physics of Fluids*, *17*, 1399.

Chen, L., Lin, Z., & White, R. (2001). On resonant heating below the cyclotron frequency, *Physics of Plasmas*, *8*, 4713.

Chiu, Y. T., & Cornwall, J. M. (1980). Electrostatic model of a quiet auroral arc. *Journal of Geophysical Research*, *85*, 543.

Chiu, Y. T., & Schulz, M. (1978). Self-consistent particle and parallel electrostatic field distributions in the magnetospheric-ionospheric auroral region. *Journal of Geophysical Research*, *83*, 629.

Chust, T., Roux, A., Kurth, W. S., Gurnett, D. A., Kivelson, M. G., & Khurana, K. K. (2005). Are Io's Alfvén wings filamented? *Galileo observations. Planetary and Space Science*, *53*(4), 395–412. https://doi.org/10.1016/j.pss.2004.09.021

Clarke, J. T., Ajello, J., Ballester, G., Jaffel, L. B., Connerney, J., Gérard, J.-C., et al. (2002). Ultraviolet auroral emissions from the magnetic footprints of Io, Ganymede, and Europa on Jupiter. *Nature*, *415*. 997.

Clarke, J. T., Ballester, G. E., Trauger, J., Evans, R., Connerney, J. E. P., Stapelfeldt, K., et al. (1996). Far-ultraviolet imaging of Jupiter's aurora and the Io "footprint". *Science*, *274*(5286), 404–409. https://doi.org/10.1126/science.274.5286.404

Clemmons, J. H., Boehm, M. H., Paschmann, G. E., & Haerendel, G. (1994). Signatures of energy-time dispersed electron fluxes observed by Freja. *Geophysical Research Letters*, *21*, 1899.

Coffin, D., Damiano, P., Delamere, P., Johnson, J., & Ng, C.-S. (2022). Broadband energization of superthermal electrons in Jupiter's inner magnetosphere. *Journal of Geophysical Research: Space Physics*, *127*, e2022JA030528. https://doi.org/10.1029/2022JA030528.

Cohen, I. J., Lessard, M. R., Kaeppler, S. R., Bounds, S. R, Kletzing, C. A., Streltsov, A. V., et al. (2013). Auroral Current and Electrodynamics Structure (ACES) observations of ionospheric feedback in the Alfvén resonator and model responses. *Journal of Geophysical Research Space Physics*, *118*, 1–9. doi:10.1002/jgra.50348

Connerney, J. E. P., Acuña, M. H., & Ness, N. F. (1981). Modeling the Jovian current sheet and inner magnetosphere. *Journal of Geophysical Research*, *86*, 8370.

Connerney, J. E. P., Adrianni, A., Allegrini, F., Bagenal, F., Bolton, S. J., Bonfond, B., et al. (2017). Jupiter's magnetosphere and aurorae observed by the Juno spacecraft during its first polar orbits. *Science*, *356*, 826.

Connerney, J. E. P., Baron, R., Satch, T., & Owen T. (1993). Images of excited $H_3^+$ at the foot of the Io flux tube in Jupiter's Atmosphere. *Science*, *262*, 1035.

Damiano, P. A., & Johnson, J. R. (2012). Electron acceleration in a geomagnetic field line resonance. *Geophysical Research Letters*, *39*, L02102. doi: 10.1029/2011GL050264

Damiano, P. A., & Wright, A. N. (2008). Electron thermal effects in standing shear Alfvén waves, *Journal of Geophysical Research, 113*, A09219. doi:10.1029/2008JA013087

Damiano, P. A., Delamere, P. A., Kim, E.-H., Johnson, J. R., & Ng, C. S. (2023). Electron energization by inertial Alfvén waves in density depleted flux tubes at Jupiter. *Geophysical Research Letters, 50*, e2022GL102467. https://doi.org/10.1029/2022GL102467

Damiano, P. A., Delamere, P. A., Stauffer, B., Ng, C.-S., & Johnson, J. R. (2019). Kinetic simulations of electron acceleration by dispersive scale Alfén waves in Jupiter's magnetosphere. *Geophysical Research Letters, 46*. https://doi.org/10.1029/2018GL081219

Damiano, P. A., Johnson, J. R., & Chaston, C. C. (2015). Ion temperature effects on magneto- tail Alfvén wave propagation and electron energization. *Journal of Geophysical Research Space Physics, 120*, 5623–5632. doi:10.1002/2015JA021074

Damiano, P. A., Johnson, J. R., & Chaston, C. C. (2016). Ion gyroradius effects on particle trap- ping in kinetic Alfvén waves along auroral field lines. *Journal of Geophysical Research Space Physics, 121*, 10,831–10,844. doi:10.1002/2016JA022566

Davisson, C. (1925). A note on Schottky's method of determining the distribution of velocities among thermionic electrons. *Physical Review, 25*, 808.

Dombeck, J., Cattell, C., Prasad, N., Meeker, E., Hanson, E., & McFadden, J. (2018). Identification of auroral electron precipitation mechanism combinations and their relationships to net downgoing energy and number flux. *Journal of Geophysical Research: Space Physics, 123*. https://doi.org/10.1029/2018JA025749

Elliott, S. S., Gurnett, D. A., Kurth, W. S., Mauk, B. H., Ebert, R. W., Clark, G., et al. (2018). The acceleration of electrons to high energies over the Jovian polar cap via whistler mode wave-particle interactions. *Journal of Geophysical Research: Space Physics, 123*, 7523–7533. https://doi.org/10.1029/2018JA025797

Elliott, S. S., Sulaiman, A. H., Kurth, W., Faden, J., Allegrini, F., Valek, P., et al. (2021). The high-latitude extension of Jupiter's Io torus: Electron densities measured by Juno Waves. *Journal of Geophysical Research: Space Physics, 126*, e2021JA029195. https://doi.org/10.1029/2021JA029195

Ergun, R. E., Carlson, C. W., McFadden, J. P., Mozer, F. S., & Strangeway, R. J. (2000). Parallel electric fields in discrete arcs. *Geophysical Research Letters, 27*, 4053.

Ergun, R. E., Andersson, L., Main, D., Su, Y.-J., Newman, D. L., Goldman, M. V., et al. (2004). Auroral particle acceleration by strong double layers: The upward current region. *Journal of Geophysical Research, 109*, A12220. doi:10.1029/2004JA010545

Ergun, R. E., Su, Y.-J., Andersson, L., Bagenal, F., Delamere, P. A., Lysak, R. L., & Strangeway, R. J. (2006). S bursts and the Jupiter ionospheric Alfvén resonator. *Journal of Geophysical Research, 111*, A06212. doi: 10.1029/2005JA011253

Evans, D. S. (1974). Precipitating electron fluxes formed by a magnetic field-aligned potential difference. *Journal of Geophysical Research, 79*, 2853.

Frank, L. A., & Ackerson, K. L. (1971). Observations of charged particle precipitation into the auroral zone. *Journal of Geophysical Research, 76*, 3612.

Fridman, M., & Lemaire, J. (1980). Relationship between auroral electron fluxes and field aligned electric potential differences. *Journal of Geophysical Research, 85*, 664.

Gershman, D. J., Connerney, J. E. P., Kotsiaros, S., DiBraccio, G. A., Martos, Y. M., Viñas, A., et al. (2019). Alfvénic fluctuations associated with Jupiter's auroral emissions. *Geophysical Research Letters, 46*. https://doi.org/10.1029/2019GL082951

Ghielmetti, A. G., Johnson, R. G., Sharp, R. D., & Shelley, E. G. (1978). The latitudinal, diurnal, and altitudinal distributions of upward flowing energetic ions of ionospheric origin. *Geophysical Research Letters, 5*, 59.

Goertz, C. K. (1979). Double layers and electrostatic shocks in space. *Reviews of Geophysics and Space Physics, 17*, 418.

Goertz, C. K. (1980). Io's interaction with the plasma torus. *Journal of Geophysical Research, 85*, 2949-2956.

Goertz, C. K., & Boswell, R. W. (1979). Magnetosphere-ionosphere coupling. *Journal of Geophysical Research, 84*, 7239.

Hasegawa, A. (1976). Particle acceleration by MHD surface wave and formation of aurora. *Journal of Geophysical Research, 81*, 5083.

Hasegawa, A., & Chen, L. (1975). Kinetic process of plasma heating due to Alfvén wave excitation. *Physical Review Letters, 35*, 370.

Hasegawa A., & Chen, L. (1976). Kinetic processes in plasma heating by resonant mode conversion of Alfvén wave. *Physics of Fluids, 19*, 1924.

Hebden, S. R., Robinson, T. R., Wright, D. M., Yeoman, T., Raita, T. & Bösinger, T. (2005). A quantitative analysis of the diurnal evolution of ionospheric Alfvén resonator magnetic resonance features and calculation of changing IAR parameters. *Annales Geophysicae, 23*, 1711.

Hess, S. L. G., Delamere, P., Dols, V., Bonfond, B., & Swift, D. (2010). Power transmission and particle acceleration along the Io flux tube. *Journal of Geophysical Research, 115*(A6), A06205. https://doi.org/10.1029/2009JA014928.

Hirano, Y., Fukunishi, H., Kataoka, R., Hasunuma, T., Nagatsuma, T., Miyake, W., & Matsuoka, A. (2005). Evidence for the resonator of inertial Alfvén waves in the cusp topside ionosphere. *Journal of Geophysical Research, 110*, A07218. doi: 10.1029/2003JA010329

Iijima, T., & Potemra, T. A. (1976). The amplitude distribution of field-aligned currents at northern high latitudes observed by TRIAD. *Journal of Geophysical Research, 81*, 2165.

Johnson, J. R., & Cheng, C. Z. (1997). Kinetic Alfvén waves and plasma transport at the magnetopause. *Geophysical Research Letters, 24*, 1423.

Johnson, J. R., & Cheng, C. Z. (2001). Stochastic ion heating at the magnetopause due to kinetic Alfvén waves. *Geophysical Research Letters, 28*, 4421–4424. doi:10.1029/2001GL013509

Johnstone, A. D., & Winningham, J. D. (1982). Satellite observations of suprathermal electron bursts. *Journal of Geophysical Research, 87*, 2321.

Keiling, A., Wygant, J. R., Cattell, C., Johnson, M., Temerin, M., Mozer, F. S., et al. (2001). Properties of large electric fields in the plasma sheet at 4-7 $R_E$ measured with Polar. *Journal of Geophysical Research, 106*, 5779.

Keiling, A., Wygant, J. R., Cattell, C., Temerin, M., Mozer, F. S., Kletzing, C. A., et al. (2000). Large Alfvén wave power in the plasma sheet boundary layer during the expansion phase of substorms. *Geophysical Research Letters, 27*, 3169.

Keiling, A., Wygant, J. R., Cattell, C., Peria, W., Parks, G., Temerin, M., et al. (2002). Correlation of Alfvén wave Poynting flux in the plasma sheet at 4–7 $R_E$ with ionospheric electron energy flux. *Journal of Geophysical Research, 107*(A7), 1132. 10.1029/2001JA900140

Kletzing, C. A. (1994). Electron acceleration by kinetic Alfvén waves. *Journal of Geophysical Research, 99*, 11,095.

Kletzing, C. A., & Hu, S. (2001). Alfvén wave generated electron time dispersion. *Geophysical Research Letters, 28*, 693.

Knight, S. (1973). Parallel electric fields. *Planetary and Space Science, 21*, 741.

Knudsen, D. J., Clemmons, J. H., & Wahlund, J.-E. (1998). Correlation between core ion energization, suprathermal electron beams, and broadband ELF plasma waves. *Journal of Geophysical Research, 103*, 4171.

Knudsen, D. J., Donovan, E. F., Cogger, L. L., Jackel, B., & Shaw, W. D. (2001). Width and structure of mesoscale optical auroral arcs. *Geophysical Research Letters, 28*, 705.

Kotsiaros, S., Connerney, J. E. P., Clark, G., Allegrini, F., Gladstone, G. R., Kurth, W., et al. (2019). Birkeland currents in Jupiter's magnetosphere observed by the polar-orbiting Juno spacecraft. *Nature Astronomy, 3*, 904-909. https://doi.org/10.1038/s41550-019-0819-7

Li, X., & Temerin, M. (1993). Ponderomotive effects on ion acceleration in the auroral zone. *Geophysical Research Letters, 20*, 13.

Lorch, C. T. S., Ray, L. C., Wilson, R. J., Bagenal, F., Crary, F., Delamere, P. A., et al. (2022). Evidence of Alfvénic activity in Jupiter's mid-to-high latitude magnetosphere. *Journal of Geophysical Research: Space Physics, 127*, e2021JA029853. https://doi.org/10.1029/2021JA029853

Lu, J. Y., Rankin, R., Marchand, R., Rae, I. J., Wang, W., Solomon, S. C., & Lei, J. (2007). Electrodynamics of magnetosphere-ionosphere coupling and feedback on magnetospheric field line resonances. *Journal of Geophysical Research, 112*, A10219. doi:10.1029/2006JA012195

Lynch, K. A., Arnoldy, R. L., Kintner, P. M., & Vago, J. L. (1994). Electron distribution function behavior during localized transverse ion acceleration events in the topside auroral zone. *Journal of Geophysical Research, 99*, 2227.

Lynch, K. A., Pietrowski, D., Torbert, R. B., Ivchenko, N., Marklund, G., & Primdahl, F. (1999). Multiple-point electron measurements in a nightside auroral arc: Auroral Turbulence II particle observations. *Geophysical Research Letters, 26*, 3361.

Lynch, K. A., Hampton, D. L., Zettergren, M., Bekkeng, T. A., Conde, M., Fernandes, P. A., et al. (2015). MICA sounding rocket observations of conductivity-gradient-generated auroral ionospheric responses: Small-scale structure with large-scale drivers. *Journal of Geophysical Research Space Physics, 120*, 9661–9682. doi:10.1002/2014JA020860

Lysak, R. L. (1991). Feedback instability of the ionospheric resonant cavity. *Journal of Geophysical Research, 96*, 1553.

Lysak, R. L. (1993). Generalized model of the ionospheric Alfven resonator. In R. L. Lysak (Ed.), *Auroral Plasma Dynamics* (p. 121). AGU Monograph 80.

Lysak, R. L. (2008). On the dispersion relation for the kinetic Alfvén wave in an inhomogeneous plasma. *Physics of Plasmas, 15*, 062901.

Lysak, R.L. (2023). Kinetic Alfvén waves and auroral particle acceleration: A review. *Reviews of Modern Plasma Phyiscs 7*, 6. https://doi.org/10.1007/s41614-022-00111-2

Lysak, R. L., & Carlson, C. W. (1981). Effect of microscopic turbulence on magnetosphere-ionosphere coupling. *Geophysical Research Letters, 8*, 269.

Lysak, R. L., & Lotko, W. (1996). On the kinetic dispersion relation for shear Alfvén waves. *Journal of Geophysical Research, 101*, 5085.

Lysak, R. L., & Song, Y. (2002). Energetics of the ionospheric feedback interaction. *Journal of Geophysical Research, 107*(A8). 10.1029/2001JA000308

Lysak, R. L., & Song, Y. (2003a). Kinetic theory of the Alfvén wave acceleration of auroral electrons. *Journal of Geophysical Research, 108*(A4), 8005. doi:10.1029/2002JA009406

Lysak, R. L., & Song, Y. (2003b). Nonlocal kinetic theory of Alfvén waves on dipolar field lines. *Journal of Geophysical Research, 108*(A8), 1327. doi:10.1029/2003JA009859.

Lysak, R. L., & Song, Y. (2005). Non-local interactions between electrons and Alfvén waves on auroral field lines. *Journal of Geophysical Research, 110*, A10S06. doi:10.1029/2004JA010803

Lysak, R. L., & Song, Y. (2020). Field line resonances in Jupiter's magnetosphere. *Geophysical Research Letters, 47*, e2020GL089473. https://doi.org/10.1029/2020GL089473

Lysak, R. L., Song, Y., Elliott, S., Kurth, W., Sulaiman, A. H., & Gershman, D. (2021). The Jovian ionospheric Alfvén resonator and auroral particle acceleration. *Journal of Geophysical Research: Space Physics, 126*, e2021JA029886. https://doi.org/10.1029/2021JA029886

Maggs, J. E., & Davis, T. N. (1968). Measurements of the thickness of auroral structures. *Planetary and Space Science, 16*, 205.

Marklund, G. T., Blomberg, L., Fälthammar, C.-G., & Lindqvist, P.-A. (1994). On intense diverging electric fields associated with black aurora. *Geophysical Research Letters, 21*, 1859.

Marklund, G. T., Karlsson, T., & Clemmons, J. (1997). On low-altitude particle acceleration and intense electric fields and their relationship to black aurora. *Journal of Geophysical Research, 102*, 17,509.

Mauk, B. H., Clark, G., Gladstone, G. R., Kotsiaros, S., Adriani, A., Allegrini, F., et al. (2020). Energetic particles and acceleration regions over Jupiter's polar cap and main aurora: A broad overview. *Journal of Geophysical Research: Space Physics, 125*, e2019JA027699. https://doi.org/10.1029/2019JA027699

Mauk, B. H., Haggerty, D. K., Paranicas, C., Clark, G., Kollman, P., Rymer, A. M., et al. (2017). Discrete and broadband electron acceleration in Jupiter's powerful aurora. *Nature, 549*, 66. doi: 10.1038/nature23648

Mauk, B. H., Haggerty, D. K., Paranicas, C., Clark, G., Kollman, P., Rymer, A. M., et al. (2018). Diverse electron and ion acceleration characteristics observed over Jupiter's main aurora. *Geophysical Research Letters*, *45*, 1277. doi: 10/1002/2017GL076901

Mauk, B. H., Szalay, J. R., Allegrini, F., Bagenal, F., Bolton, S. J., Clark, G., et al. (2023). How bi-modal are Jupiter's main aurora zones? *Journal of Geophysical Research: Space Physics*, *128*, e2022JA031237. https://doi.org/10.1029/2022JA031237

McFadden, J. P., Carlson, C. W., & Boehm, M. H. (1986). Field-aligned electron precipitation at the edge of an arc. *Journal of Geophysical Research*, *91*, 1723.

McFadden, J. P., Carlson, C. W., Boehm, M. H.Hallinan, T. J. (1987). Field-aligned electron flux oscillations that produce flickering aurora. *Journal of Geophysical Research*, *92*, 11,133.

McFadden, J. P., Carlson, C. W., & Boehm, M. H. (1990). Structure of an energetic narrow discrete arc. *Journal of Geophysical Research*, *95*, 6533.

McFadden, J. P., Carlson, C. W., Ergun, R. E., Chaston, C. C., Mozer, F. S., Temerin, M., et al. (1998). Spatial structure and gradients of ion beams observed by FAST. *Geophysical Research Letters*, *25*, 2021.

McFadden, J. P., Carlson, C. W., & Ergun, R. E. (1999). Microstructure of the auroral acceleration region as observed by FAST. *Journal of Geophysical Research*, *104*, 14,453.

McFadden, J. P., Carlson, C. W., Ergun, R. E., Klumpar, D. M., & Moebius, E. (1999). Ion and electron characteristics in aurora density cavities associated with ion beams: No evidence for cold ionospheric plasma. *Journal of Geophysical Research Space Physics*, *104*, 14,671–14,682.

Mende, S. B., Carlson, C. W., Frey, H. U., Peticolas, L. M., & Østgaard, N. (2003). FAST and IMAGE-FUV observations of substorm onset. *Journal of Geophysical Research*, *108*, 1344. doi: 10.1029/2002JA009787

Miles, D. M., Mann, I. R., Pakhotin, I. P., Burchill, J. K., Howarth, A. D., Knudsen, D. J., et al. (2018). Alfvénic dynamics and fine structuring of discrete auroral arcs: Swarm and e-POP observations. *Geophysical Research Letters*, *45*. https://doi.org/10.1002/2017GL076051

Miura, A., & Sato, T. (1980). Numerical simulation of the global formation of auroral arcs. *Journal of Geophysical Research*, *85*, 73.

Moirano, A., Mura, A., Adriani, A., Dols, V., Bonfond, B., Waite, J. H., et al. (2021). Morphology of the auroral tail of Io, Europa, and Ganymede from JIRAM L-band imager. *Journal of Geophysical Research: Space Physics*, *126*, e2021JA029450. https://doi.org/10.1029/2021JA029450

Mozer, F. S., Agapitov, O. V., Artemyev, A., Drake, J. F., Krasnoselskikh, V., Lejosne, S., & Vasko, I. (2015). Time domain structures: What and where they are, what they do, and how they are made. *Geophys. Res. Lett.*, *42*, 3627–3638, doi:10.1002/2015GL063946

Mozer, F. S., Cattell, C. A., Hudson, M. K., Lysak, R. L., Temerin, M., & Torbert, R. B. (1980). Satellite measurements and theories of auroral particle acceleration. *Space Science Reviews*, *27*, 155.

Mura, A., Adriani, A., Connerney, J. E. P., Bolton, S., Altieri, F., Bagenal, F., et al. (2018). Juno observations of spot structures and a split tail in Io-induced aurorae on Jupiter. *Science*, *361*, 774–777. https://doi.org/10.1126/science.aat1450

Neubauer, F. M. (1980). Nonlinear standing Alfvén wave current system at Io: Theory. *Journal of Geophysical Research*, *85*, 1171–1178.

Ng, C.-S., Delamere, P. A., Kaminker, V., & Damiano, P. A. (2018). Radial transport and plasma heating in Jupiter's magnetodisc. *Journal of Geophysical Research: Space Physics*, *123*. https://doi.org/10.1029/2018JA025345

Ng, C. S., Neupane, B. R., Delamere, P. A., & Damiano, P. A. (2022). A turbulent heating model combining diffusion and advection effects for giant planet magnetospheres. *Geophysical Research Letters*, *49*, e2021GL096662. https://doi.org/10.1029/2021GL096662

Pakhotin, I. P., Mann, I. R., Lysak, R. L., Knudsen, D. J., Gjerloev, J. W., Rae, I. J., et al. (2018). Diagnosing the role of Alfvén waves in magnetosphere ionosphere coupling: Swarm observations of large amplitude nonstationary magnetic perturbations during an interval of northward IMF. *Journal of Geophysical Research: Space Physics*, *123*. https://doi.org/10.1002/2017JA024713

Partamies, N., Syrjäsuo, M., Donovan, E., Connors, M., Charrois, D., Knudsen, D., & Kryzanowsky, Z. (2010). Observations of the auroral width spectrum at kilometre-scale size. *Annales Geophysicae*, *28*, 711.

Persoon, A. M., Gurnett, D. A., Peterson, W. K., Waite, J. H., Burch, J. L., & Green, J. L. (1988). Electron density depletions in the nightside auroral zone. *Journal of Geophysical Research*, *93*, 1871.

Pokhotelov, D., Lotko, W., & Streltsov, A. V. (2002). Harmonic structure of field line eigenmodes generated by ionospheric feedback instability. *Journal of Geophysical Research*, *107*(A11), 1363. doi:10.1029/2001JA000134

Polyakov, S. V., & Rapoport, V. O. (1981). Ionospheric Alfvén resonator. *Geomagnetism and Aeronomy*, *21*, 816.

Rankin, R., Frycz, P., Tikhonchuk, V. T., & Samson, J. C. (1995). Ponderomotive saturation of magnetospheric field line resonances. *Geophysical Research Letters*, *22*, 1741.

Ray, L. C., Su, Y.-J., Ergun, R. E., Delamere, P. A., & Bagenal, F. (2009). Current-voltage relation of a centrifugally confined plasma. *Journal of Geophysical Research*, *114*, A04214. doi:10.1029/2008JA013969

Robinson, R. M., Winningham, J. D., Sharber, J. R., Burch, J. L., & Heelis, R. (1989). Plasma and field properties of suprathermal electron bursts. *Journal of Geophysical Research*, *94*, 12,031.

Salveter, A., Saur, J., Clark, G., & Mauk, B. H. (2022). Jovian auroral electron precipitation budget—A statistical analysis of diffuse, mono-energetic, and broadband auroral electron distributions. *Journal of Geophysical Research (Space Physics)*, *127*(8), e30224. doi:10. 1029/2021JA030224

Saur, J., Pouquet, A., & Matthaeus, W. H. (2003). An acceleration mechanism for the generation of the main auroral oval on Jupiter. *Geophysical Research Letters*, *30*, 1260, doi: 10.1029/2002GL015761

Saur, J., Janser, S., Schreiner, A., Clark, G., Mauk, B. H., Kollman, P., et al. (2018). Wave-particle interaction of Alfvén waves in Jupiter's magnetosphere: Auroral and magnetospheric particle acceleration. *Journal of Geophysical Research: Space Physics*, *123*, 9560–9573. https://doi.org/10.1029/2018JA025948

Semeter, J., Vogt, J., Haerendel, G., Lynch, K., & Arnoldy, R. (2001). Persistent quasiperiodic precipitation of suprathermal ambient electrons in decaying auroral arcs. *Journal of Geophysical Research*, *106*(A7), 12,863–12,874.

Sharp, R. D., Johnson, R. G., & Shelley, E. G. (1979). Energetic particle measurements from within ionospheric structures responsible for auroral acceleration processes. *Journal of Geophysical Research*, *84*, 480.

Song, Y., & Lysak, R. L. (2001). The physics in the auroral dynamo regions and auroral particle acceleration. *Physics and Chemistry of the Earth*, *26*, 33.

Song, Y., & Lysak, R. L. (2006). The displacement current and the generation of parallel electric fields. *Physical Review Letters*, *96*, 145002.

Stéfant, R. J. (1970). Alfvén wave damping from finite gyroradius coupling to the ion acoustic mode. *Physics of Fluids*, *13*, 440.

Strangeway, R. J., Kepko, L., Elphic, R. C., Carlson, C. W., Ergun, R. E., McFadden, J. P., et al. (1998). FAST observations of VLF waves in the auroral zone: Evidence of very low plasma densities. *Geophysical Research Letters*, *25*, 2065.

Streltsov, A. V., & Lotko. W. (2008). Coupling between density structures, electromagnetic waves and ionospheric feedback in the auroral zone. *Journal of Geophysical Research*, *113*, A05212. doi: 10.1029/2007JA012594

Su, Y., Jones, S. T., Ergun, R. E., Bagenal, F., Parker, S. E., Delamere, P. A., & Lysak, R. L. (2006). Io-Jupiter interaction: Alfvén wave propagation and ionospheric Alfvén resonator. *Journal of Geophysical Research*, *111*, A06211. doi:10.1029/2005JA011252

Sulaiman, A. H., Hospodarsky, G. B., Elliott, S. S., Kurth, W. S., Gurnett, D. A., Imai, M., et al. (2020). Wave-particle interactions associated with Io's auroral footprint: Evidence of Alfvén, ion cyclotron, and whistler modes. *Geophysical Research Letters*, *47*, e2020GL088432. https://doi.org/10.1029/2020GL088432

Sulaiman, A. H., Mauk, B. H., Szalay, J. R., Allegrini, F., Clark, G., Gladstone, G. R., et al. (2022). Jupiter's low-altitude auroral zones: Fields, particles, plasma waves, and density depletions. *Journal of Geophysical Research: Space Physics*, *127*, e2022JA030334. https://doi.org/10.1029/2022JA030334

Sulaiman, A. H., Santolik, O., Allegrini, F., Kurth, W. S., Lysak, R. L., Elliott, S. E., et al. (2023). Electron densities in Jupiter's polar regions from the identification of Ordinary-mode wave cutoffs, submitted to *Proceedings of the Planetary Radio Emissions IX*.

Sulaiman, A. H., Szalay, J. R., Clark, G., Allegrini, F., Bagenal, F., Brennan, M. J., et al. (2023). Poynting fluxes, field-aligned current densities, and the efficiency of the Io-Jupiter electrodynamic interaction. *Geophysical Research Letters*, *50*, e2023GL103456. https://doi.org/10.1029/2023GL103456

Sydorenko, D., & Rankin, R. (2017). The stabilizing effect of collision-induced velocity shear on the ionospheric feedback instability in Earth's magnetosphere. *Geophysical Research Letters*, *44*, 6534–6542. doi:10.1002/2017GL073415

Sydorenko, D., Rankin, R., & Kabin, K. (2008). Nonlinear effects in the ionospheric Alfvén resonator. *Journal of Geophysical Research*, *113*, A10206, doi:10.1029/2008JA013579

Szalay, J. R., Bagenal, F., Allegrini, F., Bonfond, B., Clark, G., Connerney, J. E. P., et al. (2020). Proton acceleration by Io's Alfvénic interaction. *Journal of Geophysical Research: Space Physics*, *125*, e2019JA027314. https://doi.org/10.1029/2019JA027314

Szalay, J. R., Bonfond, B., Allegrini, F., Bagenal, F., Bolton, S., Clark, G., et al. (2018). In situ observations connected to the Io footprint tail aurora, *Journal of Geophysical Research: Planets*, *123*, 3061. doi: https://doi.org/10.1029/2018JE005752

Temerin, M., Cattell, C., Lysak, R., Hudson, M., Torbert, R. B., Mozer, F. S., et al. (1981). The small scale structure of electrostatic shocks. *Journal of Geophysical Research*, *86*, 11,278.

Temerin, M., McFadden, J., Boehm, M., Carlson, C. W., & Lotko, W. (1986). Production of flickering aurora and field-aligned electron flux by electromagnetic ion cyclotron waves. *Journal of Geophysical Research*, *91*, 5769.

Thompson, B. J., & Lysak, R. L. (1996). Electron acceleration by inertial Alfvén waves. *Journal of Geophysical Research*, *101*, 5359.

Tikhonchuk, V. T., & Rankin, R. (2000). Electron kinetic effects in standing shear Alfvén waves in the dipolar magnetosphere. *Physics of Plasmas*, *7*, 2630.

Trakhtengertz, V. Yu., & Feldstein, A. Ya. (1984). Quiet auroral arcs: Ionospheric effect of magnetospheric convection stratification. *Planetary and Space Science*, *32*, 127.

Vasyliunas, V. M. (1983). Plasma distribution and flow. In A. J. Dessler (Ed.), *Physics of the Jovian Magnetosphere* (p. 395). New York: Cambridge University Press.

Watanabe, T.-H., & Maeyama, S. (2018). Unstable eigenmodes of the feedback instability with collision-induced velocity shear. *Geophysical Research Letters*, *45*, 10,043–10,049. https://doi.org/10.1029/2018GL079715

Watt, C. E. J., & Rankin, R. (2009). Electron trapping in shear Alfvén waves that power the aurora. *Physical Review Letters*, *102*(4), 045002.

Watt, C. E. J., Rankin, R., Rae, I. J., & Wright, D. M. (2005). Self-consistent electron acceleration due to inertial Alfvén wave pulses. *Journal of Geophysical Research*, *110*(A10), A10S07. https://doi.org/10.1029/2004ja010877

Watt, C. E. J., Rankin, R., Rae, I. J., & Wright, D. M. (2006). Inertial Alfvén waves and acceleration of electrons in nonuniform magnetic fields. *Geophysical Research Letters*, *33*, L02106. doi:10.1029/2005GL024779

Wygant, J. R., Keiling, A., Cattell, C. A., Johnson, M., Lysak, R. L., Temerin, M., et al. (2000). Polar comparisons of intense electric fields and Poynting flux hear and within the plasma sheet-lobe boundary to UVI images: An energy source for the aurora. *Journal of Geophysical Research*, *105*, 18,675.

Wygant, J. R., Keiling, A., Cattell, C. A., Lysak, R. L., Temerin, M., Mozer, F. S., et al. (2002). Evidence for kinetic Alfvén waves and parallel electron energization at 4–6 $R_E$ altitudes in the plasma sheet boundary layer. *Journal of Geophysical Research*, *107*(A8), 1201. doi:10.1029/2001JA900113

Yelle, R. V., & Miller, S. (2004). Jupiter's thermosphere and ionosphere. In F. Bagenal, T. Dowling, & W. McKinnon (Eds.), *Jupiter: The Planet, Satellites, and Magnetosphere* (p. 185). Cambridge, UK: Cambridge University Press.

# Part III
# Earth

# 9

# Role of Alfvén Waves in Dynamic Magnetosphere–Ionosphere Coupling: New Perspectives From Satellite and Ground Observations

**Ivan P. Pakhotin[1] and Ian R. Mann[2]**

## ABSTRACT

Magnetosphere–ionosphere coupling (MIC) involves a dynamical transfer of information, momentum, and energy, with the active participation of the ionosphere. Field-aligned currents (FACs) are a major component of MIC. In the magnetohydrodynamic (MHD) paradigm, the propagation of Alfvén waves is required to establish or change FACs. Therefore, Alfvén waves play a key role in MIC processes across a range of temporal and spatial scales. This review explores the influence of Alfvén waves on various aspects of MIC: global FACs, Poynting flux energetics, and phenomena such as the ionospheric Alfvén resonator and the Alfvénic oval. Alfvén wave dynamics also imply that the ionosphere is not a passive load but can play an active role in MIC, for example, through Alfvén wave reflection and interference and Alfvén-wave-driven changes in ionospheric conductance. Interhemispheric asymmetries appear capable of preferentially redirecting incoming energy between hemispheres through Alfvén wave dynamics, affecting global energy transport and deposition into the ionosphere. Small-scale Alfvén waves including electromagnetic ion-cyclotron (EMIC) waves are also discussed within the MIC context. In summary, the characteristics of Alfvén wave generation, reflection, and interference, particularly in the context of an active ionosphere, are empathized and reviewed across a variety of temporal and spatial scales.

## 9.1. INTRODUCTION

The magnetosphere–ionosphere system is characterized by various current systems, including field-aligned currents (FACs) which are associated with the transfer of both energy and momentum between these two regions of near-Earth geospace (e.g., Ganushkina et al., 2015). While it is well known that both magnetic reconnection (e.g., Biskamp, 1996) and direct solar wind impacts and compression (e.g., Crooker et al., 1982) play important roles in the injection of energy into the magnetosphere from the solar wind, how such energy propagates and is distributed and dissipated internally within the magnetosphere is more poorly understood. In particular, the role of Alfvén waves within the dynamical magnetosphere–ionosphere continues to be an active area of research. Certainly, Alfvén waves play a systemic role in magnetosphere–ionosphere coupling (MIC) via their impacts on global FAC systems. This is, not least, because within the paradigm of single-fluid magnetohydrodynamics (MHD), the propagation of Alfvén waves along field lines is required to both establish and to dynamically change the FACs which flow between the magnetosphere and the ionosphere (see, e.g., Lysak, 2004; Song & Lysak, 2001). More fundamentally, as discussed for example by Vasyliūnas (2012), in order to describe and understand its nature, MIC requires considerations of time-dependent processes and the generation and propagation of Alfvén waves. Such wave disturbances further point to plasma flows as the fundamental and causal drivers of the electric fields, and in the case of MIC for the

[1]*Department of Physics and Astronomy, University of Calgary, Calgary, Canada*

[2]*Department of Physics, University of Alberta, Edmonton, Canada*

*Alfvén Waves Across Heliophysics: Progress, Challenges, and Opportunities, Geophysical Monograph 285*, First Edition.
Edited by Andreas Keiling.

DOI: 10.1002/9781394195985.ch09

establishment and maintenance of FACs which couple the magnetosphere and ionosphere (cf. Vasyliūnas, 2012). For this reason, and throughout this review, we implicitly consider that the dynamics of MIC on Alfvén travel times should be considered within the framework of the plasma and Maxwell equations of electrodynamics, rather than for example using models which invoke electric circuit analogies. For a further discussion on the importance of time dependence, and the fundamental role of plasma flows as the causal drivers of the electric fields at the core of magnetospheric dynamics, see also Parker (1996, 2007) and Vasyliūnas (2001, 2012).

In the coupled magnetosphere–ionosphere, dynamical changes can result in the generation and propagation of the Alfvén waves required to change or establish FACs. Moreover, such Alfvén wave propagation can also cause feedback from the ionosphere as a result, as the reflected waves interfere with incident waves in the magnetosphere. Such ionospheric feedback effects can also introduce additional complexity into the spatial and temporal dynamics. For example, changes in ionospheric conductivity due to Alfvén-wave-related energetic particle precipitation (e.g., Lysak, 1991) can affect the reflection coefficient, producing changes to wave dynamics. In extreme cases, existing plasma flow in the ionosphere can also inject wave energy into the magnetosphere, for example, as a result of ionospheric wave over-reflection (e.g., Lysak & Song, 2002a, b). In the larger magnetosphere, Alfvén waves have also been invoked to modulate existing physical processes including potentially the rate of magnetic reconnection on closed field lines at the magnetopause (e.g., Prikryl et al., 2005). Overall, the interference of incident and reflected waves (e.g., Lysak, 1991), and even their potential over-reflection (e.g., Lysak & Song, 2002a, b), can generate conditions whereby the ionosphere effectively has an active role to play in time-dependent MIC.

Clearly, the geospace system is not steadily driven, resulting in changes to the rates of energy input and the strength of MIC. Such dynamics naturally result in the generation of Alfvén waves, which both carry information about changes in the system as well as fundamentally transport energy and momentum through the system on Alfvén wave propagation timescales. For example, Wright (1996) presents an elegant description of the response of the coupled magnetosphere–ionosphere system as a result of open field-line convection in terms of Alfvén wave propagation. In one sense, the Alfvén waves are a "tip of the spear" for energy and information transfer between the two domains of geospace. However, the Alfvén waves themselves play a key role in not only establishing but also controlling the dynamics of MIC related to global FAC systems on Alfvénic timescales. For example, while Alfvén waves are observed to be embedded as small-scale filaments within larger-scale more quasi-static current systems (e.g., Pakhotin et al., 2018), they also act to serve as the agents which control the dynamics of the larger-scale FACs as well (see for example, the discussion in Pakhotin et al., 2020). Indeed, and as discussed by Vasyliūnas (2012) and others, the propagation of Alfvén waves is required in order to properly describe the magnetosphere and ionosphere response to stress imposed on the system by solar wind driving. In our view, and contrary to some assertions in the literature based for example on the fields observed in low-Earth orbit (LEO) satellite magnetic field measurements (e.g., He et al., 2012), the contribution of Alfvén waves themselves to the overall MIC energy budget also cannot be neglected. As discussed in Pakhotin et al. (2021), such Alfvén wave interactions may provide an explanation for the recently reported north–south interhemispheric asymmetry in energy input to the ionosphere. Verkhoglyadova et al. (2018) also estimated that Alfvén waves may be responsible for up to 30% of energy input in the coupled magnetosphere–ionosphere–thermosphere (MIT) system. As a result, it seems clear that understanding Alfvén wave dynamics is crucial for accurate dynamical modeling of the geospace response to variable solar wind forcing.

Ionospheric structure and dynamics can also be influenced by incident Alfvén waves. In particular, ionospheric Joule heating, a significant fraction of which is accounted for by small-scale and mesoscale Alfvén waves, can also impact the morphology of the ionosphere, while energetic particle precipitation linked to Alfvén waves can also change the conductivity. As such, magnetosphere–ionosphere coupling processes, through Alfvén waves, can impact space weather dynamics across the entire MIT domain.

Although the key role of Alfvén waves features in multiple geomagnetic energy propagation models (e.g., Lysak, 1986, 1991; Wright, 1996), a lot of past scientific literature in MIT has nonetheless tended to neglect the role of Alfvén waves, focusing rather on the more quasi-static study of larger-scale current systems as the key elements involved in MIC. A lot of this early focus on larger-scale MIC processes arose historically out of instrumental limitations, or from approaches which emphasized equilibrium relationships (cf. the exposition in Vasyliūnas, 2012). The first instruments to observe global FACs, for example, the magnetometers on the single Triad satellite which led to the first ever characterization of global FACs (Iijima & Potemra, 1976, 1978), did not have the capability to resolve spatiotemporal features or to identify smaller-scale transient Alfvén wave features. On a single satellite equipped with only magnetometers, it is in fact impossible to separate spatial disturbances from temporal disturbances, as these changes cannot be isolated from observations in the frame of the motion of the satellite.

As a result, FACs were estimated based on Ampère's law using the assumed static variations of magnetic perturbations perpendicular to the background magnetic field (see, e.g., also the discussion in Forsyth et al., 2017 and references therein). As such, in the study of MIC, focus was placed on large-scale equilibrium relationships and features which were persistent when observed by multiple satellites. As discussed in the following text, these turn out to be the large-scale and more steady-state FAC systems. In early analyses, these persistent features were superposed and characterized by latitude, magnetic local time (MLT), and geomagnetic activity, to establish their fundamental topology. It was discovered that, during southward interplanetary magnetic field (IMF) conditions, FACs organize into concentric circles around the north and south magnetic poles; in particular two rings were observed. The poleward ring was named Region 1, while the equatorward ring was labeled Region 2. Further, the cusp Region 0 system and the irregular northward Bz (NBZ) current system were also identified.

It wasn't until the advent of multi-satellite observations, and new spacecraft such as DE-2, Freja, and the Fast Auroral SnapshoT Explorer (FAST) missions which carried both magnetic and electric field instruments, that smaller-scale structures were resolved, and the role of Alfvén waves became more apparent. Knudsen et al. (1992), Ishii et al. (1992), and Lühr et al. (1994) were all able to identify Alfvén waves by utilizing combined E- and B-field data, as well as spacecraft–ground conjunctions. One of the first multi-satellite observations of intermediate-scale FACs by Ohtani et al. (1996) involved opportunistic spacecraft conjunctions. These authors used Viking and DMSP-F7 data to observe the same region of geospace and found both similarities and differences in the FAC topology inferred during the two traversals. Finally, it was possible to directly observe some aspects of the time evolution of nonstationary structures embedded within the quasi-stationary large-scale FACs. Fine-scale structure within FACs had been observed before on Triad (Iijima & Potemra, 1976; Figure 9.1(b)), but it was impossible to tell with a single satellite whether it was time-varying or not. Now, with two spacecraft observing the same region of space approximately 10 min apart, Ohtani et al. (1996) were able to infer that this was time-varying, that is, it was most likely related to a wave phenomenon.

With the advent of multi-satellite missions such as ST-5 (Slavin et al., 2008) and later Swarm (Friis-Christensen et al., 2006), it became possible to obtain systematic statistics which could be used to determine characteristic temporal and spatial scale sizes at which quasi-stationarity was violated and where temporally varying and wave-related effects took over. In a pioneering work, Gjerloev et al. (2011) demonstrated using ST-5 data that spatial and temporal scale sizes were interrelated and that magnetic disturbances on spacecraft along-track scales above approximately 200 km (corresponding to Iijima & Potemra (1978) scales for R1 and R2 field-aligned currents) were well correlated for inter-spacecraft temporal separations on the order of a minute. Meanwhile, inter-satellite correlations for smaller scale sizes dropped off sharply, indicative of non-stationarity and temporal variation which was resolvable on those smaller inter-satellite scales. These results were further confirmed and refined by Lühr et al. (2015) using Swarm data, where scale sizes of >150 km were observed to be quasi-stationary.

It was these considerations that motivated efforts to split auroral-zone electromagnetic transverse-mode disturbances into two classes: "quasi-static" and "Alfvénic." The rationale was to characterize the large-scale FAC system by targeting persistent, repeatable phenomena and largely ignoring smaller-scale disturbances, assuming that any currents carried by the smaller scales would have a smaller impact, since they could be integrated and folded into the larger-scale phenomena as a small correction. The attenuation of small scales was adopted, typically by the use of low pass filtering at 20 s, into the dual-satellite Swarm FAC density data product (Ritter et al., 2013). Multiple empirical FAC models such as the Model of FACs through Empirical Orthogonal Function analysis, or MFACE (He et al., 2012; Prokhorov et al., 2014) utilized a low pass filter at 220 km spatial scales for the same reason. The concept was also utilized in the Active Magnetosphere and Planetary Electrodynamics Response Experiment (AMPERE) dataset (Anderson et al., 2000; Waters et al., 2001). In these characterizations, small-scale Alfvén waves in global magnetosphere–ionosphere coupling were generally considered to be noise, to be largely ignored and filtered out in assessments of the larger-scale FAC systems. However, as discussed by Forsyth et al. (2017) and as we describe in detail in the following text, the stationarity required in order for such filtering of magnetic perturbations to be able to accurately reproduce FACs is often not present in the physical system especially during dynamic times.

For example, Figure 9.1 shows an example of components of the electric ($E_x$) and magnetic fields ($B_y$) (panels (a) and (b)) during an auroral-zone crossing as observed by the Swarm-A satellite during a period of southward IMF and which provides evidence supporting the hypothesis that the smaller-scale magnetic perturbations cannot be neglected within dynamical MIC, since as we discuss in the following text they have fundamental impacts not only for estimating FAC magnitudes but also for the overall energy budget. Here, electric field data from the Canadian Electric Field Instrument (C-EFI; Knudsen et al., 2017) from the Thermal Ion Imager (TII) instrument

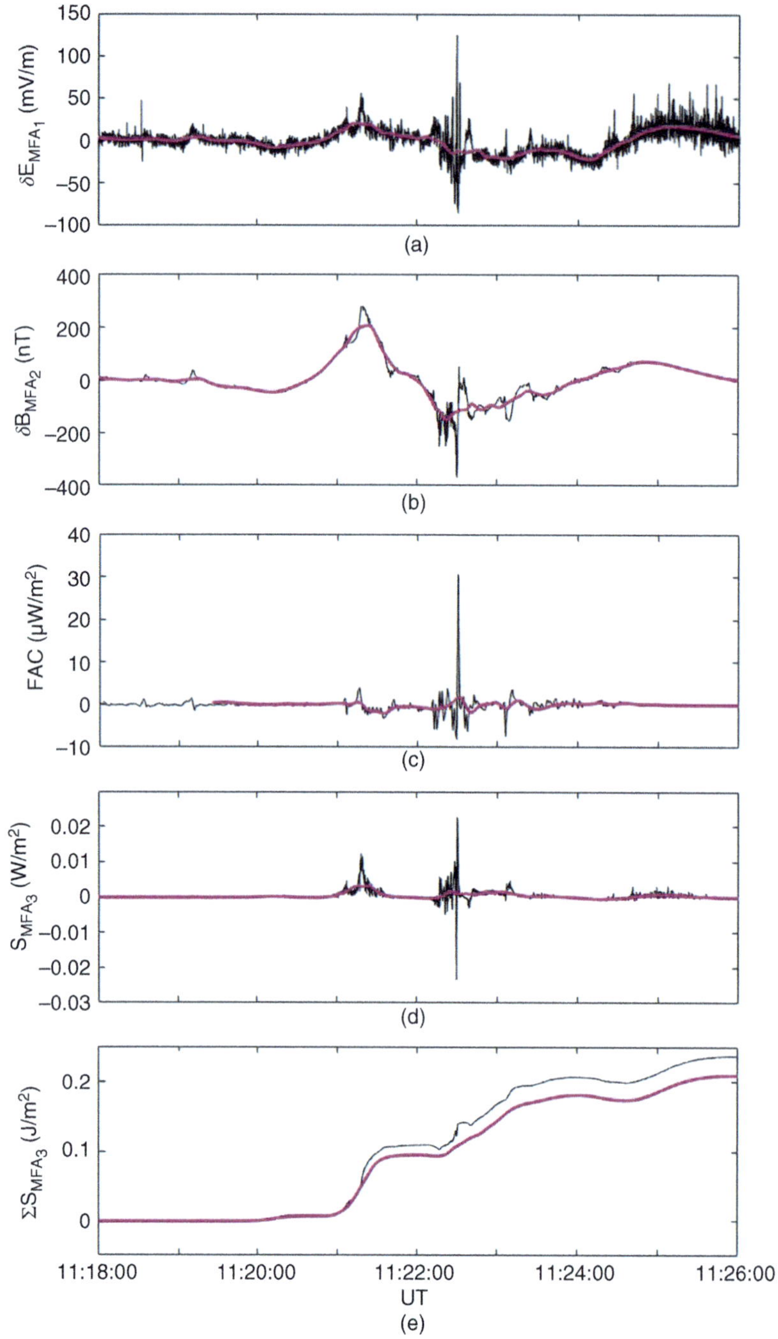


**Figure 9.1** Electric and magnetic fields, and related derived field-aligned currents (FACs) and Poynting flux, during typical auroral crossings as observed in low-Earth orbit by Swarm satellites during a period of southward IMF. The data are from Swarm-A in mean field-aligned coordinates, from 1118 to 1126 UT on 16 May 2016, and are shown at two temporal resolutions, at 50 sps and 16 sps for the magnetic and electric fields, respectively, in the raw unfiltered data (black), and the same time series after the application of a 20-s low-pass moving average filter (pink). Panel (a) shows the electric field $E_x$ component, while panel (b) shows the magnetic field $B_y$ component. Panel (c) shows the single-spacecraft (black) and dual-spacecraft (pink) FAC estimates as defined in Adapted from Ritter et al. (2013), and panel (d) shows the field-aligned Poynting flux calculated from the unfiltered and filtered time series in panels (a) and (b). Panel (e) shows the integral of the field-aligned Poynting flux from panel (d) with time. For panels (d) and (e), positive is downward. Panel (e) in particular shows the contribution of higher-frequency disturbances to the field-aligned Poynting flux when derived using data from the observing frame of the orbit of the Swarm-A satellite.

suite and magnetic field data from the vector fluxgate magnetometer (VFM; Olsen et al., 2013) are available for this crossing at raw resolutions of 16 sps (samples per second) and 50 sps, respectively (shown in black in panels (a) and (b)). A 20-s running mean filter is also applied to the data to create a lower resolution dataset (shown in pink). All data have been rotated into a mean field-aligned coordinate system where the $x$ coordinate points in the direction of the magnetic north pole perpendicular to the background magnetic field, the $z$ axis follows the mean field direction, and the $y$ axis completes the triad pointing geomagnetic east.

The auroral-zone crossing in Figure 9.1 shows the prevalence of sub-minute timescale electric field perturbations in the along-track direction around 11:22:30, with coherent signal rising above the noise in the TII measurement (the noise seen around 11:25:00). Interestingly, even on this sub-minute timescale, similar short-duration and large-amplitude magnetic field perturbations are also observed (panel (b); black line) on top of the longer timescale magnetic variations associated with larger-scale and more quasi-static Region 1 and Region 2 FAC systems. Indeed, there is also a clear association of contemporaneity and correlation between these large-amplitude magnetic- and electric-field fluctuations on timescales shorter than 20 s. Interestingly, the period with the largest high-frequency electromagnetic fluctuation amplitudes also seems to be concentrated in between the two regions of large-scale upward and downward FACs (around 11:22:20 UT) as characterized by the inflection point in the $B_y$ magnetic field (pink line) in panel (b). Panel (c) also shows that the large-amplitude short-timescale variations are likely associated with FAC perturbations on these timescales as well (black trace). Since these disturbances are also associated with similar timescale fluctuations in field-aligned Poynting flux (panel (d)), we assume that they represent large-amplitude Alfvén waves which are driven as a result of time-varying magnetosphere–ionosphere coupling. Similar small-scale electromagnetic fields and inferred FACs are also observed at the magnetic-field inflection at around 11:21:15 UT (panel (c)), again additionally associated with small-scale perturbations in field-aligned Poynting flux consistent with Alfvén waves (panel (d)).

When the field-aligned Poynting flux is integrated from the start of this interval, as shown in panel (e), the smaller-scale perturbations are revealed to have a clear influence on the overall net Poynting flux into the ionosphere when examined using the fields observed by the Swarm-A satellite along its world line as it travels along its orbit through the auroral zone. While the Poynting flux is oscillating, the integrated Poynting flux is downward as expected. The signatures of such small-scale electrodynamics can be seen in the 1-sps FAC data product (panel (c)); the 20-s filtered FAC product only registers the large-scale FACs and misses the small-scale perturbations at the FAC boundaries, and which are associated with the largest local Poynting flux. FAC products such as the dual-spacecraft Swarm FAC product, which filter at 20 s (cf. Ritter et al., 2013; and the Swarm dual-spacecraft FAC product), are clearly missing important electrodynamics associated with local finer-scale spatiotemporal structures. See the discussion in Pakhotin et al. (2020) for more details.

This review will present an overview of the role of Alfvén waves in MIC, focusing in part on recent results derived using data from the Swarm mission. This includes a discussion on the challenges of analyzing FACs and Alfvén waves invoked in MIC, with a specific exposition of the differing merits of the various approaches taken by the community with regard to diagnosing FAC density structure especially using magnetic-field measurements from low-Earth orbit satellites. This is outlined in Section 9.2. This is followed in Section 9.3 by a discussion of the effects of Alfvén waves in global MIC, including Poynting flux energetics, effects on conductance, connections to Alfvénic aurora, and in the context of an active ionosphere. In Section 9.4, the conclusions from the two preceding sections are tied together, by discussing the geophysical implications of Alfvén-wave-driven phenomena within MIT energy transport, with a focus on the key role of Alfvén waves both as initiators and as carriers of energy, momentum, and information, and the errors which may arise from attempts to exclude them from the overall MIC energy budget. Finally, Section 9.5 discusses even smaller-scale Alfvén waves, such as those related to the electromagnetic ion-cyclotron (EMIC) wave instability. In particular, we highlight recent results invoking mode conversion that introduces cross-L-shell wave propagation in the magnetosphere, in addition to the cross-L transport due to wave trapping in the ionospheric duct. Such wave propagation in the magnetosphere–ionosphere system has implications for wave transport from the near-equatorial magnetosphere domain to the ionosphere, and has particular significance from the perspective of the role of such EMIC waves in radiation-belt dynamics. Finally, a summary and concluding remarks close out the paper in Section 9.6.

## 9.2. ROLE OF ALFVÉN WAVES IN FIELD-ALIGNED CURRENT DYNAMICS

### 9.2.1. Quasi-Static Large-Scale Field-Aligned Current Topology

FACs, also known as Birkeland currents, were discovered from ground observations (Birkeland, 1908, 1913) and first mapped from space using a magnetometer onboard the Triad satellite by Iijima and Potemra (1976). FACs flow along magnetic field lines, connecting the magnetosphere with the ionosphere, and typically form two pairs of concentric rings around the north and south polar regions, with each ring typically spanning approximately 200 km in length in the north–south direction. In each hemisphere, the poleward ring, associated with convection in the outer magnetosphere, was named Region 1 or R1, while the equatorward ring, associated with the inner magnetosphere and ring current region, was labeled Region 2 or R2. Figure 9.2, reproduced from Iijima and Potemra (1978), demonstrates the large-scale FAC system. In the morning sector (approximately 2300–1100 MLT), field-aligned currents flow downward in Region 1 and upward in Region 2, while in the evening sector (approximately 1300–2300 MLT), current flow is directed upward in Region 1 and downward in Region 2. In the midnight sector (approximately 2100–0100 MLT), these regions overlap, creating the Harang discontinuity (e.g., Erickson et al., 1991), resulting also in a more complex FAC pattern which can also vary during periods of substorm activity (cf. right panel of Figure 9.2).

Interestingly, during substorms, an additional substorm current wedge (SCW) current system (e.g., Kepko et al., 2015; McPherron et al., 1973) is observed, and which is usually assumed to represent a new current system constituted by a pair of downward and upward FACs, at its eastern and western edges, respectively, connected by a westward closure current which flows in the ionosphere. Following substorm expansion phase onset, the upward FAC, associated with downgoing electrons, expands poleward and westward, being associated with auroral emissions in the westward traveling surge which is released following onset. However, the origin and cause of this SCW current system remains controversial. Recent work by Gjerloev and Hoffman (2014) argues that the usual representation of the substorm current wedge might be alternatively explained in terms of a two-wedge current system, which is indistinguishable from the point of view of ground-based equivalent currents from the prior models which assume that the SCW is a distinct current system which is created for example as a result of magnetotail dynamics associated with Earthward flows and substorm magnetic dipolarizations (see, e.g., McPherron et al., 1973). Recent observations have also used the AMPERE dataset (Anderson et al., 2000) to establish that there can be a reduction in the field-aligned current close to the onset epicenter (e.g., Murphy et al., 2012), suggesting a potentially important role for MIC in the substorm onset process. Similar work, again using AMPERE, has further demonstrated that the current wedge comprises a number of smaller-scale filamentary current systems (Murphy et al., 2013). An additional current system called the cusp current system or Region 0 (R0) is occasionally observed, mapping to the dayside cusp (e.g. Ohtani et al., 1995).

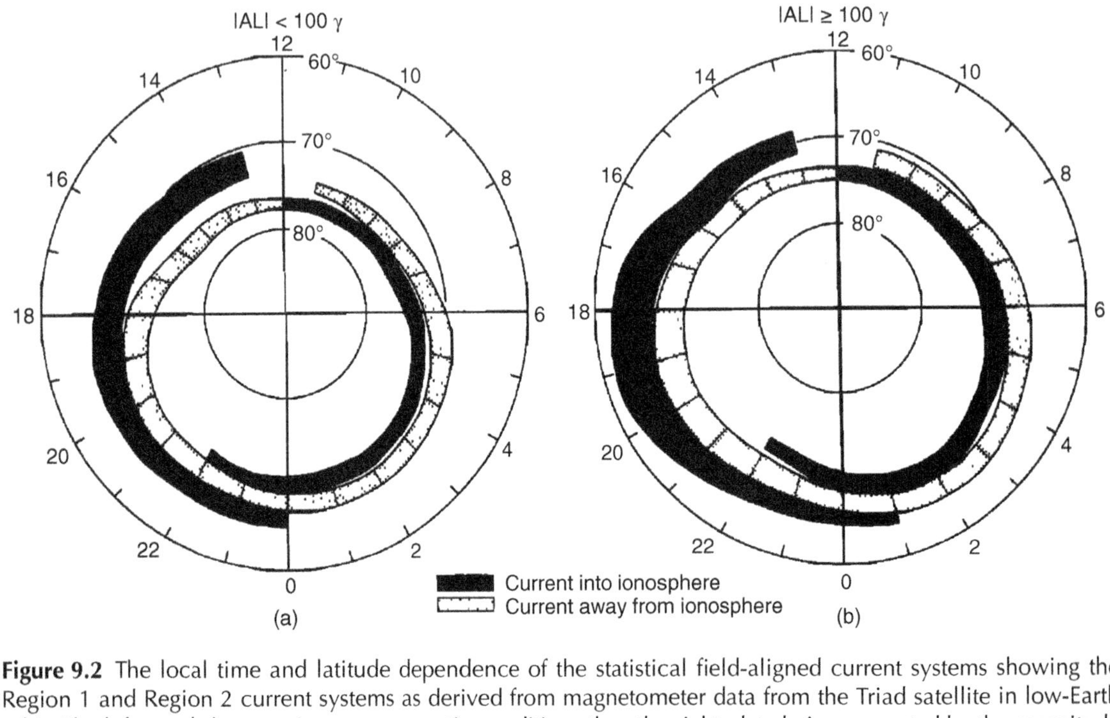


**Figure 9.2** The local time and latitude dependence of the statistical field-aligned current systems showing the Region 1 and Region 2 current systems as derived from magnetometer data from the Triad satellite in low-Earth orbit. The left panel shows quieter geomagnetic conditions than the right, data being separated by the magnitude of the auroral-electrojet lower (AL) index. Source: Iijima & Potemra (1978) / John Wiley & Sons.

During quiescent times of extended periods of northward IMF conditions, meanwhile, the rings of R1/R2 FAC structure move poleward and become more irregular and much less distinct (e.g., Iijima & Potemra, 1978). During substorms, the R1/R2 topology persists but undergoes changes as detailed in Iijima and Potemra (1978); importantly, the average latitude width increases and the centers of the R1 and R2 regions shift equatorward by approximately 2°–3° with respect to quiet-time values. In addition, the Region 1 current density is statistically larger than current density in Region 2 except during active periods and in the midnight-to-morning time sector, where the westward electrojet is most active. Region 1 and Region 2 currents also respond differently to substorms (e.g., Timoçin, 2022), suggesting that the two systems are controlled by different source regions in the magnetosphere. Region 1 FACs map to the outermost part of the magnetosphere, and are believed to be connected to magnetospheric convection, while Region 2 FACs map to regions of the plasma sheet closer to Earth (e.g., Ganushkina et al., 2015).

### 9.2.2. Methods for Inferring Quasi-Static FACs

Typically, single-spacecraft methods for inferring FAC density from low-Earth orbit have utilized the so-called curlometer method (e.g., Lühr et al., 1994), where an FAC is assumed to be a flat infinite current sheet at right angles to the orbit of the approaching spacecraft. In this case, Ampère's circuital law may be used to infer current as proportional to the curl of the magnetic field where the spatial variation is inferred from an assumed static magnetic field as measured along the track of the spacecraft:

$$\mu_0 \mathbf{j} = \nabla \times \mathbf{B}, \tag{9.1}$$

where $\mathbf{j}$ is the current density, $\nabla \times \mathbf{B}$ is the curl of the magnetic field which is caused by that current, and $\mu_0$ is the permeability of free space. For FACs, the curl of the perpendicular fields is calculated, and in the simplest case the curl of **B** is taken along the satellite track. In polar orbits, often only the cross-track fields are used. In the single-fluid MHD approximation, and for variations on Alfvén timescales, the effects of the displacement current are usually neglected. However, in order to assess a value of **j**, the current is assumed to be constant on the timescale of the FAC crossing. If **j** changes on the timescale of the satellite pass, for example, as a result of the presence of Alfvén waves, then there will be mixing of spatial and temporal magnetic field variations, which violate the stationarity assumption required. For this reason, the magnetic fields are typically low-pass-filtered, leaving only larger-scale B-field disturbances which are also assumed to be sufficiently slowly varying to be quasi-static on the timescale of the pass. Another common assumption is that the FACs are vertical, and that the current comprises a one-dimensional current sheet with a normal parallel to the orbit, that is, the satellite flies through the FAC at a 90° angle of attack. For oblique angles, solutions are possible if multi-satellite measurements are available with small along- and cross-track separation, using data from the A and C satellites in the Swarm constellation and which fly alongside each other; this is discussed as an improvement to the original single-spacecraft curlometer technique by Lühr et al. (1996).

As discussed further by Forsyth et al. (2017), FACs can have configurations which deviate from the idealized flat and one-dimensional infinite plane assumption. These authors utilized both correlation and residuals across the separation of the Swarm spacecraft as a means to quantify the degree to which these assumptions hold as a "quality check" on the validity of the calculated FAC density when derived using data from LEO such as from the Swarm constellation. In agreement with the earlier statistical results (e.g., Gjerloev et al., 2011; Lühr et al., 2015), Forsyth et al. (2017) found that small-scale FACs violate the curlometer assumptions more frequently, as both fine-structure and Alfvén wave time-varying effects become more pronounced. At discrete arc scales, simultaneous observations using magnetic-field data and discrete auroral images from the Enhanced-Polar Outflow Probe (e-POP) satellite, and magnetic fields from the Swarm satellites, confirm that both spatial and temporal assumptions can be violated (Miles et al., 2018), with the single-satellite FAC estimate being shown for this case of an oblique angle of attack to become essentially useless.

Other means of estimating FAC density involve multi-satellite approaches to construct a trapezium and to find the FAC density through this trapezium. This is the dual-spacecraft method, as applied to Swarm data by Ritter et al. (2013). This approach removes some of the assumptions inherent in the single-satellite method; for example, the algorithm can now handle irregular FAC geometry and oblique traversals. The condition of time stationarity is however still required, such that in the Ritter et al. (2013) approach the authors explicitly apply a >150-km spatial-scale low-pass filter. Similarly, the AMPERE constellation (Anderson et al., 2000; Waters et al., 2001) utilizes a constellation of Iridium satellites with magnetometers at a low resolution, with orbital tracks which have wide cross- and along-track separations. The data are then interpolated onto a spherical harmonic grid, where the curl yields a model for the large-scale field-aligned currents. A comprehensive analysis and a comparison of various methodologies for calculating FAC density may be found in Trenchi et al.

(2020), which detail the results of the European Space Agency (ESA) Field-Aligned Currents–Methodology Inter-Comparison Exercise (FAC–MICE) experiment. Note also that with larger numbers of spacecraft in orbits which are not confined to low-Earth orbit, it is possible to generalize the curlometer technique and apply it in three dimensions (see, e.g., Dunlop et al., 2015).

### 9.2.3. Spatiotemporal Dynamics in FAC Regions

The influence of time-varying electromagnetic disturbances has often been recognized as a key element of the processes which control MIC. As detailed and modeled in works such as Lysak (1986, 1991), any changes in field-aligned currents are carried by shear Alfvén waves, which propagate from the magnetosphere to the ionosphere (or vice versa). At the ionosphere, specifically at the ionospheric E-layer, a strong localized density gradient causes the Alfvén wave to reflect. This reflected wave can interfere with incident wave, setting up an interference pattern. In situations where there is a second reflection from an additional boundary, the interference of waves can create a standing wave. For example, this can occur as a result of the reflection of Alfvén waves from the ionospheric end point in conjugate hemispheres, producing a so-called magnetospheric Alfvén resonator (MAR). At smaller spatial scales, as a reflected wave travels back up the field line into the magnetosphere, it can encounter a region of strong Alfvén speed gradient at altitudes of approximately 1–2 Re, which causes it to partially reflect again back toward the ionosphere. The presence of these two reflecting surfaces (the ionospheric E-layer at the bottom and the Alfvén speed gradient at the top) forms the ionospheric Alfvén resonator (IAR; e.g. Belyaev et al., 1990), which preferentially traps waves with frequencies determined by the Alfvén speed and the length of the resonator cavity. This is a special case of the more general scenario of Alfvén wave incidence and reflection from conjugate E-layers in opposite hemispheres in the MAR.

As the electromagnetic disturbances associated with MIC occur on a variety of spatial and temporal scales, it is important to establish how the properties of Alfvén-wave-related MIC vary at different scales. In our view, many of the properties of MIC can be explained within a single-Alfvén-wave framework but which is sampled by different satellites and/or satellite constellations at different spatial and temporal scales and/or separations. For example, satellites in LEO sample the fields along their world line, such that apparent quasi-stationarity can simply occur as the result of temporal changes which occur on timescales of the order of the auroral-zone crossings. Nonetheless, even on these longer timescales, Alfvénic processes can still be crucial not least because Alfvén wave propagation is required in order for FACs to change within the MIC system. On shorter timescales, the non-stationarity becomes more obvious in LEO satellite data, and as discussed in the earlier text can still comprise an energetically significant component of MIC. With the above mentioned paradigm, it is possible to envision MIC being manifested and controlled by the propagation, reflection, and interference of a multi-scale spectrum of electromagnetic energy in the shear Alfvén wave mode, of which quasi-static FACs are the longest-timescale, largest-spatial-scale component. As such, they are going to be observed to be quasi-stationary on timescales corresponding to appropriately large multi-satellite along-track separations. Thus, rather than treating Alfvén waves and FACs as separate phenomena, it is phenomenologically possible, and physically defensible, to characterize them as different scale elements of the same phenomena: a multi-scale continuum of shear Alfvén wave energy which is dynamically exchanged between the magnetosphere and the ionosphere, and whose properties are determined by the propagation, reflection, and interference of Alfvén waves at different scales.

This does not mean that Alfvén-wave-related MIC at different scales will have identical physical properties and behavior. As detailed for example by Lysak (1991), in particular for example in Figure 11 of that paper, the observed characteristics of MIC resulting from the interference of Alfvén waves are determined by factors such as the frequency, wavelength, scale height, and the distance between the observing platform and the ionospheric E-layer. In particular, the interference patterns create characteristic frequency-dependent ratios of both the local amplitude and the phase between the orthogonal components of the perpendicular electric and magnetic fields which make up the transverse Alfvén waves. A LEO satellite observation platform such as one of the Swarm satellites (Friis-Christensen et al., 2006) is approximately 300–400 km away from the E-layer, and the full spectrum of Alfvén wave behavior may be observed, as detailed for example by Pakhotin et al. (2020). The largest-scale "Alfvén waves," corresponding to scales consistent with R1/R2 FACs, have such large wavelengths that the distance from the satellite position to the reflecting E-layer is a small fraction of the wavelength. As a result, there is little phase advance or amplitude variation over this distance. As such, the boundary condition at the ionospheric reflection will strongly control the behavior of the dynamics observed in LEO for such long-period and long field-aligned wavelength waves. As a result, and under these conditions, the satellite will see an E/B ratio controlled by

$$\frac{E}{B} = \frac{1}{\mu_0 \sum_P}, \tag{9.2}$$

where $\sum_P$ is the height-integrated Pedersen conductance of the ionosphere and $\mu_0$ is the permeability of free space. Since the orthogonal E- and B-field disturbances are in phase at the location of the ionospheric reflection, the E and B fields of these long-period disturbances will also be observed to be in phase in LEO. Since static FACs obey the same ionospheric boundary condition, under these conditions, the wave is indistinguishable from a static FAC. These considerations have been used to differentiate FACs from more classic time-varying Alfvén waves for example by Ishii et al. (1992) who established using Dynamics Explorer 2 (DE-2) data a frequency/period dependence to the E/B ratio which transitioned from one controlled by Equation (2) at large scales/long periods to one more dependent on the local Alfvén speed at smaller scales. Since such large-spatial-scale electromagnetic disturbances also have large wave periods, satellites located even tens of seconds away from each other will be able to observe the same wave incidence/interference pattern without significant changes, so long as the perpendicular scales are also large (as expected), therefore also demonstrating quasi-stationarity in LEO satellite data.

Alfvén waves with smaller scales and shorter periods will also have shorter field-aligned wavelengths. As a result, the distance between the LEO satellite and the E-layer location of ionospheric reflection can span significant fractions, or even multiple wavelengths. Under these conditions, the E/B ratio will no longer be defined by the Pedersen conductance, but will instead be a function of scale height, Pedersen conductance, and the Alfvén speed profile below. Likewise on these smaller scales, the E–B phase difference will generally no longer be zero, but will be a function of frequency indicative of the satellite observing propagating waves whose phase can also be influenced by the superposition of incident and reflected waves. At its extremes, the orthogonal E- and B-fields will be out of phase by 90°, suggesting zero time-integrated Poynting flux and perfect reflection from the E-layer. This was recently observed by Ghadjari et al. (2022) in standing Alfvén waves within equatorial plasma bubbles. Similar evidence for standing Alfvén waves in the form of field-line resonances at low latitudes was also recently presented by Lena et al. (2021) using electric-field data from the Van Allen Probes during low-altitude perigee passes along their low-inclination high-apogee orbit.

Finally, the smallest-scale Alfvén waves with frequencies above approximately 0.4 Hz are expected to have poor reflection properties and will likely be absorbed by the ionosphere rather than reflected from it (Lessard & Knudsen, 2001). In this case, only the incident wave will then be seen by the observing satellite in LEO and with properties determined by free propagation. This will have the effect of the E/B ratio being equal to the Alfvén speed, with E- and B-fields being once again in phase, making this wave scale indistinguishable from the classic Alfvén traveling wave which is observed routinely in the magnetosphere, well away from the reflecting E-layer.

It is of interest to compare the characteristics of the Alfvén waves observed in LEO during periods of magnetospheric convection and in association with different morphologies of global FACs (cf. for example Figure 9.2). Figure 9.3 taken from the study of Pakhotin et al. (2020) shows an example of global FAC structure as characterized using the AMPERE experiment (panel (a); Anderson et al., 2000; Waters et al., 2001) during a period of southward IMF, and during which a pass of Swarm-A crossed the duskside FACs. Available electric- and magnetic-field data demonstrate not only the correlation between the long period fluctuations, but also the presence of large-amplitude fluctuations at shorter timescales which are concentrated between the regions of the large-scale R1 and R2 current systems around 11:22:30 UT.

Figures 9.4 and 9.5, showcasing an event from Pakhotin et al. (2018), show a similar example from a period of northward IMF, and where the large-scale FAC map in Figure 9.4 demonstrates how the FAC patterns have contracted to high latitudes above around 80° latitude. Passes by Swarm-A and Swarm-C between 17:15 UT and 17:22 UT show large-amplitude magnetic perturbations, presented in mean field-aligned coordinates in Figure 9.5. Of significance is the difference between the $x$ and $y$ components between Swarm-A and Swarm-C, noting that the two time series have been put onto a common time base by lagging the data by the along-track separation between the satellites of 10.66 s. In particular, the bottom panel shows the difference between the $y$-component magnetic fields which reaches up to approximately 400 nT peak to peak. Interestingly, period 1 and period 2 appear to show intervals which are more and less correlated, respectively. While Swarm-A and Swarm-C are flying on polar orbits which are separated by around 1.4° in longitude, Figure 9.5 makes it clear that the magnetic perturbations seen on Swarm-A and Swarm-C are significantly different, indicative of large-amplitude and nonstationary Alfvén waves. These may also be associated with small-perpendicular scales at high latitudes, which are embedded in dynamically changing FAC patterns with irregular and small-scale features observed during periods of northward IMF.

Overall, the characteristics of Alfvén-wave-related MIC will exhibit different characteristics depending on the wave scale and the location of the observing platforms. This is also true for the condition of quasi-stationarity during auroral-zone crossings as observed from LEO, and which will be satisfied for large-scale waves but not for small-scale waves. For multiple satellite datasets, as the inter-satellite along-track separation decreases, smaller and smaller waves will meet the quasi-stationarity

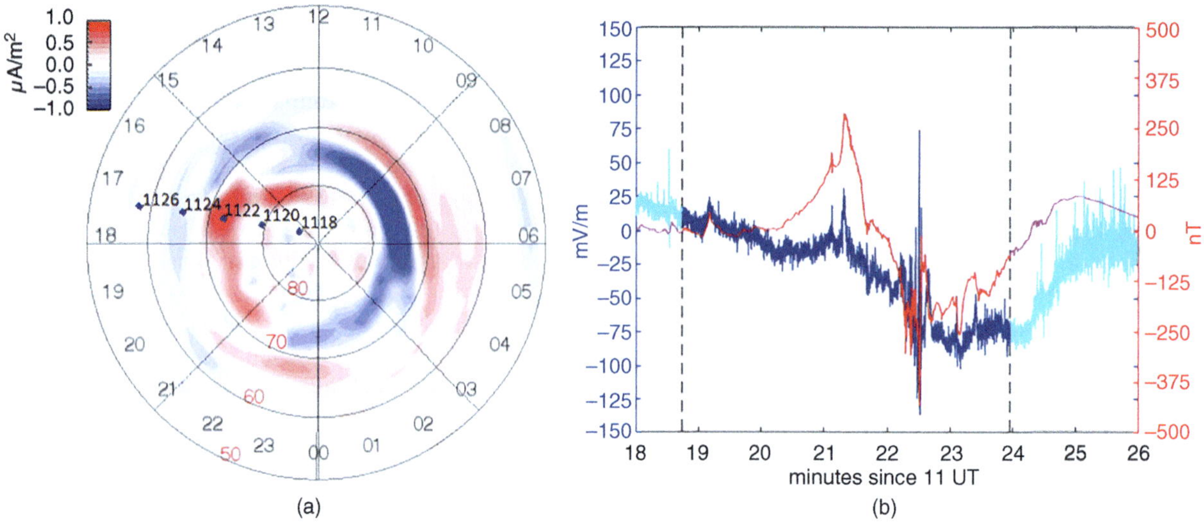


**Figure 9.3** Upward and downward R1/R2 field-aligned currents and related electric- and magnetic-field perturbations from a Swarm satellite auroral-zone pass during a period of southward IMF. (a) AMPERE FAC map, overplotted with the trajectory of Swarm-A from 11:18 UT to 11:26 UT on 16 May 2016. (b) E-field (blue) and B-field (red) for the crossing in panel (a) (E-field data outside the dashed lines plotted in cyan have larger noise). In panel (a), downward FACs are in blue, while upward FACs are in red. A section of high-frequency or small-scale, high-amplitude disturbance is seen between 11:22 and 11:23 UT, at the intersection between upward and downward FACs, similar to the behavior presented earlier in Figure 9.1 of this manuscript. Source: Pakhotin et al. (2020) /with permission of John Wiley & Sons.

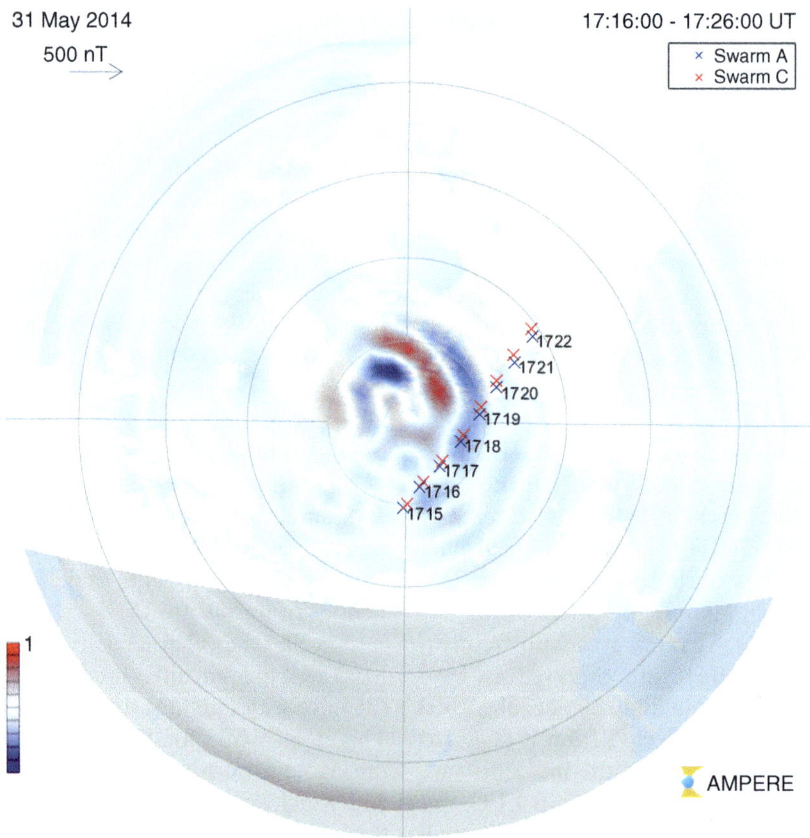


**Figure 9.4** Global FAC systems observed by AMPERE on 31 May 2014 during a period of northward IMF. Overplotted are the passes of Swarm-A (blue) and Swarm-C (red) through the high-latitude northward IMF FAC system and whose time series are shown in Figure 9.5. This is the same event as Pakhotin et al. (2018) /John Wiley & Sons / CC BY 4.0.

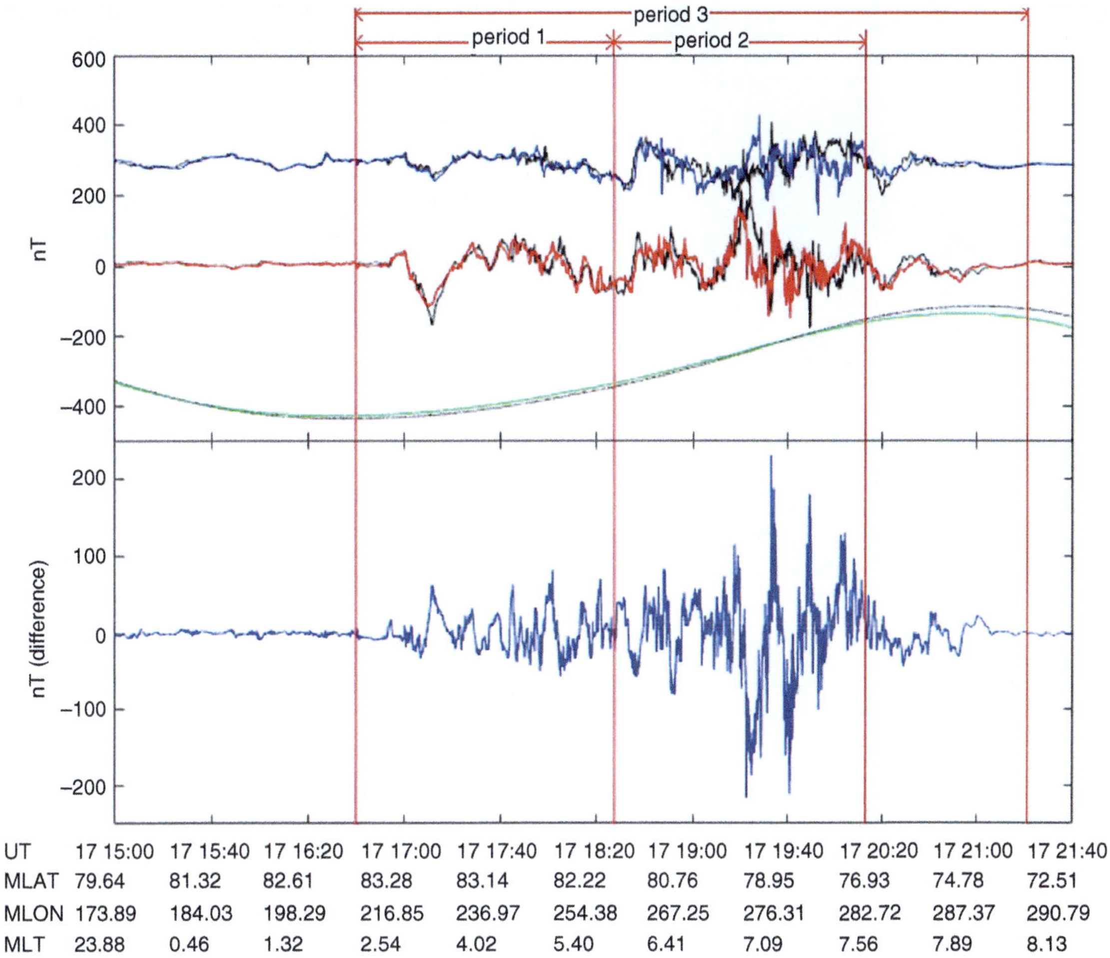


**Figure 9.5** Magnetic-field perturbations observed in mean field-aligned coordinates by Swarm-A and Swarm-C during a high-latitude pass through FAC systems during a period of northward IMF on 31 May 2014. Top: Vector magnetic field observed by Swarm-A and Swarm-C: *x* component (north–south; blue/black; offset by +300 nT), *y* component (east–west) in red/black, and field-aligned (compressional; offset by −300 nT after mean removal, green/black). Note the much smaller amplitude perturbations in the compressional compared to the transverse components. All components of the Swarm-C time series have been lagged by 10.66 s. Bottom: The difference arising from subtracting the two *y* components on Swarm-A and (lagged by 10.66 s) Swarm-C. The red bars separate Period 1, denoting good Swarm A–Swarm C correspondence, from Period 2 showing much poorer correspondence as smaller-scale disturbances dominate, and which exhibit greater variability. Here, MLAT, MLON, and MLT refer to the altitude adjusted corrected geomagnetic (AACGM) magnetic latitude, AACGM magnetic longitude, and magnetic local time, respectively. Source: Pakhotin et al. (2018) /John Wiley & Sons / CC BY 4.0.

criteria for these smaller inter-satellite separations. This result was demonstrated experimentally in Figures 7 and 8 in Gjerloev et al. (2011), the latter of which is reproduced here in our Figure 9.6. As shown in Figure 9.6, inter-satellite magnetic perturbations are only well-correlated at large spatial scales when observed by satellites with large inter-satellite separations. Conversely, closer inter-satellite along-track ST-5 satellite separations are well-correlated for both large and smaller spatial scales disturbances seen in the frame of the satellite. These data clearly suggest that there is a connection between the spatial and temporal scales of Alfvénic disturbances. It makes intuitive sense that, when the satellites are extremely close together, even the smallest-scale disturbances would appear near-identical on both spacecraft even at short along-track separations. However, the ST-5 data show that larger-scale disturbances in general vary on longer timescales. This is perfectly consistent with the Alfvén wave paradigm for MIC which we described earlier.

With regard specifically to FACs, a key question is of the energetic significance of structures at smaller scale

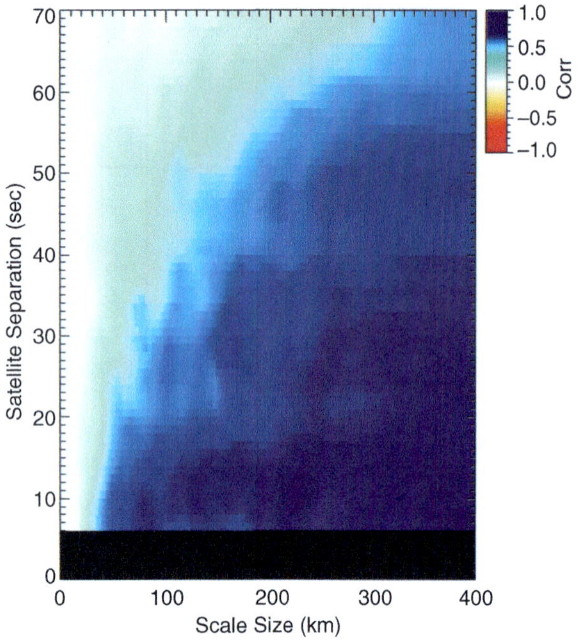


**Figure 9.6** Correlation between magnetic-field time series between different ST-5 satellite separations as a function of scale, for ST-5 inter-satellite separations less than 70 sec. Source: Gjerloev et al. (2011) /J. W. Gjerloev / CC BY 3.0.

sizes. If it could be assumed that small-scale Alfvén waves were embedded within the larger-scale FACs, in theory Ampère's law could be applied over an appropriately large area and any resultant FAC would simply be equal to the average equivalent current enclosed within the area. Thus, by selecting an appropriately large area, transient small-scale Alfvén waves could in theory be neglected, since they would be averaged out. Although the methodology is valid in principle, in practice there are several problems with this approach. First, it was found that there exist intense FACs which are small-scale and appear to differ markedly from the classic auroral-zone FACs (Rother et al., 2007). These kilometer-scale FACs (KSFACs), predominantly in the cusp/cleft region, appear to have different drivers than the large-scale auroral-zone system, and do not seem to be strongly influenced by the parameters which are known to control auroral current strengths. Thus, ignoring small-scale phenomena such as these would lead to an omission of critical physics which cannot be assumed to simply be a smaller-scale component of the larger auroral FAC system. Second, within the auroral FAC system, it was found that not all scales behave the same way. A multiscale study by McGranaghan et al. (2017) in particular found that the "FAC characteristics and dependence on controlling parameters do not map trivially across scales." These observations demand a focus on smaller scales to correctly capture the complete physics.

Third, any application of low-pass filtering invariably introduces artifacts; for example, a moving average filter may have the effect of shifting the local FAC maximum to a different location, or altering its value, which once again reduces certainty in measurements. This was one conclusion of the FAC–MICE experiment, where different techniques were employed to study various ways of calculating FACs for predetermined auroral-zone crossings (Trenchi et al., 2020). Particularly for low-correlation events, which typically correspond to active geomagnetic conditions, the violation of the non-stationarity assumption leads to processing artifacts during the attempt to preferentially extract large-scale fields. Indeed, in our view, the primary reason why smaller-scale Alfvén wave effects cannot be neglected in studies of MIC and related FAC dynamics is that the wave effects are likely to introduce non-stationarity into FAC estimates from LEO satellite data; see also the detailed discussion in Forsyth et al. (2017) and references therein. As a result, the Alfvén wave effects cannot be ignored; indeed, in our view, incorporating the Alfvénic behavior is also critical in order to correctly describe dynamical MIC and the related time-varying structure of FACs.

### 9.2.4. Interhemispheric Large-Scale FAC Asymmetry

An important aspect of both global MIC and FAC analyses is the degree to which the current systems are asymmetric between the hemispheres, and whether on a seasonally averaged basis both hemispheres see symmetric driving and energy input. Energy coupling between the solar wind and the MIT system depends on the configuration of both. Due to the Earth's nearly circular orbit, the solar energy and mass input (in the form of both extreme ultraviolet (EUV) insolation and solar wind plasma) largely affect both hemispheres equally over the course of a year, that is, the northern hemisphere is more exposed to solar insolation during northern summer than the southern hemisphere, but this reverses 6 months later during northern winter. The Earth's magnetic field, however, is offset from the Earth's rotation axis and its dipole center is northward of the Earth's center (Fraser-Smith, 1987; Koochak & Fraser-Smith, 2017). As a result, the interhemispheric asymmetry of the Earth's magnetic field can introduce an asymmetry in the FAC systems and their related MIC. Despite the seasonally averaged solar insolation and solar wind plasma being identical in each hemisphere, the asymmetry in the Earth's magnetic field can introduce an interhemispheric asymmetry by which energy input can be redistributed between the hemispheres in a seasonally asymmetric way.

A comprehensive review by Laundal et al. (2017) discusses the differences between the hemispheres and their related effects on MIT coupling. First, the geomagnetic field is generally stronger in the north between magnetically conjugate points, varying by up to a factor of 2 at 50° magnetic latitude. Second, the south-pole magnetic dipole is further offset from the Earth's rotation axis than its northern counterpart, meaning that the southern auroral oval experiences more diurnal variation relative to the northern oval. As Pedersen conductance is largely determined by EUV insolation, this parameter will therefore vary more widely in the south as the southern auroral oval will get dragged by the Earth's diurnal rotation into regions of more extreme variability than its northern counterpart. A significant result was obtained by Coxon et al. (2016) who used AMPERE data to discover that, when accounting for diurnal and seasonal variability, FACs are generally stronger in the north than in the south. Importantly, these authors compared AMPERE results with an empirical model (Milan, 2013) and found that the model, which assumes hemispheric symmetry, fails to reproduce the asymmetry observed by AMPERE data. This suggests missing physical processes and potentially an active role for the ionosphere in modulating the FAC and energy input into the ionosphere-thermosphere system on seasonally averaged basis.

More recently, Hatch et al. (2022) used the Average Magnetic field and Polar current System (AMPS) model to quantify to the degree to which it is possible to assume hemispheric symmetry with respect to Birkeland FAC currents. The study found that when averaged over IMF clock angle orientation, the total Birkeland FAC current in each hemisphere is largely symmetric except during local winter, when the northern hemisphere tends to dominate. The influences responsible for creating this asymmetry were attributed to different magnetic-field strengths in the two hemispheres, as well as to interhemispheric differences in the sunlight distribution in magnetic coordinates, which in turn affects ionospheric conductance. Alfvén wave processes are integral to ionospheric conductance dynamics and their influence on MIC energetics, and this role for Alfvén waves in impacting MIC will be covered in the next section.

## 9.3. INTERFERING ALFVÉN WAVES

### 9.3.1. Ionospheric Reflection: A Modulator of Magnetosphere–Ionosphere Coupling

Alfvén waves carry Poynting flux along magnetic field lines, therefore providing a vehicle for energy transfer between the magnetosphere and the ionosphere. Partial wave reflection arising from nonuniform Alfvén speed profiles can reduce the amount of energy reaching the ionosphere. However, interactions with the ionosphere provide the primary mechanism for MIC. This chapter addresses Alfvén wave Poynting flux, and the role of Alfvén waves in coupling the magnetosphere and ionosphere including the effects from wave incidence, absorption, reflection, and interference and related impacts on the magnetosphere–ionosphere interface.

Below the auroral acceleration region (AAR), electromagnetic Alfvén wave energy propagates in the shear mode down the field lines until it encounters the ionospheric E-layer at about 110 km height. At the E-layer, steep density gradients create a sharp gradient in Alfvén speed, which causes the wave to partially reflect back out into the magnetosphere. As discussed by Glasmeier (1984) (see also Hughes, 1974; Maltsev et al., 1974), incident Alfvén waves can be partially reflected from the ionosphere, and in the situation where one assumes a normal incidence on a uniform thin-sheet ionosphere with a height-integrated Pedersen conductivity of $\sum_P$, the reflection coefficient, R, depends on the relative size of the Alfvén impedance $\sum_A = 1/\mu_0 v_{AI}$ and $\sum_P$, where $v_{AI}$ is the Alfvén speed just above the assumed thin-sheet ionosphere (see also Newton et al., 1978) such that

$$R = \sum_A - \sum_P / \sum_A + \sum_P. \quad (9.3)$$

The strength of this reflection coefficient therefore depends on the mismatch between the Alfvén impedance, dependent on the Alfvén speed just above the layer of ionospheric reflection, and the Pedersen conductance (e.g., Knudsen et al., 1992; Lysak, 1991). In the special case that the Alfvén and Pedersen impedances match, the wave suffers no reflection and the energy is absorbed. In a realistic plasma, this condition is rarely satisfied, and partial reflections are common.

Where the reflected waves interfere with the incident waves above the ionosphere, this interference also impacts the wave structure and the characteristics of the waves' electric and magnetic fields. This varies both along magnetic-field lines and as a function of wave frequency, for a given Alfvén speed profile and ionospheric conductance structure (see Newton et al. (1978) for waves in the MAR; see also Lysak (1991) for waves in the IAR). Low-Earth orbit observations, for example, by the ESA Swarm mission (Friis-Christensen et al., 2006), orbiting at 450 km altitudes well below the AAR but also well above the E-layer, commonly observe these interference patterns in the auroral zone, manifesting as quasi-static small-scale disturbances in the E- and B-fields, whose coherency breaks down with decreasing scale size (e.g. Pakhotin et al., 2018). Knudsen et al. (1992) also provide an elegant discussion of the impacts on the local electric and magnetic fields for the case of interference from only a single incident and reflected Alfvén wave. See also Song and Lysak (2001) for a discussion of an integrated model of auroral electron acceleration and interfering Alfvén waves. Direct auroral electron acceleration can also be generated in steepening Alfvén wave fields (e.g., Watt et al., 2004). Along with powering the Alfvénic aurora (e.g., Chaston et al., 2003, 2008) and thus contributing to particle precipitation (e.g., Keiling et al., 2003), such particle precipitation can heat the ionosphere and upper thermosphere directly (Joule heating). See also Stasiewicz et al. (2000) for a discussion of Alfvén waves in the aurora. This in turn can drive thermospheric winds and uplift the topside ionosphere (e.g., Prölss, 2006), changing the density and composition of ionospheric plasma as well as its Pedersen conductance profile.

Overall, the reflection coefficient governs what happens to the remaining energy of magnetospheric origin downstream of any auroral acceleration processes: whether it is absorbed by the ionosphere, contributing to Joule heating, or reflected back up and away from the ionosphere. It is the reflection coefficient which can act to turn the ionosphere from a purely passive resistive dissipator to an active modulator of incoming magnetospheric energy: the energy and momentum arriving from the magnetosphere are approximately hemispherically symmetric, but this can be actively redistributed via hemispherically different amounts of reflection. Coupled with the fact that the reflection coefficient is scale-dependent (e.g. Forget et al., 1991; Lessard & Knudsen, 2001), the ionosphere therefore also actively modulates incoming energy at various scale sizes, preferentially absorbing or reflecting differential amounts of it depending on wavenumber.

### 9.3.2. Magnetosphere–Ionosphere–Thermosphere Coupling: Joule Heating and Poynting Flux

The Poynting flux which does not partially reflect from the E-layer is absorbed by the ionosphere, contributing to heating of the ionosphere–thermosphere. It should be noted that there is not a straightforward relationship between observed Poynting flux above the ionosphere and ionospheric heating at those specific field lines, due to three-dimensional effects (Richmond, 2010). In high-conductance regions, Poynting flux observations will tend to underestimate the energy dissipation, while in low-conductance regions the reverse is true. As such, Poynting flux alone may be insufficient to calculate the degree of heating at a specific field-line footpoint. Nevertheless, it is a very useful diagnostic tool for estimating energy input and Joule heating into the ionosphere more generally.

In a series of papers, Verkhoglyadova et al. (2016, 2017, 2018) evaluated major source and sink terms for energy transfer in the coupled MIT system. One of their findings was that the Joule heating rates appeared to be inconsistent between observations and model estimates, with these authors concluding that the likely source of the discrepancy was the omission of small-scale Alfvénic Poynting flux. Indeed, Pakhotin et al. (2018, 2020) showed that under both northward and southward IMF conditions, small-scale Alfvénic disturbances appear to have similar peak-to-peak magnitudes compared with the larger-scale FACs. In addition, these studies found that while E- and B-field perturbations often had nonzero phase differences, the Poynting flux was still directed overwhelmingly downward and did not time-integrate to zero indicating a net flow of energy into the ionosphere. As such, the evidence suggests that small-scale disturbances embedded within the larger auroral FACs carry a significant amount of Poynting flux into the ionosphere. Verkhoglyadova et al. (2018) showed that, during strong storms, Alfvén wave energy deposition may reach as much as 30% of the value of more quasi-static Joule heating. Similarly, Pakhotin et al. (2021) showed that as much as half of the Poynting flux observed by the LEO Swarm mission may be associated with events with transverse scales of less than 250 km.

This presents a key difference between analyses of FACs and Poynting flux. Whereas for Birkeland currents, it is possible to argue that in some cases the effects of small-scale currents may be averaged and folded into larger scales by integration around Ampèrian paths at larger scales to estimate the spatially averaged FAC density on that larger scale from LEO spacecraft data, it is fundamentally incorrect to attempt this when analyzing Poynting flux. Since Poynting flux is a quadratic quantity, it cannot be accurately estimated by combining two temporally averaged fields. Temporally averaging E- and B-field perturbations when analyzing data from spacecraft below the AAR where significant amounts of Poynting flux are present directly leads to systematic underestimation of the energetics associated with Alfvénic MIC—both because the time integral of the product of temporally averaged quantities is much less than the time integral of the product of the raw time series, as well as the fact that some energy may already have been converted into kinetic energy as a result of Alfvén-wave-induced field-aligned particle acceleration above the satellite in the AAR.

### 9.3.3. Pedersen Conductance Variability

The Pedersen conductance at the E-layer is affected by two main phenomena. First, direct insolation by EUV light from the Sun ionizes the neutral particles, creating more charge carriers and therefore increasing conductivity. Second, any auroral particle precipitation by accelerated electrons likewise ionizes the neutral atmosphere, increasing conductivity in the areas where the precipitation is occurring. Early studies (e.g., Moen & Brekke, 1993; Vickrey et al., 1981) utilized radar data to obtain typical values of Pedersen conductances, demonstrating clear solar zenith angle (SZA) dependences, with Pedersen conductance decreasing with increasing SZA. Meanwhile, a recent statistical study by Olifer et al. (2021) found that nightside Pedersen conductances typically range between 0 and 10 S, in agreement with the earlier Vickrey et al. (1981) and Moen and Brekke (1993) values. Olifer et al. (2021) also reported an asymmetry in Pedersen conductance values in regions of ambient upward and downward FACs on the dawnside, with regions of upward FACs exhibiting approximately 0.6 S higher conductances than regions of downward FACs. This effect was even more pronounced during active space weather conditions and was attributed to electron precipitation associated with the dawnward propagation of substorm-injected electrons. Meanwhile, the duskside Pedersen conductance values in regions of ambient upward and downward FACs did not appear to be significantly different.

The degree of insolation depends on the SZA and how much time a region of the auroral oval spends in sunlight versus in shadow. This is one of the main reasons for the observed interhemispheric asymmetry and the higher seasonally averaged intensity of both FACs (e.g., Coxon et al., 2016; Hatch et al., 2022) and Poynting flux (Cosgrove et al., 2022; Knipp et al., 2021; Pakhotin et al., 2021) observed in the northern hemisphere. Meanwhile, auroral particle precipitation can also be locally hemispherically asymmetric (e.g., Ostgaard et al., 2011), due to the southern aurora spending a different proportion of its total area in darkness during a given day as compared with its northern counterpart. Thus, both the processes responsible for determining Pedersen conductance and therefore the reflection coefficient exhibit interhemispheric asymmetry. The relationship between local Alfvén-wave-related electron precipitation can also result in localized regions of enhanced Pedersen conductance. This can create two-dimensionally inhomogeneous regions of Pedersen conductance, causing the upward propagation of additional Alfvén waves, which can change the FACs involved in MIC, and under certain circumstances can result in a positive feedback loop (e.g., Lysak, 1991, and references therein). As a result, self-consistent Alfvénic MIC can also result in the generation of spatially structured systems of upward and downward FACs and related regions of enhanced ionospheric conductivity (see Lysak & Song (2002a, b) and references therein for further discussion).

### 9.3.4. Alfvénic Aurora

Along with particle acceleration by large-scale FACs, small-scale Alfvén waves can also directly accelerate electrons, leading to the "Alfvénic aurora" (e.g., Chaston et al., 2008). Originally discovered in Polar spacecraft data (Keiling et al., 2003), its morphology was further defined by observations from above the AAR (Keiling et al., 2019), as was its dependence on the auroral electrojet (AE) index. A comprehensive summary of the Alfvénic oval and its key differences from the auroral oval are presented in Keiling et al. (2021). This phenomenon produces broadband electron acceleration, resulting in different auroral wavelengths than the classic green-line discrete arc emission.

A recent statistical study by Wu et al. (2020) explored high-frequency (in the frame of the Swarm satellite) disturbances indicative of Alfvén waves, embedded within larger-scale FAC systems. The study found that disturbances of up to 8 Hz (the Nyquist frequency of the electric field instrument) are commonly present, with roughly a third of nightside auroral-zone crossings by Swarm exhibiting Alfvénic activity. Both upward and downward FACs exhibited colocated Alfvén waves, with the cross-power spectral density or CPSD (a measure utilized by the authors as a means of evaluating the presence of Alfvénic disturbances) peaking either within downward FAC sheets (40% of all events) or on the boundary between upward and downward FACs (37%). This meant only a small minority (23%) of CPSD peaks occurred within upward FACs, suggesting only a weak relationship between Alfvén waves and auroral activity. Auroral arc systems with Alfvénic regions tended to be primarily associated with complex current sheet geometries, where 3–4 current sheets would be present; nevertheless all current sheet geometries exhibited Alfvénic activity. At these scales, many of both of the spatial and temporal assumptions underpinning FAC density calculations begin to fail. For example, as evidenced in Miles et al. (2018), their Figure 1 reproduced here in Figure 9.7, discrete arcs observed by the e-POP (Yau & James, 2015) appear to feature complex geometry and temporal quasi-periodic intensity changes, accompanied by non-time stationary transversed magnetic-field fluctuations and changing fine-scale structure FACs. The observation reported by Miles et al. (2018) occurred on the nightside when the phenomenon known as the ionospheric Alfvén resonator, or IAR, was also active (see Section 9.3.5). This was evidenced on ground stations as well as by frequency-dependent structuring of E- and B-fields and their phase relationships as observed by Swarm satellites passes, as can be seen in Figure 9.8. This event is described in more detail in the following text.

A recent review by Lysak (2023) summarized the conditions necessary for the creation of the Alfvénic aurora through small-scale Alfvén waves with kinetic effects. The Poynting flux associated with Alfvén waves provides the energy for particle energization. Small perpendicular scales are also necessary for the development of the

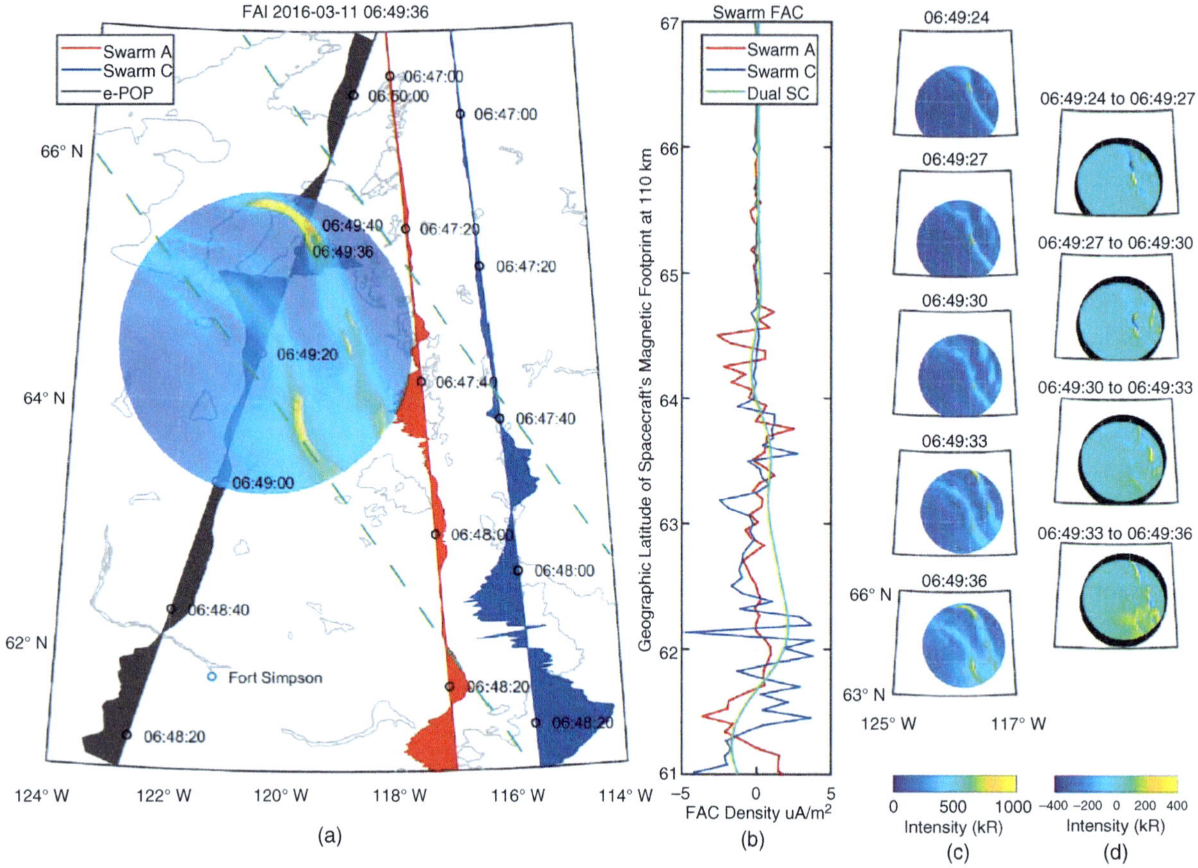


**Figure 9.7** e-POP and Swarm observations of discrete auroral arc dynamics. (a) Cross-track transverse magnetic perturbations observed by e-POP (black), Swarm-A (red), and Swarm-C (blue) overlaid with an e-POP FAI near-infrared image (color) from 06:49:36 UT showing two discrete auroral arcs. The dashed green lines extrapolate the general region of the arcs. (b) FAC estimates using single Swarm spacecraft data (A and C, red and blue) and using the Swarm dual-spacecraft technique (green), indicating how the dual-spacecraft technique cannot resolve the arc-scale FAC systems, with the single-spacecraft FACs inferred along the Swarm-A and Swarm-C trajectories being incoherent and poorly correlated suggestive of short-timescale non-stationarity. (c) and (d) The fine-scale arc structure and deviation from a planar arc and current sheet assumption are clearly visible on discrete arc scales, as is the temporal evolution showcased in (c) 3-s intervals and (d) difference plots of the e-POP FAI images. Source: Reproduced from Miles et al. (2018) / with permission of John Wiley & Sons.

parallel electric fields. These small scales are produced by phase-mixing effects during ionospheric feedback, particularly within the IAR, where multiple reflections occur and several characteristic frequencies ring simultaneously, producing the phase mixing necessary to create the kinetic Alfvén waves.

### 9.3.5. MIC and the Ionospheric Alfvén Resonator

Along with reflection from the bottom-side ionosphere at the E-layer, Alfvén waves at the magnetosphere–ionosphere interface can also reflect at the Alfvén speed gradient caused by sharp density changes in the magnetosphere at an altitude of approximately 1 Re. Under these circumstances, Alfvén wave energy can undergo a double reflection—once from the E-layer, and again at the Alfvén speed gradient out in the magnetosphere. In this case, the wave becomes trapped between the two reflecting regions, forming a phenomenon known as the IAR (e.g., Belyaev et al., 1990). It can also be observed in ground magnetometer data e.g. Canadian Array for Realtime InvestigationS of Magnetic Activity (CARISMA, Mann et al., 2008), typically on the nightside and being characterized as discrete frequency bands with a harmonic structure when plotted in dynamic spectrograms (often called spectral resonance structures, or SRS). The structure and spacing of the harmonics is dependent on the density variations inside the cavity, and from the ground

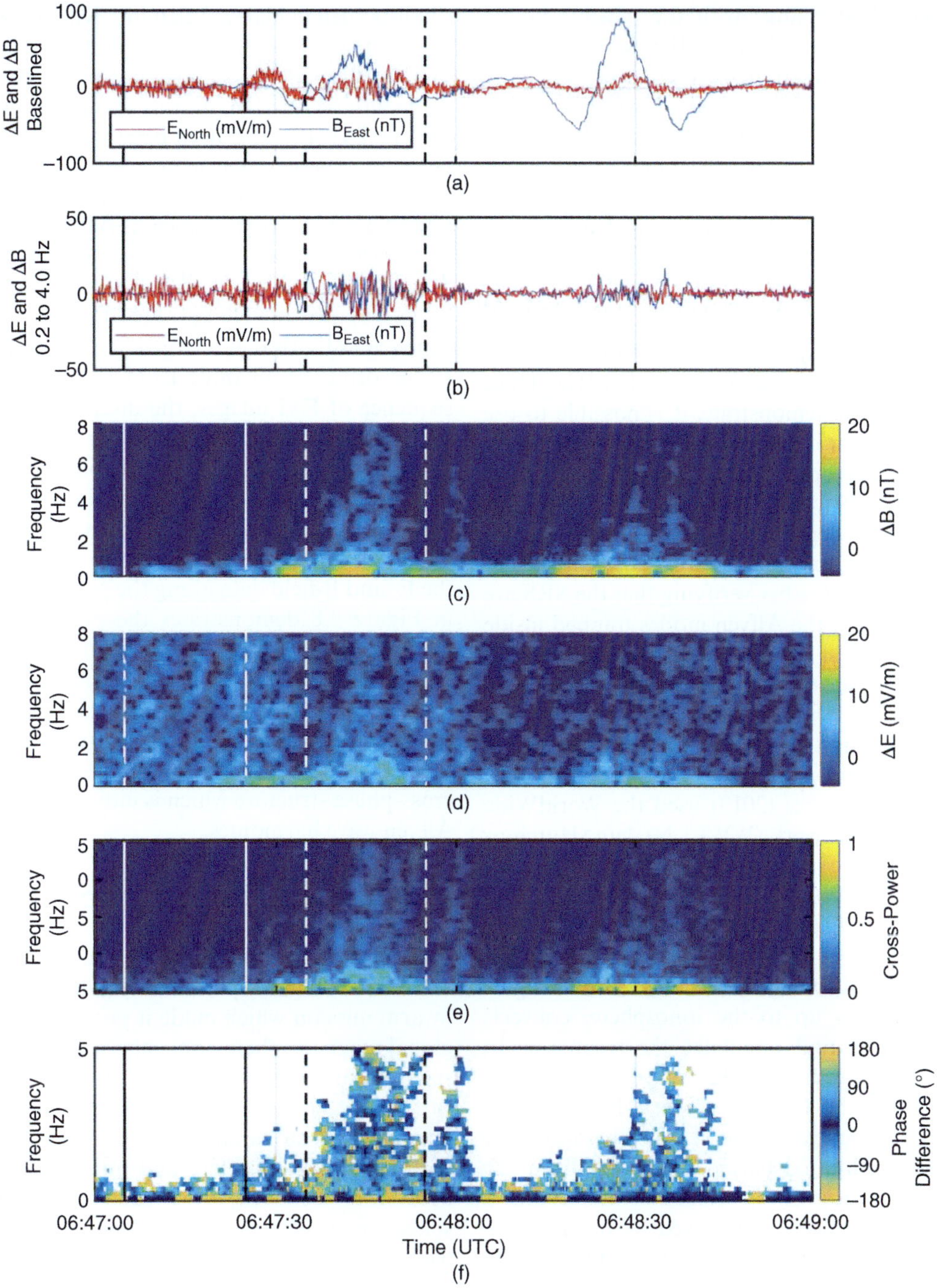


**Figure 9.8** E- and B-field time series and dynamics spectra and cross-spectra from Swarm-A during the crossing of the discrete auroral arcs shown in Figure 9.7. (a) Swarm-A northward electric and eastward magnetic fields with background removed and (b) band-pass-filtered from 0.2 to 4.0 Hz. Amplitude spectrum of the (c) northward magnetic field, (d) eastward electric field, (e) the cross-power, and (f) phase difference. The solid and dashed lines define the quiet and auroral intervals, respectively; the frequency-dependent phase difference relationship in panel (e) is suggestive of IAR harmonics. Source: Reproduced from Miles et al. (2018) / with permission of John Wiley & Sons.

these are observed to change with the passage of the ground station through nightside MLT sectors; it can also be rapidly modified by density changes occurring as a result of substorms (Parent et al., 2010). Figure 9.9, reproduced from Figure 3 of Parent et al. (2010), demonstrates the SRS, with the IAR harmonic frequencies appearing to track the F2 critical frequency f0F2, which itself is directly related to F2 maximum density (e.g., Ezquer et al., 2011):

$$N_mF_2 = 1.24e^4 * f_0F_2^2, \tag{9.4}$$

where $N_mF_2$ is the F2-layer maximum density in $cm^{-3}$, and $f_0F_2$ is the F2-layer critical frequency in MHz. Thus, as Parent et al. (2010) demonstrate, it is possible to use IAR harmonics to track the ionospheric F2-layer critical frequency and maximum density, which varies with MLT and with any substorm activity. Overall, this demonstrates that changes in density inside the IAR arising from substorm-related electron precipitation cause changes to the observed SRS, and further verifying that the SRS are harmonic signatures of the Alfvén modes trapped inside the IAR.

There is some debate as to what provides the energy source for the IAR, with some studies (e.g., Prikner et al., 2004) suggesting that IAR can contain and modulate incoming Pc1 pulsations from the magnetosphere. Meanwhile, recently Nokes (2019) used the WorldWide Lightning Location Network (WWLLN) data (Hutchins et al., 2012) to show correspondence between lightning strikes and IAR observed at multiple CARISMA ground stations to infer that lightning forms a significant energy source for the IAR (see also, e.g., Surkov et al., 2006). In this scenario, electromagnetic energy from a lightning discharge travels up to the ionosphere, converts into a plasma wave, and populates the resonator. In this paradigm, the lightning strike is a short impulsive driver which enters the IAR and reflects from the top of the resonator to be again observed as a second pulse on the ground. Additional partial reflections from the ionosphere and the top of the resonator can create a short wave train of pulses as observed by magnetometer stations on the ground. This is consistent with the proposition by Fedorov et al. (2014) who suggested that a wave train of a small number of short pulses, representing discrete reflections within the IAR cavity driven by lightning, is responsible for creating the observed SRS. That is, rather than representing a continuous ringing of the IAR, lightning excitation of the IAR cavity instead results in short pulses which appear to have a continuous structure when observed over long time spans in discrete spectrogram plots.

Very recently, Lysak (2023) suggested that when excited by Alfvén waves from above, the IAR may play a key role in the phase mixing necessary for the creation of small-perpendicular-wavelength Kinetic Alfvén Waves (KAWs), which can carry a parallel electric field and may be ultimately responsible for the Alfvénic aurora. As mentioned earlier, and as shown in Figures 9.7 and 9.8, the IAR has also been observed simultaneously with periodic, time-varying small-scale fluctuations of discrete arcs (Miles et al., 2018) which were observed using the e-POP Fast Auroral Imager (FAI) instrument (Cogger et al., 2014). Figure 9.7 shows passes of e-POP and Swarm-A and Swarm-C, during an interval characterized by two approximately parallel discrete auroral arcs and which are observed at a large angle of attack with respect to the orbits of all three LEO satellites. As seen in the sequence of FAI images, the discrete arcs evolve during the pass. The FACs inferred using single SC techniques differ significantly between the parallel Swarm-A and Swarm-C passes (panel b), consistent with the different cross-track magnetic fields shown in the feather plots along the satellite orbital tracks in panel (a). Analysis of the E- and B-field data along the pass of Swarm-A shown in Figure 9.8 demonstrates the presence of small-scale electromagnetic disturbances in the region of the arc crossing between the dashed vertical lines. The B-field (east) and E-field (north), and their cross-spectra, at frequencies between 0 and 8 Hz, show coherent waves with a strong cross-power but a frequency-dependent cross-phase structure which is indicative of potential IAR Alfvén wave harmonics.

## 9.4. GLOBAL MIC AND THE ACTIVE IONOSPHERE

In recent years, as new high-quality datasets became available for study, particularly with the advent of the Swarm mission which made it possible to map the entire global magnetosphere–ionosphere across a range of scale sizes, a paradigm shift began to develop around MIC. In the past, energy transfer was generally considered to take place mostly unidirectionally from the Sun through the outer magnetosphere and into the ionosphere, with occasional feedback loops such as through the ionospheric flywheel effect (e.g., Deng et al., 1991) or the feedback instability (e.g., Atkinson, 1970); nevertheless, the ionosphere was largely conceptualized as a passive sink for incoming energy and momentum. However, as more granular electromagnetic field and plasma density data became available, our understanding of MIC has evolved, as it became increasingly apparent that the energy flow is more complex than was hitherto realized. With the concept that ionospheric Alfvén wave reflection may play a significant role in determining the efficiency of MIC, new evidence has increasingly suggested that the ionosphere has the ability actively to modulate incoming electromagnetic wave energy, potentially redirecting incoming Alfvén wave energy from one

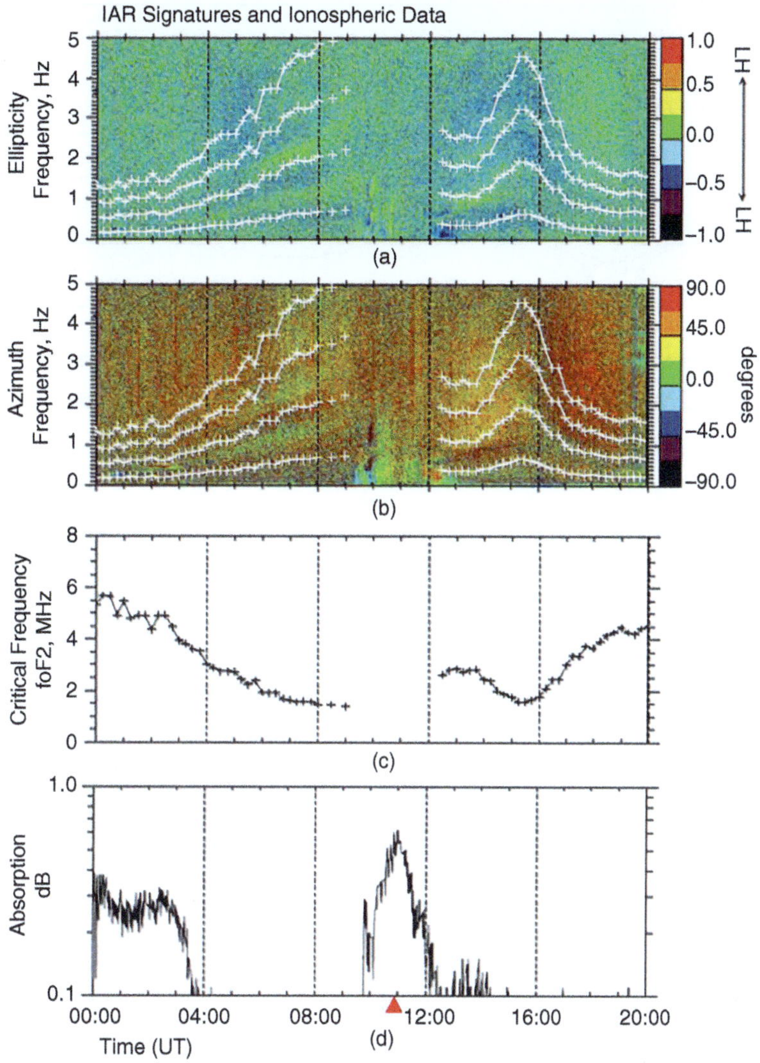


**Figure 9.9** IAR signatures visualized via (a) wave ellipticity and (b) azimuthal angle in ground-based induction-coil magnetometer data from Gakona in Alaska on 28 February 2006. Overplotted in white crosses are predicted model IAR frequencies as derived using the homogeneous IAR cavity model employed by Hebden et al. (2005), constrained by local observations of fOF2. (c) Critical frequency of the F2 layer (foF2) measured by the High-Frequency Active Auroral Research Program (HAARP) Digisonde. (d) Riometer absorption at HAARP. Local midnight is marked with a red triangle. The panels show changes in the observed IAR harmonic frequencies that can be predicted using an IAR model as a function of critical frequency fOF2. Source: Reproduced from Parent et al. (2010) / with permission of John Wiley & Sons.

hemisphere to the other (e.g., Lysak et al. (2020) and references therein). This may provide an explanation for the recently discovered interhemispheric asymmetry in incoming Poynting flux, where in a seasonally averaged sense it appears that space weather affects the northern hemisphere more than the southern one (Pakhotin et al., 2021). With recent studies of Poynting flux revealing that small scales account for a significant fraction of total energetics of MIC (Billett et al., 2022), as well as new results on upward Poynting flux and its scale dependence, the ionosphere is being re-conceptualized as an active modulator of incoming electromagnetic energy, with an increasing understanding that MIC is necessarily a two-way process that involves the ionosphere as an active element.

### 9.4.1. Scale Dependence of Poynting Flux

It was earlier demonstrated in Section 9.2 of this review that since Poynting flux is a quadratic quantity derived as the cross product of transverse magnetic- and electric-field perturbations, averaging of these fields to

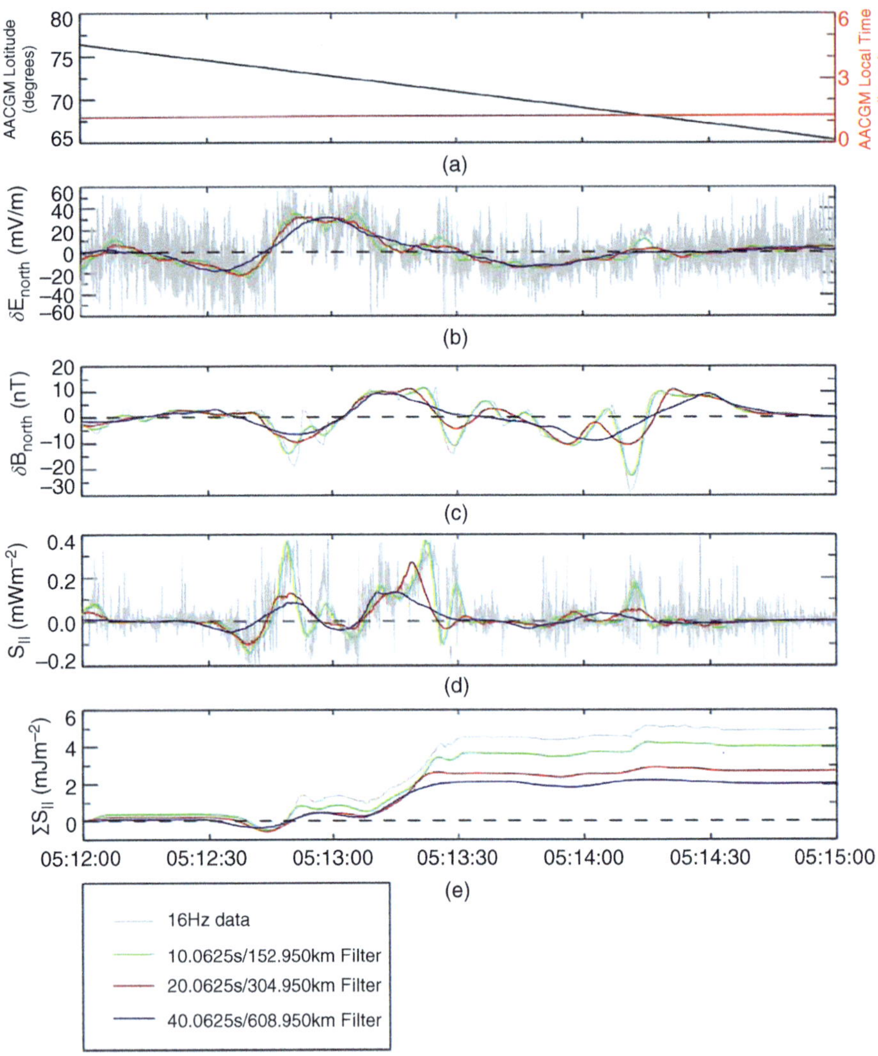


**Figure 9.10** Scale-dependence of Poynting flux during an auroral-zone crossing as observed by Swarm-A in the northern hemisphere on 17 November 2016. Illustrated is the effect of various Savitzky–Golay low-pass (SGLP) filters applied to the 16-Hz electric-/magnetic-field data and the resulting Poynting flux. (a) AACGM positional information for Swarm-A. (b) Meridional perturbation electric field. (c) Meridional perturbation magnetic field. (d) Poynting flux. (e) Time integration of the Poynting flux in (d). Note in panels (d) and (e) that positive is downward. Colors on (b)–(e) denote the unfiltered 16-Hz data product in gray, then SGLP filters of approximately 10, 20, and 40 s for green, red, and blue, respectively. The corresponding filter scale sizes (Nyquist) are also shown, assuming a satellite orbital velocity of 7.6 km/s. Source: Adapted from Billett et al. (2022); Pakhotin et al., 2021.

remove small-scale perturbations necessarily underestimates the energy flux associated with MIC. An example of a representative auroral-zone crossing is shown in Figure 9.10, reproduced from Figure 3 of Billett et al. (2022). It can be seen that as increasingly larger low-pass filter window sizes are applied to the constituent E- and B-field time series during the auroral-zone crossing, the time-integrated Poynting flux which the satellite observes along its world line during its traversal through the auroral-zone crossing decreases. In an extreme case, a Savitzky–Golay low-pass (SGLP) filter of window size 40 s (corresponding to large-scale FAC sizes) yields less than half of the along-track time-integrated Poynting flux than that obtained when using unfiltered 16-Hz data. Such band-pass filtering of transverse scale sizes of E- and B-fields also makes it possible to obtain the relative energy content in small-, meso-, and large-scale disturbances which are present in the MIC dynamics close to the magnetosphere–ionosphere interface. Overall, the amount of energy being transported into the ionosphere decreases with increasing filter window size, as smaller scales are attenuated, demonstrating the relative importance of energy transfer at smaller scales.

### 9.4.2. Poynting Flux Interhemispheric Asymmetry

The recently discovered seasonally averaged interhemispheric asymmetry in Poynting flux (Pakhotin et al., 2021) revealed that space weather has a northern preference. That is, despite arriving at the Earth's magnetosphere largely symmetrically between hemispheres in the solar wind, especially in a seasonally averaged sense, it manifests at the ionosphere in an asymmetric manner, which on a seasonally averaged basis means that more energy ends up being directed toward the northern hemisphere (NH) as opposed to southern hemisphere (SH). As discussed in Pakhotin et al. (2021), this effect was observed both on the dayside and nightside, and analyzed using Swarm noon–midnight orbit periods. Figure 9.11, a reproduction of Figure 2 from Pakhotin et al. (2021), shows the seasonally averaged time-integrated Poynting flux during electrodynamic intervals observed by Swarm. Specifically, the median and quartile Poynting fluxes at small, medium, and large spatial scales shown in Figure 9.11 clearly show the northern preference during northern winter, and how this preference evolves with season from the peak interhemispheric asymmetry in the northern summer solstice, through the equinoxes, to the northern winter solstice.

As discussed by Pakhotin et al. (2021), and reproduced here, it is interesting to note that the behavior of the interhemispheric asymmetry in electromagnetic energy flux is self-similar at small, medium, and large scales, suggesting that most likely the same physical processes are active in MIC across the entire range of spatial and temporal scales shown in Figure 9.11, and during all seasons. It is also possible that energy is transferred between scales within this system via a cascade. It can be seen that the sum totals of Poynting fluxes (north hemisphere flux plus south hemisphere flux) remain relatively similar across the seasons on the dayside (as denoted by the pink lines in Figure 9.11(a, c, e)). This suggests a relatively constant median total energy input into both hemispheres that is then redistributed differentially into the two hemispheres—but with a seasonally averaged northern preference which is especially strong near the northern summer solstice. On the nightside, the same conclusion applies—except that the northern preference is even stronger than on the dayside. Indeed, on the nightside, the northern preference is so pronounced that the median Poynting flux at Swarm altitudes is always larger in the north, and at all scales, independent of season—even near the winter solstice.

A recent study by Knipp et al. (2021) also confirmed the seasonally averaged interhemispheric asymmetry and northern preference reported by Pakhotin et al. (2021) for the dawn and dusk sectors. The collective Poynting flux results are consistent with earlier findings by Coxon et al. (2016) regarding larger field-aligned currents in the northern hemisphere. Likely both are connected to interhemispheric differences in the ionosphere. For example, the larger FACs observed in the north could be the result of higher Pedersen conductances there. The recent Poynting flux findings, meanwhile, demonstrate that incident energy is most likely being redistributed asymmetrically from a seasonally averaged symmetric source in the magnetosphere. This leads to the question of what causes the asymmetry. According to Pakhotin et al. (2021), the northern preference for Poynting flux is generated as a result of seasonally averaged asymmetries in Alfvén wave reflection.

Knipp et al. (2021) used DMSP data to confirm the interhemispheric asymmetry, determining that during the respective summer seasons of both hemispheres, co-sampled regions in the NH receive 25% more Poynting flux than their SH counterparts. The authors linked these findings with the earlier observations of stronger NH FACs by Coxon et al. (2016), arguing that stronger solar insolation and magnetic perturbations in the NH were likely responsible for the excess Poynting flux. Meanwhile, using FAST data, Cosgrove et al. (2022) found that the northern Poynting flux preference maximizes during southward IMF, in particular for local northern summer, when there can be 30% more Poynting flux flowing downward into the northern hemisphere than into the southern one. The conclusions of that paper were that the most likely sources of the asymmetry are the greater homogeneity of the geomagnetic field in the NH, which may act to produce a more pure two-cell convection pattern, and the smaller geomagnetic field magnitude (Förster & Halland, 2015) in the NH, leading to a larger ionospheric conductance (Richmond, 1995). If this conductance effect is generated by the interhemispheric asymmetry in the Earth's magnetic field, it could also be invoked within the Alfvén wave reflection paradigm advanced by Pakhotin et al. (2021).

Following the paradigm where MIC can be described through incident and reflecting Alfvén waves on a range of spatial and temporal scales, it is possible to consider the nature of the Alfvénic processes which may contribute to or explain this interhemispheric asymmetry. As discussed in the earlier text, on the nightside, discrete auroral arcs on the scale of kilometers, which have been demonstrated to contain nonstationary wave structure (e.g., Miles et al., 2018), can be responsible for significant auroral heating. Where Poynting flux is determined using measurements from LEO satellites at altitudes below the AAR (for

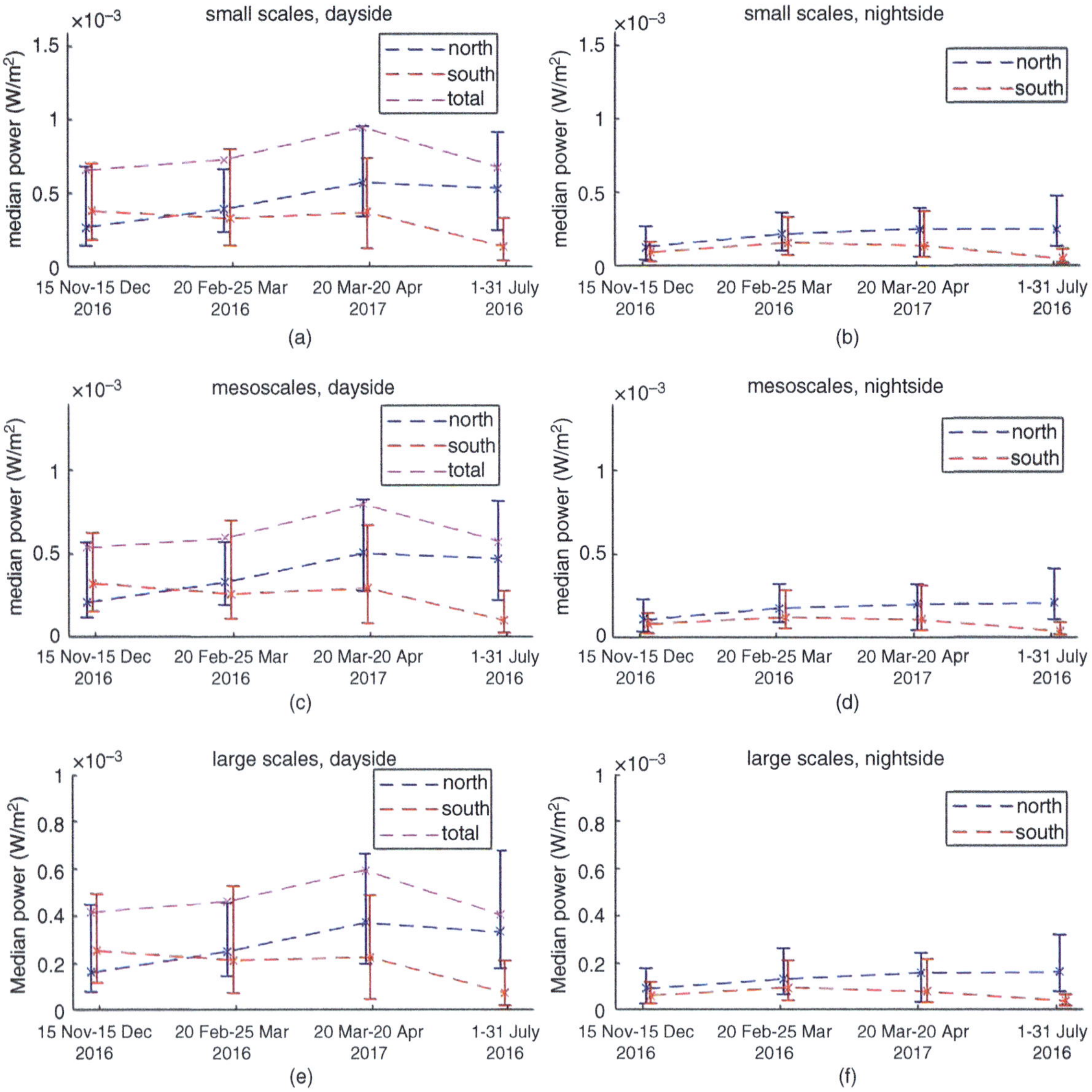


**Figure 9.11** Swarm-satellite-derived statistical dayside and nightside interhemispheric Poynting flux asymmetry for different seasons and scale sizes. (a) Small-scale dayside, (b) small-scale nightside, (c) mesoscale dayside, (d) mesoscale nightside, (e) large-scale dayside, and (f) large-scale nightside median Poynting fluxes. Small, medium, and large scales are defined to be characteristic of along-track scales of 75 km, 200 km, and 350 km, respectively, with the median power shown with a cross and the error bars indicative of quartiles in terms of magnitude. Northern hemisphere Poynting fluxes are in blue, southern hemisphere in red, while the sum of both are pink in (a), (c), and (e). As explained in Pakhotin et al. (2021), all electromagnetic power values are derived by integrating in time along the spacecraft world line during an electrodynamic event, and dividing by the duration of the event. Source: Pakhotin et al. (2021) / Springer Nature / CC BY 4.0.

example, Swarm, DMSP), such measurements represent the residual electromagnetic energy remaining after any auroral acceleration processes have taken place overhead. Importantly, due to the offset of the magnetic dipole from the center of the Earth, the southern auroral oval experiences different dwell times in darkness versus its northern counterpart. Such interhemispheric differences in solar illumination, combined for example with the fact that ionospheric feedback is hypothesized to be more effective in an ambiently dark ionosphere (e.g., Lysak & Song, 2008), may affect the amounts of auroral acceleration which occurs in each hemisphere.

In general, the Pedersen conductance in the nightside is largely driven by auroral precipitation, with accelerated electrons ionizing the upper atmosphere to form channels of enhanced Pedersen conductance. Meanwhile, the

Alfvén speed is driven by plasma density which is typically lower at night, except for the aforementioned channels of auroral heating. Therefore, both auroral acceleration and residual LEO Poynting flux, as well as wave reflection and absorption and related Joule heating, must be considered to obtain a full picture of the potentially active role or the ionosphere in creating the observed interhemispheric asymmetry. Such an approach was undertaken using Polar observations above the AAR to map the Alfvénic auroral oval, which is driven by Alfvén wave acceleration (Keiling et al., 2019, 2021).

Meanwhile on the dayside, a potentially different set of dynamical processes may dominate. The EUV insolation leads to increased plasma density outside of auroral acceleration regions, leading to ambiently higher Pedersen conductance in general. The reflection coefficient, which is primarily a function of the ratio of Alfvén impedance and the Pedersen conductance, governs the amount of reflection of incident Alfvén waves. As dayside Pedersen conductance is affected primarily by EUV insolation, an interhemispheric asymmetry in the levels of EUV insolation at given magnetic latitudes due to the offset dipole may result in, on average, hemispherically different amounts of wave reflection. In this way, even if the driver from the solar wind and magnetosphere was hemispherically symmetric in a seasonally averaged sense, the ionosphere could play an active role in modulating the transfer of this energy, potentially even redirecting it preferentially from one hemisphere to the other. Recent Swarm results, for example, Park et al. (2017) and Ivarsen et al. (2020), have examined this further, with the latter study also highlighting the importance of Hall conductance effects, which may also exhibit seasonally averaged interhemispheric asymmetry.

### 9.4.3. Upward Poynting Flux

Occasionally, observations of Poynting flux indicate an upward direction (e.g., Gary et al., 1995), representing an energy transfer from the ionosphere to the magnetosphere. These events could be due to the flywheel effect, where residual neutral wind motion from earlier activity can subsequently act as a generator for electromagnetic energy by moving charged particles in the lower ionosphere (Kelley et al., 1991). In general, upward Poynting flux events are weaker and less frequent than downward Poynting flux events, with net downward Poynting flux being the predominant state of energy transfer in MIC. An example is demonstrated in Figure 9.12 during an auroral-zone crossing, where the E- and B-fields exhibit an anti-phase relationship leading to net upgoing Poynting flux, primarily from time periods around 00:47:20 and 00:50:20 (downgoing Poynting flux is positive in Figure 9.12(d, e); upgoing is negative).

Recently, statistical maps of upward Poynting flux events have been compiled using Swarm data (Ivarsen et al., 2020), and very recently Billett et al. (2022) used high-frequency 16-Hz E- and B-field data from Swarm to determine that upward Poynting flux events have a different scale dependence compared with downward Poynting flux events. In particular, Billett et al. (2022) present evidence that upward Poynting flux energy drops off faster with decreasing scale size than downward Poynting flux events. This suggests that, for upward Poynting flux events, smaller scales contain comparatively more energy than for downward Poynting flux events. This is potentially significant as, per Lessard and Knudsen (2001), small-scale (<2 km) Alfvén waves have poor reflection properties in a realistic ionosphere; as such, Billett et al. (2022) theorize that any such small-scale waves that are observed propagating upward must come from an ionospheric source and not from ionospheric reflection. These latest results are starting to provide observational evidence in support of these earlier theoretical and modeling considerations. In general, upward Poynting flux is a relatively under-researched area of MIC, and with Swarm data it is becoming an increasingly active research topic.

## 9.5. SMALL-SCALE ALFVÉN WAVES

Having covered the topics of Alfvén waves within large-scale FAC structures and their role in global ionospheric processes through incidence, reflection, and interference, it is also important to address properties of Alfvén waves at some of the smallest scales. In this section, we focus on MIC as a result of the propagation of waves in the Pc1 range, which has periods from approximately 1–10 s, and is typically excited by the EMIC instability.

### 9.5.1. Wave Absorption at Kilometer and Sub-Kilometer Scales

An important caveat for the importance of Alfvén wave incidence and reflection at smaller scales in the ionosphere is the increased absorption of Alfvén waves at sub-kilometer perpendicular wavelengths (Borovsky, 1993). Further to this, Forget et al. (1991) demonstrated that the effective Pedersen conductance decreases sharply for wave perpendicular scales less than approximately 5 km. Lessard and Knudsen (2001), through modeling, demonstrated that under a realistic nightside ionospheric profile, reflection of Alfvén waves from the ionosphere is poor for waves >0.4 Hz in frequency and <1–2 km in perpendicular wavelength. This would place an effective lower limit on features generated as a result of wave interference; in particular, waves at these higher frequencies

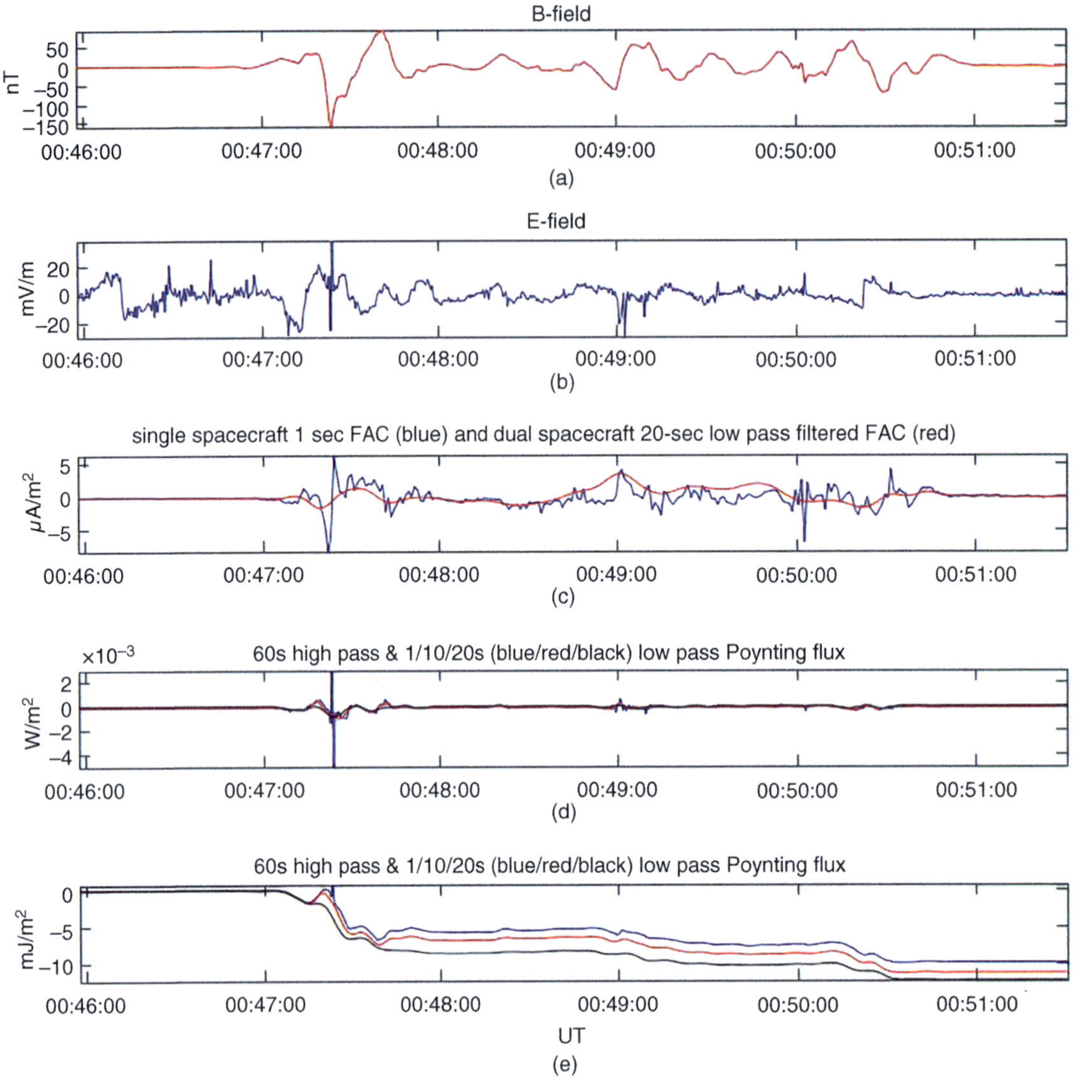


**Figure 9.12** Example of upward Poynting flux event during an auroral-zone crossing on 8 March 2016, generated using the same methodology as described in Pakhotin et al. (2021) / Springer Nature / CC BY 4.0. (a) $B_y$ (east) mean-field aligned (MFA) component. (b) $E_x$ MFA (north) component. (c) The single (blue)- and dual (red)-spacecraft FAC estimates. (d) The Poynting flux for low-pass filters of 1 s (blue), 10 s (red), and 20 s (black). (e) The time integrals of the Poynting flux from (d), indicating a net upward energy transport. Please note that in (d) and (e), downgoing Poynting flux is positive, hence upward Poynting flux is negative.

should not be able to resonate in the IAR. Miles et al. (2018) demonstrated that this agrees with the visible minimum optical features within a discrete arc observed by e-POP, which were on the order of approximately 1 km. Small-scale features within the ionosphere may still be observed (e.g., upward Poynting flux of ionospheric origin, incident EMIC waves, etc.), but these small-scale features are likely not generated as a result of ionospheric wave reflection and interference. The relative significance of incident and absorbed Pc1 waves such as EMIC waves on the overall energy budget in the ionosphere is currently unknown.

### 9.5.2. EMIC Waves

In the sub-auroral-zone, small-scale Pc1 waves on the EMIC branch are often observed. These waves originate in the magnetosphere and travel down the field lines toward the ionosphere primarily in the shear Alfvén mode, although cross-field propagation is also possible. Upon arrival at the ionosphere, they can mode-convert into the magnetosonic mode, propagating omnidirectionally from the source field line in an ionospheric duct. Simulations by Kim and Johnson (2016) outline the process and explain how the shear Alfvén wave may be ultimately observed on the ground. When observed by ground magnetometer stations (e.g., Usanova et al., 2008), the polarization of a wave can vary depending on the location of the ground station relative to the field line of arrival (cf. Mann et al., 2014). This effect is demonstrated in Figure 9.13, taken from the work of Mann et al. (2014), and where a distributed collection of magnetometer stations was seen to observe waves with different polarization. Drawing on the numerical modeling work of Woodroffe and Lysak (2012), an example of which is shown in the top right of Figure 9.13, these authors argued that the observed EMIC wave polarizations seen on the ground were indicative not only of the impact of propagation into the ionospheric duct, but also, per the modeling work of Woodroffe and Lysak (2012), that the polarization signatures could be used to diagnose the field lines which carried the EMIC wave from the magnetosphere to the ionosphere as indicated by the dashed circle in the top right panel of Figure 9.13.

EMIC waves can play an important role in MIC, but are also an important process impacting radiation-belt dynamics. A significant body of research has suggested that they may be responsible for the pitch-angle scattering of ultra-relativistic ($>\sim 2$ MeV) electrons into the loss cone on timescales of hours, leading to strong losses and dropouts of outer radiation-belt ultra-relativistic electron flux (e.g., Shprits et al. (2008) and references therein). Therefore, the study of EMIC waves, and the inference

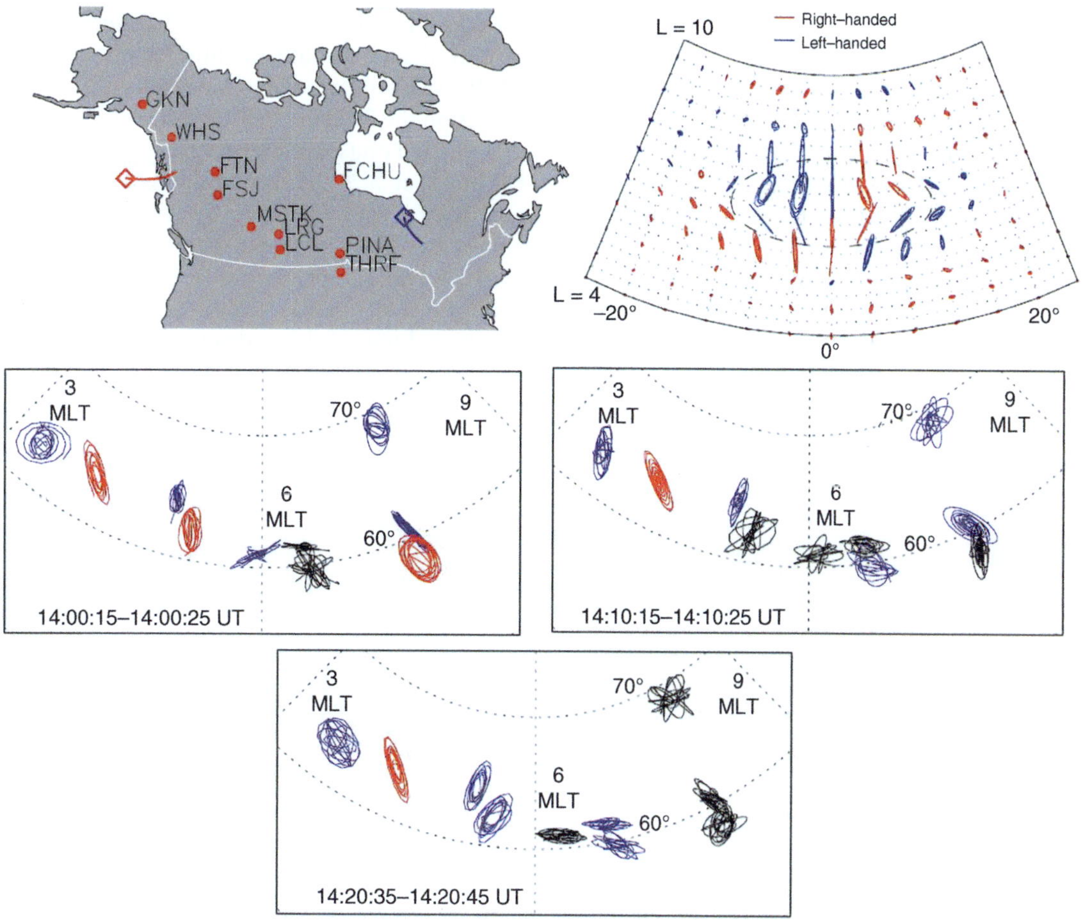


**Figure 9.13** Hodograms and polarization of EMIC waves observed at multiple ground-based induction-coil magnetometer stations. Top left shows the ground stations used, while the bottom three panels show a sequence of observed EMIC wave hodograms and polarization sense at the various ground stations at various times, demonstrating a location-dependent polarization (blue and red indicate left-hand and right-hand polarization sense in the hodograms, while black indicates an indeterminate sense of polarization rotation). The top right of the figure shows the results from a coupled magnetosphere–ionosphere simulation from Woodroffe and Lysak (2012; their Figure 7), showing the expected polarization as a function of L-shell and latitude from an EMIC wave source in the magnetosphere which is contained within the dashed oval. Source: Reproduced from Mann et al. (2014) / with permission of John Wiley & Sons.

of their points of origin, is significant for the creation of accurate radiation-belt flux forecasting systems as well as for studying the basic physics of these wave–particle interactions. In general, the fields of EMIC waves can be observed in situ by high-apogee missions such as the Van Allen Probes, THEMIS, Cluster, and MMS. Since EMIC waves are typically observed to be spatially and temporally localized phenomena, it is however often the case that EMIC waves are "missed" in the single-point measurements from a finite number of satellites at high altitudes. This can lead to EMIC wave scattering being invoked as the explanation of an observed radiation belt electron dropout, but without direct observational evidence of the EMIC wave being available. Observing EMIC waves from low-Earth orbiting platforms such as Swarm or e-POP, however, improves the return times of observations to field lines supporting EMIC waves, and therefore presents an intriguing opportunity, since a large range of L-shells can be "scanned" relatively rapidly.

For example, Swarm is in a polar orbit and has an approximately 90-min orbital period. It can therefore observe the entire range of L-shells at a particular MLT sector twice (once in each hemisphere) in approximately 45 min, which can be compared with the several hours or even days that it would take Van Allen Probes or Cluster to return to the same area. The issue with observing EMIC waves from LEO platforms is that they are often colocated within, or in close proximity to, regions of strong FACs. Disentangling the perturbations associated with the EMIC waves from those associated with the dynamical FACs, and indeed their related Alfvén wave fields, can be extremely challenging—not least because both will be associated with magnetic perturbations which are polarized primarily in the transverse east–west magnetic field component. Another way of observing the EMIC waves is to observe their magnetosonic-mode signature they generate in the ionosphere by observing the $B_z$ MFA component, or the $|B|$ modulus, as was done for example in several recent Swarm studies (e.g., Kim et al., 2018, 2020). The issue with this approach is that it does not directly yield the field line of origin, unless multi-satellite methods are employed to triangulate the location of the source.

Recent results presented by Pakhotin et al. (2022a) employed such a multi-satellite technique, observing the EMIC waves in both the transverse and the compressional modes on the Swarm satellites in LEO. These authors further utilized phase differencing (e.g., Balikhin et al., 1997; Pakhotin et al., 2013) and the known inter-spacecraft separation vector between Swarm-A and Swarm-C, to obtain the likely magnetospheric source region for the EMIC wave. The primary observational results from the Pakhotin et al. (2022a) study are shown in Figure 9.14. As shown in Figure 9.14, a primary source of EMIC waves

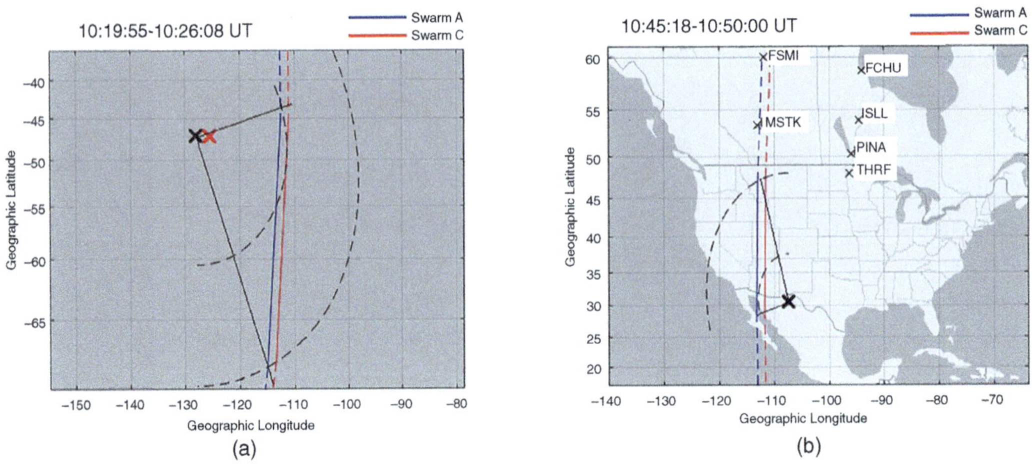


**Figure 9.14** EMIC wave observations by Swarm during a mid-latitude traversal of the American continent on 17 September 2015. In particular, (a) southern and (b) northern hemisphere tracks for Swarm-A (blue) and Swarm-C (red) are shown, with black lines denoting straight lines drawn between Swarm-A and Swarm-C inter-spacecraft separation vectors at times of maximum- and zero-phase differences. The intersection of these lines gives the approximate point of impact of the compressional wave onto the ionosphere. The blue and red traces are solid while Swarm-A and Swarm-C traversed the area between these zero-phase and maximum-phase locations, and dashed otherwise. Black dashed circles represent wavefronts assumed spreading out omnidirectionally from the source. Large black crosses denote intersection points of the phase lines, and the magnetically conjugate location (red). Panels (c) and (d) show Swarm-A and Swarm-C VFM vector magnetic fields in the VFM *y*-component, $B_y$ (labeled VFM_2), with primary and secondary EMIC signals highlighted in green and orange boxes, respectively (see text for details). Panels (e) and (f) show the Fourier power spectra for $B_y$ and the magnetic field modulus, respectively, for Swarm-A, while panels (g) and (h) show the same for Swarm-C. Source: Reproduced from Pakhotin et al. (2022a) / with permission of John Wiley & Sons.

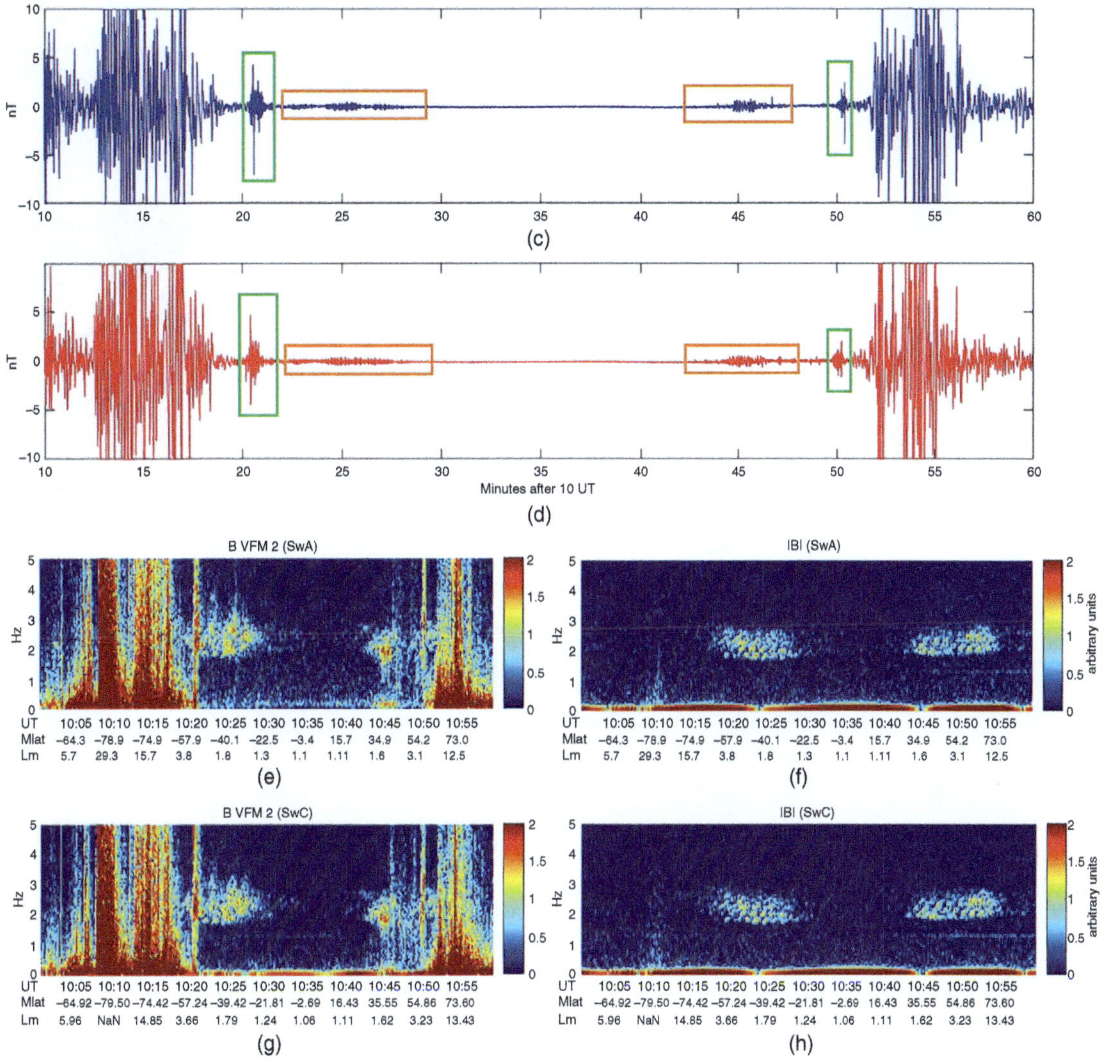


**Figure 9.14** *(continued)*

was observed to propagate from the magnetosphere and create a strong transverse signal at L ~ 3.3 as observed by Swarm-A and Swarm-C, highlighted with the green boxes in Figure 9.14(c, d), but was also associated with a second localized EMIC wave at lower L, around L ~ 1.7, highlighted with the orange boxes. Further, the principal source of the wave in the compressional mode appeared to be at this unusually low L-shell (L ~ 1.7), below that normally associated with EMIC wave generation. Although EMIC wave generation has been known to occur inward of the usual equatorial L ~ 3–6 range (e.g., Sakaguchi et al., 2013), in this case and as shown in Figure 9.15, simulations revealed another possible explanation. Namely, it is likely that the original wave originated at the equator at L ~ 3.3, and traveled in the field-aligned mode to the ionosphere where it was observed by Swarm as a strong shear mode signal. Meanwhile, the presence of multiple ion species in the magnetosphere created a Buchsbaum resonance (Buchsbaum, 1960), which acted as a reflection point where part of the wave energy was split and mode-converted, then propagating across the field lines down to low latitudes. At those low latitudes, it eventually excited a secondary field line, at L ~ 1.7, with waves from opposite hemispheres interfering to create a secondary local field-line resonance at L ~ 1.7.

Triangulation of the inferred location of the source at L ~ 1.7 was done using the projection along the lines of maximum- and zero-phase differences between the compressional waves seen on Swarm-A and Swarm-C, inferring a source location shown with a black cross in the Swarm passes in the southern and northern hemispheres shown in panels (a) and (b) of Figure 9.14. The conjugate location of the northern triangulated source is further shown in the southern hemisphere in panel (a) with a red cross, the close proximity to conjugacy of the estimates for the source location in both hemispheres providing some validation of the technique (see by Pakhotin et al., 2022a, for further details). For reasons which are not fully

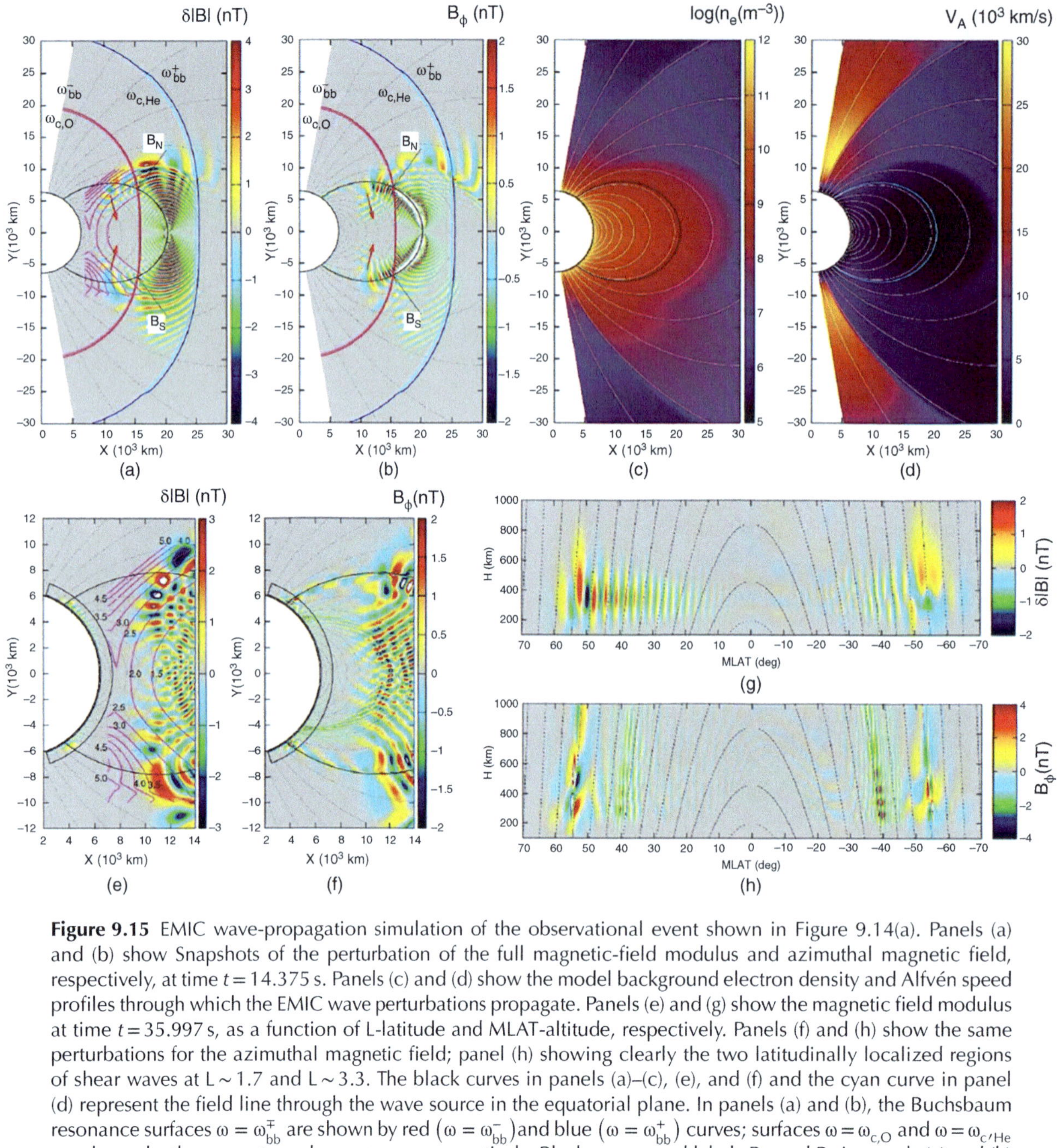


**Figure 9.15** EMIC wave-propagation simulation of the observational event shown in Figure 9.14(a). Panels (a) and (b) show Snapshots of the perturbation of the full magnetic-field modulus and azimuthal magnetic field, respectively, at time $t = 14.375$ s. Panels (c) and (d) show the model background electron density and Alfvén speed profiles through which the EMIC wave perturbations propagate. Panels (e) and (g) show the magnetic field modulus at time $t = 35.997$ s, as a function of L-latitude and MLAT-altitude, respectively. Panels (f) and (h) show the same perturbations for the azimuthal magnetic field; panel (h) showing clearly the two latitudinally localized regions of shear waves at L ~ 1.7 and L ~ 3.3. The black curves in panels (a)–(c), (e), and (f) and the cyan curve in panel (d) represent the field line through the wave source in the equatorial plane. In panels (a) and (b), the Buchsbaum resonance surfaces $\omega = \omega^{\mp}_{bb}$ are shown by red $(\omega = \omega^-_{bb})$ and blue $(\omega = \omega^+_{bb})$ curves; surfaces $\omega = \omega_{c,O}$ and $\omega = \omega_{c/He}$ are shown by the magenta and cyan curves, respectively. Black arrows and labels $B_N$ and $B_S$ in panels (a) and (b) mark the locations of Buchsbaum resonances in the northern and southern hemispheres, respectively. Red arrows in panels (a) and (b) mark equatorward propagating waves excited at locations with strongly nonuniform Alfvén speed. Source: Reproduced from Pakhotin et al. (2022a) / with permission of John Wiley & Sons.

understood, although the primary wave source at L ~ 3.3 was characterized by a strong transverse EMIC wave signatures at the altitude of Swarm, it was the secondary wave at L ~ 1.7 which was associated with the strongest compressional component as the transverse wave coupled to a compressional mode in the ionosphere and which further propagated away from the source field line in the low-altitude ionospheric duct.

In terms of the numerical simulation of the event, Figure 9.15 shows how an EMIC wave source of waves of frequency of 2 Hz, and which is locally imposed in the equatorial plane at L = 3.3, can generate multiple pathways for wave-energy propagation to the ionosphere. For example, it can be seen clearly from panels (a) and (b) in Figure 9.15 that the EMIC wave propagates in the Alfvén mode in the field-aligned direction from its equatorial source, with compressional modes additionally being ducted between the two Buchsbaum resonance surfaces $\omega = \omega_{bb}^{\mp}$(red and blue curves in panels (a) and (b) of Figure 9.15). At the locations $B_N$ and $B_S$, the waves encounter the Buchsbaum resonance, and split into two pathways, with some energy continuing down the field line to form the primary source at 50–60 MLAT (panel (h)), meanwhile a second pathway travels equatorially to lower latitudes and excites a secondary region of localized EMIC wave in the ionosphere at approximately 40 MLAT (panel (h)). In this way, heavy ion resonance effects lead a single-wave source to create pathways which reach the ionosphere being concentrated at two distinct L-shell and magnetic latitudinally localized regions at $L \sim 3.3$ and $L \sim 1.7$, respectively. See Pakhotin et al. (2022a) for more details.

Overall, the event presented by Pakhotin et al. (2022a), and summarized in Figures 9.14 and 9.15 here, showcases another pathway by which an EMIC wave generated at one L-shell may manifest on the ground as apparently having originated at a different L-shell. Importantly, the LEO Swarm mission observed both the original source wave at $L \sim 3.3$ and the secondary wave at $L \sim 1.7$, in the shear mode. This shows the utility of studying EMIC waves at the magnetosphere–ionosphere interface at LEO, where the spacecraft is close to the ground but still within the in situ plasma in the ionosphere, and where both the incident shear mode and the compressional wave excited by the interaction with the ionosphere can be observed. Moreover, the use of this polarization information can also be used to probe the pathways by which EMIC waves can propagate from the magnetosphere to the ionosphere. The case of the event studied by Pakhotin et al. (2022a) led to the discovery of the importance of the Buchsbaum resonance for introducing new pathways for EMIC wave-energy transport from the mid-latitude magnetosphere to the ionosphere. Very recent results presented by Kim and Johnson (2023) further verify the potential importance of the Buchsbaum resonance for EMIC wave energy transport from the magnetosphere to the ionosphere. These authors also further demonstrate that particularly in the presence of a tilt between the planet's dipole and the ecliptic plane (as is the case for Earth), numerous mode conversions can occur through which a wave generated at a single-source location near the equator in the magnetosphere may produce waves which arrive at the ionosphere and ground at several discrete and separated field line locations.

## 9.6. CONCLUSIONS

The review provides a summary and overview of the role of Alfvén waves at the magnetosphere–ionosphere interface and in magnetosphere–ionosphere coupling. A significant amount of new scientific literature has been generated by the community in the past decade, utilizing new high-quality datasets, in particular multi-spacecraft, multi-instrument measurements from satellite missions such as Swarm and DMSP, which have made it possible to revisit and reassess the applicability of some long-held assumptions regarding the nature of the processes associated with the MIC manifested through the propagation of Alfvén waves and the generation, evolution, and dynamics of large-scale Birkeland FAC systems. It is fair to say that, due to the work utilizing these datasets, a paradigm shift has been taking place recently within the MIC community. With an ever-increasing focus on small-scale and mesoscale processes, it is becoming increasingly clear that electromagnetic disturbances at these scales are crucial element of, and are key for a complete understanding of the nature of, the processes responsible for dynamical MIC. In turn, these same small-scale and mesoscale processes are also coming into ever-sharper focus in studies of coupling to even lower altitudes in the ionosphere–thermosphere system (e.g., Heelis & Maute, 2019).

Two new main areas of emerging consensus in the evolving understanding of MIC based on these new datasets may be identified.

Firstly, Alfvén waves are ubiquitous within the dynamical MI system and at the interface between the magnetosphere and the ionosphere. They form an energetically significant contribution to the overall energy transfer in the MIT system. In addition, they are active and can form the drivers of key change and reconfiguration processes. The impacts of these Alfvén-wave processes cannot be ignored, and filtering out smaller-scale variations is likely to lead to significant underestimates of the magnitude of the related energy transport. Indeed, in our view, any attempt to establish a distinction between quasi-static FACs and time-varying Alfvénic disturbances is largely arbitrary, based largely on the nature of the inter-spacecraft separations through which the dynamical MIC processes are being observed, especially in LEO satellite datasets. In reality, FACs, Alfvén waves, and wave modes excited within the IAR all involve the same physical processes. Each involves the propagation, reflection, interference, and interchange of shear Alfvén modes as core elements which underpin their dynamics, each representing different scale-size examples of fundamentally the same phenomenon.

Secondly, the Earth's ionosphere, far from just being a purely passive dissipator of incoming energy instead, also has an active role to play in the control and redistribution of energy in the coupled magnetosphere–ionosphere system; for example, by modulating incoming energy and displaying the capacity to differentially redirect energy back out into the magnetosphere. In particular, recent results show strong evidence for the ionosphere playing a role in generating interhemispheric asymmetries in energy transfer, both from event to event and statistically on a seasonally averaged basis. For example, on a seasonally averaged basis, a largely hemispherically symmetric energy input is affected by ionospheric interactions, being modulated by processes most likely related to the offset of the magnetic dipole from the center of the Earth in such a way as to engineer an interhemispheric asymmetry. On a seasonally averaged basis, incoming space weather energy therefore favors the northern hemisphere. The processes responsible for the active ionospheric modulation of incoming energy of solar wind origin are in our view most likely driven by Alfvénic-wave processes, such that small-scale and mesoscale processes also play a crucial role in delivering the observed interhemispheric asymmetries in energy transport, and ultimately energy absorption and dissipation in the ionosphere.

Multi-satellite LEO missions, equipped with multiple instruments, are capable of revealing hitherto unknown physics, which is leading to fundamental changes and paradigm shifts in the understanding of the processes which control magnetosphere–ionosphere coupling. The direction of travel, especially with the advent of cubesats carrying scientifically capable magnetometers (e.g., Miles et al., 2016), appears to be toward more multi-satellite, multi-instrumental missions perhaps eventually at global scales by exploiting both miniaturized cubesat platforms hosting high-fidelity and miniaturized scientific payloads. There have already been case studies showcasing how Swarm and Cluster conjunctions can be utilized to resolve field-aligned current density in three dimensions, adding more fidelity to FAC measurements (e.g., Dunlop et al., 2015).

Meanwhile, increasingly more attention is being paid to study the lower ionosphere, as understanding this understudied region of space, occasionally dubbed the "ignorosphere" and which is typically too low for satellites but too high for balloons, promises to reveal heretofore unknown and/or poorly quantified or missing physical processes required to answer some of the key remaining and unanswered scientific questions in MIT coupling. The proposed Daedalus mission (Sarris et al., 2020) is a good example of this effort. Finally, further multipoint and multiparameter in-situ satellite capabilities will be a welcome addition to a future global MIT great observatory, and a focus on smaller scales will inevitably mean an additional focus on higher sampling rates and higher signal-to-noise ratios. For example, the histogram methodology of studying E- and B-field disturbances (Grzesiak, 2000; Pakhotin et al., 2018, 2020; Wu et al., 2020) would benefit from a lower noise floor and more data points to resolve smaller-scale signals. In this context, the upcoming Geospace Dynamics Constellation (GDC) mission (Pfaff, 2016) will provide valuable multi-satellite capability to resolve multiscale MIT coupling. Meanwhile, the upcoming SMILE mission (e.g. Wang & Sun, 2022) will elucidate the influence of the near-Earth solar wind on multiscale MIC processes, providing a global picture of the magnetopause, cusps, and auroral distributions.

Additional utilization of existing datasets in new and innovative ways promises to expand our understanding of the dynamical MI system and to yield even further discoveries. In particular, one of the areas of study concerns the efficacy of MIC through the all-important Alfvén-wave reflection coefficient, and which is dependent on Alfvén speed and therefore on plasma density. It has recently been demonstrated that the current Swarm L1b plasma density measurements routinely overestimate density on the nightside, due to the prevalence of light ions (Park, 2022; Pakhotin et al., 2022b; Xiong et al., 2022). The new Swarm SLIDEM data product (Pakhotin et al., 2022b) addresses this effect, providing more robust plasma-density measurements unaffected by light ion abundances. The value of such approaches is that it allows the reprocessing of past datasets using new algorithms, yielding years of novel data without the need for a new mission.

In the upcoming immediate future, the prospect for further developments in understanding the nature of dynamical MIC remains bright. For example, Swarm operations are expected to continue such that mission data will eventually cover an entire solar cycle. This provides an opportunity to perform extensive statistical studies of the response of the active ionosphere to varying levels of solar and geomagnetic activity. Along with the fact that Swarm will operate contemporaneously with a number of other current and future ground- and space-based observatories, the environment will therefore exist to study small-scale and mesoscale disturbances at high fidelity and accuracy, utilizing opportunistic conjunctions to relax assumptions in case-study events, as well as generating high-resolution statistical datasets. When combined with recent developments of new coupled MIT models within the thermosphere modeling community, these new data promise further advances in understanding and appreciation of the role of small-scale and mesoscale Alfvén-wave processes in MIT coupling. Overall, it is clear that Alfvén waves play an absolutely critical role in MIC on a range of scales. Any hopes to be able to develop accurate future specification and forecast models of dynamical MIC will without doubt require the inclusion of the relevant Alfvén-wave physics and related physical processes to be included at their core.

## ACKNOWLEDGMENTS

Ivan P. Pakhotin is funded by the European Space Agency. The view expressed in this publication can in no way be taken to reflect the official opinion of the European Space Agency. Ian R. Mann is funded by Canadian NSERC. The authors would like to thank David Knudsen, Bob Lysak, Johnathan Burchill, and Daniel Billett for helpful discussions.

## AVAILABILITY STATEMENT

The ESA Swarm electric- and magnetic-field data needed to generate Figures 9.1 and 9.12 can be obtained from the ESA server at swarm-diss.eo.esa.int.

## REFERENCES

Anderson, B. J., Takahashi, K., & Toth, B. A. (2000). Sensing global Birkeland currents with Iridium© engineering magnetometer data. *Geophysical Research Letters*, *27*, 4045–4048. https://doi.org/10.1029/2000GL000094

Atkinson, G. (1970). Auroral arcs: Result of the interaction of a dynamic magnetosphere with the ionosphere. *Journal of Geophysical Research*, *75*, 4746–4755.

Balikhin, M. A., deWit, T. D., Alleyne, H. S. C. K., Woolliscroft, L. J. C., Walker, S. N., Krasnosel'skikh, V., et al. (1997). Experimental determination of the dispersion of waves observed upstream of a quasi-perpendicular shock. *Geophysical Research Letters*, *24*(7), 787–790. https://doi.org/10.1029/97gl00671

Belyaev, P. P., Polyakov, S. V., Rapoport, V. O., & Trakhtengertz, V. Y. (1990). The ionospheric Alfvén resonator. *Journal of Atmospheric and Terrestrial Physics*, *52*(9), 781–788. https://doi.org/10.1016/0021-9169(90)90010-K

Billett, D. D., McWilliams, K. A., Pakhotin, I. P., Burchill, J. K., Knudsen, D. J., & Martin, C. J. (2022). High-resolution Poynting flux statistics from the Swarm mission: How much is being underestimated at larger scales? *Journal of Geophysical Research: Space Physics*, *127*, e2022JA030573. https://doi.org/10.1029/2022JA030573

Biskamp, D. (1996). Magnetic reconnection in plasmas. *Astrophysics and Space Science*, *242*, 165–207. https://doi.org/10.1007/BF00645113

Birkeland, K. (1908), The Norwegian Aurora polaris expedition 1902–1903 (Vol. 1). Christiania, Norway: H. Aschelhoug.

Birkeland, K. (1913). *The Norwegian Aurora polaris expedition 1902–1903* (Vol. 2). Christiania, Norway: H. Aschelhoug.

Borovsky, J. E. (1993). Auroral arc thicknesses as predicted by various theories. *Journal of Geophysical Research*, *98*, 6101.

Buchsbaum, S. J. (1960). Ion resonance in a multicomponent plasma. *Physical Review Letters*, *5*(11), 495–497. https://doi.org/10.1103/physrevlett.5.495

Chaston, C. C., Bonnell, J. W., Carlson, C. W., McFadden, J. P., Ergun, R. E., & Strangeway, R. J. (2003). Properties of small-scale Alfvén waves and accelerated electrons from FAST. *Journal of Geophysical Research: Space Physics*, *108*(A4).

Chaston, C. C., Salem, C., Bonnell, J. W., Carlson, C. W., Ergun, R. E., Strangeway, R. J., & McFadden, J. P. (2008). The turbulent Alfvénic aurora. *Physical Review Letters*, *100*, 175003. https://doi.org/10.1103/PhysRevLett.100.175003

Cogger, L., Howarth, A., Yau, A., White, A., Enno, G., Trondsen, T., et al. (2014). Fast auroral imager (FAI) for the e-POP mission. *Space Science Reviews*, *189*(1-4), 15–25. https://doi.org/10.1007/s11214-014-0107-x

Cosgrove, R. B., Bahcivan, H., Chen, S., Sanchez, E., & Knipp, D. (2022). Violation of hemispheric symmetry in integrated Poynting flux via an empirical model. *Geophysical Research Letters*, *49*, e2021GL097329. https://doi.org/10.1029/2021GL097329

Coxon, J. C., Milan, S. E., Carter, J. A., Clausen, L. B. N., Anderson, B. J., & Korth, H. (2016). Seasonal and diurnal variations in AMPERE observations of the Birkeland currents compared to modeled results. *Journal of Geophysical Research: Space Physics*, *121*, 4027–4040. https://doi.org/10.1002/2015JA022050

Crooker, N. U., Siscoe, G. L., Mullen, P. R., Russell, C. T., & Smith, E. J. (1982). Magnetic field compression at the dayside magnetopause. *Journal of Geophysical Research*, *87*, 10,407–10,412.

Deng, W., Killeen, T. L., Burns, A. G., & Roble, R. G. (1991). The flywheel effect: Ionospheric currents after a geomagnetic storm. *Geophysical Research Letters*, *18*(10), 1845–1848.

Dunlop, M. W., Yang, J. Y., Yang, Y. Y., Xiong, C., Lühr, H., Bogdanova, Y. V., et al. (2015). Simultaneous field-aligned currents at Swarm and Cluster satellites. *Geophysical Research Letters*, *42*, 3683–3691. https://doi.org/10.1002/2015GL063738

Erickson, G. M., Spiro, R. W., & Wolf, R. A. (1991). The physics of the Harang discontinuity. *Journal of Geophysical Research*, *96*(A2), 1633–1645.

Ezquer, R. G., Cabrera, M. A., Lopez, J. L., Albornoz, M. R., Mosert, M., Marco, P., & Buresova, D. (2011). Critical frequency and maximum electrondensity of F2 region over four stations in the North American sector. *Journal of Atmospheric and Solar-Terrestrial Physics*, *73*, 420–429. doi:10.1016/j.jastp.2010.09.018

Fedorov, E., Schekotov, A., Hobara, Y., Nakamura, R., Yagova, N., & Hayakawa, M. (2014). The origin of spectral resonance structures of the ionospheric Alfvén resonator. Single high-altitude reflection or resonant cavity excitation? *Journal of Geophysical Research: Space Physics*, *119*(4), 3117–3129.

Forget, B., Cerisier, J.-C., Berthelier, A., & Berthelier, J.-J. (1991). Ionospheric closure of small-scale Birkeland Currents. *Journal of Geophysical Research*, *96*(A2), 1843–1847. https://doi.org/10.1029/90JA02376

Förster, M., & Haaland, S. (2015). Interhemispheric differences in ionospheric convection: Cluster EDI observations revisited. *Journal of Geophysical Research: Space Physics*, *120*, 5805–5823. doi:10.1002/2014JA020774.

Forsyth, C., Rae, I. J., Mann, I. R., & Pakhotin, I. P. (2017). Identifying intervals of temporally invariant field-aligned currents from Swarm: Assessing the validity of single-spacecraft methods. *Journal of Geophysical Research: Space Physics*, *122*, 3411–3419. doi:10.1002/2016JA023708.

Fraser-Smith, A. C. (1987). Centered and eccentric geomagnetic dipoles and their poles. *Reviews of Geophysics*, *25*, 1–16.

Friis-Christensen, E., Lühr, H., Knudsen, D., & Haagmans, R. (2006). Swarm—An Earth observation mission investigating geospace. *Advances in Space Research*, *41*(1), 210–216. https://doi.org/10.1016/j.asr.2006.10.008

Ganushkina, N. Y., Liemohn, M. W., Dubyagin, S., Daglis, I. A., Dandouras, I., De Zeeuw, D. L., et al. (2015). Defining and resolving current systems in geospace, *Annales Geophysicae*, *33*, 1369–1402. https://doi.org/10.5194/angeo-33-1369-2015

Gary, J. B., Heelis, R. A., & Thayer, J. P. (1995). Summary of field-aligned Poynting flux observations from DE 2. *Geophysical Research Letters*, *22*(14), 1861–1864. https://doi.org/10.1029/95gl00570

Ghadjari, H., Knudsen, D., & Skone, S. (2022). Standing Alfvén waves within equatorial plasma bubbles. *Geophysical Research Letters*, *49*, e2021GL097526. https://doi.org/10.1029/2021GL097526

Gjerloev, J. W., & Hoffman, R. A. (2014). The large-scale current system during auroral substorms. *Journal of Geophysical Research: Space Physics*, *119*, 4591–4606. doi:10.1002/2013JA019176

Gjerloev, J. W., Ohtani, S., Iijima, T., Anderson, B., Slavin, J., & Le, G. (2011). Characteristics of the terrestrial field-aligned current system. *Annales Geophysicae*, *29*, 1713–1729. https://doi.org/10.5194/angeo-29-1713-2011

Glasmeier, K. H. (1984). On the influence of ionospheres with non-uniform conductivity distribution on hydromagnetic waves. *Journal of Geophysics*, *54*(1), 125–137.

Grzesiak, M. (2000). Ionospheric Alfvén resonator as seen by Freja satellite. *Geophysical Research Letters*, *27*, 923–926. https://doi.org/10.1029/1999GL010747

Hatch, S. M., Laundal, K. M., & Reistad, J. P. (2022). Testing the mirror symmetry of Birkeland and ionospheric currents with respect to magnetic latitude, dipole tilt angle, and IMF By. *Frontiers in Astronomy and Space Sciences*, *9*, 958977. doi:10.3389/fspas.2022.958977

He, M., Vogt, J., Lühr, H., Sorbalo, E., Blagau, A., Le, G., & Lu, G. (2012). A high-resolution model of field-aligned currents through empirical orthogonal functions analysis (MFACE). *Geophysical Research Letters*, *39*, L18105. https://doi.org/10.1029/2012GL053168

Hebden, S. R., Robinson, T. R., Wright, D. M., Yeoman, T., Raita, T., & Bösinger, T. (2005). A quantitative analysis of the diurnal evolution of ionospheric Alfvén resonator magnetic resonance features and calculation of changing IAR parameters. *Annales Geophysicae*, *23*, 1711–1721.

Heelis, R. A., & Maute, A. (2020). Challenges to understanding the Earth's ionosphere and thermosphere. *Journal of Geophysical Research: Space Physics*, *125*, e2019JA027497. https://doi.org/10.1029/2019JA027497

Hughes, W. J. (1974). The effect of the atmosphere and ionosphere on long period magnetospheric micropulsations. *Planetary and Space Science*, *22*(8), 1157–1172.

Hutchins, M. L., Holzworth, R. H., Brundell, J. B., & Rodger, C. J. (2012). Relative detection efficiency of the World Wide Lightning Location Network. *Radio Science*, *47*, RS6005. doi:10.1029/2012RS005049

Iijima, T., & Potemra, T. A. (1976). The amplitude distribution of field-aligned currents at northern high latitudes observed by Triad. *Journal of Geophysical Research*, *81*, 2165–2174. https://doi.org/10.1029/JA081i013p02165

Iijima, T., & Potemra, T. A. (1978). Large-scale characteristics of field-aligned currents associated with substorms. *Journal of Geophysical Research*, *83*(A2), 599–615. https://doi.org/10.1029/JA083iA02p00599

Ishii, M., Sugiura, M., Iyemori, T., & Slavin, J. A. (1992). Correlation between magnetic and electric field perturbations in the field-aligned current regions deduced from DE 2 observations. *Journal of Geophysical Research*, *97*, 13,877–13,887. https://doi.org/10.1029/92JA00110

Ivarsen, M. F., Park, J., Kwak, Y.-S., Jin, Y., Knudsen, D. J., & Clausen, L. B. N. (2020). Observational evidence for the role of Hall conductance in Alfvén wave reflection. *Journal of Geophysical Research: Space Physics*, *125*, e2020JA028119. https://doi.org/10.1029/2020JA028119

Keiling, A. (2021). The dynamics of the Alfvénic oval. *Journal of Atmospheric and Solar-Terrestrial Physics*, *219*, 105616. https://doi.org/10.1016/J.JASTP.2021.105616

Keiling, A., Thaller, S., Wygant, J., & Dombeck, J. (2019). Global Alfvén wave power in the auroral zone in relation to the AE index. *Journal of Geophysical Research: Space Physics*, *124*, 8637–8646. https://doi.org/10.1029/2019JA026805

Keiling, A., Wygant, J. R., Cattell, C. A., Mozer, F. S., & Russell, C. T. (2003). The global morphology of wave Poynting flux: Powering the aurora. *Science*, *299*(5605), 383–386. https://doi.org/10.1126/science.1080073

Kelley, M. C., Knudsen, D. J., & Vickrey, J. F. (1991). Poynting flux measurements on a satellite: A diagnostic tool for space research. *Journal of Geophysical Research*, *96*(A1), 201–207. https://doi.org/10.1029/90ja01837

Kepko, L., McPherron, R., Amm, O., Apatenkov, S., Baumjohann, W., Birn, J., et al. (2015). Substorm current wedge revisited. *Space Science Review*, *190*, 1–46.

Kim, E.-H., & Johnson, J. R. (2016). Full-wave modeling of EMIC waves near the He+ gyrofrequency. *Geophysical Research Letters*, *43*(1), 13–21. https://doi.org/10.1002/2015GL066978

Kim, E.-H., & Johnson, J. R. (2023). Magnetic tilt effect on externally driven electromagnetic ion cyclotron (EMIC) waves. *Geophysical Research Letters*, in press, https://doi.org/10.1029/2022GL101544

Kim, H., Hwang, J., Park, J., Bortnik, J., & Lee, J. (2018). Global characteristics of electromagnetic ion cyclotron waves deduced from Swarm satellites. *Journal of Geophysical Research: Space Physics*, *123*, 1325–1336. https://doi.org/10.1002/2017JA024888

Kim, H., Shiokawa, K., Park, J., Miyoshi, Y., Hwang, J., & Kadokura, A. (2020). Modulation of Pc1 wave ducting by equatorial plasma bubble. *Geophysical Research*

*Letters*, *47*(9), e2020GL088054. https://doi.org/10.1029/2020GL088054

Knipp, D., Kilcommons, L., Hairston, M., & Coley, W. R. (2021). Hemispheric asymmetries in Poynting flux derived from DMSP spacecraft. *Geophysical Research Letters*, *48*, e2021GL094781. https://doi.org/10.1029/2021GL094781

Knudsen, D. J., Burchill, J. K., Buchert, S. C., Eriksson, A. I., Gill, R., Wahlund, J. E., et al. (2017). Thermal ion imagers and Langmuir probes in the Swarm electric field instruments. *Journal of Geophysical Research: Space Physics*, *122*, 2655–2673. https://doi.org/10.1002/2016JA022571

Knudsen, D. J., Kelley, M. C., & Vickrey, J. F. (1992). Alfvén waves in the auroral ionosphere: a numerical model compared with measurements. *Journal of Geophysical Research*, *97*, 77–90.

Koochak, Z., & Fraser-Smith, A. C. (2017). An update on the centered and accentric geomagnetic dipoles and their poles for the years 1980-2015. *Earth and Space Science*, *4*, 626–636.

Laundal, K. M., Cnossen, I., Milan, S. E., Haaland, S. E., Coxon, J., Pedatella, N. M., et al. (2017). North–South asymmetries in Earth's magnetic field. *Space Science Reviews*, *206*, 225–257.

Lena, F. R., Ozeke, L. G., Wygant, J. R., Tian, S., Breneman, A. W., & Mann, I. R. (2021). Detection of hertz frequency multiharmonic field line resonances at low-L (L= 1.1–1.5) during Van Allen probe perigee passes. *Geophysical Research Letters*, *48*(1), p. 2020GL090632.

Lessard, M. R., & Knudsen, D. J. (2001). Ionospheric reflection of small-scale Alfvén waves. *Geophysical Research Letters*, *28*, 3573–3576. https://doi.org/10.1029/2000GL012529

Lühr, H., Park, J., Gjerloev, J. W., Rauberg, J., Michaelis, I., Merayo, J. M. G., & Brauer, P. (2015). Field-aligned currents' scale analysis performed with the Swarm constellation. *Geophysical Research Letters*, *42*, 1–8. https://doi.org/10.1002/2014GL062453

Lühr, H., Warnecke, J., Zanetti, L. J., Lundquist, P. A., & Hughes, T. J. (1994). Fine structure of field-aligned current sheets deduced from spacecraft and ground-based observations: Initial Freja results. *Geophysical Research Letters*, *21*, 1883.

Lühr, H., Warnecke, J. J., & Rother, M. (1996). An algorithm for estimating field-aligned currents from single-spacecraft magnetic field measurements: A diagnostic tool applied to Freja satellite data. *IEEE Transactions on Geoscience and Remote Sensing*, *34*(6), 1369–1376. https://doi.org/10.1109/36.544560

Lysak, R. L. (1986). Coupling of the Dynamic Ionosphere to Auroral Flux Tubes. *Journal of Geophysical Research*, *91*, 7047–7056. https://doi.org/10.1029/JA091iA06p07047

Lysak, R. L. (1991). Feedback instability of the ionospheric resonant cavity. *Journal of Geophysical Research*, *96*, 1553–1568. https://doi.org/10.1029/90JA02154

Lysak, R. L. (2004). Magnetosphere-ionosphere coupling by Alfvén waves at midlatitudes. *Journal of Geophysical Research*, *109*, A07201, doi:10.1029/2004JA010454

Lysak, R. L. (2023). Kinetic Alfvén waves and auroral particle acceleration: A review. *Reviews of Modern Plasma Physics*, *7*, 6. https://doi.org/10.1007/s41614-022-00111-2

Lysak, R. L., & Song, Y. (2002a). Energetics of the ionospheric feedback interaction. *Journal of Geophysical Research*, *107*, 1160.

Lysak, R. L., & Song, Y. (2002b). A three-dimensional model of the propagation of Alfvén waves through the auroral ionosphere: First results. *Advances in Space Research*, *28*(5), 813–822.

Lysak, R. L., & Song, Y. (2008). Propagation of kinetic Alfvén waves in the ionospheric Alfvén resonator in the presence of density cavities. *Geophysical Research Letters*, *35*, L20101. https://doi.org/10.1029/2008GL035728

Lysak, R. L., Song, Y., Waters, C. L., Sciffer, M. D., & Obana, Y. (2020). Numerical investigations of interhemispheric asymmetry due to ionospheric conductance. *Journal of Geophysical Research: Space Physics*, *125*, e2020JA027866. https://doi.org/10.1029/2020JA027866

Maltsev, Y. P., Leontyev, S. V., & Lyatsky, W. B. (1974). Pi-2 pulsations as a result of evolution of an Alfvén impulse originating in the ionosphere during a brightening of aurora. *Planetary and Space Science*, *22*(11), 1519–1533.

Mann, I. R., Milling, D. K., Rae, I. J., Ozeke, L. G., Kale, A., Kale, Z. C., et al. (2008). The upgraded CARISMA magnetometer array in the THEMIS era. *Space Science Reviews*, *141*(1-4), 413–451. https://doi.org/10.1007/s11214-008-9457-6

Mann, I. R., Usanova, M. E., Murphy, K., Robertson, M. T., Milling, D. K., Kale, A., et al. (2014). Spatial localization and ducting of EMIC waves: Van Allen Probes and ground-based observations. *Geophysical Research Letters*, *41*, 785–792. doi:10.1002/2013GL058581

McGranaghan, R. M., Mannucci, A. J., & Forsyth, C. (2017). A comprehensive analysis of multiscale field-aligned currents: characteristics, controlling parameters, and relationships. *Journal of Geophysical Research: Space Physics, 122*, 11,931–11,960.

McPherron, R. L., Russell, C. T., & Aubry, M. P. (1973). Satellite studies of magnetospheric substorms on August 15, 1968: 9. Phenomenological model of substorms. *Journal of Geophysical Research*, *78*, 3131–3149. doi:10.1029/JA078i016p03131

Milan, S. E. (2013). Modeling Birkeland currents in the expanding/contracting polar cap paradigm. *Journal of Geophysical Research: Space Physics*, *118*, 5532–5542. doi:10.1002/jgra.50393.

Miles, D. M., Mann, I. R., Ciurzynski, M., Barona, D., Narod, B. B., Bennest, J. R., et al. (2016). A miniature, low-power scientific fluxgate magnetometer: A stepping-stone to cubesatellite constellation missions. *Journal of Geophysical Research: Space Physics*, *121*, 11,839–11,860. doi:10.1002/2016JA023147

Miles, D. M., Mann, I. R., Pakhotin, I. P., Burchill, J. K., Howarth, A. D., Knudsen, D. J., et al. (2018). Alfvénic dynamics and fine structuring of discrete auroral arcs: Swarm and e-POP observations. *Geophysical Research Letters*, *45*. https://doi.org/10.1002/2017GL076051

Moen, J., & Brekke, A. (1993). The solar flux influence on quiet time conductances in the auroral ionosphere. *Geophysical Research Letters*, *20*(10), 971–974. https://doi.org/10.1029/92GL02109

Murphy, K. R., Mann, I. R., Jonathan Rae, I., Waters, C. L., Anderson, B. J., Milling, D. K., et al. (2012). Reduction in field-aligned currents preceding and local to auroral substorm onset. *Geophysical Research Letters, 39*(15). https://doi.org/10.1029/2012GL052798

Murphy, K. R., Mann, I. R., Rae, I. J., Waters, C. L., Frey, H. U., Kale, A., et al. (2013). The detailed spatial structure of field-aligned currents comprising the substorm current wedge. *Journal of Geophysical Research: Space Physics, 118*(12), 7714–7727. https://doi.org/10.1002/2013JA018979

Newton, R. S., Southwood, D. J., & Hughes, W. J. (1978). Damping of geomagnetic pulsations by the ionosphere. *Planetary and Space Science, 26*(3), 201–209.

Nokes, C. D. A. (2019). *Lightning excitation of the ionospheric Alfvén resonator*, Master's thesis, University of Alberta.

Ohtani, S., Blomberg, L. G., Newell, P. T., Yamauchi, M., Potemra, T. A., & Zanetti, L. J. (1996). Altitudinal comparison of dayside field-aligned current signatures by Viking and DMSP-F7: Intermediate-scale field-aligned current systems. *Journal of Geophysical Research, 101*, 15,297–15,310. https://doi.org/10.1029/96JA00686

Ohtani, S., Potemra, T. A., Newell, P. T., Zanetti, L. J., Iijima, T., Watanabe, M., et al. (1995). Four large-scale field-aligned current systems in the dayside high-latitude region. *Journal of Geophysical Research, 100*(A1), 137–153, doi:10.1029/94JA01744

Olifer, L., Feltman, C., Ghaffari, R., Henderson, S., Huyghebaert, D., Burchill, J., et al. (2021). Swarm observations of dawn/dusk asymmetries between Pedersen conductance in upward and downward field-aligned current regions. *Earth and Space Science, 8*, e2020EA001167. https://doi.org/10.1029/2020EA001167

Olsen, N., Friis-Christensen, E., Floberghagen, R., Alken, P., Beggan, C. D., Chulliat, A., et al. (2013). The swarm satellite constellation application and research facility (SCARF) and Swarm data products. *Earth Planets and Space, 65*(11), 1189–1200. https://doi.org/10.5047/eps.2013.07.001

Østgaard, N., Humberset, B. K., & Laundal, K. M. (2011). Evolution of auroral asymmetries in the conjugate hemispheres during two substorms. *Geophysical Research Letters, 38*(3), L03101. https://doi.org/10.1029/2010GL046057

Pakhotin, I. P., Burchill, J. K., Förster, M., & Lomidze, L. (2022b). The swarm Langmuir probe ion drift, density and effective mass (SLIDEM) product. *Earth Planets, and Space, 74*, 109. https://doi.org/10.1186/s40623-022-01668-5

Pakhotin, I. P., Mann, I. R., Knudsen, D. J., Lysak, R. L., & Burchill, J. K. (2020). Diagnosing the role of Alfvén waves in global field-aligned current system dynamics during southward IMF: Swarm observations. *Journal of Geophysical Research: Space Physics, 125*, e2019JA027277. https://doi.org/10.1029/2019JA027277

Pakhotin, I. P., Mann, I. R., Lysak, R. L., Knudsen, D. J., Gjerloev, J. W., Rae, I. J., et al. (2018). Diagnosing the role of Alfvén waves in magnetosphere-ionosphere coupling: Swarm observations of large amplitude nonstationary magnetic perturbations during an interval of northward IMF. *Journal of Geophysical Research: Space Physics, 123*, 326–340. https://doi.org/10.1002/2017JA024713

Pakhotin, I. P., Mann, I. R., Sydorenko, D., & Rankin, R. (2022a). Novel EMIC wave propagation pathway through Buchsbaum resonance and interhemispheric wave interference: Swarm observations and modeling. *Geophysical Research Letters, 49*, e2022GL098249. https://doi.org/10.1029/2022GL098249

Pakhotin, I. P., Mann, I. R., Xie, K., Burchill, J. K., & Knudsen, D. J. (2021). Northern preference for terrestrial electromagnetic energy input from space weather. *Nature Communications, 12*, 199. https://doi.org/10.1038/s41467-020-20450-3

Pakhotin, I. P., Walker, S. N., Shprits, Y. Y., & Balikhin, M. A. (2013). Dispersion relation of electromagnetic ion cyclotron waves using Cluster observations. *Annales Geophysicae, 31*(8), 1437–1446. https://doi.org/10.5194/angeo-31-1437-2013

Parent, A., Mann, I. R., & Rae, I. J. (2010). Effects of substorm dynamics on magnetic signatures of the ionospheric Alfvén resonator. *Journal of Geophysical Research, 115*, A02312. doi:10.1029/2009JA014673

Park, J., Lühr, H., Knudsen, D. J., Burchill, J. K., & Kwak, Y.-S. (2017). Alfvén waves in the auroral region, their Poynting flux, and reflection coefficient as estimated from Swarm observations. *Journal of Geophysical Research: Space Physics, 122*, 2345–2360. doi:10.1002/2016JA023527.

Park, J. (2022). Ratio between oversatellite electron content and plasma density measured by Swarm: A proxy for topside scale height. *Journal of Geophysical Research: Space Physics, 127*, e2021JA030137. https://doi.org/10.1029/2021JA030137

Parker, E. N. (1996). The alternative paradigm for magnetospheric physics. *Journal of Geophysical Research: Space Physics, 101*(A5), 10587–10625. https://doi.org/10.1029/95JA02866

Parker, E. N. (2007). *Conversations on electric and magnetic fields in the cosmos*. Princeton, NJ, USA: Princeton University Press. https://doi.org/10.2307/j.ctt2111gdt

Pfaff, R-F. Jr,. (2016). An Overview of the Scientific and Space Weather Motivation for the "Notional" Geospace Dynamics Constellation Mission. *AGU Fall Meeting 2016*.

Prikner, K., Mursula, K., Kangas, J., Kerttula, R., & Feygin, F. Z. (2004). An effect of the ionospheric Alfvén resonator on multiband Pc1 pulsations. *Annales Geophysicae, 22*, 643–651. https://doi.org/10.5194/angeo-22-643-2004

Prikryl, P., Muldrew, D. B., Sofko, G. J., & Ruohoniemi, J. M. (2005). Solar wind Alfvén waves: A source of pulsed ionospheric convection and atmospheric gravity waves. *Annales Geophysicae, 23*, 401–417. https://doi.org/10.5194/angeo-23-401-2005

Prokhorov, B. E., Förster, M., He, M., Namgaladze, A. A., & Holschneider, M. (2014). Using MFACE as input in the UAM to specify the MIT dynamics. *Journal of Geophysical Research: Space Physics, 119*, 6704–6714. doi:10.1002/2014JA019981

Prölss, G. W. (2006) Ionospheric F-region Storms: Unsolved Problems. In *Characterising the Ionosphere* (pp. 10-1–10-20). Meeting Proceedings RTO-MP-IST-056, Paper 10. Neuilly-sur-Seine, France: RTO. Available from: http://www.rto.nato.int/abstracts.asp

Richmond, A. D. (1995). Ionospheric electrodynamics using magnetic apex coordinates. *Journal of Geomagnetism and Geoelectricity, 47*, 191–212

Richmond, A. D. (2010). On the ionospheric application of Poynting's theorem. *Journal of Geophysical Research, 115*, A10311. https://doi.org/10.1029/2010JA015768

Ritter, P., Lühr, H., & Rauberg, J. (2013). Determining field-aligned currents with the Swarm constellation mission. *Earth, Planets and Space, 65*(11), 1285–1294. https://doi.org/10.5047/eps.2013.09.006

Rother, M., Schlegel, K., & Lühr, H. (2007). CHAMP observation of intense kilometer-scale field-aligned currents, evidence for an ionospheric Alfvén resonator. *Annales Geophysicae, 25*(7), 1603–1615. https://doi.org/10.5194/angeo-25-1603-2007

Sakaguchi, K., Kasahara, Y., Shoji, M., Omura, Y., Miyoshi, Y., Nagatsuma, T., et al (2013). Akebono observations of EMIC waves in the slot region of the radiation belts. *Geophysical Research Letters, 40*, 5587–5591. https://doi.org/10.1002/2013gl058258

Sarris, T. E., Talaat, E. R., Palmroth, M., Dandouras, I., Armandillo, E., Kervalishvili, G., et al. (2020). Daedalus: A low-flying spacecraft for in situ exploration of the lower thermosphere–ionosphere. *Geoscientific Instrumentation, Methods and Data Systems, 9*, 153–191. https://doi.org/10.5194/gi-9-153-2020

Shprits, Y. Y., Subbotin, D. A., Meredith, N. P., & Elkington, S. R. (2008). Review of modeling of losses and sources of relativistic electrons in the outer radiation belt II: Local acceleration and loss. *Journal of Atmospheric and Solar-Terrestrial Physics, 70*, 1694–1713. doi:10.1016/j.jastp.2008.06.014, 2008.

Slavin, J. A., Le, G., Strangeway, R. J., Wang, Y., Boardsen, S. A., Moldwin, M. B., & Spence, H. E. (2008). Space technology 5 multipoint measurements of near-Earth magnetic fields: Initial results. *Geophysical Research Letters, 35*, L02107. doi:10.1029/2007GL031728

Song, Y., & Lysak, R. L. (2001). The physics in the auroral dynamo regions and auroral particle acceleration. *Physics and Chemistry of the Earth, Part C: Solar, Terrestrial & Planetary Science, 26*(1–3), pp. 33–42.

Stasiewicz, K., Bellan, P., Chaston, C., Kletzing, C., Lysak, R., Maggs, J., et al. (2000). Small scale Alfvenic Structure in the Aurora. *Space Science Reviews, 92*(3–4), 423–533. https://doi.org/10.1023/A:1005207202143

Surkov, V. V., M. Hayakawa, A. Y. Schekotov, E. N. Federov, and O. A. Molchanov (2006), Ionospheric Alfven resonator excitation due to nearby thunderstorms. *Journal of Geophysical Research, 111*, A01303, doi:10.1029/2005JA011320.

Timoçin, E. (2022). Swarm satellite observations of the effect of prompt penetration electric fields (PPEFs) on plasma density around noon and midnight side of low latitudes during the 07–08 September 2017 geomagnetic storm. *Advances in Space Research, 69*, 1335–1343. https://doi.org/10.1016/j.asr.2021.11.027, 2022

Trenchi, L., Kauristie, K., Käki, S., Vanhamäki, H., Juusola, L., Blagau, A., et al. (2020) ESA field-aligned currents methodology inter-comparison exercise. In: *Ionospheric multi-spacecraft analysis tools* (pp. 167–188). Springer, Cham. https://doi.org/10.1007/978-3-030-26732-2_8

Usanova, M. E., Mann, I. R., Rae, I. J., Kale, Z. C., Angelopoulos, V., Bonnell, J. W., et al. (2008). Multipoint observations of magnetospheric compression-related EMIC Pc1 waves by THEMIS and CARISMA. *Geophysical Research Letters, 35*, L17S25. doi:10.1029/2008GL034458.

Vasyliūnas, V. M. (2001). Electric field and plasma flow: What drives what? *Geophysical Research Letters, 28*(11), 2177–2180. https://doi.org/10.1029/2001GL013014

Vasyliūnas, V. M. (2012). The physical basis of ionospheric electrodynamics. *Annales Geophysicae, 30*(2), 357–369. https://doi.org/10.5194/angeo-30-357-2012

Verkhoglyadova, O., Meng, X., Mannucci, A. J., Tsurutani, B. T., Hunt, L. A., Mlynczak, M. G., et al. (2016). Estimation of energy budget of ionosphere-thermosphere system during two CIR-HSS events: Observations and modeling. *Journal of Space Weather and Space Climate, 6*(A20), 1–22. doi: https://doi.org/10.1051/swsc/2016013

Verkhoglyadova, O. P., Meng, X., Mannucci, A. J., Mlynczak, M. G., Hunt, L. A., Lu, G., et al. (2017). Ionosphere-thermosphere energy budgets for the ICME storms of march 2013 and 2015 estimated with GITM and observational proxies. *Space Weather, 15*, 1102–1124.

Verkhoglyadova, O. P., Meng, X., Mannucci, A. J., & McGranaghan, R. M. (2018). Semianalytical estimation of energy deposition in the ionosphere by monochromatic Alfvén waves. *Journal of Geophysical Research: Space Physics, 123*, 5210–5222. https://doi.org/10.1029/2017JA025097

Vickrey, J. F., Vondrak, R. R., & Matthews, S. J. (1981). The diurnal and latitudinal variation of auroral zone ionospheric conductivity. *Journal of Geophysical Research, 86*(A1), 65–75. https://doi.org/10.1029/JA086iA01p00065

Wang, C., & Sun, T. (2022). Methods to derive the magnetopause from soft X-ray images by the SMILE mission. *Geophysical Letters, 9*, 30. https://doi.org/10.1186/s40562-022-00240-z

Waters, C. L., Anderson, B. J., & Liou, K. (2001). Estimation of global field aligned currents using the iridium system magnetometer data. *Geophysical Research Letters, 28*, 2165–2168. https://doi.org/10.1029/2000GL012725

Watt, C. E. J., Rankin, R., & Marchand, R. (2004). Kinetic simulations of electron response to shear Alfvén waves in magnetospheric plasmas. *Physics of Plasmas, 11*(4), 1277–1284.

Woodroffe, J. R., and R. L. Lysak (2012), Ultra-low frequency wave coupling in the ionospheric Alfvén resonator: Characteristics and implications for the interpretation of ground magnetic fields. *Journal of Geophysical Research, 117*, A03223, doi:10.1029/2011JA017057.

Wright, A. N. (1996). Transfer of magnetosheath momentum and energy to the ionosphere along open field lines. *Journal of Geophysical Research, 101*, 13,169–13,178. https://doi.org/10.1029/96JA00541

Wu, J., Knudsen, D. J., Gillies, D. M., & Burchill, J. K. (2020). Swarm survey of Alfvénic fluctuations and their relation to nightside field-aligned current and auroral arc systems. *Journal of Geophysical Research: Space Physics, 125*, e2019JA027220. https://doi.org/10.1029/2019JA027220

Xiong, C., Jiang, H., Yan, R., Lühr, H., Stolle, C., Yin, F., et al (2022). Solar flux influence on the in-situ plasma density at topside ionosphere measured by Swarm satellites. *J Geophys Res Solid Earth.* https://doi.org/10.1029/2022JA030275

Yau, A. W., & James, H. G. (2015). CASSIOPE enhanced polar outflow probe (e-POP) mission overview. *Space Science Reviews, 189*, 3–14.

# 10

# Alfvén Waves in the Earth's Magnetosphere

**Andrew N. Wright[1], Michael D. Hartinger[2,3], Kazue Takahashi[4], and Tom Elsden[1]**

**ABSTRACT**

The environment surrounding the Earth that is dominated by the terrestrial magnetic field is called the magnetosphere. It comprises a highly structured magnetic field and plasma distribution. Transient events in the magnetosphere can arise from interaction with the solar wind or internal reconfigurations via reconnection, which are a natural source of magnetohydrodynamic (MHD) waves. We shall focus on the global scale waves that have wavelengths similar to the size of the magnetosphere. The different wave modes naturally couple to each other due to the structured nature of the magnetosphere. In particular, coupling of the fast magnetoacoustic mode and the Alfvén mode is common throughout the magnetosphere and is used in magneto-seismology studies. Alfvén waves can also be excited when unstable particle distributions arise. Whatever the source of the Alfvén waves, they are an agent for carrying field-aligned currents which can become particularly intense near the Earth. Consideration of these strong currents requires a two-fluid or kinetic treatment, which indicates the need for a region (at altitudes of 1–2 $R_e$) where electrons are accelerated to kiloelectron volt energies and can produce optical auroral emissions. Alfvén waves can also modify the ionosphere and induce ground currents, both of which have important implications for space weather. We review recent advances including the theory of Alfvén wave excitation in a 3D equilibrium and particle acceleration.

## 10.1. INTRODUCTION

The scope of this review is large-scale Alfvén waves within the Earth's magnetosphere with wavelengths of the order of the size of the magnetosphere itself (10 $R_e$). Specifically, we focus on advances over the last decade and cover theoretical and observational progress. For a description of developments outside of this period, we refer the reader to previous review articles and describe them briefly at the end of this section. We begin with a preamble giving a concise summary of the structure of the magnetosphere and relevant terminology for those unfamiliar with these details. Section 10.1.2 of the review by Nakariakov et al. (2016) provides an alternative overview of the magnetosphere with similar motivation. The textbook by Kivelson and Russell (1995) also provides a summary of magnetospheric structure as well as relevant plasma physics.

### 10.1.1. Structure of the Earth's Magnetosphere

Figure 10.1(a) shows a cartoon of the Earth's magnetosphere, which comprises the region surrounding the Earth that is permeated by terrestrial magnetic field lines. The magnetosphere forms a protective magnetic bubble

[1] Department of Mathematics and Statistics, University of St Andrews, St Andrews, UK
[2] Department of Earth, Planetary, and Space Sciences, University of California Los Angeles, Los Angeles, CA, USA
[3] Space Science Institute, Boulder, CO, USA
[4] The Johns Hopkins University Applied Physics Laboratory, Laurel, MD, USA

*Alfvén Waves Across Heliophysics: Progress, Challenges, and Opportunities, Geophysical Monograph 285*, First Edition.
Edited by Andreas Keiling.

DOI: 10.1002/9781394195985.ch10

that shields the Earth from the fast-flowing solar wind in which it stands. It extends about 10 $R_e$ toward the Sun. The solar wind flow is supersonic at 1 AU, so a bow shock forms upstream of the magnetosphere (not shown in Figure 10.1). The region between the bow shock and the magnetosphere contains subsonic shocked solar wind and is called the magnetosheath.

The solar wind carries magnetic field and plasma from the Sun, which bathes the magnetosphere. The boundary separating the Earth's magnetic field from that of the solar wind is called the magnetopause, and the dominant interaction between these magnetic fields is through reconnection: at the dayside magnetopause the terrestrial field is directed northward. When the solar wind field is directed southward it can reconnect with the terrestrial field on the dayside, forming 'open' field lines that have one end connected to the Earth and extend out into the solar wind.

As the field is frozen into the fast-flowing solar wind plasma, the open field lines are drawn out into a long magnetotail on the nightside. The northern half of the tail has a magnetic field directed earthward, which is opposite to that of the southern half. As the tail magnetic field is oppositely directed across the center of the tail, further reconnection can occur in the tail and lead to the formation of 'closed' field lines (that are solely connected to the Earth) and field lines that are connected to the solar wind, and exiting the tail in the anti-sunward flowing solar wind.

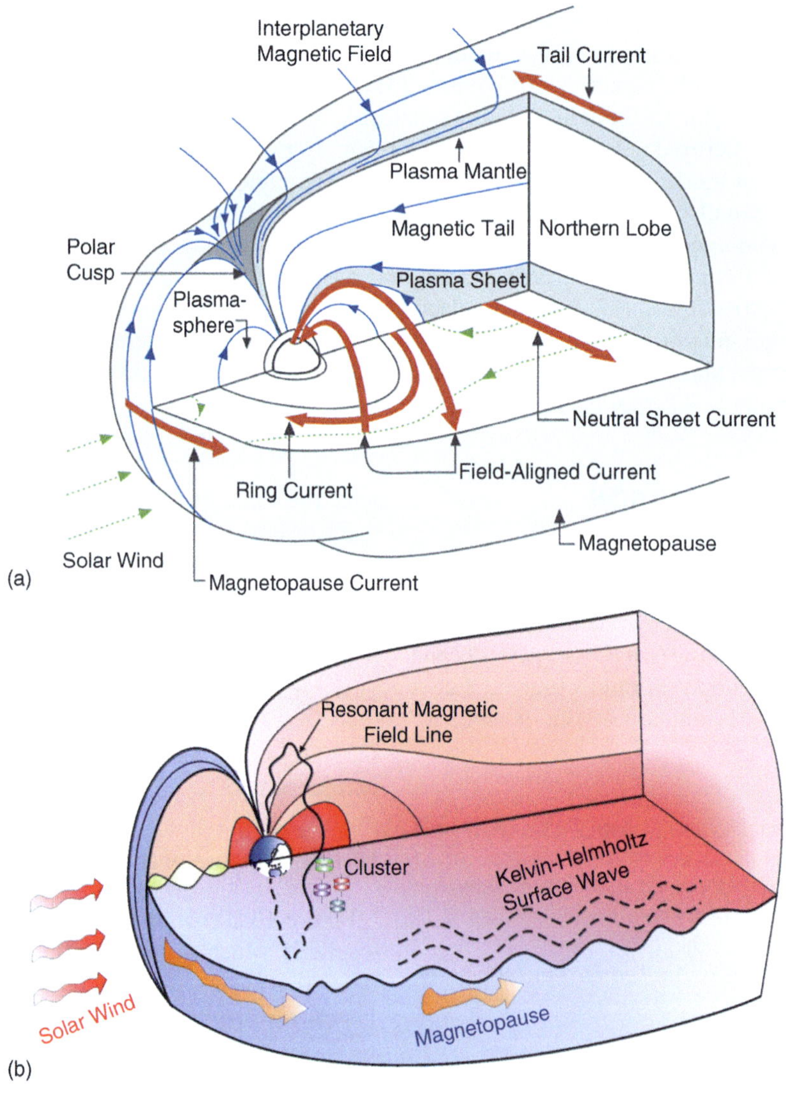


**Figure 10.1** (a) The structure of the Earth's magnetosphere (Nakariakov et al. (2016) / Springer Nature). (b) Excitation of waves in the magnetosphere by the solar wind (from the ESA/Cluster website, https://sci.esa.int/web/cluster/-/41208-cluster-helps-reveal-how-the-sun-shakes-the-earth-s-magnetic-field).

The newly formed closed magnetotail field lines move earthward under magnetic stresses and adopt a dipole-like configuration as they approach the Earth. As more closed flux is added from the tail, these field lines move around the flanks and back to the dayside, where they, once again, reconnect with the solar wind magnetic field. Thus, there is a global convection of magnetic flux and plasma within the magnetosphere, and it is important to understand the density variations this produces.

The open magnetic field lines that are stretched anti-sunward form the outer part of the magnetotail. The volume of these flux tubes increases dramatically as they are stretched, and so they have a very low plasma density. This region is known as the tail 'lobe' and has a high Alfvén speed. When reconnection occurs in the tail, the newly formed closed field lines reduce their volume as they move earthward, which leads to the center of the magnetotail having a much higher density than the tail lobes. It is known as the 'plasma sheet' and has a low Alfvén speed compared to the lobes. As we move northward from the plasma sheet to the lobe, there is a rapid change in density and Alfvén speed, and this region is termed the plasma sheet boundary layer (PSBL).

All the terrestrial field lines involved in the global convection of plasma described earlier will connect to the Earth at relatively high latitudes, as they need to map to the outer parts of the magnetosphere to undergo reconnection with the solar wind field. Field lines that connect to the Earth at low latitudes will not be part of this reconnection cycle. They will remain connected to, and co-rotate with, the Earth. They are in equilibrium with the ionospheric plasma and so have a significantly higher density and lower Alfvén speed compared to neighboring open field lines. The region of closed low-latitude field lines is known as the plasmasphere and is shown as the dark red region close to the Earth in Figure 10.1(b). The outer boundary of the plasmashere is called the plasmapause, and the Alfvén speed changes abruptly across it.

Many particles on closed dipole-like field lines gyrate about the magnetic field and undergo various guiding center drifts. Conservation of the first adiabatic invariant leads to magnetic mirroring as the particles spiral around field lines and encounter enhanced field strengths as they approach Earth. These particles are effectively trapped on the field lines and bounce back and forth across the equator unless other processes come into play. These regions of trapped particles are known as the radiation belts or Van Allen belts. Such particles will also exhibit curvature and gradient drifts. For ions, this results in westward procession, while electrons move eastward and lead to a net current encircling the Earth called the ring current. (See the westerly directed ring current in Figure 10.1(a).)

### 10.1.2. MHD Waves in the Magnetosphere

The nonuniform Alfvén speed of the magnetosphere allows for a rich variety of wave behavior. The simplest way to discuss many of the large-scale waves we observe is through magnetohydrodynamics (MHD). In particular, we shall focus on the fast magnetoacoustic and the Alfvén modes.

Historically, the waves we shall consider were referred to as Geomagnetic Micropulsations, and were first observed as fluctuations in ground magnetometer data. Jacobs et al. (1964) introduced a classification system based on whether the fluctuations have a continuous sinusoidal nature (denoted by Pc, for continuous pulsation) or an irregular nature (Pi, for irregular pulsation). Within each category there are further subdivisions to denote the period, or characteristic time scale, of the pulsation. Of particular relevance to this review is the Pi2 category. These waves have an irregular time dependence with a characteristic timescale in the range of 45–150 s, and are found to originate mainly on the nightside. The other important class of waves we consider is the Pc5 category. These are continuous oscillations with a period in the range of 150–600 s and are found most frequently outside the plasmapause. They are the lowest frequency waves of the magnetosphere and are also referred to as ULF (Ultra-Low Frequency) waves.

Much of the theory of the generation of ULF waves has assumed an axisymmetric equilibrium with wave perturbations that vary sinusoidally with the azimuthal coordinate, as characterized by the azimuthal wavenumber, $m$. We shall see that "Low-$m$" (typically $m < 10$) and "High-$m$" (typically $m > 20$) waves have very different generation mechanisms and properties. Even though modeling of these waves has progressed beyond axisymmetric equilibria (so the use of an azimuthal wavenumber has limited applicability), the terms Low-$m$ and High-$m$ are still used in an informal manner to distinguish between the two types of waves.

#### *Low-m Alfvén Waves*

The source of waves in the magnetosphere can be traced back to the solar wind. The solar wind is fast-flowing, so there is an abundance of free energy that may be tapped. For example, the flanks of the magnetopause can be susceptible to the Kelvin-Helmholtz instability (see Figure 10.1(b)). There are also plasma density variations in the solar wind that can give rise to a fluctuating dynamic pressure on the magnetopause. Sometimes this can give a sudden smack to the magnetosphere, and other times it can provide a continual buffeting of the boundary.

The fast mode is important as it can transport solar wind energy away from the magnetopause and into the magnetosphere. The most easily excited fast waves in the global magnetospheric system are the natural fast eigenmodes. Some of these can be characterized as 'cavity' modes. For example, the abrupt change in Alfvén speed across the plasmapause can lead to effective trapping of fast wave energy in the plasmaspheric cavity. It is also possible to get cavity-like fast modes on the dayside that stand radially between the magnetopause and an inner turning point or the plasmapause. This is depicted as the light green wave in Figure 10.1(b).

The fast eigenmodes on the flanks are free to propagate into the magnetotail, so they are better described as waveguide modes. Both fast cavity and waveguide modes can couple to Alfvén waves on closed field lines. The Alfvén waves have a standing structure between the two ionospheric footpoints and have natural frequencies and harmonics like those of a clamped string under tension. Given that the length of the field lines and Alfvén speeds vary throughout the magnetosphere, the frequency of the Alfvén waves will vary too.

In a simple axisymmetric model of the dipole-like field line region, coupling of the fast mode to the Alfvén mode is most efficient for low azimuthal wavenumbers ($3 < m < 6$). The Alfvén waves excited have plasma displacement, velocity, and magnetic field perturbations aligned with the azimuthal direction. They are termed 'toroidal' Alfvén waves, and their frequency $\omega_A(L)$ is a function of $L$-shell. (Here, the $L$-shell parameter $L$ is the radial distance to the field line in the equatorial plane and is measured in units of Earth radii, $R_e$.)

Resonant excitation of low-$m$ Alfvén waves occurs on the $L$-shell, where the toroidal $\omega_A(L)$ matches the fast cavity or waveguide frequency (Chen & Hasegawa, 1974; Southwood, 1974); see the resonant field line in Figure 10.1(b). These are the lowest frequency Alfvén waves of the magnetosphere and typically have periods of 150–600 s. They are classified as Ultra-Low-Frequency (ULF) waves or Pc5 waves (band 5 of the continuous pulsation classification scheme). Low-$m$ resonantly driven Alfvén waves on closed field lines are often referred to as Field Line Resonances (FLRs).

When low-$m$ Alfvén waves are excited impulsively (rather than resonantly), they can become established across a broad range of $L$-shells. After the impulse has passed, the Alfvén waves will oscillate at their natural frequency, $\omega_A(L)$. Even if the waves on all $L$-shells start with their oscillations being in-phase, as time increases, they will drift further and further out of phase with their neighbors. This process is known as phase mixing, and is characterized by the development of a wavelength across field lines, called the phase-mixing length. For an ideal axisymmetric system, the phase-mixing length in the equatorial plane in the radial direction will be

$$L_{ph} = \frac{2\pi R_e}{|d\omega_A(L)/dL|t}, \qquad (10.1)$$

so reduces with time as $t^{-1}$ (Mann et al., 1995).

### *Magnetotail and Nightside Alfvén Waves*

The northern half of the magnetotail is shown in Figure 10.2, and Alfvén waves also exist here. The structure of the magnetotail formed by a distant downtail reconnection site is shown. (The distant reconnection site is to the far right of the figure, so it is not shown.) The low Alfvén speed plasma sheet (centered on the dashed line) transitions, via the plasma sheet boundary layer (PSBL), to the high Alfvén speed lobe. Reconnection can also occur in an explosive fashion at a near-earth reconnection site, which reconnects significant tail flux in a short time. This process is known as a substorm and releases solar wind energy stored in the stretched magnetic field.

The transient nature of a substorm can cause the magnetotail to reverberate with MHD waves. Fast waves can be released (labelled (1) in Figure 10.2), and these can become trapped in the low Alfvén speed plasma sheet as they propagate earthward, in what is known as the tail waveguide model. The fast mode is able to excite Alfvén waves in the PSBL and the lobe through coupling similar to that of the low-$m$ Alfvén waves on closed field lines. Unlike the low-$m$ Alfvén waves (which have a standing structure along the closed field line), the Alfvén waves excited in the PSBL and lobe (labelled (2) in Figure 10.2) propagate along the field line. The most efficient coupling occurs in the PSBL, with smaller Alfvén waves in the lobe as the fast mode is evanescent there due to the higher Alfvén speed.

The large gradient in Alfvén speed in the PSBL leads to Alfvén wave phase mixing (3) and the generation of strong field-aligned currents there. The current in the PSBL is larger than that in the lobe (4) and is carried mainly by

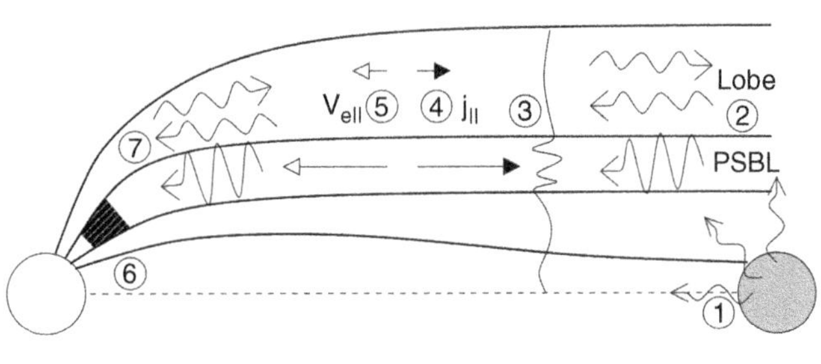


**Figure 10.2** Schematic of MHD waves in the magnetotail waveguide (Wright & Allan (2008) /John Wiley & Sons). Alfvén waves excited in the Plasma Sheet Boundary Layer (PSBL) undergo phase mixing which produces strong field-aligned currents carried by energetic electrons that can generate optical auroral emissions in the ionosphere.

the motion of electrons along field lines (5). The currents are intensified further by the converging field geometry on approaching Earth, which leads to the formation of an electron acceleration region at altitudes of 1–2 $R_e$ (6). A proper treatment of electron energization goes beyond single-fluid MHD. The acceleration of electrons can remove substantial energy from the Alfvén wave, resulting in weak reflection from the ionosphere. This is in contrast to the lobe, where the current is much weaker and allows the possibility of incident and reflected Alfvén waves superposing locally to give a standing wave (7).

The precipitation of energetic electrons associated with strong field-aligned currents in the ionosphere can give rise to optical auroral emissions modulated at the wave frequency. This is also true for Alfvén waves on closed field lines. Describing the energization of these electrons requires the use of two-fluid MHD or kinetic theory.

Besides the possibility of exciting Alfvén waves in the magnetotail through fast mode coupling, there exists the possibility of producing them more directly. A substorm will produce strong earthward flows of reconnected flux that give an impulsive push to the near-earth dipole-like field lines on the nightside. This can be an effective way of exciting the plasmapsheric cavity modes, which then couple to Alfvén waves. These waves are classified as Pi2's as they have an irregular time dependence with periods of 40–150 s.

#### *High-m Alfvén Waves*

The magnetic energy stored in the magnetotail that is released in a substorm can also energize the ions on the reconnected flux tubes as they reduce their volume when moving earthward. This results in the injection of energetic ions on dipole-like field lines that will undergo gradient and curvature drifts around the Earth and give rise to the ring current. The addition of these ions to the preexisting plasma produces an unstable particle distribution with free energy that can be converted to Alfvén waves.

The Alfvén waves produced by this mechanism are poloidal, i.e., the magnetic field, plasma displacement and velocity are in a meridional plane. Importantly, their electric field points in the azimuthal direction, as is required to exchange energy with the azimuthally drifting ion population. The resonance condition involves the poloidal Alfvén frequency (which is different to the toroidal frequency), the bounce and drift frequencies of the ions, and the azimuthal wavenumber ($m$) of the Alfvén wave. For typical values this requires $m$ to be very large ($\geq 100$). As these high-$m$ Alfvén waves on closed field lines are driven by a resonant interaction, they are also commonly referred to as Field Line Resonances (FLRs).

### 10.1.3. Other Reviews

For details of studies prior to the remit of this review, readers are referred to Wright and Mann (2006), Keiling (2009), and Keiling and Takahashi (2011). The topics covered by Wright and Mann (2006) are very similar to this review. Keiling (2009) focuses on Alfvén waves in the magetotail. Keiling and Takahashi (2011) discuss the various models of Pi2 excitation and subsequent propagation.

More recently, Nakariakov et al. (2016) have given a comparative review of MHD waves in the Earth's magnetosphere and the solar corona. Hartinger et al. (2022) have reviewed a coordinated effort to consolidate theory and observations of ULF waves in the Earth's magnetosphere. Zong et al. (2017) provide an extensive review of spacecraft observations and test particle simulations of particle flux oscillations associated with standing Alfvén waves in the Earth's magnetosphere. Duan et al. (2022) review the theory and observations of Alfvén waves in the magnetotail associated with substorms. In particular, they focus on the generation of parallel electric fields that arise in kinetic Alfvén waves when finite ion gyroradius effects and finite electron pressure play a role.

### 10.1.4. Structure of the Review

The remainder of the chapter has the following structure: Section 10.2 summarizes the developments of theoretical ideas and simulations over the last decade that are relevant to terrestrial Alfvén waves; Section 10.3 ties in these ideas with observations of low-$m$ Alfvén waves (those excited by coupling with the fast mode); Section 10.4 focuses on observations of high-$m$ Alfvén waves (driven by resonance with particles); Section 10.5 considers the importance of Alfvén waves for Space Weather; and Section 10.6 gives closing remarks on the wider application of our knowledge of terrestrial Alfvén waves.

## 10.2. THEORETICAL CONCEPTS AND SIMULATIONS

### 10.2.1. Global Magnetospheric Simulations

In this subsection, we consider Alfvén waves in global simulations such as the LFM (Lyon et al., 2004) and BATS-R-US codes (Powell et al., 1999). These nonlinear models use the magnetohydrodynamic approximation, and a key feature is that they specify the upstream solar wind conditions. The simulations naturally develop a self-consistent magnetosheath. Magnetic reconnection produces convective motions of magnetospheric flux tubes, including a magnetotail and substorm behavior. These codes were originally developed to study global magnetospheric structure and the dynamical interaction with the solar wind, rather than Alfvén waves. However, Alfvén waves are a key agent in many magnetospheric processes, such as reconnection and magnetosphere-ionosphere coupling, so it is not surprising that an accurate simulation will provide evidence of Alfvén waves.

Using the LFM code, Claudepierre et al. (2010) specified the upstream solar wind to have monochromatic pressure fluctuations and showed how these could establish toroidal Alfvén waves (Field Line Resonances, FLRs) on dayside closed magnetic field lines where the Alfvén frequency matched that of the driver. This is significant as most of the analytic theories developed at this stage were for axisymmetric equilibria, so these simulations are giving the first evidence that the resonant excitation of Alfvén waves driven by the fast mode will still operate in more general 3D equilibria.

As computer power increased, global simulations were able to adopt more realistic equilibria. A plasmasphere was included in the simulations by Claudepierre et al. (2016), and they found the magnetosphere could respond as a fast waveguide that could drive resonant Alfvén waves both inside and outside the plasmasphere. In particular, the lower Alfvén speed in the plasmasphere allowed the fast mode to penetrate deeper into the magnetosphere and hence facilitated the excitation of plasmaspheric FLRs.

Ellington et al. (2016) used the Space Weather Modeling Framework's (SWMF) code (Tóth et al., 2005) to study the effect of broadband dynamic pressure fluctuations in the solar wind on the magnetosphere. They confirmed the resonant excitation of toroidal Alfvén waves but also observed some local time asymmetries in their amplitudes. They attributed this to various possible causes, such as: coupling to Kelvin-Helmholtz magnetopause surface waves; local time differences in local ionospheric damping rates; and variations in azimuthal wavenumber. This seems reasonable as an increase in ionospheric dissipation will reduce the amplitude of FLRs. Also, it is the magnetic pressure gradient of the fast mode that acts as a driver in the Alfvén wave equation, so we would expect the Alfvén wave amplitude to be dependent on the azimuthal wavenumber. Indeed, Kivelson and Southwood (1986) show that there is an optimum azimuthal wavenumber for drving an FLR; if it is too small, the pressure gradients are also small, and coupling is weak. On the other hand, if it is too large, the fast mode is evanescent, so driving is also weak.

Simulation results for a magnetosphere subject to an impulse in solar wind dynamic pressure were presented by Archer et al. (2022) using the BATS-R-US code. This study focused on the MHD waves excited in response to the impulse. They confirmed field line resonances (FLRs) were present, and these also exhibited phase mixing. An interesting result of the highly compressed field line geometry around noon is that additional nodes of the magnetic field perturbation can be introduced (that are not present in velocity perturbations), and this can cause a reversal in the handedness of polarization ellipses.

The simulation results presented by Zhang et al. (2012) used solar wind and interplanetary magnetic field data to drive the simulation. They note that reconnection in the magnetotail produces sporadic earthward flows of plasma and is associated with Alfvén waves, particularly where information is communicated to the ionosphere. Yang et al. (2022) also found magnetotail reconnection could produce meridonally narrow flux tubes associated with explosive flows propagating earthward. They note that at the front of these tubes, low Alfvén conductances can be produced. Importantly, the conductance is relatively uniform along the flux tube, and this promotes efficient transport of Alfvénic energy along the tube as reflection from nonuniformity is minimized.

### 10.2.2. Simulations and Theory of Linear Low-*m* Alfvén Waves

The term 'low-$m$' Alfvén waves (see Section "Low-$m$ Alfvén Waves") originates from early studies that used an axisymmetric equilibrium and so could adopt an azimuthal wavenumber $m$ to facilitate the analysis. The corresponding Alfvén waves have a toroidal polarization and are driven by the fast mode. The coupling is most efficient for small values of $m$ (typically 2–6) as the fast mode becomes evanescent with increasing $m$. Low-$m$ Alfvén waves are associated with fast modes in the form of cavity and waveguide modes.

Low-$m$ Alfvén waves have a small amplitude (compared to equilibrium quantities), so linear theory can be used. The studies in this subsection have exploited this property to devise simulation codes tailored to study MHD wave behavior rather than the more general magnetospheric dynamics of the global simulation codes of the previous subsection. The linear codes do not have a self-consistent interaction with the solar wind; however, they do resolve wave behavior in the magnetosphere with great clarity and have led to substantial advances over the past decade. An important difference between the global codes and the linear codes is that the latter employ field-aligned coordinates. Indeed, this is the key to achieving clean, high resolutions results, as Alfvén waves can have a small perpendicular scale but a global parallel scale. Field-aligned coordinates allow high resolution in the perpendicular direction without unnecessary (and inefficient) over-resolution in the parallel direction.

Previous studies typically employed an axisymmetric equilibrium, used a simplified magnetic field geometry and boundaries, and idealized boundary conditions. There is also the distinction between normal mode solutions (where all variables have a prescribed $\exp(i\omega t)$ dependence) and solutions to the time-dependent equations. Progress has been made to generalize all of these details, although individual studies have tended to focus on generalizing one or two aspects.

### *Ionospheric Boundaries*

While field-aligned coordinates are central to the advances described in this subsection, the use of orthogonal field-aligned coordinates has the problem of not fitting a spherical ionospheric boundary. This aspect has been addressed by the introduction of non-orthogonal field-aligned coordinates in several studies (Degeling et al., 2010, 2018; Lysak et al., 2015, 2020; Obana et al., 2015; Takahashi, Lysak, et al., 2018). Studies where the ionospheric boundary is simply taken to be an orthogonal field-aligned coordinate have the property that the Alfvén wave magnetic field is tangential to the ionosphere, so it can easily satisfy a perfectly or partially reflecting boundary condition. For a spherical ionosphere, the Alfvén wave magnetic field is no longer guaranteed to be tangential to the ionosphere, which leads us to question whether FLRs can be excited. A study by Prokopyszyn et al. (2021) shows that FLRs can still be excited, but that an additional evanescent fast mode is present at the ionosphere as a compressional magnetic field perturbation is now required. Hence, the effect on the global FLR is not expected to be significant.

The sunlit ionosphere is generally highly conducting, and many studies adopt the perfectly reflecting limit, which gives a node of the electric field and plasma velocity (Degeling et al., 2010; Wright & Elsden, 2020). When both ionospheres are highly conducting, the standing fundamental Alfvén mode has half a wavelength between ionospheres. When the ionosphere is in darkness, the conductivity can be low, which leads to an antinode of the electric field and plasma velocity. Obana et al. (2015) used a spherical sheet ionosphere to consider an Alfvén wave excited near the terminator such that one footpoint was in a sunlit ionosphere and the other in darkness. They showed how the fundamental standing mode now had a quarter wavelength, as predicted theoretically by Allan and Knox (1979). Such modes have also been found in simulations by Lysak et al. (2020), who adopted a height resolved ionosphere (Lysak et al., 2015). They found that quarter wavelength modes result when the ratio of Pedersen conductances between the sunlit and dark ionospheres is greater than 5.

### *Outer Boundary and Waveguide/Cavity Modes*

The outer boundary for linear models is the magnetopause, as there is no solar wind or magnetosheath. Often the outer boundary is simply taken as a high latitude dipole field line $L$-shell (Elsden & Wright, 2017; Lysak et al., 2015; Obana et al., 2015), so these equilibria will not represent waves near the cusp accurately. However, they have proven to be excellent at modeling waves throughout the rest of the magnetosphere. These models have been extended to non-axisymmetric equilibria by allowing for an arbitrary plasma density distribution and adopting a realistic equatorial magnetopause to facilitate studying the flaring of the magnetopause on the flanks and into the tail (Elsden & Wright, 2022; Wright & Elsden, 2020).

The latter used orthogonal coordinates whose coordinate surfaces do not match the outer boundary. Wright and Elsden (2020) show how this can be accommodated by using some suitable interpolation. More complex compressed dipole fields that are non-axisymmetric have been adopted by Degeling et al. (2010, 2018) and, using non-orthogonal coordinates, are able to have a coordinate surface coincide with the magnetopause.

Once the outer boundary has been determined, it is necessary to apply appropriate boundary conditions. Driving by the solar wind is generally regarded as supplying a varying dynamic pressure on the magnetopause, which is modeled by specifying the magnetic pressure on the outer boundary. If the magnetopause is undriven, a node of magnetic pressure is normally applied.

Once the inner and outer boundaries are specified, we can consider fast mode propagation within the system that can carry energy from the magnetopause to locations where resonant coupling to Alfvén waves may take place. For studies that use a closed magnetopause (Degeling et al., 2010, 2018; Lysak et al., 2015, 2020; Obana et al., 2015; Takahashi, Lysak, et al., 2018), cavity modes are relevant. Other studies have mimicked the ability of the magnetotail to let energy flow out into the open tail and leave the system. They do this by introducing a dissipative buffer zone on the nightside such that any waves that enter this region will be dissipated and so will appear to have left the system (Elsden & Wright, 2017, 2019; Wright & Elsden, 2016, 2020).

In an axisymmetric cavity the fast cavity normal modes will simply propagate in the azimuthal direction. When energy is permitted to leave on the nightside, it is appropriate to consider waveguide modes, whose behavior is very different from cavity modes. Elsden and Wright (2019) used a line dipole equilibrium and found that the non-axisymmetric waveguide normal mode had a spatially standing nature in azimuth around noon, which switched to anti-sunward propagation on the flanks. They found this behavior persisted in an analytical cartesian model, including the property that these modes decay in time as they lose energy to the tail. Using an equilibrium based on a 3D dipole with flaring magnetopause on the flanks Wright and Elsden (2020) found these properties persisted even when an axisymmetric density was assumed.

### *Driver Properties*

The models in this subsection represent solar wind driving by applying a time-dependent pressure on the outer boundary (the magnetopause). Some simulations have focused on strictly monochromatic solutions and reported normal mode properties (Degeling et al., 2018; Wright & Elsden, 2016; Wright et al., 2022). Others have considered a steady monochromatic driver to focus on the time dependent evolution to a normal mode (Elsden & Wright, 2017). This study demonstrated how phase mixing persists in 3D resonant coupling and confirmed the formula for the width of the resonance as a function of time (derived in 1D by Mann et al. (1995)) remains accurate in 3D. It also validates formulae for the saturated width and amplitude of the resonance against simulation results. Lysak et al. (2015) and Degeling et al. (2010) used a decaying monochromatic driver, while others have used a broadband impulse of pressure (Elsden & Wright, 2018, 2019, 2022; Takahashi et al., 2016; Wright et al., 2018; Wright & Elsden, 2020). All these studies show clear evidence of resonant Alfvén waves, either at the driver frequency (for monochromatic driving) or the waveguide/cavity frequencies (for impulsive driving).

It is also important to consider where the driving pressure is applied. Some studies focus on the nightside outer boundary to represent the effect of substorms in the magnetotail releasing energy earthward and causing wave behavior similar to Pi2's (Lysak et al., 2015). Other studies have focused on dayside or flank driving (Degeling et al., 2010, 2018; Elsden & Wright, 2017, 2019; Wright & Elsden, 2016, 2020; Wright et al., 2018, 2022). These studies have shown the importance of the location of the energy source on the locations where FLRs are established. When the magnetopause is driven on the nightside, Alfvén resonances tend to be excited there (Lysak et al., 2015). Both Degeling and Rankin (2008) and Elsden and Wright (2019) show how varying the local time of the driver from noon toward dusk excites Alfvén waves more on the dusk flank. This raises the possibility of using the location of Alfvén waves to constrain where energy crosses the magnetopause.

### *Local Time Dependence and Plumes*

If the equilibrium and driver are symmetric about noon, the resonant Alfvén waves that are excited are also symmetric. Elsden and Wright (2018) show how the location of the Alfvén waves is relatively insensitive to the spatial and temporal scale of the driver so long as the waveguide modes, are excited. This is because it is the pressure gradients of the fast waveguide modes that actually drive the FLRs.

In reality, the magnetosphere is not symmetric about noon as the dusk flank is more dense and has lower Alfvén frequencies than at dawn. Observations show that Alfvén waves are more frequently excited, and with greater amplitude, at dawn rather than dusk (Takahashi et al., 2015). It seems plausible that this may be due to the Kelvin-Helmholz instability operating more effectively at dawn, however, Takahashi et al. (2016) show this is not the case. They also present simulations which explain the dawn/dusk Alfvén wave asymmetry in terms of the dependence of the efficiency of fast and Alfvén wave coupling on the density gradient. Wright et al. (2018) also explored the dawn/dusk asymmetry in a magnetosphere with an asymmetric density and Alfvén speed variation. Even when they applied a symmetric pressure pulse to the magnetopause, the asymmetric refraction of the fast mode was seen to excite larger amplitude Alfvén waves at dawn compared to dusk. This is consistent with other studies (Archer et al., 2015, 2017) that found fast cavity and waveguide modes may be more frequent at dawn compared to dusk due to the different local wave speed profiles.

Following substorms, there are occasions when a plasma drainage plume allows plasmaspheric material to extend into the magnetosphere beyond the nominal plasmapause (Lambour et al., 1997). This can give rise to rapid variations of density with magnetic local time (MLT) in the afternoon sector. Numerical studies of Alfvén wave FLRs with a plume present have been completed as both normal mode (Degeling et al., 2018) and impulsively driven (Wright & Elsden, 2020) calculations. Figure 10.3 is taken from the latter study and shows the depression in Alfvén speed (in the equatorial plane) associated with a dense plasma plume in the afternoon. The Alfvén speed outside the plume is axisymmetric.

Figure 10.3(b) uses $j_{\|}$ to identify where the Alfvén waves are. Several interesting features are evident in this figure. The nested colored ridges are actually contours of the Alfvén wave frequency (figure 8(b) of Elsden and Wright (2017) and figure 4 of Elsden and Wright (2018)). On the dawnside ($y < 0$), the equilibrium is locally axisymmetric, so 2D theory applies here and, as expected (Wright & Thompson, 1994), Alfvén waves sit on circles of constant $L$ (see the reference dashed circles in panel (b)). As we move radially outward the Alfvén frequency decreases giving rise to phase mixing, which manifests as successive colored ridges.

The situation is quite different in the afternoon. Tracing a ridge from noon, it moves inward and crosses $L$-shells on encountering the plume and then moves back out to larger $L$-shells on leaving the plume. The Alfvén wave current ridges in 2D theory (applicable at dawn, $y < 0$) sit on circles, but on the edges of a plume they

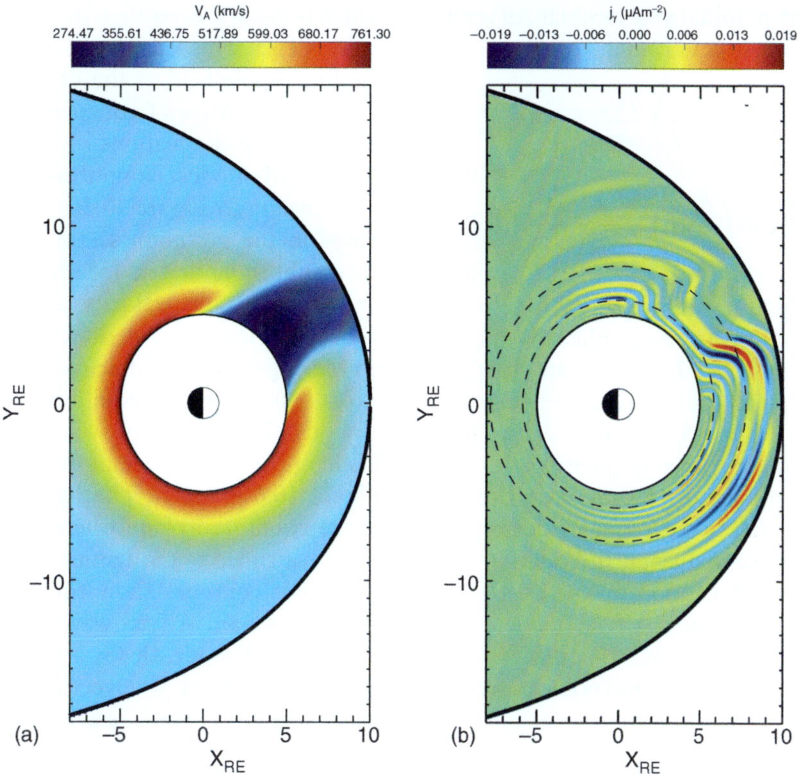


**Figure 10.3** Simulation results (Wright & Allan (2020) /John Wiley & Sons). (a) The Alfvén speed in the equatorial plane with a plasma drainage plume in the afternoon. (b) The Alfvén wave currents resulting from an impulsive pressure pulse. (The field-aligned current is calculated at the ionospheric end and then mapped along the field line and plotted in the equatorial plane.)

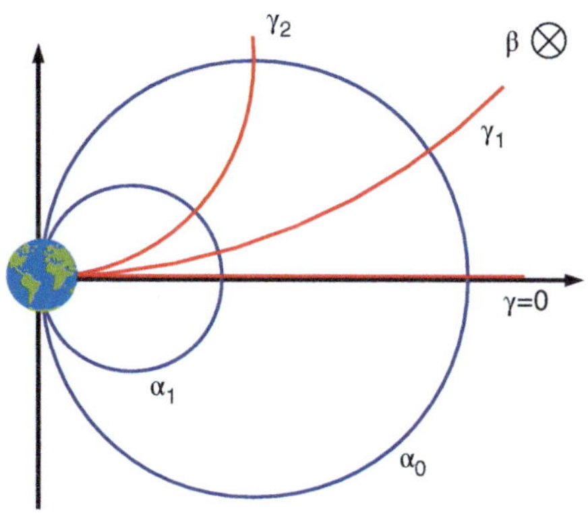


**Figure 10.4** The orthogonal field-aligned coordinates $(\alpha, \beta, \gamma)$ for a dipole field (Wright & Allan (2016) / IOP Publishing).

can be inclined. Indeed, the ridges at $(x, y) = (4.5, 6.5)$ are almost perpendicular to $L$-shells. On the edges of a plume the equilibrium is far from 2D because of the rapid variation in density with azimuth, yet resonant Alfvén waves are clearly present (Degeling et al., 2018; Wright & Elsden, 2020). These locations are ideal for studying the differences between 2D and 3D FLRs, and virtual satellite observations have been generated by Elsden and Wright (2022), which show substantial differences from 2D toroidal FLRs. The principal difference being that in 2D (axisymmetric models) the Alfvén wave velocity and field perturbations are in the azimuthal direction, while on the edges of a plume they have a polarization somewhere between poloidal and toroidal.

### *Alfvén Wave Polarization and Resonance Maps*

Before considering Alfvén waves of arbitrary polarization, we study their properties in the toroidal and poloidal limits using field-aligned coordinates $(\alpha, \beta, \gamma)$, as shown in Figure 10.4. In 2D (line dipole) and 3D dipole equilibrium magnetic fields the coordinate directions are orthogonal, with surfaces of constant $\alpha$ corresponding to nested axisymmetric surfaces of field lines (the $L$-shells defined in Section "Low-$m$ Alfvén Waves"). The invariant azimuthal coordinate is $\beta$. Along a given field line $\alpha$ and $\beta$ have constant values, so they can be used to label the field lines. The position along a field line is specified by the $\gamma$ coordinate.

The equations for the poloidal and toroidal Alfvén frequencies are given in Wright and Elsden (2016) as

$$\frac{\partial}{\partial\gamma}\left(\frac{1}{h_\gamma}\frac{\partial U_\alpha}{\partial\gamma}\right)+\frac{1}{h_\gamma}\frac{\partial}{\partial\gamma}\left(\ln\left(\frac{h_\alpha}{h_\beta}\right)\right)\frac{\partial U_\alpha}{\partial\gamma}+\frac{\omega_A^2}{V_A^2}h_\gamma U_\alpha=0, \tag{10.2}$$

$$\frac{\partial}{\partial\gamma}\left(\frac{1}{h_\gamma}\frac{\partial U_\beta}{\partial\gamma}\right)+\frac{1}{h_\gamma}\frac{\partial}{\partial\gamma}\left(\ln\left(\frac{h_\beta}{h_\alpha}\right)\right)\frac{\partial U_\beta}{\partial\gamma}+\frac{\omega_A^2}{V_A^2}h_\gamma U_\beta=0. \tag{10.3}$$

The quantities $h_\alpha$, $h_\beta$, and $h_\gamma$ are in general functions of $(\alpha, \beta, \gamma)$ and are scale factors that encapsulate the magnetic field geometry and strength ($B$). $U_\alpha$ and $U_\beta$ are simulation variables that are related to the transverse velocity components ($u_\alpha$ and $u_\beta$) by $U_\alpha = u_\alpha h_\beta B$ and $U_\beta = u_\beta h_\alpha B$. The Alfvén speed is $V_A$, and the Alfvén frequency is $\omega_A$.

Equation (10.2) has plasma displacement and velocity in the $\alpha$ (poloidal) direction, so the frequency corresponds to the poloidal Alfvén wave frequency. In Equation (10.3) the wave has a toroidal polarization (aligned with $\beta$), so $\omega_A$ here corresponds to the toroidal Alfvén frequency. The middle term of these equations shows that if $h_\alpha/h_\beta$ varies along the field line, then the toroidal and poloidal frequencies will be different.

Figure 10.5 shows the ratio of toroidal frequency ($\omega_\beta$) to poloidal frequency ($\omega_\alpha$) as a function of invariant latitude (the latitude of the field line footpoint in the ionosphere). The invariant latitude range shown of 0.5 to 1.4 rad corresponds to $\sim 28^\circ$ to $\sim 80^\circ$. The left-hand panel used a 2D line dipole, and the right-hand panel used a 3D dipole. (A 2D line dipole field is the magnetic field surrounding a pair of infinitely long straight wires that carry an equal, but opposite, current. If the wires are aligned with the $y$ direction, the magnetic field will be invariant in this direction, and only the $x$ and $z$ components of the field will be nonzero.) The black lines show the frequency ratios for the fundamental toroidal and poloidal modes which clearly have the greatest difference. Even for a modest invariant latitude of 60° ($\sim$ 1 rad) the line dipole has $\omega_\beta/\omega_\alpha \approx 1.6$, which is much greater than that of the 3D dipole where the ratio is 1.25. This led Wright and Elsden (2016) to adopt the line dipole field as the discrepancy between toroidal and poloidal Alfvén frequencies is substantial and makes it easy to see the features of 3D Alfvén resonances very clearly with modest computing power.

Returning to Figure 10.3(b), we know that the Alfvén waves in the pre-noon sector at $(x, y) = (7, -5)$ exist in an equilibrium that is locally independent of azimuth, so have a plasma velocity aligned with the azimuthal direction. However, the direction of the plasma velocity on the edges of the plume (where the Alfvén wave ridges cross $L$-shells as the equilibrium is 3D here) is not so obvious. This property was addressed by Wright and Elsden (2016) in a normal mode calculation from which Figure 10.6 is taken. The figure shows a close-up view in the equatorial plane where the resonant Alfvén wave ridge is inclined to the azimuthal direction by around 35°. The color contours show time-averaged energy density. As the resonant Alfvén wave amplitude is much greater than the fast mode, this is a good proxy for the location of the Alfvén wave. The vectors show the direction of the plasma velocity, which is evidently aligned with the path of the resonant ridge.

Resonant Alfvén waves in 3D require us to go beyond the current ideas of poloidal and toroidal Alfvén waves and to consider waves with an arbitrary polarization. To determine the Alfvén frequency for such waves on a particular field line, Wright and Elsden (2016) considered the local $\alpha$ and $\beta$ directions in the equatorial plane for this field line. These directions are then rotated by an angle $\theta$ to produce new coordinate directions, called $\alpha'$ and $\beta'$. Naively, we might expect the corresponding Alfvén wave equation to be

$$\frac{\partial}{\partial\gamma}\left(\frac{1}{h_\gamma}\frac{\partial U_{\beta'}}{\partial\gamma}\right)+\frac{1}{h_\gamma}\frac{\partial}{\partial\gamma}\left(\ln\left(\frac{h_{\beta'}}{h_{\alpha'}}\right)\right)\frac{\partial U_{\beta'}}{\partial\gamma}+\frac{\omega_A^2}{V_A^2}h_\gamma U_{\beta'}=0, \tag{10.4}$$

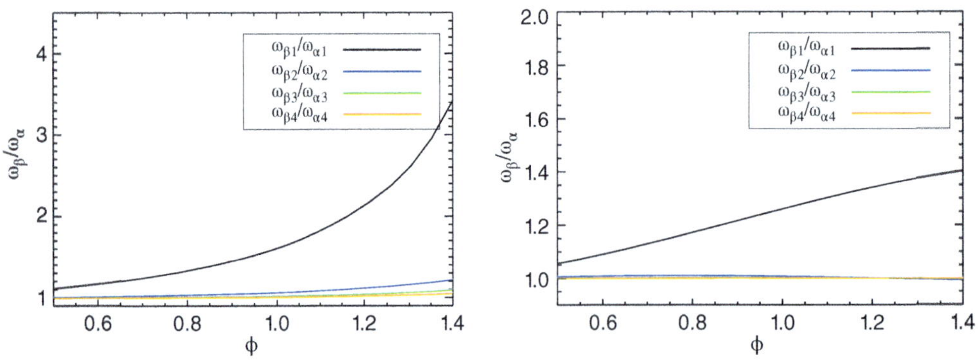


**Figure 10.5** Ratio of the toroidal ($\omega_\beta$) to poloidal ($\omega_\alpha$) Alfvén frequencies for a line dipole (left) and a 3D dipole (right) as a function of invariant latitude over the range 0.5 to 1.4 rad (28° to 80°). Black, blue, green, and brown correspond to the first four harmonics. Source: Reproduced from Elsden (2016) / University of St Andrews Library.

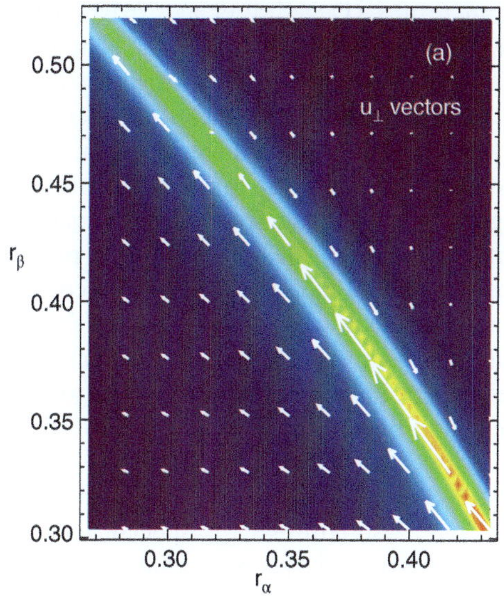


**Figure 10.6** A close-up view in the equatorial plane of time-averaged energy density, which is indicative of the location of the resonant Alfvén waves. The arrows show the plasma velocity, which is aligned with the path of the resonant Alfvén waves. Source: Reproduced from Elsden (2016) / University of St Andrews Library.

where $h_{\alpha'}$ and $h_{\beta'}$ are the scale factors for the rotated coordinate system. There are some issues with the existence of orthogonal coordinates for the rotated system (Salat & Tataronis, 2000). Nevertheless, Wright and Elsden (2016) show how the effective scale factors can be calculated for Equation (10.4) to be applicable. They provided analytical expressions for the scale factors. In more general fields, the scale factors can be calculated by tracing neighboring field lines and considering their separation (Degeling et al., 2010, 2018; Wright et al., 2022).

The result of solving Equation (10.4) is to determine the Alfvén frequency for a given polarization angle, $\omega_A(\theta)$, and such results are shown in Figure 10.7(a). The range of $\theta$ is 0 to $\pi/2$, corresponding to toroidal and poloidal polarizations, respectively. The values of $\omega_A$ in these two limits are identical to the solutions of Equations (10.3) and (10.2), so these are familiar results. What is not so well known is that $\omega_A$ can change smoothly and continuously between these two values, as shown by the $\omega_A(\theta)$ curves.

Recalling that $\alpha$ and $\beta$ have constant values on a given field line, we see that Figure 10.7(b) has five red dots that label five different field lines. (For the scale factors used in Figure 10.7, $\alpha$ and $\beta$ are equivalent to radial and azimuthal distances in the equatorial plane.) For field line (e), $\omega_A(\theta)$ is given by the top line in the left-hand panel. If the normal mode is driven with a frequency $\omega_d$ (indicated by the horizontal line in Figure 10.7(a)), then it is not possible to choose a polarization for this field line such that $\omega_A(\theta) = \omega_d$, so this field line is termed 'non-resonant.'

Assuming a simple radial monotonic variation in Alfvén speed, we can move radially out (in $\alpha$) to get to the field line (a). Curve (a) in Figure 10.7(a) shows that this field line can meet the resonant condition $\omega_A(\theta) = \omega_d$ if we choose the polarization angle to be $\theta = \pi/2$, which is the poloidal direction and is indicated by the short black line attached to the red dot labeled (a). Proceeding radially outward to field line (c), we need to choose an angle $\theta = \theta_c$ (which is intermediate to the toroidal and poloidal directions) to match the resonant condition. Moving radially further out to field line (b), we are just able to match the resonance condition with a toroidal ($\theta = 0$) polarization, but field lines beyond this will be non-resonant. This procedure can be repeated for different azimuths ($\beta$) to identify the Resonant Zone (where a resonant polarization exists) and Non-Resonant Zones (where a resonant polarization does not exist). The regions are separated by Resonant Zone Boundaries where resonance requires a

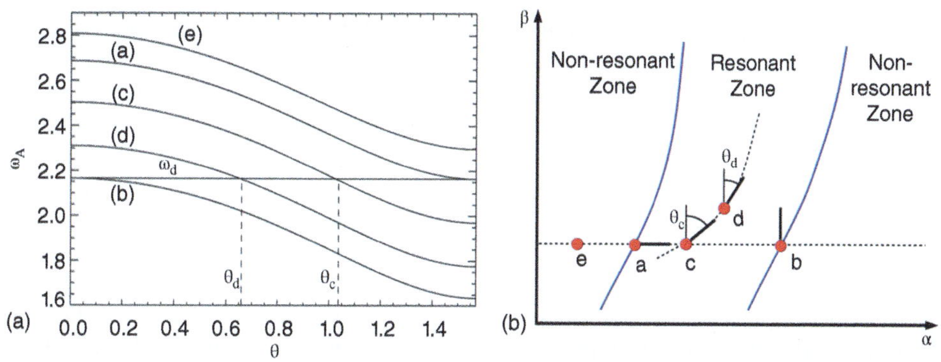


**Figure 10.7** (a) The variation of Alfvén frequency, $\omega_A$, with polarization angle, $\theta$, for five different field lines. Toroidal and poloidal polarizations correspond to $\theta = 0$ and $\pi/2$, respectively. The horizontal line denotes the driving frequency, $\omega_d$. (b) The Resonance Map showing the locations of the five field lines along with their resonant polarizations such that $\omega_A(\theta) = \omega_d$. Source: Reproduced from Elsden (2016) / University of St Andrews Library.

toroidal or poloidal polarization. Figure 10.7(b) is known as a Resonance Map (Wright & Elsden, 2016).

Wright and Elsden (2016) show how these ideas may be combined to deduce an equation for the path the resonant ridge takes in Figure 10.6 as **u** is aligned with this ridge. Consider again field line (c) in Figure 10.7(b): we require the plasma velocity to be aligned with the short black line with an angle $\theta_c$ for this field line to be resonant. Hence, the resonant Alfvén wave ridge will coincide with this line (as shown in Figure 10.6). Stepping incrementally in this direction to field line (d), which will also be on the ridge, we need to adopt a polarization angle $\theta_d$ (which is slightly different from $\theta_c$) to be resonant. Proceeding in this fashion, the entire path of the resonance can be mapped out in 3D and is denoted by the curved dashed line.

Figure 10.7(a) was constructed for a line dipole. The symmetry of this field leads to the $\omega_A(\theta)$ curves being symmetric about $\theta = \pi/2$. Thus, besides field line (c) being resonant for polarization angle $\theta = \theta_c$, it will also be resonant for $\theta = \pi - \theta_c$, so we could introduce a second resonant path in panel (b) through field line (c) that is inclined at this angle. It is also a property of the line dipole (and the 3D dipole) that the maximum $\omega_A$ is always the toroidal frequency and the minimum is always the poloidal. This is not true for more general fields such as the compressed dipole field considered by Degeling et al. (2010) and Wright et al. (2022), from which Figure 10.8 is taken.

One generalization evident in a compressed dipole field is that the maximum and minimum $\omega_A$ on a field line no longer correspond to the toroidal and poloidal directions. The Resonant Zone Boundaries are now identified as where the maximum and minimum $\omega_A$ match $\omega_d$. Figure 10.8(a) shows these boundaries as red lines, between which is the Resonant Zone. For an MLT of 0 hours (midnight, $(x, y) = (-9, 0)$) the resonant paths (black lines) approach the outer boundary with a toroidal polarization (i.e., $\omega_A(\theta_r) = \omega_d$ if $\theta_r = 0$). In contrast, the resonant paths approach the same outer boundary at noon (12 MLT, $(x, y) = (9, 0)$) with a poloidal orientation ($\theta_r = 90°$). Indeed, $\theta_r$ on the outer boundary changes continuously with MLT on both boundaries, as summarized in Figure 10.8(b). We note that at noon the poloidal frequency is greater than the toroidal, which is the opposite of the situation for an uncompressed dipole. This is due to the field lines in a meridional plane not having monotonic convergence between the equator and ionosphere and is the same geometric property that led to the unusual hodogram polarization changes reported by Archer et al. (2022) and discussed in subsection 10.2.1.

The family of resonant paths shown in Figure 10.8(a) are selected paths from an infinite number of possible solutions as every point in the Resonant Zone could potentially lie on the resonant Alfvén wave ridge. Simulation results (Degeling et al., 2010, 2018; Wright & Elsden, 2016; Wright et al., 2022) show that only one or two of these paths describe where the resonant Alfvén waves are actually found in a given simulation. Figure 10.8(c) is a simulation result with a Resonance Map like that in panel (a). The color contours are of field-aligned vorticity magnitude in the equatorial plane. Overplotted are black lines showing the possible resonance paths. Clearly, one particular path coincides with the resonant ridge, and this is shown as the green and black dashed line in panel (a). This particular resonant path seems to be favored as it corresponds to a separatrix of the resonant paths that Mager and Klimushkin (2021) explain is special as Alfvén wave energy will naturally accumulate there. Wright et al. (2022) also note that this particular resonant path is special as it meets the Resonant Zone Boundaries with a tangential alignment. This condition also applies in situations where there is only one Resonant Zone Boundary, meaning a separatrix cannot be defined (see figure 5 of Wright et al. (2022)). The examples of Resonance Maps considered so far have Resonant Zones that do not encounter the boundaries of the simulation domain (excluding periodic boundaries). When simulation boundaries are encountered, the boundary conditions there can be used to identify the resonant path that coincides with the simulation results (Wright & Elsden, 2016).

An alternative formulation of the governing equations (assuming $b_\parallel = 0$) by Degeling et al. (2010) yielded an eigenvalue problem for two Alfvén wave eigenmodes. Interestingly, these coincide with the Resonant Zone Boundaries as defined by Wright and Elsden (2016). This suggests the possibility that Degeling's eigenvalue calculation may be an efficient method of finding the Resonant Zone Boundaries, in contrast to the method of Wright and Elsden (2016), which requires iterating the shooting method to solve equation (10.4) and further iteration to determine the position where $\text{Max/Min}[\omega_A(\alpha, \beta)] = \omega_d$. It is yet to be shown mathematically that these two approaches are equivalent (Elsden et al., 2022).

Recently, Mager and Klimushkin (2021) derived analytical normal solutions in the high-$m$ limit, which means the fast mode may be neglected as it is evanescent so may not provide an effective driver of the resonant Alfvén waves (Elsden et al., 2022). They produced an analytical formula for the 3D Alfvén wave equation, which should be equivalent to Equation (10.4). Extrapolating 1D and 2D results to 3D would suggest that at the resonance $u_{\beta'}$ would have a $1/\alpha'$ singuarity and $u_{\alpha'}$ would have a logarithmic singularity. Mager and Klimushkin (2021) found the logarithmic singularity was absent and that $u_{\alpha'}$ remained finite, but with small-scale oscillations in $\alpha'$. It would be interesting to look for these properties in the simulations results.

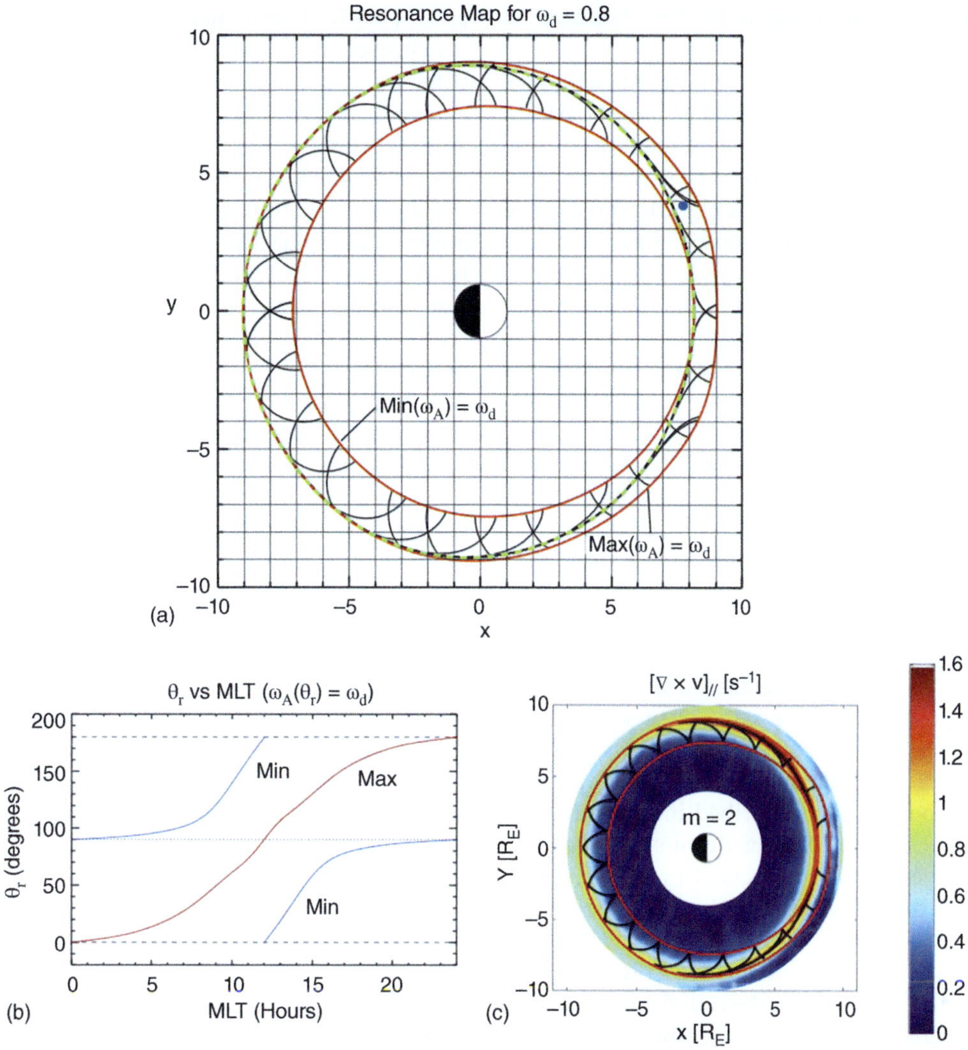


**Figure 10.8** (a) Resonance Map in the equatorial plane for the compressed dipole equilibrium used in Wright et al. (2022). The Resonant Zone Boundaries are shown in red. Black lines indicate permissible resonant Alfvén wave paths. (b) The variation of the resonant polarization angle ($\theta_r$) on the boundaries of the Resonant Zone with magnetic local time. (c) Alfvén wave amplitude (using parallel vorticity) in the equatorial plane with the resonant paths from (a) overplotted. The resonant ridge coincides with one particular path that is shown as the dashed black and green line in (a).

The approach used by Mager and Klimushkin (2021) to construct resonant paths is to consider the perpendicular group velocity of the Alfvén waves to work out the direction the wave will propagate perpendicular to the field, as this will be tangential to the paths. The Resonant Zone Boundaries are found by equating the toroidal or poloidal frequency with the normal mode frequency ($\omega_d$) in their axisymmetric magnetic field equilibrium.

Recently, Elsden et al. (2022) have provided a detailed review of the theory of 3D Alfvén resonances research and Resonance Maps. They cover both analytical work and computational studies. The review is a useful resource for readers who would like a more detailed account than given here.

### 10.2.3. Modeling High-*m* Poloidal Waves

This section concerns recent modeling of high-m poloidal waves in theory and simulation. Observations of high-m waves are reserved for Section 10.4. As briefly discussed in Section "High-*m* Alfvén Waves", high-m Alfvén waves are generally internally driven by resonance with particle populations with the appropriate drift and bounce frequencies (Southwood et al.,

1969) (the resonance condition is further discussed in Section 10.4).

Once established, Radoski (1974) first predicted, from consideration of the undriven poloidal/toroidal Alfvén wave equations in a cylindrical geometry, that the poloidal mode will decay over time. This was put on a firmer theoretical footing by Mann and Wright (1995), who showed numerically in a Cartesian box model that there is a polarization rotation in time from poloidal to toroidal. This transition is governed by phase mixing (Heyvaerts & Priest, 1983), which acts to reduce the wavelength in the direction of varying Alfvén frequency (usually the radial direction). In the case studied by Mann and Wright (1995), however, due to the straight magnetic field, the poloidal and toroidal Alfvén frequencies of any one field line are the same. Hence, this polarization rotation occurs on a single field line.

Considering the ideas developed in the preceding sections (in a dipole field) the poloidal/toroidal fundamental Alfvén frequencies differ substantially (see Figure 10.5). Therefore, a simple rotation from one to the other at a fixed location, maintaining the same frequency, is no longer viable. This idea led Elsden and Wright (2020) to consider the evolution of an initially poloidal Alfvén wave in a 2-D dipole magnetic field. Using a similar numerical model to that described in Section 10.2.2 (e.g., Wright & Elsden, 2020), a dominantly poloidal mode was initialized. The top panels of Figure 10.9 show contour plots of the initial (a) radial and (b) azimuthal velocity (proportional to azimuthal and radial electric fields, respectively). The plots are made in the equatorial plane, with $\alpha$ and $\beta$ representing the radial and azimuthal directions, respectively. Panels (c)–(f) show the evolution of the wave fields in time by plotting the magnitude of the perpendicular velocity. Phase mixing is apparent after a few wave periods (the period is $\tau = 1.342$ in dimensionless units), reducing the local radial wavelength in (c). Over time in panels (d)–(f), the region of maximum amplitude appears to travel away from the initialized region shown in (a) along well-defined curves.

It was shown that these curves are in fact contours of Alfvén frequency, as proved by the white lines overlaid in panel (g). These lines represent solutions to Equation (10.4), using a driving frequency corresponding to the poloidal Alfvén frequency at a particular location along $\beta = 0$, where the wave is initialized. These lines are therefore the Resonant Zone paths described in Section "Alfvén Wave Polarization and Resonance Maps". The perturbations travel along these paths, continuously rotating polarization from poloidal to toroidal (the wave polarization at any point is tangential to the local Alfvén frequency contour). This type of behavior had been first considered analytically in a normal mode formulation by Leonovich and Mazur (1993) and Klimushkin et al. (1995). The numerical results in Figure 10.9 give conclusive proof that this, perhaps counter-intuitive behavior, is possible: if Alfvén waves are incompressible and have a Poynting vector and group velocity strictly aligned with the equilibrium field, how can they travel perpendicular to this direction? A careful analysis of the simulation fields allowed Elsden and Wright (2022) to answer this conundrum. Leonovich et al. (2015) showed how these ideas can be used to explain the polarization features in the poloidal Alfvén wave observations reported by Dai et al. (2013), which are discussed further in Section 10.4.

In Figure 10.9, it should be stressed that the behavior of the peak wave amplitude traveling along contours of Alfvén frequency is far more pronounced for the fundamental (panel (g)) than for the second harmonic (panel (h)), due to the increased difference between the poloidal and toroidal frequencies for the fundamental (e.g., Figure 10.5).

To more realistically model waves driven by azimuthally drifting particles (drifting by gradient and curvature drifts), azimuthal phase motion has to be taken into account. Figure 10.10 is a reproduction of figure 10 from Elsden and Wright (2020), showing results from a simulation where such azimuthal phase motion is included, as well as multiple wavelengths in azimuth to genuinely model the 'high-m' regime. The figure displays the magnitude of the perpendicular velocity in the equatorial plane at four different times in the simulation, with the simulation configuration at $t = 0$ in panel (a) showing the initially dominant poloidal wave. Over time, phase mixing reduces the local wavelength in $\alpha$. The plasma velocity is aligned with the colored ridges, so it is poloidal at $t = 0$ and rotates toward toroidal as time increases. The inclination of the ridges, being aligned with the plasma velocity, can also be thought of as the polarization angle of the Alfvén waves introduced in Section "Alfvén Wave Polarization and Resonance Maps". Elsden and Wright (2020) show how the orientation of these fronts can be determined by considering how the waves will phasemix through the dependence of the Alfvén frequency on the polarization ($\theta$) and radial location ($\alpha$), such that $\omega_A = \omega_A(\alpha, \theta)$ (the medium is independent of azimuth $\beta$). Using the fact that the wave phase is constant along the phase fronts, it can be shown that these fronts must satisfy

$$\frac{d\beta}{d\alpha} = \frac{k_\alpha}{k_\beta} = -\left( \left.\frac{\partial \omega_A}{\partial \alpha}\right|_\theta + \left.\frac{\partial \omega_A}{\partial \theta}\right|_\alpha \left.\frac{\partial \theta}{\partial \alpha}\right|_t \right) \frac{t}{k_\beta}, \tag{10.5}$$

for azimuthal wavenumber $k_\beta$ (and local radial wavenumber $k_\alpha$), with the wave polarization defined by

$$\theta = \tan^{-1}\left(\frac{d\alpha}{d\beta}\right). \tag{10.6}$$

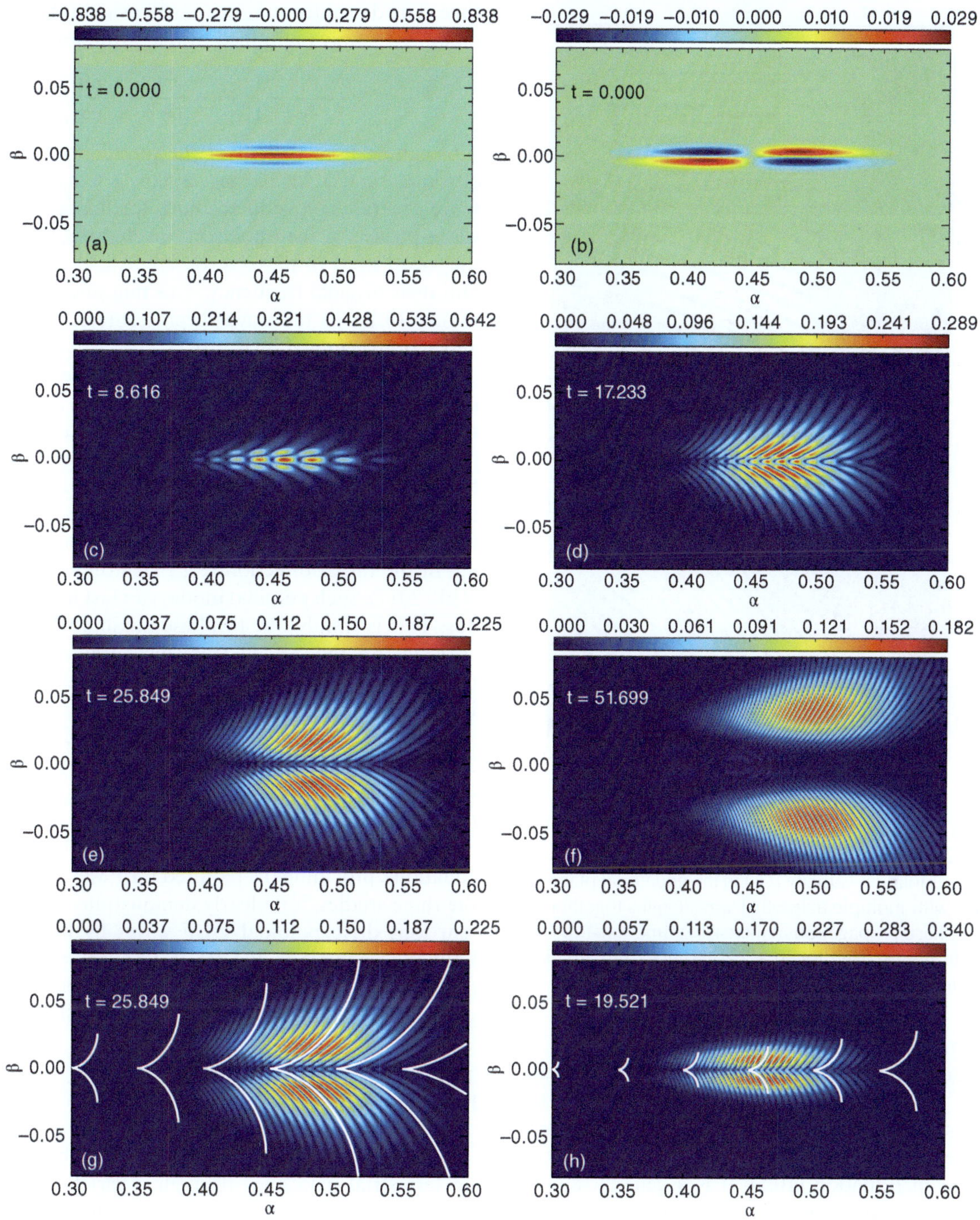


**Figure 10.9** Reproduction of figure 1 from Elsden & Wright (2020) showing the evolution of a poloidal Alfvén wave in a dipole magnetic field. α and β are the radial and azimuthal directions, respectively. In the equatorial plane: initial condition for (a) radial velocity, (b) azimuthal velocity; (c)–(f) magnitude of perpendicular velocity at four different simulation times; (g) same as panel (e) with contours of Alfvén frequency overlaid in white for the fundamental mode; (h) Color contour of $|B_\perp|$ from a separate simulation for the second field-aligned harmonic, with overlaid Alfvén frequency contours in white.

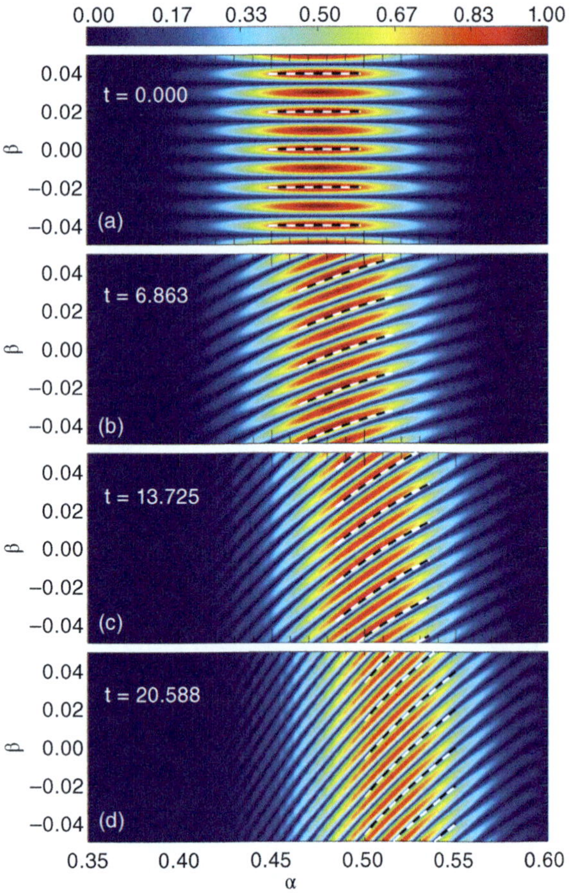


**Figure 10.10** Source: Reproduction of figure 10 by Elsden and Wright (2020). Plotted are contours of the magnitude of the perpendicular simulation velocity $|U_\perp|$ in the equatorial plane for a simulation with multiple azimuthal wavelengths together with azimuthal phase propagation at increasing times (a)–(d). Fits to the phase fronts are overplotted as dashed lines and are further described in the main text.

Equation (10.5) describes how the gradient of the phase fronts in the equatorial $(\alpha, \beta)$ plane changes in time and can be used to trace the dashed black and white lines overlaid in Figure 10.10, which clearly match well to the velocity contours. Several interesting features are evident in Equation (10.5). It is a generalization of the simpler axisymmetric toroidal phase mixing picture, which has a phase-mixing length across $L$-shells of $L_{ph} = 2\pi/(td\omega_{AT}/dL)$ for local toroidal Alfvén frequency $\omega_{AT}$. Since $L \equiv \alpha$ here, this corresponds to the local wavelength in $\alpha$, giving $k_\alpha = td\omega_{AT}/d\alpha$. Note that as $k_\beta = 0$ for axisymmetric Alfvén waves, they will remain toroidally polarized and oscillate with frequency $\omega_{AT}$. The phase fronts are inclined at an angle $\tan^{-1}(k_\beta/k_\alpha) = 0$ to the toroidal direction. Equation (10.5) shows how this generalizes for phase mixing Alfvén waves when $k_\beta \neq 0$ and the polarization angle of the waves can change in time. The dependence $k_\alpha \propto t$ is still evident; however, it is no longer simply multiplied by $d\omega_{AT}/d\alpha$, but by

$$\left.\frac{\partial\omega_A(\alpha,\theta)}{\partial\alpha}\right|_\theta + \left.\frac{\partial\omega_A(\alpha,\theta)}{\partial\theta}\right|_\alpha \left.\frac{\partial\theta}{\partial\alpha}\right|_t. \tag{10.7}$$

This can be used to recover the toroidal case by noting $\partial\theta/\partial\alpha = 0$ and $\omega_A(\alpha,\theta) = \omega_{AT}$ as $\theta(\alpha) = 0$ in this limit. The more general expression accounts for the fact that as the waves phasemix, their polarization changes, which means the waves need to move to different field lines to maintain their original frequency. This migration is evident in Figure 10.10, as the waves begin centered on $\alpha = 0.475$ and end up centered on $\alpha = 0.52$.

There have been several other recent modeling efforts to better understand high-$m$ Alfvén waves in the magnetosphere. Choi and Lee (2021) considered a high-$m$ mode in an axisymmetric dipole to provide an explanation as to how poloidal modes could be very long lasting even in the absence of a clear driving energy source. They showed that in regions where the Alfvén speed is locally not varying in the radial direction (e.g., Denton et al., 2003; Le et al., 2017), such poloidal modes can last for several periods. This is simply due to phase mixing, which depends critically on the Alfvén frequency gradient, occurring more slowly, as is evident from setting $\partial\omega_A/\partial\alpha|_\theta = 0$ in Equation (10.5).

In a series of three papers, Yamakawa et al. (2019, 2020, 2022) investigate ring current ions driving poloidal ULF waves using the drift-kinetic simulation GEMSIS-RC (Geospace Environment Modeling System for Integrated Studies-Ring Current). The model also solves Maxwell's equations, providing the required ULF wave fields. Across the three studies, it is clearly demonstrated that ring current ions drive poloidal ULF waves by both drift and bounce resonance in the Pc3-5 frequency band. A key requirement for this is found to be a positive energy gradient in the ion phase space density (PSD).

### 10.2.4. Auroral Particle Acceleration by Alfvén Waves

It is evident that Alfvén waves can be established on terrestrial field lines through a variety of processes. Specifically, the fast mode can excite Alfvén waves on closed dipole-like field lines (Section 10.2.2) and on stretched field lines in the magnetotail (Section "Magnetotail and Nightside Alfvén Waves"). Alfvén waves can also be produced from reconnection and high-$m$ Alfvén waves from wave particle interactions (Southwood et al., 1969). Whatever the origin of these Alfvén waves, they all possess field-aligned currents that are carried predominantly by electrons moving along magnetic field lines. As the currents flow toward the ionosphere, they are intensified considerably by the narrowing of the flux carrying the Alfvén wave, whose cross-sectional area scales as $1/B$. Strong currents require a large flux of electrons along the flux tube and necessitate an acceleration region at altitudes of 1-2 $R_e$ (see Figure 10.2 and Section "Magnetotail and Nightside Alfvén Waves"). The currents are so strong that the electron energy flux deposited in the ionosphere can produce a visible aurora. A proper consideration of electron energization requires a two-fluid or kinetic fomulation. However, single-fluid MHD models can be used to predict likely auroral signatures using the strength of the field-aligned current as a proxy for electron kinetic energy flux.

Wright and Allan (2008) applied this approach to nightside PSBL field lines. They found that larger amplitude Alfvén waves were excited on the PSBL field lines compared to those in the tail lobe. Moreover, the strong Alfvén speed gradient across the PSBL leads to large field-aligned currents. Working with a cartesian simulation geometry and scaling to ionospheric altitudes, they could produce realistic temporal and spatial scales for auroral arcs. These features move with the Alfvén wave phase velocity in an equatorward direction, which is determined by the direction of the Alfvén speed gradient across the PSBL.

Alfvén waves on closed dipole-like field lines can also be associated with strong field-aligned currents and optical aurora. Single-fluid MHD simulations in a dipole field with axisymmetric Alfvén speed, and the associated resonant toroidal Alfvén waves, were presented in Sakurai et al. (2022). The study was motivated by observations presented in the same paper, which found the Alfvén wave period modulated satellite electric and magnetic fields, ground magnetometer data, and optical auroral emissions. The simulation of this event aims to match the satellite and optical data and is shown in Figure 10.11. The flank magnetopause was driven by a burst of Kelvin-Helmholtz activity, which excited the fast mode in the magnetosphere that was then coupled to standing Alfvén waves. Panel (a) shows the predicted auroral emissions in a meridional plane at 6 MLT as a function of time. Note that the poleward motion of the arcs associated with phase mixing is in the opposite direction to those on PSBL field lines, as the Alfvén frequency decreases with latitude on the flank at these latitudes.

Figure 10.11(b,c) shows electric ($E$) and magnetic field ($b$) perturbations from a virtual satellite on the field line mapping to a latitude of 69.5° that sits in the center of the auroral emissions. The nodes of $b$ corresponding to maxima of the upward field-aligned current are indicated by dashed red lines that also coincide with the peak emissions in panel (a) at 69.5°. Sakurai et al. (2022) also considered properties of the four arcs, such as the initial brightening and subsequent fading, and showed how these could be accounted for in terms of the decreasing phase-mixing length and reduction of Alfvén wave amplitude in time. This could also explain the slight equatorward drift of the center of successive arcs, which was also seen in their optical data. The satellite data they reported was from a satellite on a field line that was at a higher latitude (71.68°) than the visible auroral arcs, and the corresponding virtual satellite $E$ and $b$ fields are shown in Figure 10.11(d,e). Although there are clearly Alfvén waves on this field line, there is no visible aurora at its footpoint as the thresholds of current and electron energy flux (1 $\mu Am^2$ and 1 $mWm^2$) required for the production of aurora are not met here (Chaston et al., 1091).

To consider the energetics of the process properly it is necessary to move beyond single fluid MHD, which neglects the electron mass. (This means the electrons can easily be accelerated along the field line to high speeds; however, they will have no kinetic energy, so a discussion of the energetics is not possible.) Sakurai et al. (2022) used two-fluid MHD to model the energy going into current-carrying electrons during the upward current (downgoing electron) phase and compared this with the magnetic energy of the Alfvén wave fields along the entire flux tube. They found the ratio was $\sim 0.2$, suggesting the Alfvén wave had plenty of energy to power the precipitation of the energetic electrons. However, over several cycles this could be a significant sink of Alfvén wave energy, unless the waves are continually driven by a fast wave (see Sections "Low-*m* Alfvén Waves" and 10.2.2).

Lysak and Song (2011) used a cartesian 3D two-fluid MHD simulation to focus on the electron acceleration that forms 1-2 $R_e$ above the Earth. Their motivation was the nightside aurora that often forms on the PSBL field lines during a substorm. The large Alfvén speed gradients across the PSBL naturally lead to phase mixing and strong currents that are associated with electron fluxes capable of producing optical aurora (see Figure 10.2 and Wright and Allan (2008)). Lysak and Song (2011) show how the energization is associated with the formation of a parallel electric field formed by Landau damping.

Moving beyond the two-fluid approximation, a more complete model of electron energization can be achieved using a kinetic description of the electrons. There have been several recent studies using hybrid MHD-kinetic models in order to study auroral electron acceleration by FLRs. Damiano and Johnson (2012) performed the first simulation of the upward current region of a toroidal FLR using a 2-D hybrid MHD kinetic electron model in dipolar coordinates (Damiano et al., 2007). It was found that effects due to the magnetic mirror force (Knight, 1973; Nakamura, 2000) can accelerate electrons to energies on the order of a keV. The Alfvén wave energy dissipated by this acceleration would result in completely damping the resonance within 1–2 wave periods unless otherwise strongly driven, i.e., the electron acceleration is a major sink of FLR energy (consistent with the conclusions of

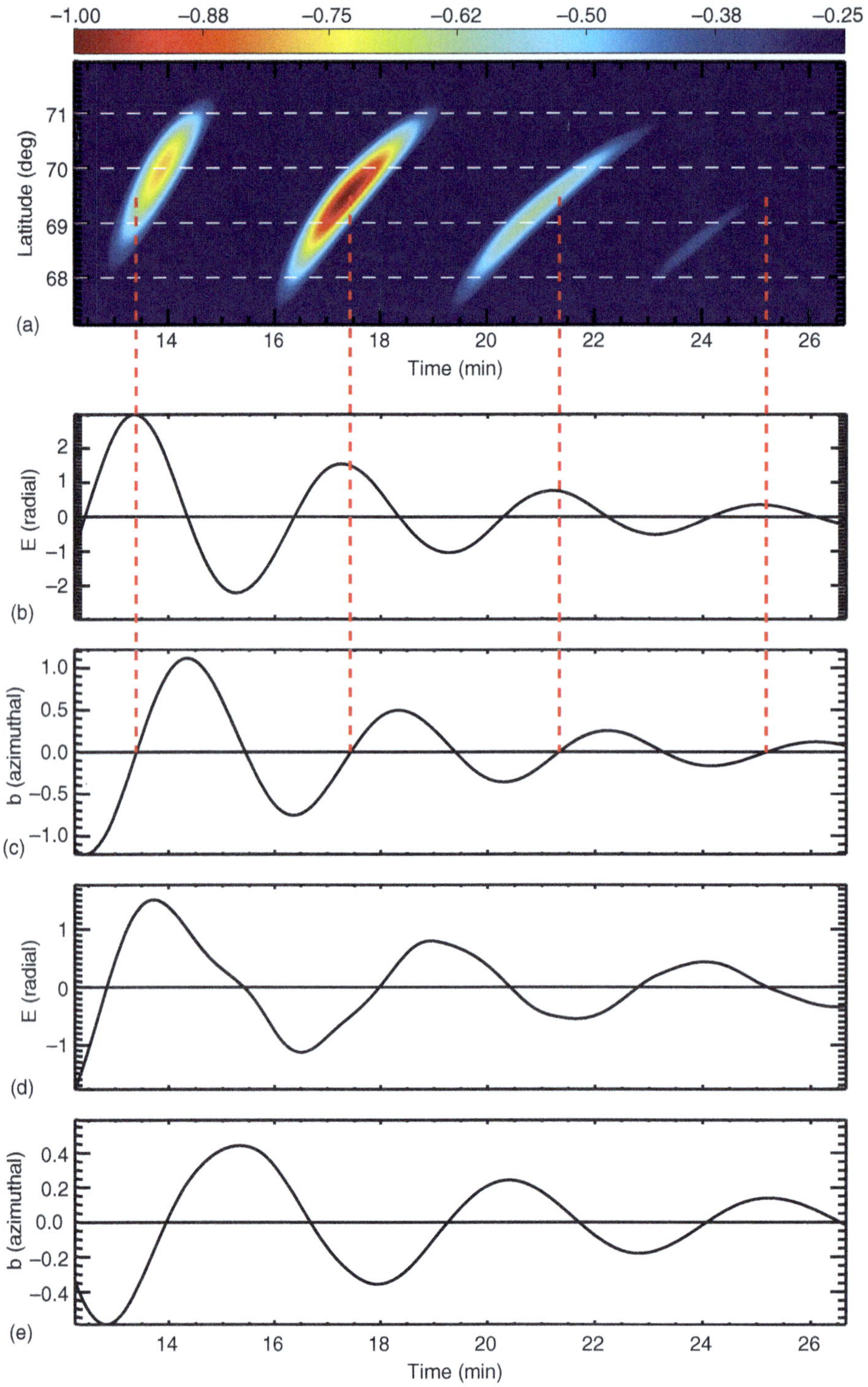


**Figure 10.11** Simulation results of auroral arcs and satellite data. (a) Predicted auroral brightness keogram in a meridian plane at 6 MLT using downgoing electron energy flux as a proxy. The scale is normalized by 4 $mWm^2$. (b) and (c) show electric and magnetic fields measured by a virtual satellite on a field line mapping to a latitude of 69.5°. The dashed red lines correspond to maxima of the upward current. (d) and (e) are the same as (b) and (c), but for a field line mapping to 71.68°. Source: After Sakurai et al. (2022).

Sakurai et al. (2022) from using two-fluid theory). Similar results were found for electrons accelerated by poloidal modes (Damiano et al., 2019).

Follow-up studies included updating the numerical model to include further physical properties (Damiano et al., 2015), e.g., ion gyroradius effects (Damiano et al., 2016). This was aimed at understanding the ability of kinetic Alfvén waves originating from the plasma sheet to trap electrons and their effect on overall auroral electron precipitation. The authors found that electrons that are trapped in the source region of the wave can be carried significantly along the field line and thereby energized by the wave. However, this is sensitive to the ion temperature, with cold ions allowing for stronger trapping by the wave. A recent review summarizing these types of studies in the broad picture of the role of Alfvén waves in auroral electron acceleration can be found in Rankin et al. (2021).

## 10.3. EXTERNALLY DRIVEN (LOW-*m*) ALFVÉN WAVES

### 10.3.1. Energy Sources for Low-*m* Alfvén Waves

For an axisymmetric magnetosphere, two solutions for standing Alfvén waves exist. One is the toroidal mode, which corresponds to $m = 0$, and the other is the poloidal mode, which corresponds to $m = \infty$. In the real magnetosphere, these idealized cases do not occur, with $m \sim 10 - 20$ being a practical boundary between toroidal-type and poloidal-type waves.

Toroidal modes are driven by a variety of energy sources upstream of the dayside magnetosphere, including pressure disturbances in the solar wind, ion foreshock, and magnetosheath. Energy from the upstream region is then processed in several ways at the magnetopause boundary, including (1) a tendency to pass energy from lower-frequency disturbances to the magnetosphere and/or (2) the excitation and growth of several types of surface waves (Plaschke, 2016). This energy is also processed in the inhomogeneous medium of the Earth's magnetosphere, where intermediate global modes serve as band-pass filters for wave energy (Hwang & Sibeck, 2016), being excited at frequencies that depend on the properties of the magnetospheric cavity (wave speed, dimensions of the cavity, etc.). Ultimately, toroidal modes can be excited by upstream energy sources via many pathways in this multi-step process, including direct excitation from impulsive upstream disturbances that do not involve significant surface wave/global mode activity to situations involving intermediate wave modes. Perhaps the most well-known example of this is the classical field line resonance mechanism, where surface waves on the magnetopause couple resonantly to toroidal waves at specific locations in the magnetosphere that depend on the properties of the surface wave and the Alfvén speed distribution in the magnetosphere (Glassmeier et al., 1999). In any of these situations, the global properties of the toroidal modes depend on the wave modes involved, including their spatial extent and locations with peak wave amplitudes.

Enabled by multi-point ground-based and in situ measurements, recent observational work has focused on understanding the effect of the orientation and spatial scale of the upstream driver on the properties of the resulting toroidal waves. For example, Oliveira et al. (2020) presented results from satellite and ground-based magnetometer measurements suggesting that the impact angle of interplanetary shocks can affect a range of toroidal mode properties, including field-aligned mode structure, frequency, and amplitude; results from this study also suggest that the MLT of the shock impact may affect the local time extent of toroidal modes as indicated in the studies discussed in Section "Driver Properties", though additional observations at a wider range of MLT are needed to quantify this effect. B. Wang et al. (2020) showed global observations of waves with frequencies of a few mHz driven by ion foreshock transients; the waves excited had a range of m-values from 5 to 23 and exhibited a dawn-dusk asymmetry consistent with the location of the pressure disturbance in the ion foreshock. B. Wang et al. (2020) suggest the results could be consistent with a running pulse on the magnetopause leading to the generation of a waveguide mode (global mode) and ultimately to toroidal modes. B. Wang et al. (2022) examined the generation of Alfvén waves and other magnetospheric Ultra-Low-Frequency waves via magnetosheath jets, finding the associated ground magnetic disturbances are more localized and transient than waves typically excited by the solar wind, as expected for the more localized, shorter duration pulses that jets represent on the magnetopause. Sarris et al. (2022) used magnetic field measurements from two satellites with different orbital inclinations to quantify ULF wave power with distance from the magnetic equator, finding results overall consistent with expectations for the fundamental mode of low-m toroidal mode waves—less wave power in the azimuthal component of the magnetic field near the magnetic equator. Their dataset and approach could be adapted to study how wave power varies during symmetric and asymmetric driving conditions.

The earlier mentioned studies mostly focused on frequencies below 10 mHz. At higher frequencies, Takahashi and Heilig (2019) used a unique network of closely spaced ground-based magnetometers to show that a global mode generated by solar wind dynamic pressure variations drove multi-harmonic toroidal modes via field line resonance that exhibited rapid spatial variations in latitude. The study demonstrates that closely spaced magnetometers can resolve the properties of Alfvén waves, including their latitude and time-dependent polarizations and phase speeds; the results also indicate that similarly closely spaced ground-based magnetometer observations can be used to separate transient from steady wave activity and examine phase mixing (Section "Driver Properties") and rapid changes in Alfvén wave polarization occurring near plumes (Section "Local Time Dependence and Plumes", Figure 10.3). Lena et al. (2021) used a unique satellite configuration (low-L perigee passes of the Van Allen Probes satellites near end of the mission) to detect multi-harmonic toroidal modes with frequencies of 0.5–3.5 Hz that were consistent with odd modes; their properties, including frequency and amplitude, varied from day to day, consistent with time-varying densities and driving conditions. Both the Takahashi and Heilig (2019) and Lena et al. (2021) studies indicate that more observations are needed at low-L values to understand the factors controlling high-frequency toroidal modes.

Though low-*m* waves tend to be excited by disturbances upstream of the dayside magnetosphere, these waves can also be excited via substorms and related phenomena on the nightside (see Sections "Magnetotail and Nightside Alfvén Waves" and "Outer Boundary and Waveguide/Cavity Modes"). For example, C.-P. Wang et al. (2022) found nightside Pi2 waves associated with plasma sheet disturbances that had polarizations consistent with toroidal modes. Takahashi et al. (2022) found nightside Pi2 waves can have low-m values or small phase shifts in longitude, consistent with a waveguide mode; furthermore, these waves often have mixed polarizations with significant wave power in the azimuthal magnetic field component away from the magnetic equator, consistent with the toroidal mode. More work is needed to understand Alfvén wave dynamics in the nightside magnetosphere, in particular the unique wave properties associated with (1) the stretched, often non-dipolar magnetic field geometry, (2) the range of possible spatial and temporal scales from bursty bulk flows, dipolarizing flux bundles, and other wave energy sources (e.g., m-values of 20–60 were reported in Liu et al., 2017), and (3) low ionospheric conductance.

Finally, whether excited by external or internal energy sources, low-*m* Alfvén waves are affected by the properties of the Alfvén speed profile (Section "Local Time Dependence and Plumes"). Recent observational work has shown that during active conditions with intense ring current, Alfvén speeds and frequencies (Sandhu et al., 2018; Wharton et al., 2020) are overall reduced. This leads to significant penetration of lower frequency wave energy, including low-*m* Alfvén waves, to the inner magnetosphere (Rae et al., 2019); this has important implications for several space weather impacts (see Section 10.5). Additionally, Sandhu et al. (2021) observed that during geomagnetic storms, localized enhancements in ULF wave power (1–15 mHz) are often observed in the dusk sector and noted these enhancements may be associated with plasmaspheric plumes trapping fast mode wave energy as simulated by Degeling et al. (2018). As discussed in Section "Local Time Dependence and Plumes", the unique Alfvén speed spatial distribution in plume regions would have also been expected to alter Alfvén wave polarizations, amplitudes, and other properties.

### 10.3.2. Magnetoseismology

Externally driven toroidal waves are a valuable tool in estimating the magnetospheric mass density, denoted $\rho_m$. Based on the fact that the frequency of the waves depends only on the background magnetic field and the $\rho_m$ distribution along it (see Equation (10.3)), we can determine $\rho_m$ if we can measure the wave frequency and have good knowledge of the magnetic field and $\rho_m$ distribution. This technique, now known as "magnetoseismology,"was conceived a long time ago (e.g., Obayashi & Jacobs, 1958). With the ever-increasing amount of data from ground magnetometers (e.g., Del Corpo et al., 2019) and spacecraft experiments (e.g., Denton et al., 2022), magnetoseismic studies have become very common and practical for producing global mass density models. This is an important research area because the total mass density is difficult to determine with particle experiments.

The success of magnetoseismology depends on the efficiency and accuracy of frequency measurements. It is often difficult to determine the wave frequency in power spectra from a single ground magnetometer because the waves may be masked by various disturbances. The recognition that wave signals can be extracted by cross-spectral analysis of magnetic field data from two stations with a small latitudinal separation (Baransky et al., 1985; Waters et al., 1991) made a major contribution to the widespread use of magnetoseismology. The situation is different with spacecraft data. Toroidal waves in spacecraft data usually produce sharp spectral lines that are easy to identify. The challenge is to determine the harmonic mode number of each spectral line. The mode numbers can be reliably determined from the amplitude and phase relationship between electric and magnetic fields, along with theoretical models of standing wave structures (Takahashi & Denton, 2021).

The success of magnetoseismology also depends on the choice of magnetic field and field line mass distribution models. As for the magnetic field, a variety of empirical models have been derived from numerous observations in space and on the ground. One can choose an appropriate model depending on the location, solar wind condition, and geomagnetic activity level. The most often used models are the dipole model, the International Geomagnetic Reference Field (IGRF) model appropriate for the inner magnetosphere, T89c (Tsyganenko, 1989) for sorting by Kp, and TS05 (Tsyganenko, 2002; Tsyganenko & Sitnov, 2005) for incorporating effects of the solar wind and the ring current. In the inner magnetosphere, the estimated mass density differs by up to 30% between the dipole and IGRF models (Berube et al., 2006). Models with no ring current effects lead to an overestimation of the mass density because the ring current reduces the equatorial magnetic field, thereby lowering the frequency of standing Alfvén waves (Vellante et al., 2014). When spacecraft data are used, this type of error can be reduced by multiplying the magnitude of the model field by the ratio of the magnitudes of the measured and model fields (Denton et al., 2004).

In contrast, there are no field line mass distribution models directly derived from particle measurements. For this reason, the ratios between the harmonics of toroidal waves have been used to infer the field line mass distribution. Most often, the distribution is assumed to have a simple functional form: $\rho_m = \rho_{eq}(r_0/r)^p$, where $r_0$ is the equatorial distance to the field line, $r$ is the distance to a point on the field line, and $\rho_{eq}$ (equatorial mass density) and $p$ (mass density power law index) are free parameters. The $p$ value varies among the studies. For example, studies

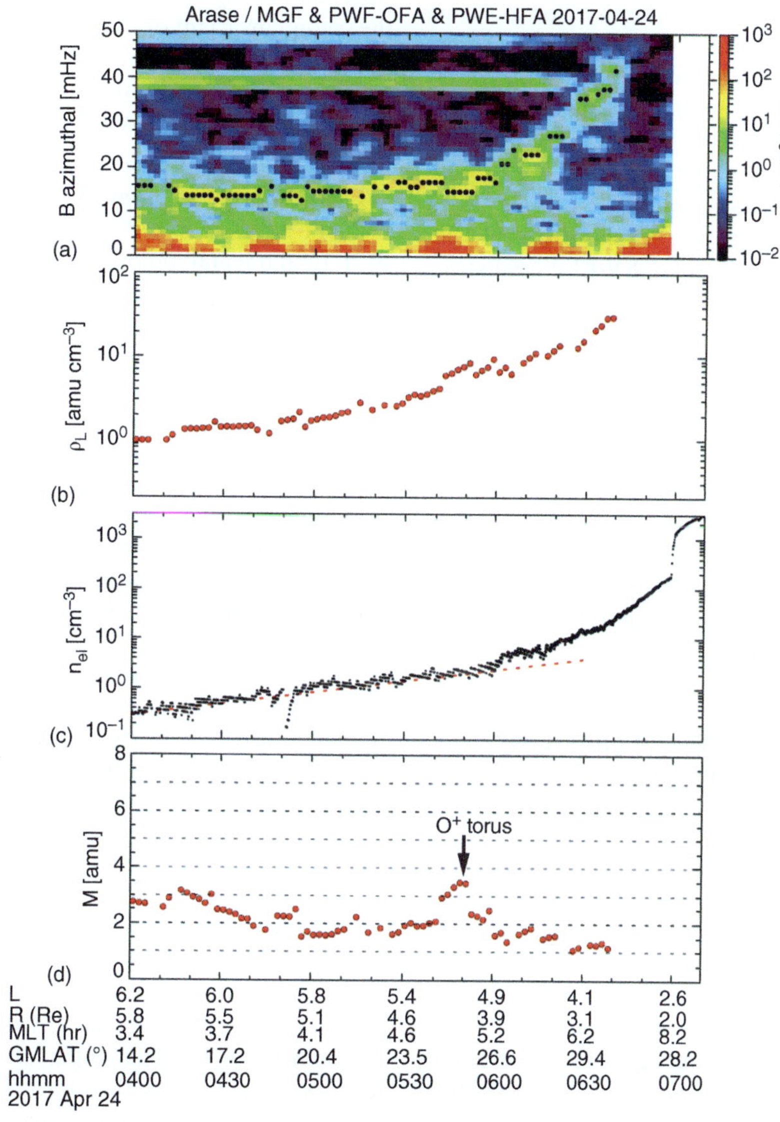


**Figure 10.12** Magnetoseismic analysis of the $O^+$ torus at Arase (Nosé et al., 2018 /John Wiley & Sons). (a) Dynamic spectra of the azimuthal component of the magnetic field, showing spectral peaks (black dots) attributed to fundamental toroidal waves. (b) Mass density derived from the wave spectra. (c) Electron density. (d) Average ion mass. The peak marked by an arrowhead is attributed to an oxygen torus.

using spacecraft data obtained $p \sim 1$ (e.g., Denton et al., 2006), whereas a study using ground magnetometer data obtained $p \sim 4$ (Wharton et al., 2018). The discrepancy might be explained by the difference in how the fundamental and second harmonic frequencies are determined and incorporated into the modeling. These harmonics are the most sensitive to the density variation near the equator. Large uncertainties in their frequencies could lead to large errors in the estimated $p$.

Once $\rho_m$ is determined, it can be combined with electron density ($n_e$) to gain information on the ion composition. $n_e$ data are routinely produced when a spacecraft operates a sensitive plasma wave experiment (e.g., Kurth et al., 2015). We simply define the average ion mass $M$ ($= \rho_m/n_e$) and use it as an indicator of heavy ion density. For a plasma consisting of $H^+$, $He^+$, and $O^+$, $M$ takes a value somewhere between 1 amu (all $H^+$ plasma) and 16 amu (all $O^+$ plasma). As the $He^+$ density is relatively steady, the variability of $M_i$ is attributed largely to the variability of the $O^+$ density (e.g., Del Corpo et al., 2022).

Variations of $\rho_m$ and $M$ occur on different time scales. A long-term variation occurs over a solar cycle (11-year period) due to the change in solar EV/EUV radiation, which controls the ionospheric feeding of $O^+$ ions into the magnetosphere (Takahashi et al., 2010). Short-time scale variations occur in association with geomagnetic storms and substorms (hours to days). When the geomagnetic activity is elevated, ionospheric $O^+$ ions are heated either in the polar region or at lower latitudes and transported to the inner magnetosphere. This process is considered to lead to the formation of an $O^+$ torus, possibly localized in longitude (Nosé et al., 2018). Figure 10.12 illustrates an example of a magnetoseismic analysis using data from the Arase spacecraft. The peak of $M$ at 0550 UT ($L = 5.1$, Figure 10.12d) is attributed to the $O^+$ torus.

## 10.4. INTERNALLY DRIVEN (HIGH-*m*) ALFVÉN WAVES

Unlike externally excited toroidal waves, which commonly exhibit several harmonics, internally excited poloidal waves tend to have wave power concentrated in a single harmonic. The harmonic mode can be readily determined when we can measure the electric ($\delta$E) and magnetic ($\delta$**B**) fields of the waves. Theory predicts how the amplitude and phase of each field component change with distance from the magnetic equator. For example, the equator is the antinode of $\delta$E and a node of $\delta$**B** of odd harmonics (symmetric waves). The node-antinode relationship switches for even harmonics (antisymmetric waves). If we have a reliable estimate of the plasma mass density, we can also determine the harmonic mode from the wave frequency.

### 10.4.1. Symmetric Waves

Symmetric poloidal waves at the fundamental harmonic are excited in the ring current regions (see Section "High-m Alfvén Waves"), with a probability much lower than that of second harmonic poloidal waves. Theory had predicted that the symmetric waves can be excited through drift resonance $\omega - m\omega_d = 0$ when an inward gradient

$$\left.\frac{\partial F}{\partial L}\right|_{M,J} < 0, \tag{10.8}$$

is established for the phase space density, $F$, of resonant ions, where $\omega$ is the wave frequency, $m$ is the azimuthal wavenumber, $\omega_d$ is the particle drift frequency, and $J$ is the second invariant of adiabatic particle motion (Chen & Hasegawa, 1991; Southwood, 1976). This circumstance requires a temporal effect in ring current dynamics because, in a steady state, the distribution is considered to be adiabatic, $\left.\frac{\partial F}{\partial L}\right|_{M,J} \approx 0$.

The first convincing evidence for the drift resonance instability was reported by Dai et al. (2013) in a study of a fundamental poloidal wave observed by the Van Allen Probe A spacecraft. Figure 10.13, taken from that study, shows that the poloidal field components ($E_\varphi$ and $B_r$) oscillate with very regular waveforms and accompany equally regular oscillations in the flux of energetic ions. The amplitude and the phase of the particle flux oscillation depend on energy, and this feature was used to determine the resonance energy (90 kev) and the azimuthal wave number ($m \sim -70$). Most importantly, the ion phase space density evaluated at the magnetic moment for the resonance ions (in the range of 11–12 MeV/G) exhibited an inward gradient (the second invariant $J$ was close to zero because the observation was made near the magnetic equator and a local pitch angle of 90° was considered). The same results for the phase space density were obtained for other fundamental poloidal wave events (Takahashi, Claudepierre, et al., 2018; Yamamoto et al., 2018).

An important finding from some of these spacecraft studies is that fundamental poloidal waves are the source of giant pulsations (Pgs) detected on the ground in magnetometer data (Takahashi, Claudepierre, et al., 2018; Yamamoto et al., 2018). Pgs have long been known for their highly monochromatic waveforms and other peculiar properties (Birkeland, 1901; Brekke et al., 1987; Sucksdorff, 1939). The origin of Pgs remained a mystery for some time, with both symmetric (Green, 1985; Takahashi et al., 1992) and antisymmetric poloidal waves (Chisham & Orr, 1991) considered as candidates. With unambiguous harmonic mode identification in space and multipoint ground magnetic field measurements around the footprints of the spacecraft (Takahashi et al.,

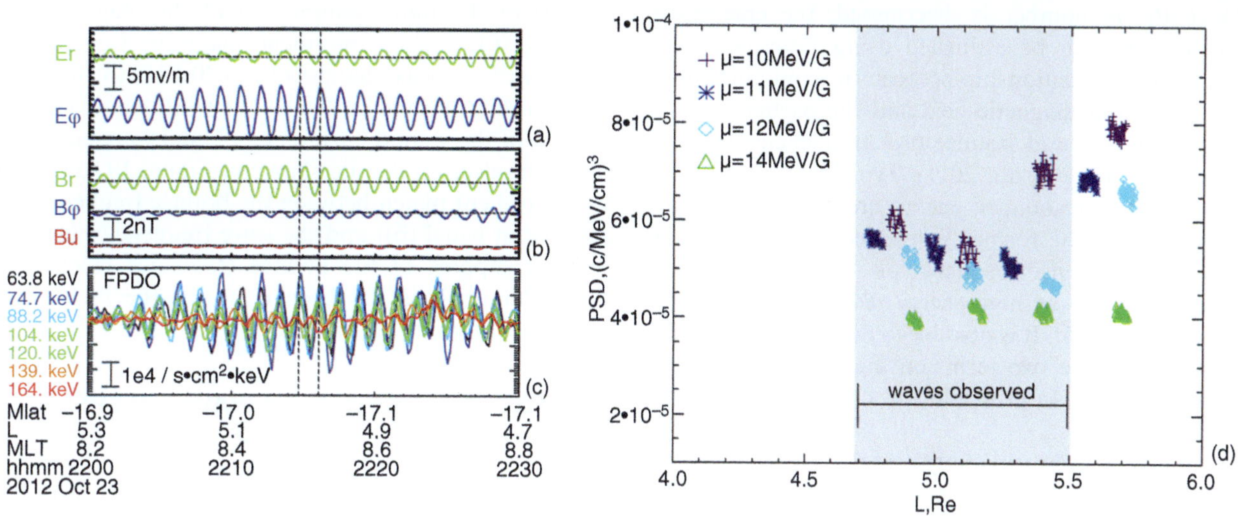


**Figure 10.13** Data from the Van Allen Probe (RBSP)—A spacecraft for a fundamental poloidal wave event (Dai et al. (2013) /John Wiley & Sons). (a) Electric field radial ($E_r$) and azimuthal ($E_\varphi$) components. (b) Magnetic field components. $B_\mu$ denotes the compressional component. (c) Perturbation of the differential energy flux of ions at seven energy steps. (d) Ion phase space density at four values of magnetic moment are plotted as a function of the $L$ value of the spacecraft.

2011), it now appears that Pgs are strictly associated with fundamental poloidal waves. However, we note that detection of fundamental poloidal waves with ground magnetometers is possible only when the magnitude of the azimuthal wave number is smaller than ~50. A fundamental poloidal wave with a large azimuthal wave number was not detected on the ground (Takahashi et al., 2013), and there are apparently many high-$m$ fundamental waves in the afternoon sector that are undetected on the ground (Brekke et al., 1987; Motoba et al., 2015).

### 10.4.2. Antisymmetric Waves

The most commonly observed poloidal waves have a second harmonic structure (Takahashi & McPherron, 1984). A theoretical prediction of the excitation of these waves has been made by considering instabilities involving the drift-bounce resonance

$$\omega - m\omega_d = N\omega_b \qquad (10.9)$$

of ring current particles with the waves, where $\omega_b$ is the particle bounce frequency and $N$ is an integer specifying the harmonic order of bounce resonance. The wave frequency is assumed to be associated with a harmonic of standing Alfvén waves. According to Southwood and Hughes (1983), the instability condition is expressed as

$$\frac{dF}{dW} = \left.\frac{\partial F}{\partial W}\right|_{M,L} + \left(\frac{dL}{dW}\right)\left.\frac{\partial F}{\partial L}\right|_{M,W} > 0, \qquad (10.10)$$

where $W$ is the particle energy, $L$ is the magnetic field shell parameter (see Section "Low-$m$ Alfvén Waves") and $M$ is the magnetic moment defined by $M = W\sin^2\alpha/B$ using the magnetic field magnitude $B$ and the pitch angle $\alpha$. An instability occurs either from energy inversion (a bump-on-tail structure), represented by the first term on the right-hand-side of the equation, or from a spatial gradient of the phase space density, represented by the second term. A bump-on-tail energetic ion distribution is commonly formed by the action of a time-dependent convection electric field (Ejiri et al., 1980). Note that the derivatives appearing on the right-hand-side of Equation (10.10) are to be evaluated for particles that are in resonance with the wave, i.e., satisfying Equation (10.9). Therefore, it is necessary to have information on the azimuthal wave number when using Equation (10.10) for instability analysis.

Recent spacecraft experiments made it possible to determine the $m$ number and the ion phase space density required for instability analysis based on Equation 10.10. The $m$ number can be determined using either multispacecraft phase delay analysis (Le et al., 2021; Takahashi, Oimatsu, et al., 2018) or ion finite Larmor radius sounding techniques (Min et al., 2017). These techniques indicate that $m$ is negative (westward-porpagating waves) and has a magnitude of the order of 100. With such high $m$ values, the ionospheric screening effect (Hughes & Southwood, 1976) is very strong, and ground magnetometers cannot detect the waves. One needs to use experiments such as a HF Doppler sounder (Wright & Yeoman, 1999) and HD radar (Shi et al., 2018; Yeoman & Wright, 2001) to detect the waves from the ground.

Once the $m$ number is determined, the energies of resonant ions can be estimated using Equation (10.9) and the known relationship between the configuration of the background magnetic field and the particle guiding center drift ($\omega_d$) and bounce ($\omega_b$) frequencies (Hamlin et al., 1961; Yang et al., 2011). Typically, the solutions for the proton resonance energy are between ~10 keV and ~100 keV. The former is referred to as low-energy resonance or bounce resonance ($\omega \sim \omega_b$), while the latter is referred to as high-energy resonance ($m\omega_d \sim \omega_b$) (Southwood, 1976). It is possible to determine the relative importance of the two terms on the right hand side of Equation (10.10). In principle, this can be done for both resonance energies.

Figure 10.14 shows an analysis of the instability condition at the low-energy resonance for a second harmonic poloidal wave event observed by Van Allen Probes (RBSP-A and RBSP-B) (Takahashi, Oimatsu, et al., 2018). The proton distribution function $F_{H+}$ exhibits a positive energy slope (Figure 10.14a) and an outward spatial gradient (Figure 10.14b). It is found that the first term has a larger magnitude and contributes to the instability. A positive energy slope ($\frac{\partial F}{\partial W} > 0$) at ~ 10 keV, resulting in a bump-on-tail energy distribution, is routinely formed by the combined action of the convection and corotation components of the global electric field in the magnetosphere (Ejiri et al., 1980). This is considered to be the reason for the ubiquitous presence of the second harmonic poloidal wave.

Figure 10.15 is a clear example of the association of a bump-on-tail energy distribution with second harmonic poloidal waves. This wave event was observed when the THEMIS-D spacecraft was moving outward in the outer plasmasphere, where a peak in the flux of ions was detected around 10 keV (Figure 10.15c). The dynamic spectra of the radial component of the magnetic field (Figure 10.15b) shows monochromatic spectral enhancement, with the peak changing from 20 mHz to 10 mHz as the spacecraft moved outward. The white trace is the frequency of the second harmonic poloidal wave computed using the electron density shown in Figure 10.15a. There is an excellent match between the bounce frequency (the red line in panel (b)) and the wave frequency (the white line) in support of the bounce resonance mechanism. A similar analysis has been done for the high-energy resonance (Oimatsu et al., 2018).

## 10.5. ROLE OF ALFVÉN WAVES IN SPACE WEATHER

Alfvén waves affect space weather in the terrestrial magnetosphere in a variety of ways, leading to potential impacts on human life and technology. This section will briefly review several ways that Alfvén waves impact space weather in order to (1) highlight gaps in our current understanding and modeling capabilities and (2) motivate future work and improvements in modeling and observational capabilities needed to account for these impacts in space weather models for nowcasts, climatology, etc.

Radiation belt and ring current particles are affected by Alfvén waves through a variety of interactions (Elkington & Sarris, 2016; Zong et al., 2017). These particles can cause unwanted charging on spacecraft in several ways, potentially leading to malfunctions and failures (Baker et al., 1998; Lanzerotti, 2001). For each wave-particle interaction, global and local effects on the particle populations depend in part on the wave amplitude, polarization, wave vector ($m$-number), and spatial extent; they also depend on the properties of the particles, including their global distribution at different energies, pitch angles,

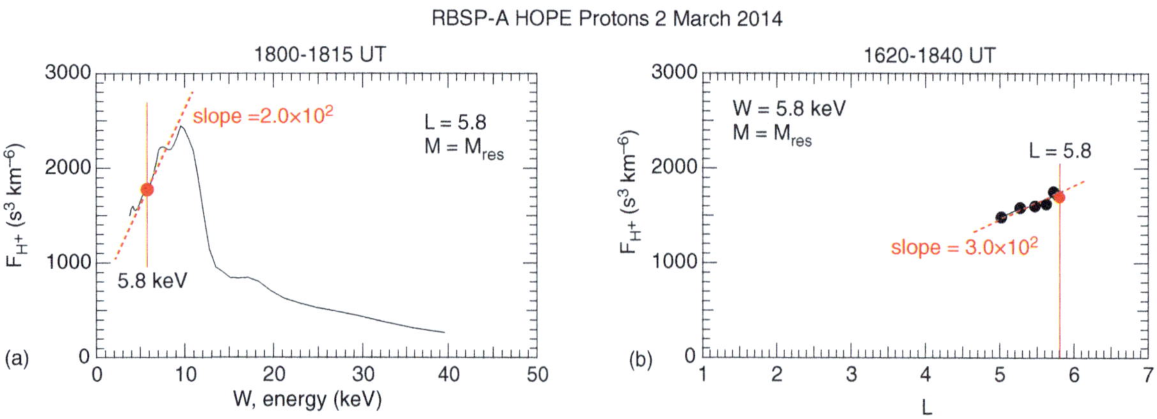


**Figure 10.14** Analysis of particle data for a second-harmonic poloidal wave event (Oimatsu, et al., 2018 /John Wiley & Sons). (a) Proton phase space density plotted as a function of energy. The phase space density was calculated from the proton differential flux measured by the HOPE instrument on RBSP-A. (b) $L$ dependence of the phase space density.

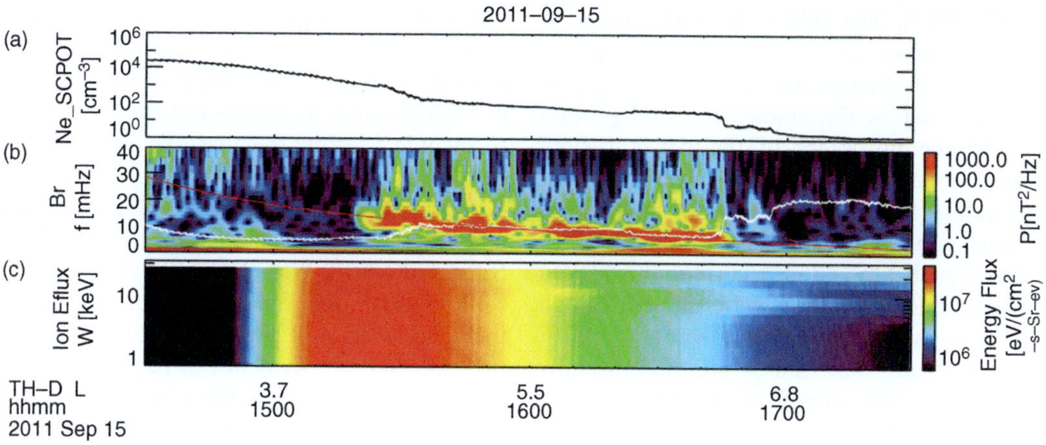


**Figure 10.15** Data from the THEMIS-D spacecraft for a second harmonic poloidal wave event (Liu et al. (2013) /John Wiley & Sons). (a) Electron density is derived from the spacecraft potential. (b) Dynamic spectra of the radial component of the magnetic field. The white trace is the model second harmonic wave frequency estimated using the electron density and assuming an all-proton plasma. The red trace is the proton bounce frequency at the energy of the bump of the ion flux shown in panel (c). (c) Ion (assumed to be protons) differential energy flux.

etc. There are many unresolved questions concerning, e.g., the nature of ULF wave-particle interactions (Kress et al., 2012) and the relative importance of ULF waves versus other wave modes in affecting particle dynamics (Drozdov et al., 2022; Lejosne et al., 2022). In the context of this review, perhaps the most important unresolved question relates to how ULF wave-particle interactions are modified by non-uniform Alfvén wave fields and, more specifically, in 3D, non-dipole geometries. A few studies have examined how non-uniform wave fields affect particle dynamics, including local-time dependence of wave amplitude (Zhao et al., 2020), rapid phase changes near a resonance (Degeling et al., 2019), and the presence of multiple wave modes (Hao et al., 2020), but this remains a mostly unexplored research area with more work needed to understand the often complex particle dynamics that are observed. One major challenge originates from the fact that particle dynamics are usually treated in coordinate systems (Schulz & Lanzerotti, 1974) that differ significantly from coordinate systems used for organizing wave fields, particularly in the outer magnetosphere; ULF waves and particles can both be treated in dipole geometries with azimuthal symmetry, but, as discussed in this review, these geometries have significant limitations due to asymmetries caused by the magnetopause (Section "Outer Boundary and Waveguide/Cavity Modes"), plasma distribution (Section "Local Time Dependence and Plumes"), and driving conditions (Section "Driver Properties").

Alfvén waves also deposit energy in the ionosphere-thermosphere system (Greenwald & Walker, 1980; Newton et al., 1978) and can significantly modulate ionospheric electron densities (Belakhovsky et al., 2016). These interactions, in turn, can cause (1) heating and thermospheric expansion (Crowley et al., 1985; Lathuillere et al., 1986; Verkhoglyadova et al., 2018) and potential loss of satellite tracking and/or unplanned, premature atmospheric reentry (Lin et al., 2022); (2) changes in ionospheric electron density that can in turn affect radio communication (Waters & Cox, 2009); and (3) changes in ionospheric electron density/conductance that in turn affect global magnetosphere-ionosphere dynamics (Buchert et al., 1999; Belakhovsky et al., 2016; Pilipenko et al., 2014). Broadly speaking, the ability of Alfvén waves to cause these effects depends on the amplitude of their ionospheric electric fields, the properties of precipitating particles associated with the Alfvén waves (Section 10.2.4), and the three-dimensional structure of the ambient ionosphere (altitude, longitude, and latitude—see Section "Ionospheric Boundaries"). There are numerous unresolved questions concerning the role of Alfvén waves in these three areas, including determining how their energy is deposited in the ionosphere-thermosphere system via electromagnetic versus kinetic (particle) energy and what conditions lead to the largest changes in ionospheric electron density/conductance. In the context of this review, it's important to determine what conditions lead to the largest magnetospheric electric field amplitudes, and thus the conditions that likely lead to the largest ionospheric electric field amplitudes.

Finally, Alfvén waves generate geomagnetic pulsations, which in turn induce geoelectric fields and can lead to geomagnetically induced currents (GIC). These GICs in turn cause disruptions in power systems that can lead to unwanted effects such as power outages and damage to transformers (Pilipenko, 2021;

Pulkkinen et al., 2017). The ability of Alfvén waves to generate large geomagnetically induced currents depends on their amplitudes, frequencies, polarizations, and spatial extents, as well as the ionospheric and ground conductivity and the properties of the power system of interest (Bedrosian & Love, 2015; Heyns et al., 2021; Yagova et al., 2021). There are still many unresolved questions concerning what conditions lead to the largest geomagnetically induced currents associated with Alfvén waves and other ULF wave modes, though there is a growing awareness that the ability of ULF waves (Hartinger et al., 2020; Shi et al., 2022) or any magnetosphere–ionosphere current system (Love et al., 2019; Lucas et al., 2020) to generate large-amplitude geomagnetically induced currents depends sensitively on the ground conductivity. In the context of this review, an important unresolved question concerns how large a GIC hazard Alfvén waves represent when compared to other ULF wave modes and other magnetosphere-ionosphere current systems. To address this, estimates of Alfvén wave amplitude, frequency, and polarization as a function of radial distance and magnetic local time should be obtained during active conditions; these are all needed as they all affect the resulting geoelectric fields and GIC. The frequency-dependent impedance related to local ground conductivity conditions should then be used to assess the hazard represented by Alfvén waves at different geographic locations (e.g., Bedrosian & Love, 2015).

Due to their space weather impacts, there is significant interest in developing global geospace models to a state where they can routinely produce hindcasts, nowcasts, and forecasts (based on upstream solar wind conditions and the current magnetospheric state) of the global properties of Alfvén waves and other Ultra-Low-Frequency waves in the Earth's magnetosphere (Hartinger et al., 2022). Achieving this goal will depend on future efforts in several areas, including increased observations needed to specify parameters for initial and boundary conditions such as Alfvén speed and ionospheric conductance, quantification of model uncertainty, identification and incorporation of missing physics (e.g., waves driven by drift-bounce resonance are not included in global MHD models), and quantification of numerical artifacts associated with grid resolution and other parameters.

## 10.6. CLOSING REMARKS

The extension of our knowledge of terrestrial Alfvén waves to other planetary magnetospheres and the solar corona provides exciting opportunities for generalizing theories to include extra physics that becomes important in these new settings and to observe their effects in data. Examples include the effects of rotation and the associated planetary disc, which provide density and current distributions not present at Earth. A review of the fast and Alfvén modes of Jupiter's magnetosphere has been given recently by Lysak (2022), who shows how the ideas of FLRs and cavity modes can be adapted to this setting to include features that have no counterparts at Earth. There are also additional sources of Alfvén waves and aurora at Jupiter associated with the Galilean moons Io and Ganymede (Bagenal & Dols, 2020; Neubauer, 1980; Saur et al., 2018; Wright & Southwood, 1987).

At Saturn, the final phase of the Cassini mission included a series of proximal orbits inside of the planetary ring system. Evidence of persistent standing Alfvén waves was found, and these were interpreted as most likely driven by fast cavity modes, as the Alfvén waves have structure, familiar from FLRs at Earth (Southwood et al., 2021). The Hermean magnetosphere also shows evidence of standing Alfvén waves, and magneto-seismology developed in the terrestrial setting has been used to infer the plasma density distribution at Mercury (James et al., 2019).

Regarding Alfvén waves at Earth and in the solar corona, there is a complimentary aspect. In the corona, it is possible to image the global oscillations of a loop or arcade of field lines and develop ideas of the global behavior of various processes directly. However, it is not possible to record *in situ* measurements of fields and plasmas. The situation is reversed at Earth, where *in situ* measurements are possible but a global visualization of waves in the magnetosphere is not. The review by Nakariakov et al. (2016) compares and contrasts MHD waves in these two settings. They show how many processes are common to both, such as the resonant excitation of Alfvén waves driven by the fast mode. This process is thought to operate in loops and arcades and play a role in heating the corona. The use of waveguide and cavity modes is also relevant to the corona, which can have highly structured density features that are ordered by the equilibrium magnetic field. On dense filamentary structures, the kink mode can play the role of a cavity mode and provide a coherent driver for resonant Alfvén waves (Goossens et al., 2014). There are also similarities between the Earth's magnetotail and helmet streamers, and it is likely that the magnetotail waveguide and processes that operate in it may play a role in understanding features in the streamer and two ribbon flares. The interested reader is referred to the review by Nakariakov et al. (2016) for an in-depth discussion of the many other examples of similarities between terrestrial Alfvén waves and those in the solar corona.

## ACKNOWLEDGMENTS

ANW was partially funded by STFC Grant ST/W001195/1. TE was funded in part by a Leverhulme Early

Career Fellowship, ECF-2019-155. KT was supported by NASA Grants 80NSSC19K0259 and 80NSSC21K0453. MDH was supported by NSF Grant AGS-2027210 and NASA Grant 80NSSC21K1683. The writing of this chapter was supported by the International Space Science Institute (ISSI) in Bern through ISSI International Team Project 483 (The Identification and Classification of 3D Alfvén Resonances). All the authors are members of this team.

## REFERENCES

Allan, W., & Knox, F. B. (1979, January). A dipole field model for axisymmetric Alfvén waves with finite ionosphere conductivities. *Planetary and Space Science*, *27*(1), 79–85. doi: 10.1016/0032-0633(79)90149-1

Archer, M. O., Hartinger, M. D., Walsh, B. M., & Angelopoulos, V. (2017, January). Magnetospheric and solar wind dependences of coupled fast-mode resonances outside the plasmasphere. *Journal of Geophysical Research (Space Physics)*, *122*, 212–226. doi: 10.1002/2016JA023428

Archer, M. O., Hartinger, M. D., Walsh, B. M., Plaschke, F., & Angelopoulos, V. (2015, December). Frequency variability of standing Alfvén waves excited by fast mode resonances in the outer magnetosphere. *Geophysical Research Letters*, *42*, 10. doi: 10.1002/2015GL066683

Archer, M. O., Southwood, D. J., Hartinger, M. D., Rastaetter, L., & Wright, A. N. (2022, February). How a realistic magnetosphere alters the polarizations of surface, fast magnetosonic, and Alfvén waves. *Journal of Geophysical Research (Space Physics)*, *127*(2), e2021JA030032. doi: 10.1029/2021JA030032

Bagenal, F., & Dols, V. (2020, May). The space environment of Io and Europa. *Journal of Geophysical Research (Space Physics)*, *125*(5), e27485. doi: 10.1029/2019JA027485

Baker, D. N., Allen, J. H., Kanekal, S. G., & Reeves, G. D. (1998, January). Disturbed space environment may have been related to pager satellite failure. *EOS Transactions*, *79*(40), 477–477. doi: 10.1029/98EO00359

Baransky, L. N., Borovkov, J. E., Gokhberg, M. B., Krylov, S. M., & Troitskaya, V. A. (1985). High-resolution method of direct measurement of the magnetic-field lines eigen frequencies. *Planetary and Space Science*, *33*(12), 1369–1374. doi: 10.1016/0032-0633(85)90112-6

Bedrosian, P. A., & Love, J. J. (2015, December). Mapping geoelectric fields during magnetic storms: Synthetic analysis of empirical United States impedances. *Geophysical Research Letters*, *42*(23), 10,160–10,170. doi: 10.1002/2015GL066636

Belakhovsky, V., Pilipenko, V., Murr, D., Fedorov, E., & Kozlovsky, A. (2016, June). Modulation of the ionosphere by Pc5 waves observed simultaneously by GPS/TEC and EISCAT. *Earth, Planets and Space*, *68*(1), 102. doi: 10.1186/s40623-016-0480-7

Berube, D., Moldwin, M. B., & Ahn, M. (2006). Computing magnetospheric mass density from field line resonances in a realistic magnetic field geometry. *Journal of Geophysical Research*, *111*(A8), A08206. doi: 10.1029/2005ja011450

Birkeland, K. (1901). Expédition norvégienne de 1899–1900 pour l'étude des aurores boréales. Résultats des recherches magnétiques. *Videnskabsselskabets Skrifter, I. Mathematisk-naturvidenskabelig klasse, Christiania*, *1*, 1–80.

Brekke, A., Feder, T., & Berger, S. (1987). Pc4 giant pulsations recorded in Tromsø, 1929–1985. *Journal of Atmospheric and Terrestrial Physics*, *49*(10), 1027–1032. doi: 10.1016/0021-9169(87)90109-7

Buchert, S. C., Fujii, R., & Glassmeier, K. H. (1999, May). Ionospheric conductivity modulation in ULF pulsations. *Journal of Geophysical Research*, *104*(A5), 10119–10134. doi: 10.1029/1998JA900180

Chaston, C. C., Peticolas, L. M., Bonnell, J. W., Carlson, C. W., Ergun, R. E., McFadden, J. P., & Strangeway, R. J. (2003, February). Width and brightness of auroral arcs driven by inertial Alfven waves. *Journal of Geophysical Research (Space Physics)*, *108*(A2), 1091. doi: 10.1029/2001JA007537

Chen, L., & Hasegawa, A. (1974, March). A theory of long-period magnetic pulsations: 1. Steady state excitation of field line resonance. *Journal of Geophysical Research*, *79*, 1024–1032. doi: 10.1029/JA079i007p01024

Chen, L., & Hasegawa, A. (1991). Kinetic-theory of geomagnetic-pulsations.1. Internal excitations by energetic particles. *Journal of Geophysical Research: Space Physics*, *96*(A2), 1503–1512. doi: 10.1029/90ja02346

Chisham, G., & Orr, D. (1991, July). Statistical studies of giant pulsations (Pgs): Harmonic mode. *Planetary and Space Science*, *39*(7), 999–1006. doi: 10.1016/0032-0633(91)90105-J

Choi, J., & Lee, D.-H. (2021). On the persistent poloidal Alfven waves. *Geophysical Research Letters*, *48*(12), e2021GL092945. Retrieved from https://agupubs.onlinelibrary.wiley.com/doi/abs/10.1029/2021GL092945 (e2021GL092945 2021GL092945) doi: https://doi.org/10.1029/2021GL092945

Claudepierre, S. G., Hudson, M. K., Lotko, W., Lyon, J. G., & Denton, R. E. (2010, November). Solar wind driving of magnetospheric ULF waves: Field line resonances driven by dynamic pressure fluctuations. *Journal of Geophysical Research (Space Physics)*, *115*, A11202. doi: 10.1029/2010JA015399

Claudepierre, S. G., Toffoletto, F. R., & Wiltberger, M. (2016, January). Global MHD modeling of resonant ULF waves: Simulations with and without a plasmasphere. *Journal of Geophysical Research (Space Physics)*, *121*(1), 227–244. doi: 10.1002/2015JA022048

Crowley, G., Wade, N., Waldock, J. A., Robinson, T. R., & Jones, T. B. (1985, August). High time-resolution observations of periodic frictional heating associated with a Pc5 micropulsation. *Nature*, *316*(6028), 528–530. doi: 10.1038/316528a0

Dai, L., Takahashi, K., Wygant, J. R., Chen, L., Bonnell, J., Cattell, C. A., et al. (2013, August). Excitation of poloidal standing Alfvén waves through drift resonance wave-particle interaction. *Geophysical Research Letters*, *40*(16), 4127–4132. doi: 10.1002/grl.50800

Damiano, P. A., & Johnson, J. R. (2012). Electron acceleration in a geomagnetic field line resonance. *Geophysical Research Letters*, *39*(2). Retrieved from https://agupubs.onlinelibrary.wiley.com/doi/abs/10.1029/2011GL050264 doi: https://doi.org/10.1029/2011GL050264

Damiano, P. A., Johnson, J. R., & Chaston, C. C. (2015). Ion temperature effects on magnetotail Alfvén wave propagation and electron energization. *Journal of Geophysical Research: Space Physics*, *120*(7), 5623–5632. Retrieved from https://agupubs.onlinelibrary.wiley.com/doi/abs/10.1002/2015JA021074 doi: https://doi.org/10.1002/2015JA021074

Damiano, P. A., Johnson, J. R., & Chaston, C. C. (2016). Ion gyroradius effects on particle trapping in kinetic Alfvén waves along auroral field lines. *Journal of Geophysical Research: Space Physics*, *121*(11), 10,831–10,844. Retrieved from https://agupubs.onlinelibrary.wiley.com/doi/abs/10.1002/2016JA022566 doi: https://doi.org/10.1002/2016JA022566

Damiano, P. A., Kim, E. H., Johnson, J. R., & Porazik, P. (2019, August). Electron energization by parallel electric fields in poloidal standing waves. *Journal of Geophysical Research (Space Physics)*, *124*(8), 6691–6700. doi: 10.1029/2019JA026849

Damiano, P. A., Wright, A. N., Sydora, R. D., & Samson, J. C. (2007, June). Energy dissipation via electron energization in standing shear Alfvén waves. *Physics of Plasmas*, *14*(6), 062904. doi: 10.1063/1.2744226

Degeling, A. W., Rae, I. J., Watt, C. E. J., Shi, Q. Q., Rankin, R., & Zong, Q. G. (2018, February). Control of ULF wave accessibility to the inner magnetosphere by the convection of plasma density. *Journal of Geophysical Research (Space Physics)*, *123*(2), 1086–1099. doi: 10.1002/2017JA024874

Degeling, A. W., & Rankin, R. (2008). Resonant drift echoes in electron phase space density produced by dayside pc5 waves following a geomagnetic storm. *Journal of Geophysical Research: Space Physics*, *113*(A10). doi: 10.1029/2008JA013254

Degeling, A. W., Rankin, R., Kabin, K., Rae, I. J., & Fenrich, F. R. (2010, October). Modeling ULF waves in a compressed dipole magnetic field. *Journal of Geophysical Research (Space Physics)*, *115*, A10212. doi: 10.1029/2010JA015410

Degeling, A. W., Rankin, R., Wang, Y., Shi, Q. Q., & Zong, Q.-G. (2019). Alteration of particle drift resonance dynamics near poloidal mode field line resonance structures. *Journal of Geophysical Research: Space Physics*, *124*(9), 7385–7401. doi: 10.1029/2019JA026946

Del Corpo, A., Vellante, M., Heilig, B., Pietropaolo, E., Reda, J., & Lichtenberger, J. (2019). Observing the cold plasma in the Earth's magnetosphere with the EMMA network. *Annals of Geophysics*, *62*(4). doi: 10.4401/ag-7751

Del Corpo, A., Vellante, M., Zhelavskaya, I. S., Shprits, Y. Y., Heilig, B., Reda, J., et al. (2022). Study of the average ion mass of the dayside magnetospheric plasma. *Journal of Geophysical Research: Space Physics*. *127*(7), A030605. doi: 10.1029/2022ja030605

Denton, R. E., Lessard, M. R., & Kistler, L. M. (2003, March). Radial localization of magnetospheric guided poloidal Pc 4-5 waves. *Journal of Geophysical Research (Space Physics)*, *108*(A3), 1105. doi: 10.1029/2002JA009679

Denton, R. E., Takahashi, K., Anderson, R. R., & Wuest, M. P. (2004). Magnetospheric toroidal Alfvén wave harmonics and the field line distribution of mass density. *Journal of Geophysical Research*, *109*(A6), A06202. doi: 10.1029/2003ja010201

Denton, R. E., Takahashi, K., Galkin, I. A., Nsumei, P. A., Huang, X., Reinisch, B. W., et al. (2006). Distribution of density along magnetospheric field lines. *Journal of Geophysical Research*, *111*(A4), A04213. doi: 10.1029/2005ja011414

Denton, R. E., Takahashi, K., Min, K., Hartley, D. P., Nishimura, Y., & Digman, M. C. (2022). Models for magnetospheric mass density and average ion mass including radial dependence. *Frontiers in Astronomy and Space Sciences*, *9*. doi: 10.3389/fspas.2022.1049684

Drozdov, A. Y., Blum, L. W., Hartinger, M., Zhao, H., Lejosne, S., Hudson, M. K., et al. (2022, February). Radial transport versus local acceleration: The long-standing debate. *Earth and Space Science*, *9*(2), e02216. doi: 10.1029/2022EA002216

Duan, S., Dai, L., & Wang, C. (2022, December). Kinetic Alfvén waves in the magnetotail during substorms. *Reviews of Modern Plasma Physics*, *6*(1), 40. doi: 10.1007/s41614-022-00100-5

Ejiri, M., Hoffman, R. A., & Smith, P. H. (1980). Energetic particle penetrations into the inner magnetosphere. *Journal of Geophysical Research*, *85*(A2), 653. doi: 10.1029/JA085iA02p00653

Elkington, S. R., & Sarris, T. E. (2016, 11). The role of Pc-5 ULF waves in the radiation belts: Current understanding and open questions. In *Waves, Particles, and Storms in Geospace: A Complex Interplay*. Oxford University Press. Retrieved from https://doi.org/10.1093/acprof:oso/9780198705246.003.0005 doi: 10.1093/acprof:oso/9780198705246.003.0005

Ellington, S. M., Moldwin, M. B., & Liemohn, M. W. (2016, March). Local time asymmetries and toroidal field line resonances: Global magnetospheric modeling in SWMF. *Journal of Geophysical Research (Space Physics)*, *121*(3), 2033–2045. doi: 10.1002/2015JA021920

Elsden, T. (2016). *Numerical Modelling of Ultra Low Frequency Waves in Earth's Magnetosphere* (Doctoral dissertation, University of St Andrews). doi: http://hdl.handle.net/10023/15663

Elsden, T., Wright, A., & Degeling, A. (2022, September). A review of the theory of 3-D Alfvén (field line) resonances. *Frontiers in Astronomy and Space Sciences*, *9*, 917817. doi: 10.3389/fspas.2022.917817

Elsden, T., & Wright, A. N. (2017, March). The theoretical foundation of 3-D Alfvén resonances: Time-dependent solutions. *Journal of Geophysical Research (Space Physics)*, *122*, 3247–3261. doi: 10.1002/2016JA023811

Elsden, T., & Wright, A. N. (2018, January). The broadband excitation of 3-D Alfvén resonances in a MHD waveguide. *Journal of Geophysical Research (Space Physics)*, *123*, 530–547. doi: 10.1002/2017JA025018

Elsden, T., & Wright, A. N. (2019, January). The effect of fast normal mode structure and magnetopause forcing on FLRs in a 3-D waveguide. *Journal of Geophysical Research (Space Physics)*, *124*(1), 178–196. doi: 10.1029/2018JA026222

Elsden, T., & Wright, A. N. (2020, August). Evolution of high-m poloidal Alfvén waves in a dipole magnetic field. *Journal of Geophysical Research (Space Physics)*, *125*(8), e28187. doi: 10.1029/2020JA028187

Elsden, T., & Wright, A. N. (2022, February). Polarization properties of 3-D field line resonances. *Journal of Geophysical Research (Space Physics)*, *127*(2), e30080. doi: 10.1029/2021JA030080

Glassmeier, K.-H., Othmer, C., Cramm, R., Stellmacher, M., & Engebretson, M. (1999, January). Magnetospheric field line resonances: A comparative planetology approach. *Surveys in Geophysics*, *20*(1), 61–109. doi: 10.1023/A:1006659717963

Goossens, M., Soler, R., Terradas, J., Van Doorsselaere, T., & Verth, G. (2014, June). The transverse and rotational motions of magnetohydrodynamic kink waves in the solar atmosphere. *Astrophysical Journal*, *788*(1), 9. doi: 10.1088/0004-637X/788/1/9

Green, C. A. (1985). Giant pulsations in the plasmasphere. *Planetary and Space Science*, *33*(10), 1155–1168. doi: 10.1016/0032-0633(85)90073-x

Greenwald, R. A., & Walker, A. D. M. (1980, October). Energetics of long period resonant hydromagnetic waves. *Geophysical Research Letters*, *7*(10), 745–748. doi: 10.1029/GL007i010p00745

Hamlin, D. A., Karplus, R., Vik, R. C., & Watson, K. M. (1961). Mirror and azimuthal drift frequencies for geomagnetically trapped particles. *Journal of Geophysical Research*, *66*(1), 1–4. doi: 10.1029/JZ066i001p00001

Hao, Y. X., Zhao, X. X., Zong, Q.-G., Zhou, X.-Z., Rankin, R., Chen, X. R., et al. (2020). Simultaneous observations of localized and global drift resonance. *Geophysical Research Letters*, *47*(17). doi: 10.1029/2020GL088019

Hartinger, M. D., Shi, X., Lucas, G. M., Murphy, B. S., Kelbert, A., Baker, J. B. H., et al. (2020, September). Simultaneous observations of geoelectric and geomagnetic fields produced by magnetospheric ULF waves. *Geophysical Research Letters*, *47*(18), e89441. doi: 10.1029/2020GL089441

Hartinger, M. D., Takahashi, K., Drozdov, A. Y., Shi, X., Usanova, M. E., & Kress, B. (2022, April). ULF wave modeling, effects, and applications: Accomplishments, recent advances, and future. *Frontiers in Astronomy and Space Sciences*, *9*, 867394. doi: 10.3389/fspas.2022.867394

Heyns, M. J., Lotz, S. I., & Gaunt, C. T. (2021, February). Geomagnetic pulsations driving geomagnetically induced currents. *Space Weather*, *19*(2), e2020SW002557. doi: 10.1029/2020SW002557

Heyvaerts, J., & Priest, E. R. (1983, January). Coronal heating by phase-mixed shear Alfven waves. *Astronomy and Astrophysics*, *117*, 220–234.

Hughes, W. J., & Southwood, D. J. (1976, July). The screening of micropulsation signals by the atmosphere and ionosphere. *Journal of Geophysical Research*, *81*(19), 3234. doi: 10.1029/JA081i019p03234

Hwang, K. J., & Sibeck, D. G. (2016, February). Role of low-frequency boundary waves in the dynamics of the dayside magnetopause and the inner magnetosphere. *Washington DC American, Geophysical Union Geophysical Monograph Senes*, *216*, 213–239. doi: 10.1002/9781119055006.ch13

Jacobs, J. A., Kato, Y., Matsushita, S., & Troitskaya, V. A. (1964, January). Classification of geomagnetic micropulsations. *Journal of Geophysical Research*, *69*, 180–181. doi: 10.1029/JZ069i001p00180

James, M. K., Imber, S. M., Yeoman, T. K., & Bunce, E. J. (2019, January). Field line resonance in the Hermean magnetosphere: Structure and implications for plasma distribution. *Journal of Geophysical Research (Space Physics)*, *124*(1), 211–228. doi: 10.1029/2018JA025920

Keiling, A. (2009, February). Alfvén waves and their roles in the dynamics of the Earth's magnetotail: A review. *Space Science Reviews*, *142*(1-4), 73–156. doi: 10.1007/s11214-008-9463-8

Keiling, A., & Takahashi, K. (2011, November). Review of Pi2 Models. *Space Science Reviews*, *161*(1-4), 63–148. doi: 10.1007/s11214-011-9818-4

Kivelson, M. G., & Russell, C. T. (1995). *Introduction to Space Physics*. 586. ISBN 0521451043. Cambridge, UK: Cambridge University Press.

Kivelson, M. G., & Southwood, D. J. (1986, April). Coupling of global magnetospheric MHD eigenmodes to field line resonances. *Journal of Geophysical Research*, *91*, 4345–4351. doi: 10.1029/JA091iA04p04345

Klimushkin, D. Y., Leonovich, A. S., & Mazur, V. A. (1995, June). On the propagation of transversally small-scale standing Alfven waves in a three-dimensionally inhomogeneous magnetosphere. *Journal of Geophysical Research*, *100*(A6), 9527–9534. doi: 10.1029/94JA03233

Knight, S. (1973, May). Parallel electric fields. *Planetary and Space Science*, *21*(5), 741–750. doi: 10.1016/0032-0633(73)90093-7

Kress, B. T., Hudson, M. K., Ukhorskiy, A. Y., & Mueller, H.-R. (2012). Nonlinear radial transport in the earth's radiation belts. In D. Summers, I. R. Mann, D. N. Baker & M. Schulz (Eds.). *Dynamics of the earth's radiation belts and inner magnetosphere* (151–160). American Geophysical Union (AGU). doi: https://doi.org/10.1029/2012GM001333

Kurth, W. S., De Pascuale, S., Faden, J. B., Kletzing, C. A., Hospodarsky, G. B., Thaller, S., & Wygant, J. R. (2015). Electron densities inferred from plasma wave spectra obtained by the waves instrument on van allen probes. *Journal of Geophysical Research: Space Physics*, *120*(2), 904–914. doi: 10.1002/2014JA020857

Lambour, R. L., Weiss, L. A., Elphic, R. C., & Thomsen, M. F. (1997, October). Global modeling of the plasmasphere following storm sudden commencements. *Journal of Geophysical Research*, *102*(A11), 24351–24368. doi: 10.1029/97JA02037

Lanzerotti, L. J. (2001, January). Space weather effects on technologies. *Washington DC American Geophysical Union Geophysical Monograph Series*, *125*, 11–22. doi: 10.1029/GM125p0011

Lathuillere, C., Glangeaud, F., & Zhao, Z. Y. (1986, February). Ionospheric ion heating by ULF Pc 5 magnetic pulsations. *Journal of Geophysical Research*, *91*(A2), 1619–1626. doi: 10.1029/JA091iA02p01619

Le, G., Chi, P. J., Strangeway, R. J., Russell, C. T., Slavin, J. A., Anderson, B., et al. (2021, May). MMS observations of field line resonances under disturbed solar wind conditions. *Journal of Geophysical Research (Space Physics)*, *126*(5), e28936. doi: 10.1029/2020JA028936

Le, G., Chi, P. J., Strangeway, R. J., Russell, C. T., Slavin, J. A., Takahashi, K., et al. (2017, April). Global observations of magnetospheric high-m poloidal waves during the 22 June 2015 magnetic storm. *Geophysical Research Letters*, *44*(8), 3456–3464. doi: 10.1002/2017GL073048

Lejosne, S., Allison, H. J., Blum, L. W., Drozdov, A. Y., Hartinger, M. D., Hudson, M. K., et al. (2022, June). Differentiating between the leading processes for electron radiation belt acceleration. *Frontiers in Astronomy and Space Sciences*, *9*, 896245. doi: 10.3389/fspas.2022.896245

Lena, F. R., Ozeke, L. G., Wygant, J. R., Tian, S., Breneman, A. W., & Mann, I. R. (2021, January). Detection of Hertz frequency multiharmonic field line resonances at Low-L (L = 1.1-1.5) during Van Allen probe perigee passes. *Geophysical Research Letters*, *48*(1), e0632. doi: 10.1029/2020GL090632

Leonovich, A. S., Klimushkin, D. Y., & Mager, P. N. (2015, July). Experimental evidence for the existence of monochromatic transverse small-scale standing Alfvén waves with spatially dependent polarization. *Journal of Geophysical Research (Space Physics)*, *120*(7), 5443–5454. doi: 10.1002/2015JA021044

Leonovich, A. S., & Mazur, V. A. (1993, Sep). A theory of transverse small-scale standing Alfvén waves in an axially symmetric magnetosphere. *Planetary and Space Science*, *41*(9), 697–717. doi: 10.1016/0032-0633(93)90055-7

Lin, D., Wang, W., Garcia-Sage, K., Yue, J., Merkin, V., Mclnerney, J. M., et al. (2022, December). Thermospheric neutral density variation during the "SpaceX" storm: Implications from physics-based whole Geospace modeling. *Space Weather*, *20*(12), e2022SW003254. doi: 10.1029/2022SW003254

Liu, J., Angelopoulos, V., Zhang, X. J., Runov, A., Artemyev, A., Plaschke, F., et al. (2017, October). Ultralow frequency waves deep inside the inner magnetosphere driven by dipolarizing flux bundles. *Journal of Geophysical Research (Space Physics)*, *122*(10), 10,112–10,128. doi: 10.1002/2017JA024270

Liu, W., Cao, J. B., Li, X., Sarris, T. E., Zong, Q. G., Hartinger, M., et al. (2013). Poloidal ULF wave observed in the plasmasphere boundary layer. *Journal of Geophysical Research: Space Physics*, *118*(7), 4298–4307. doi: 10.1002/jgra.50427

Love, J. J., Lucas, G. M., Bedrosian, P. A., & Kelbert, A. (2019, March). Extreme-value geoelectric amplitude and polarization across the northeast United States. *Space Weather*, *17*(3), 379–395. doi: 10.1029/2018SW002068

Lucas, G. M., Love, J. J., Kelbert, A., Bedrosian, P. A., & Rigler, E. J. (2020, February). A 100-year geoelectric hazard analysis for the U.S. high-voltage power grid. *Space Weather*, *18*(2), e02329. doi: 10.1029/2019SW002329

Lyon, J. G., Fedder, J. A., & Mobarry, C. M. (2004, October). The Lyon-Fedder-Mobarry (LFM) global MHD magnetospheric simulation code. *Journal of Atmospheric and Solar-Terrestrial Physics*, *66*(15-16), 1333–1350. doi: 10.1016/j.jastp.2004.03.020

Lysak, R. L. (2022, June). Field line resonances and cavity modes at Earth and Jupiter. *Frontiers in Astronomy and Space Sciences*, *9*, 913554. doi: 10.3389/fspas.2022.913554

Lysak, R. L., & Song, Y. (2011, September). Development of parallel electric fields at the plasma sheet boundary layer. *Journal of Geophysical Research (Space Physics)*, *116*, A00K14. doi: 10.1029/2010JA016424

Lysak, R. L., Song, Y., Sciffer, M. D., & Waters, C. L. (2015, January). Propagation of Pi2 pulsations in a dipole model of the magnetosphere. *Journal of Geophysical Research (Space Physics)*, *120*(1), 355–367. doi: 10.1002/2014JA020625

Lysak, R. L., Song, Y., Waters, C. L., Sciffer, M. D., & Obana, Y. (2020, July). Numerical investigations of interhemispheric asymmetry due to ionospheric conductance. *Journal of Geophysical Research (Space Physics)*, *125*(7), e27866. doi: 10.1029/2020JA027866

Mager, P. N., & Klimushkin, D. Y. (2021, January). The field line resonance in the three dimensionally inhomogeneous magnetosphere: Principal features. *Journal of Geophysical Research (Space Physics)*, *126*(1), e28455. doi: 10.1029/2020JA028455

Mann, I. R., & Wright, A. N. (1995, December). Finite lifetimes of ideal poloidal Alfvén waves. *Journal of Geophysical Research*, *100*, 23677–23686. doi: 10.1029/95JA02689

Mann, I. R., Wright, A. N., & Cally, P. S. (1995, October). Coupling of magnetospheric cavity modes to field line resonances: A study of resonance widths. *Journal of Geophysical Research*, *100*, 19441–19456. doi: 10.1029/95JA00820

Min, K., Takahashi, K., Ukhorskiy, A. Y., Manweiler, J. W., Spence, H. E., Singer, J., et al. (2017, March). Second harmonic poloidal waves observed by Van Allen Probes in the dusk-midnight sector. *Journal of Geophysical Research (Space Physics)*, *122*(3), 3013–3039. doi: 10.1002/2016JA023770

Motoba, T., Takahashi, K., Rodriguez, J. V., & Russell, C. T. (2015). Giant pulsations on the afternoonside: Geostationary satellite and ground observations. *Journal of Geophysical Research: Space Physics*, *120*. doi: 10.1002/2015ja021592

Nakamura, T. K. (2000, May). Parallel electric field of a mirror kinetic Alfvén wave. *Journal of Geophysical Research*, *105*(A5), 10729–10738. doi: 10.1029/1999JA900494

Nakariakov, V. M., Pilipenko, V., Heilig, B., Jelínek, P., Karlicky, M., Klimushkin, D. Y., et al. (2016, April). Magnetohydrodynamic oscillations in the solar corona and Earth's magnetosphere: Towards consolidated understanding. *Space Science Reviews*, *200*(1-4), 75–203. doi: 10.1007/s11214-015-0233-0

Neubauer, F. M. (1980, March). Nonlinear standing Alfvén wave current system at Io: Theory. *Journal of Geophysical Research*, *85*(A3), 1171–1178. doi: 10.1029/JA085iA03p01171

Newton, R. S., Southwood, D. J., & Hughes, W. J. (1978, March). Damping of geomagnetic pulsations by the ionosphere. *Planetary and Space Science*, *26*(3), 201–209. doi: 10.1016/0032-0633(78)90085-5

Nosé, M., Matsuoka, A., Kumamoto, A., Kasahara, Y., Goldstein, J., Teramoto, M., et al. (2018). Longitudinal structure of oxygen torus in the inner magnetosphere: Simultaneous observations by arase and van allen probe a. *Geophysical Research Letters*, *45*(19), 10,177–10,184. doi: 10.1029/2018gl080122

Obana, Y., Waters, C. L., Sciffer, M. D., Menk, F. W., Lysak, R. L., Shiokawa, K., et al. (2015, June). Resonance structure and mode transition of quarter-wave ULF pulsations around the dawn terminator. *Journal of Geophysical Research (Space Physics)*, *120*(6), 4194–4212. doi: 10.1002/2015JA021096

Obayashi, T., & Jacobs, J. A. (1958). Geomagnetic pulsations and the Earth's outer atmosphere. *Geophysical Journal International*, *1*(1), 53–63. doi: 10.1111/j.1365-246X.1958.tb00034.x

Oimatsu, S., Nose, M., Takahashi, K., Yamamoto, K., Keika, K., Kletzing, C. A., et al. (2018). Van Allen Probes observations of drift-bounce resonance and energy transfer between energetic ring current protons and poloidal Pc4 wave. *Journal of Geophysical Research: Space Physics*. *123*(5), 3421–3435.

Oliveira, D. M., Hartinger, M. D., Xu, Z., Zesta, E., Pilipenko, V. A., Giles, B. L., & Silveira, M. V. D. (2020, December). Interplanetary shock impact angles control magnetospheric ULF wave activity: Wave amplitude, frequency, and power spectra. *Geophysical Research Letters*, *47*(24), e90857. doi: 10.1029/2020GL090857

Pilipenko, V. (2021, September). Space weather impact on ground-based technological systems. *Solar-Terrestrial Physics*, *7*(3), 68–104. doi: 10.12737/stp-73202106

Pilipenko, V., Belakhovsky, V., Kozlovsky, A., Fedorov, E., & Kauristie, K. (2014). Ulf wave modulation of the ionospheric parameters: Radar and magnetometer observations. *Journal of Atmospheric and Solar-Terrestrial Physics*, *108*, 68–76. Retrieved from https://www.sciencedirect.com/science/article/pii/S1364682613003337 doi: https://doi.org/10.1016/j.jastp.2013.12.015

Plaschke, F. (2016, February). ULF waves at the magnetopause. *Washington DC American Geophysical Union Geophysical Monograph Series*, *216*, 193–212. doi: 10.1002/9781119055006.ch12

Powell, K. G., Roe, P. L., Linde, T. J., Gombosi, T. I., & De Zeeuw, D. L. (1999, September). A Solution-adaptive upwind scheme for ideal magnetohy drodynamics. *Journal of Computational Physics*, *154*(2), 284–309. doi: 10.1006/jcph.1999.6299

Prokopyszyn, A. P. K., Wright, A. N., & Hood, A. W. (2021, June). Line-tied boundary conditions can cause resonant absorption models to generate unphysically large boundary layers. *Astrophysical Journal*, *914*(1), 15. doi: 10.3847/1538-4357/abf65b

Pulkkinen, A., Bernabeu, E., Thomson, A., Viljanen, A., Pirjola, R., Boteler, D., et al. (2017, July). Geomagnetically induced currents: Science, engineering, and applications readiness. *Space Weather*, *15*(7), 828–856. doi: 10.1002/2016SW001501

Radoski, H. R. (1974, February). A theory of latitude dependent geomagnetic micropulsations: The asymptotic fields. *Journal of Geophysical Research*, *79*, 595–603. doi: 10.1029/JA079i004p00595

Rae, I. J., Murphy, K. R., Watt, C. E. J., Sandhu, J. K., Georgiou, M., Degeling, A. W., et al. (2019, October). How do ultra-low frequency waves access the inner magnetosphere during geomagnetic storms? *Geophysical Research Letters*, *46*(19), 10,699–10,709. doi: 10.1029/2019GL082395

Rankin, R., Gillies, D. M., & Degeling, A. W. (2021, June). On the relationship between shear Alfvén waves, auroral electron acceleration, and field line resonances. *Space Science Reviews*, *217*(4), 60. doi: 10.1007/s11214-021-00830-x

Sakurai, T., Wright, A. N., Takahashi, K., Elsden, T., Ebihara, Y., Sato, N., et al. (2022, August). Poleward moving auroral arcs and Pc5 oscillations. *Journal of Geophysical Research (Space Physics)*, *127*(8), e30362. doi: 10.1029/2022JA030362

Salat, A., & Tataronis, J. A. (2000, June). Conditions for existence of orthogonal coordinate systems oriented by magnetic field lines. *Journal of Geophysical Research*, *105*(A6), 13055–13062. doi: 10.1029/1999ja000221

Sandhu, J. K., Rae, I. J., Staples, F. A., Hartley, D. P., Walach, M. T., Elsden, T., & Murphy, K. R. (2021, July). The roles of the magnetopause and plasmapause in storm-time ULF wave power enhancements. *Journal of Geophysical Research (Space Physics)*, *126*(7), e29337. doi: 10.1029/2021JA029337

Sandhu, J. K., Yeoman, T. K., & Rae, I. J. (2018, November). Variations of field line eigenfrequencies with ring current intensity. *Journal of Geophysical Research (Space Physics)*, *123*(11), 9325–9339. doi: 10.1029/2018JA025751

Sarris, T. E., Li, X., Zhao, H., Papadakis, K., Liu, W., Tu, W., et al. (2022, October). Distribution of ULF wave power in magnetic latitude and local time using THEMIS and arase measurements. *Journal of Geophysical Research (Space Physics)*, *127*(10), e2022JA030469. doi: 10.1029/2022JA030469

Saur, J., Chané, E., & Hartkorn, O. (2018, January). Modeling Magneto spheric Fields in the Jupiter System. In H. Lühr, J. Wicht, S. A. Gilder, & M. Holschneider (Eds.), *Magnetic fields in the solar system*, *448*, 153–182. doi: 10.1007/978-3-319-64292-5_6

Schulz, M., & Lanzerotti, L. J. (1974). *Particle diffusion in the radiation belts*. Physics and Chemistry in Space, ISBN: 978-3-642-65677-4. Berlin: Springer. doi: 10.1007/978-3-642-65675-0

Shi, X., Baker, J. B. H., Ruohoniemi, J. M., Hartinger, M. D., Murphy, K. R., Rodriguez, J. V., et al. (2018, October). Long-lasting poloidal ULF waves observed by multiple satellites and high-latitude SuperDARN radars. *Journal of Geophysical Research (Space Physics)*, *123*(10), 8422–8438. doi: 10.1029/2018JA026003

Shi, X., Hartinger, M. D., Baker, J. B. H., Murphy, B. S., Bedrosian, P. A., Kelbert, A., & Rigler, E. J. (2022, April). Characteristics and sources of Intense geoelectric fields in the United States: Comparative analysis of multiple geomagnetic storms. *Space Weather*, *20*(4), e2021SW002967. doi: 10.1029/2021SW002967

Southwood, D. J. (1974, March). Some features of field line resonances in the magnetosphere. *Planetary Space Science*, *22*, 483–491. doi: 10.1016/0032-0633(74) 90078-6

Southwood, D. J. (1976). A general approach to low-frequency instability in the ring current plasma. *Journal of Geophysical Research*, *81*(19), 3340–3348. doi: 10.1029/JA081i019p03340

Southwood, D. J., Cao, H., Shebanits, O., Elsden, T., Hunt, G. J., & Dougherty, M. K. (2021, February). Discovery of Alfvén Waves Planetward of Saturn's Rings. *Journal of Geophysical Research (Space Physics)*, *126*(2), e28473. doi: 10.1029/2020JA028473

Southwood, D. J., Dungey, J. W., & Etherington, R. J. (1969, March). Bounce resonant interaction between pulsations and trapped particles. *Planetary and Space Science*, *17*(3), 349–361. doi: 10.1016/0032-0633(69)90068-3

Southwood, D. J., & Hughes, W. J. (1983). Theory of hydromagnetic waves in the magnetosphere. *Space Science Reviews*, *35*, 301–366. doi: 10.1007/BF00169231

Sucksdorff, E. (1939). Giant pulsations recorded at sodankylä during 1914–1938. *Journal of Geophysical Research*, *44*(2), 157–170. doi: 10.1029/TE044i002p00157

Takahashi, K., Claudepierre, S. G., Rankin, R., Mann, I. R., & Smith, C. W. (2018, June). Van Allen probes observation of a fundamental poloidal standing Alfvén wave event related to giant pulsations. *Journal of Geophysical Research (Space Physics)*, *123*(6), 4574–4593. doi: 10.1029/2017JA025139

Takahashi, K., & Denton, R. E. (2021). Nodal structure of toroidal standing Alfvén waves and its implication for field line mass density distribution. *Journal of Geophysical Research: Space Physics*, *126*(5). doi: 10.1029/2020ja028981

Takahashi, K., Denton, R. E., & Singer, H. J. (2010). Solar cycle variation of geosynchronous plasma mass density derived from the frequency of standing alfven waves. *Journal of Geophysical Research: Space Physics*, *115*, A07207. doi: ArtnA0720710.1029/2009ja015243

Takahashi, K., Glassmeier, K. H., Angelopoulos, V., Bonnell, J., Nishimura, Y., Singer, H. J., & Russell, C. T. (2011). Multisatellite observations of a giant pulsation event. *Journal of Geophysical Research: Space Physics*, *116*(A11). doi: 10.1029/2011ja016955

Takahashi, K., Hartinger, M. D., Angelopoulos, V., & Glassmeier, K.-H. (2015, August). A statistical study of fundamental toroidal mode standing Alfvén waves using THEMIS ion bulk velocity data. *Journal of Geophysical Research (Space Physics)*, *120*(8), 6474–6495. doi: 10.1002/2015JA021207

Takahashi, K., Hartinger, M. D., Angelopoulos, V., Glassmeier, K.-H., & Singer, H. J. (2013, July). Multispacecraft observations of fundamental poloidal waves without ground magnetic signatures. *Journal of Geophysical Research (Space Physics)*, *118*(7), 4319–4334. doi: 10.1002/jgra.50405

Takahashi, K., & Heilig, B. (2019, August). L versus time structures of dayside magnetic pulsations detected by the European quasi-meridional magnetometer array. *Journal of Geophysical Research (Space Physics)*, *124*(8), 6566–6584. doi: 10.1029/2019JA026796

Takahashi, K., Lee, D.-H., Merkin, V. G., Lyon, J. G., & Hartinger, M. D. (2016, October). On the origin of the dawn-dusk asymmetry of toroidal Pc5 waves. *Journal of Geophysical Research (Space Physics)*, *121*(10), 9632–9650. doi: 10.1002/2016JA023009

Takahashi, K., Lysak, R., & Vellante, M. (2022, September). Statistical analysis of Pi2 pulsations observed by Van Allen probes. *Journal of Geophysical Research (Space Physics)*, *127*(9), e30674. doi: 10.1029/2022JA030674

Takahashi, K., Lysak, R., Vellante, M., Kletzing, C. A., Hartinger, M. D., & Smith, C. W. (2018, March). Observation and numerical simulation of cavity mode oscillations excited by an interplanetary shock. *Journal of Geophysical Research (Space Physics)*, *123*(3), 1969–1988. doi: 10.1002/2017JA024639

Takahashi, K., & McPherron, R. L. (1984, November). Standing hydromagnetic oscillations in the magnetosphere. *Planetary and Space Science*, *32*, 1343–1359.

Takahashi, K., Oimatsu, S., Nosé, M., Min, K., Claudepierre, S. G., Chan, A., et al. (2018, January). Van Allen probes observations of second harmonic poloidal standing Alfvén waves. *Journal of Geophysical Research (Space Physics)*, *123*(1), 611–637. doi: 10.1002/2017JA024869

Takahashi, K., Sato, N., Warnecke, J., Lühr, H., Spence, H. E., & Tonegawa, Y. (1992). On the standing wave mode of giant pulsations. *Journal of Geophysical Research*, *97*(A7), 10717. doi: 10.1029/92ja00382

Tsyganenko, N. A. (1989). A magnetospheric magnetic field model with a warped tail current sheet. *Planetary and Space Science*, *37*(1), 5–20. doi: 10.1016/0032-0633(89)90066-4

Tsyganenko, N. A. (2002). A model of the near magnetosphere with a dawn-dusk asymmetry 2. parameterization and fitting to observations. *Journal of Geophysical Research: Space Physics*, *107*(A8), SMP 10-1–SMP 10-17. doi: 10.1029/2001ja000220

Tsyganenko, N. A., & Sitnov, M. I. (2005, March). Modeling the dynamics of the inner magnetosphere during strong geomagnetic storms. *Journal of Geophysical Research (Space Physics)*, *110*(A3), A03208. doi: 10.1029/2004JA010798

Tóth, G., Sokolov, I. V., Gombosi, T. I., Chesney, D. R., Clauer, C. R., De Zeeuw, D. L., et al. (2005). Space weather modeling framework: A new tool for the space science community. *Journal of Geophysical Research: Space Physics*, *110*(A12). Retrieved from https://agupubs.onlinelibrary.wiley.com/doi/abs/10.1029/2005JA011126 doi: https://doi.org/10.1029/2005JA011126

Vellante, M., Piersanti, M., & Pietropaolo, E. (2014). Comparison of equatorial plasma mass densities deduced from field line resonances observed at ground for dipole and IGRF models. *Journal of Geophysical Research: Space Physics*, *119*(4), 2623–2633. doi: 10.1002/2013ja019568

Verkhoglyadova, O. P., Meng, X., Mannucci, A. J., & McGranaghan, R. M. (2018). Semianalytical estimation of energy deposition in the ionosphere by monochromatic Alfvén waves. *Journal of Geophysical Research: Space Physics*, *123*(6), 5210–5222. Retrieved from https://agupubs.onlinelibrary.wiley.com/doi/abs/10.1029/2017JA025097 doi: https://doi.org/10.1029/2017JA025097

Wang, B., Liu, T., Nishimura, Y., Zhang, H., Hartinger, M., Shi, X., et al. (2020, December). Global propagation of magnetospheric Pc5 ULF waves driven by foreshock transients. *Journal of Geophysical Research (Space Physics)*, *125*(12), e28411. doi: 10.1029/2020JA028411

Wang, B., Nishimura, Y., Hietala, H., & Angelopoulos, V. (2022, November). Investigating the role of magnetosheath high-speed jets in triggering dayside ground magnetic ultra-low frequency waves. *Geophysical Research Letters*, *49*(22), e2022GL099768. doi: 10.1029/2022GL099768

Wang, C.-P., Takahashi, K., Xing, X., & Bortnik, J. (2022, May). Plasmas-pheric Pi2 pulsation enhancement in response to plasma sheet Pi2 wave source: Statistical study using Van Allen probes and THEMIS conjunctions. *Journal of Geophysical Research (Space Physics)*, *127*(5), e30399. doi: 10.1029/2022JA030399

Waters, C. L., & Cox, S. P. (2009, July). ULF wave effects on high frequency signal propagation through the ionosphere. *Annales Geophysicae*, *27*(7), 2779–2788. doi: 10.5194/angeo-27-2779-2009

Waters, C. L., Menk, F. W., & Fraser, B. J. (1991). The resonance structure of low latitude Pc3 geomagnetic-pulsations. *Geophysical Research Letters*, *18*(12), 2293–2296. doi: 10.1029/91gl02550

Wharton, S. J., Rae, I. J., Sandhu, J. K., Walach, M. T., Wright, D. M., & Yeoman, T. A. K. (2020, June). The changing eigenfrequency continuum during geomagnetic storms: Implications for plasma mass dynamics and ULF wave coupling. *Journal of Geophysical Research (Space Physics)*, *125*(6), e27648. doi: 10.1029/2019JA027648

Wharton, S. J., Wright, D. M., Yeoman, T. K., James, M. K., & Sandhu, J. K. (2018, August). Cross-phase determination of ultralow frequency wave harmonic frequencies and their associated plasma mass density distributions. *Journal of Geophysical Research (Space Physics)*, *123*(8), 6231–6250. doi: 10.1029/2018JA025487

Wright, A. N., & Allan, W. (2008, February). Simulations of Alfvén waves in the geomagnetic tail and their auroral signatures. *Journal of Geophysical Research (Space Physics)*, *113*(A2), A02206. doi: 10.1029/2007JA012464

Wright, A. N., Degeling, A. W., & Elsden, T. (2022). Resonance maps for 3D Alfvén waves in a compressed dipole field. *Journal of Geophysical Research: Space Physics*, *127*(4), e2022JA030294. doi: 10.1029/2022JA030294

Wright, A. N., & Elsden, T. (2016, December). The theoretical foundation of 3D Alfvén resonances: Normal modes. *Astrophysical Journal*, *833*, 230. doi: 10.3847/1538-4357/833/2/230

Wright, A. N., & Elsden, T. (2020, February). Simulations of MHD wave propagation and coupling in a 3-D magnetosphere. *Journal of Geophysical Research (Space Physics)*, *125*(2), e27589. doi: 10.1029/2019JA027589

Wright, A. N., Elsden, T., & Takahashi, K. (2018, August). Modeling the dawn/dusk asymmetry of field line resonances. *Journal of Geophysical Research (Space Physics)*, *123*(8), 6443–6456. doi: 10.1029/2018JA025638

Wright, A. N., & Mann, I. R. (2006, January). Global MHD Eigenmodes of the Outer Magnetosphere. In *Magnetospheric ulf waves: Synthesis and new directions* (Vol. 169, p. 51). doi: 10.1029/169GM06

Wright, A. N., & Southwood, D. J. (1987, February). Stationary Alfvénic structures. *Journal of Geophysical Research*, *92*(A2), 1167–1176. doi: 10.1029/JA092iA02p01167

Wright, A. N., & Thompson, M. J. (1994, March). Analytical treatment of Alfvén resonances and singularities in nonuniform magnetoplasmas. *Physics of Plasmas*, *1*, 691–705. doi: 10.1063/1.870815

Wright, D. M., & Yeoman, T. K. (1999). CUTLASS observations of a high-*m* ULF wave and its consequences for the DOPE HF Doppler sounder. *Annales Geo- physicae*, *17*(11), 1493–1497. doi: 10.1007/s00585-999-1493-3

Yagova, N. V., Pilipenko, V. A., Sakharov, Y. A., & Selivanov, V. N. (2021, December). Spatial scale of geomagnetic Pc5/Pi3 pulsations as a factor of their efficiency in generation of geomagnetically induced currents. *Earth, Planets and Space*, *73*(1), 88. doi: 10.1186/s40623-021-01407-2

Yamakawa, T., Seki, K., Amano, T., Miyoshi, Y., Takahashi, N., Nakamizo, A., & Yamamoto, K. (2022). Excitation of two types of storm-time Pc5 ULF waves by ring current ions based on the magnetosphere-ionosphere coupled model. *Journal of Geophysical Research: Space Physics*, *127*(8), e2022JA030486. Retrieved from https://agupubs.onlinelibrary.wiley.com/doi/abs/10.1029/2022JA030486 (e2022JA030486 2022JA030486) doi: https://doi.org/10.1029/2022JA030486

Yamakawa, T., Seki, K., Amano, T., Takahashi, N., & Miyoshi, Y. (2019). Excitation of storm time Pc5 ULF waves by ring current ions based on the drift- kinetic simulation. *Geophysical Research Letters*, *46*(4), 1911–1918. Retrieved from https://agupubs.onlinelibrary.wiley.com/doi/abs/10.1029/2018GL081573 doi: https://doi.org/10.1029/2018GL081573

Yamakawa, T., Seki, K., Amano, T., Takahashi, N., & Miyoshi, Y. (2020). Excitation of internally driven ulf waves by the drift-bounce resonance with ring current ions based on the drift-kinetic simulation. *Journal of Geophysical Research: Space Physics*, *125*(11), e2020JA028231. Retrieved from https://agupubs.onlinelibrary.wiley.com/doi/abs/10.1029/2020JA028231 (e2020JA028231 2020JA028231) doi: https://doi.org/10.1029/2020JA028231

Yamamoto, K., Nosé, M., Kasahara, S., Yokota, S., Keika, K., Matsuoka, A., et al. (2018). Giant pulsations excited by a steep earthward gradient of proton phase space density: Arase observation. *Geophysical Research Letters*. *45*(14), 6773–6781. doi: 10.1029/2018gl078293

Yang, B., Zong, Q. G., Fu, S. Y., Li, X., Korth, A., Fu, H. S., et al. (2011). The role of ULF waves interacting with oxygen ions at the outer ring current during storm times. *Journal of Geophysical Research: Space Physics*, *116*(A1). doi: 10.1029/2010ja015683

Yang, Z., Zhang, B., Lotko, W., Sorathia, K. A., Pham, K., Luan, X., et al. (2022, September). Formation of Alfvén wave ducts by magnetotail flow bursts. *Journal of Geophysical Research (Space Physics)*, *127*(9), e30841. doi: 10.1029/2022JA030841

Yeoman, T. K., & Wright, D. M. (2001). ULF waves with drift resonance and drift-bounce resonance energy sources as observed in artificially-induced HF radar backscatter. *Annales Geophysicae*, *19*(2), 159–170. doi: 10.5194/angeo-19-159-2001

Zhang, B., Lotko, W., Brambles, O., Damiano, P., Wiltberger, M., & Lyon, J. (2012, September). Magnetotail origins of auroral Alfvénic power. *Journal of Geophysical Research (Space Physics)*, *117*(A9), A09205. doi: 10.1029/2012JA017680

Zhao, X. X., Hao, Y. X., Zong, Q.-G., Zhou, X.-Z., Yue, C., Chen, X. R., et al. (2020). Origin of electron boomerang stripes: Localized ulf wave- particle interactions. *Geophysical Research Letters*, *47*(17), e2020GL087960. Retrieved from https://agupubs.onlinelibrary.wiley.com/doi/abs/10.1029/2020GL087960 (e2020GL087960 10.1029/2020GL087960) doi: https://doi.org/10.1029/2020GL087960

Zong, Q., Rankin, R., & Zhou, X. (2017, December). The interaction of ultra-low-frequency pc3-5 waves with charged particles in Earth's magnetosphere. *Reviews of Modern Plasma Physics*, *1*(1), 10. doi: 10.1007/s41614-017-0011-4

# 11

# Alfvén Waves at the Interface of Solar Wind and Magnetosphere

**Eun-Hwa Kim[1], Jay R. Johnson[2], and Peter A. Damiano[3]**


## ABSTRACT

At Earth's magnetopause, kinetic Alfvén waves play a crucial role in energy transport via mode conversion from compressional waves and mode coupling with Kelvin-Helmholtz (KH) waves. When the wave frequency is close to the ion cyclotron frequency, the compressional wave energy can transfer to the perpendicular ion cyclotron resonance (PICR), and this mode conversion effectively occurs at the magnetopause. We show that the mode-converted kinetic Alfvén waves at the PICR location radiate toward the magnetosheath, where the wave frequency ($\omega$) is higher than the local PICR frequency ($\omega_{\mathrm{PICR}}$). In the KH vortex, the PICR frequency can oscillate in the radial direction; therefore, these waves can be trapped in the KH vortex in regions where $\omega > \omega_{\mathrm{PICR}}$. When the shear flow ($V_0$) in the magnetosheath is considered, for low Alfvén Mach number, secondary KH waves having a lower frequency than the primary KH become unstable at the interface. When the shear flow transition layer exists between the magnetosheath and the magnetosphere, strong coupling between the shear Alfvén waves and the secondary KH wave occurs. As a result, significant wave energy can transfer from the boundary layer to the inner magnetosphere.


## 11.1. INTRODUCTION

The Earth's magnetopause is a transition layer that separates the magnetosheath and the magnetosphere. The transfer of solar wind mass, momentum, and energy across the magnetopause can occur through magnetic reconnection or Kelvin-Helmholtz (KH) waves (e.g., H. Hasegawa et al., 2004; Hesse & Cassak, 2020). In addition, large-amplitude Alfvén waves have been frequently observed at the magnetosphere boundary (e.g., Anderson et al., 1982; Chaston et al., 2005, 2007, 2008, 2013; Engebretson, Cahill, et al., 1991; N. Lin et al., 1991; Phan et al., 1994; Rezeau et al., 1986, 1989, 1993; Song et al., 1993; Takahashi et al., 1991; Tsurutani et al., 1982; Yao et al., 2011). A significant feature of waves related to magnetopause crossings is the abrupt polarization transition from compressional to transverse in the low-frequency wave power spectrum as satellites travel from the magnetosheath across the magnetopause into the boundary layers (e.g., Johnson & Cheng, 2003; Phan et al., 1997).

When the field-aligned phase velocity of incoming compressional waves matches with the local Alfvén speed, mode conversion occurs from the compressional wave to the shear Alfvén wave at the Alfvén resonance (Chen & Hasegawa, 1974; A. Hasegawa et al., 1983; Southwood, 1974; Radoski, 1974; Tamao, 1965; Uberoi, 1972). The waves with large amplitudes detected at the magnetopause are thought to be kinetic Alfvén waves (KAWs) resulting from mode conversion of compressional waves at this Alfvén resonance (Belmont et al., 1995; Chaston et al., 2007; Johnson et al., 2001; Johnson & Cheng, 1997;


[1] *Princeton Plasma Physics Laboratory, Princeton University, Princeton, NJ, USA*

[2] *School of Engineering, Andrews University, Berrien Springs, MI, USA*

[3] *Geophysical Institute, University of Fairbanks, Fairbanks, AK, USA*



*Alfvén Waves Across Heliophysics: Progress, Challenges, and Opportunities, Geophysical Monograph 285*, First Edition.
Edited by Andreas Keiling.


DOI: 10.1002/9781394195985.ch11

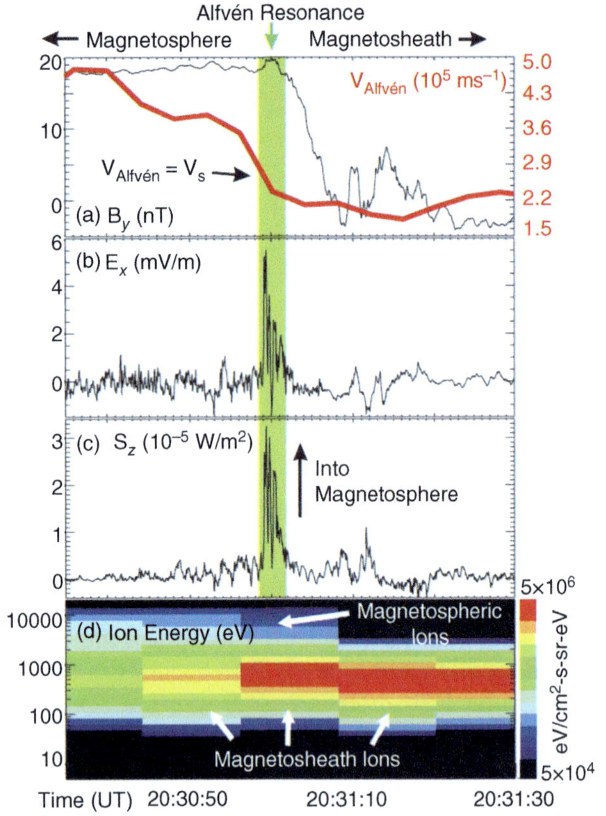


**Figure 11.1** (a) Magnetic field ($B_y$) with local Alfvén speed (red line). (b) Electric field ($E_x$). (c) The wave Poynting flux ($S_z$) normal to the magnetopause boundary in the geocentric solar magnetospheric coordinate system (GSM). (d) Ion energy spectrogram detected by Cluster 4 (Chaston et al. (2007) / The American Physical Society).

L. C. Lee et al., 1994). This mode conversion at the magnetopause leads to efficient plasma transport and entry across the boundary, leading to plasma entry into the magnetosphere (Chaston et al., 2008; Johnson & Cheng, 1997, 2001; L. C. Lee et al., 1994).

An example of kinetic Alfvén waves detected at the magnetopause is presented in Figure 11.1 (Chaston et al., 2007). As the Cluster spacecraft approached the duskside flank, it detected multiple step-like field variations resulting from the KH vortex. An enhancement in electric field amplitude occurs in each vortex. In this figure, the green shaded region denotes the location of the Alfvén resonance, and the electric field displays a sharp peak. The peak is preceded by oscillatory fluctuations that extend from the magnetosphere to the magnetopause's Alfvén resonance. The ion energy spectrogram in Figure 11.1(d) depicts how the ion population from the hot magnetosheath enters the magnetosphere, possibly in association with the strong Alfvénic fluctuations as shown in Figure 11.1(b), suggesting the waves' importance for particle transport.

Compressional modes generated in the foreshock are thought to be a source of ultralow frequency waves in the magnetosphere (Fairfield, 1969; Greenstadt et al., 1968; Odera & Stuart, 1986; Shi et al., 2017; Turc et al., 2022). The extent to which these waves penetrate the magnetopause depends on the interplanetray magnetic field (IMF) cone angle, with a quasi-parallel shock being the most efficient for penetration (Engebretson, Lin, et al., 1991). Due to the typical Parker spiral magnetic field orientation, the foreshock has a broader coverage on the dawnside than the duskside, which introduces an apparent asymmetry in the transport associated with the mode-conversion process. As a result, particle entry and heating also exhibit an asymmetry, with a higher concentration on the dawnside, as supported by a statistical study using the THEMIS spacecraft (Yao et al., 2011). This study found that the spectral energy densities of gyroradii scale waves, such as kinetic Alfvén waves, exhibit a dawn–dusk asymmetric structure and are more prominent on the dawnside of the magnetopause.

Hybrid simulations examined the mode-conversion process to kinetic Alfvén waves (Y. Lin et al., 2010, 2012; Shi et al., 2013; Wang et al., 2019). Y. Lin et al. (2010) found that the strong mode conversion occurs when wave frequency ($\omega$) is comparable to the ion cyclotron frequency ($\omega_{ci}$) at the perpendicular ion cyclotron resonance (PICR) (or Alfvén resonance) (Cross, 1988; Karney et al., 1979; Stix, 1992). When the finite $\omega/\omega_{ci}$ effects become critical, the magnetohydrodynamic (MHD) approximation is invalid, and two-fluid or multi-ion plasma effects should be considered. Like the Alfvén resonance in the MHD regime, strong resonant damping of compressional waves occurs at the PICR (Appert & Vaclavik, 1983; Karney et al., 1979; E.-H. Kim & Lee, 2003; Winglee, 1982). Y. Lin et al. (2010) also showed that the mode-converted kinetic Alfvén wave propagates toward the magnetosheath across the magnetic field line, and strong wave energy flows in the field-aligned direction.

The Cluster observations shown in Figure 11.1 are evidence of the mode conversion at the KH vortex. When considering the KH vortex, the shear flow velocity ($V_0$) effect is necessary; however, in most studies on mode-conversion processes, a low Alfvén Mach number ($M_A \equiv V_0/V_A$, where $V_A$ is the Alfvén velocity) is assumed; thus, $V_0$ is neglected. When $V_0$ is considered, coupling between KH and Alfvén waves occurs at the magnetopause (e.g., E.-H. Kim et al., 2022; Taroyan & Erdélyi, 2002; Turkakin et al., 2013).

Unstable KH waves occur at the boundary between two fluids with different velocities. These waves are responsible for mass, momentum, and energy exchange across the interface (Cowee et al., 2010; Hwang et al., 2011; Johnson et al., 2021; Matsumoto & Hoshino, 2006; Miura & Pritchett, 1982; Moore et al., 2016; Nykyri et al., 2017; Otto & Fairfield, 2000). Since KH waves are surface waves, these waves are confined near the interface and decay exponentially away from the boundary (Pu & Kivelson, 1983b; Southwood, 1968). However, because the wavelength is relatively long, the wave energy can still reach the plasma sheet and inner magnetosphere (Pu & Kivelson, 1983a) and play a critical role in generating geomagnetic pulsations and mode conversion to shear Alfvén waves (Chen & Hasegawa, 1974).

At the magnetopause, the Alfvén velocity ($V_A$) increases and the shear flow decreases in the direction from the magnetosheath to the magnetosphere, and coupling between the Alfvén surface wave and KH surface wave can occur. This coupling arises when the frequencies of two surface waves are comparable to each other at a slower flow speed. This interaction includes the resonant flow instability and negative absorption of the compressional waves, which is believed to play an important role in Earth's boundary layer and in the solar corona (Andries et al., 2000; Antolin & Van Doorsselaere, 2019; S. Kim & Kim, 2022; Mann et al., 1999; Ruderman & Wright, 1998; Taroyan & Erdélyi, 2002; Taroyan & Ruderman, 2011; Tirry et al., 1998; Turkakin et al., 2014, 2022). The presence of the shear flow can also strongly modify the eigenmode of the compressional waves in the outer magnetosphere between the magnetopause and inner magnetosphere cutoff locations (Mazur & Chuiko, 2013, 2015).

In addition to primary KH waves, secondary KH waves can also occur at lower Alfvén Mach numbers (González & Gratton, 1994; Taroyan & Erdélyi, 2002; Turkakin et al., 2013). These waves have lower wave frequencies and growth rates than primary KH waves. The secondary KH waves can carry significant wave energy into the magnetosphere, but these wave modes and related mode coupling to the Alfvén waves are less focused than the over reflection phenomena. Turkakin et al. (2013) examined characteristics of the secondary KH waves when the shear flow and the magnetic field in the magnetosheath are parallel. By comparing the Alfvén Mach number thresholds for the primary and secondary KH waves generation with the detected Alfvén Mach number from the satellites, they concluded that the secondary KH waves are dominated on the dayside flank, while the primary KH waves carry energy into the magnetosphere at the nightside flank and magnetotail.

E.-H. Kim et al. (2022) recently performed numerical simulations that showed that secondary KH waves could also arise at the boundaries when the shear velocity and background magnetic field are perpendicular if a shear transition layer exists between the magnetosphere and magnetosheath. The energy of these waves is concentrated near the shear Alfvén wave frequency of the magnetosheath, which is lower in frequency than that of primary KH waves, consistent with earlier studies. Additionally, evidence of mode conversion to the shear Alfvén waves has been discovered in these waves, suggesting efficient wave energy transfer from the boundary layer to the inner magnetosphere, similar to the detected Alfvén wave at the magnetopause in Figure 11.1.

This paper discusses how compressional MHD waves can be converted into the kinetic Alfvén waves near the PICR at the magnetopause. We examine the efficiency of the mode-conversion process and the characteristics of the inertial and kinetic Alfvén waves at the magnetopause. Furthermore, we analyze the propagation of mode-converted waves in the KH vortex and show that the kinetic Alfvén waves can be trapped in the KH vortex. We then discuss the heavy ion effects on the mode-conversion process, since the mode-converted Alfvén waves at the magnetopause can be comparable to the ion cyclotron frequency. To account for the coupling between KH and Alfvén waves, we also incorporate the effects of shear flow velocity. We reviewed the characteristics of the secondary KH waves in the dayside flank, where these waves are believed to play an important role in energy transfer. We reveal that the mode-coupling process between the secondary KH waves and Alfvén waves occurs by introducing a shear transition layer between the magnetosheath and the magnetosphere.

## 11.2. MODE CONVERSION AT THE ALFVÉN AND PERPENDICULAR ION CYCLOTRON RESONANCES

When electron inertial and Hall effects are considered in warm plasma, the MHD wave equations can be derived as,

$$\nabla \times \vec{E}_1 = -\partial_t \vec{b}_1, \tag{11.1}$$

$$\nabla \times \vec{b}_1 = \mu_0^{-1} \vec{J}_1, \tag{11.2}$$

$$\frac{m_e}{N_0 e^2}\partial_t \vec{J}_1 = \vec{E}_1 + \vec{v}_1 \times \vec{B}_0 + \frac{1}{N_0 e}\vec{J}_1 \times \vec{B}_0 - \frac{1}{N_0 e}\nabla p_{e1}, \tag{11.3}$$

$$\rho_0 \partial_t \vec{v}_1 = \vec{J}_1 \times \vec{B}_0 - \nabla p_1, \tag{11.4}$$

$$\partial_t \rho_1 = -\nabla \cdot (\rho_0 \vec{v}_1), \tag{11.5}$$

$$p_1 = \gamma (P_0/\rho_0)\rho_1. \tag{11.6}$$

Here, $\vec{B}_0$, $P_0$, $N_0$, and $\rho_0$ are the background magnetic field, pressure, number density, and mass density, and $\vec{E}_1$, $\vec{b}_1$, $\vec{J}_1$, $\vec{v}_1$, $p_{1(el)}$, and $\rho_1$ are perturbed electric field, magnetic field, current, velocity, pressure (electron pressure), and mass density, respectively, $e$ and $m_e$ are the charge and mass of the electron, and $\gamma$ is the adiabatic index.

From the linearized equations (Equations (1)–(6)), and ignoring the electron pressure gradient in Equation

(3), the wave dispersion relation can be resolved derived (Cramer, 2001) as,

$$\left[\omega^2(1+\delta_e k^2) - V_A^2 k_\parallel^2\right]$$
$$\left[\omega^2(1+\delta_e k^2)\left(\omega^2 - C_s^2 k^2\right) - V_A^2 k^2\left(\omega^2 - C_s^2 k_\parallel^2\right)\right]$$
$$= \left(\frac{\omega}{\omega_{ci}}\right)^2 V_A^2 k_\parallel^2 k^2\left(\omega^2 - C_s^2 k^2\right), \tag{11.7}$$

where $\omega$ is wave frequency, $k$ is wavevector, $k_\parallel$ is a wavevector parallel to $\vec{B}_0$, $C_s \equiv \gamma P_0/\rho_0$ is the sound speed, $\delta_e = c/\omega_{pe}$ is the electron inertial length, $\omega_{pe}$ is the electron plasma frequency, $\omega_{ci}$ is ion cyclotron frequency, and $V_A = |B_0|/\sqrt{\mu_0\rho_0}$ is the Alfvén velocity.

The wave dispersion relations using Equation (7) for $\omega \ll \omega_{ci}$ and $\omega{\sim}\omega_{ci}$ are shown in Figure 11.2. The $x$ axis represents the normalized perpendicular wavenumber ($k_\perp$) to $\delta_e$. In Figure 11.2(a), the solid lines are the wave solutions from Equation (7) for MHD waves. The red and blue lines represent two different electron thermal velocities ($V_{\text{the}}$) where $V_{\text{the}} > V_{\text{A}}$ and $V_{\text{the}} < V_{\text{A}}$, respectively. For $k_\perp{\sim}k_\parallel$, the wave solutions from both $V_{\text{the}}$ cases are almost identical to the cold plasma approximation without the electron inertial effect (represented by the green dashed line in Figure 11.2).

In this figure, the resonance occurs where $k_\perp \to \infty$ in cold plasma approximation at

$$\omega_{\text{A}} = k_\parallel V_{\text{A}}, \tag{11.8}$$

which is called the Alfvén resonance. When a fast compressional wave matches the resonance condition, the wave energy can transfer to the field-aligned propagating resonant wave mode at this location (Chen & Hasegawa, 1974; A. Hasegawa & Chen, 1975; Tamao, 1965). This resonance is also called the field-line resonance in Earth's magnetosphere and plays an important role in the magnetospheric physics of Earth (Southwood, 1974).

When we ignore the electron inertial effect in a cold plasma, the mode-converted waves localize at a resonance since $k_\perp \to \infty$. However, the electron inertial and the thermal pressure terms allow $k_\perp$ to have a finite value, enabling dispersion at the resonance and cross-field transport of kinetic-scale wave energy (e.g., Leonovich & Mazur, 1995). In Figure 11.2(a), the blue and red lines represent the IAWs for $V_{\text{the}} < V_{\text{A}}$ and KAWs for $V_{\text{the}} > V_{\text{A}}$, respectively. When the wave frequency is close to the resonant frequency ($\omega{\sim}\omega_{\text{A}}$), the wave solution can be simplified to (Cramer, 2001; Lysak, 2008)

$$\omega_{\text{A}}^2 = k_\parallel^2 V_{\text{A}}^2 \frac{1 + k_\perp^2\rho_s^2}{1 + k_\perp^2\delta_e^2}, \tag{11.9}$$

where $\rho_s = \frac{C_s}{\omega_{ci}}$ is the ion acoustic gyroradius. When electron inertial effects are ignored, this dispersion relation can also be derived from the full kinetic theory (A. Hasegawa, 1976) as,

$$\omega_{\text{A}}^2 = k_\parallel^2 V_{\text{A}}^2\left(1 + k_\perp^2\rho_i^2\left(\frac{3}{4} + \frac{T_e}{T_i}\right)\right), \tag{11.10}$$

where $\rho_i = V_{thi}/\omega_{ci}$ is the gyroradius of the thermal ions, $V_{thi}$ is the ion thermal velocity, and $T_i$ is the ion temperature.

Figure 11.2 displays black (for $V_{\text{the}} > V_{\text{A}}$) or magenta (for $V_{\text{the}} < V_{\text{A}}$) stars to represent the simplified wave dispersion relation of Equation (9). These solutions are well matched with the general solutions of red and blue lines when $k_\perp \gg k_\parallel$. It can be clearly seen that the IAW and KAW have different propagation characteristics. When $k_\perp\delta_e$ is close to 1, KAW can propagate when $\omega > \omega_{\text{A}}$, and the IAW can propagate when $\omega < \omega_{\text{A}}$. The group velocity $\partial\omega/\partial k_\perp$ is negative for IAW and positive for KAW. Hence, the IAW and KAW are backward- and forward-propagating waves, respectively.

When $\omega{\sim}\omega_{ci}$, the wave solutions in Figure 11.2(b) have almost the same trend as the MHD case. However, the resonance does not occur at $\omega = \omega_{\text{A}}$ but at

$$\omega_{\text{PICR}}^2 = k_\parallel^2 V_{\text{A}}^2\left(1 - \omega^2/\omega_{ci}^2\right). \tag{11.11}$$

This resonance condition includes finite $\omega/\omega_{ci}$ effects and is called the PICR or Alfvén resonance (Karney et al., 1979). The effect of the finite $\omega/\omega_{ci}$ on ultra-low frequency (ULF) waves becomes important near both the magnetosheath and the magnetopause (e.g., Belmont et al., 1995; Y. Lin et al., 2010; Rezeau et al., 1989), as well as in fusion devices (e.g., Appert & Vaclavik, 1983; Karney et al., 1979; Mitchell et al., 2001; Winglee, 1982). Near the resonant condition, the KAW dispersion relation is also modified to (Gekelman et al., 2011),

$$\omega_{PICR}^2 = k_\parallel^2 V_A^2 \frac{1 - \omega^2/\omega_{ci}^2 + k_\perp^2\rho_s^2}{1 + k_\perp^2\delta_e^2}. \tag{11.12}$$

Figure 11.2 shows that the wave characteristics of IAW and KAW are similar in the MHD regime. The forward-propagating KAW can propagate when $\omega > \omega_{\text{PICR}}$, while the backward-propagating IAW can propagate for $\omega < \omega_{\text{PICR}}$.

### 11.2.1. Mode-Conversion Efficiency at the Magnetopause

One crucial question is how much incoming fast compressional waves can be converted into shear Alfvén wave modes at the magnetopause. Y. Lin et al. (2010) examined the mode-conversion efficiency ($\mathcal{A}$) of compressional waves into the Alfvén resonance at the magnetopause. By assuming radial propagation across the magnetopause in the $x$ direction, they calculated the mode-conversion

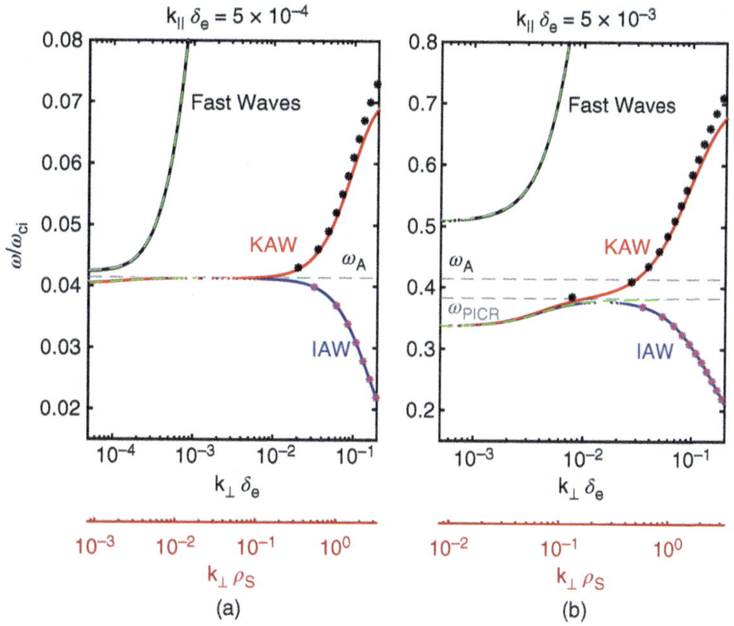


**Figure 11.2** The wave dispersion relation calculated using Equation (7) for (a) $k_\parallel\delta_e = 5 \times 10^{-4}$ and (b) $5 \times 10^{-3}$. Two different thermal velocities $V_{the}/V_A = 0.014$ and 1.4 are adopted for calculations with $V_A/c = 0.011$. The $x$ axis represents the normalized perpendicular wavenumber to $\delta_e$ and $\rho_s$ when $V_{the}/V_A = 1.4$. The black lines represent the fast compressional wave, and the red and blue lines represent the kinetic Alfvén wave (KAW) for $V_{the}/V_A = 1.4$ and the inertial Alfvén wave (IAW) for $V_{the}/V_A = 0.014$. The green dashed lines are the cold plasma wave dispersion relation ignoring the electron inertial effect. The black and magenta stars represent the approximated kinetic and inertial Alfvén waves calculated from Equation (12). The horizontal gray dashed lines represent the resonance frequencies ($\omega_A$ and $\omega_{PICR}$) from Equations (8) and (11).

efficiency as a function of the wave propagation angle $\theta = \tan^{-1}(k_y/k_\parallel)$, where $k_y$ and $k_\parallel$ represent wavenumbers along the $y$ and field-aligned $z$ directions. A set of linearized wave equations can be derived when the electron inertial effects is ignored (E.-H. Kim et al., 2011; Y. Lin et al., 2010),

$$\frac{c}{\omega}\frac{\partial Y}{\partial x} = MY, \tag{11.13}$$

where

$$Y = \begin{pmatrix} E_y \\ \frac{c}{\omega}\frac{\partial E_y}{\partial x} - in_y E_x \end{pmatrix}, \tag{11.14}$$

$$M = \begin{pmatrix} \frac{n_y D}{n_\parallel^2 - S} & 1 + \frac{n_y^2}{n_\parallel^2 - S} \\ \frac{(n_\parallel^2 - R)(n_\parallel^2 - L)}{n_\parallel^2 - S} & -\frac{n_y D}{n_\parallel^2 - S} \end{pmatrix}, \tag{11.15}$$

where $n_y$ is the refractive index in $y$ direction, $R$, $L$, $S$, and $D$ are Stix's functions (Stix, 1992). For $\omega \ll \omega_{ce}$ and $\omega_{pe}$ in two-fluid plasma, the Stix's functions can be reduced to $R(L) \cong c^2 V_A^{-2}\left(1 \pm \frac{\omega}{\omega_{ci}}\right)^{-1}$, $S \cong c^2 V_A^{-2}\left(1 - \frac{\omega^2}{\omega_{ci}^2}\right)^{-1}$, and $D \cong -S\frac{\omega}{\omega_{ci}}$. Like Lin et al. (2010), we calculate the mode-conversion efficiency by adopting a typical magnetopause profile shown in Figure 11.3(a). The mode conversion is assumed to occur at $x = 0$, and then $k_\parallel(\omega = \omega_{PICR(x=0)})$ is calculated satisfying the resonance condition. The compressional waves are launched at the boundary, and the solutions are solved using the finite difference method. The reflection ($\mathcal{R}$) and transmission ($\mathcal{T}$) coefficients can be obtained from the wave solution, and the mode-conversion efficiency becomes $\mathcal{A} = 1 - \mathcal{R} - \mathcal{T}$.

Panels (b) and (c) in Figure 11.3 present the wave solutions for $\omega/\omega_{ci} = 0.42$ when $\theta = 0°$ and $\theta = 30°$. For $\theta = 0°$ in Figure 11.3(b), the calculated mode-conversion efficiency is $\mathcal{A} = 0.27$. In this case, strong modulation of wave power is seen in the magnetosheath ($x < 0$) because most of the wave energy is reflected around $x < 0$. On the other hand, for $\theta = 30°$, $\mathcal{A} \sim 1$, wave power enhancement occurs at $x = 0$, with relatively low wave power seen in $x < 0$. Figure 11.3(d) shows the calculated mode-conversion efficiency ($\mathcal{A}$) as a function of $\omega/\omega_{ci}$ and $\theta$. At low frequencies ($\omega/\omega_{ci} \lesssim 0.1$), the MHD result is obtained with a maximum absorption of around 10%, and no absorption occurs for $\theta = 0°$. The maximum absorption occurs at $\omega/\omega_{ci}$~0.42 when $\theta$~30° at the PICR. For the higher frequency range, mode conversion also occurs for $\theta = 0°$, maximizing around $\omega/\omega_{ci} = 0.51$. Above $\omega/\omega_{ci} > 0.8$ and beyond $\theta$~52°, there is no absorption.

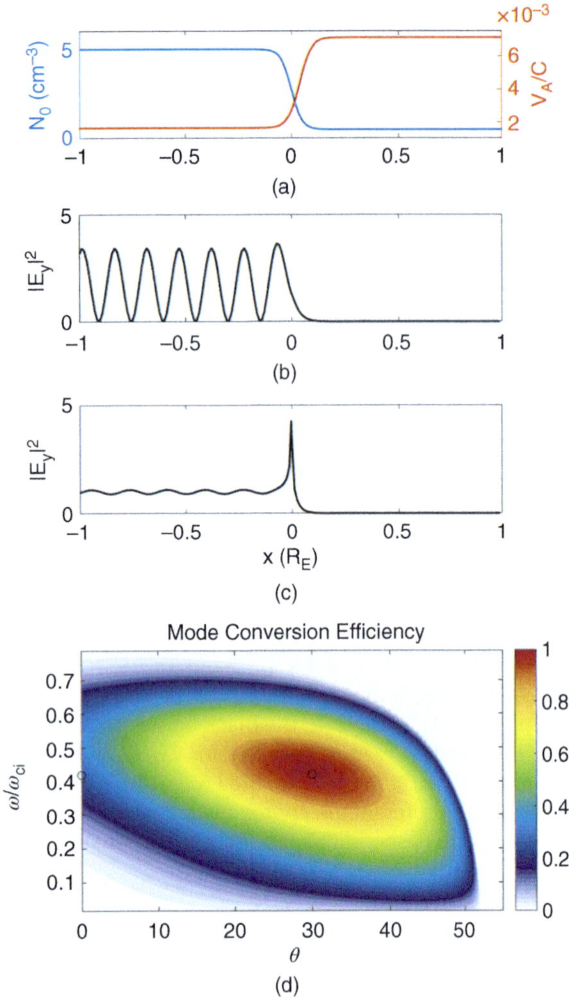


**Figure 11.3** (a) Density and Alfvén velocity profile in $x$. (b) Compressional wave energy along $x$ direction for $\theta = 0°$. (c) $\theta = 30°$. (d) The calculated mode-conversion efficiency.

These results are consistent with hybrid simulations and mode-conversion calculations from Y. Lin et al. (2010) confirm that the finite $\omega/\omega_{ci}$ becomes critical for mode conversion at the magnetopause.

### 11.2.2. Inertial and Kinetic Alfvén Wave Propagation Near the Magnetopause

The wave propagation in cold and warm plasma can be examined using a full-wave simulation. Based on a cold multi-ion plasma wave model (E.-H. Kim & Lee, 2003) and a warm electron plasma wave model (E.-H. Kim et al., 2007), we developed a two-fluid warm plasma wave model that uses the following equations,

$$\nabla \times \vec{E}_1 = -\partial_t \vec{b}_1, \tag{11.16}$$

$$\nabla \times \vec{b}_1 = \mu_0^{-1} \vec{J}_1 + c^{-2} \partial_t \vec{E}_1, \tag{11.17}$$

$$N_0 m_j \partial_t \vec{v}_{j1} = N_0 q_j (\vec{E}_1 + \vec{v}_{j1} \times \vec{B}_0) - \nabla p_{j1}, \tag{11.18}$$

$$\partial_t n_{j1} = -\nabla \cdot (N_0 \vec{v}_{j1}), \tag{11.19}$$

$$p_{j1} = \gamma n_{j1} T_{j0}, \tag{11.20}$$

where $j$ represents electron and ion. For simplicity, it is assumed that the inhomogeneity and $\vec{B}_0$ are introduced in the $x$ and $z$ directions, respectively, as shown in Figure 11.4(a). All variables are assumed to be proportional to $\exp(ik_y y + ik_\parallel z)$. We consider two cases of electron temperature, $T_e = 0$ or $T_e = T_e(x)$. Under the given conditions, the resonance frequency lies where $0.3 \le \omega_{\mathrm{PICR}}/\omega_{ci} \le 0.7$, as shown in Figure 11.4(b). An impulsive source in compressional electric field component $E_y$ with a strong center frequency of $\omega_0 = 0.42\omega_{ci}$ is launched near the left-side boundary at $x \sim -1.2\ R_E$. The center frequency is chosen from Figure 11.3(d), where the mode-conversion efficiency is maximized. The simulation domain is $-1.5\ R_E < x < 1.5\ R_E$, and we simulate for $0 < t/\tau_A < 40$, where $\tau_A = 2\pi/\omega_{A(x=0)}$.

Panels (c) and (d) in Figure 11.4 plot the time snapshot of $b_{1y}$, the resonant wave mode component, for $T_e = 0$ and $T_e = T_e(x)$. Long wavelength compressional waves propagate to the resonance layer at $t/\tau_A = 10$, and strong mode conversion occurs to the resonant wave mode at $t/\tau_A = 12 - 17$. The mode-converted waves propagate across the field line after the mode conversion occurs. For $T_e = 0$ as shown in Figure 11.4(c), waves propagate in the positive $x$ direction where $\omega_0 < \omega_{\mathrm{PICR}}$, and the wavelength becomes shorter as the wave propagates. On the other hand, waves in Figure 11.4(d) radiate to the negative $x$ direction where $\omega_0 > \omega_{\mathrm{PICR}}$ and the wavelength is slightly longer than the waves in Figure 11.4(c).

The power spectra of $b_{1y}$ obtained through the fast Fourier transform in $x$ direction demonstrate the phase velocity direction. When the compressional waves reach the mode-conversion layer, waves propagate with a positive $k_x$. However, both short-wavelength waves in Figure 11.4(c, d) have a negative $k_x$. Therefore, for $T_e = 0$, the short wavelength waves' group and phase velocity directions are opposite, which is backward-propagating IAWs. The forward-propagating KAWs appear for the case of $T_e = T_e(x)$ with the same group and phase velocity directions. Therefore, the IAW and KAW propagating characteristics near the ion cyclotron frequency are identical to those in the MHD regime (Lysak, 2008) and hybrid simulation (Y. Lin et al., 2010) (see Figure 11.5).

### 11.2.3. KAW Propagation within a KH Vortex

As shown in Figure 11.1, the mode conversion at the Alfvén resonance occurs within KH vortices (Chaston

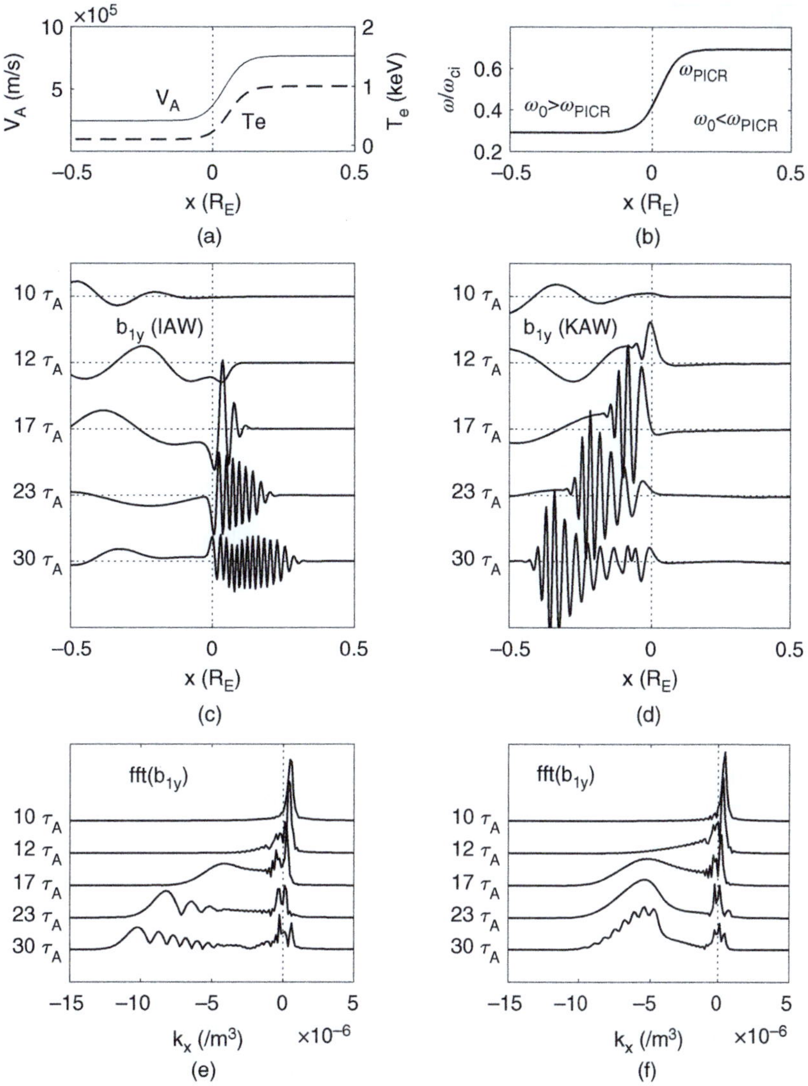


**Figure 11.4** (a) Adopted Alfvén velocity ($V_A$) and electron temperature ($T_e$). (b) Calculated resonance frequency ($\omega_{PICR}$). Time snapshot and the power spectra of $b_{1y}$ of (c, e) inertial Alfvén wave (IAW) and (d, f) kinetic Alfvén wave (KAW).

et al., 2007). Considering wave propagation across the magnetopause interface in the presence of KH vortices as illustrated in Figure 11.6(a), the $V_A$ profile oscillates, and wave energy can be trapped in the region of minimum $V_A$. We assumed that $\omega_{PICR}$ oscillates along $x$ as shown in Figure 11.6(b). Similar to Section 11.2.2, we also examined the mode-conversion process at the KH vortex by assuming two center frequencies of $\omega_0/\omega_{PICR} = 0.42$ and 0.63. Panels (c) and (d) of Figure 11.6 show the time snapshot of $b_{1y}$ at $t/\tau_A$=23. The gray-shaded regions in this figure represent the magnetosheath where $\omega_0 > \omega_{PICR}$.

When the compressional waves from the magnetosheath reach the first mode-conversion location near $x\sim -0.1$, the first mode conversion occurs, and the mode-converted KAW propagates toward the magnetosheath. Some of the compressional wave energy is reflected, but the rest of the waves can tunnel through the $V_A$ bump and reach the second and third mode-conversion locations near $x\sim 0$ and $x\sim 0.15$, respectively. The mode-converted KAW can propagate toward the region where $\omega_0 > \omega_{PICR}$ and the wavelength is much shorter to tunnel through the $V_A$ bump; thus, these waves are trapped in the region of the local minimum of $V_A$.

Unlike the monotonically increasing resonance condition shown in Figure 11.4, multiple resonances at different locations can occur. The mode-converted KAW propagating toward the magnetosheath where $x < 0$ shows a wave packet. For example, waves with $\omega_0/\omega_{PICR} = 0.63$ have the second power node near $x\sim -0.15$. This behavior results because the compressional waves are reflected

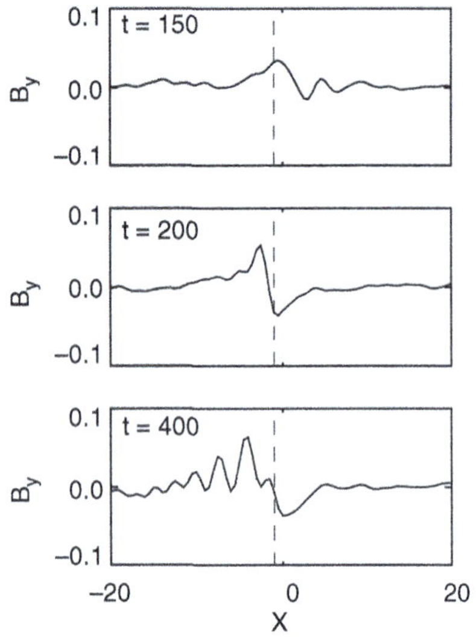


**Figure 11.5** Spatial cuts of the transverse component ($B_y$) of a hybrid simulation as a function of $x$ at different simulation time steps. Strong KAWs with larger $k_\perp$ radiate into the magnetosheath on the left side (Lysak & Lotko, 2020 / with permission of John Wiley & Sons).

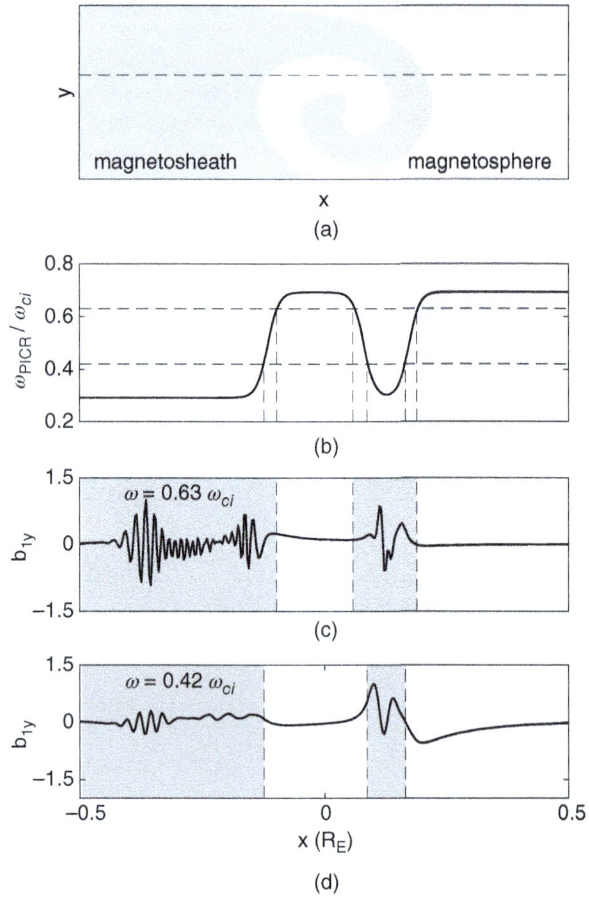


**Figure 11.6** (a) Illustration of KH vortex. The gray-shaded area is the magnetosheath. The profile of $V_A$ can oscillate in $x$ providing a potential well that can trap Alfvén wave energy inside the KH vortex. (b) Adopted perpendicular ion cyclotron resonance frequency ($\omega_{PICR}$) for the simulations. Time snapshots of the perturbed magnetic-field component in transverse direction ($b_{1y}$) for (c) $\omega = 0.63\ \omega_{ci}$ and (d) $\omega = 0.42\ \omega_{ci}$ at $t/\tau_A$=23. The gray-shaded area represents the magnetosheath where $\omega > \omega_{PICR}$.

inside the KH vortex. The long-wavelength reflected compressional waves near the second or third resonance layer inside the KH vortex could tunnel through the $V_A$ bump and reach the first mode-conversion location. Another mode-conversion occurs from the reflected compressional waves to the KAW. Therefore, there is a time delay between the first and the second power peak of the mode-converted KAW in the magnetosheath area due to the transit time of the reflected compressional wave between the first and the third mode-conversion location.

### 11.2.4. Heavy Ion Effects on Mode-Conversion Process

The solar wind contains protons as its primary constituent but also comprises a notable amount of $He^{++}$ ions (Gloeckler & Geiss, 1989). The $He^{++}$ ions are also detected at the magnetopause and the ratio between $He^{++}$ ions and protons is about 6–7% (Delano et al., 2019). The ionosphere is a major source of $O^+$ ions (Horwitz et al., 1984; Shelley et al., 1972; Yau & Andre, 1997), which can be the main contributor to the mass density in the magnetopause (Bouhram et al., 2005; S. H. Lee et al., 2016). The $O^+$ ions account for 30% and 5% of the mass density on the dusk and the dawn flank, respectively (Bouhram et al., 2005), and the ratio of $O^+$ to $H^+$ at the magnetopause can reach up to 40% (Zeng et al., 2020).

The presence of heavy ions can have a profound impact on plasma entry and momentum transport at the magnetopause. For example, shear-driven KH instabilities (KHI) would be more unstable because the increased mass density reduces the effect of restoring magnetic tension force; thus, heavy ions can modify the threshold shear flow velocity and the growth rate of the KHI (Delamere et al., 2011). Including heavy ions increases the mass density, which can modify the KAW properties in the MHD regime, such as wave polarization and the electric field to magnetic field amplitude ratio (Moya et al., 2022). Heavy ions also can strongly affect the efficiency of mode conversion of compressional waves to transverse Alfvén waves (E.-H. Kim et al., 2011).

In multi-ion plasmas, the dispersion relation of fast compressional wave can be derived as (Stix, 1992)

$$n_\perp^2 \cong \left(R - n_\parallel^2\right)\left(L - n_\parallel^2\right) \Big/ \left(S - n_\parallel^2\right), \qquad (11.21)$$

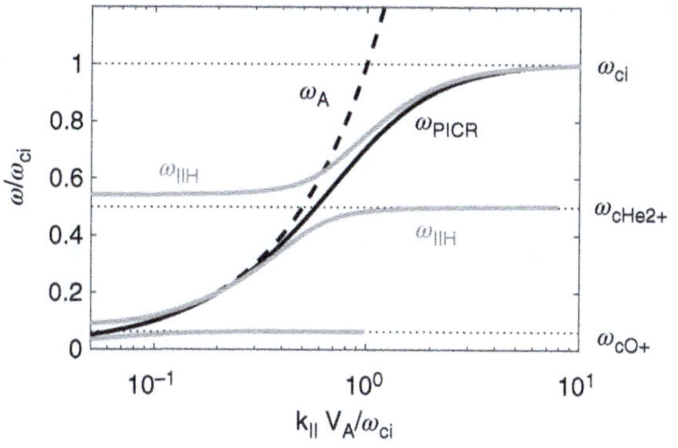


**Figure 11.7** Alfvén resonance ($\omega_A$) from Equation (8), perpendicular ion cyclotron resonance in two-fluid plasma ($\omega_{PICR}$) from Equation (11), and ion–ion hybrid resonance ($\omega_{IIH}$) in 5% of $He^{++}$ and 5% of $O^+$ plasma from Equation (22). Here, $\omega_{ci(cHe2+,cO+)}$ is proton ($He^{++}$ or $O^+$) cyclotron frequency.

where $n_{\perp(\parallel)}$ is wave refractive index perpendicular (parallel) to $B_0$, and the resonance condition is shown at

$$n_{\parallel}^2 = S. \tag{11.22}$$

This resonance occurs between each set of two ion cyclotron frequencies and is called the ion–ion hybrid (IIH) resonance (Buchsbaum, 1960). The IIH resonance is reduced to the Alfvén or PICR resonances in two-fluid plasmas. It is believed that the IIH resonance is critical in field-line resonances at Mercury (e.g., E.-H. Kim et al., 2008; E.-H. Kim, Johnson, Valeo, et al., 2015; Othmer et al., 1999), electromagnetic ion cyclotron waves at Earth (e.g., E.-H. Kim, Johnson, Kim, et al., 2015; Klimushkin et al., 2010; D.-H. Lee et al., 2008; Mikhailova et al., 2019), and tokamak plasmas (e.g., Bertelli et al., 2017; Jaeger et al., 2003).

Figure 11.7 plots the resonance condition of Alfvén resonance in the MHD regime ($\omega_A$), the PICR ($\omega_{PICR}$) in two-fluid plasma, and the IIH resonance ($\omega_{IIH}$) for electron–proton–$He^{++}$–$O^+$ plasma as a function of $k_{\parallel}$. There are several resonance branches in different frequencies due to heavy ions, even for the case of minimal $k_{\parallel}$. The mode-conversion efficiency to the IIH resonance depends on various parameters, such as the heavy-ion-density ratio and wave propagation angle $\theta$ (E.-H. Kim et al., 2011). Thus, the strong mode conversion may occur at the lower frequency below the heaviest ion cyclotron frequency. As discussed, compressional waves in the Pc3 frequency range from the foreshock can be sources of the mode conversion. Since Pc3 waves have a frequency ranging from 20–80 mHz, which is comparable to $O^+$ cyclotron frequency near the nose of the magnetopause approximately 60 mHz, the heavier ion effects are expected to be very important at the magnetopause.

The mode-converted waves at the IIH resonance can also propagate across the field line similar to the IAWs or KAWs, as electron inertial waves (or electrostatic ion cyclotron waves) in cold plasma (E.-H. Kim et al., 2019; D.-H. Lee et al., 2008; Ono, 1979) and ion Bernstein waves in warm/hot plasmas (Bertelli et al., 2017; Y. Lin et al., 2004; Phillips et al., 1996; Vincena et al., 2013; Wukitch et al., 2005). The ion Bernstein waves are strongly damped, and the power carried by the mode-converted waves results in direct electron heating in fusion devices (Y. Lin et al., 2003; Phillips et al., 2000), which is similar to electron heating by KAWs (Damiano et al., 2016; Hull et al., 2022; Watt & Rankin, 2009; Wygant et al., 2000) in Earth's magnetosphere. Therefore, the mode-conversion process and associated wave–particle interaction at the multi-ion magnetopause should be considered in the future.

## 11.3. COUPLING BETWEEN ALFVÉN AND KELVIN-HELMHOLTZ WAVES

The previous section discusses how compressional waves can convert to kinetic Alfvén waves at the magnetopause, and how these waves propagate across the field line and can be trapped in the KH vortex. When $V_0$ is considered in the magnetosheath, both $V_A$ and $V_0$ have a discontinuity at the magnetopause. This discontinuity leads to the coupling of field-aligned Alfvén waves and compressional KH waves. In this section, we will discuss the effects of the coupling of Alfvén waves and KH waves in detail.

If we ignore the electron inertial effect and assume $\omega/\omega_{ci} \to 0$, the linearized MHD wave equations in warm plasma can be derived as follows (Taroyan et al., 2002),

$$\frac{\partial \vec{b}_1}{\partial t} = \nabla \times (\vec{v}_1 \times \vec{B}_0) + \nabla \times (\vec{V}_0 \times \vec{b}_1), \tag{11.23}$$

$$\frac{\partial \rho_1}{\partial t} = -\nabla \cdot (\rho_0 \vec{v}_1 + \rho_1 \vec{V}_0), \tag{11.24}$$

$$\rho_0 \left[ \left( \frac{\partial}{\partial t} + \vec{V}_0 \cdot \nabla \right) \vec{v}_1 + \vec{V}_1 \cdot \nabla \vec{V}_0 + \vec{V}_0 \cdot \nabla \vec{V}_1 \right] + \rho_1 \vec{V}_0 \cdot \nabla \vec{V}_0 = \frac{1}{\mu_0} [(\nabla \times \vec{b}_1) \times \vec{B}_0 + (\nabla \times \vec{B}_0) \times \vec{b}_1] - \nabla p_1, \tag{11.25}$$

$$\left[ \frac{\partial}{\partial t} + \vec{V}_0 \cdot \nabla \right] p_1 - C_s^2 \left[ \frac{\partial}{\partial t} + \vec{V}_0 \cdot \nabla \right] \rho_1 = -\frac{1}{\rho_0} [\rho_1 \vec{V}_0 \cdot \nabla P_0 - \gamma p_1 \vec{V}_0 \cdot \nabla \rho_0] - \vec{v}_1 \cdot \nabla P_0 - \vec{V}_0 \cdot \nabla p_1 + C_s^2 \vec{v}_1 \cdot \nabla \rho_0. \tag{11.26}$$

Assuming that the plasma exhibits inhomogeneity along the $x$ direction, the perturbation value ($f$) can be expressed as a function of the location in $x$, frequency ($\omega$), and wavenumbers along the $y$ and $z$ directions ($k_{y,z}$); thus, $f(x,y,z,t) \rightarrow f(x,k_y,k_z,\omega)\exp(ik_y y + ik_z z - \omega t)$. With this assumption, Equations (23)–(26) can be simplified into the two coupled first-order differential equations (e.g., S. Kim & Kim, 2022; Taroyan et al., 2002),

$$C_0 \frac{\partial \xi_x}{\partial x} = -C_1 P, \tag{11.27}$$

$$\frac{\partial P}{\partial x} = C_2 \xi_x, \tag{11.28}$$

where $\xi_x$ is the Lagrangian displacement, $P$ is the Eulerian of total pressure, $C_{0,1,2}$ is constant in the homogeneous regions, $C_0(x) = \rho_0 \left(C_S^2 + V_A^2\right)\left(\omega^2 - \omega_T^2\right)\left(\omega^2 - \omega_A^2\right)$, $C_1(x) = \left(\omega^2 - \omega_{cut1}^2\right)\left(\omega^2 - \omega_{cut2}^2\right)$, $C_2(x) = \rho_0\left(\omega^2 - \omega_A^2\right)$, $\omega_{cut1,2}$ are the slow and fast cutoff frequencies, $\omega_A = k_\parallel V_A$ and $\omega_T = \omega_A C_S/\sqrt{V_A^2 + C_S^2}$, respectively. Eigenmode analysis using Equations (27) and (28) (e.g., S. Kim & Kim, 2022, Tirry et al., 1998) and time-dependent simulations using Equations (23)–(26) (E.-H. Kim et al., 2022; Turkakin et al., 2022) examined the properties of KH waves.

### 11.3.1. Characteristics of the Secondary KH Waves

When the wavelength is much longer than the thickness of the boundary layer, the boundary layer can be treated as a zero-thickness surface (e.g., Mills & Wright, 1999; Pu & Kivelson, 1983a; Taroyan & Ruderman, 2011; Turkakin et al., 2022). Considering the zero-thickness magnetopause, the boundary layer can be illustrated as shown in Figure 11.8(a) (Turkakin et al., 2013). Here, the magnetosheath and magnetosphere are represented by Regions I and III, respectively, and the magnetosphere in Region III is assumed to have a width of $d_{\mathrm{III}}$. At the magnetopause, the interface between Regions I and III, the Alfvén velocity increases from the magnetosheath to the magnetosphere ($V_{\mathrm{AI}} < V_{\mathrm{AIII}}$), while the shear flow velocity decreases to zero at the magnetosphere ($V_{0\mathrm{III}} \rightarrow 0$). This figure shows the shear flow along the $y$ direction ($\vec{V}_0 = V_0\hat{y}$) in the magnetosheath. The magnetic field can be assumed to be parallel or perpendicular to $\vec{V}_0$. For instance, Turkakin et al. (2013) considered that the magnetic field in the magnetosheath is parallel to shear velocity ($\vec{B}_{0I} \parallel \vec{V}_0$), and magnetic fields in the magnetosheath and magnetosphere are perpendicular ($\vec{B}_{0\mathrm{I}} \perp \vec{B}_{0\mathrm{III}}$) in the dayside flank. The magnetosphere can be assumed to be a semi-infinite plasma or have a finite width ($d_{\mathrm{III}}$). E.-H. Kim et al. (2022) used the magnetic field in the magnetosheath to be perpendicular to the shear flow ($\vec{B}_{0\mathrm{I}} \perp \vec{V}_0$) and the magnetic field in the magnetopause lies in the same direction as the magnetosheath ($\vec{B}_{0\mathrm{I}} \parallel \vec{B}_{0\mathrm{III}}$).

In a semi-infinite homogeneous plasma in Regions I and III, $\xi_x$ and $P$ in Equations (27) and (28) are set to $e^{-i\kappa_{\mathrm{I}}x}$ and $e^{-i\kappa_{\mathrm{III}}x}$, respectively, and must be continuous at the interface. Then, the wave dispersion relation becomes (Turkakin et al., 2013)

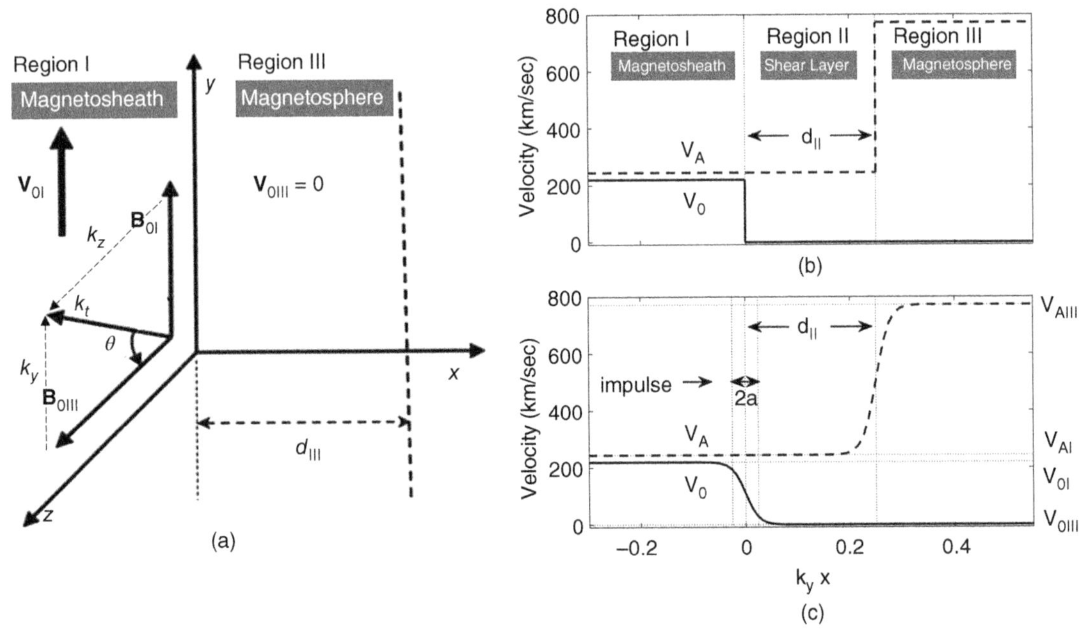


**Figure 11.8** (a) Illustration of the adopted background plasma profiles near the magnetopause (Adapted from Turkakin et al. (2013) and E.-H. Kim et al. (2022). Regions I and III represent the magnetosheath and the magnetosphere, respectively, and Region II in panels (b) and (c) is the shear transition layer. The zero-thickness magnetopause is adopted in panels (a) and (b).

$$\left[\tilde{\omega}_{\mathrm{I}}^2 - (\vec{V}_{AI}\cdot\vec{k}_t)^2\right]\frac{\rho_{0I}}{\kappa_I} + \left[\tilde{\omega}_{III}^2 - (\vec{V}_{AIII}\cdot\vec{k}_t)^2\right]\frac{\rho_{0III}}{\kappa_{III}} = 0, \tag{11.29}$$

where $\tilde{\omega}_{\mathrm{I,III}} = \omega - \vec{V}_{0I,0III}\cdot\vec{k}_t$ is the Doppler shifted frequency at Region I or III and $\kappa_{I,III}^2 = \tilde{\omega}_{I,III}^4 / \left[\tilde{\omega}_{I,III}^2\left(V_{AI,AIII}^2 + C_{SI,SIII}^2\right) - (\vec{V}_{AI,AIII}\cdot\vec{k}_t)^2 C_{SI,SIII}^2\right] - k_t^2$ is a squared wavenumber along the $x$ direction.

If Region III has a finite width ($d_{\mathrm{III}}$) and the inner boundary is assumed to be a perfect reflector (such as a plasmapause), then the solution becomes

$$i\tan(\kappa_{III}d_{III})\left[\tilde{\omega}_I^2 - (\vec{V}_{AI}\cdot k_t)^2\right]\frac{\rho_{0I}}{\kappa_I} + \left[\tilde{\omega}_{III}^2 - (\vec{V}_{AIII}\cdot k_t)^2\right]\frac{\rho_{0III}}{\kappa_{III}} = 0. \tag{11.30}$$

The real wave frequency ($\omega_{\mathrm{r}}$) and the growth rate ($\omega_i$) can be calculated using Equation (29) or (30) as a function of $V_0 V_A$, and the wave propagation angle $\theta_2 = 90° - \mathrm{atan}\left(\frac{k_z}{k_y}\right)$ respectively.

Turkakin et al. (2013) investigated the unstable KH waves on the dayside and nightside flanks of the magnetosphere and magnetotail and discovered that secondary KH waves are most prevalent on the dayside flank. Panels (a) and (b) of Figure 11.9 show characteristics of the secondary KH waves in the dayside flank from eigenmode analysis. The figure shows $\omega_{\mathrm{r}}$ and $\omega_i$ as a function of the Alfvén Mach number normalized to the Alfvén velocity in the magnetosphere ($M_{AIII} = V_{0\mathrm{I}}/V_{\mathrm{AIII}}$) when $\vec{V}_{0\mathrm{I}}(x) = V_{0\mathrm{I}}(x)\hat{y}$, $\vec{B}_{0\mathrm{I}} = B_{0\mathrm{I}}\hat{y}$, and $\vec{B}_{0\mathrm{III}} = B_{0\mathrm{III}}\hat{z}$.

For $\theta_2 = 90°$, thus $\vec{k}_t = k_y$ in Figure 11.9(a), two fast wave modes are found: forward- and backward-propagating modes with positive and negative values at $M_{AIII} = 0$ (e.g., Taroyan et al., 2002). The fast waves are stable when $M_{AIII} < M_{As} \sim 0.7$, where $M_{As}$ is the Alfvén Mach number of the KH instability threshold. However, for $M_{AIII} > M_{As}$, unstable primary KH waves are generated due to wave interaction between modes 3 and 4 in this figure, similar to the typical KH waves (Johnson et al., 2014).

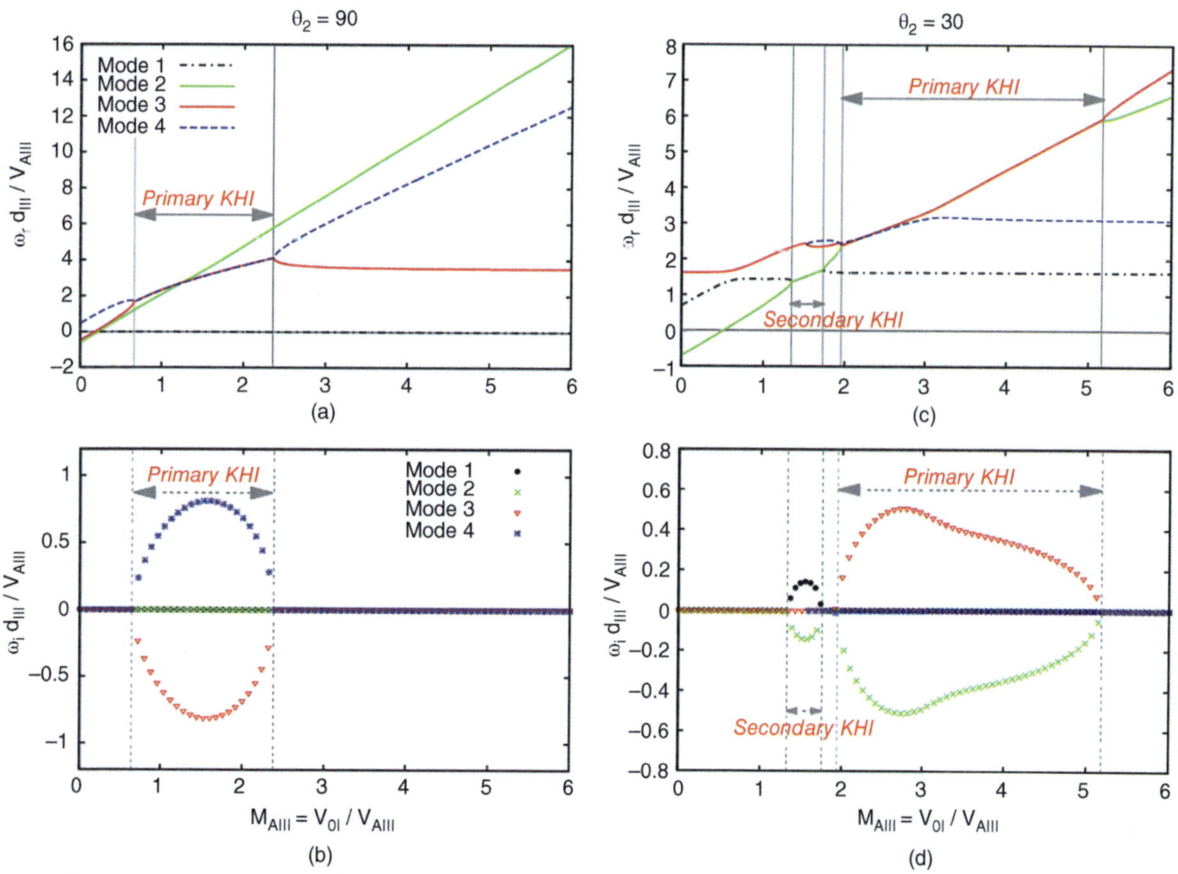


**Figure 11.9** Real frequencies and growth rates at the dayside magnetopause for a semi-infinite magnetosphere (Turkakin et al. (2013) / John Wiley & Sons). (a) and (b) For $\theta_2 = 90°$, only primary KH waves appear. (c) and (d) The secondary KH waves are also generated for $\theta_2 = 30°$.

On the other hand, for the oblique propagation of $\theta_2 = 30°$ in Figure 11.9(c,d), the secondary KH waves appear for $1.3 \leq M_{AIII} \leq 1.8$ as result of wave interaction between modes 1 and 2. Here, it is clearly seen that the secondary KH waves are generated with lower frequency, lower growth rate, and lower shear flow velocity and Alfvén Mach number threshold than the primary KH waves appearing at $M_{AIII} \geq 2$.

This analysis was expanded to cover a broad range of angles of propagation $\theta_2$. The total phase velocity of each wave mode in the magnetosheath and magnetosphere can also be calculated from the eigenmode analysis. Figure 11.10 shows the Friedrichs phase speed diagram of the fast and slow modes and the primary and secondary KH waves for $M_{AIII} = 2.3$. In this figure, the gray- and yellow-shaded regions represent the primary and secondary KH waves, respectively. As shown in Figure 11.9, the primary KH waves are generated due to wave coupling between modes 3 and 4 marked in this figure, and the secondary waves are due to interaction between modes 1 and 2. The figure also shows that the primary KH waves propagate at broad angles of $\theta_2 \cong 26° - 90°$, while the secondary KH waves appear at narrow angles of $\theta_2 \cong 19° - 24°$. This suggests that the secondary KH waves only propagate in a very narrow range of angles in the magnetosphere and magnetosheath. In the group velocity diagram shown in Figure 11.10(c), the primary and secondary KH waves can propagate energy to the magnetosphere. While the primary KH waves can carry the energy across the field line in the magnetosheath, the secondary KH waves can only propagate energy at a very narrow range of propagation angles.

Secondary KH waves can be generated in the nightside flank or magnetotail, where the magnetic field in the magnetosheath ($\vec{B}_{0I}$) is parallel to the magnetic field in the magnetosphere ($\vec{B}_{0III}$) (e.g., Taroyan et al., 2002). These unstable wave modes can also be developed when the shear flow velocity ($\vec{V}_{0I}$) is perpendicular to the magnetic field in the magnetosheath ($\vec{B}_{0I}$) (E.-H. Kim et al., 2022).

For $\vec{V}_0(x) = V_0(x)\hat{y}$ and $\vec{B}_0 = B_0\hat{z}$, the coupled MHD equations, Equations (27) and (28), in cold plasma are reduced to the following equations,

$$\frac{\partial p}{\partial x} = \mu_0\rho_0\left(\tilde{\omega}^2 - k_\parallel^2 V_A^2\right)\xi_x, \tag{11.31}$$

$$\mu_0\rho_0\frac{\partial \xi_x}{\partial x} = -\left(\frac{\tilde{\omega}^2 - k_y^2 V_A^2 - k_\parallel^2 V_A^2}{\tilde{\omega}^2 - k_\parallel^2 V_A^2}\right)\frac{P}{B_0^2}. \tag{11.32}$$

Since the Alfvén velocity has a large gradient typically wider than the velocity shear layer (Paschmann et al., 1993), E.-H. Kim et al. (2022) introduced a shear transition layer in Region II, as shown in Figure 11.8(b,c). Unlike Figure 11.8(a), the interfaces between I and II ($V_0$ interface) and between II and III ($V_A$ interface) are not located at the same location. Therefore, if there is no coupling between Alfvén and KH waves, the KH instability is driven at the $V_0$ interface, while the Alfvén resonance occurs at the $V_A$ interface. In this case, the wave dispersion becomes

$$(H_I + H_{II})(H_{II} + H_{III}) - (H_I - H_{II})(H_{II} - H_{III}) \times \exp(-2\kappa_{II} d_{II}) = 0, \tag{11.33}$$

where $H_J = \mu_0\rho_{0J}\left(\tilde{\omega}^2 - k_\parallel^2 V_{AJ}^2\right)/\kappa_J$ with $J$ = I, II, and III, and $\kappa_J^2 = \frac{\tilde{\omega}^2}{V_{AJ}^2} - k_y^2 - k_\parallel^2$. The $\omega_r$ and $\omega_i$ for various

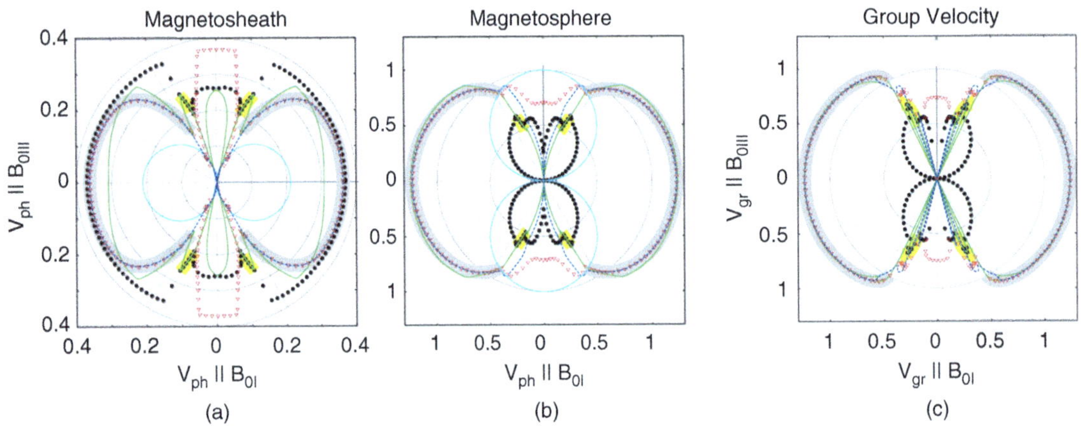


**Figure 11.10** (a) and (b) Friedrichs phase speed (normalized by $V_{AIII}$) diagrams at shear flow velocity $M_{AIII} = 2.3$ at flow for dayside unbounded magnetosphere in the magnetosheath and magnetosphere, and (c) group velocity diagram. Here, $V_{ph}$ and $V_{gr}$ are phase and group velocities. Black dots, green blue lines, red triangle, and blue dashed lines represented different wave modes, and light blue line is Alfvén wave. Yellow shaded areas mark the secondary KH waves and gray shaded areas mark the primary KH waves (Turkakin et al. (2013) / John Wiley & Sons).

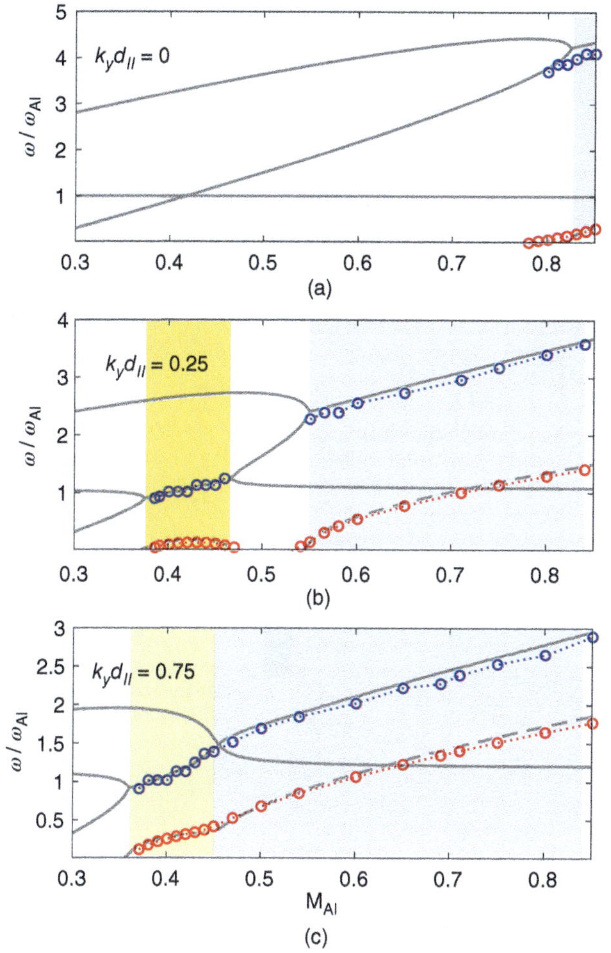


**Figure 11.11** Unstable wave frequency (solid black line) and growth rate (dashed black line) at the $V_0$ and $V_A$ interfaces from eigenmode analysis for $\vec{V}_0 \perp \vec{B}_0$ when the finite-width ($k_y d_{||}$) shear transition layer exists between the magnetosheath and magnetosphere (E.-H. Kim et al (2022) / Frontiers Media S.A / CC BY 4.0). Blue and red circles are the real frequency and growth rate, respectively, from time-dependent simulations. The gray-shaded area is where the primary KH waves are developed, the yellow-shaded area in panel (b) is where the secondary KH waves appear, and the light-yellow area in panel (c) shows the quasi-secondary KH waves.

widths of the shear transition layer ($d_{II}$) are calculated using Equation (33), as shown in Figure 11.11.

Figure 11.11 shows the real ($\omega_r$, solid black lines) and imaginary ($\omega_i$, dashed black lines) frequencies as a function of the Alfvén Mach number ($M_{AI} \equiv V_{0I}/V_{AI}$). When the shear transition layer is ignored in Figure 11.11(a) ($k_y d_{II} = 0$), a forward and a backward fast wave propagating mode exist for $M_{AI} < M_{As}$ and unstable primary KH waves appear for $M_{AI} > M_{As}$, similar to Figure 11.9(a,b). We also found a shear Alfvén wave mode ($\omega = \omega_{AI}$), and near $M_{AI}$~0.45, there is a crossing point between the fast wave and shear Alfvén wave modes, but there is no coupling between the two modes.

A finite width of the shear transition layer (Region II) can significantly change the wave dispersion relations. For $k_y d_{II} = 0.25$ in Figure 11.11(b), when the wave frequency is close to the Alfvén wave frequency ($\omega \to \omega_{AI} = k_{\|} V_{AI}$), unstable secondary KH waves appear where $0.355 \leq M_A \leq 0.4$, which is shaded in yellow in Figure 11.11(b). The dispersion relation clearly shows that this secondary KH wave mode is due to the coupling between the backward-propagating fast and the shear Alfvén waves. In this case, the secondary KH waves are clearly separated from the primary KH waves, consistent with Figure 11.9(c,d).

When the $V_A$ interface is further away from the $V_0$ interface ($k_y d_{II} = 0.75$), as shown in Figure 11.11(c), semi-secondary KH waves appear in the lightly shaded region. The wave characteristics are similar to the secondary KH waves in Figure 11.11(b), but this wave mode has been merged with the primary KH waves. When $d_{II}$ becomes much wider, the two interfaces cannot interact with each other, and $V_A$ profile at the magnetopause boundary can be estimated as a constant.

### 11.3.2. Mode-Conversion between the Secondary KH and the Alfvén Waves

The eigenmode analysis could adequately provide the eigenmode frequency and growth rate; however, the mode-conversion process cannot be examined. On the other hand, time-dependent wave simulations can examine the mode conversion as well as wave generation and propagation near the boundary layer (E.-H. Kim et al., 2022; Turkakin et al., 2022).

E.-H. Kim et al. (2022) performed numerical simulations solving the MHD equations, Equations (23)–(26), in cold plasma as an initial-valued problem in slab coordinates. In this case, they assumed that $\vec{V}_0 \perp \vec{B}_0$ and that $V_A$ and $V_0$ have hyperbolic tangent profiles, as shown in Figure 11.8(c). The real and imaginary parts of the frequency ($\omega_r$ and $\omega_i$) can be obtained from the time histories of the wave solution at each grid point. Figure 11.11 shows the $\omega_r$ (red circles) and $\omega_i$ (blue circles) from the time-dependent simulations and the results show excellent agreement with the eigenmode analysis (black solid and dashed lines), even though the boundary conditions at each interface are different for the eigenmode calculations and the simulations, as shown in Figure 11.8(b,c).

The detailed characteristics of secondary KH waves for the case of $k_y d_{II} = 0.25$ are presented in Figure 11.12. For a lower flow velocity of $M_{AI} = 0.45$ in Figure 11.12(a), the enhancement of wave power is shown at $\omega/\omega_{AI}$~1.2 in both the transverse ($b_{1y}$) and compressional ($b_{1z}$) components.

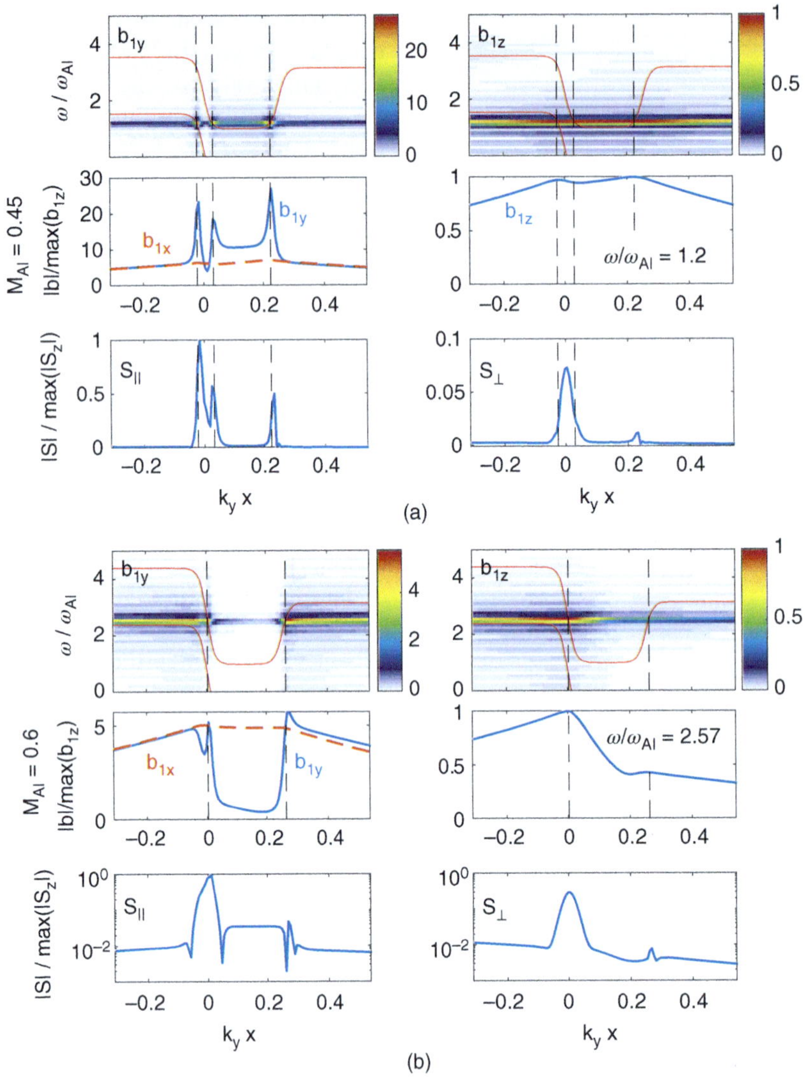


**Figure 11.12** Time-dependent simulation results showing the primary and secondary KH waves for (a) $M_{AI} = 0.45$ and (b) $M_{AI} = 0.6$ and when $k_y d_{\parallel} = 0.25$. Upper panels are perturbed magnetic field compressional ($b_{1z}$) and transverse ($b_{1y}$) components, middle panels are the spatial structure of the peak frequency, and lower panels are the Poynting flux parallel ($S_{\parallel}$) and perpendicular ($S_{\perp}$) to the $B_0$ (E.-H. Kim et al (2022) / Frontiers Media S.A / CC BY 4.0).

The transverse component $b_{1y}$ has three power enhancements near the $V_0$ (at $k_y x = -0.022$ and 0.033) and $V_A$ interfaces (at $k_y x = 0.022$). At these three locations, the Doppler-shifted wave frequency matches the Alfvén resonance condition,

$$\omega^{\pm} - k_y V_0 = \pm k_{\parallel} V_A . \tag{11.34}$$

Unlike the discontinuous $V_0$ profile in Figure 11.8(a), the time-dependent simulation used a finite width for the interface near $x = 0$. In this figure, two amplitude peaks appear near $x = 0$ because $\omega^-$ can also be positive near the $V_0$ interface when $k_y V_0 > k_{\parallel} V_A$. On the other hand, $b_{1z}$ decays away from the $V_A$ and $V_0$ interfaces, exhibiting the characteristics of KH waves. The wave power of $b_{1y}$ is significantly stronger than that of $b_{1z}$, as shown in the middle panel of Figure 11.12(b), and the field-aligned Poynting flux in the bottom panel is dominated. Therefore, wave energy significantly flows along the field line at each interface. Because the waves are initially launched in the compressional component, and the growing KH waves are also compressional waves, the intense field-aligned energy flow is clear evidence of mode conversion from the secondary KH waves to the shear Alfvén wave. This energy transport at the boundary layer is similar to observations (e.g., Chaston et al., 2007).

In Figure 11.11(b) for the higher $M_{AI}$ case ($M_{AI} = 0.6$), the maximum value of $b_{1z}$ is observed at the $V_0$ interface, whereas the secondary power peak at the $V_A$ interface is much weaker compared to the $M_{AI} = 0.45$ case. Like the lower $M_{AI}$ case, $b_{1y}$ exhibits power enhancement at each interface, but the amplitude ratio between $b_{1y}$ and $b_{1z}$ is significantly reduced. Although the field-aligned Poynting flux ($S_{\parallel}$) at the $V_A$ and $V_0$ interfaces remains stronger than the flux across the field line ($S_{\perp}$), it is not dominant. This implies that the mode-conversion process from the primary KH waves occurs less effectively than in the secondary KH waves.

The simulation results also showed that the secondary KH wave properties in the magnetosphere and magnetosheath are similar to those of the primary KH waves; thus, perpendicular and parallel Poynting flux are comparable, $S_{\parallel} \sim S_{\perp}$. On the other hand, Turkakin et al. (2013) demonstrated that the secondary KH waves propagate almost along the field line with relatively small propagation angle to the magnetic field as shown in Figure 11.10(c). The disparity is likely due to the magnetic field direction, the rotation of the magnetic field at the magnetopause, and the shear transition layer. Therefore, further investigation is necessary to understand the secondary KH waves and the process of mode conversion.

## 11.4. SUMMARY

This paper reviews the mode-conversion and mode-coupling processes of the kinetic Alfvén waves at the magnetopause. When the wave frequency is close to the ion cyclotron frequency, the resonance occurs at the perpendicular ion cyclotron resonance, while the resonance occurs at the Alfvén resonance in the MHD frequency regime. At the magnetopause, the maximum mode-conversion efficiency occurs for the higher frequency range at the perpendicular ion cyclotron resonance, and the mode converted kinetic Alfvén waves propagate across the field line toward the location where the $V_A$ is minimized. Thus, the kinetic Alfvén waves can be trapped into KH vortex across the field-line direction. Since heavy ions play an important role in plasma transportation, strong absorption of the compressional wave to the shear Alfvén waves occurs near the ion cyclotron frequency at the magnetopause. We also consider the heavy ion effects on the mode conversion, and the heavy ions from the ionospheric outflow (such as $O^+$) and the solar wind (such as $He^{++}$) are expected to significantly affect the mode-conversion process at the magnetopause.

When considering the shear flow velocity, wave–wave interaction can lead to the development of the secondary KH waves with a lower frequency and growth rate than the primary KH waves. The orientation of the background magnetic field may be critical to the growing wave frequency, growth rate, and propagation characteristics in both the magnetosheath and magnetosphere. Furthermore, the inclusion of the shear transition layer between the shear velocity interface and the Alfvén velocity interface results in strong mode conversion from the secondary KH waves to shear Alfvén waves, which transfers significant energy to the magnetosphere along the field line. Consequently, energy exchanges occur in the magnetopause through the mode conversion of compressional waves to Alfvén waves, either in the absence of shear velocity, or through the conversion of secondary KH waves when shear flows are slower than the typical threshold for the primary KH instability.

## ACKNOWLEDGMENTS

This material is based upon work supported by the U.S. Department of Energy, Office of Science, Office of Fusion Energy Sciences under contract DE-AC02-09CH11466. The authors acknowledge support from NASA grants 80HQTR19T0076, 80NSSC19K0843, 80NSSC21K2009, 80NSSC21K1678, 80NSSC20K0355, 80NSSC20K0704, 80NSSC22K0515, and NSF AGS grant 2131013.

## AVAILABILITY STATEMENT

The raw data were saved in MATLAB files and published in the Data Space of Princeton University (https://doi.org/10.34770/m8rz-5718).

## REFERENCES

Anderson, R. R., Harvey, C. C., Hoppe, M. M., & Tsurutani, B. T. (1982). Plasma waves near the magnetopause. *Journal of Geophysical Research*, *87*, 2087.

Andries, J., Tirry, W. J., & Goossens, M. (2000). Modified Kelvin-Helmholtz instabilities and resonant F low Instabilities in a One-dimensional Coronal Plume Model: Results for Plasma beta=0. *The Astrophysical Journal*, *531*(1), 561–570. 10.1086/308430

Antolin, P., & Van Doorsselaere, T. (2019). Influence of resonant absorption on the generation of the Kelvin-Helmholtz Instability. *Frontiers in Physics*, *7*, 85. 10.3389/fphy.2019.00085

Appert, K., & Vaclavik, J. (1983). Effects of finite ion cyclotron frequency on the absorption of magnetohydrodynamic waves at the spatial Alfven resonance. *Plasma Physics*, *25*(5), 551–561. 10.1088/0032-1028/25/5/007

Belmont, G., Reberac, F., & Rezeau, L. (1995). Resonant amplification of magnetosheath MHD fluctuations at the magnetopause. *Geophysical Research Letters*, *22*, 295–298.

Bertelli, N., Valeo, E. J., Green, D. L., Gorelenkova, M., Phillips, C. K., Podestà, M., et al. (2017). Full-wave simulations of ICRF heating regimes in toroidal plasma with non-Maxwellian distribution functions. *Nuclear Fusion*, *57*(5), 056035. https://doi.org/10.1088/1741-4326/aa66db

Bouhram, M., Klecker, B., Paschmann, G., Haaland, S., Hasegawa, H., Blagau, A., et al. (2005). Survey of energetic O+ ions near the dayside mid-latitude magnetopause with cluster. *Annales Geophysicae*, *23*, 1281–1294. https://doi.org/10.5194/angeo-23-1281-2005

Buchsbaum, S. J. (1960). Resonance in a plasma with two ion species. *Physics of Fluids*, *3*, 418.

Chaston, C. C., Bonnell, J., McFadden, J. P., Carlson, C. W., Cully, C., Le Contel, O., et al. (2008). Turbulent heating and cross-field transport near the magnetopause from THEMIS. *Geophysical Research Letters*, *35*, 17. https://doi.org/10.1029/2008GL033601

Chaston, C. C., Phan, T. D., Bonnell, J. W., Mozer, F. S., Acuńa, M., Goldstein, M. L., et al. (2005). Drift-kinetic Alfvén waves observed near a reconnection X line in the Earth's magnetopause. *Physical Review Letters*, *95*(6), 065002. https://doi.org/10.1103/PhysRevLett.95.065002

Chaston, C. C., Wilber, M., Mozer, F. S., Fujimoto, M., Goldstein, M. L., Acuna, M., et al. (2007). Mode conversion and anomalous transport in Kelvin-Helmholtz vortices and Kinetic Alfvén waves at the Earth's magnetopause. *Physical Review Letters*, *99*(17), 175004. https://doi.org/10.1103/PhysRevLett.99.175004

Chaston, C. C., Yao, Y., Lin, N., Salem, C., & Ueno, G. (2013). Ion heating by broadband electromagnetic waves in the magnetosheath and across the magnetopause. *Journal of Geophysical Research: Space Physics*, *118*(9), 5579–5591. https://doi.org/10.1002/jgra.50506

Chen, L., & Hasegawa, A. (1974). Plasma heating by spatial resonance of Alfven wave. *Physics of Fluids*, *17*(7), 1399–1403.

Cowee, M. M., Winske, D., & Gary, S. P. (2010). Hybrid simulations of plasma transport by Kelvin-Helmholtz instability at the magnetopause: Density variations and magnetic shear. *Journal of Geophysical Research: Space Physics*, *115*(A6), A06214. https://doi.org/10.1029/2009JA015011

Cramer, N. F. (2001). *The physics of Alfvén waves*. Wiley-VCH, Berlin, Germany.

Cross, R. (1988). *An introduction to Alfven waves*. 1st edn. Adam Hilger, Bristol, England.

Damiano, P. A., Johnson, J. R., & Chaston, C. C. (2016). Ion gyroradius effects on particle trapping in kinetic Alfvén waves along auroral field lines. *Journal of Geophysical Research: Space Physics*, *121*(11), 10,831-10,844. https://doi.org/10.1002/2016JA022566

Delamere, P. A., Wilson, R. J., & Masters, A. (2011). Kelvin-Helmholtz Instability at Saturn's Magnetopause: Hybrid Simulations. *Journal of Geophysical Research*, *116*, A10222. doi:10.1029/2011JA016724

Delano, K., Fuselier, S. A., Elliott, H. A., Burch, J. L., Mukherjee, J., Petrinec, S., et al. (2019). The He++/H+ density ratio across earth's subsolar magnetopause and its implications for the presence of a mass-dependent reflection coefficient. *Journal of Geophysical Research: Space Physics*, *124*(12), 9893–9903. https://doi.org/10.1029/2019JA027182

Engebretson, M. J., Cahill, L. J., Jr., Arnoldy, R. L., Anderson, B. J., & Rosenberg, T. J. (1991). The role of the ionosphere in coupling upstream ULF wave power into the dayside magnetosphere. *Journal of Geophysical Research*, *96*, 1527–1542.

Engebretson, M. J., Lin, N., Baumjohann, W., Luehr, H., & Anderson, B. J. (1991). A comparison of ULF fluctuations in the solar wind, magnetosheath, and dayside magnetosphere. I - Magnetosheath morphology. II—Field and plasma conditions in the magnetosheath. *Journal of Geophysical Research*, *96*, 3441–3464.

Fairfield, D. H. (1969). Bow shock associated waves observed in the far upstream interplanetary medium. *Journal of Geophysical Research*, *74*, 3541–3553.

Gekelman, W., Vincena, S., Van Compernolle, B., Morales, G. J., Maggs, J. E., Pribyl, P., & Carter, T. A. (2011). The many faces of shear Alfvén wavesa. *Physics of Plasmas*, *18*(5), 055501. https://doi.org/10.1063/1.3592210

Gloeckler, G., & Geiss, J. (1989). The abundances of elements and isotopes in the solar wind. *AIP Conference Proceedings*, *183*(1), 49–71. https://doi.org/10.1063/1.37985

González, A. G., & Gratton, J. (1994). The role of a density jump in the Kelvin-Helmholtz instability of compressible plasma. *Journal of Plasma Physics*, *52*(2), 223–244. https://doi.org/10.1017/S0022377800017888

Greenstadt, E. W., Green, I. M., Inouye, G. T., Hundhausen, A. J., Bame, S. J., and Strong, I. B. (1968), Correlated magnetic field and plasma observations of the Earth's bow shock, *Journal of Geophysical Research*, *73*(1), 51–60, doi:10.1029/JA073i001p00051

Hasegawa, A. (1976). Particle acceleration by MHD surface wave and formation of aurora. *J. Geophys. Res.*, *81*( 28), 5083–5090, doi:10.1029/JA081i028p05083.

Hasegawa, A., & Chen, L. (1975). Kinetic process of the plasma heating by Alfvén wave. *Physical Review Letters*, *35*, 370.

Hasegawa, A., Tsui, K. H., & Assis, A. S. (1983). A theory of long period magnetic pulsation III—Local field line oscillations. *Geophysical Research Letters*, *10*, 765.

Hasegawa, H., Fujimoto, M., Phan, T.-D., Rème, H., Balogh, A., Dunlop, M. W., et al. (2004). Transport of solar wind into Earth's magnetosphere through rolled-up Kelvin-Helmholtz vortices. *Nature*, *430*(7001), 755–758. https://doi.org/10.1038/nature02799

Hesse, M., & Cassak, P. A. (2020). Magnetic reconnection in the space sciences: Past, present, and future. *Journal of Geophysical Research: Space Physics*, *125*(2), e2018JA025935. https://doi.org/10.1029/2018JA025935

Horwitz, J. L., Comfort, R. H., & Chappell, C. R. (1984). Thermal ion composition measurements of the formation of the new outer plasmasphere and double plasmapause during storm recovery phase. *Geophysical Research Letters*, *11*, 701–704.

Hull, A. J., Damiano, P. A., Chaston, C. C., Johnson, J. R., & Reeves, G. D. (2022). Electron energization signatures in traveling kinetic Alfvén waves at storm time injection fronts. *Geophysical Research Letters*, *49*(15), e2022GL099318. https://doi.org/10.1029/2022GL099318

Hwang, K.-J., Kuznetsova, M. M., Sahraoui, F., Goldstein, M. L., Lee, E., & Parks, G. K. (2011). Kelvin-Helmholtz waves under southward interplanetary magnetic field. *Journal of Geophysical Research: Space Physics*, *116*(A8), A08210. https://doi.org/10.1029/2011JA016596

Jaeger, E. F., Berry, L. A., Myra, J. R., Batchelor, D. B., D'Azevedo, E., Bonoli, P. T., et al. (2003). Sheared poloidal flow driven by mode conversion in tokamak plasmas. *Physical Review Letters*, *90*, 195001.

Johnson, J. R., & Cheng, C. Z. (1997). Kinetic Alfvén waves and plasma transport at the magnetopause. *Geophysical Research Letters*, *24*(11), 1423–1426. https://doi.org/10.1029/97GL01333

Johnson, J. R., & Cheng, C. Z. (2001). Stochastic ion heating at the magnetopause due to kinetic Alfvén waves. *Geophysical Research Letters*, *28*(23), 4421–4424. https://doi.org/10.1029/2001GL013509

Johnson, J. R., & Cheng, C. Z. (2003). Kinetic Alfvén waves at the magnetopause mode conversion, transport and formation of LLBL. In P.T. Newell & T. Onsager (Eds.), Earth's low-latitude boundary layer. American Geophysical Union. https://doi.org/10.1029/133GM21

Johnson, J. R., Cheng, C. Z., & Song, P. (2001). Signatures of mode conversion and kinetic Alfvén waves at the magnetopause. *Geophysical Research Letters*, *28*(2), 227–230. https://doi.org/10.1029/2000GL012048

Johnson, J. R., Wing, S., & Delamere, P. A. (2014). Kelvin Helmholtz instability in planetary magnetospheres. *Space Science Reviews*, *184*(1), 1–31. https://doi.org/10.1007/s11214-014-0085-z

Johnson, J. R., Wing, S., Delamere, P., Petrinec, S., & Kavosi, S. (2021). Field Aligned Currents in Auroral Vortices. *Journal of Geophysical Research: Space Physics*, *126*(2), e28583. https://doi.org/10.1029/2020JA028583

Karney, C. F. F., Perkins, F. W., & Sun, Y.-C, (1979). Alfven resonance effects on magnetosonic modes in large tokamaks. *Physical Review Letters*, *24*, 1621.

Kim, E.-H., & Lee, D.-H. (2003). Resonant absorption of ULF waves near the ion cyclotron frequency: A simulation study. *Geophysical Research Letters*, *30*, 2240.

Kim, E.-H., Cairns, I. H., & Robinson, P. A. (2007). Extraordinary-mode radiation produced by linear-mode conversion of Langmuir waves. *Physical Review Letters*, *99*(1), 015003. https://doi.org/10.1103/PhysRevLett.99.015003

Kim, E.-H., Johnson, J. R., & Lee, D.-H. (2008). Resonant absorption of ULF waves at Mercury's magnetosphere. *Journal of Geophysical Research*, *113*, A11207.

Kim, E.-H., Johnson, J. R., & Lee, K.-D. (2011). ULF wave absorption at Mercury. *Geophysical Research Letters*, *38*, L16111. doi:10.1029/2011GL048621.

Kim, E.-H., Johnson, J. R., Kim, H., & Lee, D.-H. (2015). Inferring magnetospheric heavy ion density using EMIC waves. *Journal of Geophysical Research: Space Physics*, *120*, doi:10.1002/2015JA021092.

Kim, E.-H., Johnson, J. R., & Lee, D.-H. (2019). Electron inertial effects on linearly polarized electromagnetic ion cyclotron waves at Earth's magnetosphere. *Journal of Geophysical Research: Space Physics*, *124*(4), 2643–2655. https://doi.org/10.1029/2019JA026532

Kim, E.-H., Johnson, J. R., & Nykyri, K. (2022). Coupling between Alfvén wave and Kelvin–Helmholtz waves in the low latitude boundary layer. *Frontiers in Astronomy and Space Sciences*, *8*, 785413. https://doi.org/10.3389/fspas.2021.785413

Kim, E.-H., Johnson, J. R., Valeo, E., & Phillips, C. K. (2015). Global modeling of ULF waves at Mercury. *Geophysical Research Letters*, *42*, 5147–5154, doi:10.1002/2015GL064531.

Kim, S., & Kim, K. (2022). Giant overreflection of magnetohydrodynamic waves from inhomogeneous plasmas with nonuniform shear flows. *Physics of Fluids*, *34*(12), 127108. https://doi.org/10.1063/5.0127265

Klimushkin, D. Y., Mager, P. N., & Marilovtseva, O. S. (2010). Parallel structure of Pc1 ULF oscillations in multi-ion magnetospheric plasma at finite ion gyrofrequency. *Journal of Atmospheric and Solar-Terrestrial Physics*, *72*, 1327. doi:10.1016/j.jastp.2010.09.019.

Leonovich, A. S., & Mazur, V. A. (1995) Linear transformation of the standing Alfven wave in an axisymmetric magnetosphere. *Planetary and Space Science*, *43*, 885. https://doi.org/10.1016/0032-0633(94)00207-8

Lee, D.-H., Johnson, J. R., Kim, K., & Kim, K.-S. (2008). Effects of heavy ions on ULF wave resonances near the equatorial region. *Journal of Geophysical Research*, *113*, A11212, doi:10.1029/2008JA 013088.

Lee, L. C., Johnson, J. R., & Ma, Z. W. (1994). Kinetic Alfven waves as a source of plasma transport at the dayside magnetopause. *Journal of Geophysical Research*, *99*(A9), 17405.

Lee, S. H., Zhang, H., Zong, Q.-G., Otto, A., Rème, H., & Liebert, E. (2016). A statistical study of plasmaspheric plumes and ionospheric outflows observed at the dayside magnetopause. *Journal of Geophysical Research: Space Physics*, *121*(1), 492–506. https://doi.org/10.1002/2015JA021540

Lin, N., Engebretson, M. J., McPherron, R. L., Kivelson, M. G., Baumjohann, W., Luehr, H., et al. (1991). A comparison of ULF fluctuations in the solar wind, magnetosheath, and dayside magnetosphere. 2. Field and plasma conditions in the magnetosheath. *Journal of Geophysical Research*, *96*, 3455–3464.

Lin, Y., Wukitch, S. J., Bonoli, P. T., Marmar, E., Mossessian, D., Nelson-Melby, E., et al. (2003). Ion cyclotron range of frequencies mode conversion electron heating in deuterium-hydrogen plasmas in the Alcator C-Mod tokamak. *Plasma Physics and Controlled Fusion*, *45*(6), 1013–1026. https://doi.org/10.1088/0741-3335/45/6/313

Lin, Y., Wukitch, S., Bonoli, P., Nelson-Melby, E., Porkolab, M., Wright, J. C., et al. (2004). Investigation of ion cyclotron range of frequencies mode conversion at the ion-ion hybrid layer in Alcator C-Mod. *Physics of Plasmas*, *11*(5), 2466–2472. https://doi.org/10.1063/1.1651489

Lin, Y., Johnson, J. R., & Wang, X. Y. (2010). Hybrid simulation of mode conversion at the magnetopause. *Journal of Geophysical Research: Space Physics*, *115*, A04208. https://doi.org/10.1029/2009JA014524

Lin, Y., Johnson, J. R., & Wang, X. (2012). Three-Dimensional Mode Conversion Associated with Kinetic Alfvén Waves. *Physical Review Letters*, *109*(12), 125003. https://doi.org/10.1103/PhysRevLett.109.125003

Lysak, R. L. (2008). On the dispersion relation for the kinetic Alfvén wave in an inhomogeneous plasma. *Physics of Plasmas*, *15*(6), 062901. https://doi.org/10.1063/1.2918742

Mann, I. R., Wright, A. N., Mills, K. J., & Nakariakov, V. M. (1999). Excitation of magnetospheric waveguide modes by magnetosheath flows. *Journal of Geophysical Research: Space Physics*, *104*(A1), 333–353. https://doi.org/10.1029/1998JA900026

Matsumoto, Y., & Hoshino, M. (2006). Turbulent mixing and transport of collisionless plasmas across a stratified velocity shear layer. *Journal of Geophysical Research: Space Physics*, *111*(A5), A05213. https://doi.org/10.1029/2004JA010988

Mikhailova, O. S., Mager, P. N., & Klimushkin, D. Y. (2019). Two modes of ion-ion hybrid waves in magnetospheric plasma. *Plasma Physics and Controlled Fusion*, *62*(2), 025026. https://doi.org/10.1088/1361-6587/ab5b32

Mills, K. J., & Wright, A. N. (1999). Azimuthal phase speeds of field line resonances driven by Kelvin-Helmholtz unstable waveguide modes. *Journal of Geophysical Research*, *104*(A10), 22667–22678. https://doi.org/10.1029/1999JA900280

Mitchell, C., Vincena, S., Maggs, J., & Gekelman, W. (2001). Laboratory observation of Alfvén resonance. *Geophysical Research Letters*, *28*, 923–926. https://doi.org/10.1029/2000GL012165

Miura, A., & Pritchett, P. L. (1982). Nonlocal stability analysis of the MHD Kelvin-Helmholtz instability in a compressible plasma. *Journal of Geophysical Research: Space Physics*, *87*(A9), 7431–7444. https://doi.org/10.1029/JA087iA09p07431

Moore, T. W., Nykyri, K., & Dimmock, A. P. (2016). Cross-scale energy transport in space plasmas. *Nature Physics*, *12*, 1164–1169. doi:10.1038/NPHYS3869. https://doi.org/DOI:10.1038/NPHYS3869

Moya, P. S., Zenteno-Quinteros, B., Gallo-Méndez, I., & Pinto, V. A. (2022). The role of O+ and He+ in the propagation of kinetic Alfvén waves in the earth's inner magnetosphere. *The Astrophysical Journal*, *933*(1), 32. https://doi.org/10.3847/1538-4357/ac713b

Mazur, V. A., & Chuiko, D. A. (2013) Kelvin-Helmholtz instability on the magnetopause, magnetohydrodynamic waveguide in the outer magnetosphere, and Alfvén resonance deep in the magnetosphere. *Plasma Physics Report*, *39*, 488–503. https://doi.org/10.1134/S1063780X13060068

Mazur, V. A., & Chuiko, D. A. (2015) Azimuthal inhomogeneity in the MHD waveguide in the outer magnetosphere. *Journal of Geophysical Research: Space Physics*, *120*, 4641–4655. doi: 10.1002/2014JA020819

Nykyri, K., Ma, X., Dimmock, A., Foullon, C., Otto, A., & Osmane, A. (2017). Influence of velocity fluctuations on the Kelvin-Helmholtz instability and its associated mass transport. *Journal of Geophysical Research: Space Physics*, *122*(9), 9489–9512. https://doi.org/10.1002/2017JA024374

Odera, T. J., & Stuart, W. F. (1986). Morphology of low-frequency waves in the solar wind and their relation to ground pulsations. *Journal of Atmospheric and Terrestrial Physics*, *48*, 83–95.

Ono, M. (1979). Cold, electrostatic, ion-cyclotron waves and ion-ion hybrid resonances. *Physical Review Letters*, *42*, 1267–1270. https://doi.org/10.1103/PhysRevLett.42.1267

Othmer, C., Glassmeier, K.-H., & Cramm, R. (1999). Concerning field line resonances in Mercury's magnetosphere. *Journal of Geophysical Research*, *104*(A5), 10369–10378. https://doi.org/10.1029/1999JA900009

Otto, A., & Fairfield, D. H. (2000). Kelvin-Helmholtz instability at the magnetotail boundary: MHD simulation and comparison with Geotail observations. *Journal of Geophysical Research*, *105*(A9), 21,175-21,190. https://doi.org/10.1029/1999JA000312

Paschmann, G., Baumjohann, W., Sckopke, N., Phan, T.-D., & Lühr, H. (1993). Structure of the dayside magnetopause for low magnetic shear. *Journal of Geophysical Research: Space Physics*, *98*(A8), 13409–13422. https://doi.org/10.1029/93JA00646

Phan, T.-D., Paschmann, G., Baumjohann, W., Sckopke, N., & Luehr, H. (1994). The magnetosheath region adjacent to the dayside magnetopause: AMPTE/IRM observations. *Journal of Geophysical Research*, *99*, 121–141.

Phan, T.-D., Larson, D., McFadden, J., Lin, R. P., Carlson, C., Moyer, M., et al. (1997). Low-latitude dusk flank magnetosheath, magnetopause, and boundary layer for low magnetic shear: Wind observations. *Journal of Geophysical Research: Space Physics*, *102*(A9), 19883–19895. https://doi.org/10.1029/97JA01596

Phillips, C. K., Hosea, J. C., Ignat, D., Majeski, R., Rogers, J. H., Schilling, G., & Wilson, J. R. (1996). Majority ion heating near the ion-ion hybrid layer in tokamaks. In The 11th topical Conference on radio frequency power in plasmas (Vol. 355, pp. 289–292). https://doi.org/10.1063/1.49530

Phillips, C. K., Bell, M. G., Bell, R. E., Bernabei, S., Bettenhausen, M., Bush, C. E., et al. (2000). ICRF heating and profile control techniques in TFTR. *Nuclear Fusion*, *40*(3Y), 461–466. https://doi.org/10.1088/0029-5515/40/3Y/304

Pu, Z.-Y., & Kivelson, M. G. (1983a). Kelvin-Helmholtz Instability at the magnetopause: Energy flux into the magnetosphere. *Journal of Geophysical Research: Space Physics*, *88*(A2), 853–861. https://doi.org/10.1029/JA088iA02p00853

Pu, Z.-Y., & Kivelson, M. G. (1983b). Kelvin-Helmholtz instability at the magnetopause: Solution for compressible plasmas. *Journal of Geophysical Research: Space Physics*, *88*(A2), 841–852. https://doi.org/10.1029/JA088iA02p00841

Radoski, H. R. (1974), A theory of latitude dependent geomagnetic micropulsations: The asymptotic fields. *Journal of Geophysical Research*, *79*(4), 595–603, doi:10.1029/JA079i004p00595.

Rezeau, L., Perraut, S., & Roux, A. (1986). Electromagnetic fluctuations in the vicinity of the magnetopause. *Geophysical Research Letters*, *13*, 1093–1096.

Rezeau, L., Morane, A., Perraut, S., & Roux, A. (1989). Characterization of Alfvénic fluctuations in the magnetopause boundary layer. *Journal of Geophysical Research*, *94*, 101.

Rezeau, L., Roux, A., & Russell, C. T. (1993). Characterization of small-scale structures at the magnetopause from ISEE measurements. *Journal of Geophysical Research*, *98*, 179–186.

Ruderman, M. S., & Wright, A. N. (1998). Excitation of resonant Alfvén waves in the magnetosphere by negative energy surface waves on the magnetopause. *Journal of Geophysical Research: Space Physics*, *103*(A11), 26573–26584. https://doi.org/10.1029/98JA02296

Shelley, E. G., Johnson, R. G., & Sharp, R. D. (1972). Satellite observations of energetic heavy ions during a geomagnetic storm. *Journal of Geophysical Research*, *77*, 6104, doi:10.1029/JA077i031p06104.

Shi, F., Cheng, L., Lin, Y., & Wang, X. (2017). Foreshock wave interaction with the magnetopause: Signatures of mode conversion. *Journal of Geophysical Research: Space Physics*, *122*(7), 7057–7076. https://doi.org/10.1002/2016JA023114

Shi, F., Lin, Y., & Wang, X. (2013). Global hybrid simulation of mode conversion at the dayside magnetopause. *Journal of Geophysical Research: Space Physics*, *118*, 6176–6187. https://doi.org/10.1002/jgra.50587

Song, P., Russell, C. T., Strangeway, R. J., Wygant, J. R., Cattell, C. A., Fitzenreiter, R. J., & Anderson, R. R. (1993). Wave properties near the subsolar magnetopause—Pc 3-4 energy coupling for northward interplanetary magnetic field. *Journal of Geophysical Research*, *98*, 187–196.

Southwood, D. J. (1968). The hydromagnetic stability of the magnetospheric boundary. *Planetary and Space Science*, *16*(5), 587–605. https://doi.org/10.1016/0032-0633(68)90100-1

Southwood, D. J. (1974). Some features of field line resonances in the magnetosphere. *Planetary and Space Science*, *22*, 483–491.

Stix, T. H. (1992). *Waves in plasmas*. New York: American Institute of Physics.

Takahashi, K., Sibeck, D. G., Newell, P. T., & Spence, H. E. (1991). ULF waves in the low-latitude boundary layer and their relationship to magnetospheric pulsations—A multisatellite observation. *Journal of Geophysical Research*, *96*, 9503–9519.

Tamao, T. (1965). Transmission and coupling resonance of hydromagnetic disturbances in the nonuniform Earth's magnetosphere. *Science Reports of Tohoku University*, *5*, 43.

Taroyan, Y., & Erdélyi, R. (2002). Resonant and Kelvin–Helmholtz instabilities on the magnetopause. *Physics of Plasmas*, *9*(7), 3121–3129. https://doi.org/10.1063/1.1481746

Taroyan, Y., & Ruderman, M. S. (2011). MHD Waves and Instabilities in Space Plasma Flows. *Space Science Reviews*, *158* (2–4), 505–523. https://doi.org/10.1007/s11214-010-9737-9

Tirry, W. J., Cadez, V. M., Erdelyi, R., & Goossens, M. (1998). Resonant flow instability of MHD surface waves. *Astronomy and Astrophysics*, *332*, 786–794.

Tsurutani, B. T., Smith, E. J., Anderson, R. R., Ogilvie, K. W., Scudder, J. D., Baker, D. N., & Bame, S. J. (1982). Lion roars and nonoscillatory drift mirror waves in the magnetosheath. *Journal of Geophysical Research*, *87*, 6060–6072.

Turc, L., Zhou, H., Tarvus, V., Ala-Lahti, M., Battarbee, M., Pfau-Kempf, Y., et al. (2022). A global view of Pc3 wave activity in near-Earth space: Results from hybrid-Vlasov simulations. *Frontiers in Astronomy and Space Sciences*, *9*, 989369. https://doi.org/10.3389/fspas.2022.989369

Turkakin, H., Mann, I. R., & Rankin, R. (2014). Kelvin-Helmholtz unstable magnetotail flow channels: Deceleration and radiation of MHD waves. *Geophysical Research Letters*, *41*(11), 3691–3697. https://doi.org/10.1002/2014GL060450

Turkakin, H., Mann, I. R., & Rankin, R. (2022). Linear and nonlinear Kelvin–Helmholtz instability and magnetohydrodynamic wave emission in sheared astrophysical plasma flows. *The Astrophysical Journal*, *939*(1), 30. https://doi.org/10.3847/1538-4357/ac9404

Turkakin, H., Rankin, R., & Mann, I. R. (2013). Primary and secondary compressible Kelvin-Helmholtz surface wave instabilities on the Earth's magnetopause. *Journal of Geophysical Research: Space Physics*, *118*(7), 4161–4175. https://doi.org/10.1002/jgra.50394

Uberoi, C. (1972). Alfven waves in inhomogeneous magnetic fields. *Physics of Fluids*, *15*, 1673.

Vincena, S. T., Farmer, W. A., Maggs, J. E., & Morales, G. J. (2013). Investigation of an ion-ion hybrid Alfvén wave resonator. *Physics of Plasmas*, *20*, 012111. https://doi.org/10.1063/1.4775777

Wang, H., Lin, Y., Wang, X., & Guo, Z. (2019). Generation of kinetic Alfvén waves in dayside magnetopause reconnection: A 3-D global-scale hybrid simulation. *Physics of Plasmas*, *26*(7), 072102. https://doi.org/10.1063/1.5092561

Watt, C. E. J., & Rankin, R. (2009). Electron trapping in shear Alfvén waves that power the aurora. *Physical Review Letters*, *102*(4), 045002. https://doi.org/10.1103/PhysRevLett.102.045002

Winglee, R. M. (1982). Finite frequency effects on magnetosonic wave mode conversion. *Plasma Physics*, *24*, 1161.

Wukitch, S. J., Lin, Y., Parisot, A., Wright, J. C., Bonoli, P. T., Porkolab, M., et al. (2005). Ion cyclotron range of frequency mode conversion physics in Alcator C-Mod: Experimental measurements and modeling. *Physics of Plasmas*, *12*(5), 056104.

Wygant, J. R., Keiling, A., Cattell, C. A., Lysak, R. L., Temerin, M., Mozer, F. S., et al. (2000). Evidence for kinetic Alfven waves and parallel electron energization at 4-6 $R_E$ altitudes in the plasma sheet boundary layer. *Journal of Geophysical Research*, *107*(A8), doi:10.1029/2001JA900113.

Yao, Y., Chaston, C. C., Glassmeier, K.-H., & Angelopoulos, V. (2011). Electromagnetic waves on ion gyro-radii scales across the magnetopause. *Geophysical Research Letters*, *38*(9), L09102. https://doi.org/10.1029/2011GL047328

Yau, A. W., & Andre, M. (1997). Sources of ion outflow in the high latitude ionosphere. *Space Science Reviews*, *80*, 1–25.

Zeng, C., Wang, C., Duan, S., Dai, L., Fuselier, S. A., Burch, J. L., et al. (2020). Statistical study of oxygen ions abundance and spatial distribution in the dayside magnetopause boundary layer: MMS observations. *Journal of Geophysical Research: Space Physics*, *125*, e2019JA027323.

# 12

# Laboratory-Space Comparisons of Alfvén Waves

**James W. R. Schroeder**

**ABSTRACT**

Laboratory studies of Alfvén waves are providing increasingly detailed measurements of linear and nonlinear behaviors relevant to space plasma. Spacecraft measurement, remote observation, simulation, and theory are indispensable amid efforts to understand the contributions of Alfvén waves to the heliosphere; however, lab experiments add measurements of the underlying physics with unmatched detail. After a brief summary of linear Alfvén wave theory relevant to many space and lab plasmas, experimental results are reviewed where the unique level of control in the lab enabled thorough tests of dispersion and damping under conditions relevant to space plasmas. Some experiments have explored the properties of Alfvén waves in inhomogeneous, multi-ion, and partially ionized plasmas, providing access in the lab to processes believed to be at work in dynamic portions of the heliosphere including the solar corona, magnetospheres, and magnetosphere–ionosphere coupling. More recent lab experiments summarized here have measured nonlinear Alfvén wave physics. The increasing focus on nonlinearities involving Alfvén waves in the lab has created an avenue for progress in the grand challenges of turbulence, instabilities, and particle acceleration. Some experiments show a departure from standard theory, which points to opportunities for collaboration between experiment, theory, and simulation.

## 12.1. INTRODUCTION

Alfvén waves are magnetohydrodynamic waves below the ion cyclotron frequency that propagate in the direction of the background magnetic field $\mathbf{B}_0$ (Alfvén, 1942). They are now understood to be present in magnetized plasmas across the Universe. Almost immediately after their predicted existence, there was recognized potential for Alfvén waves to account for significant energy transfer from region to region and scale to scale required by space and astrophysical plasmas for heating, particle acceleration, and turbulence. A paradigm developed, still under consideration today, where Alfvén waves propagate outward from the photosphere (Alfvén, 1945; Mathioudakis et al., 2013; Morton et al., 2013), deposit a fraction of their energy into heating the $10^6$ K corona (Alfvén, 1947; Hahn & Savin, 2014; Tomczyk et al., 2007), provide pressure to drive the solar wind (De Pontieu et al., 2007; Hollweg & Johnson, 1988; Kasper et al., 2019), and seed fluctuations of the turbulent cascade found within the solar wind (Bale et al., 2019; Belcher et al., 1969; C. H. K. Chen, 2016; Parker, 1957). It is impossible to study solar physics today without considering the central role played by Alfvén waves.

Closer to home, Alfvén waves contribute significantly to the flow of energy through geospace. Amid the solar wind flow, Earth's magnetosphere-ionosphere system can be viewed as a resonant cavity with waves found throughout. Variable compression of Earth's magnetic field by the solar wind can excite magnetoacoustic waves that mode convert to Alfvén waves across inhomogeneities (Hasegawa, 1976). Additionally, Alfvén waves

*Department of Physics and Engineering, Wheaton College, Wheaton, IL, USA*

*Alfvén Waves Across Heliophysics: Progress, Challenges, and Opportunities, Geophysical Monograph 285*, First Edition.
Edited by Andreas Keiling.

DOI: 10.1002/9781394195985.ch12

can be generated by unstable beams of ions and electrons (L. Chen et al., 2014; Hong et al., 2012; Yin et al., 2007), produced by magnetic reconnection at the dayside magnetopause (Øieroset et al., 2014; Yao et al., 2011), and are found in the diffusion and exhaust regions of magnetotail reconnection (C. C. Chaston et al., 2009, 2012). Alfvén waves have been detected in the near-Earth magnetotail (Duan et al., 2016) and the plasma sheet boundary layer (Dombeck et al., 2005). Global standing Alfvén wave modes, called field-line resonances, are known to be associated with variations of auroras (Samson et al., 1991). As visualized in Figure 12.1, Alfvén waves are commonly found in the magnetotail, excited by magnetic disturbances such as changes in magnetic topology produced by magnetic reconnection (C. C. Chaston et al., 2009). Hasegawa (1976) predicted that Alfvén waves generated in the outer regions of the magnetosphere would propagate inward to auroral regions. Indeed, Alfvén waves traveling earthward are often found above auroras (C. C. Chaston et al., 2007; Keiling et al., 2003; Keiling et al., 2020) and have been shown to accelerate the electrons that cause auroras (Schroeder et al., 2021). Alfvén waves have even been inferred in Earth's interior, traversing the liquid metal outer core over the course of years with a regularity corresponding to slight length of day variations detected in the 20th century (Gillet et al., 2010).

With Alfvén waves playing consequential roles throughout the heliosphere (Nakariakov et al., 2016), and with society's growing dependence on the electromagnetic properties of geospace (Hapgood, 2012), the space physics community has prioritized increased understanding of Alfvén wave physics as it relates to space plasma (DPP, 2020; NRC, 2013). These reports affirm that an effective research community studies a complex system like the heliosphere with many approaches including theory, simulation, remote sensing, spacecraft measurements, and lab experiments. Because accurate understanding of the heliosphere and actionable space weather predictions require a thorough grasp of the fundamental plasma physics at work (Howes, 2018), there is good reason for lab experiments to continue as part of this community effort. Spacecraft missions provide invaluable insight while sampling one or perhaps a few points in space, but determining a unique description of phenomena often requires more control. Lab experiments have leveraged control and repeatability to answer questions that elude space-based measurement, all while offering a cost-effective complement to building, launching, and operating spacecraft (Howes, 2018).

Other chapters of this monograph review Alfvén wave physics in various regions of the heliosphere. The present work aims to provide a complementary perspective by reviewing results from lab experiments and the insights they have generated for understanding Alfvén waves in space. Section 12.2 presents Alfvén wave theory relevant to many of the lab experiments discussed here. Sections 12.3 and 12.4 present linear and nonlinear Alfvén wave experiments, respectively. This grouping allows work to be organized into approximately chornological order as experiments in the last decade have transitioned to largely focus on nonlinear physics. The more recent results in Section 12.3 benefit from the context of some earlier work, although we make no claims that the review of 20th century Alfvén wave experiments given here is complete. A more thorough treatment of these earlier works has been published under separate cover (Gekelman, 1999). Finally, Section 12.5 offers some concluding remarks.

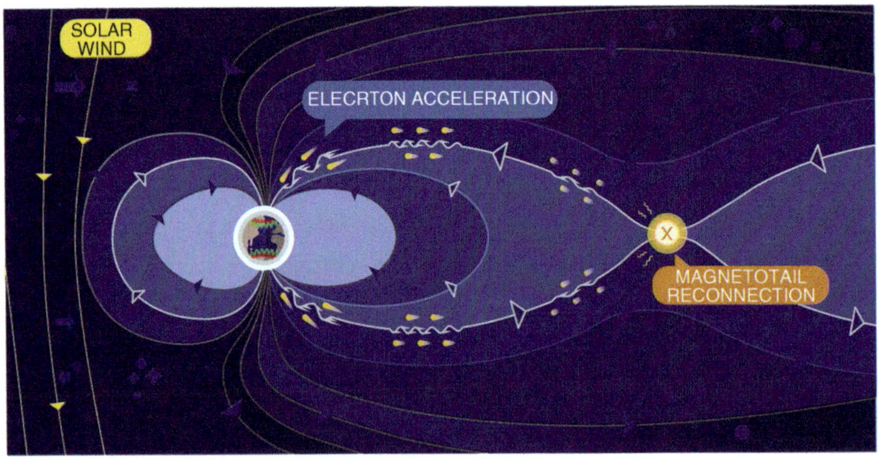


**Figure 12.1** Drawing of Earth's magnetosphere with Alfvén waves generated by magnetic reconnection in the magnetotail. Alfvén waves descend Earthward and are often found above auroras. Lab experiments have demonstrated Alfvén waves can accelerate the electrons that cause auroras. Image credit: Austin Montelius, College of Liberal Arts and Sciences, University of Iowa.

## 12.2. LINEAR ALFVÉN WAVE THEORY IN IDEAL MHD, INERTIAL, AND KINETIC LIMITS

Because of the potentially dizzying array of literature sources with different limits and applications in mind, it is worth articulating theory relevant to many of the lab results discussed below. An advisor once suggested, responding to confusion over the zoo of sometimes inconsistent plasma wave terminology, that dispersion relations are a reliable and unambiguous way to identify plasma waves. In light of this advice, we include here a concise derivation of Alfvén wave dispersion in limits relevant to many lab experiments (Schroeder, 2017).

An Alfvén wave dispersion relation with cold ions and an arbitrary distribution of electrons will allow us to recover the ideal MHD limit as well as the common results of dispersive Alfvén waves in the so-called inertial $v_A \gg v_{te}$ and kinetic $v_A \ll v_{te}$ limits, where $v_A$ is the Alfvén speed and $v_{te} = \sqrt{kT_e/m_e}$ is the electron thermal speed. In the intermediate $v_A \approx v_{te}$ regime, Landau damping is strong (Landau, 1946), and so the Alfvén wave is typically found as a propagating wave mode in either the inertial or kinetic limits. Using a geometry where $\mathbf{B}_0 = B_0\hat{\mathbf{z}}$ and perpendicular variations of phase fronts $k_\perp$ are in $\hat{\mathbf{x}}$, then plane wave solutions have the form $e^{i(k_z z + k_\perp x - \omega t)}$. A solution can be built up by linearly combining multiple plane wave solutions as needed. This geometry assumes the oscillating wave quantities have the polarization $\mathbf{B}_1 = (0, B_y, 0)$ and $\mathbf{E}_1 = (E_x, 0, E_z)$.

Using the plane wave assumption specified above, Faraday's law and Ampére's law give

$$k_z E_x - k_\perp E_z = \omega B_y, \tag{12.1}$$

$$-ik_z B_y = \mu_0 J_x, \tag{12.2}$$

$$ik_\perp B_y = \mu_0 J_z, \tag{12.3}$$

where $\mu_0$ is the permeability of free space. Using the plasma dielectric tensor $\mathbf{K}$ and $\mathbf{J} = -i\omega\epsilon_0\mathbf{K}\cdot\mathbf{E}$, where $\epsilon_0$ is the permittivity of free space,

$$\mu_0 J_x = -\frac{i\omega}{c^2}K_{xx}E_x. \tag{12.4}$$

$K_{xy}$ has been neglected in Equation 12.4 using a low frequency assumption $\omega/\omega_{ci} \ll 1$. To specify $K_{xx}$, the cold ion assumption applicable to many lab experiments is introduced. The cold, two-fluid expression for $K_{xx}$ is given by Gurnett and Bhattacharjee (2005),

$$K_{xx} = 1 - \sum_s \frac{\omega_{ps}^2}{\omega^2 - \omega_{cs}^2}, \tag{12.5}$$

and it highlights that contributions to this entry of the dielectric scale with species mass. Consequently, the ion contribution to $K_{xx}$ dominates, and the electron term can be discarded. This is equivalent to the well-known conclusion that the ion polarization drift carries the perpendicular Alfvén wave current. Accordingly, keeping only the ion term in the low-frequency limit, $K_{xx}$ simplifies to

$$K_{xx} \approx \frac{c^2}{v_A^2}. \tag{12.6}$$

Combining Equations 12.4 and 12.6 gives

$$\mu_0 J_x = -\frac{i\omega}{v_A^2}E_x. \tag{12.7}$$

Next, the system of equations above is manipulated to eliminate $E_x$ and $B_y$ in favor of $J_z$ and $E_z$. Combining Equations 12.2 and 12.7 gives

$$E_x = \frac{v_A^2 k_z}{\omega}B_y. \tag{12.8}$$

Equations 12.1 and 12.8 give

$$k_\perp E_z = \left(\frac{v_A^2 k_z^2}{\omega} - \omega\right)B_y. \tag{12.9}$$

The combination of Equations 12.3 and 12.9 produces

$$\frac{q_e}{m_e}E_z = \frac{i\mu_0\omega}{k_\perp^2}\left(1 - \frac{v_A^2 k_z^2}{\omega^2}\right)\frac{q_e}{m_e}J_z, \tag{12.10}$$

where both sides have been multiplied by the electron charge-to-mass ratio $q_e/m_e$.

The current $J_z$ can be written in terms of the perturbation of the parallel electron velocity distribution $g_{e1}(v_z)$

$$J_z = q_e n_e \int_{-\infty}^{\infty} v_z g_{e1} dv_z. \tag{12.11}$$

Substituting this into Equation 12.10 gives

$$\frac{q_e}{m_e}E_z = \frac{i\omega}{k_\perp^2\delta_e^2}\left(1 - \frac{v_A^2 k_z^2}{\omega^2}\right)\int_{-\infty}^{\infty} v_z g_{e1}(v_z) dv_z. \tag{12.12}$$

Using the linearized gyroaveraged Boltzmann equation, $g_{e1}(v_z)$ is

$$-i\omega g_{e1} + ik_z v_z g_{e1} + \frac{q_e}{m_e}E_z\frac{\partial g_{e0}}{\partial v_z} = -\nu g_{e1}, \tag{12.13}$$

where $\nu$ is the velocity-dependent Coulomb collision rate included as part of a rudimentary but analytically-tractable collision operator (Schroeder et al., 2017). Collisions are often necessary for accurate modeling of Alfvén wave propagation in lab plasmas (Gekelman et al., 1994; Kletzing et al., 2010; S. Vincena et al., 2001), and some lab-relevant theory departs at this point from our derivation by including more sophisticated collision operators (Thuecks et al., 2009). Solving for $g_{e1}$ produces

$$g_{e1} = \frac{i}{k_z v_z - \omega - i\nu} \frac{\partial g_{e0}}{\partial v_z} \frac{q_e}{m_e} E_z. \tag{12.14}$$

Replacing $(q_e/m_e)E_z$ in Equation 12.14 using equation 12.12 gives

$$g_{e1} = \frac{1}{k_z v_z - \omega - i\nu} \frac{\partial g_{e0}}{\partial v_z} \frac{i\omega}{k_\perp^2 \delta_e^2} \left( \frac{v_A^2 k_z^2 - 1}{\omega^2} \right) \int_{-\infty}^{\infty} v_z g_{e1} dv_z. \tag{12.15}$$

Multiplying the previous equation by $q_e v_z n_e$ and integrating over $v_z$ gives

$$q_e n_e \int_{-\infty}^{\infty} v_z g_{e1} dv_z = \frac{v_A^2 k_z^2 - \omega^2}{\omega k_\perp^2 \delta_e^2} \left( q_e n_e \int_{-\infty}^{\infty} v_z g_{e1} dv_z \right) \times \left( \int_{-\infty}^{\infty} \frac{v_z(\partial g_{e0}/\partial v_z)}{k_z v_z - \omega - i\nu} dv_z \right). \tag{12.16}$$

In the preceding equation, the left-hand side and the quantity in the first set of parentheses on the right-hand side are both recognized as $J_z$, yielding

$$J_z \left( 1 - \frac{v_A^2 k_z^2 - \omega^2}{\omega k_\perp^2 \delta_e^2} \int_{-\infty}^{\infty} \frac{v_z(\partial g_{e0}/\partial v_z)}{k_z v_z - \omega - i\nu} dv_z \right) = 0. \tag{12.17}$$

The quantity in parentheses is the Alfvén wave dispersion relation for cold ions and an arbitrary distribution of electrons $g_{e0}(v_z)$. This quantity must be zero for non-trivial solutions to exist. The condition is written below:

$$\frac{v_A^2 k_z^2 - \omega^2}{\omega k_\perp^2 \delta_e^2} \int_{-\infty}^{\infty} \frac{v_z(\partial g_{e0}/\partial v_z)}{k_z v_z - \omega - i\nu} dv_z = 1. \tag{12.18}$$

Once the distribution of electrons $g_{e0}(v_z)$ is known, values $(\omega, k_\perp, k_z)$ that satisfy this equation can be determined numerically, if not analytically. This equation includes the effects of collisionless and collisional damping as well as finite $k_\perp$ for all values of $v_A/v_{te}$ ranging from the kinetic to the inertial limit, although finite frequency effects have not been retained. Allowing for an arbitrary electron velocity distribution makes equation 12.18 useful in circumstances where electrons are not in thermodynamic equilibrium, which includes many lab plasmas.

For the sake of connecting to more familiar theory, equation 12.18 can be evaluated in collisionless inertial and kinetic limits. Assuming the background distribution is Maxwellian,

$$g_{e0} = \frac{1}{\sqrt{2\pi} v_{te}} e^{-v_z^2/(2v_{te}^2)}, \tag{12.19}$$

where $v_{te}^2 = kT_e/m_e$. With the variable substitution $z = v_z/(\sqrt{2}v_{te})$, $\zeta = \omega/(\sqrt{2}k_z v_{te})$, and assuming the collisionless scenario $\nu = 0$, then the integral in Equation 12.18 becomes

$$\int_{-\infty}^{\infty} \frac{v_z(\partial g_{e0}/\partial v_z)}{k_z v_z - \omega - i\nu} dv_z = -\frac{\sqrt{2}}{v_{te}\sqrt{\pi} k_z} \int_{-\infty}^{\infty} z^2 \frac{e^{-z^2}}{z - \zeta} dz, \tag{12.20}$$

which can be rewritten so that

$$\int_{-\infty}^{\infty} z^2 \frac{e^{-z^2}}{z - \zeta} dz = \int_{-\infty}^{\infty} \left( z + \zeta + \frac{\zeta^2}{z - \zeta} \right) e^{-z^2} dz. \tag{12.21}$$

The first term in parentheses is an odd function integrated over even limits. The remaining Guassian integrals can be evaluated to find

$$\int_{-\infty}^{\infty} z^2 \frac{e^{-z^2}}{z - \zeta} dz = \sqrt{\pi} \left( \zeta + \zeta^2 Z(\zeta) \right), \tag{12.22}$$

where $Z(\zeta)$ is the plasma dispersion function (Fried & Conte, 1961; Huba, 2013). Using 12.22 and 12.20, Equation 12.18 can be written as

$$1 = \frac{2\zeta^2 - v_A^2/v_{te}^2}{k_\perp^2 \delta_e^2} (1 + \zeta Z(\zeta)). \tag{12.23}$$

This is the collisionless version of the dispersion relation given in Equation 12.18. No assumptions about the relative size of $v_A$ and $v_{te}$ have yet been made.

In the inertial limit, $v_A \gg v_{te}$ and $\zeta \gg 1$. Using the large argument expansion of the plasma dispersion function (Huba, 2013) gives

$$\frac{\omega}{k_z} = \frac{v_A}{\sqrt{1 + k_\perp^2 \delta_e^2}}, \tag{12.24}$$

where $\delta_e = c/\omega_{pe}$ is the electron skin depth, and $\omega_{pe}$ is the electron plasma frequency. This is the well-known dispersion relation for the inertial Alfvén wave, where the parallel phase velocity decreases as $k_\perp^2$ increases (Goertz & Boswell, 1979; Stasiewicz et al., 2000).

In the kinetic limit, $v_A \ll v_{te}$ and $\zeta \ll 1$. The small argument expansion of the plasma dispersion function is appropriate (Huba, 2013). This yields

$$\frac{\omega}{k_z} = v_A \sqrt{1 + k_\perp^2 \rho_s^2}, \tag{12.25}$$

where $\rho_s = C_s/\omega_{ce}$ is the ion acoustic gyroradius, and $C_s = \sqrt{T_e/m_i}$ is the ion acoustic speed. This is the dispersion relation for the kinetic Alfvén wave, where the parallel phase velocity increases with $k_\perp^2$ (Hasegawa, 1976; Stasiewicz et al., 2000).

In the limit of infinite conductivity associated with ideal MHD, Ohm's law dictates that $E_z$ must be zero. Returning to Equation 12.9, this reproduces the well-known dispersion relation for ideal MHD Alfvén waves

$$\frac{\omega}{k_z} = v_A, \tag{12.26}$$

where Alfvén waves of all frequencies propagate with the same group and phase velocities.

## 12.3. LINEAR ALFVÉN WAVE PHYSICS

A significant body of work exists that identifies the linear behaviors of Alfvén waves based on spacecraft measurement, remote observation, and laboratory experiment. This constellation of efforts has produced conclusions about where Alfvén waves are found in space plasmas and the properties of their propagation in various conditions. This work provides an essential foundation upon which nonlinear Alfvén wave physics, including turbulence, instabilities, and particle acceleration, can be studied.

The work of studying Alfvén waves in space and lab plasma begins, of course, with their identification. Fluctuations of the magnetic field, electric field, and plasma current are associated with Alfvén waves, and single-point measurements typically identify Alfvén waves by comparing the magnitude and phase of these fluctuating quantities with expectations from theory. See, e.g., Section 12.2 for relationships between wave fields associated with shear Alfvén waves. When data are available at multiple locations, phase differences between points of measurement are compared to the predicted propagation delay.

### 12.3.1. Alfvén Wave Identification in Space

An early report of Alfvén waves suspected in measurements of solar wind plasma was provided by Coleman (1966). Low-frequency variations ranging from 1–50 cycles per day in 3D magnetic field data and radial velocity yield correlations that are consistent with Alfvén and fast waves. The addition of density fluctuations measured by Mariner 5 enabled a more definitive identification of Alfvén waves (Belcher et al., 1969; Belcher & Davis, 1971). Specifically, the absence of density perturbations, which are expected with fast and slow magnetoacoustic waves but not with Alfvén waves, led to the conclusion that wave activity measured by Mariner 5 was dominated by Alfvén waves. More recently, single-point solar wind measurements have been decomposed into MHD fast, slow, and Alfvén modes using a spectral filtering technique where the orientation of the wavevector along the background magnetic field is determined from fluctuations of the magnetic field, velocity, and density (Glassmeier et al., 1995). Time series data are interpreted as spatial variations advected past the spacecraft by the high-speed solar wind flow and filtered into constituent MHD wave modes. This technique has been applied to Parker Solar Probe data, leading several authors to conclude that fluctuations in the solar wind are generally Alfvénic (C. C. Chaston et al., 2020; Zhu et al., 2020).

Direct evidence of Alfvén waves associated with sunspot structures and the solar corona was obtained from Doppler velocity perturbations and non-thermal broadening of spectroscopic measurements (De Pontieu et al., 2007; Tomczyk et al., 2007). More recently, the advancement of solar observations has allowed improved understanding of the character of oscillations in the chromosphere and corona (Jess et al., 2015), and it has been suggested that much of the wave energy in this region is carried by Alfvén waves (Mathioudakis et al., 2013; Morton et al., 2013).

In Earth's magnetosphere, Gurnett et al. (1984) find persistent oscillations of the magnetic and electric fields above the auroral oval in data from Dynamics Explorer 1 (DE-1) orbiting between altitudes of 0.1 and 3.65 Earth radii. Analysis shows the ratio of electric to magnetic field is inconsistent with expectations for propagating Alfvén waves, e.g., as given by equation 12.8. Gurnett suggests Alfvén waves reflected by the conducting ionosphere produce interference and cause the measured wavefields to depart from their expected values for isolated propagating waves. Later calculations support the plausibility of this scenario (Lysak, 1998). An example of an unambiguous identification of propagating Alfvén waves in the lower magnetosphere comes from Boehm et al. (1990). Using electric and magnetic field measurements made aboard a 1985 rocket flight from Greenland into the dayside cusp at an altitude of 600 km, the electric-to-magnetic field ratio matched the expected value for propagating Alfvén waves. More recently, the ratio of electric to magnetic fields is regularly used with data from missions like FAST and Polar to identify propagating Alfvén waves. Following predictions for dispersive kinetic and inertial Alfvén waves, a range of $E_\perp/B_\perp$ values is typically deemed consistent with propagating Alfvén waves (Andersson et al., 2002; C. C. Chaston et al., 2007; Keiling et al., 2003; C. Keiling et al., 2020; Wygant et al., 2002).

The expansion of multi-point spacecraft missions has enabled simultaneous measurements at several locations to be compared with predicted propagation and dispersion (Neubauer & Glassmeier, 1990). This technique has been commonly applied in the terrestrial magnetosphere, e.g., using the Magnetospheric Multiscale (MMS) Mission (He et al., 2020; Narita et al., 2016). In the solar wind, Cluster and MMS data have been used in this way to identify Alfvén waves (Sahraoui et al., 2010; S. Q. Zhao et al., 2022). It has also been used to suggest that a significant fraction of what has previously been categorized as propagating Alfvén waves in the solar wind may in fact be nonpropagating structures in the solar wind frame advected past spacecraft by the solar wind flow (L. L. Zhao et al., 2023).

### 12.3.2. Alfvén Wave Generation and Dispersion in the Lab

Coinciding with significant interest in Alfvén waves from the space physics community in the mid-20th century, laboratory experiments began studying fundamental Alfvén wave properties, including propagation and damping. While a significant portion of laboratory studies were motivated by space physics applications, interest in lab studies of Alfvén waves also came from the fusion community (Fisch & Karney, 1981). A review by Gekelman (1999) highlights early lab work that details the propagation of Alfvén waves in various plasma conditions and geometries.

The first laboratory measurements of Alfvén waves used a 15-cm-long cylindrical vessel of liquid mercury placed in a 13 kG magnetic field (Lundquist, 1949). A fan at the bottom of the vessel generated velocity perturbations perpendicular to the imposed magnetic field at frequencies between 0.1 Hz and 10 Hz. A mirror floating atop the mercury oscillated as the wave arrived at the surface. By recording light reflected off the floating mirror onto a ruler, it was determined that the phase of the oscillating mirror relative to the driving fan is consistent with the delay expected for Alfvén wave propagation. Researchers at Tufts provided the first detection of Alfvén waves in laboratory plasma (Bostick & Levine, 1952). A Langmuir probe in a toroidal discharge shows 100 μs oscillations of ion and electron current. By reducing the magnetic field, oscillations vanish. Data are interpreted as spontaneous Alfvén standing waves, with the Alfvén speed and the toroidal length of the device used to determine the standing wave frequency. Additional work by Wilcox et al. (1960) finds Alfvén waves follow predictions of the phase speed as a function of the background magnetic field.

Tsushima et al. (1982) excited shear and compressional Alfvén waves using loops spaced along the background magnetic field with oscillating currents driven 180° out of phase in adjacent loops. The plasma diameter and length were 10 cm and 200 cm, respectively, though there existed a strong radial density gradient with the density falling by 50% only 3 cm from the central axis. Using B-dot coils, the radial position of the maximum wave amplitude is found to vary with wavelength and frequency, interpreted as the result of waves spreading radially across the density gradient until they reach a critical radius beyond which they become evanescent, effectively creating a pileup of wave energy. Predictions are made for the critical radius of the shear and compressional Alfvén waves and found to coincide with the measured radius of peak Alfvén wave power.

Using the same plasma device, Amagishi et al. (1989) report a test of the Alfvén wave dispersion relation. Waves were driven by a helical antenna wrapped around the 12-cm-diameter plasma column, driven by a resonant circuit delivering short bursts of power to the antenna at frequencies below the ion cyclotron frequency. The helical antenna excites MHD surface wave eigenmodes of the device. Similar to the mode-conversion predicted by Hasegawa (1976), evidence was found for the conversion of the MHD surface wave to shear Alfvén waves across the radial density gradient. Using the parallel wave number $k_z$ determined from measurements at multiple locations, results for the phase speed at several antenna driving frequencies are found to agree with predictions from the Alfvén wave dispersion relation.

Borg et al. (1985) studied Green's function solutions for waves driven by delta functions of antenna current oriented in various directions relative to $\mathbf{B}_0$. This approach predicts that antenna currents perpendicular to $\mathbf{B}_0$ produce compressional Alfvén waves, while currents aligned with $\mathbf{B}_0$ preferentially generate shear Alfvén waves. Following predictions, Alfvén waves were generated using field-aligned current elements in hydrogen plasma with $n_e = 10^{13}$ cm$^{-3}$, $T_e = 10$ eV, and $4 < B_0 < 10$ kG, where the inertial limit $v_A > v_{te}$ is relevant. The measured profile of shear Alfvén wave magnetic fields is reproduced with a model based on inertial Alfvén wave dispersion and by allowing waves in the model to encircle the toroidal device multiple times.

Gekelman et al. (1994) performed experiments using helium plasma in UCLA's Large Plasma Device (LAPD) with an electron density of $n_e = 10^{12}$ cm$^{-3}$, electron temperature $T_e$ of a few eV, and a background magnetic field $B_0 = 1100$ G to produce $v_A > v_{te}$. Although upgraded several times since the work published by Gekelman et al. (1994), the LAPD, shown in Figure 12.2(a), remains a cylindrical magnetized device where primary electrons are accelerated from a cathode to collisionally generate plasma in the main volume of the experiment. The LAPD provided the first laboratory environment for Alfvén wave studies that was not collisionally dominated. Alfvén waves were excited using an exposed metal mesh disk immersed in the plasma. Oscillating voltage applied to the disk collected current flowing along $\mathbf{B}_0$ that produced Alfvén waves. Because of the small size, the disk antenna produced inertial Alfvén waves that are significantly dispersive with $|k_\perp \delta_e| \approx 1$. The inertial Alfvén wave dispersion relation predicts a group velocity at a slight angle to $\mathbf{B}_0$, and so the circular ring of Alfvén waves produced by the disk exciter expands downfield into a cone as waves propagate away from the antenna. Accompanying theory from Morales et al. (1994) finds the cone angle is dependent on several factors including $k_\perp$, collisions, and Landau damping. Gekelman concludes that the need for Landau damping in their model is evidence for the field-aligned electric field of inertial Alfvén waves.

The trapping of dispersive Alfvén waves on magnetic field lines conjugate to auroras (A. Streltsov & Lotko,

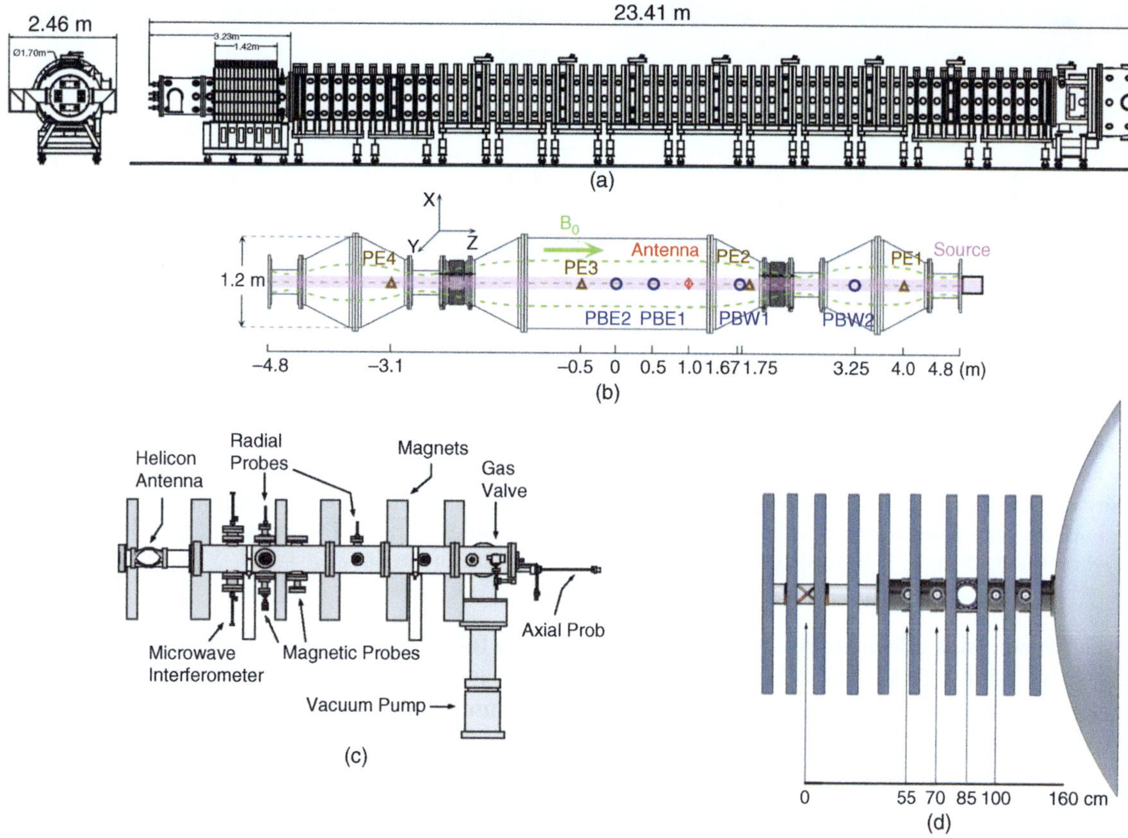


**Figure 12.2** Plasma devices that have been used for Alfvén wave experiments. (a) The Large Plasma Device (LAPD) is a collaborative research facility at UCLA. (b) The Keda Mirror with AXisymmetricity (KMAX) at the University of Science and Technology of China. (c) The Hot hELIcon eXperiment (HELIX) at West Virginia University. (d) The Auburn Linear Experiment for Space Plasma Investigations (ALESPI). Schematic (a) provided by Tomas Sketchley, private communication. Source: Schematics (b)–(d) adapted from Shi, Yang, et al. (2019), Hanna & Watts (2001) /American Institute of Physics and Houshmandyar and Scime (2011) / American Institute of Physics, respectively.

1995; A. V. Streltsov et al., 1998) has been explored in the LAPD by S. Vincena et al. (2004). Experiments found a pileup of Alfvén wave energy at a critical radius outside the primary Alfvén wave cone. Plasma conditions within 15 cm of the central axis were fairly uniform and within the kinetic limit $v_A < v_{te}$. Outside the core of the plasma column, electron density and temperature both decreased so that $v_A > v_{te}$, corresponding to the inertial limit. An analysis of the Alfvén wave group velocity and its local variations with the plasma density and temperature reveal a radius where the perpendicular group velocity goes to zero. This is because the perpendicular group velocity and phase velocities of the kinetic Alfvén wave are aligned, while the two velocities are anti-aligned in the inertial limit. Consequently, in the transition between kinetic and inertial Alfvén waves, the perpendicular group velocity goes to zero, and Alfvén waves entering from the kinetic region of the plasma are stopped, producing the pileup of wave energy detected experimentally. As seen in the left panel of Figure 12.3, wave energy density accumulates in time near where plasma conditions give $\overline{\beta}_e = 1$, the nominal transition from kinetic to inertial Alfvén waves. Curve (a) on the right of Figure 12.3 shows the radial peak of wave energy density determined experimentally as a function of wave frequency. Measurements are found to be consistent with overlapping theory curves (b) and (c) that include finite frequency effects.

A detailed test of Alfvén wave dispersion was performed by Kletzing et al. (2003). An electrostatic antenna was used that, like the disk exciters of Gekelman et al. (1994) and others, generated current in the plasma using biased metal meshes. While the disk antennas used by Gekelman et al. (1994) and others at UCLA produced cones of Alfvén waves described most naturally by the superposition of multiple Bessel eigenmodes, the antenna used for this work produce Alfvén waves approximately described by monochromatic plane waves. This is accomplished using an array of mesh grids spaced along $\hat{\mathbf{x}}$ and uniform in $\hat{\mathbf{y}}$ that generate current channels along $\mathbf{B_0} = B_0\hat{\mathbf{z}}$. The adjustable phase and amplitude of currents

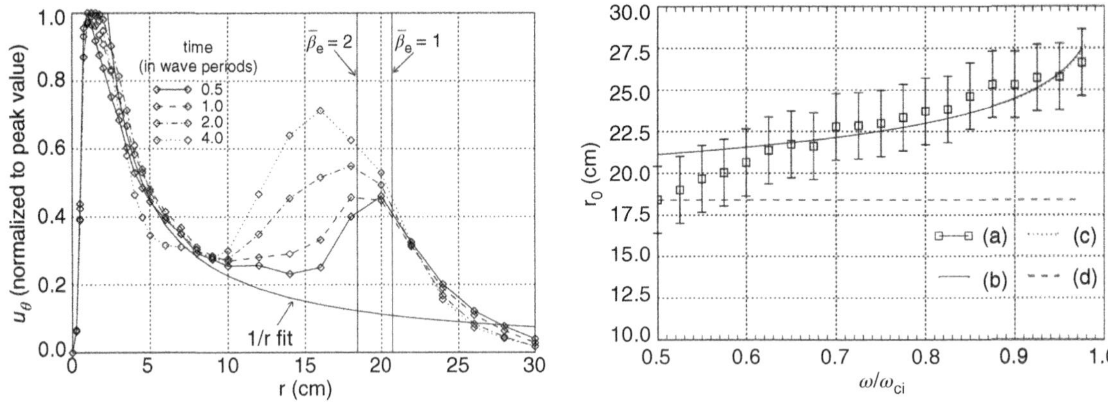


**Figure 12.3** A pileup of energy density is caused by dispersive Alfvén waves that spread across magnetic field lines and travel into a region where the perpendicular group velocity is zero. (Left) The accumulation of energy density is resolved over time. (Right) The measured radial peak location of wave energy density (line a) varies with frequency and compares favorably with overlapping theory curves (lines b and c). Theory curve (d) neglects finite frequency effects. Source: S. Vincena et al. (2004) / American Institute of Physics.

collected by adjacent grids allow tuning of the antenna to select the $k_\perp$ of launched Alfvén waves. By correlating the arrival of phase fronts between two probes, the propagation delay and phase speed were determined for $0.1 < k_\perp \rho_s < 1$. As expected for the $v_A \approx v_{te}$ conditions of the experiment, the phase speed increased at higher values of $k_\perp \rho_s$.

Thuecks et al. (2009) and Kletzing et al. (2010) continued experiments of Alfvén wave dispersion in the inertial $v_A \gg v_{te}$ and kinetic $v_A \ll v_{te}$ limits. Experiments were performed using the LAPD and an upgraded antenna, named the Arbitrary Spatial Waveform antenna, based on a design similar to Kletzing et al. (2003). Figure 12.4, adapted from Thuecks et al. (2009), shows the parallel phase velocity and damping as a function of $k_\perp$. Measured results are compared with predictions for several collision operators, including non-conserving Krook (N.C.), number conserving (Num), energy conserving (Energy), and pitch-angle scattering (Lorentz). In the kinetic limit, shown in Figure 12.4(a,b), measurements are well described by collisionless theory. Meanwhile, the inertial limit results in (c,d) are consistent with the non-conserving operator. It is noteworthy that collisionless damping is prominent in (b), indicating wave-particle interactions increasing with $k_\perp$ as expected due to the parallel electric field increasing with $k_\perp$ as well. Also noteworthy are the increasing and decreasing phase velocities seen in (a) and (c) which are characteristic of the kinetic and inertial limits, respectively.

Follow-up work by Nielson et al. (2010) compared the results of Thuecks et al. (2009) with a gyrokinetic simulation. While modeling performed by Thuecks did not find a single collision operator capable of describing Alfvén wave behavior in both the kinetic and inertial limits, a fully-conservative collision operator in the simulation was able to reproduce both sets of measurements. Determining the role of collisions is critical for understanding Alfvén waves in the lab and the relationship of lab experiments to Alfvén waves in space plasma where collisions may be more or less important.

Alfvén waves have been recognized as an important signature of fast 3D reconnection (J. F. Drake, 1995; Rogers et al., 2001). Spacecraft data from Cluster and MMS support the hypothesis that the diffusion and exhaust regions of guide field reconnection contain kinetic Alfvén waves with effects on the rate of reconnection and particle energization (C. C. Chaston et al., 2009; Gershman et al., 2017). In the lab, the first report of Alfvén waves produced by magnetic reconnection comes from the Keda Mirror with AXisymmetricity (KMAX) (Shi, Huang, et al., 2019). As shown in Figure 12.2(b), KMAX is used to generate guide field reconnection between adjacent flux ropes created within a helicon plasma. As seen in Figure 12.5, wave fields are recorded at three $z$ locations along $\mathbf{B}_0$. Magenta lines trace the measured magnetic field in the $x$-$y$ plane perpendicular to $\mathbf{B}_0$, while the color scale indicates the calculated parallel electric field. 'O' points of the magnetic field pattern are seen to rotate both in time and $z$ in a manner consistent with $m = -2$ kinetic Alfvén wave phase fronts rotating in the same sense as gyro-orbiting ions. As predicted, measurements find that reconnection rate and Alfvén wave amplitude are well-correlated.

The overwhelming majority of Alfvén wave measurements in the lab use B-dot probes that produce signals proportional to the changing magnetic flux through a loop of wire. An array of such probes provides spatial resolution to determine the wavenumber, phase speed,

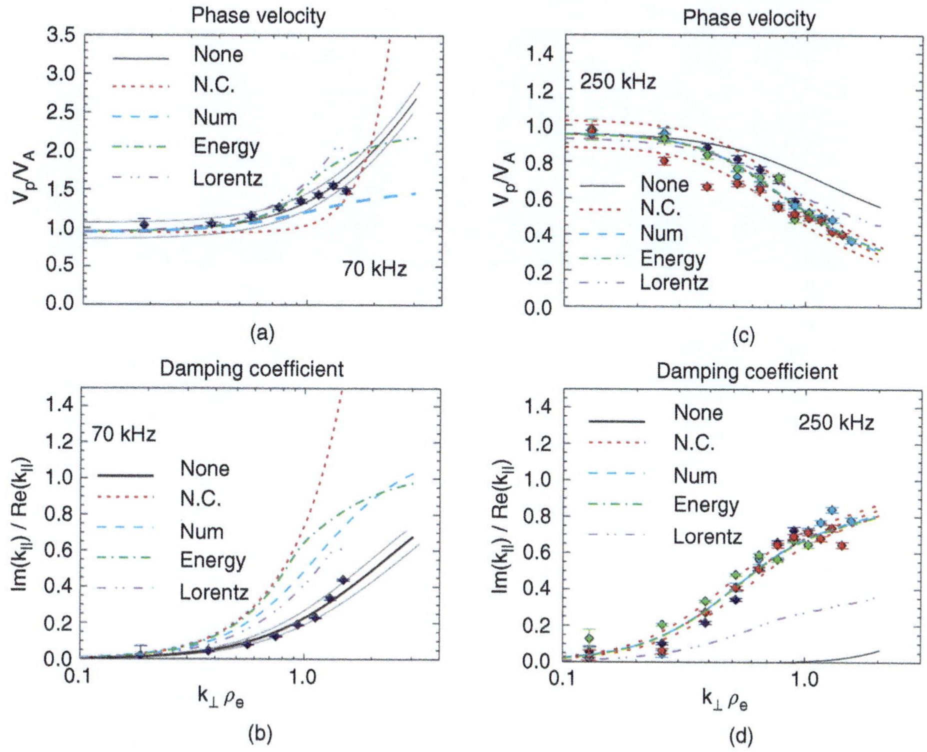


**Figure 12.4** Dispersive Alfvén wave parallel phase speed and damping measured in the LAPD. (a), (b) Kinetic Alfvén waves are well-described by the collisionless dispersion and damping. (c), (d) A non-conserving Krook collision operator faithfully describes the behavior of inertial Alfvén waves in experiments. Source: Thuecks et al. (2009) / American Institute of Physics.

and damping, enabling comparison to dispersion relations. In contrast, Alfvén waves in space plasma are most often identified from single-point measurements using the ratio of electric and magnetic fields. This approach has been brought to the lab with probes presented by D. J. Drake et al. (2011). Called Elsässer probes, they combine B-dot probes to measure magnetic fluctuations and double probes to record potential differences between two points and infer the electric field. Limited bandwidth of electric probes in lab plasmas typically arises from the combined impedance of the plasma sheath around the probe tip with the capacitance of typical lengths of cable leading to external signal amplifiers and digitizers. Powered amplifiers integrated into the probe tip alleviate these concerns and enable calibrated measurements of the Alfvén wave electric field. D. J. Drake et al. (2011) present the theory needed to relate probe outputs to the Elsässer variables $\mathbf{z}^{\pm} = \mathbf{v} \pm \delta\mathbf{B}/\sqrt{\mu_0\rho_0}$ used for a symmetric expression of incompressible MHD, where $\mathbf{v}$ and $\delta\mathbf{B}$ are the velocity and magnetic perturbations, respectively (Elsässer, 1950). Initial Elsässer probe data collected in the LAPD are used to calculate Poynting flux and the ratio of electric and magnetic wave fields, both of which are consistent with traveling Alfvén waves.

Elsässer probes provide a single-point diagnostic to identify Alfvén waves in the lab without requiring an array of measurements along $\mathbf{B}_0$ to determine phase variations. This diagnostic is useful for Alfvén wave experiments where integration between probes is complicated, for example, by inhomogeneous plasma.

### 12.3.3. Alfvén Waves in Inhomogeneous, Multi-Ion, and Partially-Ionized Lab Plasmas

Electron Landau damping and ion-cyclotron damping are investigated in experiments performed by S. Vincena et al. (2001) using a magnetic beach geometry (Stix, 1992). Alfvén waves produced by disk exciters propagate into a flared background magnetic field. Near the antenna, $v_{te} \approx v_A$ so significant Landau damping is expected. Away from the antenna, $B_0$ decreased so that $\omega \approx \Omega_{ci}$, and ion cyclotron damping is expected to dominate in this region. This experiment provided the first laboratory study of Alfvén waves propagating through a transition region where different dissipation mechanisms dominate. WKB solutions are produced for models including combinations of ion cyclotron damping, damping from Coulomb collisions, and electron Landau damping. Comparison of measurements and the model solutions shows that

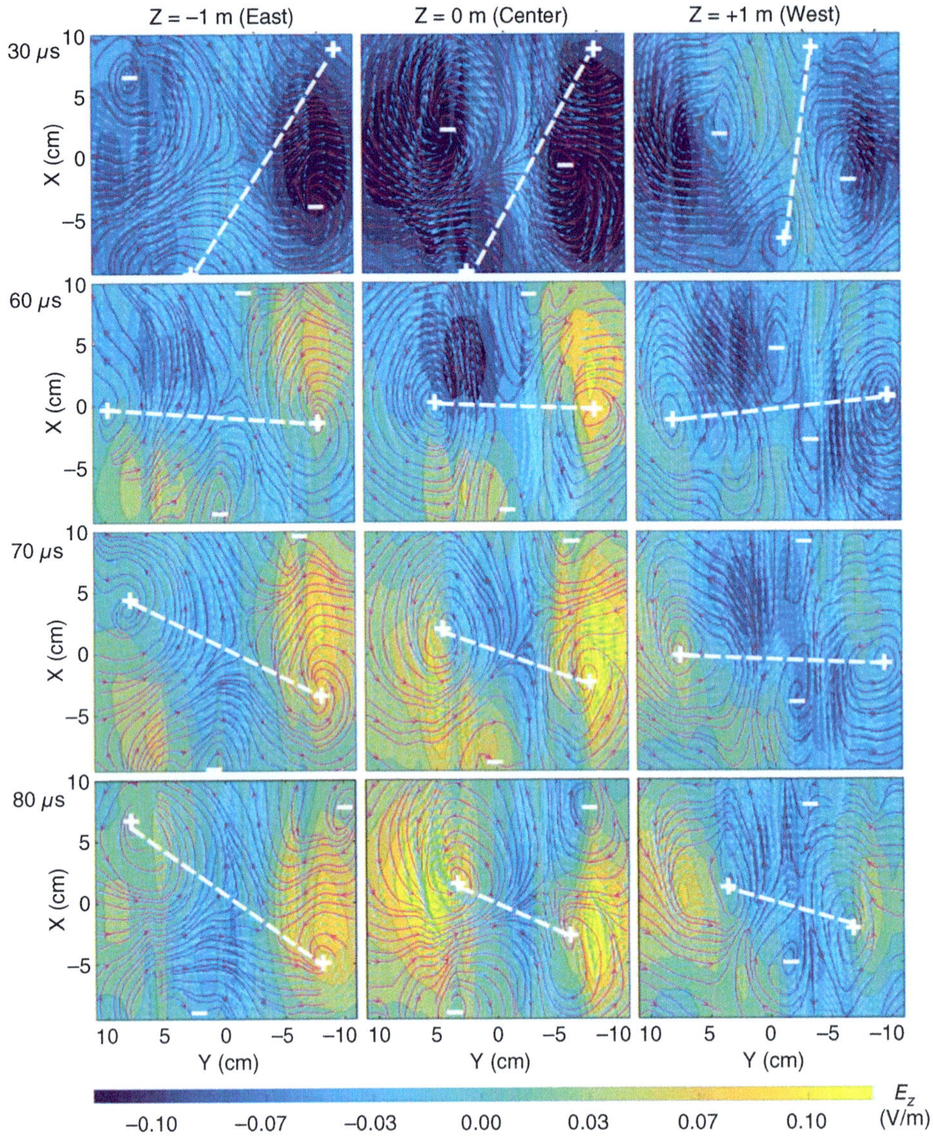


**Figure 12.5** The first report of Alfvén waves generated by magnetic reconnection in the lab. Magenta lines trace the measured magnetic field in the *x*-*y* plane perpendicular to $\mathbf{B}_0$, while color represents the calculated parallel electric field. Rotation of magnetic field 'o' points in both time and *z*, seen with the changing slope of the dashed white line, is consistent with the dispersion of kinetic Alfvén waves. Source: Shi et al. (2019) / IOP Publishing Ltd.

all three damping mechanisms are needed to accurately describe the phase and amplitude of the Alfvén wave along a ray path. Furthermore, all three damping mechanisms contribute approximately equally to the integrated results of the WKB model.

Bose et al. (2019) use lab experiments to explore the dynamics of Alfvén waves in coronal holes, regions where the solar magnetic field expands into interplanetary space. It is known that coronal holes are a source of fast solar wind, leading to questions about the energy budget and the role of Alfvén waves (Cranmer, 2009; McIntosh et al., 2011). Observations of significant Alfvén wave damping in the corona (Bemporad & Abbo, 2012; Hahn et al., 2012) support the hypothesis that Alfvén waves cause heating of coronal plasma and drive acceleration of the solar wind. Carried out in UCLA's LAPD, experiments used a magnetic beach with a $\mathbf{B}_0$ gradient stronger than earlier work by S. Vincena et al. (2001). The flared magnetic field is characterized by a scale length over which the Alfvén speed changes and is adjusted to be relevant to the inhomogeneities of the corona. Alfvén waves are used with low frequency and low $k_\perp$ that are mostly consistent with an ideal MHD description. Damping of waves through the magnetic beach is observed to be significantly increased compared to the collisional damping that occurs in homogeneous LAPD plasma. The lost Alfvén wave energy could not be detected in the generation of reflected waves, mode conversion, or ion-cyclotron resonance. Authors conclude that the most likely explanation is that Alfvén waves are damped by heating the plasma, although follow-up tests are needed.

Several experiments explore shear Alfvén wave propagation in multi-ion lab plasma. S. T. Vincena et al. (2010) document propagation characteristics; Rahbarnia et al. (2010) study the superposition of R and L modes generated by the addition of a second ion species; and Robertson et al. (2020) explores dispersion as a means for diagnosing the relative density of each species. In a multi-ion plasma, each additional ion species contributes a new cyclotron resonance and cutoff frequency, separating the Alfvén wave into distinct propagating bands. Alfvén waves around the cyclotron resonances, also called electromagnetic ion cyclotron (EMIC) waves (Young et al., 1981), contribute to the dynamics of Earth's radiation belts (Summers & Thorne, 2003). The ion-ion-hybrid frequency in multi-ion plasmas is a resonance for compressional Alfvén waves and a cutoff for shear Alfvén waves (Buchsbaum, 1960; Smith & Brice, 1964). As the fusion community moves toward regular use of deuterium-tritium plasmas, the propagation of Alfvén waves in multi-ion plasmas will become increasingly relevant to energy research, with initial tokamak experiments by Borg and Cross (1987) and Intrator et al. (1996). Alfvén eigenmodes of tokamaks are known to enhance transport of energetic particles (Heidbrink et al., 1993), and the effects of multiple ion species on torodial Alfvén eigenmodes is an area for additional study. The effect of an ion-ion resonator in ITER is considered by Farmer and Morales (2014), where high ion temperatures are found to complicate matters by enabling mode coupling to the fast wave, and it is predicted that energetic alpha particles can amplify resonant ion-ion Alfvén wave modes.

In a two-ion plasma, propagating Alfvén waves are found in two distinct frequency bands. The upper band is bound at high frequency by the cyclotron resonance of the lighter ion and at low frequency by the ion-ion-hybrid cutoff. Because the ion-ion-hybrid frequency increases with $B_0$, it is possible to have Alfvén waves in the upper band propagate from lower $B_0$ toward higher $B_0$ until they match the ion-ion-hybrid frequency and reflect. A magnetic well would have two such reflection points, resulting in a resonant cavity for ion-ion Alfvén waves. This effect is relevant to Earth's magnetosphere by allowing shear Alfvén waves to become trapped between reflection points (Guglielmi et al., 2000). Alfvén wave properties consistent with multi-ion effects and reflections from the ion-ion-hybrid cutoff have been detected in GEOS and S3-3 satellite data (Perraut et al., 1984; Temerin & Lysak, 1984). The ion-ion Alfvén wave resonator in the magnetosphere has been studied via ray tracing as well (Rauch & Roux, 1982).

S. T. Vincena et al. (2011, 2013) report experiments and theory for an ion-ion Alfvén wave resonantor using a magnetic well in the LAPD. Figure 12.6(a,b), taken from S. T. Vincena et al. (2013), shows the lower and upper Alfvén wave bands in the magnetic well for two different ratios $\eta$ of the hydrogen to helium density. Green and red portions of the upper band are expected to be trapped by reflections. Color-coded transparent bars on (c) and (d), corresponding to (a) and (b), respectively, indicate the range of trapped frequencies. Peaks in the normalized wave field amplitudes, indicated with arrows in (c) and (d), show resonant modes within the trapping region and are found to correspond to predicted resonant frequencies. The resonant eigenmode frequencies shift with the ratio of hydrogen to helium ions $\eta$, supporting the interpretation that reflection is produced by the ion-ion-hybrid frequency and not other inhomogeneities associated with the $\mathbf{B}_0$ mirror profile.

Neutral particles influence the behavior of Alfvén waves in many partially-ionized space and astrophysical plasma environments ranging from the solar photosphere and chromosphere, the coupling of Earth's magnetosphere and ionosphere, and interstellar clouds (De Pontieu et al., 2001; Goyal & Sharma, 2018; Haerendel, 1992; Leake et al., 2005; Sciffer & Waters, 2002). The interaction of ions participating in Alfvén waves with surrounding neutral particles can lead to energy transfer, wave damping, coupled motions of the ion and neutral fluids, and heating. Such interactions are relevant to the dynamics of chromospheric heating (De Pontieu et al., 2001), spicules (de Pontieu & Haerendel, 1998; De Pontieu et al., 2007; Haerendel, 1992), and the dissipation of Alfvén wave turbulence (Krishan & Gangadhara, 2008). The effect of neutral particles on Alfvén wave propagation, energy flow, cutoff, and resonance has been considered by several authors (Carbonell et al., 2010; Li et al., 2014; Vranjes et al., 2008; Vranjes & Kono, 2014; Woods, 1962). One effect of ion-neutral collisions described by Woods (1962) is a complex effective ion mass that varies with the collision rate. As might be expected, when the collision rate is much larger than the wave frequency, ions and neutrals are strongly coupled in their motion. At higher wave frequencies, there starts to be differential motion between ions and neutrals, or "slip." This frequency-dependent complex effective ion mass alters the Alfvén wave dispersion relation, changing the wavelength and introducing collisional damping.

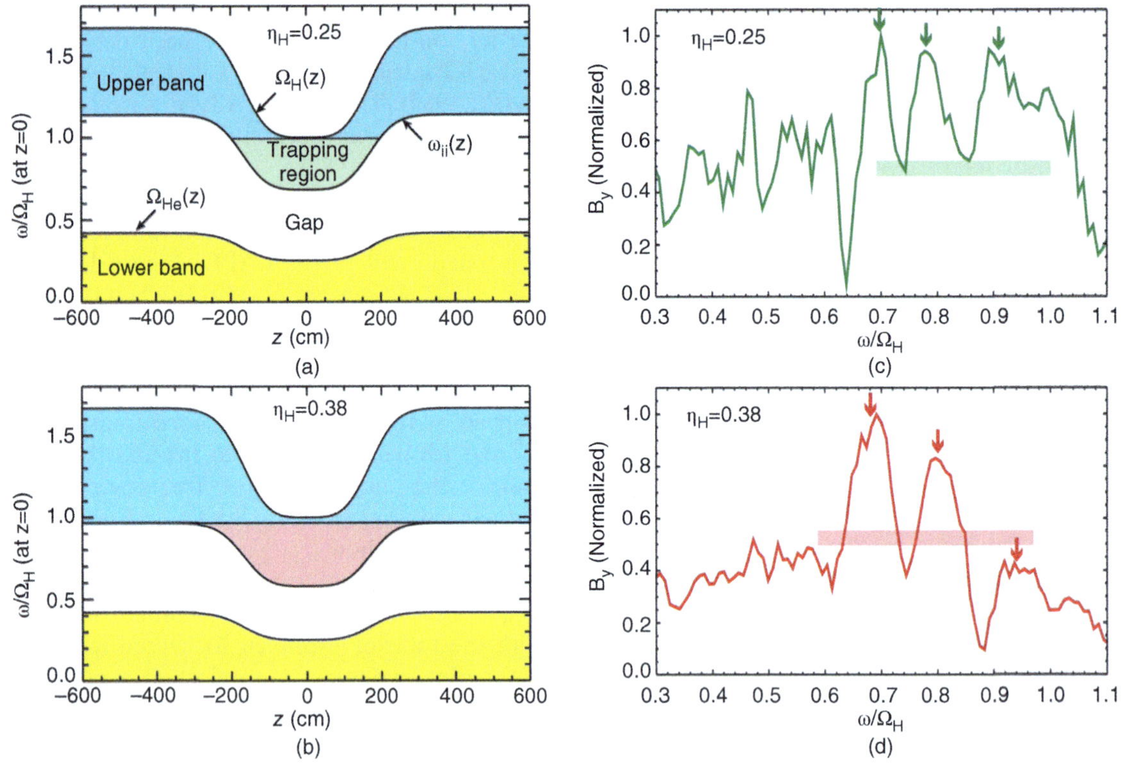


**Figure 12.6** An ion-ion Alfvén wave resonator in the LAPD. (a,b) The upper and lower bands in the magnetic well, with trapped frequencies in green and red for hydrogen to helium ion ratios $\eta = 0.25$ and $\eta = 0.38$, respectively. (c) Resonant modes measured within the trapped frequency range of (a). (d) The resonant modes corresponding to (b). Source: Vincena et al. (2013) / American Institute of Physics.

An early test of Alfvén wave behavior in partially-ionized plasma comes from Mueller (1974). Plasma generated by collisional ionization from current driven through a neutral fill gas produced ion densities of $10^{15}$ $cm^{-3}$ in a 5 cm diameter, 1 m long cylindrical device. Experiments scan the Alfvén wave frequency and use magnetic pickup coils at several axial locations to determine the complex wave number. Neutral density is not measured directly but rather adjusted in the model to achieve the best fit between measured and predicted dispersion. This process produces a good agreement between measured and predicted wave numbers as a function of frequency with the resulting neutral density similar to the ion density. A series of follow-up experiments have been conducted in plasma generated by helicon sources. It is known that helicon sources have several modes of operation that lead to different levels of ionization. The Auburn Linear Experiment for Space Plasma Investigations (ALESPI), shown in Figure 12.2(c), produced results consistent with fully ionized plasma (Hanna & Watts, 2001). Results from follow-up ALESPI experiments are shown in Figure 12.7. The left panel of Figure 12.7 shows results from the high-ionization mode of the helicon source. Measured Alfvén wavenumber versus frequency, shown with black squares, follows the dispersion curve for a fully ionized plasma. On the right, experiments in low-ionization plasma show that interactions with neutrals modify the Alfvén wave, especially near resonance. The ionization fraction varies with distance from the core of the plasma. Results show good agreement between the measured dispersion and damping when the neutral fraction is adjusted for best fit. Though not measured directly, results suggest an ionization fraction of only a few percent. Measurements and results of Alfvén wave experiments in the Hot hELIcon eXperiment (HELIX) at West Virginia University, shown in Figure 12.2(d), are similar, except neutral density is determined directly by laser absorption spectroscopy (Houshmandyar & Scime, 2011).

With lab results documenting the linear physics of Alfvén waves in homogenious, inhomogenious, multi-ion, and partially-ionized conditions, experiments exploring nonlinear effects can be pursued to generate insight into Alfvén wave instabilities, turbulence, and particle acceleration believed to be at work in many space and astrophysical plasmas. Lab experiments have largely confirmed linear theory, but several important questions remain, including those reviewed above from L. L. Zhao et al. (2023) and Bose et al. (2019).

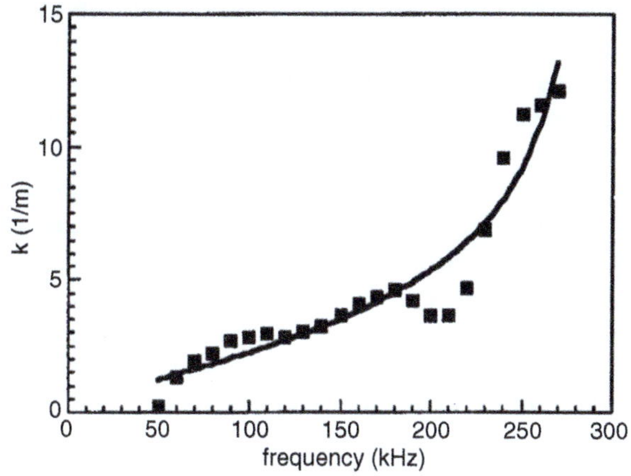


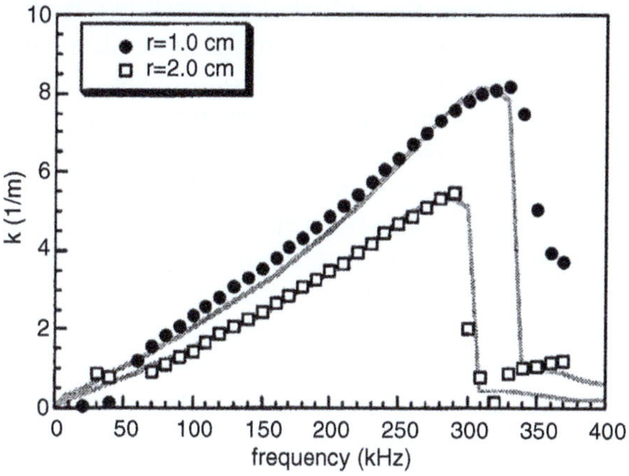


**Figure 12.7** Alfvén wave dispersion in the Auburn Linear Experiment for Space Plasma Investigations (ALESPI). (left) Wavenumber versus frequency in a high-efficiency helicon discharge characterized by a high ionization fraction. Measured dispersion, shown with black squares, follows the prediction for a fully-ionized plasma. (right) Low ionization modifies Alfvén wave dispersion, especially near resonance. Ionization fraction varies with the radial distance from the core. Source: Watts & Hanna (2004) / American Institute of Physics.

## 12.4. NONLINEAR ALFVÉN WAVE PHYSICS

Efforts to understand the origin of wave spectra in space plasmas, e.g., in the solar corona and solar wind, require a study of the nonlinear transfer of Alfvén wave energy from scale to scale and mode to mode. The central process of Alfvénic turbulence is the nonlinear interaction of counterpropagating Alfvén waves that generates a third Alfvén wave with a wavelength smaller than either of the original waves (Boldyrev, 2006; Goldreich & Sridhar, 1995; Kraichnan, 1965). Parametric beat, modulational, and decay instabilities have also been a focus of theory (Derby, 1978; Goldstein, 1978; Hollweg, 1994). Parametric instabilities are considered important in establishing the turbulent fluctuations in the solar corona and solar wind (Del Zanna et al., 2001a; Goldstein, 1978; Hollweg et al., 1993; Inhester, 1990; Wong & Goldstein, 1986). Alfvén wave instabilities could also contribute to the thermodynamics of space and astrophysical plasmas in cases where acoustic modes are dissipated more readily than Alfvén waves (Spangler, 1991).

### 12.4.1. Parametric Instabilities

Parametric instabilities can be studied by considering the unperturbed plasma and a lone large-amplitude "pump" Alfvén wave as the background upon which the roots of the dispersion relation can have positive growth rates. In general, parametric instabilities generate additional Alfvén waves and acoustic modes. Of the parametric instabilities, decay is most commonly cited because it is predicted to produce counterpropagating Alfvén waves needed for Alfvén wave turbulence. Complicating matters, Hollweg (1994) identifies that solar wind conditions are often in a regime where fluid models indicate the decay instability would not arise, although this conclusion is tempered by the acknowledgement that kinetic effects are known to change the character of parametric instabilities. Regardless, the point stands that study is needed to determine the properties of parametric instabilities and their relevance to space and astrophysical plasma. While theory and numerical simulation support the expectation that these processes are widespread and consequential in space plasmas, relatively few observations detect them; notable examples come from Spangler et al. (1997) and Narita et al. (2007). Lab experiments have recently stepped in to determine the properties of parametric instabilities and explore possible signatures relevant to observation.

An early report suggesting nonlinear Alfvén wave physics in laboratory plasma comes from Borg et al. (1993). In this experiment, Alfvén waves are launched by RF antennas in a tokamak plasma, and the resulting spectrum of waves includes several harmonics. While the authors cannot exclude the possibility of harmonics being excited by currents in the sheath, they also note that it is possible that harmonics arise from nonlinear Alfvén wave physics such as wave steepening, a nonlinear effect where large-amplitude waves create significant distortions of the plasma, causing wave troughs and crests to begin traveling at separate speeds and distorting the original waveform (Davidson, 1972).

The first lab experiment to record the spatial structure of nonlinear Alfvén wave interactions relevant to parametric instabilities comes from Carter et al. (2006). Using UCLA's LAPD, co-propagating Alfvén waves were generated using the resonant cavity formed by the cathode and mesh anode. In the LAPD, primary electrons are accelerated from the cathode to a mesh anode that allows primary electrons to pass through and collisionally ionize the neutral fill gas. The cathode-anode segment of the plasma column forms a resonant cavity that, for some discharge settings, spontaneously generates a lower-frequency shear Alfvén wave described by the $m = 0$ Bessel mode and a higher-frequency $m = 1$ mode. Alfvén waves are transmitted through the mesh anode into the main volume of the experiment, where magnetic fluctuations are recorded by a movable probe immersed in the plasma. Because the phase of spontaneously-generated Alfvén waves varies from shot to shot, a stationary probe is used to provide a phase reference. Langmuir probe measurements find oscillations of the ion saturation current $I_{sat}$ at a frequency matching the beat frequency between the two Alfvén modes. Additionally, when $I_{sat}$ oscillations are present, a higher-frequency Alfvén sideband is detected, separated from the $m = 1$ mode by the beat frequency. It is hypothesized that the crests of the spontaneous $m = 0$ and $m = 1$ modes slide past each other with differing phase speeds described by kinetic Alfvén wave dispersion; this generates a nonlinear $I_{sat}$ beat mode, and the beat mode scatters the original Alfvén waves to produce the upper sideband. To test this hypothesis, the plasma discharge is operated in such a way as to spontaneously produce only the $m = 1$ mode, while the $m = 0$ mode is excited by driving an oscillating current between the cathode and anode on top of the primary discharge current. This approach allows control of the frequency separation of the $m = 0$ and $m = 1$ modes and reveals that the $I_{sat}$ and upper Alfvén sidebands follow the frequency separation. It is determined that $I_{sat}$ fluctuations are a nonlinear pseudomode that, in the absence of the driving interaction, would not be a solution to the dispersion relation. Pseudomodes are known to occur between copropagating Alfvén waves during modulational instability (Hollweg, 1994). Spatial mapping of the $I_{sat}$ oscillations gives measured values for the wavenumber and phase speed that meet three-wave matching conditions. While these experiments do not produce a parametric instability, they do generate the three-wave interaction believed to be at the heart of the modulational instability.

Beat-wave experiments demonstrating the three-wave interaction of the decay instability are reported by Dorfman and Carter (2013). In this work, linearly polarized counterpropagating Alfvén waves are generated by antennas on opposite ends of the LAPD. A beat wave, again detected with Langmuir probe measurements of $I_{sat}$, arises at the difference frequency between the two launched Alfvén waves. After shutting off the Alfvén wave antennas, the beat wave persists momentarily after the last Alfvén wave passes the probe location. This observation informs the hypothesis that, unlike Carter et al. (2006), the three-wave interaction in these experiments excites a propagating wave. The magnitude of the beat wave, seen in Figure 12.8(a), is recorded while varying the difference frequency. This process reveals a resonance, interpreted to be a natural mode of the ion acoustic wave in the LAPD. Using the kinetic Alfvén wave dispersion relation, the ion acoustic wave dispersion relation, and three-wave matching conditions, it is predicted that the resonant ion acoustic wave frequency that varies with the Alfvén wave frequency and the ion cyclotron frequency. While the model shown in (a) does not recreate the entire response curve, it does capture the resonant frequency. The prediction guides the design of follow-up experiments that scan a range of Alfvén wave frequencies and background magnetic field values in both hydrogen and helium plasmas. As seen in Figure 12.8(b), the resonant frequency

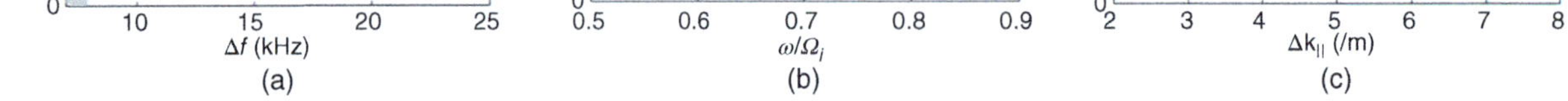


**Figure 12.8** Counterpropagating Alfvén waves in the LAPD generate an ion acoustic wave consistent with the parametic decay instability. (a) The normalized response of the beat wave varies while scanning the frequency separation of the launched Alfvén waves, revealing a resonance. (b) A model for the resonant frequency is consistent with data in hydrogen and helium plasmas. (c) The frequency and wavenumber of the beat wave follows the ion acoustic wave dispersion relation. Source: Dorfman & Carter, 2013 / American Physical Society.

measured in all trials is consistent with predictions. In (c), analysis finds the frequency and wavenumber of the beat wave agree with the ion acoustic wave dispersion relation (Dorfman & Carter, 2015). While the decay instability is the result of a single Alfvén wave that produces an ion acoustic wave and a counterpropagating Alfvén wave, the growth rate of the decay instability in this experiment is comparable to the transit time of Alfvén waves through the LAPD. Therefore, consistent with these experiments, it is not expected that the decay instability itself will be detected, but the three-wave interaction of this instability is still produced by launching counterpropagating Alfvén waves.

Experiments using higher-amplitude Alfvén waves were performed in the LAPD in an effort to achieve growth rates large enough to generate parametric instabilities in the device (Dorfman & Carter, 2016). Using a single antenna, a pump Alfvén wave above a threshold amplitude is seen to generate three child waves, including upper and lower Alfvén sidebands, and a nonresonant mode at the beat frequency of the pump wave with the child sidebands. Figure 12.9(a) shows the spectrum of driven and spontaneous fluctuations below the threshold (blue). Above the threshold (black), sidebands M- and M+ are seen, along with the lower-frequency M1 nonresonant mode. Following predictions, the data in Figure 12.9(b) show that increasing the pump wave amplitude increases the beat frequency, altering the frequency of all three child waves. Aligning with predictions for modulational instability (Hollweg, 1994), Figure 12.9(c) shows all three child waves copropagating with the pump wave. M+ and M- are consistent with Alfvén wave dispersion. M1 is inconsistent with Alfvén and acoustic waves and is identified as a nonresonant wave, a solution of the dispersion relation only when the driving instability is present. However, unlike predictions of modulational instability, the nonresonant wave includes a magnetic signature instead of being exclusively a density perturbation. A composite measurement of the nonresonant child wave spanning the plane perpendicular to $\mathbf{B_0}$, stitched together from many shots using a movable probe, reveals the nonresonant mode peaks along the perpendicular gradient of the pump wave. Meanwhile, theory predicts that the nonresonant mode will be concentrated at the peak of the pump wave. Despite these inconsistencies, the modulational instability, with its prediction that all child waves copropagate with the pump wave, is most consistent with the experimental results. This experiment indicates the theory of parametric Alfvén wave instabilities may be underdeveloped. Not only are the measured child waves partially inconsistent with modulational instability theory, but the modulational instability itself is not predicted to be the dominant instability in these conditions. Results from Wong and Goldstein (1986) and Hollweg (1994) predict the decay instability growth rate will be three orders of magnitude higher than the modulational instability. Dorfman and Carter (2016) suggest collisions or the perpendicular structure of the pump wave, both present in the experiment but neglected in theory, may be the source of discrepancies.

Alfvén waves launched with a solenoid immersed in the Keda Mirror with AXisymmetricity (KMAX), a linear tandem mirror machine at the University of Science and Technology of China, spontaneously generate forward and backward propagating waves (Shi, Yang, et al., 2019). Data reported come from a suite of three-axis magnetic probes and Langmuir probes. The parent and two child waves satisfy frequency and wavenumber matching conditions. The launched wave is found to have magnetic but

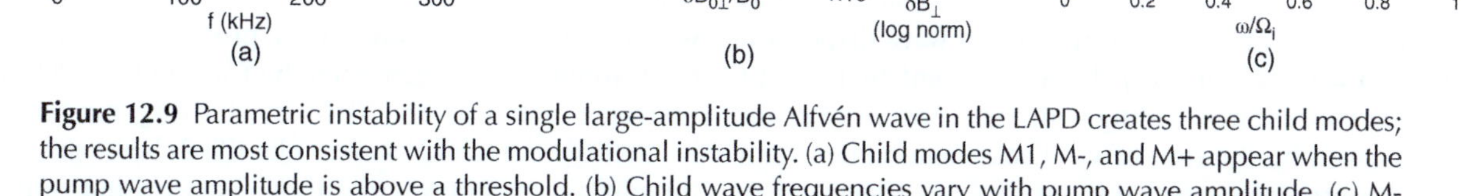


**Figure 12.9** Parametric instability of a single large-amplitude Alfvén wave in the LAPD creates three child modes; the results are most consistent with the modulational instability. (a) Child modes M1, M-, and M+ appear when the pump wave amplitude is above a threshold. (b) Child wave frequencies vary with pump wave amplitude. (c) M- and M+ are Alfvén waves, while M1 is a nonresonant mode that meets three-wave matching conditions. Source: S. Dorfman & Carter (2016) / American Institute of Physics.

not density fluctuations, be, linearly polarized, and agree with the Alfvén wave dispersion relation of Hall MHD, which includes the finite frequency of the wave compared to the ion cyclotron frequency. The spontaneously generated waves contain both magnetic and density fluctuations that are in phase for the copropagating child, as expected with the fast magnetosonic wave. Meanwhile, magnetic and density fluctuations are $\approx 180^o$ out of phase for the counterpropagating child wave, consistent with the slow magnetosonic wave. Interpretation is complicated by the order of magnitude variation in density over the length of the central chamber as well as the launched wavelength being longer than the central chamber. However, variations in the background plasma are relevant to the physics of Alfvén waves space plasma, and Alfvén wave instabilities in inhomogeneous plasma have not been thoroughly explored in the lab. The exploration of parametric instabilities in inhomogeneous plasma is an area for continued study where laboratory experiments can bridge theory and space.

### 12.4.2. Alfvén Wave Collisions and Turbulence

Turbulent fluctuations of Alfvén waves are believed to play consequential roles in the heating of the solar corona and acceleration of the solar wind (McIntosh et al., 2011), as well as the dynamics of geospace (Sundkvist et al., 2005; Zimbardo et al., 2010). As seen with the Elsässer equation, ideal MHD predicts the nonlinear interactions between Alfvén waves required for a turbulent cascade to occur between counterpropagating waves (Elsässer, 1950; Kraichnan, 1965). However, the nature of Alfvén wave interactions, particularly in the case of strong turbulence, is unresolved (Boldyrev, 2006; Goldreich & Sridhar, 1995). Because spacecraft data are not well-suited for studying the interaction of individual Alfvén wave modes, it is necessary for lab experiments to test and verify theory.

A series of papers applies the theory of Alfvén wave collisions to the design of LAPD experiments (D. J. Drake et al., 2013; Howes & Nielson, 2013; Howes, Nielson, Drake, et al., 2013; Howes, Nielson, Schroeder, et al., 2013). Alfvén wave interactions contributing to a turbulent cascade must generate propagating child waves. It was believed the three-wave interaction would be unable to achieve this result due to three requirements: (1) the three-wave matching conditions $\omega_1 + \omega_2 = \omega_3$ and $\mathbf{k_1} + \mathbf{k_2} = \mathbf{k_3}$; (2) the ideal MHD Alfvén wave dispersion relation $\omega = |k_\parallel| v_A$; and (3) the requirement that parent waves are counterpropagating with the direction of propagation given by the sign of $k_\parallel$. These requirements can only be nontrivially satisfied by either $k_{\parallel 1} = 0$ or $k_{\parallel 2} = 0$. Researchers initially interpreted these requirements to mean one of the parent waves is nonpropagating, the three-wave interaction is zero, and therefore the lowest-order nonlinearity in which parent Alfvén waves generate a propagating child wave is the four-wave interaction (Sridhar & Goldreich, 1994). However, there is no reason to expect Alfvén wave packets in space plasma to have the symmetry required for the $k_\parallel = 0$ mode to be absent. For example, Alfvén waves propagating through spreading or wandering magnetic field lines, common in space plasmas, would include an effective $k_\parallel = 0$ component. This allows three-wave interactions to be the lowest-order nonlinearity of weak plasma turbulence (Galtier et al., 2000; Howes & Nielson, 2013; Ng & Bhattacharjee, 1996).

To produce a three-wave interaction in the laboratory that results in a propagating child wave, counteprop-agating Alfvén waves are launched from either end of the LAPD (D. J. Drake et al., 2013; Howes, Nielson, Schroeder, et al., 2013). One of the parent waves was driven at a low frequency to produce a wavelength longer than the LAPD itself. Similar to the $k_\parallel = 0$ mode present in Alfvén waves propagating through a wandering magnetic field, this low-frequency wave contains an effective $k_\parallel = 0$ component in the Fourier transform over the length of the experiment. Magnetic fluctuations were recorded across the plane perpendicular to $\mathbf{B_0}$ by combining multiple shots from movable probes. Measurements of each parent wave driven individually and together allow the subtraction of data sets, revealing a propagating child wave that satisfies the three requirements enumerated above (Howes et al., 2012). The child wave, seen in Figure 12.10(d), has a perpendicular wave vector $(k_x, k_y)$ that is the sum of the two parent waves, Figure 12.10(b,c). The frequency of the child wave matches the parent wave without the $k_\parallel = 0$ component. These characteristics are consistent with expectations that Alfvén wave turbulence transfers energy preferentially to smaller scales perpendicular to the background magnetic field (Goldreich & Sridhar, 1995). These results verify that the three-wave interaction is the lowest-order nonlinearity that underpins the turbulence of ideal MHD Alfvén waves.

Non-propagating child waves produced by Alfvén wave collisions were studied in follow-up experiments by Drake et al. (2016). Data were collected during LAPD experiments using counterpropagating Alfvén waves with the same antennas and diagnostics as Howes et al. (2012). Analysis focuses on the $k_\parallel \neq 0$ components of the parent waves and the child waves they produce. Similar to the nonresonant modes generated by Carter et al. (2006) and Dorfman and Carter (2016) in studying modulational instability, the child waves identified here satisfy the three-wave matching conditions but are not solutions of the dispersion relation in the absence of the driving interaction of parent waves. Measurements compare

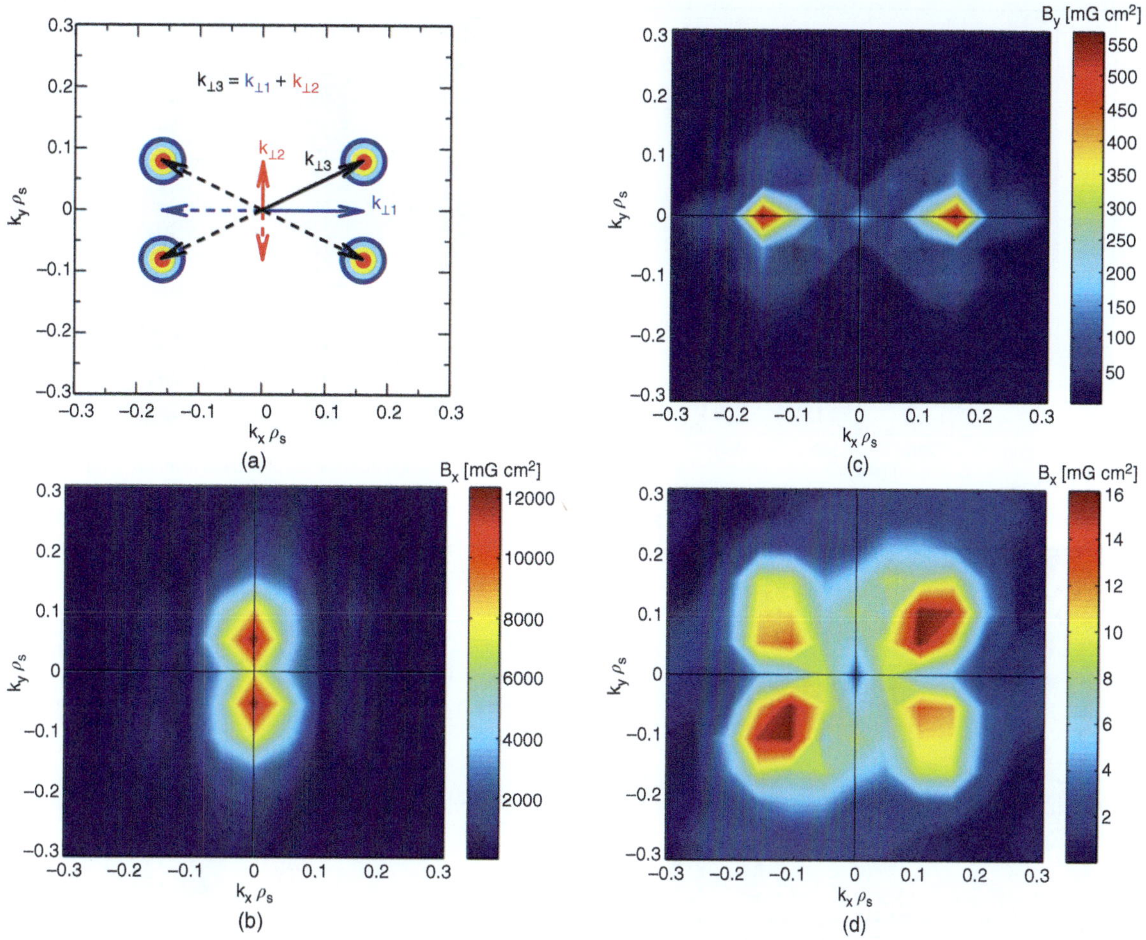


**Figure 12.10** Predicted and measured $\mathbf{k}_\perp$ of the lowest-order nonlinear interaction responsible for ideal MHD Alfvén wave turbulence. (a) Every possible combination of the parent wave perpendicular wave vectors $\mathbf{k}_{\perp 1}$ and $\mathbf{k}_{\perp 2}$ is expected, resulting in four contributions to the child wave vector $\mathbf{k}_{\perp 3}$. (b) and (c) show each individual parent wave after Fourier transform in the *x–y* plane. (d) The child wave. Source: Howes et al. 2012 / American Institute of Physics.

favorably with the three expectations from theory. First, as seen in Figure 12.11(a), child waves are identified at frequencies matching the sum and difference of the parent wave fundamental frequencies. Second, consistent with the nonlinear term of the Elsässer equation (Elsässer, 1950), beat waves in the experiment (d) are spatially concentrated where the extrema of one parent wave (c) overlaps with the gradient of the other (b). Third, measured child wave amplitudes scale with the product of the parent amplitudes, as expected. Because non-propagating child waves vanish as soon as the parent waves are turned off, they are not the recipients of any net energy transfer. However, they can participate in subsequent three-wave interactions to produce additional child waves that may be propagating. The combined effect of two three-wave interactions is equivalent to a four-wave interaction (Howes, Nielson, Schroeder, et al., 2013).

### 12.4.3. Alfvén Wave-Particle Interactions

While plasma waves can transport energy over large distances, wave-particle interactions contribute to the local energetics and thermodynamics of space and astrophysical plasmas. The interactions of Alfvén waves with ions and electrons remove energy from turbulence and cause collisionless damping. The result is the heating and acceleration of particles in the solar corona and wind (Velli, 2003), planetary magnetospheres (C. Chaston et al., 2008; Saur, 2004), the interstellar medium (Minter & Spangler, 1997), and accretion disks (Quataert, 1998). In these systems, the transfer of energy from waves to particles occurs at length scales shorter than the collisional mean free path (Howes, 2010), implicating wave-particle interactions that require a kinetic description. Depending on conditions, ions or electrons can be preferentially energized by interactions with Alfvén waves,

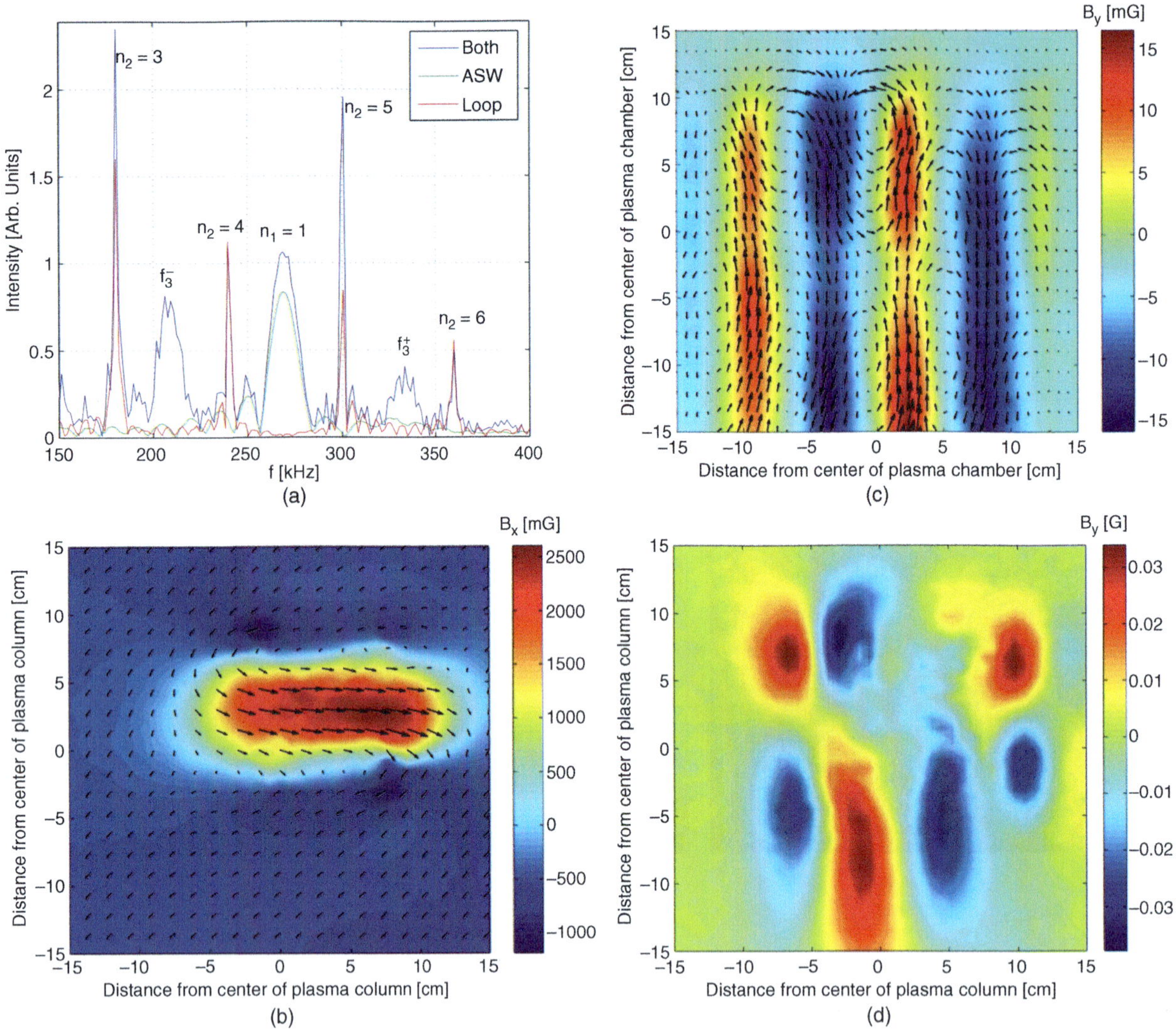


**Figure 12.11** Nonresonant beat wave interactions between counterpropagating Alfvén waves. (a) Frequency spectrum of magnetic fluctuations present when two antennas (ASW, Loop) operate individually (green, red) and together (blue). The fundamental modes $n_1 = 1$ and $n_2 = 1$ (not shown) interact to produce beat waves with frequencies $f_3^-$ and $f_3^+$. Parent waves are launched by the loop antenna (b) and ASW antenna (c). (d) Beat wave detected using a frequency bandpass filter around $f_3^-$. Source: D. J. Drake et al. (2016) / American Institute of Physics.

including Landau damping and transit-time damping, a wave-particle interaction similar to Landau damping where the magnetic moment of particles interacts with the wave magnetic field (Quataert, 1998). Despite being "one of the critical unsolved problems in the study of space and astrophysical turbulence" (NASEM, 2021), few lab experiments have probed the dissipation of turbulence and Alfvén wave-particle interactions. The advancement of diagnostics providing access to velocity space in lab plasmas is starting to change this.

One area where measurements of the electron velocity distribution in lab plasma have provided insight is the interaction of Alfvén waves with electrons in conditions relevant to the auroral magnetosphere. Dispersive Alfvén waves have long been recognized as a candidate for accelerating auroral electrons via Landau resonance (Goertz & Boswell, 1979; Hasegawa, 1976; Kletzing, 1994). Test particle simulations have supported this hypothesis, showing Landau resonant interactions mediated by the parallel electric field of inertial Alfvén waves (C. C. Chaston, 2006; Kletzing & Hu, 2001). Polar orbiting satellites have provided statistical understanding of the common correspondence between auroras and Alfvén waves descending Earthward on conjugate magnetic field lines. More than one author has concluded that Alfvén waves provide energy input to the auroral magnetosphere sufficient for about ~ 1/3 of all auroras (C. C. Chaston et al., 2003; Keiling et al., 2003), although some recent estimates are lower (Keiling et al., 2020). Notably, ascending Alfvén waves are less common, suggesting Alfvén wave energy is absorbed in the auroral magnetosphere (C. C. Chaston et al., 2007). At some locations and times, this fraction has been documented to be significantly higher (C. C. Chaston et al., 2007). An analysis of data by geomagnetic activity reveals Alfvén waves are especially prevalent above auroras during geomagnetic storms, when auroras are commonly bright and active (Keiling et al., 2019, 2020).

While theory and satellite data point to Alfvén waves as likely contributors to the generation of auroras, the wave-particle interaction believed to be at work went untested for decades. This has been attributed to the difficulty of isolating the interaction of individual wave modes with electrons in satellite data, while at the same time, lab measurements were unable to provide the precision and resolution needed to detect changes to the electron distribution caused by Alfvén waves (Howes, 2018). A diagnostic was developed for use on the LAPD that leverages the resonant absorption of a small-amplitude whistler-mode wave to determine the suprathermal parallel electron velocity distribution (Thuecks et al., 2012; Schroeder et al., 2017). This diagnostic technique, called wave absorption, has been previously deployed on MIT's Versator II to explore the role of suprathermal electrons, electrons above the thermal speed $v_{te}$, in carrying plasma current (Skiff et al., 1993, 1994). Earlier wave absorption measurements were reported by Arunasalam et al. (1968) and Kirkwood et al. (1990). The use of wave absorption to measure the background distribution of electrons in the LAPD is reported by Thuecks et al. (2012), with results giving values of electron density $n_e$ and temperature $T_e$ consistent with separate Langmuir probe measurements.

Resolution of the electron velocity distribution's interaction with Alfvén waves in the LAPD is gained using multiple phase-shifted data sets, where the Alfvén wave trigger is shifted by fractions of an Alfvén wave period with respect to the timing of the wave absorption diagnostic (Schroeder et al., 2016). Inertial Alfvén waves are launched from one end of the LAPD, while the wave absorption diagnostic is used to simultaneously record the parallel suprathermal electron distribution (Schroeder et al., 2021). Figure 12.12 shows results from this approach with the background distribution in (a); (b) shows measured perturbations to the electron distribution resolved in Alfvén wave phase $\phi_{Ez}$ referenced to the wave's parallel electric field $E_z$; and (c) gives modeled linear perturbations to the electron distribution that compare favorably with measurements. Wave absorption is known to access only the suprathermal tails of the parallel electron distribution, while the bulk of the distribution remains opaque to this diagnostic technique (Thuecks et al., 2012). This limitation is acceptable for these experiments because the phase velocity of inertial Alfvén waves falls in the tail of the electron distribution by definition.

Measurements of the perturbed electron distribution are analyzed with the field-particle correlation technique to reveal net changes to the energy content in different regions of phase space (Klein & Howes, 2016). Key to interpreting these results is that the energy content of the distribution at a given velocity can only increase (decrease) if the number of particles at that velocity increases (decreases). Figure 12.13(a) shows the results of the field-particle correlation applied to experimental data. Values consistent with zero are interpreted to be velocities where the electron distribution is not gaining or losing energy, while negative (positive) values indicate velocities where the electron distribution is losing (gaining) net energy. The zero crossing from negative to positive around 18 eV indicates a loss of electrons at low velocity, causing the distribution below the zero crossing

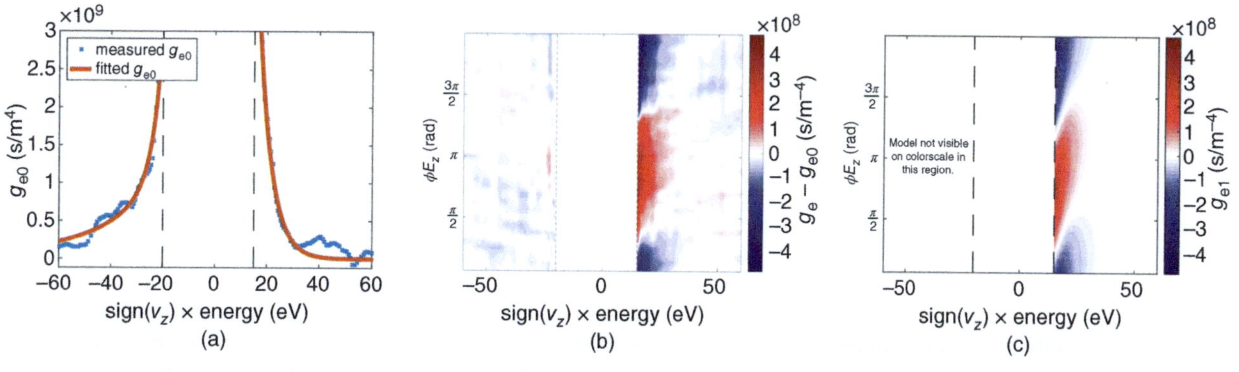


**Figure 12.12** The parallel suprathermal electron velocity distribution measured by wave absorption. (a) The background distribution. (b) Measured perturbations to the background distribution are coherent with the phase of the launched inertial Alfvén wave. Phase $\phi_{Ez}$ is referenced to $E_z$, the parallel electric of Alfvén waves in the experiment. (c) Modeled linear perturbations are consistent with measurements. Source: Schroeder et al. (2021) / Springer Nature / CC BY 4.0.

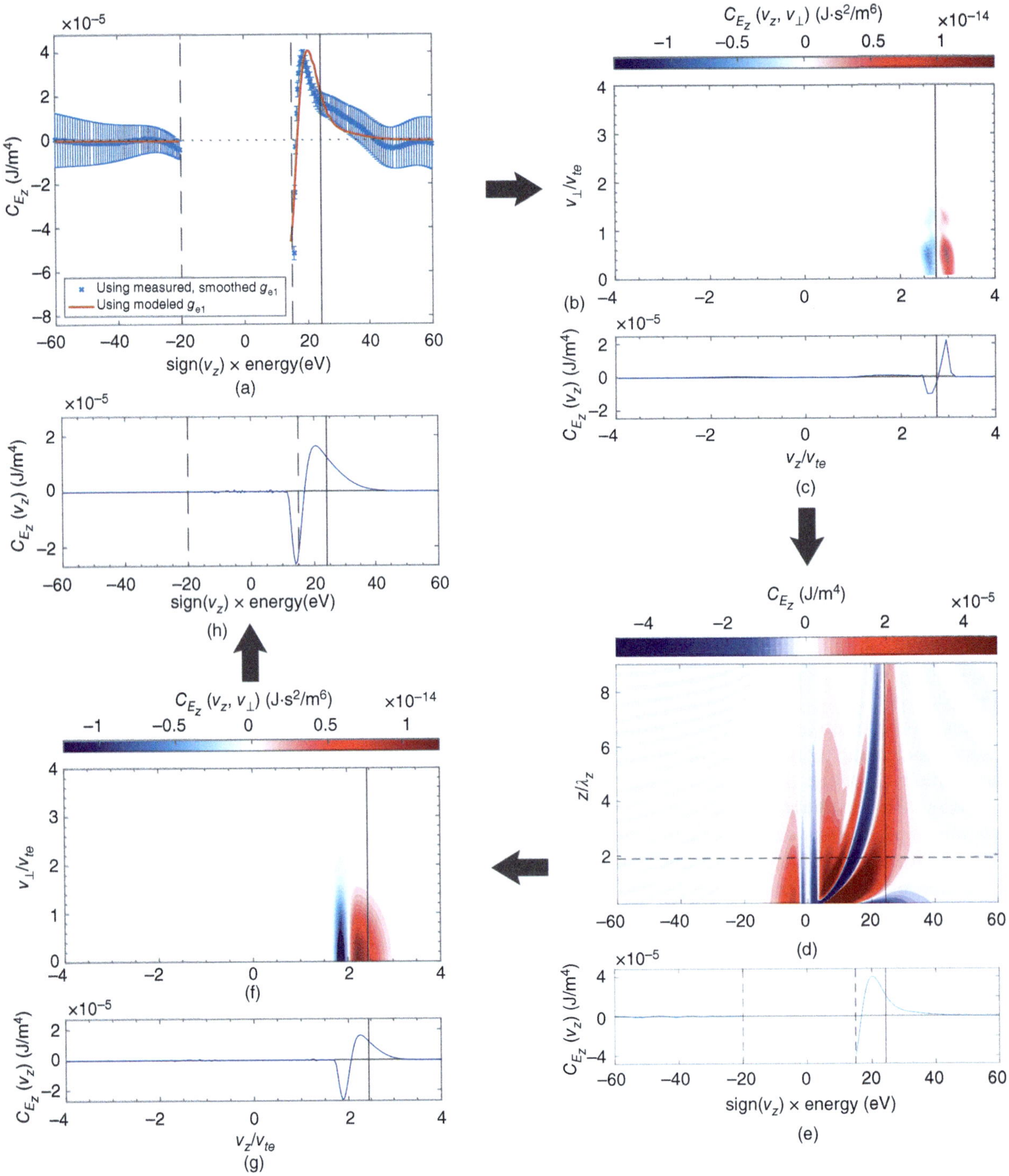


**Figure 12.13** Application of the field-particle correlation technique to Alfvénic perturbations of the electron distribution. Black arrows are included to guide comparison. (a) Unlike traditional expectations for Landau resonance, experimental results produce an acceleration signature that crosses from negative to positive below the phase velocity (solid vertical line). (b) and (c) Gyrokinetic results show the Landau resonance acceleration signature centered on the phase velocity. (d) Analytical theory incorporates the finite length of the experiment and reveals that a short interaction length, measured by $z/\lambda_z$, causes the acceleration signature to cross zero below the phase velocity. (e) A slice of (d) at $z/\lambda_z \approx 2$, corresponding to the interactions length of the experiment. (f) and (g) Results from a separate test particle simulation that matches the length of the experiment show the same qualitative features as (e). (h) Test particle calculation results are converted to the coordinates of experimental data for direct comparison with (a). The model curve in (a) is taken directly from (e). Source: Figure from Schroeder et al. (2021).

to lose net energy, followed by a gain of electrons at higher velocity causing the distribution above the zero crossing to gain net energy. In other words, the field-particle correlation is showing the promotion of electrons to higher velocities by Alfvén waves in the experiment. The vertical black line in (a) shows the Alfvén phase speed. Common interpretations of the Landau resonance would lead to expectations that acceleration would be centered on the phase velocity, while data show an acceleration signature below the phase velocity. Gyrokinetic simulation with a periodic domain in (b) and (c) show the Landau resonant acceleration of electrons centered on the phase velocity as expected. However, (d) shows an analytical Laplace-Fourier solution of the Boltzmann equation that incorporates the finite length of the experiment. This model indicates that an acceleration signature centered on the phase speed arises after many wavelengths of wave-particle interaction $z/\lambda_z \gg 1$, where $z$ is distance from the antenna and $\lambda_z$ is the parallel Alfvén wavelength. For $z/\lambda_z \approx 2$ in the experiment, the acceleration signature in this model is shifted to lower velocity, interpreted physically to mean electrons starting near the antenna have not yet reached their maximum velocity. A cut of the Laplace–Fourier model at $z/\lambda_z \approx 2$ is shown in (e) and also plotted in red in (a). (f) and (g) give the results of a separate test particle calculation where electrons are mapped through two wavelengths of inertial Alfvén wave fields, which produces qualitative features similar to the experimental results. The comparison to experimental data is completed in (h) by converting test particle calculation results to the same coordinates as (a). This final comparison supports the conclusion that inertial Alfvén waves accelerate the electrons that cause auroras.

## 12.5. DISCUSSION AND CONCLUSION

Central goals of heliophysics include (1) characterizing the interactions between various regions of the heliosphere, like the Sun, solar wind, and magnetospheres, and (2) developing predictive models that will allow more accurate anticipation of space weather effects on our increasingly technological society (Howes, 2018). As Alfvén waves are found throughout the heliosphere carrying significant amounts of energy (Belcher & Davis, 1971; Gosling et al., 2009; Hahn & Savin, 2014; Keiling et al., 2003), accomplishing the goals of heliophysics requires understanding the processes that govern how Alfvén waves transport energy and when Alfvén waves deposit energy. In situ spacecraft measurements of plasmas throughout the heliosphere are indispensable, setting boundaries on possible explanations. However, the nature of spacecraft measurements at a single point, or even at several points, means a unique understanding of the underlying physics is not always guaranteed. This is where lab experiments have made—and offer opportunities to continue making—unique and needed contributions to heliophysics. The ability to tune background plasma conditions, precisely vary the perturbations that generate waves, and capture measurements separating spatial and temporal evolution mean lab investigations of Alfvén wave physics enable comparisons to theory at a level of detail that is difficult to achieve elsewhere.

The dispersion of Alfvén waves in MHD as well as inertial and kinetic limits has been thoroughly verified by lab experiments (Gekelman et al., 1994; Kletzing et al., 2003, 2010; Nielson et al., 2010; Thuecks et al., 2009). In addition, modifications to Alfvén wave propagation due to collisions, collisionless damping, and ion-neutral interactions have been studied in the lab (Hanna & Watts, 2001; Houshmandyar & Scime, 2011; Nielson et al., 2010; Thuecks et al., 2009; S. T. Vincena et al., 2010; Watts & Hanna, 2004). Confirmation of these essential foundations of Alfvén wave physics aids in the identification of wave modes in space plasma, a nontrivial task. In some studies, collisions and neutrals are an obstacle to comparisons with collisionless space plasmas. In others, collisions and neutrals in the lab provide access to space-relevant processes. For example, coupling of the magnetosphere, ionosphere, and thermosphere is an emphasis of three missions proposed by the National Research Council (NRC, 2013), and Alfvén waves and neutrals are critical contributors to the flow of energy between these regions.

Essential processes that are a focus of ongoing study in heliopheric plasma include turbulence, instabilities, and particle acceleration. Alfvén waves are implicated in notable examples of these processes, from the solar chromosphere down to Earth's ionosphere. The participation of Alfvén waves in these processes is likely key to answering long-standing questions like the heating of the solar corona, the acceleration of the solar wind and origin of its tangled magnetic character, the interaction of the solar wind with magnetospheres through magnetic reconnection, and the energization of auroral particles. The lab experiments presented here have advanced understanding of all of these essential processes as they relate to Alfvén waves and are positioned to make many additional advances in the near future. In line with this, the last decade has seen a shift toward experiments increasingly focused on the nonlinearities of Alfvén waves contributing to turbulence, instabilities, and particle acceleration. There are examples where the contribution of lab experiments has been to verify theories, while others have advanced the discussion between competing theories. Still, other experiments reviewed here have shown discrepancies with theory, including the growth rates of parametric instabilities and the properties of nonlinearly generated child waves. It is likely that such discrepancies will continue to be uncovered as studies of nonlinear

Alfvén wave physics in the lab progress. This makes the lab fertile ground for collaboration between experimentalists, theorists, and modelers; surely an intermediate step on the path to reliable models of the heliosphere must be the ability to accurately model nonlinear Alfvén wave physics in the lab.

Significant opportunity exists for lab work to continue enabling essential tests of Alfvén waves in space plasma theory and simulation. Opportunities include continued progress in studying the parametric instabilities of Alfvén waves and Alfvén wave-particle interactions. Diagnostic advances that provide access to velocity space open up a new set of experiments that study resonant and non-resonant wave-particle interactions and the dissipation of turbulence. Hopefully, this era of studying the nonlinearities of Alfvén wave turbulence, instabilities, and particle acceleration in the lab will also be an era of increased collaboration between lab experimentalists and space plasma theorists and modelers.

## ACKNOWLEDGMENTS

In memory of Professors Craig A. Kletzing and Donald A. Gurnett. Their love for space plasma physics and passion for mentoring proved formative for many, including the author. Craig and I agreed to write this chapter together, but I ended up having to write it alone. His absence is palpable, not only in this chapter but also in our community.

This material is based upon work supported by the National Science Foundation under NSF CAREER Award AGS-2238191.

## REFERENCES

Alfvén, H. (1942, October). Existence of electromagnetic-hydrodynamic waves. *Nature*, *150*, 405–406. doi: 10.1038/150405d0

Alfvén, H. (1945, January). Magneto-hydrodynamic waves and sunspots. I, II. *Monthly Notices of the Royal Astronomical Society*, *105*, 382. doi: 10.1093/mnras/105.6.382

Alfvén, H. (1947, January). Magneto hydrodynamic waves, and the heating of the solar corona. *Monthly Notices of the Royal Astronomical Society*, *107*, 211. doi: 10.1093/mnras/107.2.211

Amagishi, Y., Saeki, K., & Donnelly, I. J. (1989). Excitation of mhd surface waves propagating with shear alfven waves in an inhomogeneous cylindrical plasma. *Plasma Physics and Controlled Fusion*, *31*(5), 675. Retrieved from http://stacks.iop.org/0741-3335/31/i=5/a=001

Andersson, L., Ergun, R. E., Newman, D. L., McFadden, J. P., Carlson, C. W., & Su, Y. J. (2002, August). Characteristics of parallel electric fields in the downward current region of the aurora. *Physics of Plasmas*, *9*(8), 3600–3609. doi: 10.1063/1.1490134

Arunasalam, V., Meservey, E. B., Gurnee, M. N., & Davidson, R. C. (1968, May). Absorption in a hot plasma near the second harmonic of the electron cyclotron frequency. *Physics of Fluids*, *11*(5), 1076–1084. doi: 10.1063/1.1692046

Bale, S. D., Badman, S. T., Bonnell, J. W., Bowen, T. A., Burgess, D., Case, A. W., et al. (2019, December). Highly structured slow solar wind emerging from an equatorial coronal hole. *Nature*, *576*(7786), 237–242. doi: 10.1038/s41586-019-1818-7

Belcher, J. W., & Davis, L., Jr. (1971, January). Large-amplitude Alfvén waves in the interplanetary medium, 2. *Journal of Geophysical Research*, *76*(16), 3534. doi: 10.1029/JA076i016p03534

Belcher, J. W., Davis, L., Jr., & Smith, E. J. (1969, January). Large-amplitude Alfvén waves in the interplanetary medium: Mariner 5. *Journal of Geophysical Research*, *74*(9), 2302. doi: 10.1029/JA074i009p02302

Bemporad, A., & Abbo, L. (2012, June). Spectroscopic signature of Alfvén waves damping in a polar coronal hole up to 0.4 solar radii. *Astrophysical Journal*, *751*(2), 110. doi: 10.1088/0004-637X/751/2/110

Boehm, M. H., Carlson, C. W., McFadden, J. P., Clemmons, J. H., & Mozer, F. S. (1990, August). High-resolution sounding rocket observations of large-amplitude Alfven waves. *Journal of Geophysical Research*, *95*, 12157–12171. doi: 10.1029/JA095iA08p12157

Boldyrev, S. (2006, March). Spectrum of magnetohydrodynamic turbulence. *Physical Review Letters*, *96*(11), 115002. doi: 10.1103/PhysRevLett.96.115002

Borg, G. G., Brennan, M. H., Cross, R. C., Giannone, L., & Donnelly, I. J. (1985, October). Guided propagation of Alfven waves in a toroidal plasma. *Plasma Physics and Controlled Fusion*, *27*, 1125–1149. doi: 10.1088/0741-3335/27/10/004

Borg, G. G., & Cross, R. C. (1987, June). Guided propagation of Alfven and ion-ion hybrid waves in a plasma with two ion species. *Plasma Physics and Controlled Fusion*, *29*(6), 681–696. doi: 10.1088/0741-3335/29/6/001

Borg, G. G., Lister, J. B., Piazza, D., & Martin, Y. (1993, June). An experimental study of the harmonics generated during Alfven wave heating at TCA. *Nuclear Fusion*, *33*(6), 841–848. doi: 10.1088/0029-5515/33/6/I01

Bose, S., Carter, T., Hahn, M., Tripathi, S., Vincena, S., & Savin, D. W. (2019, September). Measured reduction in Alfvén wave energy propagating through longitudinal gradients scaled to match solar coronal holes. *Astrophysical Journal*, *882*(2), 183. doi: 10.3847/1538-4357/ab2fe0

Bostick, W. H., & Levine, M. A. (1952, August). Experimental demonstration in the laboratory of the existence of magneto-hydrodynamic waves in ionized Helium. *Physical Review*, *87*, 671–671. doi: 10.1103/PhysRev.87.671

Buchsbaum, S. J. (1960, May). Resonance in a plasma with two ion species. *Physics of Fluids*, *3*(3), 418–420. doi: 10.1063/1.1706052

Carbonell, M., Forteza, P., Oliver, R., & Ballester, J. L. (2010, June). The spatial damping of magnetohydrodynamic waves in a flowing partially ionised prominence plasma. *Astronomy & Astrophysics*, *515*, A80. doi: 10.1051/0004-6361/200913024

Carter, T. A., Brugman, B., Pribyl, P., & Lybarger, W. (2006, April). Laboratory observation of a nonlinear interaction between shear Alfvén waves. *Physical Review Letters*, *96*(15), 155001–+. doi: 10.1103/PhysRevLett.96.155001

Chaston, C., Bonnell, J., McFadden, J. P., Carlson, C. W., Cully, C., Le Contel, O., et al. (2008, May). Turbulent heating and cross-field transport near the magnetopause from THEMIS. *Geophysical Research Letters*, *35*, 17–+. doi: 10.1029/2008GL033601

Chaston, C. C. (2006). ULF waves and Auroral electrons. In K. Takahashi, P. J. Chi, R. E. Denton, & R. L. Lysak (Eds.), *Magnetospheric ulf waves: Synthesis and new directions* (Vol. 169, p. 239). Washington, DC: American Geophysical Union.

Chaston, C. C., Bonnell, J. W., Bale, S. D., Kasper, J. C., Pulupa, M., Dudok de Wit, T., et al. (2020, February). MHD mode composition in the inner heliosphere from the Parker solar probe's first perihelion. *Astrophysical Journal, Supplement*, *246*(2), 71. doi: 10.3847/1538-4365/ab745c

Chaston, C. C., Bonnell, J. W., Carlson, C. W., McFadden, J. P., Ergun, R. E., & Strangeway, R. J. (2003, April). Properties of small-scale Alfvén waves and accelerated electrons from FAST. *Journal of Geophysical Research*, *108*, 8003. doi: 10.1029/2002JA009420

Chaston, C. C., Bonnell, J. W., Clausen, L., & Angelopoulos, V. (2012, September). Energy transport by kinetic-scale electromagnetic waves in fast plasma sheet flows. *Journal of Geophysical Research (Space Physics)*, *117*(A9), A09202. doi: 10.1029/2012JA017863

Chaston, C. C., Carlson, C. W., McFadden, J. P., Ergun, R. E., & Strangeway, R. J. (2007, April). How important are dispersive Alfven waves for auroral particle acceleration? *Geophysical Research Letters*, *34*, 7101. doi: 10.1029/2006GL029144

Chaston, C. C., Johnson, J. R., Wilber, M., Acuna, M., Goldstein, M. L., & Reme, H. (2009, January). Kinetic Alfvén wave turbulence and transport through a reconnection diffusion region. *Physical Review Letters*, *102*(1), 015001. doi: 10.1103/PhysRevLett.102.015001

Chen, C. H. K. (2016, December). Recent progress in astrophysical plasma turbulence from solar wind observations. *Journal of Plasma Physics*, *82*(6), 535820602. doi: 10.1017/S0022377816001124

Chen, L., Wu, D. J., Zhao, G. Q., Tang, J. F., & Huang, J. (2014, September). Excitation of kinetic Alfvén waves by fast electron beams. *Astrophysical Journal*, *793*(1), 13. doi: 10.1088/0004-637X/793/1/13

Coleman, P. J. (1966, July). Hydromagnetic waves in the interplanetary plasma. *Physical Review Letters*, *17*(4), 207–211. doi: 10.1103/PhysRevLett.17.207

Cranmer, S. R. (2009, September). Coronal holes. *Living reviews in solar physics*, *6*(1), 3. doi: 10.12942/lrsp-2009-3

Davidson, R. C. (1972). *Methods in nonlinear plasma theory.* New York and London: Academic Press.

de Pontieu, B., & Haerendel, G. (1998, October). Weakly damped Alfven waves as drivers for spicules. *Astronomy & Astrophysics*, *338*, 729–736.

De Pontieu, B., Martens, P. C. H., & Hudson, H. S. (2001, September). Chromospheric Damping of Alfvén waves. *Astrophysical Journal*, *558*(2), 859–871. doi: 10.1086/322408

Del Zanna, L., Velli, M., & Londrillo, P. (2001a, February). Parametric decay of circularly polarized Alfven waves: Multidimensional simulations in periodic and open domains. *Astronomy & Astrophysics*, *367*, 705–718. doi: 10.1051/0004-6361:20000455

Del Zanna, L., Velli, M., & Londrillo, P. (2001b, February). Parametric decay of circularly polarized Alfven waves: Multidimensional simulations in periodic and open domains. *Astronomy & Astrophysics*, *367*, 705–718. doi: 10.1051/0004-6361:20000455

De Pontieu, B., McIntosh, S., Carlsson, M., Hansteen, V., Tarbell, T., Schrijver, C., et al. (2007). Chromospheric Alfvénic waves strong enough to power the solar wind. *Science*, *318*(5856), 1574–1577.

Derby, N. F., Jr. (1978, September). Modulational instability of finite-amplitude, circularly polarized Alfvén waves. *Astrophysical Journal*, *224*, 1013–1016. doi: 10.1086/156451

Dombeck, J., Cattell, C., Wygant, J. R., Keiling, A., & Scudder, J. (2005, December). Alfvén waves and Poynting flux observed simultaneously by Polar and FAST in the plasma sheet boundary layer. *Journal of Geophysical Research*, *110*(A12), A12S90. doi: 10.1029/2005JA011269

Dorfman, S., & Carter, T. A. (2013, May). Nonlinear excitation of acoustic modes by large-amplitude Alfvén waves in a laboratory plasma. *Physical Review Letters*, *110*(19), 195001. doi: 10.1103/PhysRevLett.110.195001

Dorfman, S., & Carter, T. A. (2015, May). Non-linear Alfvén wave interaction leading to resonant excitation of an acoustic mode in the laboratorya). *Physics of Plasmas*, *22*(5), 055706. doi: 10.1063/1.4919275

Dorfman, S., & Carter, T. A. (2016, May). Observation of an Alfvén wave parametric instability in a laboratory plasma. *Physical Review Letters*, *116*(19), 195002. doi: 10.1103/PhysRevLett.116.195002

DPP. (2020). American Physical Society – Division of Plasma Physics – Community Planning Process. Retrieved from https://sites.google.com/pppl.gov/dpp-cpp/home?authuser=0

Drake, D. J., Howes, G. G., Rhudy, J. D., Terry, S. K., Carter, T. A., Kletzing, C. A., et al. (2016). Measurements of the nonlinear beat wave produced by the interaction of counterpropagating Alfvén waves. *Physics of Plasmas*, *23*(2), 022305. Retrieved from http://scitation.aip.org/content/aip/journal/pop/23/2/10.1063/1.4941977. doi: http://dx.doi.org/10.1063/E4941977

Drake, D. J., Kletzing, C. A., Skiff, F., Howes, G. G., & Vincena, S. (2011, October). Design and use of an Elsässer probe for analysis of Alfvén wave fields according to wave direction. *Review of Scientific Instruments*, *82*(10), 103505. doi: 10.1063/1.3649950

Drake, D. J., Schroeder, J. W. R., Howes, G. G., Kletzing, C. A., Skiff, F., Carter, T. A., & Auerbach, D. W. (2013, July). Alfvén wave collisions, the fundamental building block of plasma turbulence. IV. Laboratory experiment. *Physics of Plasmas*, *20*(7), 072901. doi: 10.1063/1.4813242

Drake, J. F. (1995, January). Magnetic reconnection: A kinetic treatment. *Geophysical Monograph Series*, *90*, 155. doi: 10.1029/GM090p0155

Duan, S. P., Dai, L., Wang, C., Liang, J., Lui, A. T. Y., Chen, L. J., et al. (2016, May). Evidence of kinetic Alfvén eigenmode in the near-Earth magnetotail during substorm expansion phase. *Journal of Geophysical Research (Space Physics)*, *121*(5), 4316–4330. doi: 10.1002/2016JA022431

Elsässer, W. M. (1950, July). The hydromagnetic equations. *Physical Review*, *79*, 183–183. doi: 10.1103/PhysRev.79.183

Farmer, W. A., & Morales, G. J. (2014, June). The ion-ion hybrid Alfvén resonator in a fusion environment. *Physics of Plasmas*, *21*(6), 062507. doi: 10.1063/1.4882662

Fisch, N. J., & Karney, C. F. F. (1981, January). Current generation with low-frequency waves. *Physics of Fluids*, *24*, 27–39. doi: 10.1063/1.863243

Fried, B. D., & Conte, S. D. (1961). *The plasma dispersion function*. New York and London: Academic Press.

Galtier, S., Nazarenko, S. V., Newell, A. C., & Pouquet, A. (2000, June). A weak turbulence theory for incompressible magnetohydrodynamics. *Journal of Plasma Physics*, *63*(5), 447–488. doi: 10.1017/S0022377899008284

Gekelman, W. (1999, July). Review of laboratory experiments on Alfvén waves and their relationship to space observations. *Journal of Geophysical Research*, *104*(A7), 14417–14436. doi: 10.1029/98JA00161

Gekelman, W., Leneman, D., Maggs, J., & Vincena, S. (1994, December). Experimental observation of Alfvén wave cones. *Physics of Plasmas*, *1*, 3775–3783. doi: 10.1063/1.870851

Gershman, D. J., F-Vmas, A., Dorelli, J. C., Boardsen, S. A., Avanov, L. A., Bellan, P. M., et al. (2017, March). Wave-particle energy exchange directly observed in a kinetic Alfvén-branch wave. *Nature Communications*, *8*, 14719. doi: 10.1038/ncomms14719

Gillet, N., Jault, D., Canet, E., & Fournier, A. (2010, May). Fast torsional waves and strong magnetic field within the Earth's core. *Nature*, *465*(7294), 74–77. doi: 10.1038/nature09010

Glassmeier, K. H., Motschmann, U., & von Stein, R. (1995, January). Mode recognition of MHD wave fields at incomplete dispersion measurements. *Annales Geophysicae*, *13*(1), 76–83. doi: 10.5194/angeo-13-76-1995

Goertz, C. K., & Boswell, R. W. (1979, December). Magnetosphere-ionosphere coupling. *Journal of Geophysical Research*, *84*, 7239–7246. doi: 10.1029/JA084iA12p07239

Goldreich, P., & Sridhar, S. (1995, January). Toward a theory of interstellar turbulence. II. Strong Alfvenic turbulence. *Astrophysical Journal*, *438*, 763. doi: 10.1086/175121

Goldstein, M. L. (1978, January). An instability of finite amplitude circularly polarized Afvén waves. *Astrophysical Journal*, *219*, 700–704. doi: 10.1086/155829

Gosling, J. T., McComas, D. J., Roberts, D. A., & Skoug, R. M. (2009, April). A one-sided aspect of Alfvenic fluctuations in the solar wind. *Astrophysical Journal Letters*, *695*(2), L213–L216. doi: 10.1088/0004-637X/695/2/L213

Goyal, R., & Sharma, R. P. (2018, March). Effect of ion-neutral collisions on the evolution of kinetic Alfvén waves in plasmas. *Plasma Physics and Controlled, Fusion*, *60*(3), 035002. doi: 10.1088/1361-6587/aaa118

Guglielmi, A. V., Potapov, A. S., & Russell, C. T. (2000, September). The ion cyclotron resonator in the magnetosphere. *Soviet Journal of Experimental and Theoretical Physics Letters*, *72*(6), 298–300. doi: 10.1134/1.1328441

Gurnett, D. A., & Bhattacharjee, A. (2005). *Introduction to plasma physics*. Cambridge, UK: Cambridge University Press.

Gurnett, D. A., Huff, R. L., Menietti, J. D., Burch, J. L., Winningham, J. D., & Shawhan, S. D. (1984, October). Correlated low-frequency electric and magnetic noise along the auroral field lines. *Journal of Geophysical Research*, *89*, 8971–8985. doi: 10.1029/JA089iA10p08971

Haerendel, G. (1992, November). Weakly damped Alfvén waves as drivers of solar chromospheric spicules. *Nature*, *360*(6401), 241–243. doi: 10.1038/360241a0

Hahn, M., Landi, E., & Savin, D. W. (2012, July). Evidence of wave damping at low heights in a polar coronal hole. *Astrophysical Journal*, *753*(1), 36. doi: 10.1088/0004-637X/753/1/36

Hahn, M., & Savin, D. W. (2014, November). Evidence for wave heating of the quiet-Sun corona. *Astrophysical Journal*, *795*(2), 111. doi: 10.1088/0004-637X/795/2/111

Hanna, J., & Watts, C. (2001, September). Alfvén wave propagation in a helicon plasma. *Physics of Plasmas*, *8*(9), 4251–4254. doi: 10.1063/1.1386801

Hapgood, M. (2012, April). Astrophysics: Prepare for the coming space weather storm. *Nature*, *484*(7394), 311–313. doi: 10.1038/484311a

Hasegawa, A. (1976, October). Particle acceleration by MHD surface wave and formation of aurora. *Journal of Geophysical Research*, *81*, 5083–5090. doi: 10.1029/JA081i028p05083

He, J., Zhu, X., Verscharen, D., Duan, D., Zhao, J., & Wang, T. (2020, July). Spectra of diffusion, dispersion, and dissipation for kinetic Alfvénic and compressive turbulence: Comparison between kinetic theory and measurements from MMS. *Astrophysical Journal*, *898*(1), 43. doi: 10.3847/1538-4357/ab9174

Heidbrink, W. W., Strait, E. J., Chu, M. S., & Turnbull, A. D. (1993, August). Observation of beta-induced Alfvén eigenmodes in the DIII-D tokamak. *Physical Review Letters*, *71*(6), 855–858. doi: 10.1103/PhysRevLett.71.855

Hollweg, J. V. (1994, December). Beat, modulational, and decay instabilities of a circularly polarized Alfven wave. *Journal of Geophysical Research*, *99*, 23431–+. doi: 10.1029/94JA02185

Hollweg, J. V., Esser, R., & Jayanti, V. (1993, March). Modulational and decay instabilities of Alfvén waves: Effects of streaming $He^{++}$. *Journal of Geophysical Research*, *98*(A3), 3491–3500. doi: 10.1029/92JA02347

Hollweg, J. V., & Johnson, W. (1988, September). Transition region, corona, and solar wind in coronal holes: Some two-fluid models. *Journal of Geophysical Research*, *93*(A9), 9547–9554. doi: 10.1029/JA093iA09p09547

Hong, M. H., Lin, Y., & Wang, X. Y. (2012, July). Generation of kinetic Alfvén waves by beam-plasma interaction in non-uniform plasma. *Physics of Plasmas*, *19*(7), 072903. doi: 10.1063/1.4736988

Houshmandyar, S., & Scime, E. E. (2011, November). Ducted kinetic Alfvén waves in plasma with steep density gradients. *Physics of Plasmas*, *18*(11), 112111. doi: 10.1063/1.3662113

Howes, G. G. (2010). A prescription for the turbulent heating of astrophysical plasmas. *Monthly Notices of the Royal Astronomical Society*, *409*, L104–L108. doi: 10.1111/j.1745-3933.2010.00958.x

Howes, G. G. (2018, May). Laboratory space physics: Investigating the physics of space plasmas in the laboratory. *Physics of Plasmas*, *25*(5), 055501. doi: 10.1063/1.5025421

Howes, G. G., Drake, D. J., Nielson, K. D., Carter, T. A., Kletzing, C. A., & Skiff, F. (2012, December). Toward astrophysical turbulence in the laboratory. *Physical Review Letters*, *109*(25), 255001. doi: 10.1103/PhysRevLett.109.255001

Howes, G. G., & Nielson, K. D. (2013, July). Alfvén wave collisions, the fundamental building block of plasma turbulence. I. Asymptotic solution. *Physics of Plasmas*, *20*(7), 072302. doi: 10.1063/1.4812805

Howes, G. G., Nielson, K. D., Drake, D. J., Schroeder, J. W. R., Skiff, F., Kletzing, C. A., & Carter, T. A. (2013, July). Alfvén wave collisions, the fundamental building block of plasma turbulence. III. Theory for experimental design. *Physics of Plasmas*, *20*(7), 072304. doi: 10.1063/1.4812808

Howes, G. G., Nielson, K. D., Schroeder, J. W. R., Drake, D. J., Skiff, F., Kletzing, C. A., & Carter, T. A. (2013). Alfven wave collisions, the fundamental building block of plasma turbulence V: Simulations and magnetic shear interpretation. *Physics of Plasmas*. (in preparation)

Huba, J. D. (2013). *NRL PLASMA FORMULARY Supported by The Office of Naval Research*. Washington, DC: Naval Research Laboratory. Retrieved from http://wwwppd.nrl.navy.mil/nrlformulary/

Inhester, B. (1990, July). A drift-kinetic treatment of the parametric decay of large- amplitude Alfven waves. *Journal of Geophysical Research*, *95*, 10525–10539. doi: 10.1029/JA095iA07p10525

Intrator, T., Vukovic, M., Elfimov, A., Probert, P. H., & Winz, G. (1996, March). Discrete spectrum of Alfvén ion-ion hybrid waves. *Physics of Plasmas*, *3*(3), 1054–1063. doi: 10.1063/1.871761

Jess, D. B., Morton, R. J., Verth, G., Fedun, V., Grant, S. D. T., & Giagkiozis, I. (2015, July). Multiwavelength studies of MHD waves in the solar chromosphere. An overview of recent results. *Space Science Reviews*, *190*(1-4), 103–161. doi: 10.1007/s11214-015-0141-3

Kasper, J. C., Bale, S. D., Belcher, J. W., Berthomier, M., Case, A. W., Chandran, B. D. G., et al. (2019, December). Alfvénic velocity spikes and rotational flows in the near-Sun solar wind. *Nature*, *576*(7786), 228–231. doi: 10.1038/s41586-019-1813-z

Keiling, A., Thaller, S., Dombeck, J., & Wygant, J. (2020, April). Temporal evolution of substorm-driven global Alfvén wave power above the auroral acceleration region. *Journal of Geophysical Research*, *125*(4), e27444. doi: 10.1029/2019JA027444

Keiling, A., Thaller, S., Wygant, J., & Dombeck, J. (2019, June). Assessing the global Alfvén wave power flow into and out of the auroral acceleration region during geomagnetic storms. *Science Advances*, *5*(6), eaav8411. doi: 10.1126/sciadv.aav8411

Keiling, A., Wygant, J. R., Cattell, C. A., Mozer, F. S., & Russell, C. T. (2003, January). The Global Morphology of Wave Poynting Flux: Powering the Aurora. *Science*, *299*, 383–386. doi: 10.1126/science.1080073

Kirkwood, R., Hutchinson, I., Luckhardt, S., & Squire, J. (1990). Measurement of suprathermal electrons in tokamaks via electron cyclotron transmission. *Nuclear Fusion*, *30*(3), 431. Retrieved from http://stacks.iop.org/0029-5515/30/i=3/a=005

Klein, K. G., & Howes, G. G. (2016, August). Measuring collisionless damping in heliospheric plasmas using field-particle correlations. *Astrophysical Journal Letters*, *826*, L30. doi: 10.3847/2041-8205/826/2/L30

Kletzing, C. A. (1994, June). Electron acceleration by kinetic Alfvén waves. *J. Geophys. Res.*, *99*, 11095–11104. doi: 10.1029/94JA00345

Kletzing, C. A., Bounds, S. R., Martin-Hiner, J., Gekelman, W., & Mitchell, C. (2003, January). Measurements of the shear Alfvén wave dispersion for finite perpendicular wave number. *Physical Review Letters*, *90*(3), 035004. doi: 10.1103/PhysRevLett.90.035004

Kletzing, C. A., & Hu, S. (2001, February). Alfvén wave generated electron time dispersion. *Geophysical Research Letters*, *28*, 693–696. doi: 10.1029/2000GL012179

Kletzing, C. A., Thuecks, D. J., Skiff, F., Bounds, S. R., & Vincena, S. (2010, March). Measurements of inertial limit Alfvén wave dispersion for finite perpendicular wave number. *Physical Review Letters*, *104*(9), 095001. doi: 10.1103/PhysRevLett.104.095001

Kraichnan, R. H. (1965, July). Inertial-range spectrum of hydromagnetic turbulence. *Physics of Fluids*, *8*(7), 1385–1387. doi: 10.1063/1.1761412

Krishan, V., & Gangadhara, R. T. (2008, April). Mean-field dynamo in partially ionized plasmas - I. *Monthly Notices of the Royal Astronomical Society*, *385*(2), 849–853. doi: 10.1111/j.1365-2966.2007.12824.x

Landau, L. D. (1946). On the vibrations of the electronic plasma. *Journal of Physics*, *10*, 25.

Leake, J. E., Arber, T. D., & Khodachenko, M. L. (2005, November). Collisional dissipation of Alfvén waves in a partially ionised solar chromosphere. *Astronomy & Astrophysics*, *442*(3), 1091–1098. doi: 10.1051/0004-6361:20053427

Li, B., Chen, L., & Wu, D. (2014). Effects of inelastic collisions on alfvén waves in partially ionized plasmas. *Chinese Science Bulletin*, *59*, 740–746.

Lundquist, S. (1949, July). Experimental Demonstration of Magneto-hydrodynamic Waves. *Nature*, *164*, 145–146. doi: 10.1038/164145a0

Lysak, R. L. (1998). The relationship between electrostatic shocks and kinetic Alfvén waves. *Geophys. Res. Lett.*, *25*, 2089–2092. doi: 10.1029/98GL00065

Mathioudakis, M., Jess, D. B., & Erdelyi, R. (2013). Alfvén waves in the solar atmosphere: From theory to observations. *Space Science Reviews*, *175*, 1–27.

McIntosh, S. W., de Pontieu, B., Carlsson, M., Hansteen, V., Boerner, P., & Goossens, M. (2011, July). Alfvénic waves with sufficient energy to power the quiet solar corona and fast solar wind. *Nature*, *475*(7357), 477–480. doi: 10.1038/nature10235

Minter, A. H., & Spangler, S. R. (1997, August). Heating of the Interstellar Diffuse Ionized Gas via the Dissipation of Turbulence. *Astrophysical Journal*, *485*(1), 182–194. doi: 10.1086/304396

Morales, G. J., Loritsch, R. S., & Maggs, J. E. (1994, December). Structure of Alfvén waves at the skin-depth scale. *Physics of Plasmas*, *1*, 3765–3774. doi: 10.1063/1.870850

Morton, R., Verth, G., Fedun, V., Shelyag, S., & Erdélyi, R. (2013). Evidence for the photospheric excitation of incompressible chromospheric waves. *The Astrophysical Journal*, *768*(1), 17.

Mueller, G. (1974, September). Experimental study of torsional Alfven waves in a cylindrical partially ionized magnetoplasma. *Plasma Physics*, *16*(9), 813–822. doi: 10.1088/0032-1028/16/9/003

Nakariakov, V. M., Pilipenko, V., Heilig, B., Jelínek, P., Karlický, M., Klimushkin, D. Y., et al. (2016, April). Magnetohydrodynamic oscillations in the solar corona and Earth's magnetosphere: Towards consolidated understanding. *Space Sci. Rev.*, *200*(1-4), 75–203. doi: 10.1007/s11214-015-0233-0

Narita, Y., Glassmeier, K.-H., Fränz, M., Nariyuki, Y., & Hada, T. (2007, July). Observations of linear and nonlinear processes in the foreshock wave evolution. *Nonlinear Processes in Geophysics*, *14*, 361–371. doi: 10.5194/npg-14-361-2007

Narita, Y., Plaschke, F., Nakamura, R., Baumjohann, W., Magnes, W., Fischer, D., et al. (2016, May). Wave telescope technique for MMS magnetometer. *Geophysical Research Letters*, *43*(10), 4774–4780. doi: 10.1002/2016GL069035

NASEM. (2021). *Plasma science: Enabling technology, sustainability, security, and exploration.* Washington, DC: The National Academies Press. (https://doi.org/10.17226/25802)

Neubauer, F. M., & Glassmeier, K. H. (1990, November). Use of an array of satellites as a wave telescope. *Journal of Geophysical Research*, *95*(A11), 19115–19122. doi: 10.1029/JA095iA11p19115

Ng, C. S., & Bhattacharjee, A. (1996, July). Interaction of Shear-Alfven wave packets: Implication for weak magnetohydrodynamic turbulence in astrophysical plasmas. *Astrophysical Journal*, *465*, 845. doi: 10.1086/177468

Nielson, K. D., Howes, G. G., Tatsuno, T., Numata, R., & Dorland, W. (2010, February). Numerical modeling of Large Plasma Device Alfvén wave experiments using AstroGK. *Physics of Plasmas*, *17*(2), 022105. doi: 10.1063/1.3309486

NRC. (2013). *Solar and space physics: A science for a technological society*. The National Academies Press on behalf of the Committee on a Decadal Strategy for Solar and Space Physics (Heliophysics); Space Studies Board; Aeronautics and Space Engineering Board; Division of Earth and Physical Sciences; National Research Council. Retrieved from http://www.nap.edu/openbook.php?record_id=13060

Øieroset, M., Sundkvist, D., Chaston, C. C., Phan, T. D., Mozer, F. S., McFadden, J. P., et al. (2014, August). Observations of plasma waves in the colliding jet region of a magnetic flux rope flanked by two active X lines at the subsolar magnetopause. *Journal of Geophysical Research (Space Physics)*, *119*(8), 6256–6272. doi: 10.1002/2014JA020124

Parker, E. N. (1957, December). Sweet's mechanism for merging magnetic fields in conducting fluids. *Journal of Geophysical Research*, *62*, 509–520. doi: 10.1029/JZ062i004p00509

Perraut, S., Gendrin, R., Roux, A., & de Villedary, C. (1984, January). Ion cyclotron waves: Direct compariosn between ground-based measurements and observations in the source region. *Journal of Geophysical Research*, *89*(A1), 195–202. doi: 10.1029/JA089iA01p00195

Quataert, E. (1998, June). Particle heating by Alfvénic turbulence in hot accretion flows. *Astrophysical Journal*, *500*, 978–991. doi: 10.1086/305770

Rahbarnia, K., Ullrich, S., Sauer, K., Grulke, O., & Klinger, T. (2010, March). Alfvén wave dispersion behavior in single- and multicomponent plasmas. *Physics of Plasmas*, *17*(3), 032102. doi: 10.1063/1.3322852

Rauch, J. L., & Roux, A. (1982, October). Ray tracing of ULF waves in a multicomponent magnetospheric plasma: Consequences for the generation mechanism of ion cyclotron waves. *Journal of Geophysical Research*, *87*(A10), 8191–8198. doi: 10.1029/JA087iA10p08191

Robertson, J., Carter, T. A., & Vincena, S. (2020, December). Propagation of shear Alfvén waves in a two-ion plasma and application as a diagnostic for the ion density ratio. *Journal of Plasma Physics*, *86*(6), 905860613. doi: 10.1017/S0022377820001403

Rogers, B. N., Denton, R. E., Drake, J. F., & Shay, M. A. (2001, November). Role of dispersive waves in collisionless magnetic reconnection. *Physical Review Letters*, *87*(19), 195004. doi: 10.1103/PhysRevLett.87.195004

Sahraoui, F., Goldstein, M. L., Belmont, G., Canu, P., & Rezeau, L. (2010, September). Three Dimensional Anisotropic k Spectra of Turbulence at Subproton Scales in the Solar Wind. *Phys. Rev. Lett.*, *105*(13), 131101. doi: 10.1103/PhysRevLett.105.131101

Samson, J. C., Hughes, T. J., Creutzberg, F., Wallis, D. D., Greenwald, R. A., & Ruohoniemi, J. M. (1991, September). Observations of a detached, discrete arc in association with field line resonances. *Journal of Geophysical Research*, *96*(A9), 15683–15695. doi: 10.1029/91JA00796

Saur, J. (2004, February). Turbulent heating of Jupiter's middle magnetosphere. *Astrophysical Journal Letters*, *602*, L137–L140. doi: 10.1086/382588

Schroeder, J. W. R. (2017). *Exploring the Alfvén-wave acceleration of auroral electrons in the laboratory* (Unpublished doctoral dissertation). University of Iowa.

Schroeder, J. W. R., Howes, G. G., Kletzing, C. A., Skiff, F., Carter, T. A., Vincena, S., & Dorfman, S. (2021, January). Laboratory measurements of the physics of auroral electron acceleration by Alfvén waves. *Nature Communications*, *12*, 3103. doi: 10.1038/s41467-021-23377-5

Schroeder, J. W. R., Skiff, F., Howes, G. G., Kletzing, C. A., Carter, T. A., & Dorfman, S. (2017, March). Linear theory and measurements of electron oscillations in an inertial Alfvén wave. *Physics of Plasmas*, *24*(3), 032902. doi: 10.1063/1.4978293

Schroeder, J. W. R., Skiff, F., Kletzing, C. A., Howes, G. G., Carter, T. A., & Dorfman, S. (2016, May). Direct measurement of electron sloshing of an inertial Alfvén wave. *Geophysical Research Letters*, *43*, 4701–4707. doi: 10.1002/2016GL068865

Sciffer, M. D., & Waters, C. L. (2002, October). Propagation of ULF waves through the ionosphere: Analytic solutions for oblique magnetic fields. *Journal of Geophysical Research (Space Physics)*, *107*(A10), 1297. doi: 10.1029/2001JA000184

Shi, P., Huang, K., Lu, Q., & Sun, X. (2019, December). Experimental observation of kinetic Alfvén wave generated by magnetic reconnection. *Plasma Physics and Controlled Fusion, 61*(12), 125010. doi: 10.1088/1361-6587/ab4f9c

Shi, P., Yang, Z., Luo, M., Wang, R., Lu, Q., & Sun, X. (2019, March). Observation of spontaneous decay of Alfvénic fluctuations into co- and counter-propagating magnetosonic waves in a laboratory plasma. *Physics of Plasmas, 26*(3), 032105. doi: 10.1063/1.5082802

Skiff, F., Boyd, D. A., & Colborn, J. A. (1993, July). Measurements of electron parallel-momentum distributions using cyclotron wave transmission*. *Physics of Fluids B, 5*, 2445–2450. doi: 10.1063/1.860729

Skiff, F., Boyd, D. A., & Colborn, J. A. (1994, August). Measurements of electron dynamics during lower hybrid current drive. *Plasma Physics and Controlled Fusion, 36*, 1371–1379. doi: 10.1088/0741-3335/36/8/009

Smith, R. L., & Brice, N. (1964, December). Propagation in multicomponent plasmas. *Journal of Geophysical Research, 69*(23), 5029–5040. doi: 10.1029/JZ069i023p05029

Spangler, S. R. (1991, August). The dissipation of magnetohydrodynamic turbulence responsible for interstellar scintillation and the heating of the interstellar medium. *Astrophysical Journal, 376*, 540. doi: 10.1086/170303

Spangler, S. R., Leckband, J. A., & Cairns, I. H. (1997, March). Observations of the parametric decay instability of nonlinear magnetohydrodynamic waves. *Physics of Plasmas, 4*(3), 846–855. doi: 10.1063/1.872183

Sridhar, S., & Goldreich, P. (1994, September). Toward a theory of interstellar turbulence. I. Weak Alfvenic turbulence. *Astrophysical Journal, 432*, 612. doi: 10.1086/174600

Stasiewicz, K., Bellan, P., Chaston, C., Kletzing, C., Lysak, R., Maggs, J., et al. (2000, May). Small scale Alfvénic structure in the Aurora. *Space Science Reviews, 92*, 423–533.

Stix, T. H. (1992). *Waves in plasmas*. New York: Springer.

Streltsov, A., & Lotko, W. (1995, October). Dispersive field line resonances on auroral field lines. *Journal of Geophysical Research, 100*, 19457–19472. doi: 10.1029/95JA01553

Streltsov, A. V., Lotko, W., Johnson, J. R., & Cheng, C. Z. (1998, November). Small-scale, dispersive field line resonances in the hot magnetospheric plasma. *Journal of Geophysical Research, 103*, 26559–26572. doi: 10.1029/98JA02679

Summers, D., & Thorne, R. M. (2003, April). Relativistic electron pitch-angle scattering by electromagnetic ion cyclotron waves during geomagnetic storms. *Journal of Geophysical Research (Space Physics), 108*(A4), 1143. doi: 10.1029/2002JA009489

Sundkvist, D., Krasnoselskikh, V., Shukla, P. K., Vaivads, A., Andre, M., Buchert, S., & Rème, H. (2005, August). In situ multi-satellite detection of coherent vortices as a manifestation of Alfvénic turbulence. *Nature, 436*, 825–828. doi: 10.1038/nature03931

Temerin, M., & Lysak, R. L. (1984, May). Electromagnetic ion cyclotron mode (ELF) waves generated by auroral electron precipitation. *Journal of Geophysical Research, 89*(A5), 2849–2860. doi: 10.1029/JA089iA05p02849

Thuecks, D. J., Kletzing, C. A., Skiff, F., Bounds, S. R., & Vincena, S. (2009, May). Tests of collision operators using laboratory measurements of shear Alfvén wave dispersion and damping. *Physics of Plasmas, 16*(5), 052110. doi: 10.1063/1.3140037

Thuecks, D. J., Skiff, F., & Kletzing, C. A. (2012, August). Measurements of parallel electron velocity distributions using whistler wave absorption. *Review of Scientific Instruments, 83*(8), 083503. doi: 10.1063/1.4742766

Tomczyk, S., McIntosh, S., Keil, S., Judge, P., Schad, T., Seeley, D., & Edmondson, J. (2007). Alfvén waves in the solar corona. *Science, 317*(5842), 1192–1196.

Tsushima, A., Amagishi, Y., & Inutake, M. (1982, April). Observations of spatial Alfven resonance. *Physics Letters A, 88*, 457–460. doi: 10.1016/0375-9601(82)90540-0

Velli, M. (2003, December). MHD turbulence and the heating of astrophysical plasmas. *Plasma Physics and Controlled Fusion, 45*(12A), A205–A216. doi: 10.1088/0741-3335/45/12A/014

Vincena, S., Gekelman, W., & Maggs, J. (2001, September). Shear Alfvén waves in a magnetic beach and the roles of electron and ion damping. *Physics of Plasmas, 8*, 3884–3896. doi: 10.1063/1.1389092

Vincena, S., Gekelman, W., & Maggs, J. (2004, September). Shear Alfvén wave perpendicular propagation from the kinetic to the inertial regime. *Physical Review Letters, 93*(10), 105003. doi: 10.1103/PhysRevLett.93.105003

Vincena, S. T., Farmer, W. A., Maggs, J. E., & Morales, G. J. (2011, June). Laboratory realization of an ion-ion hybrid Alfvén wave resonator. *Geophysical Research Letters, 38*(11), L11101. doi: 10.1029/2011GL047399

Vincena, S. T., Farmer, W. A., Maggs, J. E., & Morales, G. J. (2013, January). Investigation of an ion-ion hybrid Alfvén wave resonator. *Physics of Plasmas, 20*(1), 012111. doi: 10.1063/1.4775777

Vincena, S. T., Morales, G. J., & Maggs, J. E. (2010, May). Effect of two ion species on the propagation of shear Alfvén waves of small transverse scale. *Physics of Plasmas, 17*(5), 052106. doi: 10.1063/1.3422549

Vranjes, J., & Kono, M. (2014, January). On the Alfvén wave cut-off in partly ionized collisional plasmas. *Physics of Plasmas, 21*(1), 012110. doi: 10.1063/1.4862781

Vranjes, J., Poedts, S., Pandey, B. P., & de Pontieu, B. (2008, February). Energy flux of Alfvén waves in weakly ionized plasma. *Astronomy & Astrophysics, 478*(2), 553–558. doi: 10.1051/0004-6361:20078274

Watts, C., & Hanna, J. (2004, April). Alfvén wave propagation in a partially ionized plasma. *Physics of Plasmas, 11*(4), 1358–1365. doi: 10.1063/1.1649993

Wilcox, J. M., Boley, F. I., & De Silva, A. W. (1960, January). Experimental study of Alfvén-wave properties. *Physics of Fluids, 3*(1), 15–19. doi: 10.1063/1.1705996

Wong, H. K., & Goldstein, M. L. (1986, May). Parametric instabilities of the circularly polarized Alfven waves including dispersion. *Journal of Geophysical Research, 91*, 5617–5628. doi: 10.1029/JA091iA05p05617

Woods, L. C. (1962, January). Hydromagnetic waves in a cylindrical plasma. *Journal of Fluid Mechanics, 13*, 570–586. doi: 10.1017/S0022112062000944

Wygant, J. R., Keiling, A., Cattell, C. A., Lysak, R. L., Temerin, M., Mozer, F. S., et al. (2002, August). Evidence for kinetic Alfvén waves and parallel electron energization at

4-6 $R_E$ altitudes in the plasma sheet boundary layer. *Journal of Geophysical Research (Space Physics)*, *107*, 1201. doi: 10.1029/2001JA900113

Yao, Y., Chaston, C. C., Glassmeier, K. H., & Angelopoulos, V. (2011, May). Electromagnetic waves on ion gyro-radii scales across the magnetopause. *Geophysical Research Letters*, *38*(9), L09102. doi: 10.1029/2011GL047328

Yin, L., Winske, D., Daughton, W., & Bowers, K. J. (2007, June). Kinetic Alfvén waves and electron physics. I. Generation from ion-ion streaming. *Physics of Plasmas*, *14*(6), 062104. doi: 10.1063/1.2734950

Young, D. T., Perraut, S., Roux, A., de Villedary, C., Gendrin, R., Korth, A., et al. (1981, August). Wave-particle interactions near $\Omega_{He+}$ observed in GEOS 1 and 2 1. Propagation of ion cyclotron waves in $He^+$ -rich plasma. *Journal of Geophysical Research*, *86*(A8), 6755–6772. doi: 10.1029/JA086iA08p06755

Zhao, L. L., Zank, G. P., Nakanotani, M., & Adhikari, L. (2023, February). Observations of waves and structures by frequency-wavenumber spectrum in solar wind turbulence. *Astrophysical Journal*, *944*(1), 98. doi: 10.3847/1538-4357/acb33b

Zhao, S. Q., Yan, H., Liu, T. Z., Liu, M., & Wang, H. (2022, October). Multispacecraft analysis of the properties of magnetohydrodynamic fluctuations in sub-Alfvénic solar wind turbulence at 1 au. *Astrophys. J.*, *937*(2), 102. doi: 10.3847/1538-4357/ac822e

Zhu, X., He, J., Verscharen, D., Duan, D., & Bale, S. D. (2020, September). Wave composition, propagation, and polarization of magnetohydrodynamic turbulence within 0.3 au as observed by Parker solar probe. *Astrophysical Journal Letters*, *901*(1), L3. doi: 10.3847/2041-8213/abb23e

Zimbardo, G., Greco, A., Sorriso-Valvo, L., Perri, S., Vörös, Z., Aburjania, G., et al. (2010, October). Magnetic turbulence in the geospace environment. *Space Science Reviews*, *156*(1-4), 89–134. doi: 10.1007/s11214-010-9692-5

# INDEX




*Alfvén Waves Across Heliophysics: Progress, Challenges, and Opportunities, Geophysical Monograph 285*, First Edition.
Edited by Andreas Keiling.



DOI: 10.1002/9781394195985.bindex